AF594965

Phytochemical Antioxidants and Health

Phytochemical Antioxidants and Health

Editor

Baojun Xu

Basel • Beijing • Wuhan • Barcelona • Belgrade • Novi Sad • Cluj • Manchester

Editor
Baojun Xu
Beijing Normal
University-Hong Kong
Baptist University United
International College
Zhuhai, China

Editorial Office
MDPI
St. Alban-Anlage 66
4052 Basel, Switzerland

This is a reprint of articles from the Special Issue published online in the open access journal *Antioxidants* (ISSN 2076-3921) (available at: https://www.mdpi.com/journal/antioxidants/special_issues/Phytochemical_Antioxidants_Health).

For citation purposes, cite each article independently as indicated on the article page online and as indicated below:

Lastname, A.A.; Lastname, B.B. Article Title. *Journal Name* **Year**, *Volume Number*, Page Range.

ISBN 978-3-0365-8764-6 (Hbk)
ISBN 978-3-0365-8765-3 (PDF)
doi.org/10.3390/books978-3-0365-8765-3

Contents

About the Editor

Baojun Xu

Dr. Xu is a Chair Professor at Beijing Normal University-Hong Kong Baptist University United International College (UIC, a full English teaching college in China), Fellow of the Royal Society of Chemistry, Zhuhai Scholar Distinguished Professor, Department Head of the Department of Life Sciences, Program Director of the Food Science and Technology Program, and author of over 300 peer-reviewed papers. Dr. Xu received Ph.D. in Food Science from Chungnam National University, South Korea. He conducted postdoctoral research work at North Dakota State University (NDSU), Purdue University, and the Gerald P. Murphy Cancer Foundation in the USA from 2005 to 2009. He conducted short-term visiting research at NDSU in 2012 and the University of Georgia in 2014, followed by visiting research during his sabbatical leave (7 months) at Pennsylvania State University in the USA in 2016. Dr. Xu serves as the Associate Editor-in-Chief of *Food Science and Human Wellness*, the Associate Editor of *Food Research International*, the Associate Editor of *Food Frontiers*, and is an Editorial Board Member of around 10 international journals. He received the inaugural President's Award for Outstanding Research of UIC in 2016 and the President's Award for Outstanding Service of UIC in 2020. Dr. Xu was listed in the world's top 2% of scientists by Stanford University in 2020, 2021, and 2022, and has been listed in the Best Scientists in the world in the field of Biology and Biochemistry at Research.com in 2023.

Article

Comparative Study on Phytochemical Profiles and Antioxidant Capacities of Chestnuts Produced in Different Geographic Area in China

Ziyun Xu [1,2,†], Maninder Meenu [1,†], Pengyu Chen [1,†] and Baojun Xu [1,*]

1 Food Science and Technology Programme, Beijing Normal University-Hong Kong Baptist University United International College, Zhuhai 519087, China; ziyun.xu@mail.mcgill.ca (Z.X.); meenu_maninder@yahoo.com (M.M.); l630013005@mail.uic.edu.hk (P.C.)

2 Department of Food Science and Agricultural Chemistry, McGill University, Quebec, QC H9X 3V9, Canada

* Correspondence: baojunxu@uic.edu.hk; Tel.: +86-7563620636; Fax: +86-7563620882

† These authors contributed equally to the article as the first authors.

Received: 27 January 2020; Accepted: 22 February 2020; Published: 25 February 2020

Abstract: This study aimed to systematically assess the phenolic profiles and antioxidant capacities of 21 chestnut samples collected from six geographical areas of China. All these samples exhibit significant differences ($p < 0.05$) in total phenolic contents (TPC), total flavonoids content (TFC), condensed tannin content (CTC) and antioxidant capacities assessed by DPPH free radical scavenging capacity (DPPH), ABTS free radical scavenging capacities (ABTS), ferric reducing antioxidant power (FRAP), and 14 free phenolic acids. Chestnuts collected from Fuzhou, Jiangxi (East China) exhibited the maximum values for TPC (2.35 mg GAE/g), CTC (13.52 mg CAE/g), DPPH (16.74 μmol TE/g), ABTS (24.83 μmol TE/g), FRAP assays (3.20 mmol FE/100 g), and total free phenolic acids (314.87 μg/g). Vanillin and gallic acids were found to be the most abundant free phenolic compounds among other 14 phenolic compounds detected by HPLC. Overall, the samples from South China revealed maximum mean values for TPC, CTC, DPPH, and ABTS assays. Among the three chestnut varieties, *Banli* presented prominent mean values for all the assays. These finding will be beneficial for production of novel functional food and developing high-quality chestnut varieties.

Keywords: chestnuts; *Castanea mollissima*; phenolic properties; antioxidant capacities

---

1. Introduction

Chestnuts (*Castanea* spp.), belonging to family Fagaceae, are extensively cultivated in Asian countries. China is the largest producer of chestnut followed by Bolivia, Turkey, Korea, and Italy [1]. Chestnuts were mainly produced from four economically important species, namely *Castanea mollissima* (Chinese chestnut), *C. crenata* (Japanese chestnut), *C. dentate* (American chestnut) and *C. sativa* (European chestnut) [2]. Chinese chestnut variety is preferred by its high yielding and easy cultivation [3]. It is a rich source of carbohydrates, fiber and minerals [4]. Fresh Chinese chestnut fruits exhibit significant amount of water (52.0%), carbohydrates (42.2%), proteins (4.2%), and lipids (0.7%) [5]. According to a prehistoric encyclopedia of China Compendium of Materia Medica (Ben Cao Gang Mu) from Ming Dynasty (A.D. 1590), Chinese chestnuts improve kidney functioning [6]. Thus, chestnuts are popular among the Chinese population from ancient times due to its nutritional value as well as the health benefits attributed to the presence of various antioxidant compounds [7]. Antioxidants, such as phenolic acids and their derivatives, are the group of naturally occurring functional substances in plant-based foods, especially in fruits, vegetables, and nuts. Chestnuts presented abundant antioxidant content (4.7 mmol Fe^{2+}/100 g) compared to many legumes (0.11–1.97 mmol Fe^{2+}/100 g), fruits (0.4–2.4 mmol Fe^{2+}/100 g), and grain products (0.5–1.3 mmol Fe^{2+}/100 g) [8]. Phenolic compounds present in chestnuts are responsible for free radical scavenging properties that in turn exhibit protective effects against coronary heart disease, cancer, neurodegenerative diseases and osteoporosis [9]. As a typical group of phenolic compounds, phenolic acids accounted for about 1/3 of phenolic compounds in plant-derived food, and most of them were derivatives of benzoic acid and cinnamic acid, existed in form of both free and bound [10].

Based on climatic characteristics of China, chestnut is mainly produced in five different geographical areas, namely North China, East China, Central China, South China, and Southwest China. Chinese chestnuts (*C. mollissima*) are classified into three subgroups, i.e., *Banli, Youli,* and *Maoli* based on the different morphological features. Among these varieties, *Maoli* is the smallest in size and contains a comparatively higher level of sugar and glutinous starch content [11]. Banli is the most common variety of chestnut and the fruit is flattened on one or two sides. Youli variety of chestnut exhibits round shape, darker color and lustrous outer shell.

Various researchers have studied the phenolic profiles and antioxidant activities of chestnuts collected from various geographical regions and also explored the impact of different processing techniques on phenolic content and antioxidant capacities of chestnuts [9,12–14]. A previous study reported the variation in phenolic content and flavonoids content of Chinese chestnut collected from North China to South region of China [15].

Although many researchers have investigated the antioxidant activities of various chestnut species from several geographical regions, phenolic profile in terms of total phenolic contents (TPC), total flavonoids content (TFC), condensed tannin contents (CTC), and antioxidant properties of three sub-varieties (*Banli, Youli,* and *Maoli*) of Chinese chestnuts (*C. mollissima*) collected from different geographic regions are still unexplored. Besides, phenolic profile in terms of 14 free phenolic acids of Chinese chestnuts were also unknown. Thus, the present study was carried out with an aim to systematically assess the phytochemical profiles as well as antioxidant capacities of twenty one raw chestnut fruits grown in five geographic areas of China.

2. Materials and Methods

2.1. Chestnuts Produced in Different Parts of China

The chestnut samples were collected from five different geographic areas in China in 2016. All the chestnut samples were identified as *Castanea mollissima* Blume by Professor Jingzheng Zhang from Chestnut Research Center, Hebei Normal University of Science and Technology, Hebei, China. All the collected samples were further classified as *Banli, Maoli,* and *Youli* based on their morphological features. Soil source for all chestnuts was sandy loam soil (pH 5.5–6.5). Harvested chestnuts were stored in specialized refrigerator (2–6 °C). The information regarding the common name, size, the specific growing area, average temperature, and monthly sunlight duration from April to September [16] is summarized in Table 1. The Supplemental Figure S1. is presenting the sampling geographical regions in China. The Supplemental Figure S2. is presenting the morphological appearance of twenty-one chestnut samples explored in this study.

2.2. Chemicals and Reagents

Folin-Ciocalteu reagent and 2, 2′-azino-bis (3-ethylbenzothiazoline-6-sulfonic acid) (ABTS) were purchased from Shanghai Yuanye Biological Technology Co., Ltd. (Shanghai, China). (+)-Catechin, 2,4, 6-tri(2-pyridyl)-s-triazine (TPTZ), and 2,2-diphenyl-1-picrylhydrazyl (DPPH) were obtained from Sigma-Aldrich (Shanghai, China). Absolute ethanol, 6-hydroxy-2,5,7,8-tetramethylchromane-2-carboxylic acid (Trolox), acetone, and methanol were provided by Tianjin Fuyu Fine Chemical Co., Ltd. (Tianjin, China). Trifluroacetic acid (TFA), butylated hydroxytoluene (BHT), and methanol (HPLC grade) were purchased from Sigma-Aldrich Co., Ltd. (Shanghai, China). All chemicals employed in this study were of analytical grade.

2.3. Sample Preparation

Chestnut samples were peeled with a chestnut peeler (550 W, Kenong Technology Co., Ltd., Jiangsu, China) and stored overnight at −80 °C. Samples were then freeze-dried using freeze-dryer (Freezone Benchtop, Labconco Corporation, Kansas City, MO, USA) and ground into fine flours. The percentage yield of dried chestnut flours was calculated by dividing dried chestnut flour weight by the weight of the fresh chestnut fruit.

Table 1. Information of chestnut samples in different parts of China.

Code	Variety	Common Name	Size	Growing Area	Region	Average Temperature (°C)	Average Sunlight Duration (h/month)
1	*Castanea mollissima*	*Banli*	Medium	Huairou, Beijing	North China	22.33	213.12
2	*Castanea mollissima*	*Youli*	Small	Jixian, Tianjin			
3	*Castanea mollissima*	*Banli*	Medium	Tangshan, Hebei			
4	*Castanea mollissima*	*Banli*	Medium	Xingtai, Hebei			
5	*Castanea mollissima*	*Banli*	Medium	Qinhuangdao, Hebei			
6	*Castanea mollissima*	*Maoli*	Small	Shangluo, Shanxi	Northwest China		
7	*Castanea mollissima*	*Banli*	Medium	Wuxi, Jiangsu	East China	23.72	182.82
8	*Castanea mollissima*	*Youli*	Medium	Wenzhou, Zhejiang			
9	*Castanea mollissima*	*Banli*	Medium	Anqing, Anhui			
10	*Castanea mollissima*	*Youli*	Medium	Nanping, Fujian			
11	*Castanea mollissima*	*Maoli*	Small	Fuzhou, Jiangxi			
12	*Castanea mollissima*	*Maoli*	Small	Taian, Shandong			
13	*Castanea mollissima*	*Maoli*	Small	Linyi, Shandong			
14	*Castanea mollissima*	*Youli*	Medium	Xinyang, Henan	Central China	23.82	180.84
15	*Castanea mollissima*	*Banli*	Medium	Xiangyang, Hubei			
16	*Castanea mollissima*	*Youli*	Large	Yangjiang, Guangdong	South China	27.26	185.05
17	*Castanea mollissima*	*Banli*	Medium	Guilin, Guangxi			
18	*Castanea mollissima*	*Banli*	Medium	Liuzhou, Guangxi			
19	*Castanea mollissima*	*Maoli*	Small	Haikou, Hainan			
20	*Castanea mollissima*	*Maoli*	Small	Zhaotong, Yunnan	Southwest China	21.21	162.02
21	*Castanea mollissima*	*Banli*	Medium	Kunming, Yunnan			

2.4. Determination of Moisture Content and Color Attributes

The moisture content of dried chestnut flours was determined by the fast water content analyzer (MA150, Sartorius Corporation, Goettingen, Germany). Colorimeter (CR-410, Konica Minolta, Japan) was used to measure the color of all the chestnut samples. The color was expressed based on a three-axis color system $L^*a^*b^*$; here L^* denotes lightness, a^* represents red (+) or green (−), and b^* represents yellow (+) or blue (−). The colorimeter was calibrated with a standard white background plate before measurement.

2.5. Extraction of Total Phenolics from Chestnut Samples

For the extraction of phenolics, 0.5 g of chestnut powder was extracted with 5 mL of extraction solvent (acetone/water/acetic acid: 70:29.5:0.5, *v*/*v*/*v*) according to the previously mentioned procedure [17].

2.6. Determination of Total Phenolic Content (TPC)

TPC was determined by employing Folin–Ciocalteu assay as described by Xu and Chang [17]. Gallic acid was used as an external standard. The absorbance of the reaction mixture was measured at 765 nm using UV-Vis Spectrophotometer (UT-1901). TPC values of chestnut samples were expressed as milligram gallic acid equivalents per gram freeze-dried sample (mg GAE/g).

2.7. Determination of Total Flavonoid Content (TFC)

The TFC of chestnut samples was determined using colorimetric assay as described previously [17]. (+)-Catechin was used as an external standard and the absorbance of the reaction mixture was determined at 510 nm using UV-Visible spectrophotometer. TFC values were expressed as milligram (+)-catechin equivalents per gram of freeze-dried sample (mg CAE/g).

2.8. Determination of Condensed Tannin Content (CTC)

The CTC was determined according to the method described by Xu and Chang [17] with slight modifications. Catechin was used as an external standard and the absorbance of the resultant reaction mixture was measured at 500 nm using a UV-visible spectrophotometer. The CTC of chestnut samples was expressed as (+)-catechin equivalents per gram of freeze-dried sample (mg CAE/g).

2.9. DPPH Free Radical Scavenging Activity (DPPH) Assay

DPPH values of samples were performed using Trolox as external standard according to the previously described procedure [17]. The absorbance of the resultant reaction mixture was measured using a UV-visible spectrophotometer at 517 nm against the ethanol blank. Results were expressed as micromole of Trolox equivalents per gram of freeze-dried samples (μmol TE/g).

2.10. Ferric-Reducing Antioxidant Power (FRAP) Assay

A colorimetric reaction assay was used to determine the FRAP values of chestnut samples according to the method described by Xu and Chang [17]. The absorbance of the reaction mixture was measured at 593 nm using a UV-visible spectrophotometer. The FRAP value was expressed as millimoles of Fe^{2+} equivalent (FE) per 100 g freeze-dried samples (mmol FE/100 g).

2.11. ABTS Free Radical Scavenging Assay

ABTS free radical scavenging capacities of samples were performed according to the method reported by Xu and Chang [17] with slight modifications. Trolox was used as an external standard and the absorbance of the reaction mixture was measured at 734 nm using a UV-visible spectrophotometer against the ethanol blank. Results were expressed as micromole of Trolox equivalents per gram of freeze-dried samples (μmol TE/g).

2.12. HPLC Analysis of Free Phenolic Acids

The free phenolic acid contents of chestnut samples were determined by HPLC (High Performance Liquid Chromatography) according to the method described by Xu and Chang [18]. Briefly, 0.5 g of ground sample was extracted with 5 mL extraction solvent (methanol/water/acidic acid/BHT = 85:15:0.5:0.2, *v*/*v*) twice. The mixture was filtered through Whatman no. 42 filter paper and the supernatant was evaporated at 40 °C until dryness. The residue was dissolved in 2.5 mL methanol (25%, *v*/*v*) and 20 μL of the extract was subjected to HPLC system (Waters, e2695 Separations Modulek, Milford, MA, USA) equipped with a photodiode array detector. A reverse phase Zorbax C18 column (5 μm, 250 × 4.6 mm) was employed at temperature of 40 °C. Mobile phase for analysis include solvent A (0.1% acetic acid in water) and solvent B (methanol). The flow rate was set at 0.7 mL/min and the working wavelength of the detector was set at 262 nm. The chromatograms of 14 phenolic acids were extracted at different maximum absorption wavelength from 210 nm to 320 nm. The contents of 14 free phenolic acids were expressed as microgram free phenolic acid per gram sample (g/g sample) on dry weight basis. The regressive equations and correlation coefficients for phenolic acid standards are provided in the Supplemental Table S1.

2.13. Statistical Analysis

All experiments were performed in triplicates and the data were expressed as mean ± standard deviation. The significant differences among mean values were analyzed using One-Way ANOVA. Duncan test was performed to determine the significant differences ($p < 0.05$) among the mean values of different samples. Statistical analysis was performed by using IBM SPSS Statistics version 22 (IBM Corporation, New York, USA).

3. Results

3.1. *Yield of Chestnut Flours and Moisture Content*

Among all the chestnut samples under investigation, the flour yield ranged from 39.11% in case of samples from Jixian, Tianjin (North China) to 61.17% in case of samples from Huairou (North China) as shown in Table 2. Moisture content values of all the chestnut samples exhibit a significant difference ($p < 0.05$) as shown in Table 2. Among all the samples under investigation, the highest moisture content was recorded as 13.14% in samples from Jixian, Tianjin, (North China) while the lowest value (5.02%) was found in samples collected from Anqing, Anhui, (East China).

Table 2. Yield, moisture content, and color value of chestnuts from different geographic areas.

Code	Region	Growing Area	Yield	Moisture Content (%)	Color Value *L*	*a**	*b**
1	North China	Huairou, Beijing	61.17%	$9.15 \pm 0.00^{d,e}$	88.81^{g}	-1.53^{q}	$11.39^{d,e}$
2		Jixian, Tianjin	39.11%	13.14 ± 0.00^{a}	$91.04^{e,f}$	-0.83^{k}	$10.94^{g,h}$
3		Tangshan, Hebei	46.92%	10.21 ± 0.00^{c}	$91.04^{e,f}$	-0.36^{d}	9.03^{o}
4		Xingtai, Hebei	58.38%	$9.17 \pm 0.01^{d,e}$	$90.56^{e,f}$	-0.28^{c}	11.72^{c}
5		Qinhuangdao, Hebei	56.55%	7.24 ± 0.00^{h}	$92.57^{a,b,c}$	-0.95^{m}	$10.76^{h,i}$
6	Northwest China	Shangluo, Shanxi	42.23%	$7.54 \pm 0.00^{g,h}$	$90.77^{e,f}$	-0.89^{l}	$11.09^{f,g}$
7	East China	Wuxi, Jiangsu	44.86%	$5.76 \pm 0.00^{i,j}$	$91.62^{c,d,e}$	-1.00^{n}	9.76^{n}
8		Wenzhou, Zhejiang	49.55%	$9.56 \pm 0.01^{c,d}$	93.51^{a}	-0.06^{b}	7.63^{q}
9		Anqing, Anhui	48.68%	5.02 ± 0.00^{j}	93.39^{a}	-0.44^{e}	8.57^{p}
10		Nanping, Fujian	51.90%	$8.95 \pm 0.01^{d,e,f}$	$90.98^{e,f}$	-0.70^{i}	$11.14^{e,f,g}$
11		Fuzhou, Jiangxi	54.70%	$8.37 \pm 0.01^{e,f,g}$	85.77^{h}	0.50^{a}	10.16^{m}
12		Taian, Shandong	47.16%	10.30 ± 0.01^{c}	$91.55^{c,d,e}$	-0.42^{e}	$10.70^{h,i,j}$
13		Linyi, Shandong	52.06%	$8.24 \pm 0.00^{e,f,g}$	$92.17^{b,c,d}$	-1.25^{p}	$10.28^{l,m}$
14	Central China	Xinyang, Henan	49.97%	11.54 ± 0.01^{b}	$92.57^{a,b,c}$	-1.81^{r}	13.00^{b}
15		Xiangyang, Hubei	53.16%	6.21 ± 0.01^{i}	$90.81^{e,f}$	-1.04^{o}	11.46^{d}
16	South China	Yangjiang, Guangdong	47.03%	$5.87 \pm 0.00^{i,j}$	$92.95^{a,b}$	-2.52^{s}	14.94^{a}
17		Guilin, Guangxi	51.27%	$7.66 \pm 0.00^{g,h}$	$90.80^{e,f}$	-0.74^{j}	$10.59^{i,j,k}$
18		Liuzhou, Guangxi	47.19%	$8.16 \pm 0.00^{f,g,h}$	90.00^{f}	-0.58^{g}	$11.41^{d,e}$
19		Haikou, Hainan	49.78%	$7.43 \pm 0.01^{g,h}$	$91.40^{d,e}$	-0.65^{h}	$10.46^{j,k,l}$
20	Southwest China	Zhaotong, Yunnan	54.65%	$8.69 \pm 0.00^{d,e,f}$	90.10^{f}	-0.91^{l}	$11.35^{d,e,f}$
21		Kunming, Yunnan	50.76%	$8.99 \pm 0.01^{d,e,f}$	90.22^{f}	-0.561^{f}	$10.35^{k,l,m}$

Values are expressed as the mean of triplicates ± standard deviation. Means in the same column with unlike superscripts ($^{a–s}$) differ significantly. ($p < 0.05$).

3.2. *Color Value*

The color values of all the twenty-one chestnut samples are mentioned in Table 2. Among all the samples, the significant differences ($p < 0.05$) were observed in their color parameters L*a *b. The lightness value (L) of the samples were ranged from 85.77 in case of samples collected from Fuzhou, Jiangxi (East China) to 93.51 in chestnut samples from Wenzhou, Zhejiang (East China). The a* value was varied from −2.52 in samples from Yangjiang, Guangdong (South China) to 0.50 in case of samples collected from Fuzhou, Jiangxi (East China). The b* value was ranged from 7.63 in chestnut samples from Wenzhou, Zhejiang (East China) to 14.94 in case of samples belong to Yangjiang, Guangdong (South China).

3.3. *Phenolic Profiles and Antioxidant Capacities of Chestnut Samples*

The phenolic profiles in terms of TPC, TFC, and CTC values along with the antioxidant activity of all the chestnut samples as assessed by DPPH, FRAP, and ABTS assay are presented in Table 3. The samples collected from different geographical areas exhibited a wide range of variation among their TPC, TFC, CTC values, and antioxidant capacities.

Table 3. Phenolic profiles (total phenolic contents (TPC), total flavonoids content (TFC), and condensed tannin content (CTC)), and antioxidant capacities (DPPH free radical scavenging capacity (DPPH), ferric reducing antioxidant power (FRAP), and ABTS free radical scavenging capacities (ABTS)) of chestnut samples.

Code	Region	Growing Area	TPC (mg GAE/g)	TFC (mg CAE/g)	CTC (mg CAE/g)	DPPH (μmol TE/g)	FRAP (mmol FE/100 g)	ABTS (μmol TE/g)
1	North China	Huairou, Beijing	$1.18 \pm 0.11^{k,l}$	$0.60 \pm 0.03^{j,k}$	4.99 ± 0.49^{i}	$8.60 \pm 0.26^{g,h}$	0.82 ± 0.06^{g}	9.45 ± 0.84^{h}
2		Jixian, Tianjin	$1.53 \pm 0.14^{h,i}$	$0.65 \pm 0.06^{h,i,j,k}$	7.79 ± 0.41^{f}	$9.86 \pm 0.48^{d,e,f}$	1.97 ± 0.04^{d}	$12.36 \pm 0.41^{e,f}$
3		Tangshan, Hebei	$1.59 \pm 0.09^{f,g,h,i}$	$0.78 \pm 0.03^{e,f,g}$	$6.81 \pm 0.20^{g,h}$	$10.18 \pm 0.39^{d,e}$	2.13 ± 0.18^{d}	14.31 ± 0.46^{d}
4		Xingtai, Hebei	$1.58 \pm 0.13^{f,g,h,i}$	$0.68 \pm 0.01^{g,h,i,j,k}$	$7.42 \pm 0.42^{f,g}$	$10.02 \pm 0.10^{d,e}$	1.40 ± 0.12^{e}	$13.08 \pm 0.43^{d,e}$
5		Qinhuangdao, Hebei	$1.92 \pm 0.15^{c,d,e}$	$0.84 \pm 0.05^{d,e}$	11.15 ± 0.83^{c}	12.81 ± 0.06^{c}	3.10 ± 0.29^{a}	17.69 ± 1.49^{c}
21	Northwest China	Shangluo, Shanxi	$1.50 \pm 0.09^{h,i,j}$	$0.75 \pm 0.07^{e,f,g,h}$	5.27 ± 0.41^{i}	$10.00 \pm 0.20^{d,e}$	1.48 ± 0.04^{e}	$11.72 \pm 0.60^{e,f,g}$
	Mean ± SD		1.55 ± 0.24	0.72 ± 0.09	7.24 ± 2.22	10.25 ± 1.38	1.82 ± 0.78	13.10 ± 2.77
6	East China	Wuxi, Jiangsu	$1.38 \pm 0.11^{i j k}$	$0.68 \pm 0.05^{g,h,i,j}$	6.33 ± 0.49^{h}	$9.12 \pm 0.43^{f,g}$	0.73 ± 0.05^{g}	$10.44 \pm 0.66^{g,h}$
7		Wenzhou, Zhejiang	$1.80 \pm 0.18^{d,e,f}$	$0.97 \pm 0.06^{b,c}$	$9.97 \pm 0.68d^{e}$	$10.12 \pm 0.48^{d,e}$	1.55 ± 0.04^{e}	$12.53 \pm 1.20^{e,f}$
8		Anqing, Anhui	$1.97 \pm 0.11^{b,c,d}$	$0.91 \pm 0.08^{c,d}$	12.12 ± 0.66^{b}	$10.16 \pm 0.24^{d,e}$	2.68 ± 0.15^{b}	16.82 ± 0.87^{c}
9		Nanping, Fujian	$1.29 \pm 0.08^{j,k}$	0.57 ± 0.03^{k}	3.43 ± 0.34^{j}	$7.97 \pm 0.58^{h,i}$	1.07 ± 0.02^{f}	$10.67 \pm 0.51^{g,h}$
10		Fuzhou, Jiangxi	2.35 ± 0.22^{a}	1.01 ± 0.08^{b}	13.52 ± 1.10^{a}	16.74 ± 0.92^{a}	3.20 ± 0.26^{a}	24.83 ± 0.19^{a}
11		Taian, Shandong	$1.68 \pm 0.13^{e,f,g,h}$	$0.82 \pm 0.06^{d,e,f}$	$6.76 \pm 0.54^{g,h}$	10.42 ± 0.35^{d}	2.05 ± 0.15^{d}	16.71 ± 1.63^{c}
12		Linyi, Shandong	$1.40 \pm 0.01^{i,j,k}$	$0.65 \pm 0.08^{h,i,j,k}$	9.43 ± 0.91^{e}	$9.38 \pm 0.13^{e,f}$	0.66 ± 0.03^{g}	$12.55 \pm 0.18^{e,f}$
	Mean ± SD		1.70 ± 0.38	0.80 ± 0.17	8.79 ± 3.52	10.56 ± 2.85	1.71 ± 0.98	14.94 ± 5.07
13	Central China	Xinyang, Henan	$1.19 \pm 0.06^{k,l}$	$0.63 \pm 0.05^{j,k}$	$4.26 \pm 0.33^{i,j}$	$7.49 \pm 0.27^{i,j}$	1.38 ± 0.09^{e}	9.51 ± 0.72^{h}
14		Xiangyang, Hubei	$1.92 \pm 0.19^{c,d,e}$	$0.85 \pm 0.08^{d,e}$	12.07 ± 0.23^{b}	10.45 ± 0.66^{d}	2.38 ± 0.08^{c}	19.86 ± 0.85^{b}
	Mean ± SD		1.56 ± 0.52	0.74 ± 0.16	8.17 ± 5.52	8.97 ± 2.09	1.88 ± 0.71	14.69 ± 7.32
15	South China	Yangjiang, Guangdong	$1.54 \pm 0.14^{g,h,i}$	$0.70 \pm 0.04^{g,h,i,j}$	5.26 ± 0.48^{i}	$10.10 \pm 0.57^{d,e}$	1.36 ± 0.12^{e}	$11.26 \pm 0.69^{f,g}$
16		Guilin, Guangxi	$2.19 \pm 0.10^{a,b}$	1.03 ± 0.03^{b}	12.33 ± 0.33^{b}	13.87 ± 0.70^{b}	3.24 ± 0.12^{a}	$17. 92 \pm 0.59^{c}$
17		Liuzhou, Guangxi	$2.12 \pm 0.19^{a,b,c}$	$0.71 \pm 0.04^{g,h,i,j}$	13.58 ± 0.57^{a}	13.62 ± 0.27^{b}	2.77 ± 0.13^{b}	19.61 ± 1.01^{b}
18		Haikou, Hainan	$1.72 \pm 0.17^{d,e,f,g,h}$	$0.74 \pm 0.04^{f,g,h,i}$	$6.48 \pm 0.31^{g,h}$	$9.44 \pm 0.33^{e,f}$	2.12 ± 0.11^{d}	$13.16 \pm 1.32^{d,e}$
	Mean ± SD		1.89 ± 0.31	0.80 ± 0.16	9.41 ± 4.15	11.76 ± 2.31	2.37 ± 0.82	14.68 ± 8.16
19	Southwest China	Zhaotong, Yunnan	1.03 ± 0.05^{l}	$0.63 \pm 0.05^{i,j,k}$	4.57 ± 0.27^{i}	7.08 ± 0.15^{j}	0.84 ± 0.06^{g}	6.53 ± 0.35^{i}
20		Kunming, Yunnan	$1.79 \pm 0.14^{d,e,f,g}$	1.13 ± 0.07^{a}	$10.69 \pm 0.59^{c,d}$	$9.52 \pm 0.14^{e,f}$	1.13 ± 0.11^{f}	16.51 ± 1.23^{c}
	Mean ± SD		1.41 ± 0.54	0.88 ± 0.35	7.63 ± 4.33	8.30 ± 1.73	0.99 ± 0.21	11.52 ± 7.06

Values are expressed as the mean of triplicates ± standard deviation. Means in the same column with unlike superscripts ($^{a–l}$) differ significantly ($p < 0.05$).

Among all the chestnut samples, TPC values were ranged from 1.03 mg GAE/g in the samples collected from Zhaotong, Yunnan (Southwest China) to 2.35 mg GAE/g in case of samples belonging to Fuzhou, Jiangxi (East China). Overall, chestnut samples from South China exhibited higher mean TPC values (1.89 mg GAE/g) compared to the samples from other regions, whereas, the samples from Southwest China revealed minimum mean TPC values (1.41 mg GAE/g). It was also interesting to observe that the TPC values of chestnuts samples collected from Guilin (2.19 mg GAE/g) and Liuzhou (2.12 mg GAE/g) cities of Guangxi province were comparatively higher than samples collected from other regions in South China.

TFC values were ranged from 0.57 mg CAE/g in case of samples collected from Nanping, Fujian (East China) to 1.13 mg CAE/g in case of samples procured from Kunming, Yunnan (Southwest China). Overall, the samples from Southwest region of China demonstrated the highest mean value of TFC (0.88 mg CAE/g) and samples from North China exhibited least mean value for TPC (0.72 mg CAE/g) compared to samples from other regions. It was also observed that the TFC content of chestnut samples collected from different cities of the same province exhibit a significant difference ($p < 0.05$). In this study, TFC values of samples from Tangshan (0.78 mg CAE/g), Xingtai (0.68 mg CAE/g) and Qinhuangdao (0.84 mg CAE/g) cities of Hebei province presented significant differences ($p < 0.05$).

The chestnut samples procured from different geographical areas had also presented a wide range of variation in their CTC values. The highest CTC value (13.58 mg CAE/g) was observed in the case of samples collected from Liuzhou, Guangxi (South China). In general, chestnut samples from South China exhibit maximum mean value for CTC (9.41 mg CAE/g) and the samples from North China presented minimum mean value for CTC (7.24 mg CAE/g). Alike the TFC values, significant differences ($p < 0.05$) were also observed among the CTC values of samples collected from different regions of the same province.

DPPH values of chestnut samples under investigation were ranged from 7.08 μmol TE/g in case of samples procured from Zhaotong, Yunnan (Southwest China) to 16.74 μmol TE/g in case of samples from Fuzhou, Jiangxi (East China). Overall, among all the geographical regions, the highest mean DPPH value was exhibited by the samples from South China (11.76 μmol TE/g) and the lowest mean DPPH value was presented by the samples procured from Southwest China (8.30 μmol TE/g). Alike TFC and CTC values, DPPH values of chestnut samples collected from different cities of same province also exhibits a significant difference ($p < 0.05$) except the samples procured from Tangshan (10.18 μmol TE/g) and Xingtai (10.02 μmol TE/g) cities of Hebei province and the samples from Guilin (13.87 μmol TE/g) and Liuzhou (13.62 μmol TE/g) from the Guangxi Province.

The FRAP values of chestnut samples under investigation varied from 0.66 mmol FE/100 g in case of samples from Linyi, Shandong (East China) to 3.24 mmol FE/100 g in case of samples collected from Guilin, Guangxi (South China). Overall, the samples from South China presented higher mean FRAP value (2.37 mmol FE/100 g), whereas, the chestnut samples from Southwest China (0.99 mmol FE/100 g) exhibit lower mean FRAP values compared to the samples collected from other regions. It was also observed that the chestnut samples collected from different cities of the same province also exhibit significant differences ($p < 0.05$) in their FRAP values.

The ABTS values of chestnut samples collected from various regions of China also exhibit a wide range of variation. Among all the samples, chestnut procured from Zhaotong, Yunnan (Southwest China) exhibit the lowest ABTS value (6.53 μmol TE/g). Whereas, the samples from Fuzhou, Jiangxi (East China) exhibit higher ABTS value (24.83 μmol TE/g). While comparing the mean ABTS values presented by samples from five different geographical regions, it was observed that samples collected from East China exhibit the highest mean ABTS value (14.94 μmol TE/g) followed by the samples from Central China (14.69 μmol TE/g), South China (14.68 μmol TE/g), and North China (13.10 μmol TE/g). The lowest mean ABTS value (11.52 μmol TE/g) was presented by the samples collected from Southwest China.

3.4. Free Phenolic Acids Contents of Chestnut Samples

The chromatograms of 14 phenolic acids (including gallic acid, protocatechuic acid, 2,3,4-trihydroxybenzoic acid, protocatechualdehyde, *p*-hydroxybenzoic acid, gentisic acid, chlorogenic acid, vanillic and caffeic acid, syringic acid, vanillin, *p*-coumaric and syringaldehyde, ferulic acid, sinapic acid, and salicylic acid) determined by HPLC were shown in Supplemental Figure S3. Chemical structure of each phenolic compound was presented in Table 4. The contents of these phenolic acids in chestnut samples among different geographical areas were significantly different ($p < 0.05$), and values were summarized in Table 5.

Table 4. Chemical structures of phenolic compounds.

Name	Structure	Name	Structure
Gallic acid	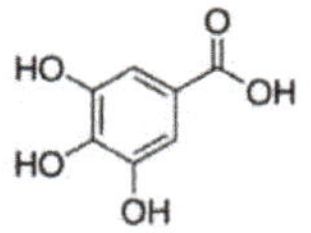	Vanillic acid	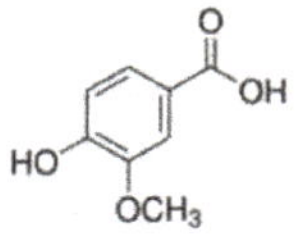
Protocatechuic acid	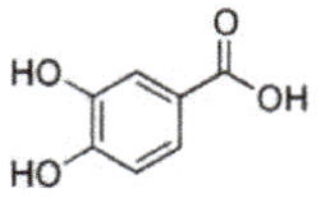	Caffeic acid	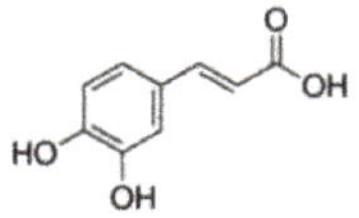
2,3,4-Trihydroxybenzoic acid	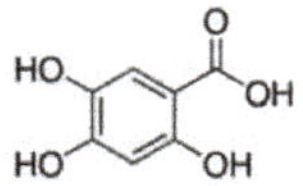	Syringic acid	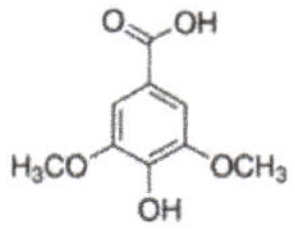
Protocatechualdehyde	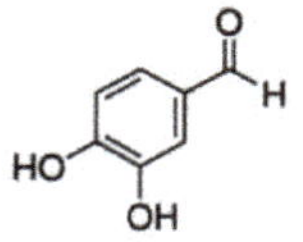	Vanillin	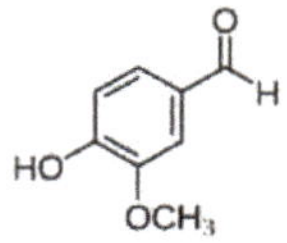
p-Hydroxybenzoic acid	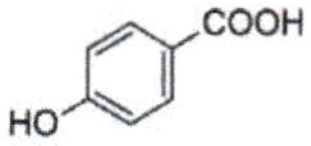	*p*-Coumaric acid	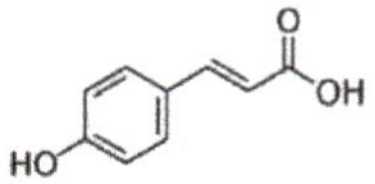
Gentisic acid	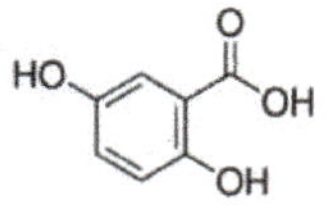	Ferulic acid	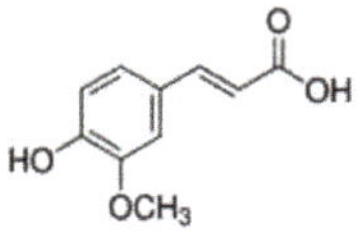
Chlorogenic acid	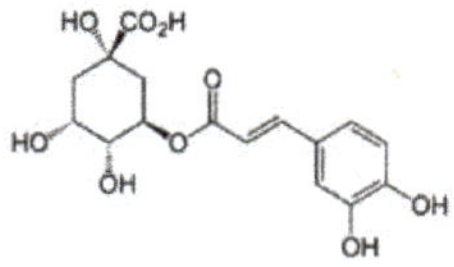	Sinapic acid	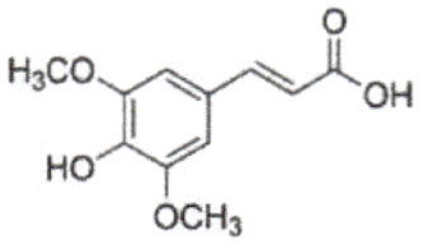

Table 5. Phenolic acids content of 21 chestnut samples.

Region	Gallic Acid	Protocatechuic Acid	2,3,4-Trihydroxybenzoic Acid	Protocatechualdehyde	*p*-Hydroxybenzoic Acid	Gentisic Acid	Chlorogenic Acid	Vanillic Acid + Caffeic Acid	
Huairou, Beijing	19.47 ± 0.61 ^{m}	11.87 ± 0.47 ^{h}	7.46 ± 0.28 ^{c}	1.85 ± 0.14 ^{d}	7.05 ± 0.46 ^{h,i}	11.77 ± 0.53 ^{k}	13.07 ± 0.16 ^{c}	22.64 ± 0.40 ^{h,i}	North China
Jixian, Tianjin	62.61 ± 0.38 ^{a}	21.82 ± 0.25 ^{c}	6.47 ± 0.02 ^{e,f}	1.44 ± 0.07 ^{g}	2.00 ± 0.08 ^{l}	9.16 ± 0.32 ^{m,n}	10.50 ± 0.14 ^{d,e}	5.04 ± 0.07 ^{p}	
Tangshan, Hebei	28.96 ± 1.44 ^{k,l}	11.97 ± 0.62 ^{g,h}	ND	1.38 ± 0.03 ^{g,h}	7.71 ± 0.17 ^{g,h}	8.95 ± 0.57 ^{n}	4.14 ± 0.13 ^{n}	28.93 ± 0.45 ^{f}	
Xingtai, Hebei	40.44 ± 2.67 ^{e}	12.70 ± 1.01 ^{g,h}	6.13 ± 0.02 ^{h}	1.63 ± 0.01 ^{f}	7.92 ± 0.10 ^{g,h}	16.90 ± 1.59 ^{h,i}	8.57 ± 0.30 ^{g}	30.12 ± 0.96 ^{e}	
Qinhuangdao, Hebei	34.43 ± 0.61 ^{h}	19.02 ± 0.06 ^{d}	ND	1.19 ^{j,k}	11.62 ± 0.32 ^{d,e}	9.52 ± 0.33 ^{m,n}	4.42 ± 0.01 ^{n}	20.57 ± 0.47 ^{j}	
Shangluo, Shanxi	26.97 ± 1.04 ^{l}	12.37 ± 0.26 ^{g,h}	6.41 ± 0.01 ^{e,f,g}	2.10 ± 0.04 ^{b}	6.43 ± 0.24 ^{i,j}	15.34 ± 0.71 ^{j}	7.85 ± 0.28 ^{h,i}	24.46 ± 0.13 ^{g}	Northwest China
Mean ± SD	35.48 ± 1.13	14.96 ± 0.45	4.41 ± 0.08	1.60 ± 0.06	7.12 ± 0.23	11.94 ± 0.68	8.09 ± 0.17	21.96 ± 0.41	
Wuxi, Jiangsu	35.56 ± 0.63 ^{f}	17.19 ± 1.32 ^{e,f}	8.00 ± 0.25 ^{b}	1.79 ± 0.05 ^{d,e}	5.90 ± 0.54 ^{j}	20.05 ± 0.61 ^{d,e}	7.33 ± 0.36 ^{i,j}	23.68 ± 0.49 ^{g,h}	East China
Wenzhou, Zhejiang	46.00 ± 0.84 ^{d}	19.31 ± 0.62 ^{d}	5.71 ^{k}	2.30 ± 0.01 ^{a}	8.14 ± 0.54 ^{g}	18.96 ± 0.43 ^{e,f}	6.31 ± 0.111 ^{m}	35.48 ± 0.86 ^{b,c}	
Anqing, Anhui	40.39 ± 0.21 ^{e}	26.48 ± 0.70 ^{b}	6.32 ± 0.02 ^{g}	1.74 ± 0.02 ^{e}	12.70 ± 0.58 ^{b,c}	21.50 ± 0.81 ^{c}	22.86 ± 0.10 ^{a}	37.98 ± 0.53 ^{a}	
Nanping, Fujian	39.64 ± 1.26 ^{e,f}	8.94 ± 0.30 ^{i}	6.35 ± 0.04 ^{f,g}	1.95 ± 0.07 ^{c}	10.31 ± 0.40 ^{f}	11.05 ± 0.23 ^{k,l}	6.79 ± 0.18 ^{k,l}	30.68 ± 1.36 ^{e}	
Fuzhou, Jiangxi	55.08 ± 0.57 ^{b}	51.78 ± 2.19 ^{a}	5.83 ± 0.05 ^{j,k}	2.21 ± 0.10 ^{a}	27.01 ± 1.31 ^{a}	35.15 ± 0.13 ^{a}	10.23 ± 0.83 ^{e}	36.57 ± 1.20 ^{b}	
Taian, Shandong	39.98 ± 1.94 ^{e}	7.17 ± 0.25 ^{j,k}	5.97 ± 0.03 ^{i,j}	0.60 ± 0.03 ^{m}	5.57 ± 0.34 ^{j}	9.72 ± 0.51 ^{m,n}	6.17 ± 0.27 ^{m}	22.81 ± 1.01 ^{h,i}	
Linyi, Shandong	35.18 ± 0.41 ^{h}	16.15 ± 0.35 ^{f}	ND	1.97 ± 0.01 ^{c}	8.64 ± 0.17 ^{g}	18.63 ± 0.56 ^{f,g}	7.37 ± 0.36 ^{i,j}	33.10 ± 0.42 ^{d}	
Mean ± SD	41.69 ± 0.84	21.00 ± 0.82	5.45 ± 0.07	1.80 ± 0.04	11.18 ± 0.55	19.30 ± 0.47	9.58 ± 0.32	31.47 ± 0.84	
Xinyang, Henan	48.42 ± 3.14 ^{c}	5.97 ± 0.14 ^{k}	ND	1.12 ± 0.04 ^{k}	ND	17.58 ± 0.80 ^{g,h}	7.05 ± 0.22 ^{j,k}	17.14 ± 0.60 ^{m}	Central China
Xiangyang, Hubei	37.51 ± 1.55 ^{f}	13.40 ± 0.45 ^{g}	6.55 ± 0.05 ^{e}	1.25 ± 0.05 ^{i,j}	11.09 ± 0.56 ^{e,f}	11.45 ± 1.14 ^{k}	6.33 ± 0.31 ^{l,m}	18.49 ± 0.60 ^{l}	
Mean ± SD	42.97 ± 2.34	9.68 ± 0.30	3.28 ± 0.05	1.19 ± 0.05	5.55 ± 0.56	14.52 ± 0.97	6.69 ± 0.26	17.82 ± 0.60	
Yangjiang, Guangdong	39.12 ± 0.93 ^{e,f}	8.35 ± 0.12 ^{i,j}	6.10 ± 0.05 ^{h,i}	1.56 ± 0.06 ^{f}	3.62 ± 0.18 ^{k}	27.61 ± 0.89 ^{b}	7.31 ± 0.28 ^{i,j,k}	22.24 ± 0.62 ^{i}	South China
Guilin, Guangxi	31.97 ± 0.37 ^{i,j}	16.95 ± 1.66 ^{e,f}	7.26 ± 0.13 ^{d}	0.91 ± 0.01 ^{l}	13.57 ± 1.31 ^{b}	11.78 ± 0.26 ^{k}	10.81 ± 0.07 ^{d}	15.88 ± 0.47 ^{n}	
Liuzhou, Guangxi	29.19 ± 2.15 ^{k,l}	18.01 ± 1.73 ^{d,e}	ND	1.31 ± 0.09 ^{h,i}	7.94 ± 0.78 ^{g,h}	18.34 ± 0.85 ^{f,g}	8.30 ± 0.53 ^{g,h}	13.98 ± 0.91 ^{o}	
Haikou, Hainan	30.24 ± 0.60 ^{j,k}	16.87 ± 0.45 ^{e,f}	9.45 ± 0.08 ^{a}	2.31 ± 0.03 ^{a}	12.17 ± 0.34 ^{c,d}	10.12 ± 0.55 ^{l,m}	16.64 ± 0.50 ^{b}	18.92 ± 0.38 ^{k,l}	
Mean ± SD	32.63 ± 1.01	15.05 ± 0.99	5.70 ± 0.09	1.52 ± 0.05	9.33 ± 0.65	16.96 ± 0.64	10.77 ± 0.34	17.76 ± 0.59	
Zhaotong, Yunnan	33.99 ± 1.10h ^{i}	11.68 ± 0.38 ^{h}	5.99 ± 0.04 ^{h,i}	2.22 ± 0.03 ^{a}	6.15 ± 0.26 ^{i,j}	20.36 ± 0.31 ^{d}	9.43 ± 0.35 ^{f}	34.85 ± 0.74 ^{c}	Southwest China
Kunming, Yunnan	39.21 ± 1.08 ^{e,f}	15.74 ± 0.64 ^{f}	ND	1.60 ± 0.06 ^{f}	10.18 ± 0.74 ^{f}	16.36 ± 0.27 ^{i,j}	3.40 ± 0.08 ^{o}	19.94 ± 0.52 ^{j,k}	
Mean ± SD	36.60 ± 1.09	13.71 ± 0.51	3.00 ± 0.04	1.91 ± 0.04	8.16 ± 0.50	18.36 ± 0.29	6.42 ± 0.21	27.40 ± 0.63	

Region	Syringic Acid	Vanillin	*p*-Coumaric Acid + Syringaldehyde	Ferulic Acid	Sinapic Acid	Total	
Huairou, Beijing	2.80 ± 0.25 ^{n,o}	60.11 ± 4.26 ^{d}	1.35 ± 0.05 ^{n}	3.75 ± 0.18 ^{o}	1.25 ^{o}	164.43 ± 7.81	North China
Jixian, Tianjin	3.18 ± 0.07 ^{m,n}	67.07 ± 0.97 ^{b,c}	2.07 ± 0.03 ^{l}	11.50 ± 0.18 ^{b}	3.57 ± 0.02 ^{a}	206.44 ± 2.60	
Tangshan, Hebei	3.52 ± 0.07 ^{m}	52.99 ± 2.13 ^{e}	1.63 ± 0.02 ^{m}	2.84 ± 0.13 ^{q}	1.64 ± 0.02 ^{l}	159.30 ± 5.79	
Xingtai, Hebei	10.05 ± 0.51 ^{a}	59.90 ± 0.93 ^{d}	3.18 ± 0.08 ^{g,h}	7.78 ± 0.12 ^{f}	2.76 ± 0.02 ^{e}	208.07 ± 8.31	
Qinhuangdao, Hebei	5.51 ± 0.24 ^{f,g,h}	63.45 ± 1.79 ^{c,d}	2.42 ± 0.03 ^{j,k}	4.50 ± 0.07 ^{n}	1.39 ± 0.03 ^{m,n}	182.67 ± 3.96	
Shangluo, Shanxi	5.74 ± 0.29 ^{e,f,g}	65.28 ± 2.10 ^{c}	2.33 ± 0.06 ^{k}	4.90 ± 0.26 ^{l,m}	2.03 ± 0.06 ^{h}	182.21 ± 5.48	Northwest China
Mean ± SD	5.13 ± 0.24	61.47 ± 2.03	2.16 ± 0.04	5.88 ± 0.15	2.11 ± 0.03	183.85 ± 5.66	
Wuxi, Jiangsu	4.80 ± 0.21 ^{j,k}	44.26 ± 4.13 ^{h}	1.19 ± 0.04 ^{n}	3.40 ± 0.13 ^{p}	1.07 ± 0.03 ^{p}	174.22 ± 8.78	East China
Wenzhou, Zhejiang	5.37 ± 0.29 ^{f,g,h,i}	50.51 ± 0.88 ^{e,f}	5.76 ± 0.02 ^{a}	9.36 ± 0.19 ^{c}	2.50 ± 0.06 ^{f}	215.72 ± 4.86	
Anqing, Anhui	7.25 ± 0.56 ^{d}	49.50 ± 0.64 ^{e,f}	4.42 ± 0.07 ^{e}	5.40 ± 0.13 ^{j,k}	1.33 ^{n,o}	237.67 ± 4.38	
Nanping, Fujian	8.91 ± 0.19 ^{b}	65.12 ± 2.02 ^{c}	3.32 ± 0.14 ^{g}	7.35 ± 0.27 ^{g}	3.44 ± 0.13 ^{b}	203.86 ± 6.59	
Fuzhou, Jiangxi	7.86 ± 0.65 ^{c}	70.72 ± 1.34 ^{b}	3.58 ± 0.04 ^{f}	6.90 ± 0.24 ^{h}	1.96 ± 0.06 ^{h,i}	314.87 ± 8.71	
Taian, Shandong	5.20 ± 0.27 ^{h,i,j}	26.46 ± 1.03 ^{j}	3.63 ± 0.17 ^{f}	9.15 ± 0.44 ^{c,d}	1.85 ± 0.08 ^{i,j}	144.28 ± 6.36	
Linyi, Shandong	6.05 ± 0.26 ^{e}	45.08 ± 0.76 ^{g,h}	3.09 ± 0.08 ^{h,i}	6.23 ± 0.16 ^{i}	2.85 ± 0.07 ^{e}	188.95 ± 3.62	
Mean ± SD	6.49 ± 0.35	50.24 ± 1.54	3.54 ± 0.08	6.83 ± 0.22	2.14 ± 0.06	211.37 ± 6.19	

Table 5. *Cont.*

Region	Gallic Acid	Protocatechuic Acid	2,3,4-Trihydroxybenzoic Acid	Protocatechualdehyde	*p*-Hydroxybenzoic Acid	Gentisic Acid	Chlorogenic Acid	Vanillic Acid + Caffeic Acid	
Xinyang, Henan	4.23 ± 0.43 ^{l}	21.23 ± 1.13 ^{k}		5.73 ± 0.34 ^{a}	5.71 ± 0.26 ^{j}	3.55 ± 0.16 ^{a,b}		142.35 ± 7.27	Central China
Xiangyang, Hubei	5.84 ± 0.38 ^{e,f}	67.00 ± 3.84 ^{c}		4.98 ± 0.24 ^{c}	8.38 ± 0.25 ^{e}	3.00 ± 0.16 ^{d}		195.27 ± 9.58	
Mean ± SD	5.03 ± 0.40	44.12 ± 2.49		5.36 ± 0.29	7.04 ± 0.26	3.28 ± 0.16		168.81 ± 8.43	
Yangjiang, Guangdong	5.23 ± 0.22 ^{g,h,i,j}	35.88 ± 0.76 ^{i}		2.41 ± 0.06 ^{j,k}	5.22 ± 0.08 ^{k,l}	1.73 ± 0.04 ^{k,l}		166.37 ± 4.30	South China
Guilin, Guangxi	4.61 ± 0.21 ^{kl,}	61.02 ± 4.11 ^{d}		1.69 ± 0.02 ^{m}	9.06 ± 0.27 ^{c,d}	1.26 ± 0.03 ^{o}		186.78 ± 8.94	
Liuzhou, Guangxi	2.43 ± 0.09 ^{o}	47.95 ± 2.67 ^{f,g}		2.91 ± 0.08 ^{i}	4.82 ± 0.21 ^{m,n}	1.48 ± 0.05 ^{m}		161.28 ± 10.13	
Haikou, Hainan	6.78 ± 0.25 ^{d}	99.33 ± 1.50 ^{a}		2.57 ± 0.06 ^{j}	13.27 ± 0.25 ^{a}	1.76 ± 0.05 ^{j,k}		240.43 ± 5.04	
Mean ± SD	4.76 ± 0.19	61.05 ± 2.26		2.40 ± 0.05	8.09 ± 0.20	1.56 ± 0.04		188.72 ± 7.10	
Zhaotong, Yunnan	4.87 ± 0.40 ^{i,j,k}	52.08 ± 1.59 ^{e}		4.77 ± 0.11 ^{d}	8.84 ± 0.09 ^{d}	3.16 ± 0.03 ^{c}		198.37 ± 5.42	Southwest China
Kunming, Yunnan	3.31 ± 0.15 ^{m,n}	27.31 ± 1.23 ^{j}		5.36 ± 0.16 ^{b}	4.77 ± 0.07 ^{m,n}	2.30 ± 0.03 ^{g}		154.11 ± 5.03	
Mean ± SD	4.09 ± 0.28	39.70 ± 1.41		5.06 ± 0.13	6.80 ± 0.08	2.73 ± 0.03		176.24 ± 5.23	

Data were expressed as mean ± standard deviation (n = 3). Data in the same column marked with different superscript letters ($^{a-p}$) differed significantly ($p \leq 0.05$).

Overall, the highest total phenolic content (314.87 μg/g) was observed in samples collected from Fuzhou, Jiangxi. Among five geographical regions, samples collected from East China were shown to have the highest mean value of total phenolic acids (211.37 μg/g), followed by South China (188.72 μg/g) and North China (183.85 μg/g).

In terms of individual phenolic compounds, vanillin was the most abundant phenolic compound and the contents were ranged from 21.23 μg/g in the sample collected from Xinyang, Henan to 99.33 μg/g in the sample collected from Haikou, Hainan. Overall, samples from North China contained the highest mean value of vanillin among all five different regions, whereas samples from Central China contained the lowest amount. Besides, although high amount of vanillin was found in North China region, no significant differences ($p < 0.05$) were found in chestnut samples between different cities. However, significant differences of vanillin content were found in samples collected from cities in South China.

Gallic acid was found to be the second dominant phenolic compound in chestnut samples. Contents of gallic acid were ranged from 19.47 μg/g in the sample collected from Huairou, Beijing to 62.61 μg/g in the sample collected from Jixian, Tianjin. Overall, chestnut samples from Central China contained relatively higher mean value of gallic acid than the samples from other regions. However, chestnut samples collected from South China contained the minimum mean value of gallic acid.

With regard to gentisic acid, levels ranged from 8.95 μg/g in samples collected from Tangshan, Hebei to 35.15 μg/g in samples collected from Fuzhou, Jiangxi. The highest mean value of gentisic acid was found in East China, and samples collected from Fuzhou, Jiangxi were shown as much higher value of gentisic acid (35.15 μg/g) than other cities in the group of East China. The similar tendency was also observed in vanillic acid and caffeic acid. In terms of 2,3,4-trihydroxybenzoic acid, contents were ranged from 5.71 μg/g in samples collected from Wenzhou, Zhejiang to 9.45 μg/g in samples collected from Haikou, Hainan. No results were detected in samples collected from several cities, including Tangshan and Qinhuangdao (North China), Linyi (East China), Xinyang (Central China), Liuzhou (South China), and Kunming (Southwest China).

Contents of protocatechuic acid were ranged from 5.97 μg/g in the sample collected from Xinyang, Henan to 51.78 μg/g in the sample collected from Fuzhou, Jiangxi. Overall, samples collected from East China contained highest average value of protocatechuic acid while that from Central China contained the lowest value. The similar tendency was also observed in *p*-hydroxybenzoic acid. The content of *p*-hydroxybenzoic acid were ranged from 2 μg/g in the samples collected from Jixian, Tianjin to 27.01 μg/g in the samples collected from Fuzhou, Jiangxi. For chlorogenic acid, contents were ranged from 3.40 μg/g in samples collected from Kunming, Yunnan to 22.86 μg/g in the samples collected from Anqing, Anhui.

In terms of protocatechualdehyde, syringic acid, *p*-coumaric acid and syringaldehyde, ferulic acid, and sinapic acid, all detected contents were pretty low (typically below 10 μg/g). The highest level of protocatechualdehyde and ferulic acids were found in Southern regions. However, more synaptic acid and *p*-coumaric acid were observed in Central China. The content of syringic acid ranged from 2.43 μg/g to 10.05 μg/g, and was observed most in East China. Overall, contents of different phenolic compounds were significantly different ($p < 0.05$) from different regions.

4. Discussion

4.1. Phenolic and Antioxidant Properties among 21 Raw Chestnut Samples

This study has revealed that all the chestnut samples under investigation are rich sources of TPC, TFC, and CTC. These samples have also presented high antioxidant activities as assessed by DPPH, FRAP, and ABTS assay. It has been well reported in the literature that phenolic compounds prevent cardiovascular diseases, cataract development, oxidative injury caused by heat stress, lower the incidence of influenza infection, reduce fat absorption, and enhance energy expenditure [9,19]. Overall, significant differences ($p < 0.05$) were observed among the phenolic profiles and antioxidant activities of all chestnut samples procured from five different geographic regions of China. As mentioned in

Table 3, raw chestnut samples from in South China presented relatively higher values of TPC, CTC, and antioxidant activities in terms of DPPH, FRAP, and ABTS values compared to samples collected from other regions. However, the samples from Southwest China presented relatively higher TFC values but contain lower level of TPC, DPPH, FRAP, and ABTS values compared to the samples from other geographical regions.

In particular, samples from Fuzhou (East China) exhibit higher values corresponding to TPC (2.35 mg GAE/g), DPPH (16.74 μmol TE/g), FRAP (3.20 mmol FE/100 g), and ABTS (24.83 μmol TE/g) assays. Previously, a study has reported 2.84 mg GAE/g of TPC in chestnut fruits collected from Tenerife, Spain (C. sativa Mill) [20] Whereas, Otles and Selek [21] mentioned even higher TPC values from 5.00 to 32.82 mg GAE/g in Turkish chestnuts (*C. sativa* Mill). The relatively lower phenolic contents of chestnuts investigated in this study may be attributed to the certain degree of oxidation of raw chestnuts peeled by chestnut peeler. Earlier, chestnuts from Italy reported to exhibit lower values for ABTS (4.77 to 8.15 μmol TE/g) compared to the present study [22]. The samples from Kunming (Southwest China) presented the highest values for TFC (1.13 mg CAE/g). However, earlier 2.62 mg CAE/g of TFC was reported in Spanish chestnuts (*C. sativa* Mill) [6].

Furthermore, in spite of differences among the phenolic profile of samples from different geographic regions, inter-provincial and intra-provincial disparities among the TPC, TFC, CTC, and antioxidant values of all the chestnut samples were also observed. The samples collected from Qinhuangdao, Hebei presented relatively higher values for phenolics and antioxidant capacities compared to the samples procured from Tangshan and Xingtai cities of Hebei Province. However, samples from Tangshan and Xingtai of Hebei Province presented less difference among the TPC, TFC, CTC, DPPH and ABTS values except for the FRAP value (2.13 mmol FE/100 g in case of samples from Tangshan and 1.40 mmol FE/100 g in case of samples from Xingtai). A similar phenomenon was also presented by the samples collected from South China. The samples from Guilin and Liuzhou of Guangxi Province exhibit comparatively less difference in TPC, TFC, DPPH, and FRAP values. However, as shown in Table 3, a considerable difference was observed in CTC and ABTS values of samples collected from these two different regions of Guangxi Province. Nevertheless, the samples from Zhaotong and Kunming of Yunnan Province exhibit significant differences among the values of all the assays. The TFC (1.13 mg CAE/g) and CTC (10.69 mg CAE/g) values of samples procured from Kunming were about two times compared to TFC (0.63 mg CAE/g) and CTC (4.57 mg CAE/g) values presented by samples from Zhaotong. However, the ABTS value of samples from Kumming was observed to be approximately three times higher than the ABTS value presented by samples from Zhaotong. The samples procured from Taian and Linyi region of Shandong Province also revealed significant differences among all the values related to phenolics and antioxidant assays. It was also observed that the samples from Taian exhibit significantly higher ABTS values (16.71 μmol TE/g) compared to samples collected from Linyi region (12.55 μmol TE/g).

Based on these findings, it may be concluded that differences among the phenolic profiles and antioxidant activities of chestnut samples significantly depend on the geographical factors, such as temperature and sunlight exposure. The correlation between average temperature and sunlight with phenolic contents and antioxidant capacities were shown in Supplemental Figure S4. Higher average temperature during growing period (27.26 °C in South China) of chestnut helps present higher phenolic contents (in terms of TPC and CTC) and antioxidant activities. This may due to the elevated temperature could facilitate photosynthetic capacity of plants to produce more secondary metabolites, such as phenolic acids [23]. Chestnuts grown in Fuzhou, Jiangxi, although belong to East China (average temperature = 23.72 °C), its geographical location is very near the South China and the average temperature reaches to 26.33 oC during growing period [16]. On the other side, more sunlight duration (213.12 h/month in North China) also helps improve antioxidant activities in terms of DPPH and FRAP. i.e., Chestnut samples from North China (213.12 h/month) exhibited higher average values than Central China (180.84 h/month) and Southwest China (162.02 h/month). In response to high levels of sunlight, plants are able to adapt to the circumstances and release various secondary metabolites including phenolic compounds and triterpenoids, which have well-known antioxidant properties [24,25]. However, TFC value is less affected by these geographical factors. This finding is in agreement with results of the previous study that reported a significant difference in flavonoids and phenolic content of Chinese chestnut collected from various ecological regions [15]. Compared with other chestnut varieties, *Castanea sativa* Mill (European variety) were found to present the highest performance in net photosynthesis with higher temperature (26 °C) in September [23]. This was also in agreement with results from Almeida et al. [26], in which the optimal temperature for the highest rates of net photosynthesis of chestnut (*C. sativa* Mill) were in a range of 31 to 33.5 °C.

4.2. Analysis of Phenolic and Antioxidant Contents Based on Morphological Features

The mean values of TPC, TFC, CTC, and antioxidant activities of three identified sub-groups of chestnut, *Banli*, *Maoli*, and *Youli* are described in Table 6. Overall, the phenolic profile and antioxidant activities exhibit significant differences ($p < 0.05$) among these three sub-groups. It was also observed that *Banli* presented the highest values for TPC, TFC, CTC, and antioxidant capacities as assessed by DPPH, FRAP, and ABTS assays, followed by *Maoli* and *Youli*.

Table 6. Results of phenolic profiles and antioxidant values based on morphological properties.

Mean	TPC (mg GAE/g)	TFC (mg CAE/g)	CTC (mg CAE/g)	DPPH (μmol TE/g)	FRAP (mM FE/100 g)	ABTS (μmol TE/g)
Banli	1.76 ± 0.33 ^{a}	0.82 ± 0.17 ^{a}	9.75 ± 3.05 ^{a}	10.84 ± 1.89 ^{a}	2.04 ± 0.95 ^{a}	15.57 ± 3.63 ^{a}
Maoli	1.61 ± 0.44 ^{a,b}	0.77 ± 0.14 ^{a,b}	7.67 ± 3.31 ^{b}	10.51 ± 3.26 ^{a,b}	1.73 ± 0.94 ^{b}	14.25 ± 6.13 ^{a}
Youli	1.47 ± 0.24 ^{b}	0.70 ± 0.16 ^{b}	6.14 ± 2.69 ^{b}	9.11 ± 1.27 ^{b}	1.47 ± 0.33 ^{c}	11.27 ± 1.25 ^{b}

Values are expressed as the mean of triplicates ± standard deviation. Means in the same column with unlike superscripts ($^{a–c}$) differ significantly ($p < 0.05$).

Specifically, among all *Banli* varieties, chestnuts from Guilin and Liuzhou of Guangxi Province (South China) contributed considerably towards the higher level of phenolics and antioxidant activities. The *Banli* samples collected from Kunming (Southwest China) observed to impose a major impact on the overall high TFC level of *Banli* variety. The *Banli* samples from Xiangyang, Hubei (Central China) presented the highest ABTS value (19.86 μmol TE/g) compared to other samples.

Amongst *Maoli* varieties, samples from Fuzhou, Jiangxi (East China) exhibit a major contribution towards the high mean values of all the assays employed to determine the phenolic profile and antioxidant capacities. On the other hand, chestnuts from Zhaotong, Yunnan (Southwest China) exhibited the lowest values in TPC, TFC, CTC, DPPH, and ABTS assays.

In case of *Youli* varieties, samples from Wenzhou, Zhejiang (East China) presented relatively higher values for TPC, TFC, CTC, DPPH, and ABTS assays. However, *Youli* samples from Xinyang, Henan (Central China) contain relatively lower values for TPC, TFC, CTC and antioxidant activities determined by DPPH, FRAP, and ABTS assay. These results are also in agreement with conclusion that phenolic profiles and antioxidant activities are largely depend on other geographical factors not limited to sample varieties.

4.3. Correlation among Phenolic Contents, Antioxidant Activities, and Color Values

The correlation coefficient (r) between phenolic compounds, antioxidant capacities, and color values has also been established and presented in Table 7. The strong and positive correlations were observed in phenolic profiles of chestnut samples in terms of TPC, TFC, and CTC. The highest correlation coefficient value was found between TPC and CTC ($r = 0.834$, $p < 0.01$), followed by between TPC and TFC ($r = 0.762$, $p < 0.01$) and between TFC and CTC ($r = 0.708$, $p < 0.01$). Additionally, the stronger positive correlation was also exhibited between three antioxidant assays. The highest correlation coefficient (r) was determined as 0.875 ($p < 0.01$) between DPPH and ABTS values, followed by 0.819 ($p < 0.01$) between FRAP and DPPH as well as between FRAP and ABTS values. All the parameters related to phenolic contents and antioxidant capacities have presented a positive linear correlation with each other. The highest correlation value was shown as 0.884 ($p < 0.01$) between TPC and ABTS. A comparatively low and positive correlation was found between TFC and FRAP (r = 0.540, $p < 0.05$). For color values, no significant correlations were found between lightness (L) and other phenolic and antioxidant parameters.

Table 7. Correlation analysis between phenolic contents, antioxidant activities, and color value.

Correlation Coefficient (r)		TPC	TFC	CTC	DPPH	FRAP	ABTS
TPC		-	-	-	-	-	-
TFC		0.762 **	-	-	-	-	-
CTC		0.834 **	0.708 **	-	-	-	-
DPPH		0.821 **	0.575 **	0.782 **	-	-	-
FRAP		0.866 **	0.540 *	0.719 **	0.819 **	-	-
ABTS		0.884 **	0.684 **	0.866 **	0.875 **	0.819 **	-
Color	L	−0.214	−0.114	−0.182	−0.444 *	−0.162	−0.374
	a	0.539 *	0.515 *	0.483 *	0.471 *	0.423	0.539 *
	b	−0.344	−0.502 *	−0.363	−0.180	−0.167	−0.209

(Sample size: $N = 21$, $p < 0.05$ was recorded as *; $p < 0.01$ was recorded as **).

4.4. Analysis of Phenolic Acid Profile Based on Geographic Regions

Overall, chestnut samples from all five regions of China were abundant with phenolic acids. The beneficial function of phenolic acid has been illustrated in this article, including preventing cancer, heart disease, and cardiovascular disease [27]. In all 14 phenolic acids detected in this study, gallic acid and vanillin were two most predominant phenolic acids found in chestnuts, which in accordance with the results from the research conducted by Otles and Selek [21]. However, three phenolic acids, 2,3,4-trihydroxybenzoic acid, protocatechualdehyde and sinapic acid, were found the least values in chestnut samples collected in China.

Based on different geographic areas of China, chestnut samples collected from East China contained the highest total phenolic acids (211.37 μg/g), followed by South China (188.72 μg/g) and North China (183.85 μg/g). Central China contained the fewest overall phenolic acids (168.81 μg/g). Phenolic acids in samples collected from different regions varied significantly, mainly attributed to both geographical factors and some human factors. From the perspective of geographical factors, adequate exposure to sunlight and moderate precipitation contributed to higher value of phenolic acid inside plants [25]. Especially, Fuzhou, Jiangxi (East China) exhibited overwhelmingly high contents of vanillin (70.72 μg/g), gallic acid (55.08 μg/g) and protocatechuic acid (51.78 μg/g). The higher contents of chestnuts collected from Fuzhou may attributed to warmer temperature (26.33 °C) and adequate precipitation (1600 mm) during growing seasons. The results observed by HPLC were also in accordance with the previous colorimetric assays.

Taking a deeper look at the different types of phenolic acids, vanillin was the most abundant one among other phenolic acids observed in chestnuts samples collected from China. The higher contents of vanillin were found in South China (61.05 μg/g) and North China (61.47 μg/g), due to the warmer average temperature (27.26 °C in South China) and sufficient sunlight exposure (213.12 h/month in North China). It is delightful to observe higher amount of vanillin in chestnut samples as vanillin has been proved to possess potent antioxidant capacity [28]. According to Clemens et al. [29], vanillin is also shown to have some beneficial health effects to human, such as inhibiting lipid oxidation, preventing DNA damage from exposure to excessive sunlight, preventing the forming of cancer, etc. Besides, Sanz et al. [30] has proved that toasting will lead to the degradation of lignin, and promote releasing of low-molecular weight phenolic compound, such as vanillin. Therefore, higher contents of vanillin found in chestnut samples (99.33 μg/g) collected from Haikou, Hainan could be explained by sufficient sunlight exposure and higher average temperature during the growing season.

In terms of gallic acid, it is known as having anti-inflammatory, anti-microbial and radical scavenging activities which can be very helpful in treating diseases including cancer, asthma, Alzheimer, and so on according to [31]. In contrast to vanillin, the highest contents of gallic acid were found in Central China (42.97 μg/g) and East China (41.69 μg/g). A reasonable explanation for this can also be found in research conducted by Sanz et al. [30], which illustrated that gallic acid was very sensitive to heat, thus decomposition of gallic acid may occur with higher temperature.

4.5. Analysis of Phenolic Acid Profile Based on Morphological Features

Mean values of 14 phenolic acids of three types of chestnuts based on morphological features were described in Table 8. Overall, the total phenolic acid content among three chestnut varieties were significantly differed ($p < 0.05$), and *Youli* presented the highest value (206.53 μg/g), followed by *Banli* (193.43 μg/g) and *Maoli* (176.78 μg/g).

Specifically, in terms of *Youli* varieties, the extremely high value of total phenolic content (314 μg/g) was observed in chestnut samples collected from Fuzhou, Jiangxi, which becomes the key factor for higher value of this variety. However, there is only two out of five of *Youli* variety chestnut samples exceeded the average total phenolic acid value (206.53 μg/g). The lowest value of phenolic acid among 21 chestnuts was observed in samples collected from Xinyang, Henan (142.35 μg/g).

With regard to *Banli* varieties, five out of ten chestnut samples exceeded the mean value (193.43 μg/g). The major contributors to higher phenolic content of *Banli* variety are samples collected from Haikou, Hainan (240.43 μg/g) and Anqing, Anhui (237.67 μg/g). The lowest total phenolic acid content among *Banli* varieties was observed in Kunming, Yunnan (154.11 μg/g).

In terms of *Maoli* varieties, the mean phenolic acid content was the lowest among three varieties which is mainly due to the sample collected from Taian, Shandong (144.28 μg/g). Based on the above-mentioned findings, it could be seen that no observable relationships between phenolic acid content and morphological features were found.

Table 8. Phenolic acid contents of chestnut based on morphological features.

Variety	Gallic Acid	Protocatechuic Acid	2,3,4-Trihydroxybenzoic Acid	Protocatechualdehyde	*p*-Hydroxybenzoic Acid	Gentisic Acid	Chlorogenic Acid	Vanillic Acid + Caffeic Acid
Youli	41.30 ± 1.34^{a}	20.47 ± 0.78^{a}	5.11 ± 0.10^{b}	1.75 ± 0.07^{a}	10.66 ± 0.72^{a}	18.98 ± 0.61^{a}	$8.60 \pm 0.33^{a,b}$	26.06 ± 0.73^{a}
Banli	38.14 ± 1.08^{ab}	16.59 ± 0.74^{b}	4.27 ± 0.07^{c}	1.71 ± 0.04^{a}	8.55 ± 0.39^{b}	15.11 ± 0.64^{b}	9.59 ± 0.27^{a}	24.24 ± 0.61^{a}
Maoli	34.41 ± 1.00^{b}	12.59 ± 0.45^{c}	5.29 ± 0.05^{a}	1.43 ± 0.03^{b}	7.83 ± 0.44^{b}	15.72 ± 0.50^{b}	7.67 ± 0.21^{b}	23.47 ± 0.57^{a}

Variety	Syringic Acid	Vanillin	*p*-Coumaric Acid + Syringaldehyde	Ferulic Acid	Sinapic Acid	Total Phenolic Content
Youli	5.22 ± 0.40^{a}	53.91 ± 2.29^{a}	4.28 ± 0.14^{a}	6.82 ± 0.22^{a}	2.45 ± 0.09^{a}	206.53 ± 7.65^{a}
Banli	5.63 ± 0.24^{a}	55.85 ± 1.70^{a}	2.95 ± 0.08^{b}	6.73 ± 0.17^{a}	2.22 ± 0.04^{a}	$193.43 \pm 6.03^{a,b}$
Maoli	5.19 ± 0.27^{a}	50.69 ± 1.90^{a}	2.87 ± 0.07^{b}	6.95 ± 0.20^{a}	1.90 ± 0.04^{b}	176.78 ± 5.74^{b}

Values are expressed as the mean ± standard deviation. Means in the same column with unlike superscripts ($^{a\text{–}c}$) differ significantly ($p < 0.05$).

5. Conclusions

The phenolic profile and antioxidant activity of chestnut from five different geographical areas of China have been explored in this study. All the samples from different regions and varieties exhibit significant difference ($p < 0.05$) in TPC, TFC, CTC, DPPH, FRAP, and ABTS values. It was observed that the chestnut samples from Fuzhou, Jiangxi (East China) exhibited the higher level of TPC (2.35 mg GAE/g) and CTC (13.52 mg CAE/g) and antioxidant activity among all the chestnut samples, and also exhibited the highest total phenolic acid content (314.87 µg/g). However, the samples collected from Kunming, Yunnan (Southwest China) presented the highest level of TFC (1.13 mg CAE/g). Among the five geographical regions, samples from South China revealed maximum mean values for TPC (1.89 mg GAE/g), CTC (9.41 mg CAE/g), DPPH (11.76 µmol TE/g), and FRAP (2.37 mmol FE/100 g). Whereas the samples from Southwest China exhibit minimum mean values for TPC (1.41 mg GAE/g), DPPH (8.30 µmol TE/g), FRAP (0.99 mmol FE/100 g) and ABTS (11.52 µmol TE/g). Among 14 free phenolic compounds, vanillin and gallic acid were found to be most abundant. The content of vanillin is more in warmer regions because high temperature may lead to decomposition of lignin and release more phenolic compounds. Higher temperature and more sunlight exposure during growing period of chestnuts help to improve phenolic profiles and antioxidant activities of chestnut samples. Among three varieties of chestnut, *Banli* presented higher mean values for TPC, TFC, CTC, and antioxidant capacities, followed by *Maoli* and *Youli*. However, no observable relationships between phenolic acid content and morphological features were found. Overall, chestnuts samples exhibit a considerable number of phenolic compounds and potent antioxidant activities. The significant variations in phenolic compounds and antioxidant activity were observed based on the geographical regions and varieties of chestnuts. The findings of this study will have a major importance for the consumers, food scientists, plant breeders and commercial chestnut growers for the better selection of specific chestnut variety from a particular geographical region for maximum health benefits, production of functional food, developing high-value chestnut varieties and selection of geographical site for further cultivation of chestnut plants. In future study, the effect on the thermal processing of chestnuts from different geographical areas will be further investigated.

Supplementary Materials: The following are available online at http://www.mdpi.com/2076-3921/9/3/190/s1, Table S1: The maximum detection wavelengths, retention time, regressive equations and correlation coefficient of 14 phenolic acids, Figure S1: Chestnut sampling regions in China (marked by stars), Figure S2: Twenty one raw chestnut samples investigated in this study. Figure S3: Chromatogram of 14 phenolic acids. Figure S4: Correlation coefficient (r) between temperature and sunlight with phenolic contents and antioxidant activities.

Author Contributions: Conceptualization, B.X.; methodology, B.X.; Z.X.; P.C.; software, Z.X.; P.C.; validation, Z.X.; P.C.; formal analysis, Z.X. and P.C.; investigation, Z.X.; M.M.; P.C.; resources, B.X.; data curation, Z.X.; M.M.; P.C.; writing—original draft preparation, Z.X. and M.M.; writing—review and editing, Z.X.; M.M.; B.X.; visualization, B.X.; supervision, B.X.; project administration, B.X.; funding acquisition, B.X. All authors have read and agreed to the published version of the manuscript.

Funding: The work was jointly supported by two grants R201714 and R201914 from Beijing Normal University-Hong Kong Baptist University United International College, Zhuhai, Guangdong, China.

Conflicts of Interest: The authors declare no conflict of interest.

References

1. FAO. Food and Agricultural Organization of United Nations. FAOSTAT. Available online: http://www.fao.org/faostat/en/#data/QC (accessed on 23 March 2019).
2. Tan, Z.L.; Wu, M.C.; Wang, Q.Z.; Wang, C. Effect of calcium chloride on chestnut. *J. Food Process. Preserv.* **2007**, *31*, 298–307. [CrossRef]
3. Hunt, K.; Gold, M.; Reid, W.; Warmund, M. Growing Chinese chestnuts in Missouri. In *Agroforestry in Action*; AF1007-2009; University of Missouri Center for Agroforestry: Columbia, MO, USA, 2009; pp. 1–16.
4. Borges, O.P.; Carvalho, J.S.; Correia, P.R.; Silva, A.P. Lipid and fatty acid profiles of *Castanea sativa* Mill. Chestnuts of 17 native Portuguese cultivars. *J. Food Compos. Anal.* **2007**, *20*, 80–89. [CrossRef]

5. Yang, Y.X.; Pan, X.C.; Wang, G.Y. *China Food Composition*, 2nd ed.; Peking University Medical Press: Beijing, China, 2009; p. 80.
6. Li, Q.; Shi, X.; Zhao, Q.; Cui, Y.; Ouyang, J.; Xu, F. Effect of cooking methods on nutritional quality and volatile compounds of Chinese chestnut (*Castanea mollissima* Blume). *Food Chem.* **2016**, *201*, 80–86. [CrossRef] [PubMed]
7. De Vasconcelos, M.C.B.M.; Bennett, R.N.; Rosa, E.A.; Ferreira-Cardoso, J.V. Composition of European chestnut (*Castanea sativa* Mill.) and association with health effects: Fresh and processed products. *J. Sci. Food Agric.* **2010**, *90*, 1578–1589. [CrossRef] [PubMed]
8. Carlsen, M.H.; Halvorsen, B.L.; Holte, K.; Bohn, S.K.; Dragland, S.; Sampson, L.; Willey, C.; Senoo, H.; Umezono, Y.; Sanada, C.; et al. The total antioxidant content of more than 3100 foods; beverages; spices; herbs and supplements used worldwide. *Nutr. J.* **2010**, *9*, 3. [CrossRef] [PubMed]
9. Antonio, A.L.; Fernandes, A.; Barreira, J.C.M.; Bento, A.; Botelho, M.L.; Ferreira, I.C.F.R. Influence of gamma irradiation in the antioxidant potential of chestnuts (*Castanea sativa* Mill.) fruits and skins. *Food Chem. Toxicol.* **2011**, *49*, 1918–1923. [CrossRef]
10. Jiang, P.; Xu, G.; Liu, D.; Chen, J.; Ye, X. Determination of phenolic acid in 15 citrus peels. *Food Ferment. Ind.* **2008**, *6*, 142–146.
11. Luo, F.L. Present situation of chestnut products processing in China. *Food Mach.* **2004**, *1*, 3–8.
12. Barros, A.I.; Nunes, F.M.; Gonçalves, B.; Bennett, R.N.; Silva, A.P. Effect of cooking on total vitamin C contents and antioxidant activity of sweet chestnuts (*Castanea sativa* Mill.). *Food Chem.* **2011**, *128*, 165–172. [CrossRef]
13. Zhu, F. Effect of Processing on Quality Attributes of Chestnut. *Food Bioprocess Technol.* **2016**, *9*, 1429–1443. [CrossRef]
14. Zhang, F.; Liu, F.; Abbasi, A.; Chang, X.; Guo, X. Effect of steaming processing on phenolic profiles and cellular antioxidant activities of *Castanea mollissima*. *Molecules* **2019**, *24*, 703. [CrossRef] [PubMed]
15. Yang, F.; Liu, Q.; Pan, S.; Xu, C.; Xiong, Y.L. Chemical composition and quality traits of Chinese chestnuts (*Castanea mollissima*) produced in different ecological regions. *Food Biosci.* **2015**, *11*, 33–42. [CrossRef]
16. National Bureau of Statistics. Monthly average temperature of major cities 2016. In *China Statistical Yearbook 2017*; Sheng, L., Ed.; China Statistic Press: Beijing, China, 2017; p. 8.
17. Xu, B.J.; Chang, S.K.C. A comparative study on phenolic profiles and antioxidant activities of legμmes as affected by extraction solvents. *J. Food Sci.* **2007**, *72*, 159–166. [CrossRef] [PubMed]
18. Xu, B.J.; Chang, S.K. Total phenolics, phenolic acids, isoflavones, and anthocyanins and antioxidant properties of yellow and black soybeans as affected by thermal processing. *J. Agric. Food Chem.* **2008**, *56*, 7165–7175. [CrossRef]
19. Meenu, M.; Kamboj, U.; Sharma, A.; Guha, P.; Mishra, S. Green method for determination of phenolic compounds in mung bean (*Vigna radiata* L.) based on near-infrared spectroscopy and chemometrics. *Int. J. Food Sci. Technol.* **2016**, *51*, 2520–2527. [CrossRef]
20. Suárez, M.H.; Galdón, B.R.; Mesa, D.R.; Romero, C.D.; Rodríguez, E.R. Sugars, organic acids and total phenols in varieties of chestnut fruits from Tenerife (Spain). *Food Nutr. Sci.* **2012**, *3*, 705–715. [CrossRef]
21. Otles, S.; Selek, I. Phenolic compounds and antioxidant activities of chestnut (*Castanea sativa Mill*) fruits. *Qual. Assur. Saf. Crop.* **2012**, *4*, 199–205. [CrossRef]
22. Neri, L.; Dimitri, G.; Sacchetti, G. Chemical composition and antioxidant activity of cured chestnuts from three sweet chestnut (*Castanea sativa* Mill.) ecotypes from Italy. *J. Food Compos. Anal.* **2010**, *23*, 23–29. [CrossRef]
23. Gomes-Laranjo, J.; Peixoto, F.; Sang, H.W.W.F.; Torres-Pereira, J. Study of the temperature effect in three chestnut (*Castanea sativa* Mill.) cultivars' behaviour. *J. Plant. Physiol.* **2006**, *163*, 945–955. [CrossRef]
24. Boardman, N.K. Comparative photosynthesis of sun and shade plants. *Ann. Rev. Plant Physiol.* **1977**, *28*, 355–377. [CrossRef]
25. Yang, L.; Wen, K.; Ruan, X.; Zhao, Y.; Wei, F.; Wang, Q. Response of plant secondary metabolites to envionmental factors. *Molecules* **2018**, *23*, 762. [CrossRef] [PubMed]
26. Almeida, P.; Dinis, L.; Coutinho, J.; Pinto, T.; Anjos, R.; Ferreira-Cardoso, J.; Pimentel-Pereira, M.; Peixoto, F.; Gomes-Laranjo, J. Effect of temperature and radiation on photosynthesis productivity in chestnut populations (*Castanea sativa* Mill. Cv. Judia). *Acta Agron. Hung.* **2007**, *55*, 193–203. [CrossRef]
27. Liu, Q.; Dong, H.; Liu, P.; Zhang, L.; Wang, F. Study on photosynthetic characteristics of Chinese chestnut. *J. Fruits Sci.* **2005**, *4*, 34–37.

28. Burri, J.; Graf, M.; Lambelet, P.; Loliger, J. Vanillin: More than a flavouring agent—A potent antioxidant. *J. Sci. Food Agric* **1989**, *48*, 49–56. [CrossRef]
29. Clemens, R. Vanillin: More than just a favorite flavoring. *Food Technol.* **2015**, *69*, 20.
30. Sanz, M.; Cadahía, E.; Esteruelas, E.; Muñoz, A.; Fernández de Simón, B.; Hernández, T.; Estrella, I. Phenolic compounds in chestnut (*Castanea sativa* Mill.) heartwood effect of toasting at cooperage. *J. Agric. Food Chem.* **2010**, *58*, 9631–9640. [CrossRef]
31. Kim, S.H.; Jun, C.D.; Suk, K.; Choi, B.J.; Lim, H.; Park, S.; Lee, S.H.; Shin, H.Y.; Kim, D.K.; Shin, T.Y. Gallic acid inhibits histamine release and pro-inflammatory cytokine production in mast cells. *Toxicol. Sci.* **2006**, *91*, 123–131. [CrossRef]

Article

The Health-Promoting Potential of *Salix* spp. Bark Polar Extracts: Key Insights on Phenolic Composition and In Vitro Bioactivity and Biocompatibility

Patrícia A. B. Ramos [1,2], Catarina Moreirinha [1], Sara Silva [3], Eduardo M. Costa [3], Mariana Veiga [3], Ezequiel Coscueta [3], Sónia A. O. Santos [1], Adelaide Almeida [4], M. Manuela Pintado [3], Carmen S. R. Freire [1], Artur M. S. Silva [1,2] and Armando J. D. Silvestre [1,*]

1 CICECO—Aveiro Institute of Materials, Department of Chemistry, University of Aveiro, 3810-193 Aveiro, Portugal; patriciaaramos@ua.pt (P.A.B.R.); catarina.fm@ua.pt (C.M.); santos.sonia@ua.pt (S.A.O.S.); cfreire@ua.pt (C.S.R.F.); artur.silva@ua.pt (A.M.S.S.)

2 QOPNA & LAQV-REQUIMTE, Department of Chemistry, University of Aveiro, 3810-193 Aveiro, Portugal

3 CBQF—Centro de Biotecnologia e Química Fina—Laboratório Associado, Escola Superior de Biotecnologia, Universidade Católica Portuguesa, Rua de Diogo Botelho 1327, 4169-005 Porto, Portugal; snsilva@porto.ucp.pt (S.S.); emcosta@porto.ucp.pt (E.M.C.); mveiga@porto.ucp.pt (M.V.); ecoscueta@gmail.com (E.C.); mpintado@porto.ucp.pt (M.M.P.)

4 Biology Department and CESAM—Centre for Environmental and Marine Studies, University of Aveiro, 3810-193 Aveiro, Portugal; aalmeida@ua.pt

* Correspondence: armsil@ua.pt; Tel.: +351-234-370-711

Received: 27 September 2019; Accepted: 26 November 2019; Published: 30 November 2019

Abstract: *Salix* spp. have been exploited for energy generation, along with folk medicine use of bark extracts for antipyretic and analgesic benefits. Bark phenolic components, rather than salicin, have demonstrated interesting bioactivities, which may ensure the sustainable bioprospection of *Salix* bark. Therefore, this study highlights the detailed phenolic characterization, as well as the in vitro antioxidant, anti-hypertensive, *Staphylococcus aureus* growth inhibitory effects, and biocompatibility of *Salix atrocinerea* Brot., *Salix fragilis* L., and *Salix viminalis* L. bark polar extracts. Fifteen phenolic compounds were characterized by ultra-high-performance liquid chromatography-ultraviolet detection-mass spectrometry analysis, from which two flavan-3-ols, an acetophenone, five flavanones, and a flavonol were detected, for the first time, as their bark components. *Salix* bark extracts demonstrated strong free radical scavenging activity (5.58–23.62 μg mL^{-1} IC_{50} range), effective inhibition on angiotensin-I converting enzyme (58–84%), and *S. aureus* bactericidal action at 1250–2500 μg mL^{-1} (6–8 log CFU mL^{-1} reduction range). All tested *Salix* bark extracts did not show cytotoxic potential against Caco-2 cells, as well as *S. atrocinerea* Brot. and *S. fragilis* L. extracts at 625 and 1250 μg mL^{-1} against HaCaT and L929 cells. These valuable findings can pave innovative and safer food, nutraceutical, and/or cosmetic applications of *Salix* bark phenolic-containing fractions.

Keywords: *Salix* spp. bark polar extracts; phenolic compounds; antioxidant activity; anti-hypertensive potential; antibacterial effect; bioeconomy-based value chain

1. Introduction

Presently, the population's growing rate, the climate change, and the ecosystem degradation have aroused society's awareness and political decisions for the utmost importance to consume and produce chemicals, energy, and materials in a more ecological and sustainable way. The European Commission launched in 2012 the bioeconomy strategy for addressing the conversion of biomass into bioenergy, food and feed ingredients, fine chemicals, and biomaterials, in order to boost the modernization of economic primary and secondary activities, contributing to reduce fossil fuel dependency and respecting the ecological world's boundaries [1]. In this context, biorefinery-based industrial plants are attracting broad interest, but the biomass demand has risen the attention to the potential stress on agricultural land use, environment, and ecosystem [2]. In addition to the by-products and wastes of the agriculture, forestry, and food industries, energy crops and short-rotation woody crops can be valuable biomass sources for biorefinery-based plants. *Salix* (Salicaceae), commonly known as willow, is among the most promising short-rotation woody crops, since it grows quickly and can provide high commercial biomass yields, generally reaching 8–10 dry t ha^{-1} $year^{-1}$ in European countries [3]. Additionally, it can be cultivated in abandoned soils, and not necessarily in agricultural fertile fields, leading to a positive impact on biodiversity and rural income [3,4].

Willow has traditionally been used in basket manufacturing and for ornamental aspects, and more recently, for thermal and electricity generation [3]. Furthermore, *Salix* spp. bark extracts are well-known in folk medicine, since the ancient Egyptian, Greek, and Roman civilizations, owing to their analgesic and antipyretic actions which are mainly ascribed to the physiological oxidation of salicin to salicylic acid [5]. In fact, salicin-standardized extracts of *Salix fragilis* L., *Salix purpurea* L., and *Salix daphnoides* Vill. barks are also recommended for lower back pain [6]. Moreover, *Salix* spp. bark polar extracts and phenolic-enriched fractions have exhibited anti-inflammatory, antioxidant, and tumor antiproliferative effects, which have been related with the presence of catechin and procyanidins, instead of the extracts' marker component salicin [7–9]. Other phenolic compounds, namely, acetophenones (e.g., picein), chalcones (e.g., isosalipurposide), and flavanones (e.g., naringenin 7-*O*-glucoside), have also been reported in several *Salix* spp. bark extracts, including commercial ones [10–13]. These phenolic compounds have also demonstrated anti-hypertensive [14,15], cytoprotective [16], and antimicrobial [17] effects.

Considering the vast set of biological activities of *Salix* spp. bark extracts and their phenolic constituents, along with society's increasing interest for natural components rather than synthetic ones, innovative food, nutraceutical, and cosmetic purposes can be envisaged. Actually, phenolic compounds have been increasingly used in the food industry as natural additives [18], as well as in the cosmetic field, including sunscreen and anti-aging cream formulations [19]. All of these applications are normally associated with their antioxidant activity, since phenolic compounds can disrupt the cascade oxidation reactions, either in food matrixes, allowing longer shelf life [18], or in dermatological preparations, preventing the oxidation of the other ingredients [19]. At the same time, the oral or topical administration of phenolic compounds can promote human wellbeing [18–20]. In this sense, alternative or complementary natural-based therapeutics have been researched for tackling current worldwide health problems, like hypertension [14,21] and multidrug-resistant bacterial infections [22,23].

Hypertension affects ca. 1.13 billion people, and is associated with premature mortality and disability [24]. Synthetic inhibitors of angiotensin-I converting enzyme (ACE) are the most used anti-hypertensive drugs; however, they can cause skin rashes, cough, angioedema, hypotension, renal disfunction, and other disturbing side effects [25]. Among phenolic compounds, flavan-3-ols, in particular procyanidins, have shown active ACE inhibition, being promising natural anti-hypertensive agents or coadjuvants [14].

Additionally, multidrug-resistant bacterial infections are a serious threat to public health, with an increased risk of morbidity and mortality, and financial burden on healthcare systems. Despite colonizing the skin of healthy humans, *Staphylococcus aureus* represents one of the leading causes of bacteremia, in addition to skin, soft tissue, and bone infections [26]. This Gram-positive bacterium can also lead to gastrointestinal illness, which comes from food contaminated by one of the 20 staphylococcal enterotoxins [27]. Phenolic compounds, like acetophenones and hydroxycinnamic acids, have exhibited anti-*S. aureus* effect [17,22,28].

Although *Salix* cultivation is more expanded in Northern Europe, several species of this genera are disseminated in Continental Portugal, namely *Salix atrocinerea* Brot., *S. fragilis* L., and *Salix viminalis* L., assuming a huge importance for the biodiversity and the soil stability in humid zones, within riparian ecosystems. Few works have evidenced the presence of phenolic constituents in the bark of these *Salix* species [29–32], but it is still missing a systematic approach, integrating the detailed phenolic composition, bioactivity, and biocompatibility of polar extracts of the individual species in question. Given the adequate edaphoclimatic conditions of Portugal for the selected *Salix* spp., this knowledge can boost their sustainable exploitation in Southern Europe, preserving the riparian ecosystem, enhancing the biodiversity, and contributing to rural development, in the context of the bioeconomy concept.

In the scope of our interest in bioprospecting *Salix* spp. bark [32], the present work aims to characterize the phenolic composition of *S. atrocinerea* Brot., *S. fragilis* L., and *S. viminalis* L. barks by ultra-high-performance liquid chromatography-diode array-tandem mass spectrometry (UHPLC-DAD-MS^n), as well as to evaluate three key in vitro biological activities of their polar extracts, such as: (1) antioxidant activity, using two *in chimico* assays; (2) anti-hypertensive via ACE inhibitory effect; and (3) antibacterial effect against *S. aureus*. The cytotoxicity of the studied *Salix* spp. bark phenolic-containing extracts is also approached in three mammalian cell lines, namely, Caco-2, HaCaT, and L929 cell lines, towards potential safe food, nutraceutical, and cosmetic usages.

2. Materials and Methods

2.1. Chemicals

Dichloromethane (p.a., ≥99%), methanol (p.a., ≥99.8%), HPLC-grade methanol and acetonitrile were supplied by Fisher Scientific (Pittsburgh, PA, USA). Before UHPLC analysis, mobile phase solvents were previously filtered via a Solvent Filtration Apparatus 58061 from Supelco (Bellefonte, PA, USA). Acetic acid glacial (p.a., ≥99.5%) was purchased from Labkem (Madrid, Spain). Sodium carbonate (p.a., ≥99.9%) was obtained from Panreac AppliChem ITW Reagents (Barcelona, Spain). Gallic acid (≥97.5%), Folin–Ciocalteu's phenol reagent (2 N), HPLC-grade water, formic acid (≥98%), catechin (>99%), eriodictyol (≥98%), naringenin (98%), procyanidin B1 (≥90%), procyanidin B2 (≥90%), quercetin (>98%), quercetin 3-*O*-galactoside (≥97%), 2,2-diphenyl-1-picrylhydrazyl free radical (DPPH•), 2,2′-azino-bis(3-ethyl-benzothiazoline-6-sulfonic acid) diammonium salt (ABTS), ascorbic acid (≥99.5%), angiotensin-I converting enzyme (ACE) (peptidyl-di-peptidase A, EC 3.4.15.1, 5.1 U mg^{-1}), MEM non-essential amino acid solution, phenazine methosulfate, and 2,3-bis-(2-methoxy-4-nitro-5-sulfophenyl)-2*H*-tetrazolium-5-carboxanilide (XTT) were supplied by Sigma-Aldrich (Merck, Darmstadt, Germany). The intramolecularly quenched fluorescent tripeptide *o*-aminobenzoylglycyl-*p*-nitro-L-phenylalanyl-L-proline [Abz–Gly–Phe(NO_2)–Pro] was purchased from Bachem Feinchemikalien (Bubendorf, Switzerland). Tris [tri(hydromethyl) aminomethane] was afforded by Fluka (Gmbh, Germany). Dulbecco's Modified Eagle Medium (DMEM) high glucose and Penicillin-Streptomycin mixture were obtained from Lonza (Basel, Switzerland). Fetal bovine serum (FBS) was purchased from Biowest (Nuaillé, France). Piceol (≥98%), *m*-hydroxybenzoic acid (≥99%), tryptic soy broth and tryptic soy agar were afforded by Merck (Darmstadt, Germany). *p*-Hydroxybenzoic acid (>99%) was purchased from Fisher Scientific (Thermo Fisher Scientific Inc., Waltham, MA, USA). Naringenin 7-*O*-glucoside (≥99%) was supplied by Extrasynthese (Lyon, France).

2.2. Sampling of Salix spp. Barks

Branches from 8-year-old trees of *S. atrocinerea* Brot., *S. fragilis* L., and *S. viminalis* L. were collected nearby Aveiro (GPS coordinates 40°41′54.78″ N, 8°36′3.23″ W), from an industrial experimental plantation of The Navigator Company, in October 2017, and air-dried at room temperature until the biomass weight was stable [32]. Bark samples were hand-separated and ground using a hammer mill, in order to select the fraction with a granulometry lower than 1 mm.

2.3. Extraction of Phenolic Compounds

The lipophilic components were previously removed from the milled barks of the three *Salix* spp., as earlier reported [32]. Then, 2 g of lipophilic component free-dry bark was submitted to methanol/water/acetic acid (49.5:49.5:1) extraction, by stirring at 900 r.p.m. for 24 h, in the dark at room temperature, following a previously described approach [33]. After vacuum filtration through a glass filter of porosity 3 to separate the extract from the biomass, methanol was removed at 37 °C using a rotative evaporator, whilst water was removed by freeze-drying. *Salix* spp. extracts were prepared in triplicate, and the respective extractive yield (EY) was expressed as the percentage of dry bark. The extracts were then kept at room temperature and protected from the light, until the chemical analysis and the biological activity assays were performed.

2.4. Total Phenolic Content

The total phenolic content (TPC) of *Salix* spp. barks was determined using the Folin–Ciocalteu reagent, according to procedures carried out elsewhere [34,35], with some alterations. In a 96-well plate, 150 µL of Folin–Ciocalteu reagent previously diluted 1:10 (*v*/*v*) with water, and 120 µL of 75 g L^{-1} sodium carbonate aqueous solution were added to 30 µL of *Salix* spp. bark extracts, previously dissolved in methanol/water (1:1, *v*/*v*) and diluted with water, corresponding to 0.2 mg mL^{-1} of extract. After 60 min of incubation at room temperature and in the dark, the absorbance was recorded at 750 nm, against a blank containing 30 µL of water instead of the sample volume, in a Thermo Scientific Multiskan™ FC microplate reader (Thermo Fisher Scientific Inc., Waltham, MA, USA). TPC was determined as gallic acid equivalents (GAE) using the linear regression equation ($y = 0.0103x - 0.0276$; $r^2 = 0.9995$) obtained from the standard curve of gallic acid (5–100 µg mL^{-1}), and expressed as grams of GAE per kilogram of dry bark and milligrams of GAE per gram of extract, according to Equations (1) and (2), as follows:

$$\text{TPC (g GAE kg}^{-1}\text{ of dry bark)} = \text{TPC (g GAE kg}^{-1}\text{ of extract)} \times [\text{EY (kg of extract kg}^{-1}\text{ of dry bark)/100}] \quad (1)$$

$$\text{TPC (mg GAE g}^{-1}\text{ of extract)} = [\text{TPC (µg GAE mL}^{-1}) \times \text{dilution factor}]/[\text{extract concentration (g L}^{-1}) \times 0.001] \quad (2)$$

All of the assays were performed three times, each one in triplicate ($n = 9$).

2.5. Identification of Phenolic Compounds by UHPLC-DAD-MS^n Analysis

Salix spp. bark extracts were first dissolved in methanol/water (1:1, *v*/*v*), at 10 mg mL^{-1} and filtered using PTFE filters with 0.2 µm pore diameter. Extracts (10 µL) were injected in the UHPLC system equipped with an Accela 600 LC pump, an Accela autosampler (set at 16 °C), and an Accela 80 Hz photo diode array detector (DAD) (Thermo Fisher Scientific, San Jose, CA, USA). The separation of the extract components was developed in a Hypersil Gold RP C18 column (100 × 2.1 mm; 1.9 µm particle size) afforded by Thermo Fisher Scientific (San Jose, CA, USA), preceded by a C18 pre-column (2.1 mm i.d.) supplied by Thermo Fisher Scientific (San Jose, CA, USA), and both were kept at 45 °C. The binary mobile phase included (A) water/acetonitrile (99:1, *v*/*v*) and (B) acetonitrile, both containing 0.1% (*v*/*v*) formic acid. A gradient elution program was applied at a flow rate of 0.45 mL min^{-1}, as follows: 1% B kept from 0 to 3 min; 1–31% B from 3 to 30 min; 31–100% B from 30 to 32 min, and 100–1% B from 32 to 36 min, keeping 1% B from 36 to 40 min for column re-equilibration. The chromatograms were recorded at 235, 280, and 370 nm and UV-Vis spectra from 210 to 600 nm.

The UHPLC system was coupled to a LCQ Fleet ion trap mass spectrometer (ThermoFinnigan, San Jose, CA, USA), equipped with an electrospray ionization (ESI) source. The ESI-MS was operated under the negative ionization mode with a spray voltage of 5 kV and capillary temperature of 320 °C. The flow rate of nitrogen sheath and auxiliary gas were 40 and 5 (arbitrary units), respectively. The capillary and tube lens voltages were set at −44 and −225 V, respectively. CID-MS^n experiments were executed on mass-selected precursor ions in the range of *m*/*z* 100–2000. The isolation width of precursor ions was 1.0 mass units. The scan time was 100 ms and the collision energy was 35 arbitrary units, using helium as collision gas. The data acquisition was carried out by using Xcalibur® data system (Thermo Finnigan, San Jose, CA, USA).

2.6. Quantification of Phenolic Compounds by UHPLC-UV Analysis

Standard curves were obtained through the UHPLC injection of catechin, *m*-hydroxybenzoic acid, naringenin, piceol, and quercetin standard solutions in HPLC grade methanol/water (1:1, *v*/*v*), with six concentrations ranging from 0.10 to 30.89 μg mL^{-1}. The quantification of individual phenolic compounds was determined by using the linear regression equation (Table 1), obtained with the most similar standard compound. The limit of detection (LOD) and the limit of quantification (LOQ) were approached for each standard curve (Table 1), based on Equations (3) and (4), respectively, as follows:

$$\text{LOD} = (\text{standard deviation of the ordinate intercept/slope of the linear regression}) \times 3 \quad (3)$$

$$\text{LOQ} = (\text{standard deviation of the ordinate intercept/slope of the linear regression}) \times 10 \quad (4)$$

Table 1. Standard data used for the HPLC-UV quantification of phenolic compounds present in methanol/water/acetic acid (49.5:49.5:1) extracts of *Salix* spp. bark.

Standard Compound	λ (nm) [A]	Concentration Range (μg mL^{-1})	Linear Regression Equation [B]	r^2	LOD (μg mL^{-1})	LOQ (μg mL^{-1})
Catechin	280	0.10–30.29	$y = 93621x + 17212$	0.9998	0.52	1.74
m-Hydroxybenzoic acid	235	0.51–30.89	$y = 245747x + 909936$	0.9929	3.42	11.40
Naringenin	280	0.11–21.17	$y = 398130x + 61541$	0.9990	0.87	2.89
Piceol	280	0.30–18.23	$y = 765733x + 59082$	0.9992	0.68	2.25
Quercetin	370	0.10–19.21	$y = 320421x - 99949$	0.9989	0.85	2.83

[A] Wavelength used in the quantitative analysis; [B] y = peak area, x = concentration in μg mL^{-1}. LOD, limit of detection; LOQ, limit of quantification.

The quantitative analysis was performed in triplicate for each sample ($n = 3$).

2.7. Antioxidant Activity

2.7.1. DPPH Free Radical Scavenging Effect

The DPPH• scavenging effect of *Salix* spp. bark extracts was measured according to a former method [34], with slight modifications for 96-well microplate scale. Ascorbic acid was used as the natural antioxidant reference. Briefly, stock solutions of extracts and ascorbic acid were firstly prepared in methanol/water (1:1, *v*/*v*). Then, 30 μL of 1 mM DPPH• methanolic solution was added to 75 μL of sample and 195 μL of methanol, in each microwell. The control was constituted by 270 μL of methanol and 30 μL of 1 mM DPPH• methanolic solution. The concentrations of extracts and ascorbic acid were tested in the 1–40 μg mL^{-1} and 0.5–20 μg mL^{-1} range, respectively. After a gentle mixing, the microplate was kept in the dark for 30 min, and the absorbance at 520 nm was thereafter read against the blank (methanol), using a Thermo Scientific Multiskan™ FC microplate reader. The DPPH• scavenging effect percentage was calculated according to Equation (5):

$$\text{DPPH• scavenging effect (\%)} = [(A_{control} - A_{sample})/A_{control}] \times 100 \quad (5)$$

where $A_{control}$ and A_{sample} are the absorbances at 520 nm of control and sample, respectively. The inhibitory concentration of extracts and ascorbic acid able to scavenge 50% of DPPH• (IC_{50}) was calculated through the graph of scavenging effect percentage against concentration logarithm.

To compare the obtained results with the literature, the Antioxidant Activity Index (AAI) was determined according to Equation (6) [36]:

$$\text{AAI} = \text{DPPH• final concentration } (\mu g\ mL^{-1})/IC_{50}\ (\mu g\ mL^{-1}) \quad (6)$$

where DPPH• final concentration was 61.874 μg mL^{-1}. All of the assays were performed three times, each one in triplicate ($n = 9$).

2.7.2. ABTS Radical Cation Scavenging Effect

The ABTS radical cation ($ABTS^{•+}$) scavenging effect of *Salix* spp. bark extracts was assayed based on the methodology reported elsewhere [34,37], which was adapted to the 96-well microplate scale.

Ascorbic acid was used as the reference antioxidant. The ABTS•$^+$ was first generated by mixing 7 mM ABTS and 2.45 mM potassium persulfate, and keeping the reactional mixture in the dark at room temperature for 16 h. Then, the ABTS•$^+$ solution was diluted with methanol, in order to reach the absorbance value of 0.700 at 750 nm. Meanwhile, stock solutions of extracts and ascorbic acid were prepared in methanol/water (1:1, *v*/*v*). In each microwell, 250 μL of diluted ABTS•$^+$ solution was added to 50 μL of sample, obtaining the 0.5–40 μg mL^{-1} and 0.5–16 μg mL^{-1} range for extracts and ascorbic acid, respectively. The control contained 250 μL of diluted ABTS•$^+$ solution and 50 μL of methanol. Then, the microplate was kept in the dark for 30 min, and the absorbance at 750 nm was read against the blank (methanol) using the Thermo Scientific Multiskan™ FC microplate reader. The ABTS•$^+$ scavenging effect percentage was determined according to Equation (7):

$$\text{ABTS}\bullet^{+}\text{ scavenging effect (\%)} = [(A_{control} - A_{sample})/A_{control}] \times 100 \quad (7)$$

where $A_{control}$ and A_{sample} are the absorbances at 750 nm of control and sample, respectively. The IC_{50} of extracts and ascorbic acid was determined from the scavenging effect percentage versus logarithm of concentration. All of the assays were performed three times, each one in triplicate (n = 9).

2.8. Angiotensin-I Converting Enzyme Inhibitory Activity

The ACE-inhibitory activity of *Salix* spp. bark extracts, at 625 μg mL^{-1}, was measured by fluorescence using the method of Sentandreu and Toldrá [38], with some modifications [21]. The method consists in the ACE-catalyzed hydrolysis of a specific substrate [ABz–Gly–Phe(NO_2)–Pro] to the fluorescent *o*-aminobenzoylglycine. Commercial ACE was diluted in 5 mL of 50% (*v*/*v*) glycerol aqueous solution, which was kept at −20 °C until use. Thereafter, the ACE solution was diluted (1:24) with 150 mM Tris buffer solution pH 8.3, containing 1 μM zinc chloride, for a final concentration of 42 mU mL^{-1}. Then, 40 μL of ultrapure water or ACE working solution was added to each microplate well, and the volume was thereafter adjusted to 80 μL by adding ultrapure water to blank, control, or samples. A sample blank was also made. The enzymatic reaction was started by adding 160 μL of substrate solution (0.45 mM ABz–Gly–Phe(NO_2)–Pro prepared in 150 mM Tris buffer pH 8.3, and containing 1.125 M sodium chloride), and then the mixture was incubated at 37 °C. The generated fluorescence was measured at 30 min using a Multidetection plate reader (Synergy H1, BioTek Instruments, Winooski, VT, USA). The assay was performed in a black 96-well microplate (Thermo Scientific Nunc, Roskilde, Denmark). Excitation and emission wavelengths were 350 and 420 nm, respectively. The inhibitory activity was calculated as the percentage decrease of ACE activity compared with the maximum ACE activity (control). All of the assays were performed two times, each one in duplicate (n = 4).

2.9. Inhibitory Effect Against Staphylococcus aureus Growth

The inhibitory effects of *Salix* spp. bark extracts were evaluated against the growth of a Gram-positive *S. aureus* strain (ATCC® 6538). This bacterium was aseptically inoculated in tryptic soy broth, and grown at 37 °C under 120 r.p.m. for 24 h. Before the antibacterial test, the *S. aureus* density was adjusted to 0.5 McFarland in phosphate-buffered saline (PBS) solution, corresponding to 10^8–10^9 colony forming units (CFUs) mL^{-1}. Then, the bacterial inoculum was incubated with the aqueous solutions of *Salix* spp. bark extracts at 37 °C for 24 h, obtaining the final concentrations of 625, 1250, and 2500 μg mL^{-1}. The control containing only bacterial inoculum in PBS was also performed. Thereafter, the *S. aureus* bacterial density was determined by plating serial dilutions in tryptic soy agar. After 24 h of incubation at 37 °C, the antibacterial effect was assayed by determining the logarithm units of CFU mL^{-1} and comparing it with that of growth control group. In this study, the bacteriostatic and bactericidal effects were considered as the decrease of <3-log and ≥3-log in CFU mL^{-1}, respectively, in comparison with the control inoculum [39]. All of the assays were performed three times, each one in duplicate (n = 6).

2.10. In Vitro Biocompatibility

2.10.1. Mammalian Cell Lines

Three different cell lines were considered throughout this work, namely, Caucasian colon adenocarcinoma cells—Caco-2 (86010202, Sigma-Aldrich, St. Louis, MO, USA); human keratinocyte—HaCaT (300493, CLS, Eppelheim, Germany); and mouse fibroblast cells—L929 (NCTC) (ECACC 85103115). Caco-2 cells were maintained in DMEM high glucose supplemented with 10% (*v*/*v*) FBS, 1% (*v*/*v*) penicillin-streptomycin, and MEM non-essential amino acid solution. HaCaT and L929 cells were maintained using DMEM high glucose supplemented with 10% (*v*/*v*) FBS and 1% (*v*/*v*) penicillin-streptomycin. All of cell lines were incubated at 37 °C in a 5% (*v*/*v*) CO_2 humidified atmosphere.

2.10.2. Metabolic Inhibition via XTT Assay

Cells were detached using TrypLE Express (Thermo Scientific, Waltham, MA, USA), seeded (1×10^4 cells/well) into 96-well Nunclon Delta microplates (Thermo Scientific, Waltham, MA, USA), and incubated for 24 h. Afterwards, the culture media were carefully removed and replaced with *Salix* spp. bark extracts at 625, 1250, and 2500 µg mL^{-1} (sterile filtered). After incubation for 24 h, the cytotoxicity of the samples was evaluated using the XTT assay. Immediately before use, 10 µL of 10 mM phenazine methosulfate solution was added to 4 mL of 1 mg mL^{-1} XTT solution prepared in DMEM. Then, 25 µL of this mixture was added to each well, and the plates were, once again, incubated at 37 °C. After 2 h, the optical density at 485 nm was measured using a microplate reader (Synergy H1, Biotek Instruments, Winooski, VT, USA). Cells in culture medium were used as control, and wells without cells were used as blanks. The metabolic inhibition was determined according to the following Equation (8):

$$\text{Metabolic inhibition (\%)} = [(A_{control} - A_{sample})/A_{control}] \times 100 \quad (8)$$

where $A_{control}$ and A_{sample} are the absorbances at 485 nm of control and sample, respectively. Five replicates for each condition were performed ($n = 5$).

2.11. Statistical Analysis

The statistical analysis was performed using the IBM® SPSS® Statistics Version 25 (IBM Corporation, New York, NY, USA). The EY, TPC, and the in vitro bioactivity assay data were analyzed through the one-way analysis of variance (ANOVA). Where differences existed, the source of the differences at $p < 0.05$ of significance level was identified by all pairwise multiple comparison procedures, through the Tukey's honestly significant difference (HSD) post-hoc test. The Pearson's correlation r values between TPC or phenolic compound abundances and the antioxidant activity IC_{50} values were also determined using the aforementioned software.

3. Results

3.1. Extractive Yield and Total Phenolic Content

In the present work, a methanol/water/acetic acid (49.5:49.5:1) solution was used for the extraction and chemical analysis of phenolic compounds in the studied *Salix* spp. barks, as it has proven to be suitable for the removal of these type of bioactive compounds from crops' biomass [33].

The EY and TPC of *S. atrocinerea* Brot., *S. fragilis* L., and *S. viminalis* L. barks are summarized in Table 2.

S. atrocinerea Brot. bark showed the highest EY (15.1% of dry bark (*w*/*w*)), being significantly higher than EYs of *S. fragilis* L. and *S. viminalis* L. barks ($p < 0.05$). Considering the TPC determined using the Folin–Ciocalteu reagent, *S. atrocinerea* Brot. bark revealed the highest TPC, accounting for 44.47 g GAE kg^{-1} dry weight (dw). In terms of TPC expressed in mg g^{-1} of extract, *S. atrocinerea* Brot. bark extract also demonstrated the highest TPC (293.36 mg GAE g^{-1} of extract), but it was not statistically different from TPC of *S. viminalis* L. bark extract ($p > 0.05$). *S. fragilis* L. bark extract also presented considerable TPC, reaching 17.47 g kg^{-1} dw and 179.06 mg GAE g^{-1} of extract.

Table 2. Extractive yield (EY) and total phenolic content (TPC) of methanol/water/acetic acid (49.5:49.5:1) extracts of *Salix atrocinerea* Brot., *Salix fragilis* L., and *Salix viminalis* L. barks.

Salix spp.	EY (% of Dry Bark, *w/w*)	TPC (g GAE kg^{-1} of Dry Bark)	TPC (mg GAE g^{-1} of Extract)
Salix atrocinerea Brot.	15.1 ± 1.7 ^{b}	44.47 ± 6.68 ^{b}	293.36 ± 19.52 ^{b}
Salix fragilis L.	9.7 ± 0.3 ^{a}	17.47 ± 3.19 ^{a}	179.06 ± 30.64 ^{a}
Salix viminalis L.	10.1 ± 0.8 ^{a}	24.76 ± 0.82 ^{a}	246.44 ± 16.58 ^{a,b}

The results represent the mean ± standard deviation. Means with different superscript minor case letters (a, b) within the same column are statistically different (one-way ANOVA, followed by Tukey's HSD test, $p < 0.05$). GAE, gallic acid equivalents.

3.2. Phenolic Composition

3.2.1. Identification of Phenolic Compounds

Figure 1 depicts the UHPLC-UV chromatograms of methanol/water/acetic acid extracts from *S. atrocinerea* Brot., *S. fragilis* L., and *S. viminalis* L. barks.

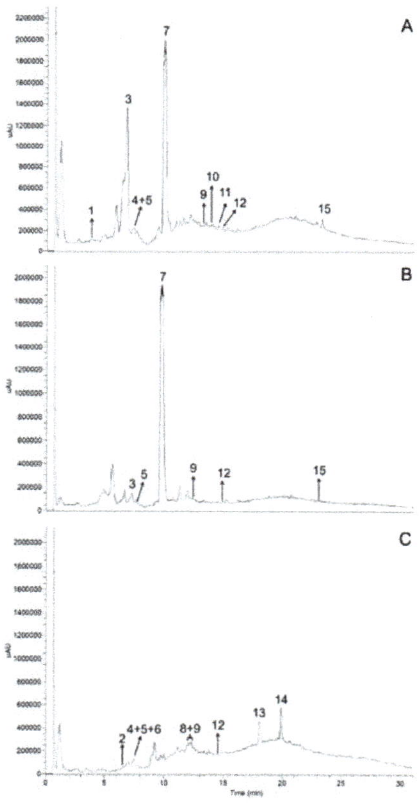

Figure 1. UHPLC-UV chromatograms of methanol/water/acetic acid (49.5:49.5:1) extracts, derived from (**A**) *Salix atrocinerea* Brot., (**B**) *Salix fragilis* L., and (**C**) *Salix viminalis* L. barks, recorded at 280 nm. The peak numbers correspond to those represented in Tables 3 and 4 and Figure 2.

Table 3. UHPLC-DAD-MS^n data of phenolic compounds detected in methanol/water/acetic acid (49.5:49.5:1) extracts of *Salix atrocinerea* Brot., *Salix fragilis* L. and *Salix viminalis* L. barks.

No.	RT (min)	Compound	λ_{max} (nm)	$[M-H]^-$ (*m/z*)	MS^n Product Ions (*m/z*) [B]	Id.
1	3.79	(Epi)gallocatechin-(epi)catechin dimer isomer	233, 273	593	MS^2: 575, 525, 467, 441, 425, 423, 407, 303, <u>289</u>, 245 MS^3: 245	[40,41]
2	6.68	B-type procyanidin dimer isomer 1	237, 277	577	MS^2: 559, 451, 425, 407, <u>289</u>, 287, 245 MS^3: 245, 229, 205	[41]
3	6.77	Picein	229, 264	343 [A]	MS^2: 297, 135, 120	[11]
4	7.19	Procyanidin B1	236, 278	577	MS^2: 559, 451, 425, 407, <u>289</u>, 287, 245 MS^3: 245	Co
5	7.37	Catechin	235, 278	289	MS^2: 271, 245, 205, 203, 179	Co
6	7.61	B-type procyanidin dimer isomer 2	237, 278	577	MS^2: 559, 451, 425, 407, <u>289</u>, 287, 245 MS^3: 245, 205	[41]
7	10.10	Piceol	229, 274	135	MS^2: 93	Co
8	12.24	B-type procyanidin dimer isomer 3	241, 279	577	MS^2: 559, 451, 425, 407, <u>289</u> MS^3: 289, 245	[41,42]
9	12.27	Salicylic acid	241, 299	137	MS^2: 93	[11]
10	13.86	Naringenin-*O*-hexoside isomer 1	241, 277	433	MS^2: 433, 416, 365, 313, <u>271</u>, 151 MS^3: 151	[11]
11	14.52	Naringenin-*O*-hexoside isomer 2	241, 274	433	MS^2: 313, <u>271</u>, 251, 151 MS^3: <u>151</u>, 107	[11]
12	14.88	Quercetin 3-*O*-galactoside	241, 268, 346	463	MS^2: 417, 395, 379, 343, <u>301</u>, 300, 271, 179, 151 MS^3: 179, 151	Co
13	18.09	Eriodictyol-*O*-hexoside isomer	238, 282, 330sh	449	MS^2: 431, 413, 403, 381, 297; <u>287</u>, 269, 175, 151, 135 MS^3: 287, 269, 151, 135, 125, 107	[33]
14	19.88	Eriodictyol	238, 284, 330sh	287	MS^2: 287, 151, 135, 125, 107	Co
15	23.23	Naringenin	237, 279	271	MS^2: 227, 177, 151, 119, 107	Co

[A] Compound 3 was detected as a formate adduct ($[M+HCOO]^-$ ion). [B] *m/z* underlined was subjected to MS^n analysis. The numbers (No.) of phenolic compounds correspond to the chromatographic peaks assigned in Figure 1, and the proposed chemical structures illustrated in Figure 2. Co, co-injection of a commercial standard; Id., identification; RT, retention time; sh, shoulder wavelength.

Fifteen phenolic compounds were detected in the studied *Salix* spp. bark polar extracts by HPLC-DAD-MS^n analysis, as listed in Table 3 and explained thoroughly below.

Flavan-3-ols

Compound 1 was tentatively assigned as a prodelphinidin dimer isomer, or (epi)gallocatechin-(epi)catechin dimer isomer (Figure 2), based on its UV spectrum (Figure S1A), and on the detection of the $[M-H]^-$ ion at *m/z* 593 and MS^n fragmentation (Table 3). The MS^2 spectrum of the $[M-H]^-$ ion showed the base peak at *m/z* 425, resulting from the retro-Diels-Alder fission of the C ring in the upper subunit ($[M-H-168]^-$), as well as the product ion at *m/z* 407 given the sequent loss of a water molecule ($[M-H-168-H_2O]^-$) (see mass fragmentation 1 in Figure S2, in Supplementary Material) [40,41]. Moreover, the MS^2 spectrum presented two product ions at *m/z* 303 and *m/z* 289 formed by the cleavage of the interflavanic linkage, corresponding to the quinone methide of the upper unit residue ($[(\text{epi})\text{gallocatechin}-3H]^-$) and the deprotonated ion of the lower unit residue ($[(\text{epi})\text{catechin}-H]^-$), respectively (see mass fragmentation 2 in Figure S2, in Supplementary Material). The product ion at *m/z* 289 can also be generated from the interflavanic fission of the product ion at *m/z* 467, after the C ring fission and the loss of a phloroglucinol moiety ($[M-H-126]^-$) (see mass fragmentation 3 in Figure S2, in Supplementary Material) [43]. Additionally, the MS^3 spectrum of the ion at *m/z* 289 showed the product ion at *m/z* 245, which is common to the mass fragmentation of catechin and epicatechin [42].

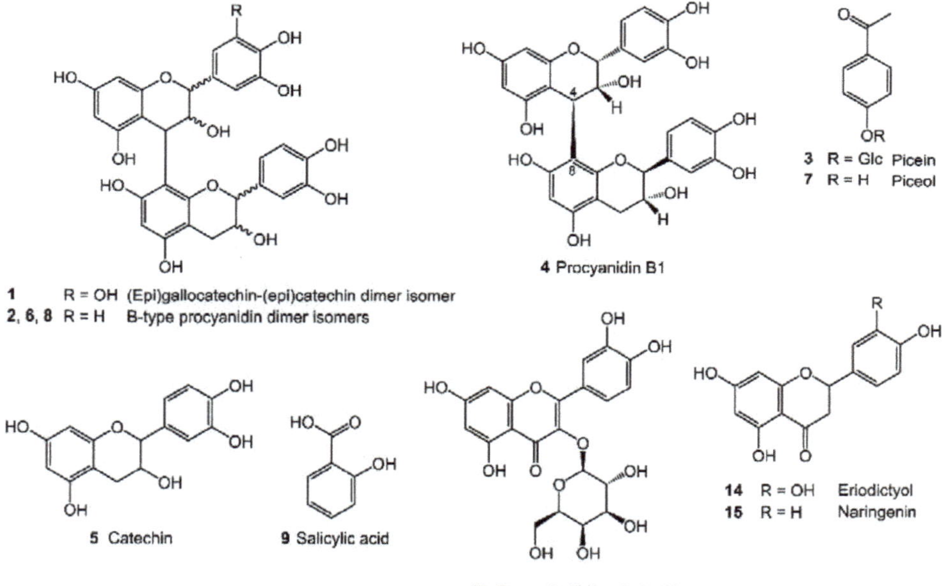

Figure 2. Proposed chemical structures for main phenolic compounds detected in the *Salix atrocinerea* Brot., *Salix fragilis* L., and *Salix viminalis* L. barks. Glc, glucosyl.

Compound 4 was identified as procyanidin B1 ((−)-epicatechin-(4β-8)-(+)-catechin) (Figure 2). The retention time, UV spectrum, the detection of the $[M-H]^-$ ion at *m/z* 577, and the MS^2 and MS^3 fragmentations (Table 3) are in agreement with that of commercial standard, injected in the UHPLC-DAD-MS system, under the same experimental conditions.

Compounds 2, 6, and 8 were tentatively identified as B-type procyanidin dimer isomers (Figure 2) formed by two (epi)catechin units, due to their characteristic UV spectra (Figure S1B, in Supplementary Material), the detection of the $[M-H]^-$ ion at *m/z* 577, and the MS^n data (Table 3). The MS^2 fragmentation of the [M−H]$^-$ ion originated the product ion at *m/z* 425 (base peak) from the retro-Diels-Alder fission of the C ring ([M−H−152]$^-$), which afforded the product ion at *m/z* 407 after a water molecule loss ([M−H−152−H_2O]$^-$) [43]. Furthermore, four characteristic product ions were detected, namely, at *m/z* 559 (loss of a water molecule), *m/z* 451 (heterocyclic C ring fission with the phloroglucinol moiety loss), as well as at *m/z* 289 and *m/z* 287, which resulted from the quinone methide fission of the interflavanic linkage between C and D rings [41,43]. Moreover, the MS^3 spectrum of the ion at *m/z* 289 presented the characteristic product ions of catechin or epicatechin [42]. It was not possible to attribute the chemical structures of compounds 2, 6, and 8 to procyanidin B2, since none of their retention times were coincidental with the corresponding commercial standard. Since these compounds are procyanidin dimers of (epi)catechin units, there are six hypotheses of B-type procyanidins, namely, procyanidins B4, B5, and B8 [41], in addition to procyanidins B3, B6, and B7 detected earlier in *Salix* species [31], which can be suggested for their identification.

Compound 5 was identified as catechin (Figure 2 and Table 3) based on its UV spectrum, the detection of the [M−H]$^-$ ion at *m/z* 289, and the characteristic MS^2 data of the ion at *m/z* 289 [42], in addition to the injection of the commercial standard, at the same experimental conditions.

Acetophenones

Compound 3 presented a similar UV spectrum (Figure S1A, in Supplementary Material) to that of the picein (Figure 2), and afforded the ion at *m/z* 343, under the negative ionization [11], which corresponds to the formate adduct ion of that acetophenone ([M+HCOO]$^-$) (Table 3). In addition to the [M−H]$^-$ ion at *m/z* 297, two product ions were detected in the MS^2 spectrum of the ion at *m/z* 343, namely, the base peak at *m/z* 135 resulting from the loss of a hexosyl unit of the [M−H]$^-$ ion ([M−H−162]$^-$), and the anion radical at *m/z* 120 which may be originated by homolytic fission of the methyl group from the aglycone ion ([M−H−162−CH_3]$^-$) [44]. In this sense, compound 3 was most likely assigned as picein, although the mass spectrometry analysis did not allow to discriminate the position of the *O*-glycosyl substituent. Nevertheless, the elution order of compound 3 relative to the commercial standard of piceol is in agreement with literature data [45].

Compound 7 was identified as piceol (Figure 2). The retention time, the UV spectrum, the detection of the [M−H]$^-$ ion at *m/z* 135, and the MS^2 fragmentation of this ion (Table 3), yielding the product ion at *m/z* 93 from the ketene loss ([M−H−42]$^-$) [44], were concordant with that of commercial standard injected under the same experimental conditions. Although the product ion at *m/z* 120 would be expected in the mass fragmentation of the [M−H]$^-$ ion at *m/z* 135 of compound 7, by comparison with the MS data of compound 3 and with literature [44], it was not found in the MS^2 spectrum of the [M−H]$^-$ ion obtained from the studied *Salix* extracts, or from the corresponding commercial standard. However, this fact does not hamper its unambiguous identification, since it has been corroborated with the retention time and MS data of the commercial standard.

Hydroxybenzoic Acids

Compound 9 was identified as *o*-hydroxybenzoic acid, commonly known as salicylic acid (Figure 2), presenting a UV spectrum (Figure S1B, in Supplementary Material) similar to that of salicylic acid [11], and the [M−H]$^-$ ion at *m/z* 137 (Table 3). Additionally, the product ion at *m/z* 93 was found in the MS^2 spectrum of the [M−H]$^-$ ion, due to the CO_2 loss ([M−H−44]$^-$) [46], being concordant with the MS/MS data of salicylic acid, under the negative ionization mode [11]. Moreover, the retention time of compound 9 was different from those of commercial standards of *m*- and *p*-hydroxybenzoic acids injected in the HPLC-UV-MS system, under the same experimental conditions, thus being assigned as salicylic acid.

Flavanones

Compounds 10 and 11 were tentatively identified as two naringenin-*O*-hexoside isomers 1 and 2, respectively, whilst compound 15 was identified as naringenin (Figure 2) in the studied *Salix* spp. bark extracts (Table 3). Naringenin-*O*-hexoside isomers were assigned considering their UV spectra (Figure S1A, in Supplementary Material), the detection of the $[M-H]^-$ ion at *m/z* 433, and the characteristic MS^n fragmentation [11]. Indeed, the base peak of the MS^2 spectrum of the aforementioned $[M-H]^-$ ion was noted at *m/z* 271, which evidenced the loss of a hexosyl residue (−162 amu). Additionally, the MS^3 spectrum of the ion at *m/z* 271 presented the characteristic product ions of naringenin [47]. Despite similar MS data, none of these compounds might be the chalcone isosalipurposide, since their absorption maxima wavelengths (274 and 277 nm) are completely different from that of the latter (370 nm) [11]. Furthermore, the retention times of compounds 10 and 11 were not concordant with that of commercial standard of naringenin 7-*O*-glucoside. Several hypotheses of naringenin-*O*-hexoside isomers can be proposed for their identification, including (+)- and (−)-naringenin 5-*O*-glucoside, as these have previously been found in *S. daphnoides* bark [11,12]. Nevertheless, the UV and MS data did not allow to distinguish the chemical structure of the glycosyl substituent and its position in the naringenin. Therefore, it was not possible to unequivocally identify compounds 10 and 11, only by UV spectra and MS data, being compounds' isolation and chemical structure elucidation by NMR needed for their unambiguous identification. Compound 15 was identified as naringenin, based on the UV spectrum, the detection of the $[M-H]^-$ ion at *m/z* 271, and the MS^2 spectrum (Table 3). Furthermore, its identification was confirmed by running a commercial standard in the UHPLC-UV-MS system, at the same experimental conditions.

Compounds 13 and 14 were tentatively assigned as eriodictyol-*O*-hexoside isomer and eriodictyol (Figure 2), respectively, considering their UV spectra (Figure S1A, in Supplementary Material), the detection of the $[M-H]^-$ ions at, respectively, *m/z* 449 and *m/z* 287, and the respective MS^n fragmentation (Table 3) [33,48]. Regarding compound 13, the MS^2 spectrum of the ion at *m/z* 449 showed the base peak at *m/z* 287, as a consequence of the hexosyl unit loss (−162 amu), whose MS^3 spectrum demonstrated the characteristic product ions of the eriodictyol mass fragmentation [33]. Also, the earlier elution order of eriodictyol-*O*-hexoside in comparison to naringenin 7-*O*-glucoside is concordant with that reported in the literature [33]. Furthermore, the identification of eriodictyol was confirmed by comparing its retention time, molecular absorption UV spectrum, and MS data with that of a commercial standard.

Flavonols

Compound 12 was identified as quercetin 3-*O*-galactoside (Figure 2), taking into account its UV spectrum, the detection of the $[M-H]^-$ ion at *m/z* 463, and the MS^n data (Table 3) [49]. In fact, the MS^2 spectrum of the $[M-H]^-$ ion presented the base peak at *m/z* 301, whose product ion resulted from a hexosyl unit loss ($[M-H-162]^-$. Furthermore, the MS^3 spectrum of the ion at *m/z* 301 was concordant with that of quercetin [47]. The identification of compound 12 was unambiguously confirmed with the injection of the commercial standard of quercetin 3-*O*-galactoside in the HPLC-UV-MS system, under the same experimental conditions.

3.2.2. Quantification of Identified Phenolic Compounds by UHPLC-UV Analysis

Table 4 depicts the contents of phenolic compounds present in the studied *Salix* spp. methanol/water/acetic acid extracts, expressed in mg kg^{-1} of dry weight (dw) and in mg g^{-1} of extract.

The total contents of identified phenolic compounds ranged from 490 mg kg^{-1} dw in *S. viminalis* L. bark (4.83 mg g^{-1} of extract) to 2871 mg kg^{-1} dw in *S. atrocinerea* Brot. bark (19.18 mg g^{-1} of extract).

Acetophenones represented the predominant phenolic compounds identified in *S. atrocinerea* Brot. bark extracts, accounting for 2155 mg kg^{-1} dw (14.42 mg g^{-1} of extract), as well as in *S. fragilis* L. bark extracts (1564 mg kg^{-1} dw and 16.15 mg g^{-1} of extract), mainly represented by piceol (**7**). Picein (**3**) was the second major phenolic compound identified in *S. atrocinerea* Brot. bark extracts, while it was present in *S. fragilis* L. bark extracts at a much lower content (up to 29-fold).

Table 4. Abundance of phenolic compounds in the methanol/water/acetic acid extracts (49.5:49.5:1) of *Salix atrocinerea* Brot., *Salix fragilis* L., and *Salix viminalis* L. barks.

No.	Compound	λ (nm)	mg kg^{-1} of Dry Weight			mg g^{-1} of Extract		
			Salix atrocinerea Brot.	*Salix fragilis* L.	*Salix viminalis* L.	*Salix atrocinerea* Brot.	*Salix fragilis* L.	*Salix viminalis* L.
1	(Epi)gallocatechin-(epi)catechin dimer isomer [A]	280	213	–	–	1.40	–	–
2	B-type procyanidin dimer isomer 1 [A]	280	–	–	19	–	–	0.19
4	Procyanidin B1[A]	280	404 [F(4+5)]	–	159 [F(4+5+6)]	2.70 [F(4+5)]	–	1.55 [F(4+5+6)]
5	Catechin [A]	280	[F(4+5)]	146	[F(4+5+6)]	[F(4+5)]	1.51	[F(4+5+6)]
6	B-type procyanidin dimer isomer 2 [A]	280	–	–	[F(4+5+6)]	–	–	[F(4+5+6)]
8	B-type procyanidin dimer isomer 3 [B]	235	–	–	[F(8+9), G]	–	–	[F(8+9), G]
	Σ Flavan-3-ols		**617**	**146**	**178**	**4.10**	**1.51**	**1.73**
3	Picein [C]	280	797	27	–	5.32	0.28	–
7	Piceol [C]	280	1358	1537	–	9.10	15.87	–
	Σ Acetophenones		**2155**	**1564**	**–**	**14.42**	**16.15**	**–**
9	Salicylic acid [B]	235	traces	58	200 [F(8+9), G]	traces	0.59	2.00 [F(8+9), G]
	Σ Hydroxybenzoic acids		**traces**	**58**	**200**	**traces**	**0.59**	**2.00**
10	Naringenin-*O*-hexoside isomer 1 [D]	280	6	–	–	0.04	–	–
11	Naringenin-*O*-hexoside isomer 2 [D]	280	13	–	–	0.09	–	–
13	Eriodictyol-*O*-hexoside isomer [D]	280	–	–	51	–	–	0.50
14	Eriodictyol [D]	280	–	–	52	–	–	0.51
15	Naringenin [D]	280	44	5	–	0.30	0.05	–
	Σ Flavanones		**64**	**5**	**103**	**0.43**	**0.05**	**1.00**
12	Quercetin 3-*O*-galactoside [E]	370	35	6	10	0.23	0.06	0.09
	Σ Flavonols		**35**	**6**	**10**	**0.23**	**0.06**	**0.09**
	TOTAL		**2871**	**1779**	**490**	**19.18**	**18.37**	**4.83**

The results represent the means obtained from *Salix* spp. bark extracts injected in triplicate (standard deviation less than 5%). Standard curves used for the quantification of phenolic compounds: [A] catechin; [B] *m*-hydroxybenzoic acid; [C] piceol; [D] naringenin; [E] quercetin. [F] The abundance of co-eluting compounds 4, 5, and 6 was determined at 280 nm, using the calibration curve of catechin. [G] The abundance of co-eluting compounds 8 and 9 was assayed at 235 nm, through the calibration curve of *m*-hydroxybenzoic acid, as the maximum absorbance was higher at 235 nm than at 280 nm.

S. atrocinerea Brot. bark also contained the highest flavan-3-ol content, representing 617 mg kg^{-1} dw (4.10 mg g^{-1} of extract), and being up to 4-fold higher than in *S. fragilis* L. bark. In particular, procyanidin B1 (**4**) and catechin (**5**) were the major flavan-3-ols present in *S. atrocinerea* Brot. bark extracts. Salicylic acid (**9**) was majorly present in *S. viminalis* L. bark extracts (200 mg kg^{-1} dw and 2.00 mg g^{-1} of extract). Additionally, *S. viminalis* L. bark extracts demonstrated the highest flavanone content (103 mg kg^{-1} dw and 1.00 mg g^{-1} of extract), being 21-fold higher relative to *S. fragilis* L., whereas *S. atrocinerea* Brot. bark extracts showed the highest flavonol content (up to 6-fold higher when compared with *S. fragilis* L. bark extracts).

Taking in account the differentiated phenolic composition of *Salix* spp. bark extracts, in addition to the reported antioxidant, anti-hypertensive, and antimicrobial effects of the identified phenolic compounds [9,14,15,17], three in vitro biological activities were evaluated, as follows: (1) antioxidant activity, via scavenging effects against DPPH• and ABTS•$^+$ free radicals; (2) anti-hypertensive potential, via the inhibitory effect on ACE enzymatic activity; and (3) antibacterial action via inhibitory effect against *S. aureus*. Finally, to ensure that the extracts can be safely used, for instance in food, nutraceutical, or cosmetic formulations, the in vitro biocompatibility of *Salix* spp. bark extracts was also conducted in three mammalian cell lines, namely, Caco-2, HaCaT, and L929 cell lines.

3.3. In Vitro Bioactivity of Salix spp. Bark Polar Extracts

3.3.1. Antioxidant Activity

The antioxidant activity of *Salix* spp. bark polar extracts was assessed through the DPPH• and ABTS•$^+$ scavenging effect assays, as denoted in Table 5.

Table 5. Antioxidant activity of methanol/water/acetic acid extracts (49.5:49.5:1) of *Salix atrocinerea* Brot., *Salix fragilis* L., and *Salix viminalis* L. bark, through DPPH• and ABTS•$^+$ scavenging effects.

Salix spp. Bark Extract/Reference	DPPH• Scavenging Effect			ABTS•$^+$ Scavenging Effect
	IC_{50} (µg mL^{-1})	IC_{50} (mg AAE g^{-1} of Dry Bark)	AAI	IC_{50} (µg mL^{-1})
Salix atrocinerea Brot.	10.98 ± 0.77 ^{a,b}	54.41 ± 8.22 ^{b}	5.64	5.58 ± 0.72 ^{a,b}
Salix fragilis L.	23.62 ± 4.82 ^{c}	16.79 ± 3.54 ^{a}	2.62	10.24 ± 1.54 ^{c}
Salix viminalis L.	14.06 ± 1.73 ^{b}	28.63 ± 4.34 ^{a}	4.40	7.82 ± 0.45 ^{b,c}
Ascorbic acid	3.92 ± 0.08 ^{a}	-	-	3.37 ± 0.06 ^{a}

The results represent the mean ± standard deviation ($n = 9$). Means with different superscript minor case letters (a–c) within the same column are statistically different (one-way ANOVA, followed by Tukey's HSD test, $p < 0.05$). AAE, ascorbic acid equivalents; AAI, antioxidant activity index; IC_{50}, inhibitory concentration at 50%.

S. atrocinerea Brot. bark extracts were the most active in scavenging the DPPH• and ABTS•$^+$ (IC_{50} of 10.98 and 5.58 µg mL^{-1}, respectively), although their IC_{50} were not statistically different from *S. viminalis* L. extracts ($p > 0.05$), when using Tukey's HSD test for pairwise multiple comparison procedure. On the other hand, *S. fragilis* L. bark extracts presented the lowest antioxidant effect, with IC_{50} of 23.62 and 10.24 µg mL^{-1}, respectively in the DPPH• and ABTS•$^+$ scavenging activities. Particularly in what concerns to the ABTS•$^+$ scavenging effect, the IC_{50} of *S. fragilis* L. extracts was not statistically different from that of *S. viminalis* L. extracts ($p > 0.05$), using Tukey's HSD test. Comparing the DPPH• and ABTS•$^+$ scavenging effects of *Salix* spp. bark extracts with that of ascorbic acid, *S. atrocinerea* Brot. was 2.8- and 1.7-fold less effective than the natural antioxidant standard in the respective assays, but with no statistical differences using the Tukey's HSD test ($p > 0.05$) were observed. Furthermore, the antioxidant activity of *S. fragilis* L. and *S. viminalis* L. bark extracts was significantly weaker than that of ascorbic acid ($p < 0.05$) in both assays. According to the AAI rank suggested by Scherer and Godoy [36], all *Salix* spp. bark extracts presented very strong antioxidant activity in the DPPH• assay.

3.3.2. Angiotensin-I Converting Enzyme Inhibitory Activity

The inhibitory effects of *Salix* spp. bark extracts were assessed at 625 μg mL^{-1} against the enzymatic activity of ACE, as illustrated in Figure 3.

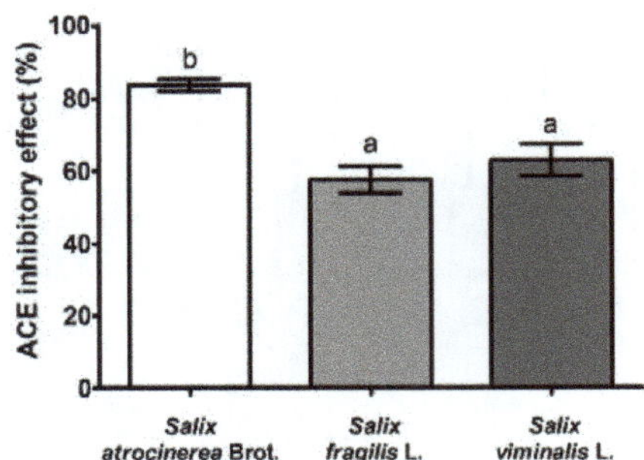

Figure 3. Inhibitory effect of 625 μg mL^{-1} *Salix atrocinerea* Brot., *Salix fragilis* L., and *Salix viminalis* L. bark extracts against the angiotensin I-converting enzyme (ACE). Each column and bar represents the mean and the standard deviation, respectively (n = 4). Columns with different minor case letters (a, b) are statistically different (one-way ANOVA, followed by Tukey's HSD test, $p < 0.05$).

Hence, *S. atrocinerea* Brot. bark polar extracts largely decreased the enzymatic activity of ACE (84 ± 2% of inhibition), being 1.5- and 1.3-fold more active than *S. fragilis* L. (58 ± 4%) and *S. viminalis* L. (63 ± 4%) bark extracts, respectively ($p < 0.05$). To the best of our knowledge, the inhibitory effect of *Salix* spp. bark extracts on ACE was evaluated herein for the first time, showing the promising potential for the anti-hypertensive purpose.

3.3.3. Inhibitory Effect against *S. aureus* Growth

The inhibitory effects of *S. atrocinerea* Brot., *S. fragilis* L., and *S. viminalis* L. bark extracts were tested for 24 h against the growth of the Gram-positive bacterium *S. aureus*, as depicted in Figure 4.

All the 24 h-treatments reduced the bacterial growth in a concentration-dependent manner, but statistical differences were not found with 625 μg mL^{-1} *Salix* spp. bark extracts, when compared with the growth control group ($p > 0.05$). Notwithstanding, all *Salix* spp. bark extracts tested at 1250 and 2500 μg mL^{-1} reduced significantly the bacterial growth regarding to the control group ($p < 0.05$), exhibiting bactericidal effects, as caused ≥3 log CFU mL^{-1} decrease. More specifically, 1250 μg mL^{-1} *S. atrocinerea* Brot. led to 6 log CFU mL^{-1} decrease, whereas *S. fragilis* L. and *S. viminalis* L. extracts to 7 log CFU mL^{-1} reduction. It is noteworthy to highlight that no bacterial colonies were detected after the treatments with 2500 μg mL^{-1} *S. atrocinerea* Brot. and *S. fragilis* L. extracts (8 log CFU mL^{-1} reduction), whilst *S. viminalis* L. extracts decreased significantly 7 log CFU mL^{-1}, at the same concentration, in comparison with the control ($p < 0.05$).

3.4. In Vitro Biocompatibility of Salix spp. Bark Polar Extracts

The cytotoxicity of *S. atrocinerea* Brot., *S. fragilis* L., and *S. viminalis* L. bark extracts was assayed at the 625–2500 μg mL^{-1} range for 24 h in Caco-2, HaCaT, and L929 cells (Figure 5).

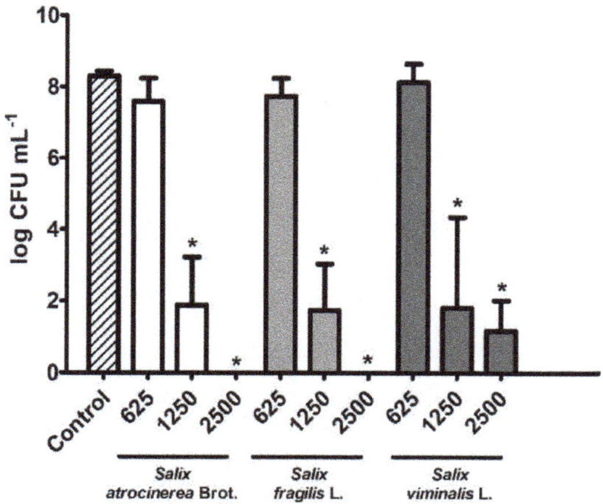

Figure 4. Bacterial density expressed as log CFU mL^{-1} of *Staphylococcus aureus* ATCC® 6538, after 24 h of incubation with 625, 1250, and 2500 µg mL^{-1} of *Salix atrocinerea* Brot., *Salix fragilis* L., and *Salix viminalis* L. bark polar extracts. Growth bacterial control is also depicted. Each column and bar represents the mean and the standard deviation, respectively ($n = 6$). Columns with the symbol * are statistically different from the growth control (one-way ANOVA, followed by Tukey's HSD test, $p < 0.05$). CFU, colony forming unit.

According to the international standard ISO 10993-5 for the biological evaluation of medical devices (part 5: Tests for in vitro cytotoxicity), the threshold value for a sample to be cytotoxic is a metabolic inhibition of 30%. As such, as can be seen in Figure 5A, none of the tested *Salix* spp. bark extracts exerted a cytotoxic effect against Caco-2 cells. In fact, some of the concentrations appeared to stimulate the mitochondrial metabolism of this cell line. For HaCaT cells (Figure 5B), none of the tested concentrations of *S. fragilis* L. bark extracts exhibited a cytotoxic effect against this cell line, with the *S. atrocinerea* Brot bark extracts at 625 and 1250 µg mL^{-1} demonstrating the same effect. In fact, only the highest concentration of *S. atrocinerea* Brot. exhibited a clear cytotoxic effect, as well as the two lower concentrations of *S. viminalis* L. (625 and 1250 µg mL^{-1}) resulted in a metabolic inhibition which is close to the threshold value, therefore requiring further studies (namely, to study the production of apopototic markers), particularly as *S. viminalis* L. bark extract at 2500 µg mL^{-1} did not exert a cytotoxic effect. The mouse fibroblast L929 cells (Figure 5C) appeared to be more susceptible to the presence of the extracts than the remaining tested cell lines, with the highest concentration of *S. atrocinerea* Brot. and *S. fragilis* L. extracts, along with all concentrations of *S. viminalis* L. extracts, resulting in metabolic inhibitions above 30%. Overall, it is important to mark that, at 625 and 1250 µg mL^{-1}, *S. atrocinerea* Brot. and *S. fragilis* L. extracts did not exert a cytotoxic effect against any of the tested cell lines.

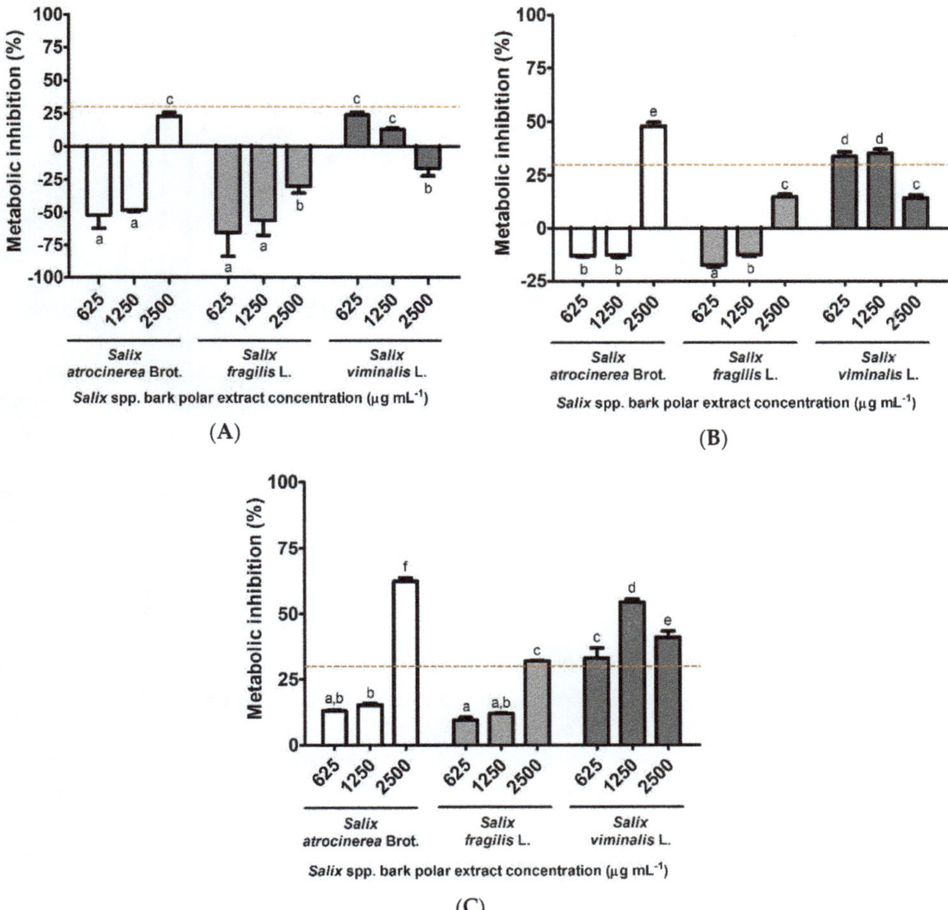

Figure 5. Metabolic inhibition of *Salix atrocinerea* Brot., *Salix fragilis* L., and *Salix viminalis* L. bark polar extracts at 625, 1250, and 2500 μg mL^{-1} for 24 h against three mammalian cell lines, namely: (**A**) human colorectal adenocarcinoma Caco-2 cells; (**B**) human keratinocyte HaCaT cells; and (**C**) mouse fibroblast L929 cells. Each column and bar represents the mean and the standard deviation, respectively ($n = 5$). Columns with different minor case letters (a–e) are statistically different (one-way ANOVA, followed by Tukey's HSD test, $p < 0.05$).

4. Discussion

The present work describes, for the first time, the detailed phenolic characterization, as well as the in vitro bioactivity and biocompatibility, of *S. atrocinerea* Brot., *S. fragilis* L., and *S. viminalis* L. bark polar extracts, aiming at their sustainable and safer bioprospection towards novel and innovative food, nutraceutical, and/or cosmetic applications.

Using methanol/water/acetic acid (49.5:49.5:1) solution for the phenolic compounds' extraction from the studied *Salix* spp. barks, the EY ranged from 9.7% in *S. fragilis* L. to 15.1% (*w*/*w*) in *S. atrocinerea* Brot. barks (Table 2). Comparing with the literature data for the distinct *Salix* species and extraction solvents, the EY of *S. atrocinerea* Brot. bark was 1.2-fold higher than that of 70% (*v*/*v*) acetone extract of *S. psammophila* bark, but 1.8- and 2.0-fold lower than the same kind of extracts of *S. sachalinensis* and *S. pet-susu* bark, respectively [50].

Moreover, *S. atrocinerea* Brot. bark showed the highest TPC (Table 2), accounting for 44.47 g GAE kg^{-1} dw and 293.36 mg GAE g^{-1} of extract, but with no statistical significance ($p > 0.05$) when comparing its TPC expressed in mg GAE g^{-1} of extract with that of *S. viminalis* L. bark. *S. fragilis* L. bark also demonstrated considerable TPC, i.e., 17.47 g GAE kg^{-1} dw and 179.06 mg GAE g^{-1} of extract. TPC of *S. atrocinerea* Brot. bark was 2.8-fold higher relative to *S. psammophila* bark [50], but lower than *S. subserrata* Willd. (up to 1.8-fold) [51], *S. aegyptiaca* L. (up to 4.8-fold) [52], *S. sachalinensis* (2.3-fold), and *S. pet-susu* (2.5-fold) barks [50]. Nevertheless, some caution should be taken in these comparisons, since *Salix* barks from different species and geographical origins were used, in addition to the different extraction media and methodologies applied, obviously affecting EY and TPC.

Fifteen phenolic compounds were found in bark polar extracts of the three *Salix* spp. in study, by UHPLC-UV-MS^n (Figure 1 and Table 3), namely six flavan-3-ols (**1**, **2**, **4–6**, and **8**), two acetophenones (**3** and **7**), a hydroxybenzoic acid (**9**), five flavanones (**10**, **11**, **13–15**), and a flavonol (**12**) (Figure 2). Regarding to flavan-3-ols, procyanidin B1 (**4**) and catechin (**5**) have been previously detected in *S. viminalis* L. bark [31,53]. In addition to procyanidin B1 (**4**), three other B-type procyanidin dimer isomers (**2**, **6**, and **8**) were also detected in *S. viminalis* L. bark. However, no B-type procyanidin dimer isomers were herein identified in *S. fragilis* L. bark, contrarily to what reported by Pobłocka-Olech and Krauze-Baranowska [31], which may be related not only with the extraction methodology and analytical techniques, but also with the geographical origin, climatic conditions, season, plant age, and genotype-phenotype associations. Still, it is remarkable the number of B-type procyanidins besides procyanidin B1 that have already been identified in several *Salix* spp., namely, procyanidins B3, B6, and B7 [31], which may potentiate interesting applications of this biomass in the food and health fields, due to their vast biological effects, including antioxidant.

This work also evidences, for the first time, the identification of catechin (**5**) in *S. atrocinerea* Brot. and *S. fragilis* L. barks, as well as a prodelphinidin dimer isomer (**1**) and procyanidin B1 (**4**) in *S. atrocinerea* Brot. bark. Considering the acetophenones, picein (**3**) was identified here, for the first time, in *S. atrocinerea* Brot. and *S. fragilis* L. barks. Furthermore, this phenolic compound has been described in the bark of other *Salix* species, namely, *S. daphnoides* [11], *S. purpurea* [13], and willow hybrid "Karin" [10]. In the same sense, piceol (**7**) and salicylic acid (**9**) were also found in the studied extracts of the two aforementioned *Salix* species, being recently identified in the respective lipophilic fractions [32].

In what concerns flavanones, two naringenin-*O*-hexoside isomers (**10** and **11**) were herein identified for the first time as phenolic constituents of the three analyzed *Salix* spp. barks. Naringenin (**15**) has recently been detected in *S. fragilis* L. bark [54], but it is revealed, for the first time in this work, as a phenolic component of *S. atrocinerea* Brot. bark. It is worth underlining that (+)- and (−)-naringenin 5-*O*-glucoside, naringenin 7-*O*-glucoside, and naringenin (**15**) have also been found in the barks of *S. daphnoides* and *S. purpurea* [11–13,30]. Eriodictyol-7-*O*-glucoside and eriodictyol (**14**) have been isolated from a commercial willow bark extract, with *S. fragilis* L. bark included in the formulation [55,56]. However, to the best of our knowledge, an eriodictyol-*O*-hexoside isomer (**13**) and eriodictyol (**14**) were reported herein for the first time in *S. atrocinerea* Brot. and *S. viminalis* L. barks. Regarding flavonols, only quercetin 3-*O*-galactoside (**12**) was found in *S. atrocinerea* Brot., *S. fragilis* L., and *S. viminalis* L. barks, being described in these raw materials for the first time.

Considering their quantitative analysis (Table 4), the total contents of identified phenolic compounds varied between 490 mg kg^{-1} dw in *S. viminalis* L. bark and 2871 mg kg^{-1} dw in *S. atrocinerea* Brot. bark. Comparing these results with TPC (Table 2), not only they did not follow the same trend as TPC, but they also corresponded to a minor part of TPC (ca. 2–10%), similar to what has been observed with other shrubs [57]. This may be explained by an array of extracts' components other than phenolic compounds that can react with the Folin–Ciocalteu's reagent in alkaline medium, including sugars and organic acids, among others [58].

Acetophenones were the main phenolic constituents of *S. atrocinerea* Brot. and *S. fragilis* L. barks, accounting for 2155 and 1564 mg kg^{-1} dw, respectively. Piceol (**7**) abundance is clearly higher in both *Salix* bark polar extracts than in the respective lipophilic fractions [32], while picein (**3**) content is up to 10.5-fold higher than in *S. caprea* L. bark [59], but up to 5.7-fold lower than that described for *S. phylicifolia* L., *S. myrsinifolia* Salisb., and *S. pentandra* L. barks [60], which may be related to the aforementioned factors.

Flavan-3-ols were also present at considerable contents in *S. atrocinerea* Brot. bark, accounting for 617 mg kg^{-1} dw, followed by *S. viminalis* L. and *S. fragilis* L. barks, but being 4.5-fold lower than the one reported for *S. viminalis* L. bark [53]. Minor abundances of flavanones and flavonols were detected in the studied *Salix* species.

Due to the antioxidant [9,61], anti-hypertensive [14,15], and antimicrobial [17] effects exhibited by the analyzed phenolic compounds, *Salix* spp. bark polar extracts were evaluated for these biological activities.

The antioxidant activity of *Salix* spp. bark polar extracts was assessed through the scavenging activity against DPPH• and $ABTS\bullet^{+}$ radicals (Table 5). Ascorbic acid was used as a natural antioxidant reference. Indeed, *S. atrocinerea* Brot. extracts were more active in scavenging DPPH• and $ABTS\bullet^{+}$ radicals, although with no statistical difference ($p > 0.05$) when using Tukey's HSD test, in comparison with *S. viminalis* L. extracts. Despite evidencing higher IC_{50} values, the antioxidant activity of *S. atrocinerea* Brot. extracts was not significantly different from ascorbic acid in both assays ($p > 0.05$), using Tukey's HSD test. Nevertheless, the DPPH• scavenging effect of *Salix* spp. bark extracts can be considered as very strong, according to the AAI (Table 5) [36]. Yet, taking the AAI in consideration, *S. atrocinerea* Brot. bark extracts are 8.5- and 2.3-fold stronger than, respectively, *S. alba* L. bark 70% methanol [62] and *S. aegyptiaca* L. bark ethanol extracts [52], but slightly weaker (1.4-fold) than *S. subserrata* Willd. bark 80% methanol extracts [51]. Comparing the $ABTS\bullet^{+}$ scavenging effect of the studied *Salix* spp. bark extracts with other species, all are considerably much stronger than the water extracts of *S. myrsinifolia* and *S. purpurea* barks (IC_{50} values of 7 and 20 mg mL^{-1}, respectively) [63]. Bridging the phenolic composition with the antioxidant activity of the studied *Salix* spp. bark extracts, the strongest significant correlation in each assay was found between flavan-3-ol content and DPPH• scavenging effect (Pearson's correlation, $r = -0.637$; $p < 0.033$), and between flavan-3-ol abundance and $ABTS\bullet^{+}$ scavenging effect (Pearson's correlation, $r = -0.669$; $p < 0.024$). Actually, flavan-3-ols like procyanidins have demonstrated strong DPPH• and $ABTS\bullet^{+}$ scavenging effects [64,65]. A significant correlation was also achieved between the flavonol abundance and $ABTS\bullet^{+}$ scavenging effect (Pearson's correlation, $r = -0.647$; $p < 0.030$). Moreover, flavanone and flavonol contents could be slightly correlated with DPPH• scavenging effect (Pearson's correlation, r values of −0.580 and −0.543 respectively), but they were not statistically significant ($p > 0.05$). TPC was also significantly correlated with the DPPH• scavenging effect (Pearson's correlation, $r = -0.665$; $p < 0.025$). Although there was a smooth correlation between TPC and $ABTS\bullet^{+}$ scavenging effect (Pearson's correlation, $r = -0.546$), it was not significant ($p > 0.05$). The hypothesis of synergisms occurring between flavan-3-ols and other phenolic compounds, or even other extracts' components, should indeed be placed.

Salix spp. bark phenolic-containing extracts were investigated for their anti-hypertensive potential (Figure 3), through the inhibitory effects against ACE. Hence, all *Salix* spp. bark extracts at 625 µg mL^{-1} diminished the enzymatic activity of ACE, ranging from 58 to 84% of inhibition, with *S. atrocinerea* Brot. bark extracts as the most effective ($p < 0.05$). To the best of our knowledge, the inhibitory effect of *Salix* spp. bark extracts on ACE was evaluated herein for the first time. In fact, the ACE inhibitory effect has been poorly approached for polar extracts of woody plants, as assayed with 70% ethanol extracts of *Populus tremula* L. (Salicaceae) bark, *Betula pendula* Rot. (Betulaceae) buds, and *Quercus robur* L. (Fagaceae) bark, at 100 µg mL^{-1}, and ranging from 11 to 28% of inhibition, respectively [66]. Flavan-3-ols may be strongly involved in the inhibitory effect of *S. atrocinerea* Brot. bark extracts, since these phenolic compounds have shown an interesting ACE inhibitory activity [14,15].

The inhibitory effect of *Salix* spp. bark polar extracts was also evaluated against the growth of the bacterium *S. aureus* ATCC® 6538 (Figure 4). Thus, the 24 h-treatments with *S. atrocinerea* Brot., *S. fragilis* L., and *S. viminalis* L. bark polar extracts reduced *S. aureus* growth, in a concentration-dependent manner (Figure 4), but statistical differences were not found at 625 μg mL^{-1} *Salix* spp. bark extracts, when compared with the control group ($p > 0.05$). On the other hand, *Salix* spp. bark extracts at 1250 and 2500 μg mL^{-1} significantly decreased the *S. aureus* growth compared to the control group ($p < 0.05$), with a 6–8 log CFU mL^{-1} range reduction, meaning that all extracts were bactericidal for this microorganism. Previous studies have demonstrated the anti-*S. aureus* potential of other *Salix* spp. bark extracts, including *S. mucronate* L. bark ethyl acetate (minimum inhibitory concentration (MIC) of 3125 μg mL^{-1}) [67] and *S. capense* extracts (5–1000 μg mL^{-1} MIC range) [68]. Phenolic compounds, like picein (**3**) and piceol (**7**), have also exhibited inhibitory effect against *S. aureus* growth, with MICs of 650 and 900 μg mL^{-1}, respectively [17]. Synergisms between phenolic and other extracts' constituents may have occurred, but a bioactive-guided fractionation should be conducted.

For future safe usage of *Salix* spp. bark phenolic-containing extracts, especially in what regards to food, nutraceutical, or cosmetic applications, their in vitro biocompatibility was addressed in human colorectal adenocarcinoma Caco-2 cells, human keratinocyte HaCaT cells, and mouse fibroblast L929 cells, analyzing the 24 h-inhibitory effects on cell metabolism (Figure 5). Globally, *S. atrocinerea* Brot. and *S. fragilis* L. bark extracts did not present cytotoxic effects at 625 and 1250 μg mL^{-1} against the three cell lines, as the metabolic inhibition was lower than 30%. Regarding to *S. viminalis* L. bark extracts, they were not cytotoxic against Caco-2 cells at all tested doses (Figure 5A), and against HaCaT cells at the highest concentration (Figure 5B), although all the tested concentrations suppressed the metabolism of L929 cells (Figure 5C) more than 30%. The cytotoxic potential of these *Salix* spp. bark extracts may be associated with some of the identified phenolic compounds, such as naringenin (**15**) and catechin (**5**), which have previously shown mild cytotoxic effect against H1299 human lung cancer cells after 24 h-incubation [69]. In this sense, the proposed applications of *Salix* spp. bark polar extracts should be tuned based on their non-cytotoxic concentrations.

In summary, *Salix* spp. bark polar extracts evidenced strong antioxidant activity, promising anti-hypertensive potential and effective antibacterial action against *S. aureus*. Notwithstanding, some attention should be paid to the non-cytotoxic concentrations of these extracts, being necessary to plan applications of *S. atrocinerea* Brot. and *S. fragilis* L. extracts for non-cytotoxic doses, and to better understand the cytotoxic effect of *S. viminalis* L. extracts. Moreover, an activity-guided fractionation is further needed in order to clarify the main bioactive constituents of *Salix* spp. bark extracts. Alternative extraction methodologies and solvents, like ultrasound and microwave-assisted extractions, and deep eutectic solvents should be considered for the extraction of *Salix* spp. bark phenolic compounds, intending their promising applicability in food, nutraceutical, and dermatological fields, towards the sustainable exploitation of this biomass and, at the same time, contributing for the biodiversity and rural profits.

5. Conclusions

The present study evidences, for the first time, the detailed phenolic characterization of three Portuguese *Salix* spp. bark samples, namely, *S. atrocinerea* Brot., *S. fragilis* L., and *S. viminalis* L., as well as the in vitro health-promoting potential of these polar extracts, such as antioxidant, anti-hypertensive, and antibacterial effects, and biocompatibility. Fifteen phenolic compounds were revealed in *Salix* spp. barks, by UHPLC-UV-MS^n, being two flavan-3-ols, an acetophenone, five flavanones, and a flavonol, detected for the first time in the studied *Salix* spp. barks. *S. atrocinerea* Brot. extracts demonstrated the highest total content of identified phenolic compounds (2871 mg kg^{-1} dw and 19.18 mg g^{-1} of extract), including acetophenones (2155 mg kg^{-1} dw and 14.42 mg g^{-1} of extract) and flavan-3-ols (617 mg kg^{-1} dw and 4.10 mg g^{-1} of extract). In what concerns the *in vitro* biological activity, *Salix* spp. bark extracts exhibit strong DPPH• and $ABTS•^+$ free radical scavenging effects (5.58–23.62 μg mL^{-1} IC_{50} range) and ACE inhibitory effects (58–84% of inhibition). Moreover, all extracts at 1250–2500 μg mL^{-1} exhibited bactericidal activity (6–8 log CFU mL^{-1} reduction) against *S. aureus*. The three in vitro biological activities may be mainly related to the presence of flavan-3-ols and acetophenones, but synergism effects may occur between these compounds and other extracts' phenolic subclasses or constituents.

Indeed, a bioactivity-guided fractionation should be further performed to clearly elucidate the bioactive component(s). Nonetheless, some caution should be taken in the safe use of these extracts, considering their non-cytotoxic doses. Overall, these promising insights can foster the economic valorization of the three studied Portuguese *Salix* spp., as raw materials of phenolic-containing extracts with an array of biological activities, towards innovative and novel food, nutraceutical, or cosmetic applications, along with the energy generation, being in line with the biorefinery concept.

Supplementary Materials: The following are available online at http://www.mdpi.com/2076-3921/8/12/609/s1, Figure S1: UHPLC-UV chromatograms of methanol/water/acetic acid (49.5:49.5:1) extracts, from (**A**) *Salix atrocinerea* Brot. and (**B**) *Salix viminalis* L. barks, recorded at 280 nm. The peak numbers correspond to compounds **1–6**, **8–11** and **13**. The molecular absorption UV spectra of these compounds are also depicted; Figure S2: Mass fragmentation of a prodelphinidin dimer isomer, under negative ionization mode.

Author Contributions: P.A.B.R. conceptualized and performed the experimental assays, analyzed the data, and prepared the original paper draft; S.A.O.S. carried out the UHPLC-UV-MS^n analysis and analyzed the respective data; C.M. and A.A. developed the antibacterial assays and analyzed the respective data; S.S., E.M.C., M.V., E.C., and M.M.P. conducted the ACE and cytotoxic assays and analyzed the respective data; S.A.O.S., A.A., M.M.P., C.S.R.F., A.M.S.S., and A.J.D.S. acquired funding and contributed to data interpretation; all authors greatly contributed to the critical review and editing of the paper.

Funding: The authors acknowledge FCT/MCTES for the financial support to the CICECO—Aveiro Institute of Materials (UID/CTM/50011/2019), to the QOPNA (UID/QUI/00062/2019), to the CBQF/ESB-UCP (UID/Multi/50016/2019), and to the CESAM (UID/MAR/LA0017/2019) through national funds; co-financing, where applicable, came from FEDER within the PT2020, and also to the Portuguese NMR Network. P.A.B.R., S.S., and E.C. acknowledge the "MultiBiorefinery" project (POCI-01-0145-FEDER-016403) for their post-doctoral grants. C.M. and S.A.O.S. thank the "AgroForWealth" project (CENTRO-01-0145-FEDER-000001) funded by Centro2020, through FEDER and PT2020 for the post-doctoral grant and the contract, respectively. C.S.R.F. also thanks FCT/MCTES for her contract under the Stimulus of Scientific Employment 2017 (CEECIND/00464/2017).

Acknowledgments: The authors are grateful to The Navigator Company and RAIZ for kindly supplying *Salix* spp. samples.

Conflicts of Interest: The authors declare no conflicts of interest.

References

1. EC A Sustainable Bioeconomy for Europe. *Strengthening the Connection between Economy, Society and the Environment—Updated Bioeconomy Strategy*; Publications Office of the European Union: Brussels, Belgium, 2018; pp. 4–16.
2. Parajuli, R.; Knudsen, M.T.; Dalgaard, T. Multi-criteria assessment of yellow, green, and woody biomasses: Pre-screening of potential biomasses as feedstocks for biorefineries. *Biofuels Bioprod. Biorefining* **2015**, *9*, 545–566. [CrossRef]
3. Wickham, J.; Rice, B.; Finnan, J.; McConnon, R. *A Review of Past and Current Research on Short Rotation Coppice in Ireland and Abroad: Report prepared for COFORD and Sustainable Energy Authority of Ireland*; COFORD, National Council for Forest Research and Development: Dublin, Ireland, 2010; pp. 1–36.
4. Djomo, S.N.; Kasmioui, O.E.L.; Ceulemans, R. Energy and greenhouse gas balance of bioenergy production from poplar and willow: A review. *Glob. Chang. Biol. Bioenergy* **2011**, 3, 181–197. [CrossRef]
5. Setty, A.R.; Sigal, L.H. Herbal medications commonly used in the practice of rheumatology: Mechanisms of action, efficacy, and side effects. *Semin. Arthritis Rheum.* **2005**, *34*, 773–784. [CrossRef]
6. EMA European Union herbal monograph on Salix (various species including *S. purpurea* L., *S. daphnoides* Vill., *S. fragilis* L.), cortex; London, 2017. Available online: https://www.ema.europa.eu/documents/herbal-monograph/final-european-union-herbal-monograph-salix-various-species-including-s-purpurea-l-s-daphnoides-vill_en.pdf (accessed on 5 July 2019).
7. Bonaterra, G.A.; Heinrich, E.U.; Kelber, O.; Weiser, D.; Metz, J.; Kinscherf, R. Anti-inflammatory effects of the willow bark extract STW 33-I (Proaktiv®) in LPS-activated human monocytes and differentiated macrophages. *Phytomedicine* **2010**, *17*, 1106–1113. [CrossRef]
8. Bonaterra, G.A.; Kelber, O.; Weiser, D.; Metz, J.; Kinscherf, R. In vitro anti-proliferative effects of the willow bark extract STW 33-I. *Arzneimittelforschung* **2010**, *60*, 330–335. [CrossRef]

9. Agnolet, S.; Wiese, S.; Verpoorte, R.; Staerk, D. Comprehensive analysis of commercial willow bark extracts by new technology platform: Combined use of metabolomics, high-performance liquid chromatography–solid-phase extraction–nuclear magnetic resonance spectroscopy and high-resolution radical scavenging assay. *J. Chromatogr. A* **2012**, *1262*, 130–137.
10. Dou, J.; Xu, W.; Koivisto, J.J.; Mobley, J.K.; Padmakshan, D.; Kögler, M.; Xu, C.; Willför, S.; Ralph, J.; Vuorinen, T. Characteristics of hot water extracts from the bark of cultivated willow (*Salix* sp.). *ACS Sustain. Chem. Eng.* **2018**, *6*, 5566–5573. [CrossRef]
11. Kammerer, B.; Kahlich, R.; Biegert, C.; Gleiter, C.H.; Heide, L. HPLC-MS/MS analysis of willow bark extracts contained in pharmaceutical preparations. *Phytochem. Anal.* **2005**, *16*, 470–478. [CrossRef] [PubMed]
12. Krauze-Baranowska, M.; Poblocka-Olech, L.; Glod, D.; Wiwart, M.; Zielinski, J.; Migas, P. HPLC of flavanones and chalcones in different species and clones of *Salix*. *Acta Pol. Pharm.* **2013**, *70*, 27–34. [PubMed]
13. Sulima, P.; Krauze-Baranowska, M.; Przyborowski, J.A. Variations in the chemical composition and content of salicylic glycosides in the bark of *Salix purpurea* from natural locations and their significance for breeding. *Fitoterapia* **2017**, *118*, 118–125. [CrossRef]
14. Actis-Goretta, L.; Ottaviani, J.I.; Keen, C.L.; Fraga, C.G. Inhibition of angiotensin converting enzyme (ACE) activity by flavan-3-ols and procyanidins. *FEBS Lett.* **2003**, *555*, 597–600. [CrossRef]
15. Tsutsumi, Y.; Shimada, A.; Miyano, A.; Nishida, T.; Mitsunaga, T. In vitro screening of angiotensin I-converting enzyme inhibitors from Japanese cedar (*Cryptomeria japonica*). *J. Wood Sci.* **1998**, *44*, 463–468. [CrossRef]
16. Han, X.; Pan, J.; Ren, D.; Cheng, Y.; Fan, P.; Lou, H. Naringenin-7-*O*-glucoside protects against doxorubicin-induced toxicity in H9c2 cardiomyocytes by induction of endogenous antioxidant enzymes. *Food Chem. Toxicol.* **2008**, *46*, 3140–3146. [CrossRef] [PubMed]
17. Zajdel, S.M.; Graikou, K.; Sotiroudis, G.; Glowniak, K.; Chinou, I. Two new iridoids from selected *Penstemon* species-antimicrobial activity. *Nat. Prod. Res.* **2013**, *27*, 2263–2271. [CrossRef] [PubMed]
18. Veiga, M.; Costa, E.M.; Silva, S.; Pintado, M. Impact of plant extracts upon human health: A review. *Crit. Rev. Food Sci. Nutr.* **2018**, 1–14. [CrossRef]
19. Kusumawati, I.; Indrayanto, G. Chapter 15—Natural Antioxidants in Cosmetics. In *Studies in Natural Products Chemistry*, 1st ed.; Atta-ur-Rahman, Ed.; Elsevier: Amsterdam, the Netherlands, 2013; Volume 40, pp. 485–505.
20. Kilfoyle, B.E.; Kaushik, D.; Terebetski, J.L.; Bose, S.; Michniak-Kohn, B.B. The use of quercetin and curcumin in skin care consumer products. In *Formulating, Packaging, and Marketing of Natural Cosmetic Products*; John Wiley & Sons, Ltd.: Hoboken, NJ, USA, 2011; pp. 259–286.
21. Coscueta, E.R.; Campos, D.A.; Osório, H.; Nerli, B.B.; Pintado, M. Enzymatic soy protein hydrolysis: A tool for biofunctional food ingredient production. *Food Chem. X* **2019**, *1*, 100006. [CrossRef]
22. Alves, M.J.; Ferreira, I.C.F.R.; Froufe, H.J.C.; Abreu, R.M.V.; Martins, A.; Pintado, M. Antimicrobial activity of phenolic compounds identified in wild mushrooms, SAR analysis and docking studies. *J. Appl. Microbiol.* **2013**, *115*, 346–357. [CrossRef]
23. Parreira, P.; Soares, B.I.G.; Freire, C.S.R.; Silvestre, A.J.D.; Reis, C.A.; Martins, M.C.L.; Duarte, M.F. *Eucalyptus* spp. outer bark extracts inhibit *Helicobacter pylori* Growth: In vitro studies. *Ind. Crop. Prod.* **2017**, *105*, 207–214. [CrossRef]
24. WHO Hypertension. Available online: https://www.who.int/news-room/fact-sheets/detail/hypertension (accessed on 5 July 2019).
25. Puchalska, P.; Alegre, M.L.M.; López, M.C.G. Isolation and characterization of peptides with antihypertensive activity in foodstuffs. *Crit. Rev. Food Sci. Nutr.* **2015**, *55*, 521–551. [CrossRef]
26. ECDC Surveillance of antimicrobial resistance in Europe 2017. In *Annual Report of the European Antimicrobial Resistance Surveillance Network (EARS-Net)*; ECDC: Stockholm, Sweden, 2018; pp. 54–56.
27. CDC Staphylococcal (Staph) Food Poisoning. Available online: https://www.cdc.gov/foodsafety/diseases/staphylococcal.html (accessed on 5 July 2019).
28. Masika, P.J.; Sultana, N.; Afolayan, A.J.; Houghton, P.J. Isolation of two antibacterial compounds from the bark of *Salix capensis*. *S. Afr. J. Bot.* **2005**, *71*, 441–443. [CrossRef]
29. Pobłocka-Olech, L.; Krauze-Baranowska, M.; Głód, D.; Kawiak, A.; Łojkowska, E. Chromatographic analysis of simple phenols in some species from the genus *Salix*. *Phytochem. Anal.* **2010**, *21*, 463–469. [CrossRef] [PubMed]

30. Pobłocka-Olech, L.; van Nederkassel, A.-M.; Vander Heyden, Y.; Krauze-Baranowska, M.; Glód, D.; Baczek, T. Chromatographic analysis of salicylic compounds in different species of the genus *Salix*. *J. Sep. Sci.* **2007**, *30*, 2958–2966. [CrossRef] [PubMed]
31. Pobłocka-Olech, L.; Krauze-Baranowska, M. SPE-HPTLC of procyanidins from the barks of different species and clones of *Salix*. *J. Pharm. Biomed. Anal.* **2008**, *48*, 965–968. [CrossRef] [PubMed]
32. Ramos, P.A.B.; Moreirinha, C.; Santos, S.A.O.; Almeida, A.; Freire, C.S.R.; Silva, A.M.S.; Silvestre, A.J.D. Valorisation of bark lipophilic fractions from three Portuguese *Salix* species: A systematic study of the chemical composition and inhibitory activity on *Escherichia coli*. *Ind. Crop. Prod.* **2019**, *132*, 245–252. [CrossRef]
33. Ramos, P.A.B.; Santos, S.A.O.; Guerra, Â.R.; Guerreiro, O.; Freire, C.S.R.; Rocha, S.M.; Duarte, M.F.; Silvestre, A.J.D. Phenolic composition and antioxidant activity of different morphological parts of *Cynara cardunculus* L. var. *altilis* (DC). *Ind. Crop. Prod.* **2014**, *61*, 460–471. [CrossRef]
34. Santos, S.A.O.; Vilela, C.; Domingues, R.M.A.; Oliveira, C.S.D.; Villaverde, J.J.; Freire, C.S.R.; Neto, C.P.; Silvestre, A.J.D. Secondary metabolites from *Eucalyptus grandis* wood cultivated in Portugal, Brazil and South Africa. *Ind. Crop. Prod.* **2017**, *95*, 357–364. [CrossRef]
35. Singleton, V.L.; Rossi, J.A. Colorimetry of total phenolics with phosphomolybdic-phosphotungstic acid reagents. *Am. J. Enol. Vitic.* **1965**, *16*, 144–158.
36. Scherer, R.; Godoy, H.T. Antioxidant activity index (AAI) by the 2, 2-diphenyl-1-picrylhydrazyl method. *Food Chem.* **2009**, *112*, 654–658. [CrossRef]
37. Re, R.; Pellegrini, N.; Proteggente, A.; Pannala, A.; Yang, M.; Rice-Evans, C. Antioxidant activity applying an improved ABTS radical cation decolorization assay. *Free Radic. Biol. Med.* **1999**, *26*, 1231–1237. [CrossRef]
38. Sentandreu, M.Á.; Toldrá, F. A rapid, simple and sensitive fluorescence method for the assay of angiotensin-I converting enzyme. *Food Chem.* **2006**, *97*, 546–554. [CrossRef]
39. Belley, A.; Neesham-Grenon, E.; Arhin, F.F.; McKay, G.A.; Parr, T.R.; Moeck, G. Assessment by time-kill methodology of the synergistic effects of oritavancin in combination with other antimicrobial agents against *Staphylococcus aureus*. *Antimicrob. Agents Chemother.* **2008**, *52*, 3820–3822. [CrossRef]
40. Friedrich, W.; Eberhardt, A.; Galensa, R. Investigation of proanthocyanidins by HPLC with electrospray ionization mass spectrometry. *Eur. Food Res. Technol.* **2000**, *211*, 56–64. [CrossRef]
41. Teixeira, N.; Azevedo, J.; Mateus, N.; de Freitas, V. Proanthocyanidin screening by LC-ESI-MS of Portuguese red wines made with teinturier grapes. *Food Chem.* **2016**, *190*, 300–307. [CrossRef] [PubMed]
42. Santos, S.A.O.; Vilela, C.; Camacho, J.F.; Cordeiro, N.; Gouveia, M.; Freire, C.S.R.; Silvestre, A.J.D. Profiling of lipophilic and phenolic phytochemicals of four cultivars from cherimoya (*Annona cherimola* Mill.). *Food Chem.* **2016**, *211*, 845–852. [CrossRef] [PubMed]
43. Cheynier, V. Flavonoids in wine. In *Flavonoids: Chemistry, Biochemistry, and Applications*; Andersen, Ø.M., Markham, K.R., Eds.; CRC Press: Boca Raton, FL, USA, 2006; pp. 263–318.
44. Attygalle, A.B.; Ruzicka, J.; Varughese, D.; Bialecki, J.B.; Jafri, S. Low-energy collision-induced fragmentation of negative ions derived from *ortho-*, *meta-*, and *para-*hydroxyphenyl carbaldehydes, ketones, and related compounds. *J. Mass Spectrom.* **2007**, *42*, 1207–1217. [CrossRef]
45. Mageroy, M.H.; Parent, G.; Germanos, G.; Giguère, I.; Delvas, N.; Maaroufi, H.; Bauce, É.; Bohlmann, J.; Mackay, J.J. Expression of the β-glucosidase gene *Pgβglu-1* underpins natural resistance of white spruce against spruce budworm. *Plant J.* **2015**, *81*, 68–80. [CrossRef]
46. Hossain, M.B.; Rai, D.K.; Brunton, N.P.; Martin-Diana, A.B.; Barry-Ryan, C. Characterization of phenolic composition in Lamiaceae spices by LC-ESI-MS/MS. *J. Agric. Food Chem.* **2010**, *58*, 10576–10581. [CrossRef]
47. Fabre, N.; Rustan, I.; de Hoffmann, E.; Quetin-Leclercq, J. Determination of flavone, flavonol, and flavanone aglycones by negative ion liquid chromatography electrospray ion trap mass spectrometry. *J. Am. Soc. Mass Spectrom.* **2001**, *12*, 707–715. [CrossRef]
48. Vallverdú-Queralt, A.; Jáuregui, O.; Medina-Remón, A.; Andrés-Lacueva, C.; Lamuela-Raventós, R.M. Improved characterization of tomato polyphenols using liquid chromatography/electrospray ionization linear ion trap quadrupole Orbitrap mass spectrometry and liquid chromatography/electrospray ionization tandem mass spectrometry. *Rapid Commun. Mass Spectrom.* **2010**, *24*, 2986–2992. [CrossRef]
49. Chang, Q.; Wong, Y.-S. Identification of flavonoids in Hakmeitau beans (*Vigna sinensis*) by high-performance liquid chromatography–electrospray mass spectrometry (LC-ESI/MS). *J. Agric. Food Chem.* **2004**, *52*, 6694–6699. [CrossRef]

50. Kubo, S.; Hashida, K.; Makino, R.; Magara, K.; Kenzo, T.; Kato, A. Aorigele Chemical composition of desert willow (*Salix psammophila*) grown in the Kubuqi desert, inner Mongolia, China: Bark extracts associated with environmental adaptability. *J. Agric. Food Chem.* **2013**, *61*, 12226–12231. [CrossRef] [PubMed]
51. Tawfeek, N.; Sobeh, M.; Hamdan, D.I.; Farrag, N.; Roxo, M.; El-Shazly, A.M.; Wink, M. Phenolic compounds from *Populus alba* L. and *Salix subserrata* Willd. (Salicaceae) counteract oxidative stress in *Caenorhabditis elegans*. *Molecules* **2019**, *24*, 1999. [CrossRef] [PubMed]
52. Enayat, S.; Banerjee, S. Comparative antioxidant activity of extracts from leaves, bark and catkins of *Salix aegyptiaca* sp. *Food Chem.* **2009**, *116*, 23–28. [CrossRef]
53. Pobłocka-Olech, L.; Krauze-Baranowska, M.; Wiwart, M. HPTLC determination of catechins in different clones of the genus *Salix*. *JPC-J. Planar Chromatogr.-Mod. TLC* **2007**, *20*, 61–64.
54. Gligorić, E.; Igić, R.; Suvajdžić, L.; Grujić-Letić, N. Species of the genus *Salix* L.: Biochemical screening and molecular docking approach to potential acetylcholinesterase inhibitors. *Appl. Sci.* **2019**, *9*, 1842.
55. Freischmidt, A.; Jürgenliemk, G.; Kraus, B.; Okpanyi, S.N.; Müller, J.; Kelber, O.; Weiser, D.; Heilmann, J. Contribution of flavonoids and catechol to the reduction of ICAM-1 expression in endothelial cells by a standardised willow bark extract. *Phytomedicine* **2012**, *19*, 245–252. [CrossRef]
56. Piazzini, V.; Bigagli, E.; Luceri, C.; Bilia, A.R.; Bergonzi, M.C. Enhanced solubility and permeability of salicis cortex extract by formulating as a microemulsion. *Planta Med.* **2018**, *84*, 976–984. [CrossRef]
57. Touati, R.; Santos, S.A.O.; Rocha, S.M.; Belhamel, K.; Silvestre, A.J.D. Phenolic composition and biological prospecting of grains and stems of *Retama sphaerocarpa*. *Ind. Crop. Prod.* **2017**, *95*, 244–255. [CrossRef]
58. Prior, R.L.; Wu, X.L.; Schaich, K. Standardized methods for the determination of antioxidant capacity and phenolics in foods and dietary supplements. *J. Agric. Food Chem.* **2005**, *53*, 4290–4302. [CrossRef]
59. Julkunen-Tiitto, R.; Tahvanainen, J. The effect of the sample preparation method of extractable phenolics of Salicaceae species. *Planta Med.* **1989**, *55*, 55–58. [CrossRef]
60. Meier, B.; Julkunen-Tiitto, R.; Tahvanainen, J.; Sticher, O. Comparative high-performance liquid and gas-liquid chromatographic determination of phenolic glucosides in Salicaceae species. *J. Chromatogr. A* **1988**, *442*, 175–186. [CrossRef]
61. Ishikado, A.; Sono, Y.; Matsumoto, M.; Robida-Stubbs, S.; Okuno, A.; Goto, M.; King, G.L.; Blackwell, T.K.; Makino, T. Willow bark extract increases antioxidant enzymes and reduces oxidative stress through activation of Nrf2 in vascular endothelial cells and *Caenorhabditis elegans*. *Free Radic. Biol. Med.* **2013**, *65*, 1506–1515. [CrossRef] [PubMed]
62. Zaiter, A.; Becker, L.; Petit, J.; Zimmer, D.; Karam, M.-C.; Baudelaire, É; Scher, J.; Dicko, A. Antioxidant and antiacetylcholinesterase activities of different granulometric classes of *Salix alba* (L.) bark powders. *Powder Technol.* **2016**, *301*, 649–656. [CrossRef]
63. Durak, A.; Gawlik-Dziki, U.; Sugier, D. Coffee enriched with willow (*Salix purpurea* and *Salix myrsinifolia*) bark preparation—interactions of antioxidative phytochemicals in a model system. *J. Funct. Foods* **2015**, *18*, 1106–1116. [CrossRef]
64. Janceva, S.; Lauberte, L.; Dizhbite, T.; Krasilnikova, J.; Telysheva, G.; Dzenis, M. Protective effects of proanthocyanidins extracts from the bark of deciduous trees in lipid systems. *Holzforschung* **2017**, *71*, 675–680. [CrossRef]
65. Villaño, D.; Fernández-Pachón, M.S.; Moyá, M.L.; Troncoso, A.M.; García-Parrilla, M.C. Radical scavenging ability of polyphenolic compounds towards DPPH free radical. *Talanta* **2007**, *71*, 230–235. [CrossRef] [PubMed]
66. Ivanov, S.A.; Garbuz, S.A.; Malfanov, I.L.; Ptitsyn, L.R. Screening of Russian medicinal and edible plant extracts for angiotensin I-converting enzyme (ACE I) inhibitory activity. *Russ. J. Bioorgan. Chem.* **2013**, *39*, 743–749. [CrossRef]
67. Eldeen, I.M.S.; Elgorashi, E.E.; van Staden, J. Antibacterial, anti-inflammatory, anti-cholinesterase and mutagenic effects of extracts obtained from some trees used in South African traditional medicine. *J. Ethnopharmacol.* **2005**, *102*, 457–464. [CrossRef]

68. Masika, P.J.; Afolayan, A.J. Antimicrobial activity of some plants used for the treatment of livestock disease in the Eastern Cape, South Africa. *J. Ethnopharmacol.* **2002**, *83*, 129–134. [CrossRef]
69. Jeon, S.H.; Chun, W.; Choi, Y.J.; Kwon, Y.S. Cytotoxic constituents from the bark of *Salix hulteni*. *Arch. Pharm. Res.* **2008**, *31*, 978–982. [CrossRef]

Article

Effects of *Lespedeza Bicolor* Extract on Regulation of AMPK Associated Hepatic Lipid Metabolism in Type 2 Diabetic Mice

Younmi Kim [1,†], Heaji Lee [1,†], Sun Yeou Kim [2] and Yunsook Lim [1,*]

1 Department of Food and Nutrition, Kyung Hee Univerity, 26 Kyung Hee-Daero, Dongdamun-Gu, Seoul 02447, Korea; younme_810@khu.ac.kr (Y.K.); ji3743@khu.com (H.L.)

2 Gachon Institute of Pharmaceutical Science, Gachon University, #191, Hambakmoero, Yeonsu-gu, Incheon 21936, Korea; sunnykim@gachon.ac.kr

* Correspondence: ylim@khu.ac.kr; Tel.: 82-2-961-0262; Fax: 82-961-0260

† These authors contributed equally to this work.

Received: 30 September 2019; Accepted: 27 November 2019; Published: 29 November 2019

Abstract: *Lespedeza bicolor* (LB) is one of the ornamental plants used for the treatment of inflammation caused by oxidative damage. However, its beneficial effects on hyperglycemia-induced hepatic damage and the related molecular mechanisms remain unclear. We hypothesized that *Lespedeza bicolor* extract (LBE) would attenuate hyperglycemia-induced liver injury in type 2 diabetes mellitus (T2DM). Diabetes was induced by a low dosage of streptozotocin (STZ) injection (30 mg/kg) with a high fat diet in male C57BL/6J mice. LBE was administered orally at 100 mg/kg or 250 mg/kg for 12 weeks. LBE supplementation regardless of dosage ameliorated plasma levels of hemoglobin A1c (HbA1c) in diabetic mice. Moreover, both LBE supplementations upregulated AMP-activation kinase (AMPK), which may activate sirtuin1 (SIRT) associated pathway accompanied by decreased lipid synthesis at low dose of LBE supplementation. These changes were in part explained by reduced protein levels of oxidative stress (nuclear factor erythroid 2-related factor 2 (Nrf2) and catalase), inflammation (nuclear factor kappa B (NF-κB), interleukin-1β (IL-1β), interleukin-6 (IL-6), and nitric oxide synthases (iNOS)), and fibrosis (α-smooth muscle actin (α-SMA) and protein kinase C (PKC)) in diabetic liver. Taken together, LBE might be a potential nutraceutical to ameliorate hepatic damage by regulation of AMPK associated pathway via oxidative stress, inflammation, and fibrosis in T2DM.

Keywords: *Lespedeza bicolor*; type 2 diabetes; AMPK; lipid metabolism; inflammation; oxidative stress; fibrosis

1. Introduction

With around more than 500 million prevalent cases in 2018, type 2 diabetes mellitus (T2DM) is one of the most frequent metabolic syndromes in the world. In T2DM, insulin resistance (IR) caused by hyperglycemia leads to various diabetic complications [1–3]. As an insulin-sensitive tissue, the liver is susceptible to hyperglycemia-induced oxidative stress, which can cause hepatic damage [4,5].

Oxidative stress causes an imbalance between free radicals and antioxidants and reduces proliferation of mature hepatocytes [6–9]. As a result, chronic oxidative stress proliferates hepatic stellate cells (HSCs), which play a key role in the progression of hepatic fibrosis [10]. Nuclear factor erythroid 2-related factor 2 (Nrf2), a major transcription factor, regulates cellular resistance to oxidant exposure [11]. It modulates its downstream antioxidant defense mediators such as catalase, glutathione peroxidase (GPx), NAD(P)H dehydrogenase quinone 1 (NQO1), and superoxide dismutase (SOD) which eliminate excessive reactive oxygen species (ROS) [11]. Furthermore, oxidative stress triggers nuclear factor kappa B (NF-κB), which modulates various inflammatory mediators including interleukin-1β (IL-1β), interleukin-6 (IL-6), and tumor necrosis factor α (TNFα), causing chronic inflammation [9,11].

On the other hand, there is substantial evidence suggesting that adenosine monophosphate -activation kinase (AMPK) is dysregulated in metabolic syndrome such as obesity and T2DM [12]. More specifically, sirtuin1 (SIRT1) and AMPK have clinical relevance with regard to type 2 diabetes because of their effects on various cellular metabolism such as energy turnover, glucose metabolism, and lipotoxicity [12]. When the level of AMP is increased, AMPK is activated to maintain energy homeostasis. SIRT1 is an NAD+-dependent protein deacetylase which acts as a major regulator of energy homeostasis in response to nutrient availability. AMPK relies on SIRT1 activity to regulate lipid metabolism related pathway [12,13]. Moreover, AMPK downregulates lipogenesis related factors such as sterol regulatory element-binding protein1 (SREBP1), SREBP2, and fatty acid synthase (FAS) [12]. AMPK activation also decreases CAAT box/enhancer binding protein alpha (C/EBPα), which upregulates adipocyte differentiation [12–14]. Therefore, activation of AMPK associated pathway would be a therapeutic mechanism to reduce hepatic lipid accumulation in diabetes.

Lespedeza bicolor (LB) is a perennial deciduous shrub belonging to the leguminosae and is cultivated for ornamental purposes throughout Asia [15]. LB has been used traditionally for the treatment of inflammation. LB contains various compounds such as genistein, quercetin, daidzein, catechin, rutin, luteolin, and naringin [16]. These natural phytoconstituents affluent in *Lespedeza bicolor* extract (LBE) have been confirmed to exert antioxidants, decreasing the blood glucose level and anti-inflammatory activity. Especially, genistein, quercetin, and naringin have antioxidant activities such as electron donating and ROS scavenging activity [17]. Importantly, the previous study has shown that LBE ameliorated endothelial dysfunction induced methylglyoxal glucotoxicity in vitro [18]. Furthermore, LB attenuated methylglyoxal (MGO)-induced diabetic renal damage in vitro and in vivo [19]. These results suggest that LB had a potential for preventing or curing diabetic complications related to hyperglycemia.

However, no research has focused on the effect of LB on hyperglycemia-induced hepatic damage and its molecular mechanism. In this study, we hypothesized isoflavones and quercetin enriched LBE would ameliorative the effect on hyperglycemia-induced hepatic lipid metabolism by regulation of lipid metabolism in T2DM.

2. Materials and Methods

2.1. Preparation of Lespedeza Bicolor Extracts (LBE)

The aerial parts of LB were purchased from Jayeonchunsa Co. (Damyang, Korea). The preprocessing of LB was described in our previous research [18]. Briefly, LB was extracted with 70% ethanol at room temperature overnight. Afterwards, the extract was filtered, evaporated, and freeze-dried. The extract was dissolved in distilled water at a concentration of 100 and 250 mg/kg body weight (BW), respectively. The concentration of each stock solution was 25 mg/ml (low dosage of LBE, LL) and 62.5 mg/ml (high dosage of LBE, HL), respectively.

2.2. Animals and Experimental Design

C57BL/6J male (n = 40; 5-weeks-old) mice were provided (Raon Bio, Gyeonggi-do, Korea) and lodged in a room at 22 ± 1 °C, 50 ± 5% suitable humidity, and 12 h dark/light cycle. In a constant environment (12 h light/dark cycle, 21 ± 1 °C, and 50 ± 3% humidity), food and distilled water were supplied ad libitum. A randomly allocated diabetic group were fed with 40% kcal high-fat diet, while a non-diabetic control group (NC) was fed with 10% kcal control diet (AIN-93G). After 4 weeks, the diabetic group was intraperitoneally injected with 30 mg/kg body weight (BW) of streptozotocin (STZ) twice to induce diabetes [20]. The normal control mice were injected with only a citric acid buffer. One week after the second injection, mice with fasting blood glucose (FBG) levels higher than 140.4 mg/dl were included in the diabetic group. After induction of diabetes for 9 weeks, all mice were divided into 4 experimental groups (n = 10 per group) as follows: normal control, NC; diabetes mellitus control, DMC; low dosage of LBE, LL; high dosage of LBE, HL. The treatment groups were administrated with 100 mg/kg BW (LL) and 250 mg/kg BW (HL) by oral gavage every day for 12 weeks.

LBE was freshly suspended in distilled water. At the same time, LBE untreated groups, the NC and DMC groups were treated with identical volumes of distilled water. During the treatment period, body weight, food intake, and fasting blood glucose (FBG) levels from the tail vein were measured once a week. At the end of treatment for 12 weeks, the animals were anesthetized by inhalation with diethyl ether (Duksan, Seoul, Korea). Blood sample was collected by heparin-coated (Sigma Aldrich, St. Louis, MO, USA) syringe from cardiac puncture and centrifuged at 845 g at 4 °C for 10 min to obtain plasma. The hepatic tissue was weighed and washed by saline. For protein extraction, part of the hepatic tissue was frozen in liquid nitrogen, and stored at −80 °C before experiments. Other parts of the hepatic tissue were fixed in 10% formaldehyde for paraffin embedding. All experiments were approved by Kyung Hee University for animal welfare (KHUASP(SE)-16-001) and were performed in accordance with the guidelines.

2.3. *Hemoglobin A1c (HbA1c)*

HbA1c levels were measured according to commercial reagent methods (Crystal Chem., Downers Grove, Elk Grove Village, IL, USA).

2.4. *Plasma Glutamate Oxaloacetate Transaminase (GOT) and Glutamate Pyruvate Transaminase (GPT)*

Plasma GOT and GPT were measured using commercial detection kits (Bio Clinical System, Gyeonggi-do, Korea).

2.5. *Lipid Profile Analysis*

Hepatic triglyceride (TG) and total cholesterol (TC) concentrations were measured using commercial kits (Bio-Clinical System, Gyeonggi-do, Korea) according to the manufacturer's recommendation.

2.6. *Histological Analysis*

Hepatic tissue was fixed in 10% buffered formalin and embedded in paraffin wax. Histological sections (4 µm) of hepatic tissue were stained with hematoxylin and eosin (H&E) for conventional morphological evaluation using an optical microscope (Olympus BX51; Olympus Optical, Tokyo, Japan).

2.7. *Western Blot Analysis*

Hepatic tissue was homogenized in lysis buffer (20 mM Tris-Hcl, 150 mM NaCl, pH 7.5, 1% NP40, 0.5% Na-deoxycholate stock, 1mM ethylene diamineteraacetic acid (EDTA) and 0.1% sodium dodecyl sulfate (SDS)) and then centrifuged at 9000 g at 4 °C for 30 min. The supernatants were used for hepatic cytosol protein extract, the pelleted nuclei remnants were resuspended in a hypertonic buffer containing glycerol, 10 mM 4-(2-hydroxyethyl)-1-piperazineethanesulfonic acid (HEPES), 4 mM NaCl, 1 mM $MgCl_2$, 500 mM EDTA, 1 mM dithiothreitol (DTT), phenylmethylsulfonyl fluoride (PMSF), 1 mM benzamidine, pepstatin, leupeptin, aprotinin, and distilled water. The lysed nuclei were stored at −80 °C until used for nuclear analysis. The hepatic extract was separated by 10% sodium dodecyl sulphate-polyacrylamide gel electrophoresis (SDS-PAGE) and transferred to a polyvinylidene fluoride (PVDF) membrane (Millipore, Billerica, MA, USA). The membranes were blocked with 3% bovine serum albumin (BSA) and incubated overnight at 4 °C with the primary antibodies: NF-κB, monocyte chemoattractant protein-1 (MCP-1), α-smooth muscle actin (α-SMA), catalase, C/EBPα, p-AMPK, AMPK, peroxisome proliferator-activated receptor-γ (PPARγ) (Cell Signaling Technology, Inc., Danvers, MA, USA, 1:500); Nrf2, SIRT1, nSREBP1, peroxisome proliferator-activated receptor-α (PPARα), FAS, GPx, NQO1, c-reactive protein (CRP), receptor AGE (RAGE) (Abcam, Cambridge, MA, USA, 1:1000); TNF-α, IL-1β, IL-6, MnSOD, transforming growth factor β (TGF-β), protein kinase C (PKC), CuZnSOD, peroxisome proliferator-activated receptor gamma coactivator 1-α (PGC1α), protein kinase C-βII (PKCβII), β-actin (Santa Cruz Biotechnology, Santa Cruz, CA, USA, 1:200); nitric oxide synthases (iNOS), cyclooxygenase-2 (COX2) (Stressgen, 1:1000); heme oxygenase-1 (HO-1), proliferating cell nuclear antigen (PCNA) (Enzolife science, 1:1000), 4-hydroxynonenal (4-HNE) (R&D system, Inc.). After washing by phosphate-buffered saline supplemented with Tween (PBS-T) three times, the membrane was then incubated with the relative secondary antibodies (Santa Cruz Biotechnology, Santa Cruz, CA, USA). After being washed an additional three times by PBS-T, the membranes were developed using the enhanced chemiluminescence (ECL) luminol reagent (Biorad, Hercules, CA, USA). The luminescent signal was recorded and quantified with the Syngene G box (Syngene, Cambridge, MA, USA).

2.8. Statistical Analysis

Results were presented as means ± SEM. The significance of difference was analyzed by one-way ANOVA followed by Tukey's test. A probability level of $p < 0.05$ was considered statically significant. All statistical analysis used SPSS software (version 20.0 K for windows, Armonk, NY, USA) and Graphpad Prism (Version 5.0, San Diego, CA, USA).

3. Results

3.1. Effects of LBE Supplementation on Body Weight, Food Intake, and FBG Level in T2DM

After diabetes was induced, the body weight and FBG level of the DMC group was significantly higher compared to that in the NC group. However, LBE supplementation did not change body weight, food intake, and FBG levels in the diabetic mice (Table 1).

Table 1. Effects of *Lespedeza bicolor* extract (LBE) supplementation on body weight and food intake in type 2 diabetes (T2DM) mice.

Group	NC	DMC	LL	HL
Body weight (g)				
Before treatment	26.58 ± 1.35 [a]	32.04 ± 3.29 [b]	30.94 ± 2.40 [b]	32.51 ± 4.13 [b]
After treatment	30.42 ± 1.83 [a]	43.93 ± 4.97 [b]	39.02 ± 5.15 [b]	40.34 ± 6.35 [b]
Gain	3.85 ± 0.33 [a]	8.89 ± 0.70 [b]	8.09 ± 1.34 [b]	7.82 ± 1.72 [b]
Food intake (g/day)	2.40 ± 0.731 [a]	3.36 ± 0.89 [a,b]	3.56 ± 1.70 [b]	4.16 ± 1.27 [b]
Fasting blood glucose (FBG) (mg/dL)	122.2 ± 16.78 [a]	173.8 ± 31.97 [b]	135.6 ± 13.28 [a,b]	170.8 ± 28.31 [b]

Values are means ± SEM ($n = 10$). Mean values with different letters ([a] and [b]) were significantly different. ($p < 0.05$).

3.2. Effects of LBE Supplementation on Glycation Products in T2DM Mice

HbA1c and advanced glycation end products receptor (RAGE) expression in plasma were used to estimate advanced glycation end products (AGE) formation. As shown in Figure 1A, HbA1c was significantly higher in the DMC group than that in the NC group. However, LBE supplementation lowered the HbA1c level in the diabetic mice regardless of dose.

The protein levels of plasma RAGE in the DMC group were significantly higher compared to that in the NC group. The HL group showed significantly lower levels of RAGE than that in the DMC group. However, the protein level of RAGE in the LL group was not significantly different from that in the DMC group (Figure 1B).

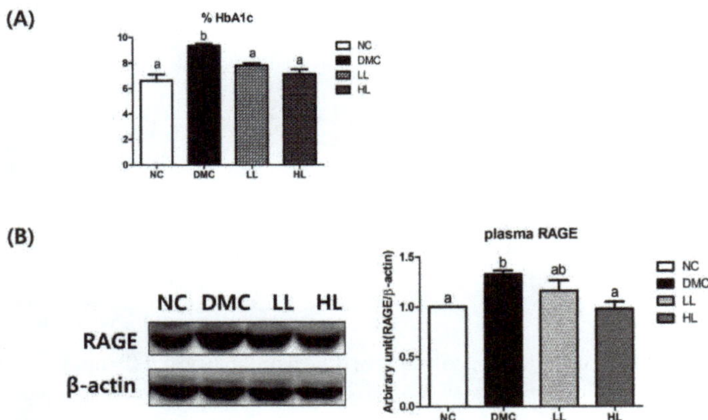

Figure 1. Effects of LBE supplementation on glycation products in T2DM mice. (**A**) %HbA1c and (**B**) plasma advanced glycation end products receptor (RAGE). Values are means ± SEM (n = 6). Mean values with different letters ([a] and [b]) were significantly different. ($p < 0.05$). NC: non-diabetic control group; DMC: diabetes mellitus control; LL: low dosage of LBE; HL: high dosage of LBE.

3.3. Effects of LBE Supplementation on Plasma GOT and GPT in T2DM Mice

Plasma GOT and GPT levels were measured as biomarkers of liver injury. GOT and GPT levels were significantly higher in the DMC group than those in the NC group. Low dosage of LBE supplementation significantly lowered GOT and GPT levels compared to the DMC group whereas a high dosage of LBE supplementation did not (Figure 2).

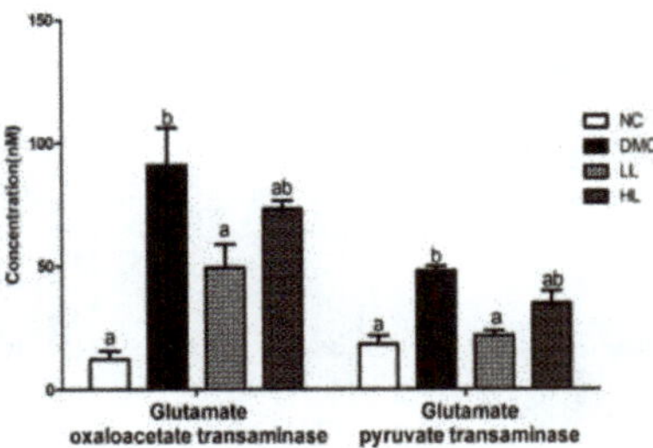

Figure 2. Effects of LBE supplementation on glutamate oxaloacetate transaminase (GOT) and glutamate pyruvate transaminase (GPT) in T2DM mice. Values are means ± SEM (n = 10). Mean values with different letters ([a] and [b]) were significantly different. ($p < 0.05$).

3.4. Effects of LBE on Hepatic Morphology and Lipid Profiles in T2DM Mice

Figure 3A shows the hepatic histology in each group. The white area estimated by fat deposition in the liver was increased in the DMC group compared to that in the NC group. However, in particular, the LL group showed a decrease in white areas compared to the DMC group. These findings could represent less fat deposition after the LBE treatment.

Moreover, TG and TC levels were significantly higher in the DMC group compared to those in the NC group. However, TG and TC levels in the LBE treatment groups were significantly lower than those in the DMC group (Figure 3B).

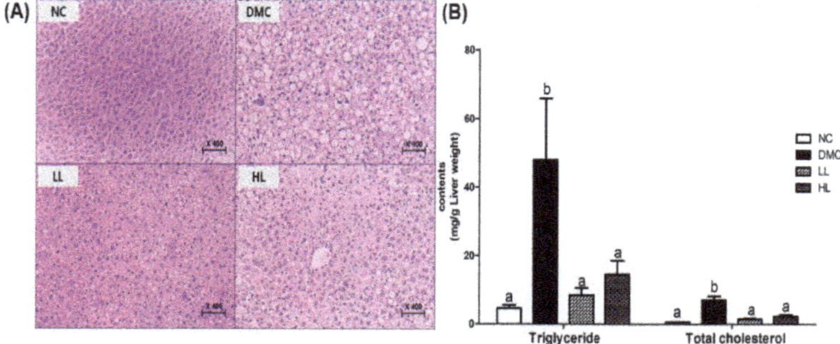

Figure 3. Effects of LBE supplementation on (**A**) hepatic morphology and (**B**) lipid profiles in T2DM mice. Levels of triglyceride (TG) and total cholesterol (TC) were measured in hepatic tissues. Values are means ± SEM (n = 6). Mean values with different letters ([a] and [b]) were significantly different. ($p < 0.05$).

3.5. Effects of LBE Supplementation on Hepatic Protein Levels of Lipid Metabolism Related Markers in T2DM Mice

The protein levels of nSREBP1, C/EBPα, PPARγ, and FAS in the DMC group were significantly higher than those in the NC group. Only the LL group showed normalized lipid metabolism related markers compared to the DMC group. The protein level of PPARα in the DMC group was significantly lower than those in the NC group. The LL group showed a significantly higher level of PPARα than that in the DMC group (Figure 4). The protein level of FAS was increased in the DMC group compared to that of the NC group, but it was lowered by a low dose of LBE supplementation.

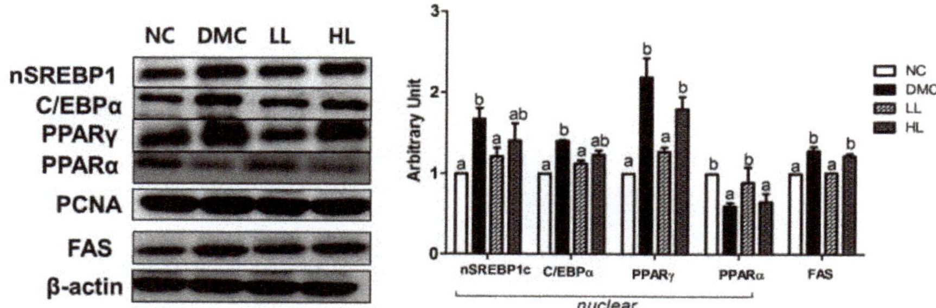

Figure 4. Effects of LBE supplementation on hepatic protein levels of lipid metabolism related markers: nuclear sterol regulatory element-binding protein1 (nSREBP1), CAAT box/enhancer binding protein alpha (C/EBPα), peroxisome proliferator-activated receptor-γ (PPARγ), peroxisome proliferator-activated receptor-α (PPARα), and fatty acid synthase (FAS) in T2DM mice. The hepatic protein was measured by Western blot. The bands show the intensity of the bands that were densitometrically measured and normalized to the band levels of proliferating cell nuclear antigen (PCNA) (nucleus) or β-actin (cytosol). Data are presented as means ± SEM (n = 6). Values with the same superscript letter ([a] and [b]) are not significantly different. ($p < 0.05$).

3.6. Effects of LBE Supplementation on Hepatic Protein Levels of Energy Metabolism Related Markers in T2DM Mice

The protein levels of energy metabolism related markers including AMPK, p-AMPK, nuclear PGC1α, and SIRT1 were measured. The protein levels of AMPK, P-AMPK, *nuclear* PGC1α, and SIRT1 were decreased in the DMC group compared to those in the NC group. However, the protein levels of AMPK and p-AMPK in the LB treatment groups were significantly higher compared to those in the DMC group. The LL group showed a significantly higher level of SIRT1 than that in the DMC group. However, the protein level of PGC1α was not normalized by LBE supplementation (Figure 5).

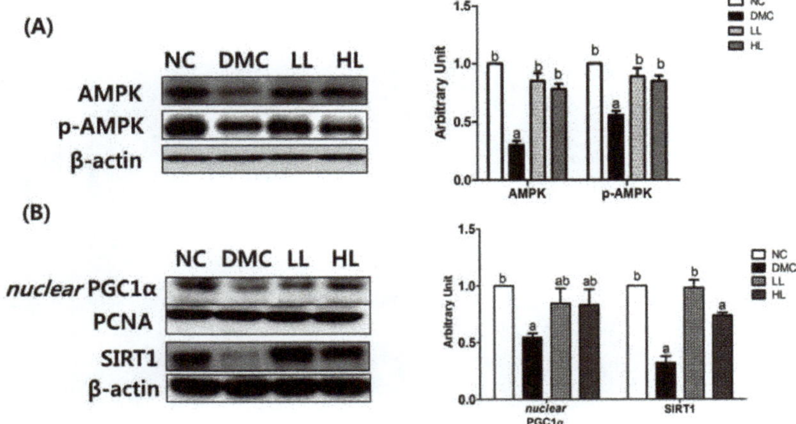

Figure 5. Effects of LBE supplementation on hepatic protein levels of energy metabolism related markers in T2DM mice. The hepatic protein was measured by Western blot. Representative band images of (**A**) adenosine monophosphate activation kinase (AMPK) phosphorylation and (**B**) nuclear peroxisome proliferator-activated receptor gamma coactivator 1-α (PGC1α) and Sirtuin1 (SIRT1) activation. The bands show the intensity of the bands that were densitometrically measured and normalized to the band levels of PCNA (nucleus) or β-actin (cytosol). Data are presented as means ± SEM (n = 6). Values with the same superscript letter ([a] and [b]) are not significantly different. ($p < 0.05$).

3.7. Effects of LBE Supplementation on Plasma and Hepatic Protein Levels of Oxidative Stress Markers T2DM Mice

4-HNE and protein carbonyls were used as markers for oxidative stress in plasma (Figure 6A). The protein levels of plasma 4-HNE and protein carbonyls in the DMC group were significantly higher than those in the NC group. Both LL and HL groups showed significantly lower levels of 4-HNE than the DMC group. The level of protein carbonyls in the HL group was significantly lowered compared to that in the DMC group. The protein levels of nuclear Nrf2 and cytosolic CuZnSOD, MnSOD, HO-1, catalase, and NQO1 were significantly higher in the DMC group compared to those in the NC group. The protein levels of Nrf2 and catalase in the LBE supplementation groups were significantly lowered compared to those in the DMC group (Figure 6B). The protein levels of GPx were not different among the groups.

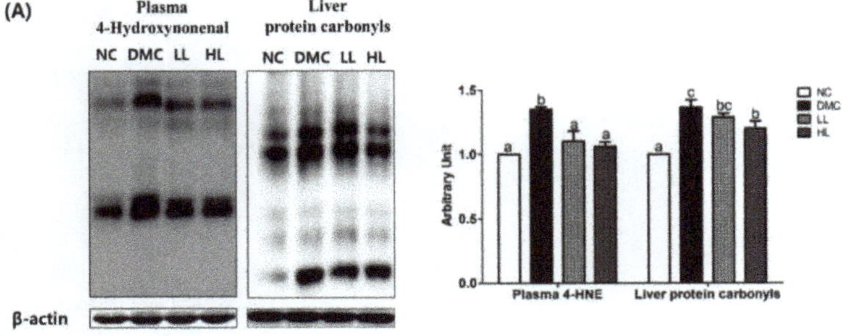

Figure 6. *Cont.*

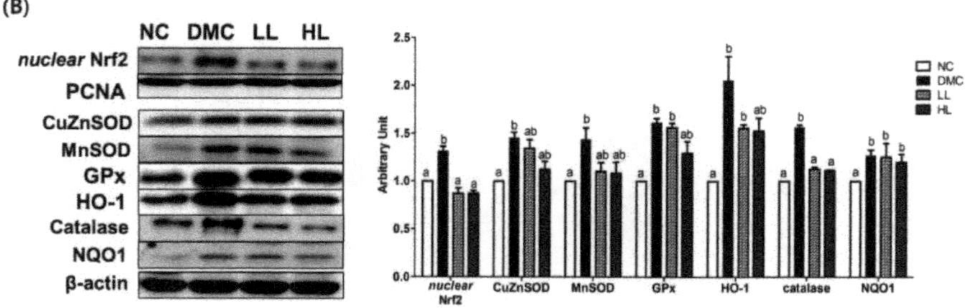

Figure 6. Effects of LBE supplementation on plasma and hepatic protein levels of oxidative stress markers in T2DM mice. (**A**) Plasma 4-hydroxynonenal (4-HNE) and liver protein carbonyls and (**B**) nuclear factor erythroid 2-related factor 2 (Nrf2) associated antioxidant defense markers: nuclear factor erythroid 2-related factor 2 (Nrf2), copper-zinc-superoxide dismutase (SOD), manganese superoxide dismutase (SOD), glutathione peroxidase (GPx), heme oxygenase-1 (HO-1), catalase, and NAD(P)H dehydrogenase quinone 1 (NQO1). The bands show the intensity of the bands that were densitometrically measured and normalized to the band levels of PCNA (nucleus) or β-actin (cytosol). Data are presented as means ± SEM ($n = 6$). Values with the same superscript letter (^{a} and ^{b}) are not significantly different. ($p < 0.05$).

3.8. Effects of LBE Supplementation on Hepatic Protein Levels of Inflammatory Response Related Markers in T2DM Mice

The protein levels of inflammatory response related markers were measured by Western blot in hepatic tissue (Figure 7). The protein levels of nuclear factor kappa B (NF-κB) and its related inflammatory genes including TNF-α, IL-1β, IL-6, iNOS, MCP-1, and CRP were significantly higher in the DMC group than those in the NC group. However, the levels of NF-κB, IL-1β, IL-6, and iNOS in both LBE treated groups were significantly lowered compared to those in the DMC group. Furthermore, the protein levels of COX2 and MCP-1 were lowered in the HL group compared to the DMC group. The protein levels of TNF-α and CRP were not reduced in both LBE supplementation groups.

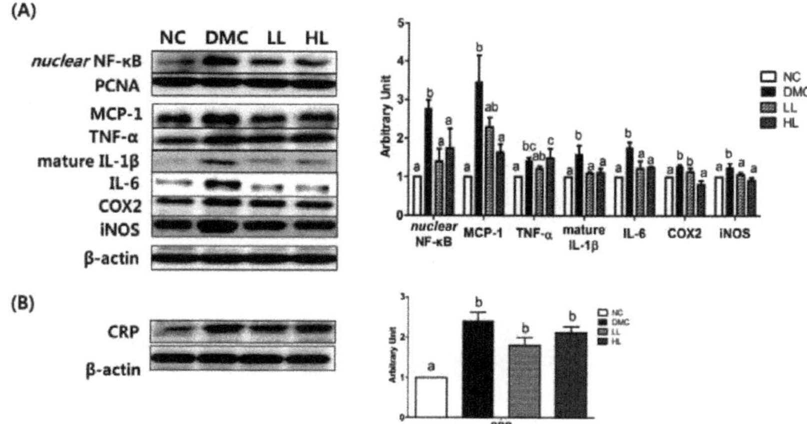

Figure 7. Effects of LBE supplementation on hepatic protein levels of inflammatory response related markers in T2DM mice. (**A**) Nuclear factor kappa B (NF-κB)-related markers: monocyte chemoattractant protein-1 (MCP-1), tumor necrosis factor α (TNFα), interleukin-1β (IL-1β), interleukin-6 (IL-6), cyclooxygenase-2 (COX2), and nitric oxide synthases (iNOS) and (**B**) inflammatory proteins: c-reactive protein (CRP). The hepatic protein was measured by Western blot. The bands show the intensity of the bands that were densitometrically measured and normalized to the band levels of PCNA (nucleus) or β-actin (cytosol). Data are presented as means ± SEM ($n = 6$). Values with the same superscript letter (^{a} and ^{b}) are not significantly different. ($p < 0.05$).

3.9. *Effects of LBE Supplementation on Hepatic Fibrosis in T2DM Mice*

The protein levels of fibrosis-related markers including α-SMA, TGF-β, PKC, and PKCβII were significantly higher in the DMC group compared to those in the NC group. The protein levels of α-SMA and PKC in both LBE treatment groups regardless of dose were significantly lower than those in the DMC group. Furthermore, the protein level of PKCβII only in the HL group was significantly lower than that in the DMC group (Figure 8). The protein level of TGF-β was not significantly reduced in both LB supplementation groups.

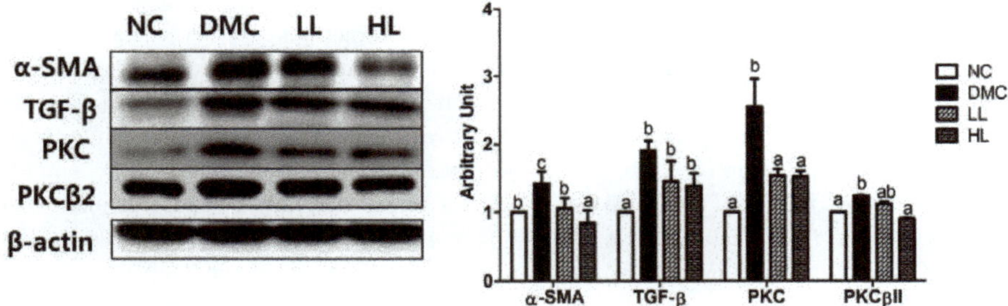

Figure 8. Effects of LBE supplementation on hepatic fibrosis markers: α-smooth muscle actin (α-SMA), transforming growth factor β (TGF-β), protein kinase C (PKC), and protein kinase C-βII (PKCβII) in T2DM mice. The hepatic protein was measured by Western blot. The bands show the intensity of the bands that were densitometrically measured and normalized to the band levels of β-actin (cytosol). Data are presented as means ± SEM ($n = 6$). Values with the same superscript letter (^{a}, ^{b} and ^{c}) are not significantly different. ($p < 0.05$).

4. Discussion

In the present study, we investigated the effect of LBE on hyperglycemia-induced hepatic damage in diabetes. Consequently, the results demonstrated that LBE effectively attenuated hepatic damage by regulation of lipogenesis associated with oxidative stress, inflammation, and fibrosis in T2DM.

According to the HPLC analysis previously reported by our group, the concentration of genistein, daidzein, quercetin, and naringenin in LBE were determined as about 0.053 mg/g, 0.165 mg/g, 0.853 mg/g, and 0.08 mg/g, respectively [18]. These natural compounds showed antioxidant effects and exerted anti-diabetic and anti-lipogenic potentials in in vitro studies [21–25]. In the current study, a high dose of LBE supplementation reduced the levels of HbA1c along with RAGE, which was considered as an index of chronic hyperglycemic states [26], although a low dose of LBE treatment decreased only HbA1c, which is a more useful clinical biomarker of diabetes. Therefore, one can conclude that LBE is beneficial for attenuating the hyperglycemic condition in T2DM.

Hyperglycemia is a key contributor of hepatic damage in T2DM. We measured plasma levels of GOT and GPT which are sensitive clinical markers of hepatic damage [27]. Our data showed that a low dosage of LBE ameliorated hepatic damage by reducing GOT and GPT levels. Furthermore, LBE decreased hepatic fat droplets by reducing hepatic TG and TC levels in diabetes. These results are well in accordance with our histological observation of hepatic tissue. Therefore, LBE can be a potential nutrient attenuating hepatic lipid accumulation without extra hepatic burden.

We examined how LBE influences hyperglycemia-induced abnormal hepatic lipid metabolism. AMPK, which has a major role in lipid metabolism is known to decrease in diabetes [28]. Especially, AMPK downregulates the expression of SREBP1, which is a major transcription factor of fatty acid synthesis [29]. SREBP1 leads to an increase in the expression of lipogenic enzymes such as ACC and FAS [29]. Moreover, AMPK can change the NAD+/NADH ratio and accordingly stimulate SIRT1 expression in the hepatocyte [28,30]. SIRT1 activation also leads to an increase fatty acid oxidation via PPARα and PGC1α as well as a decrease in inflammatory response via NF-κB regulation [31]. Previous studies have shown that polyphenols in natural products activated

AMPK-SIRT1 signaling pathway, which triggers lipogenesis and β-oxidation [31–34]. For the first time, the present study revealed that LBE treatment regardless of dosage increased the protein level of hepatic AMPK, which may increase NAD+ production by β-oxidation and lead to SIRT1 activation in T2DM mice.

In addition, LBE supplementation ameliorated hyperglycemia-induced oxidative stress in our study. Hyperglycemia-induced AGEs formation leads to RAGE production in the cell membrane [35]. Activated RAGE increases ROS, subsequently, leading to chronic oxidative stress [36]. As mentioned before, LBE supplementation at a high dose reduced the levels of RAGE expression in diabetic mice. A previous study also showed the ameliorative effects of LBE on MGO-induced RAGE expression in kidney tissue, which subsequently reduced AGE-RAGE interactions [37]. These results suggested that LBE reduced glycation products which can cause hepatic complications in diabetes. Furthermore, LBE has been known to have an antioxidant effect by scavenging nitrite in normal mice [38]. Our data demonstrated that the 4-HNE and protein carbonyls, representative biomarker of oxidative stress [39], were significantly increased in the DMC group compared to those in the NC. LBE supplementation regardless of dosage decreased the levels of 4-HNE but only a high dose of LBE treatment reduced protein carbonyl level in the diabetic mice. In addition, both LBE treatment reduced oxidative stress by regulation of Nrf2 and its downstream enzymes including catalase in diabetic liver.

Increased oxidative stress directly contributes to inflammation via activation of NF-κB, which regulates expression of inflammatory mediators [40]. In the current study, both LBE treatments significantly ameliorated NF-κB activation and its related inflammatory proteins including IL-1β, IL-6, and iNOS in diabetic mice and only a high dose of LBE treatment reduced levels of COX-2 and MCP-1, suggesting that LBE ameliorated hepatic hyper-inflammation related to NF-κB activation under diabetic condition. On the contrary, LBE treatment did not attenuate hepatic protein levels of CRP, which is not directly/indirectly regulated by NF-κB activation, in T2DM mice. A previous study also reported that LBE is capable of inhibiting NO production by inhibiting NF-κB in vitro [15]. As mentioned previously, genistein, quercetin, daidzein, and naringenin are affluent in LBE, and previous research showed that these compounds inhibited NF-κB activation along with decreased iNOS expression and NO production in vitro [41–43] and in vivo [44–46]. Therefore, it might be inferred that LBE supplementation might reduce hepatic oxidative stress along with NFκB associated inflammatory responses in T2DM.

Oxidative stress and inflammation resulting from chronic hyperglycemia also promote hepatic fibrosis in diabetes [46]. HSCs transform into proliferative and fibrogenic myofibroblasts, which express α-SMA in response to ROS [47]. Oxidative stress also induces the production of TGF-β and PKC which can cause cell death and activate collagen synthesis resulting in hepatic fibrosis [19,48]. The current study showed that LBE treatment regardless of dose attenuated hepatic protein levels of α-SMA and PKC II and only a high dose of LBE treatment reduced PKCβ level in T2DM mice. In the previous study, the level of increased fibrotic collagen in MGO-induced renal damage was attenuated by LBE in diabetic nephropathy [19]. Therefore, it can be concluded that LBE has an ameliorative effect on fibrosis related mediators along with reduced oxidative stress and inflammation in hyperglycemia-induced damaged tissue.

5. Conclusions

The present study demonstrated that LBE supplementation attenuated hyperglycemia-induced hepatic damage by regulation of AMPK associated lipogenesis in T2DM. Furthermore, LBE ameliorated hepatic oxidative stress, inflammation, and fibrosis although some molecular markers were selectively ameliorated at different treatment dosage of LBE in T2DM mice. Conclusively, LBE could be considered as a potential nutraceutical to ameliorate hyperglycemia-induced diabetic damage in T2DM.

Author Contributions: Conceptualization, Y.L.; data curation, Y.K. and H.L.; formal analysis, Y.L. and Y.K.; investigation, Y.L. and S.Y.K.; funding acquisition, Y.L.; methodology, Y.L., Y.K., and H.L.; supervision, Y.L.; writing, review, and editing, Y.K., H.L., and Y.L.

Funding: This research was supported by a grant (2018R1D1A1B07046778) funded by the Ministry of Education, Science and Technology, Republic of Korea.

Conflicts of Interest: The authors declare no conflicts of interest.

Abbreviations

The following abbreviations are used in this manuscript:

AGE	Advanced glycation end products
AMPK	adenosine monophosphate activation kinase
AUC	Area under the curve
BSA	Bovine serum albumin
BW	Body weight
C/EBPα	CAAT box/enhancer binding protein alpha
DM	Diabetes mellitus
FAS	Fatty acid synthase
FBG	Fasting blood glucose
4-HNE	Four-hydroxynonenal
GOT	Glutamate oxaloacetate transaminase
GPT	Glutamate pyruvate transaminase
GPx	Glutathione peroxidase
H&E	Hematoxylin and eosin
HSCs	Hepatic stellate cells
IL-1β	Interleukin-1β
IL-6	Interleukin-6
LB	Lespedeza Bicolor
MGO	Methylglyoxal
NC	Normal control
NF-κB	nuclear factor kappa B
NQO1	NAD(P)H degydrogenase quinine 1
Nrf2	Nuclear factor erythroid2-related factor 2
OGTT	Oral glucose tolerance test
PVDF	Polyvinylidene fluoride
SIRT1	Sirtuin1
SOD	Superoxide dismutase
STZ	Streptozotocin
RAGE	Advanced glycation end products receptor
ROS	Reactive oxygen species
RNS	Reactive nitrogen species
SREBP1	Sterol regulatory element-binding protein 1
T2DM	Type 2 diabetes mellitus
TC	Total cholesterol
TG	Triglyceride
TNFα	Tumor necrosis factor α

References

1. Blair, M. Diabetes Mellitus Review. *Urol. Nurs.* **2016**, *36*, 27–36. [CrossRef] [PubMed]
2. Pedersen, O.B. Diabetes mellitus and malfunctions of insulin secretion. *Ugeskr. Laeger.* **1997**, *159*, 7118–7119. [PubMed]
3. DeFronzo, R.A.; Ferrannini, E.; Groop, L.; Henry, R.R.; Herman, W.H.; Holst, J.J.; Hu, F.B.; Kahn, C.R.; Raz, I.; Shulman, G.I.; et al. Type 2 diabetes mellitus. *Nat. Rev. Dis. Primers* **2015**, *1*, 15019. [CrossRef] [PubMed]
4. Leite, N.C.; Villela-Nogueira, C.A.; Cardoso, C.R.; Salles, G.F. Non-alcoholic fatty liver disease and diabetes: From physiopathological interplay to diagnosis and treatment. *World J. Gastroenterol.* **2014**, *20*, 8377–8392. [CrossRef]

5. Stefan, N.; Haring, H.U. The metabolically benign and malignant fatty liver. *Diabetes* **2011**, *60*, 2011–2017. [CrossRef]
6. Fabbrini, E.; Sullivan, S.; Klein, S. Obesity and nonalcoholic fatty liver disease: Biochemical, metabolic, and clinical implications. *Hepatology* **2010**, *51*, 679–689. [CrossRef]
7. Cohen, J.C.; Horton, J.D.; Hobbs, H.H. Human fatty liver disease: Old questions and new insights. *Science* **2011**, *332*, 1519–1523. [CrossRef]
8. Williams, K.H.; Shackel, N.A.; Gorrell, M.D.; McLennan, S.V.; Twigg, S.M. Diabetes and nonalcoholic Fatty liver disease: A pathogenic duo. *Endocr. Rev.* **2013**, *34*, 84–129. [CrossRef]
9. Marchesini, G.; Marzocchi, R. Metabolic syndrome and NASH. *Clin. Liv. Dis.* **2007**, *11*, 105–117. [CrossRef]
10. Tsuchida, T.; Friedman, S.L. Mechanisms of hepatic stellate cell activation. *Nat. Rev. Gastroenterol. Hepatol.* **2017**, *14*, 397–411. [CrossRef]
11. Zhang, H.; Davies, K.J.A.; Forman, H.J. Oxidative stress response and Nrf2 signaling in aging. *Free Radic. Biol. Med.* **2015**, *88*, 314–336. [CrossRef] [PubMed]
12. Li, J.; Liu, M.; Yu, H.; Wang, W.; Han, L.; Chen, Q.; Ruan, J.; Wen, S.; Zhang, Y.; Wang, T. Mangiferin Improves Hepatic Lipid Metabolism Mainly Through Its Metabolite-Norathyriol by Modulating SIRT-1/AMPK/SREBP-1c Signaling. *Front. Pharmacol.* **2018**, *9*, 201. [CrossRef] [PubMed]
13. Rodriguez-Ramiro, I.; Vauzour, D.; Minihane, A.M. Polyphenols and non-alcoholic fatty liver disease: Impact and mechanisms. *Proc. Nutr. Soc.* **2016**, *75*, 47–60. [CrossRef] [PubMed]
14. Mira, L.; Fernandez, M.T.; Santos, M.; Rocha, R.; Florencio, M.H.; Jennings, K.R. Interactions of flavonoids with iron and copper ions: A mechanism for their antioxidant activity. *Free Radic. Res.* **2002**, *36*, 1199–1208. [CrossRef] [PubMed]
15. Lee, S.J.; Hossaine, M.D.; Park, S.C. A potential anti-inflammation activity and depigmentation effect of Lespedeza bicolor extract and its fractions. *Saudi J. Boil. Sci.* **2016**, *23*, 9–14. [CrossRef] [PubMed]
16. Maximov, O.B.; Kulesh, N.I.; Stepanenko, L.S.; Dmitrenok, P.S. New prenylated isoflavanones and other constituents of Lespedeza bicolor. *Fitoterapia* **2004**, *75*, 96–98. [CrossRef]
17. Miyase, T.; Sano, M.; Yoshino, K.; Nonaka, K. Antioxidants from Lespedeza homoloba (II). *Phytochemistry* **1999**, *52*, 311–319. [CrossRef]
18. Do, M.H.; Lee, J.H.; Wahedi, H.M.; Pak, C.; Lee, C.H.; Yeo, E.J.; Lim, Y.; Ha, S.K.; Choi, I.; Kim, S.Y. Lespedeza bicolor ameliorates endothelial dysfunction induced by methylglyoxal glucotoxicity. *Phytomed. Int. J. Phytother. Phytopharm.* **2017**, *36*, 26–36. [CrossRef]
19. Do, M.H.; Lee, J.H. Therapeutic Potential of Lespedeza bicolor to Prevent Methylglyoxal-Induced Glucotoxicity in Familiar Diabetic Nephropathy. *J. Clin. Med.* **2019**, *8*, 1138. [CrossRef]
20. Zhang, M.; Lv, X.Y.; Li, J.; Xu, Z.G.; Chen, L. The characterization of high-fat diet and multiple low-dose streptozotocin induced type 2 diabetes rat model. *Exp. Diabetes Res.* **2008**, *2008*, 704045. [CrossRef]
21. Lee, C.W.; Seo, J.Y.; Lee, J.; Choi, J.W.; Cho, S.; Bae, J.Y.; Sohng, J.K.; Kim, S.O.; Kim, J.; Park, Y.I. 3-O-Glucosylation of quercetin enhances inhibitory effects on the adipocyte differentiation and lipogenesis. *Biomed. Pharmacother. Biomed. Pharmacother.* **2017**, *95*, 589–598. [CrossRef] [PubMed]
22. Chen, S.; Jiang, H. Therapeutic Effects of Quercetin on Inflammation, Obesity, and Type 2 Diabetes. *Mediat. Inflamm.* **2016**, *2016*, 9340637. [CrossRef] [PubMed]
23. Guo, T.L.; Germolec, D.R.; Zheng, J.F.; Kooistra, L.; Auttachoat, W.; Smith, M.J.; White, K.L.; Elmore, S.A. Genistein protects female nonobese diabetic mice from developing type 1 diabetes when fed a soy- and alfalfa-free diet. *Toxicol. Pathol.* **2015**, *43*, 435–448. [CrossRef] [PubMed]
24. Shin, J.H.; Jung, J.H. Non-alcoholic fatty liver disease and flavonoids: Current perspectives. *Clin. Res. Hepatol. Gastroenterol.* **2017**, *41*, 17–24. [CrossRef]
25. Al-Dosari, D.I.; Ahmed, M.M.; Al-Rejaie, S.S.; Alhomida, A.S.; Ola, M.S. Flavonoid Naringenin Attenuates Oxidative Stress, Apoptosis and Improves Neurotrophic Effects in the Diabetic Rat Retina. *Nutrients* **2017**, *9*, 1161. [CrossRef]
26. Do, M.H.; Hur, J.; Choi, J.; Kim, Y.; Park, H.Y. Spatholobus suberectus Ameliorates Diabetes-Induced Renal Damage by Suppressing Advanced Glycation End Products in db/db Mice. *Int. J. Mol. Sci.* **2018**, *19*, 2774. [CrossRef]
27. Ann, J.Y.; Eo, H.; Lim, Y. Mulberry leaves (*Morus alba* L.) ameliorate obesity-induced hepatic lipogenesis, fibrosis, and oxidative stress in high-fat diet-fed mice. *Genes. Nutr.* **2015**, *10*, 46. [CrossRef]

28. He, L.; Zhou, X.; Huang, N.; Li, H.; Tian, J.; Li, T.; Yao, K.; Nyachoti, C.M.; Kim, S.W.; Yin, Y. AMPK Regulation of Glucose, Lipid and Protein Metabolism: Mechanisms and Nutritional Significance. *Curr. Protein Pept. Sci.* **2017**, *18*, 562–570. [CrossRef]
29. Li, Y.; Xu, S.; Mihaylova, M.M.; Zheng, B.; Hou, X.; Jiang, B. AMPK phosphorylates and inhibits SREBP activity to attenuate hepatic steatosis and atherosclerosis in diet-induced insulin-resistant mice. *Cell Metab.* **2011**, *13*, 376–388. [CrossRef]
30. Thirupathi, A.; Souza, C.T. Multi-regulatory network of ROS: The interconnection of ROS, PGC-1 alpha, and AMPK-SIRT1 during exercise. *J. Physiol. Biochem.* **2017**, *73*, 487–494. [CrossRef]
31. Shang, J.; Chen, L.L.; Xiao, F.X.; Sun, H.; Ding, H.C.; Xiao, H. Resveratrol improves non-alcoholic fatty liver disease by activating AMP-activated protein kinase. *Acta Pharmacol. Sin.* **2008**, *29*, 698–706. [CrossRef] [PubMed]
32. Joven, J.; Espinel, E.; Rull, A.; Aragones, G.; Rodriguez-Gallego, E.; Camps, J.; Micol, V.; Herranz-Lopez, M.; Menendez, J.A.; Borras, I.; et al. Plant-derived polyphenols regulate expression of miRNA paralogs miR-103/107 and miR-122 and prevent diet-induced fatty liver disease in hyperlipidemic mice. *Biochim. Biophys. Acta* **2012**, *1820*, 894–899. [CrossRef] [PubMed]
33. Huang, C.; Qiao, X.; Dong, B. Neonatal exposure to genistein ameliorates high-fat diet-induced non-alcoholic steatohepatitis in rats. *Br. J. Nutr.* **2011**, *106*, 105–113. [CrossRef]
34. Wang, L.M.; Wang, Y.J.; Cui, M.; Luo, W.J.; Wang, X.J.; Barber, P.A.; Chen, Z.Y. A dietary polyphenol resveratrol acts to provide neuroprotection in recurrent stroke models by regulating AMPK and SIRT1 signaling, thereby reducing energy requirements during ischemia. *Eur. J. Neurosci.* **2013**, *37*, 1669–1681. [CrossRef] [PubMed]
35. Salil, G.; Nithya, R.; Nevin, K.G.; Rajamohan, T. Dietary coconut kernel protein beneficially modulates NFkappaB and RAGE expression in streptozotocin induced diabetes in rats. *J. Food Sci. Technol.* **2014**, *51*, 2141–2147. [CrossRef] [PubMed]
36. Maki, D.G. Review: HbA1c has low accuracy for prediabetes; lifestyle programs and metformin reduce progression to T2DM. *Ann. Int. Med.* **2017**, *166*, Jc41. [CrossRef] [PubMed]
37. Lee, Y.S.; Chang, Z.; Park, S.C.; Lim, N.R.; Kim, N.W. Antioxidanat activity and lrritation response of Lespedeza bicolor. *J. Toxicol.* **2005**, *21*, 115–119.
38. Dong, X.G.; An, Z.M.; Guo, Y.; Zhou, J.L.; Qin, T. Effect of triptolide on expression of oxidative carbonyl protein in renal cortex of rats with diabetic nephropathy. *J. Huazhong Univ. Sci. Technol. Med. Sci. Hua Zhong Ke Ji Da Xue Xue Bao. YI Xue Ying De Wen Ban Huazhong Keji Daxue Xuebao. Yixue Yingdewen Ban.* **2017**, *37*, 25–29, (Chinese). [CrossRef] [PubMed]
39. Pieper, C.M.; Roza, A.M.; Henderson, J.D., Jr.; Zhu, Y.R.; Lai, C.S. Spatial distribution and temporal onset of NF-kB activation and inducible nitric oxide synthase within pancreatic islets in the pre-diabetic stage of genetic, diabetic-prone BB rats: Attenuation by drug intervention decreases inflammatory cell infiltration and incidence of diabetes. *Inflammation* **2004**, *53*, 22–30.
40. Alderton, W.K.; Cooper, C.E.; Knowles, R.G. Nitric oxide synthases: Structure, function and inhibition. *Biochem. J.* **2001**, *357*, 593–615. [CrossRef]
41. Bogdan, C. Nitric oxide and the immune response. *Nat. Immunol.* **2001**, *2*, 907–916. [CrossRef] [PubMed]
42. Hamalainen, M.; Nieminen, R.; Vuorela, P.; Heinonen, M.; Moilanen, E. Anti-inflammatory effects of flavonoids: Genistein, kaempferol, quercetin, and daidzein inhibit STAT-1 and NF-kappaB activations, whereas flavone, isorhamnetin, naringenin, and pelargonidin inhibit only NF-kappaB activation along with their inhibitory effect on iNOS expression and NO production in activated macrophages. *Med. Inflamm.* **2007**, *2007*, 45673.
43. Zhang, X.; Li, H.; Feng, H.; Xiong, H.; Zhang, L.; Song, Y.; Yu, L.; Deng, X. Valnemulin downregulates nitric oxide, prostaglandin E2, and cytokine production via inhibition of NF-kappaB and MAPK activity. *Int. Immunopharmacol.* **2009**, *9*, 810–816. [CrossRef] [PubMed]
44. Szliszka, E.; Skaba, D.; Czuba, Z.P.; Krol, W. Inhibition of inflammatory mediators by neobavaisoflavone in activated RAW264.7 macrophages. *Molecules* **2011**, *16*, 3701–3712. [CrossRef]
45. Guo, Z.; Xu, H.Y.; Xu, L.; Wang, S.S.; Zhang, X.M. In vivo and in vitro immunomodulatory and anti-inflammatory effects of total flavonoids of astragalus. *Afr. J. Tradit. Complement. Altern. Med. AJTCAM* **2016**, *13*, 60–73. [CrossRef]
46. Richter, K.; Kietzmann, T. Reactive oxygen species and fibrosis: Further evidence of a significant liaison. *Cell Tissue Res.* **2016**, *365*, 591–605. [CrossRef]

47. Xu, P.; Zhang, Y.; Liu, Y.; Yuan, Q.; Song, L.; Liu, M.; Liu, Z.; Yang, Y.; Li, J.; Li, D.; et al. Fibroblast growth factor 21 attenuates hepatic fibrogenesis through TGF-beta/smad2/3 and NF-kappaB signaling pathways. *Toxicol. Appl. Pharmacol.* **2016**, *290*, 43–53. [CrossRef]
48. Fan, H.N.; Wang, H.J.; Yang-Dan, C.R.; Ren, L.; Wang, C.; Li, Y.F. Protective effects of hydrogen sulfide on oxidative stress and fibrosis in hepatic stellate cells. *Mol. Med.* **2013**, *7*, 247–253. [CrossRef]

Article

Functional Analysis of Macromolecular Polysaccharides: Whitening, Moisturizing, Anti-Oxidant, and Cell Proliferation

Chien-Jen Kao [1], Hsin-Yu Chou [2,3], Yu-Chen Lin [3,4], Qinghong Liu [5,*] and Hui-Min David Wang [3,6,7,*]

1. Department of Internal medicine of Gangshan Branch of Kaohsiung Armed Forces General Hospital, National Defense Medical Center, Kaohsiung 82049, Taiwan; kaochienjen@yahoo.com.tw
2. Program in Tissue Engineering and Regenerative Medicine, National Chung Hsing University, Taichung 402, Taiwan; s9412105@gmail.com
3. Graduate institute of biomedical engineering, Chung Hsing University, Taichung 402, Taiwan; 114eric0110@gmail.com
4. Department of Chemical Engineering, National Chung Hsing University, Taichung 402, Taiwan
5. Department of Vegetable, College of Horticulture, China Agricultural University, Beijing 100193, China
6. Graduate Institute of Medicine, College of Medicine, Kaohsiung Medical University, Kaohsiung 807, Taiwan
7. Department of Medical Laboratory Science and Biotechnology, China Medical University, Taichung 404, Taiwan

* Correspondence: qhliu@cau.edu.cn (Q.L.); davidw@dragon.nchu.edu.tw (H.-M.D.W.); Tel.: +886-935753718 or +886-4-22840733-651 (H.-M.D.W.)

Received: 1 October 2019; Accepted: 4 November 2019; Published: 7 November 2019

Abstract: In this research we utilized extracts from two different nature products, *Achatina fulica* and *Heimiella retispora*, to enhance skin moisturizing abilities, anti-oxidative properties, and cell proliferations. It was observed that two polysaccharides with anti-oxidative effects by chelating metal ions reduced oxidative stress and further blocked the formation of reactive oxygen species syntheses. To detect whether there was a similar effect within the cellular mechanism, a flow cytometry was applied for sensing the oxidative level and it was found that both materials inhibited the endogenous oxidative stress, which was induced by phorbol-12-myristate-13-acetate (PMA). Both polysaccharides also stimulated the production of collagen to maintain skin tightness and a moisturizing effect. In summary, we developed two macromolecular polysaccharides with potential applications in dermal care.

Keywords: *Achatina fulica*; *Heimiella retispora*; reactive oxygen species (ROS), collagen; moisturizing

---

1. Introduction

The skin is the body's largest organ. It consists of three layers: The epidermis, dermis, and the subcutaneous tissue. The outer layer of the epidermis consists of dead skin cells, natural oils, and lipids to protect the skin from irritants and toxins and to prevent the loss of water, biomolecules, and electrolytes [1]. When the outer layer of the epidermis is exposed to some detergent and detergent ingredients, these protective elements on the surface of the skin are peeled off. Once these irritants penetrate the outer layers of the skin, they can cause dry skin and skin health problems. One of the most important functions of the skin is to provide a barrier to prevent excessive transepidermal water loss [2,3]. Constant water movement on the skin plays an important role in the epidermal repair process. Thus, skin moisturization is very important [4]. The base layer of the epidermis is attached to the dermis, wherein the basal layer contains melanocytes. The synthesis of melanin causes the skin, eyes, and hair to darken, which is composed of colored biopolymers in the melanosomes of melanocytes.

Melanin is a protective mechanism against UV damage [5,6]. A copper-containing enzyme plays a key role in melanin production. Known as tyrosinase, tyrosinase catalyzes the hydroxylation of L-tyrosine to L-3,4-dihydroxyphenylalanine (L-DOPA) and then oxidizes L-DOPA to dopaquinone [7]. Several tyrosinase inhibitors are currently used as agents for epidermal hyperpigmentation. Many skin lightening agents have been developed, such as hydroquinone, kjoic acid, and amla fruit extract powder. Ultraviolet radiation (UVR) can cause light irritation, photoaging, and carcinogenesis to induce skin inflammation. Enhanced endogenous protective mechanisms of oxidative damage are promising strategies for reducing skin damage [5,7–11]. Skin cells are formulated with antioxidants to eliminate ROS to maintain a balance of pro-oxidant/antioxidants. However, the proliferation of reactive oxygen species (ROS) leads to the consumption of antioxidants and the further formation of reaction products, leading to oxidative stresses [12,13]. When the skin is physically damaged, a wound is created, and the wound is repaired by inflammation, cell proliferation, and tissue remodeling. This study focuses on cell proliferation because cell proliferation is very important in the speed of wound repair.

The study of polysaccharides has been very popular in recent years. Depending on the corresponding chemical structure, polysaccharides and their derivatives have some special biological characteristics, such as biological response modifiers, anti-inflammatory, hypolipidemic, and anticoagulant. However, no one has specifically applied it to skin maintenance. In this century, snails have become available through aquaculture. Due to its traditional sensory qualities and particularly high nutritional value in Europe, especially in France, Spain, the Netherlands, Belgium, and Portugal, the use of snail-related products and the snails themselves are still considered to be extravagant [14]. The main species of edible snails belong to two families: *Helicidae* and *Achantinide*. The *Helicidae* is mainly found in Europaen countries, and *Achantinide* is usually found in African and Asian countries [15–17]. In the Taiwan market *Achatina fulica* exists in a traditional dish that is delicious and nutritious, called hot-fried snails. The research on *A. fulica* in the scientific field is limited, and the main research is biology [14]. Snails are rich in beneficial ingredients, the polysaccharide derivative isolated from *A. fulica* can selectively block angiogenesis in an inflammatory model induced by VEGF (vascular endothelial growth factor), and it is speculated the bioactive polysaccharide may have some health promoting activity [14]. A rot fungus is called "*Heimiella retispora*" by the Chinese, and it has been used to improve human health and longevity in the past millennium [18]. Many studies have confirmed polysaccharides isolated from *H. retispora* possess many special characteristics, such as improving insulin sensitivity and having anti-inflammatory, immunomodulatory anti-inflammatory, and anti-tumor properties [18–21]. *H. retispora* have flourished in the food and pharmaceutical industries in recent years. However, data on skin maintenance and repair of *H. retispora* is limited.

Polysaccharides have been researched and found to have many beneficial activities, but research on polysaccharides in skin care is limited. This research performed a series of biofunctional tests for two polysaccharides enriched extracts—*A. fulica* extracts and *H. retispora* extracts. Due to its moisturizing and skin-repairing properties, we discovered these two extracts have potential for the application in skin care products.

2. Materials and Methods

2.1. Chemicals and Reagents

Ascorbic acid (vitamin C), 3-(4,5-dimethylthiazol-2-yl)-2,5-diphenyltetrazolium bromide (MTT), L-3,4-dihydroxyphenylalanine (L-DOPA), dimethyl sulfoxide (DMSO), 1,1-diphenyl-2-picrylhydrazyl (DPPH), ethanol, ethylenediaminetetraacetic acid (EDTA), ferrous chloride ($FeCl_2{\cdot}4H_2O$), ferric chloride ($FeCl_3$), kojic acid, methanol, potassium ferricyanide ($K_3Fe(CN)_6$), 3-tert-butyl-4-hydroxyanisole (BHA), 5-hydroxy-2-hydroxymethyl-4-pyrone (kojic acid), 12-O-Tetradecanoylphorbol-13-acetate (PMA), and L-tyrosine were purchased from Sigma-Aldrich Company (St. Louis, MO, USA). Dulbecco's modified Eagle's medium (DMEM) and fetal bovine serum (FBS) were obtained from Gibco BRL

(Gaithersburg, MD, USA). Other chemical buffers and reagents were purchased at the highest available purity and quality.

2.2. Extraction from Giant African Land Snail (Achatina Fulica) and Heimiella Retispora

Giant African land snails were purchased from a local snail farm, and the fruiting bodies of *H. retispora* were from the mushroom market in Kunming, Yunnan Province in China. The snails were put into a box with a sieve at the bottom, and the box was vibrated by an ultrasonic wave instrument to stimulate the snails' secretions. The secretions were collected and diluted with reverse osmosis (RO) water 1000 times. Ammonium sulfate was added into the snail secreted solution until 80% saturation was reached. After standing at 4 °C for 12 h, the sediments were collected by centrifugation at 40 *g* × 30 min, at 4 °C. The sediments were dialyzed with molecular weight cut-off (MWCO) of 10 kDa against water. After dialyzing, the solution was centrifuged at 300 *g* × 30 min, at 4 °C to discharge the sediments and collected the supernatant. After the supernatant was lyophilized, the snail secretion powder was stored at −20 °C.

The fruiting bodies of *H. retispora* (500 g) were homogenized with a waring blender. The homogenized sample was heated in a water boiler at 90 °C for 4 h. The supernatant were collected by a centrifugation at 300 *g* × 30 min, at 4 °C. Methanol was added to the supernatant until the concentration reached 70% (*v/v*). After standing at room temperature for 10 h, the sediment was collected by a centrifugation at 500 *g* × 30 min, at 4 °C. The sediments were dialyzed with MWCO 35 kDa against water. After dialyzing, the solution was centrifuged at 300 *g* × 30 min, at 4 °C to discharge the sediments and to collect the supernatant. After the supernatant were lyophilized, the snail secretion powder was stored at −20 °C.

2.3. Evaporation Rate of A. fulica Extracts and H. retispora Extracts

A simple method was used to estimate the evaporation rate of *A. fulica* extracts and *H. retispora* extracts. Samples mixed with 1 mL water were applied to a 4-cm in diameter glass dish. This amount was chosen so the sample fluid covered the entire filter glass at all times during the experiment. The dish was placed on a scale in a draft-free environment at 32 °C for 2 h (the skin surface is typically about 32 °C). The weight of the samples was recorded at time zero to determine the exact amount applied and two hours later to determine the amount that had evaporated during that period, and the evaporation rate was computed as:

$$\text{Evaporation rate } (\%) = \frac{(\Delta M_{sample})}{(\Delta M_{control})} \times 100\% \quad (1)$$

where ΔM_{sample} means weight change of sample; $\Delta M_{control}$ means weight change of control.

2.4. Determination of 1,1-Diphenyl-2-Picrylhydrazyl (DPPH) Radical Scavenging Capacity

DPPH has stable free radicals and it is an anti-oxidant assay to detect the ability of anti-oxidants to scavenge free radicals [22]. It is a purple reagent, which transforms into yellow if the hydrogen of DPPH transfers to anti-oxidants. Correction concentration samples were added to DPPH (60 µM), and the DPPH became a bright color at 517 nm because the optical absorbance reduced. The percentages of remaining DPPH and the sample were used to calculate the amount of anti-oxidant required. Scavenging activity (%) was calculated according to:

$$\text{Scavenging activity } (\%) = \frac{(A_{sample} - A_{blank})}{A_{control}} \times 100\% \quad (2)$$

where A_{sample} means absorption of samples at 595 nm wavelength; A_{blank} means absorption of blank at 595 nm wavelength; $A_{control}$ means absorption of control at 595 nm wavelength.

2.5. Metal Chelating Activity

Metal ion can cause lipid peroxidation, especially ferrous ion which is pro-oxidant. The samples were filled into 10 μL $FeCl_2{\cdot}4H_2O$ (2 mM) and then mixed in 20 μL ferrozine (5 mM). The admixture was shaken and held at 25 °C for 10 min. The absorbance of the sample solution was observed at 562 nm. EDTA acted as a positive control, and the chelating power calculation formula was based on Equation (2).

2.6. Reducing Power

The determination of the reduction force is based on the method of Oyaizu. Briefly, samples were mixed with phosphate-buffered saline buffer (85 μL, 67 mM, pH6.8) and potassium ferricyanide ($K_3Fe(CN)_6$) (2.5 μL, 20%). Then, the reaction was carried out at 50 °C for 20 min. Following this, 160 μL of trichloro acetic acid (10%) was mixed with the reaction and 20 min was spent to centrifuge 300×. The optical density was determined at 700 nm through a 96-well plate after the solution was mixed with 2% FeC13 (25 μL). Our positive control was based on butylated hydroxyanisole (BHA), and a higher reducing performance had the property of higher light absorption.

2.7. Cytotoxicity Examinations

A total of 5% CO_2 at 37 °C was used to cultivate human dermal fibroblasts cell line HS68 (ATCC® CRL-1635™) (ATCC, Manassas, VA, USA) cell at a consistent monolayer culture of Dulbecco's modified Eagle medium (DMEM) for 24 h. Fetal bovine serum (FBS) (10%), penicillin (100 U/mL), streptomycin (100 mg/mL), amphotericin B (0.25 μg/mL), and amphotericin B (0.25 μg/mL) were the ingredients of DMEM. *A. fulica* mucus were dissolved in adding dimethyl sulfoxide (DMSO) at different concentrations without any impurity, and the DMSO concentration was less than 1.0% compared to the final working volume. The influences of testing samples on cell development were estimated with 3-(4,5-dimethylthiazol-2-yl)-2,5-diphenyltetrazolium bromide (MTT) assay. Cells were seeded at 1×10^4 cells/well in 96-well plates and allowed to hatch for 24 h before adding the extracts. After 24 h, the MTT solution was dispensed into each well. After another two hours, the culture medium was discarded, and DMSO was added to each well. The absorbance of the formazan salt was 595 nm, and the cell viability was computed as Equation (3).

$$\text{Cell viability } (\%) = \frac{(A_{sample} - A_{blank})}{(A_{control} - A_{blank})} \times 100\% \tag{3}$$

2.8. B16-F10 Cellular Tyrosinase Activity

According to the previous assay, the tyrosinase activity was turned on the dopachrome formation rate [23]. Melanoma B16-F10 cells (10^5 cells/well) were added to 1000 μL of medium and seeded in a 12-well plate. During the next 24 h, they were treated by assigned concentrations of graphene oxide nanoribbons. B16-F10 cells were lysed with 1% Triton X-100/ phosphate buffered saline (PBS) buffer after PBS washing, and then, 50 μL of 2 mM L-tyrosine was added. The mixture was incubated in darkness for 3 h at 37 °C. The optical absorbance was spectrophotometrically monitored at 490 nm. The tyrosinase activity evaluation formula was similar to Equation (2).

2.9. Detection of ROS by 2′,7′-dichlorodihydrofluorescein Diacetate (DCFDA) Stain

HS68 cells were seeded in a 6-well micro-plate at a density of 1.2×10^5 cells/well as described [24]. After 24 h, the cells were treated with 10 and 25 mg/L of samples. Cells were then washed twice with PBS buffer, suspended with 1 mL trypsin, and loaded with DCFDA (5 μM) for 30 min at 37 °C in DMEM without phenol red. Individual cells were suspended by gentle pipetting up and down three times prior to flow cytometry analyses. DCFDA was illuminated with a 488 nm laser and detected at 535 nm.

2.10. Statistical Analysis

Three of each concentration for the standard and the samples were used. Using Student's *t*-test, the results were statistically compared and were expressed using the average of the mean values ± standard deviation (SD).

3. Results and Discussion

3.1. Moisturizing Activities

The moisture of the skin is strongly correlated with skin maintenance. The moist environment can decrease the rate of skin aging, promote wound repair, and reduce scar production. Therefore, excellent moisturizing ability is very important for the application of the material in the skin. In the article, we used *A. fulica* extracts and *H. retispora* extracts to measure the moisturizing activities of the skin. We found *A. fulica* extracts decreased the evaporation rates by 12.8% and 14.3% at a concentration of 10 mg/mL and 25mg/mL, and the *H. retispora* extracts had a better effect on moisturizing function; it decreased 47.1% and 77.5% evaporation at a concentration of 10 mg/L and 25 mg/L. Moreover, marketed essence has only 8% moisturizing power (Figure 1). Some previous reports explored the use of natural products in moisturizing research, but most of them found indirect evidence, including the production of hyaluronic acid and collagen in skin cells [25,26]. However, the outermost layer of the skin is the stratum corneum rather than the cells, and the water does not directly evaporate from the cells. Therefore, we made a measurement of moisture retention, making sure that it reduced water evapotranspiration. From the results of Figure 1, both *A. fulica* extracts and *H. retispora* extracts have the effects of inhibiting the evaporation of water directly, and the potential to be utilized in skin moisturizing.

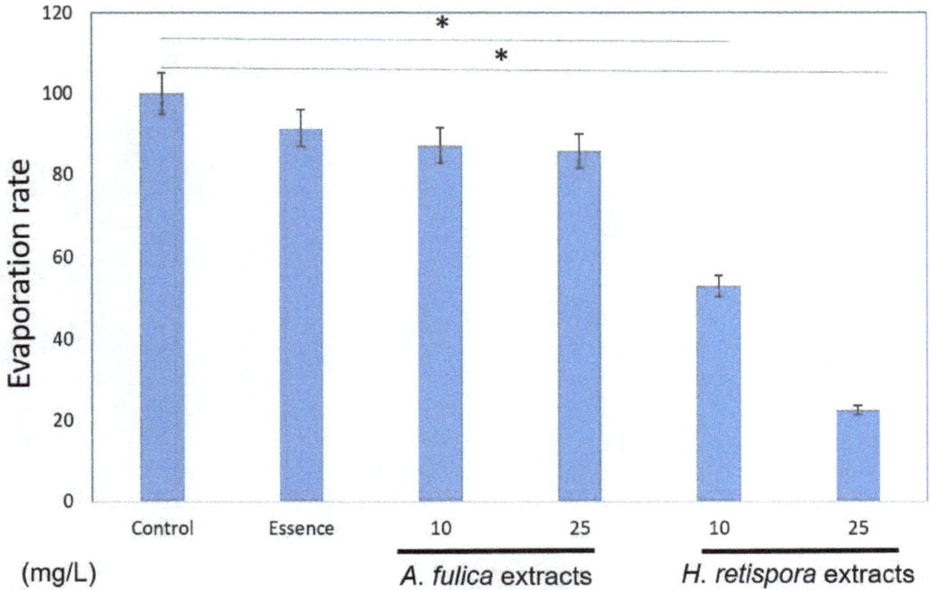

Figure 1. *Achatina fulica* extracts and *Heimiella retispora* extracts showed potential moisturizing activities. (Data represents mean ± S.D of three independent experiments performed. * $p < 0.01$).

3.2. Anti-Oxidative Properties of the A. fulica Extracts and H. retispora Extracts

Antioxidant properties are much more abundant in the reporting of natural substances than in moisturizing [22,27], and the use of antioxidant properties has a good effect in areas such as inflammation and whitening. In this research, we used multiple methods to determine these two natural substances, including DPPH, chucking, reducing power tests intracellular oxidative stress analysis. We found these extracts were particularly effective on the chelating test.

3.2.1. *A. fulica* Extracts and *H. retispora* Extracts Had No DPPH Free Radical Scavenging Activity

DPPH free radical scavenging activity is a common antioxidant activity method. However, the two extracts had no significant effect in the DPPH test. We speculate *A. fulica* extracts and *H. retispora* extracts do not work in DPPH. Since the material properties may vary, different antioxidant detection methods will have different reactions (Table 1).

Table 1. Antioxidant activities of *A. fulica* extracts and *H. retispora* extracts, including reducing power, 1,1-diphenyl-2-picrylhydrazyl (DPPH) free radical scavenging activity, and ferrous ion chelating power. Data represents mean ± S.D of three independent experiments performed.

Samples	Concentration (mg/L)	DPPH (%)	Chelating (%)	Reducing Power (OD700)
Vitamin C	100 μM	87.4 ± 0.1	-	-
EDTA	100 μM	-	85.6 ± 0.1	-
BHA	100 μM	-	-	0.67 ± 0.02
A. fulica extracts	10	21.83 ± 0.06	N/A	0.25 ± 0.06
	25	40.15 ± 0.03	N/A	0.24 ± 0.02
H. retispora extracts	10	14.48 ± 0.05	N/A	0.17 ± 0.01
	25	28.03 ± 0.04	N/A	0.18 ± 0.01

N/A: Unable to measure valid values.

3.2.2. Ferrous Ions Chelating Capacity Measurements

In the Ferrous ions chelating capacity measurements, ferrozine reacts with iron ions and turns into a dark red color. Once the analyte can be chelated with iron ions, which causes a reduction reaction, the Fe^{2+} complex is destroyed and the color is lightened. EDTA was used as the positive control at a concentration of 100 μM. In Table 1, *A. fulica* extracts and *H. retispora* extracts had chelating properties at 25 mg/L (40.15 ± 0.05%, 28.03 ± 0.04%), while the EDTA reached the same condition at 100 μM (87.38 ± 0.08%). It can be observed from the experimental results that *A. fulica* extracts and *H. retispora* extracts tend to achieve antioxidant effects by means of chelating iron ions.

3.2.3. Ferric Reducing Antioxidant Power (FRAP) Index Assessments

This study quantifies the Fe (III)-ferricyanide complex. The reduction reaction will change the complex from yellow to blue. Table 1 shows the reducing powers of *A. fulica* extracts and *H. retispora* extracts were OD 0.25 and 0.17 at 10 mg/L. From three different aspects of antioxidant response testing, we can observe *A. fulica* and *H. retispora* extracts have antioxidant properties in vitro. We further test the intracellular antioxidant assay to examine the antioxidant ability in the skin cells.

3.2.4. *A. fulica* Extracts and *H. retispora* Extracts Inhibit Intracellular ROS Accumulation

Previous reports revealed ROS caused damage on cellular biological morphological structures, including cell membranes, organelles, DNA, and protein configurations [28–30]. Meanwhile, a high level oxidative stress induced massive productions of melanin. Therefore, reducing oxidative stress is one good guideline to decrease melanin production. We used a cell-permeant DCFDA stain assay to test whether *A. fulica* extracts and *H. retispora* extracts treatments diminished intracellular ROS levels. Chemical fluorescent DCFDA staining is often applied to measure oxidative stress, which can be defined as the absence of oxidation. Typically, DCFDA is introduced into target cells via a small amount of aqueous solution, and then rapidly diffuses through the cell membrane as a colorless probe. Once the two acetate groups are cleaved by esterases within the cell, DCFDA fluorescence is detectable. A valuable property of DCFDA is that it cannot exit the cell while it has been cleaved within the cell. This increasing period of time during DCFDA reactions can be determined as an oxidative stress indicator. We used phorbol 12-myristate 13-acetate (PMA, 20 ng/mL), which is an inducer for endogenous superoxide production, as a negative control to induce oxidative stress, and then treated *A. fulica* extracts and *H. retispora* extracts for 24 h [31]. When the solution was comprised of 20 ng/mL of PMA, it enhanced the relative expressions of ROS by 106%, and adding PMA enlarged the ROS expression while AFMPS and GLNPS lessened the relative expression. There was a parallel trend among *A. fulica* extracts and *H. retispora* extracts with PMA, with the latter of the two displaying a larger difference compared to the solution containing only PMA. We observed both of these two materials decreased ROS levels, and the anti-oxidative effect of *H. retispora* was better than *A. fulica* (Figure 2). It showed similar results on in vitro antioxidant tests in Table 1.

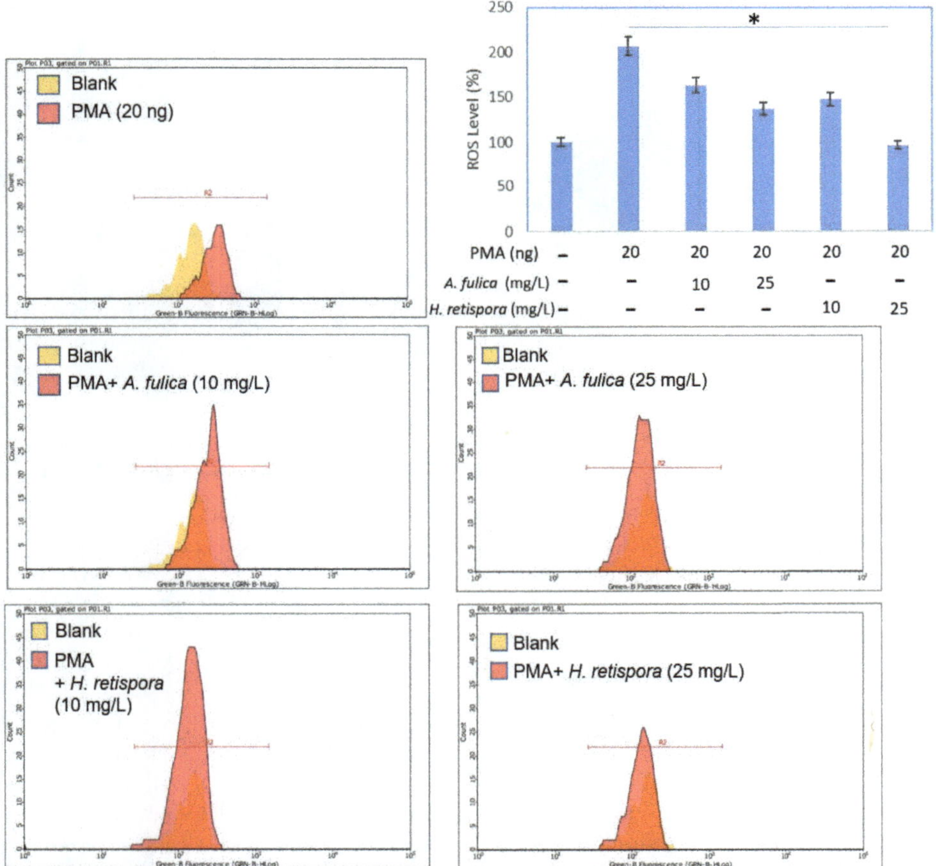

Figure 2. The 2′,7′-dichlorodihydrofluorescein diacetate (DCFDA) assay results showing that *A. fulica* extracts and *H. retispora* extracts treatment decreased ROS production in HS68 cells. The phorbol-12-myristate-13-acetate (PMA) was used as negative control to increase the oxidative level. Data represents mean ± S.D of three independent experiments performed. (Data represents mean ± S.D of three independent experiments performed. * $p < 0.01$).

3.3. Cell Growth of A. fulica Extracts and H. retispora Extracts Treated in Human Fibroblasts

Figure 3 shows the *A. fulica* extracts and *H. retispora* extracts increased the cell growth rate. The *A. fulica* extracts increased the growth rate by 118% 24 h after the addition of the extract, while *H. retispora* extracts increased by 146% at a concentration 25 mg/L. The chart shows as the length of concentration after the addition of *A. fulica* extracts and *H. retispora* extracts into the medium increased, the cell viability of fibroblasts was enhanced as well. In a previous study, both *A. fulica* extracts and *H. retispora* extracts were found to contain polysaccharides [14], Polysaccharides have the effect of promoting growth and can be applied to cell repair, skin care, and scar removal. This has also been verified in this experiment, especially in GLNPS with better promotion of proliferation.

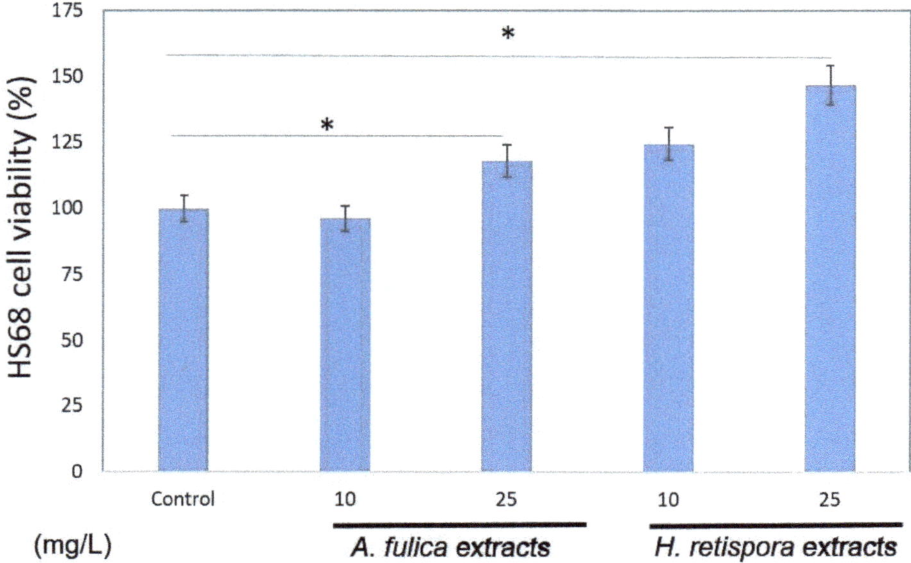

Figure 3. *A. fulica* extracts and *H. retispora* extracts effects on human cell viability with various doses. Fibroblasts were seeded in a 96-well micro titer plate which had a density of about 1×10^4 cells/well and treated with 1, 5, and 10 mg/L of *A. fulica* extracts and *H. retispora* extracts for 24 h. The cell viability of fibroblasts was measured by MTT assay 24 h after compound treatment. (Data represents mean ± S.D of three independent experiments performed. * $p < 0.01$).

3.4. Collagen Productions in Sirius Red Assays

The increased oxidative stress not only destroys the inside of the cell, but also affects the extracellular matrix. The change in the extracellular matrix is also one of the reasons for inducing cell carcinogenesis and tumor metastasis [32,33]. In previous experiments, we observed *A. fulica* extracts and *H. retispora* extracts have antioxidant properties, so we did a test of collagen in the cell content. PMA was used as a negative control, which induced oxidative stress and inhibited the production of collagen in the cells. After adding *A. fulica* extracts and *H. retispora* extracts, we observed a significant response of collagen, which was restored at a concentration of 25 mg/L. 44.9% and 55.4% (Figure 4), meaning *A. fulica* extracts and *H. retispora* extracts contribute to the repair of extracellular matrices.

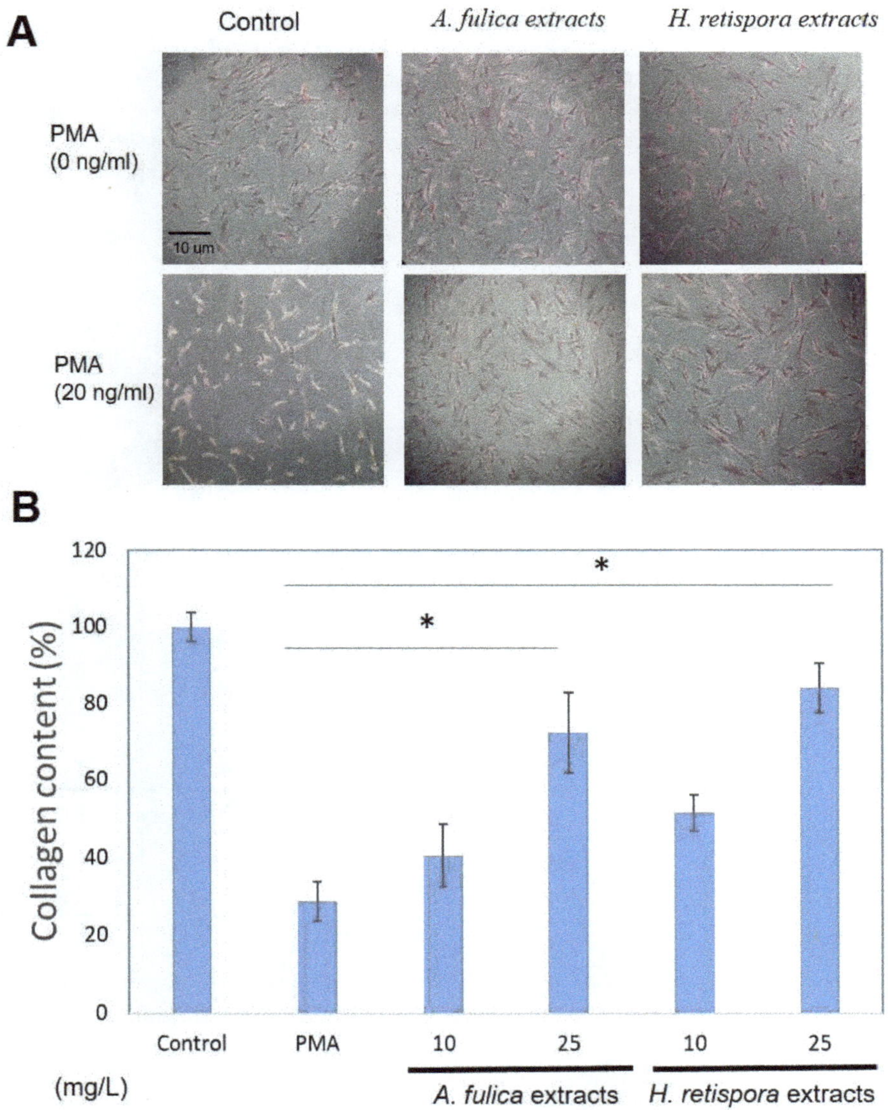

Figure 4. **(A)** HS68 cells collagen production with *A. fulica* extracts and *H. retispora* extracts treatments in Sirius red assay. PMA is used as negative control. (**B**) The quantitative data of collagen production; Data represents mean ± S.D of three independent experiments performed. (Data represents mean ± S.D of three independent experiments performed. * $p < 0.01$).

3.5. *A. fulica Extracts and H. retispora Extracts on B16-F10 Cellular Tyrosinase Activity*

Tyrosinase, a rate limiting enzyme, plays a critical role in pigment biosynthesis reactions. Other downstream enzymes influence the differences in color types for the syntheses of eumelanin and phenomelanin. To observe whether the four nanocarbons reduced melanin synthesis by down-regulating tyrosinase, in vitro tyrosinase activity was examined in the melanoma cell B16-F10 type. The data showed *A. fulica* extracts and *H. retispora* extracts inhibited 17.1 ± 2.9% and 12.6 ± 3.5% of the tyrosinase activity at a concentration of 25 mg/L, with both of them being in a dose-dependent manner from 10–25 mg/L. We observed *A. fulica* extracts and *H. retispora* extracts had better properties to suppress tyrosinase activity (Figure 5). Melanin is the source of skin color, which absorbs UV radiation to avoid UV-induced DNA damage and mutation. Although melanin can protect the skin from UV damage, the over expression of melanin also causes some skin disorders [34]. The experimental results suggested *A. fulica* extracts and *H. retispora* extracts could inhibit the synthesis of melanin. Table 1 shows the effect of chelating metal ion *A. fulica* extracts is better than the *H. retispora* extracts, meaning *A. fulica* extracts could be more positively used in cosmetic production as whitening agents than *H. retispora* extracts.

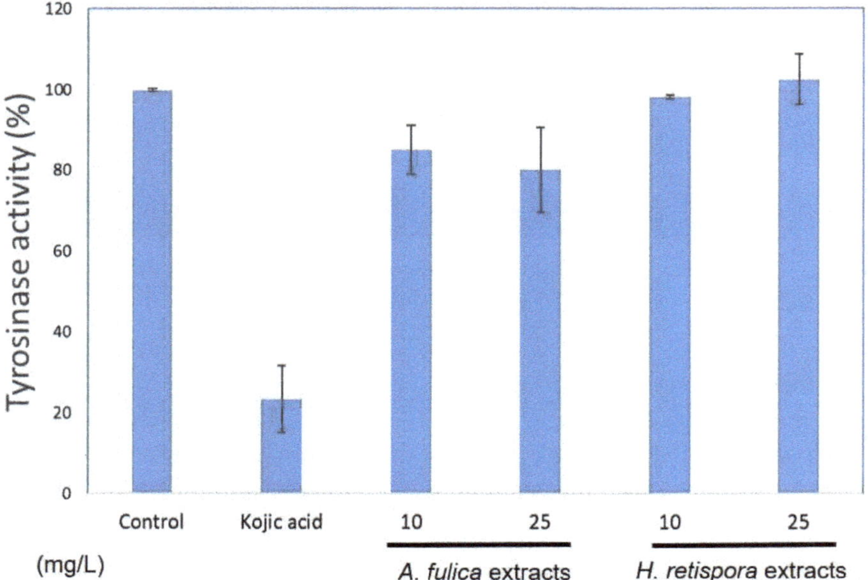

Figure 5. The inhibitory effects of tyrosinase activity. Treated with 10 and 25 μg/mL of *A. fulica* extracts and *H. retispora* extracts.

4. Conclusions

To sum up, we tested *A. fulica* and *H. retispora* macromolecular polysaccharide extracts as potential raw materials for cosmeceutical applications because of the multiple biofunctional properties. The experiments showed that these two extracts played roles as antioxidant ingredients and electron donors to stop free radical chain reactions. At the same time, *A. fulica* extracts inhibited tyrosinase activity, and *H. retispora* extracts promoted cell proliferation.

Author Contributions: H.-M.D.W., H.-Y.C. and C.-J.K. conceived and designed the experiments; H.-Y.C., Q.L. and H.-M.D.W. performed the experiments and analyzed the data; Q.L. and Y.-C.L. contributed the reagents, materials, and analysis tools; H.-Y.C. and H.-M.D.W. wrote the manuscript.

Funding: This research was supported by grants from the Ministry of Science and Technology (MOST107-2221-E-005-063) and National Key R&D Program of China (Project No. 2018YFD0400200). We also thank the projects of Research Center for Sustainable Energy and Nanotechnology, NCHU 107S0203B and China Agriculture Research System (No. CARS-20-08B).

Conflicts of Interest: The authors declare no conflict of interest

Data Availability Statement: The data used to support the findings of this study are available from the corresponding author upon request.

References

1. Li, P.H.; Liu, L.H.; Chang, C.C.; Gao, R.; Leung, C.H.; Ma, D.L.; David Wang, H.M. Silencing stem cell factor gene in fibroblasts to regulate paracrine factor productions and enhance c-Kit expression in melanocytes on melanogenesis. *Int. J. Mol. Sci.* **2018**, *19*, 1475. [CrossRef] [PubMed]
2. Zhang, Q.; Murawsky, M.; LaCount, T.; Kasting, G.B.; Li, S.K. Transepidermal water loss and skin conductance as barrier integrity tests. *Toxicol. Vitr.* **2018**, *51*, 129–135. [CrossRef] [PubMed]
3. Alexander, H.; Brown, S.; Danby, S.; Flohr, C. Research techniques made simple: Transepidermal water loss measurement as a research tool. *J. Investig. Dermatol.* **2018**, *138*, 2295–2300.e1. [CrossRef] [PubMed]
4. Grubauer, G.; Elias, P.M.; Feingold, K.R. Transepidermal water loss: The signal for recovery of barrier structure and function. *J. Lipid Res.* **1989**, *30*, 323–333. [PubMed]
5. Wang, Y.-C.; Haung, X.-Y.; Chiu, C.-C.; Lin, M.-Y.; Lin, W.-H.; Chang, W.-T.; Tseng, C.-C.; Wang, H.-M.D. Inhibitions of melanogenesis via phyllanthus emblica fruit extract powder in B16F10 cells. *Food Biosci.* **2019**, *28*, 177–182. [CrossRef]
6. Wu, P.Y.; Lin, T.Y.; Hou, C.W.; Chang, Q.X.; Wen, K.C.; Lin, C.Y.; Chiang, H.M. 1,2-Bis[(3-Methoxyphenyl) Methyl]Ethane-1,2-Dicarboxylic Acid Reduces UVB-Induced Photodamage In Vitro and In Vivo. *Antioxidants* **2019**, *8*, 452. [CrossRef]
7. Seo, S.Y.; Sharma, V.K.; Sharma, N. Mushroom tyrosinase: Recent prospects. *J. Agric. Food Chem.* **2003**, *51*, 2837–2853. [CrossRef]
8. Jimbow, K.; Obata, H.; Pathak, M.A.; Fitzpatrick, T.B. Mechanism of depigmentation by hydroquinone. *J. Investig. Dermatol.* **1974**, *62*, 436–449. [CrossRef]
9. Omarsdottir, S.; Olafsdottir, E.S.; Freysdottir, J. Immunomodulating effects of lichen-derived polysaccharides on monocyte-derived dendritic cells. *Int. Immunopharmacol.* **2006**, *6*, 1642–1650. [CrossRef]
10. Nishigori, C.; Yarosh, D.B.; Donawho, C.; Kripke, M.L. The immune system in ultraviolet carcinogenesis. *J. Investig. Dermatol. Symp. Proc.* **1996**, *1*, 143–146. [CrossRef]
11. Pillai, S.; Oresajo, C.; Hayward, J. Ultraviolet radiation and skin aging: Roles of reactive oxygen species, inflammation and protease activation, and strategies for prevention of inflammation-induced matrix degradation—A review. *Int. J. Cosmet. Sci.* **2005**, *27*, 17–34. [CrossRef] [PubMed]
12. Svobodova, A.R.; Galandakova, A.; Sianska, J.; Dolezal, D.; Ulrichova, J.; Vostalova, J. Acute exposure to solar simulated ultraviolet radiation affects oxidative stress-related biomarkers in skin, liver and blood of hairless mice. *Biol. Pharm. Bull.* **2011**, *34*, 471–479. [CrossRef] [PubMed]
13. Pereira-Marostica, H.V.; Castro, L.S.; Goncalves, G.A.; Silva, F.M.S.; Bracht, L.; Bersani-Amado, C.A.; Peralta, R.M.; Comar, J.F.; Bracht, A.; Sa-Nakanishi, A.B. Methyl Jasmonate Reduces Inflammation and Oxidative Stress in the Brain of Arthritic Rats. *Antioxidants* **2019**, *8*, 485. [CrossRef] [PubMed]
14. Liu, J.; Shang, F.; Yang, Z.; Wu, M.; Zhao, J. Structural analysis of a homogeneous polysaccharide from Achatina fulica. *Int. J. Biol. Macromol.* **2017**, *98*, 786–792. [CrossRef]
15. Paszkiewicz, W.; Kozyra, I.; Rzeżutka, A. A refinement of an international standard method (ISO/TS 15216–2: 2013) to allow extraction and concentration of human enteric viruses from tissues of edible snail species. *Food Anal. Method* **2015**, *8*, 799–806. [CrossRef]
16. Kim, Y.S.; Jo, Y.Y.; Chang, I.M.; Toida, T.; Park, Y.; Linhardt, R.J. A new glycosaminoglycan from the giant African snail Achatina fulica. *J. Biol. Chem.* **1996**, *271*, 11750–11755. [CrossRef]
17. Lv, S.; Zhang, Y.; Liu, H.X.; Hu, L.; Yang, K.; Steinmann, P.; Chen, Z.; Wang, L.Y.; Utzinger, J.; Zhou, X.N. Invasive snails and an emerging infectious disease: Results from the first national survey on Angiostrongylus cantonensis in China. *PLoS Negl. Trop. Dis.* **2009**, 3, e368. [CrossRef]
18. Zhang, L.; Zhang, Y. Structure and immunological activity of a novel polysaccharide from the spores of Ganoderma lucidum. *Afr. J. Biotechnol.* **2011**, *10*, 10923–10929. [CrossRef]
19. Zhang, K.; Liu, Y.; Zhao, X.; Tang, Q.; Dernedde, J.; Zhang, J.; Fan, H. Anti-inflammatory properties of GLPss58, a sulfated polysaccharide from Ganoderma lucidum. *Int. J. Biol. Macromol.* **2018**, *107*, 486–493. [CrossRef]
20. Xu, S.; Dou, Y.; Ye, B.; Wu, Q.; Wang, Y.; Hu, M.; Ma, F.; Rong, X.; Guo, J. Ganoderma lucidum polysaccharides improve insulin sensitivity by regulating inflammatory cytokines and gut microbiota composition in mice. *J. Funct. Foods* **2017**, *38*, 545–552. [CrossRef]

21. Zhao, X.; Zhou, D.; Liu, Y.; Li, C.; Zhao, X.; Li, Y.; Li, W. Ganoderma lucidum polysaccharide inhibits prostate cancer cell migration via the protein arginine methyltransferase 6 signaling pathway. *Mol. Med. Rep.* **2018**, *17*, 147–157. [CrossRef] [PubMed]
22. Lin, L.C.; Chen, C.Y.; Kuo, C.H.; Lin, Y.S.; Hwang, B.H.; Wang, T.K.; Kuo, Y.H.; Wang, H.D. 36H: A Novel Potent Inhibitor for Antimelanogenesis. *Oxidative Med. Cell. Longev.* **2018**, *2018*, 6354972. [CrossRef] [PubMed]
23. Li, P.H.; Chiu, Y.P.; Shih, C.C.; Wen, Z.H.; Ibeto, L.K.; Huang, S.H.; Chiu, C.C.; Ma, D.L.; Leung, C.H.; Chang, Y.N.; et al. Biofunctional Activities of Equisetum ramosissimum Extract: Protective Effects against Oxidation, Melanoma, and Melanogenesis. *Oxidative Med. Cell. Longev.* **2016**, *2016*, 2853543. [CrossRef] [PubMed]
24. Panchuk, R.R.; Skorokhyd, N.R.; Kozak, Y.S.; Lehka, L.V.; Moiseenok, A.G.; Stoika, R.S. Tissue-protective activity of selenomethionine and D-panthetine in B16 melanoma-bearing mice under doxorubicin treatment is not connected with their ROS scavenging potential. *Croat. Med. J.* **2017**, *58*, 171. [CrossRef] [PubMed]
25. Lorz, L.R.; Yoo, B.C.; Kim, M.-Y.; Cho, J.Y. Anti-Wrinkling and Anti-Melanogenic Effect of Pradosia mutisii Methanol Extract. *Int. J. Mol. Sci.* **2019**, *20*, 1043. [CrossRef] [PubMed]
26. Kim, Y.; Kim, D.; Park, C.; Park, T.; Park, B. Anti-Inflammatory and Skin-Moisturizing Effects of a Flavonoid Glycoside Extracted from the Aquatic Plant Nymphoides indica in Human Keratinocytes. *Molecules* **2018**, *23*, 2342. [CrossRef] [PubMed]
27. Boo, Y.C. Human Skin Lightening Efficacy of Resveratrol and Its Analogs: From in Vitro Studies to Cosmetic Applications. *Antioxidants* **2019**, *8*, 332. [CrossRef]
28. Yin, H.; Xu, L.; Porter, N.A. Free radical lipid peroxidation: Mechanisms and analysis. *Chem. Rev.* **2011**, *111*, 5944–5972. [CrossRef]
29. Tung, C.-H.; Chang, J.-H.; Hsieh, Y.-H.; Hsu, J.-C.; Ellis, A.V.; Liu, W.-C.; Yan, R.-H. Comparison of hydroxyl radical yields between photo-and electro-catalyzed water treatments. *J. Taiwain Inst. Chem. Eng.* **2014**, *45*, 1649–1654. [CrossRef]
30. Ohshima, H.; Yoshie, Y.; Auriol, S.; Gilibert, I. Antioxidant and pro-oxidant actions of flavonoids: Effects on DNA damage induced by nitric oxide, peroxynitrite and nitroxyl anion. *Free Radic. Biol. Med.* **1998**, *25*, 1057–1065. [CrossRef]
31. Chou, H.Y.; Lee, C.; Pan, J.L.; Wen, Z.H.; Huang, S.H.; Lan, C.W.; Liu, W.T.; Hour, T.C.; Hseu, Y.C.; Hwang, B.H.; et al. Enriched Astaxanthin Extract from Haematococcus pluvialis Augments Growth Factor Secretions to Increase Cell Proliferation and Induces MMP1 Degradation to Enhance Collagen Production in Human Dermal Fibroblasts. *Int. J. Mol. Sci.* **2016**, *17*, 955. [CrossRef] [PubMed]
32. Erdogan, B.; Webb, D.J. Cancer-associated fibroblasts modulate growth factor signaling and extracellular matrix remodeling to regulate tumor metastasis. *Biochem. Soc. Trans.* **2017**, *45*, 229–236. [CrossRef] [PubMed]
33. Yuzhalin, A.E.; Lim, S.Y.; Kutikhin, A.G.; Gordon-Weeks, A.N. Dynamic matrisome: ECM remodeling factors licensing cancer progression and metastasis. *Biochim. Biophys. Acta Rev. Cancer* **2018**, *1870*, 207–228. [CrossRef] [PubMed]
34. Xiao, L.; Matsubayashi, K.; Miwa, N. Inhibitory effect of the water-soluble polymer-wrapped derivative of fullerene on UVA-induced melanogenesis via downregulation of tyrosinase expression in human melanocytes and skin tissues. *Arch. Dermatol. Res.* **2007**, *299*, 245–257. [CrossRef] [PubMed]

Article

Maltol Improves APAP-Induced Hepatotoxicity by Inhibiting Oxidative Stress and Inflammation Response via NF-κB and PI3K/Akt Signal Pathways

Zi Wang [1,2], Weinan Hao [1], Junnan Hu [1], Xiaojie Mi [1], Ye Han [1], Shen Ren [1], Shuang Jiang [1,2], Yingping Wang [1,2], Xindian Li [1,*] and Wei Li [1,2,*]

1 College of Chinese Medicinal Materials, Jilin Agricultural University, Changchun 130118, China; wangzi8020@126.com (Z.W.); 15764380497@163.com (W.H.); junnanhu005@126.com (J.H.); njx006@yahoo.com.cn (X.M.); jiruoxiao@126.com (Y.H.); rs0109@163.com (S.R.); jiangshuang0503@hotmail.com (S.J.); yingpingw@126.com (Y.W.)

2 National & Local Joint Engineering Research Center for Ginseng Breeding and Development, Changchun 130118, China

* Correspondence: xdli2018@126.com (X.L.); liwei7727@126.com (W.L.); Tel.: +86-0431-84533308 (X.L.); +86-0431-84533304 (W.L.); Fax: 86-0431-84538012 (X.L.); 86-0431-84538011 (W.L.)

Received: 25 June 2019; Accepted: 5 September 2019; Published: 12 September 2019

Abstract: Maltol, a food-flavoring agent and Maillard reaction product formed during the processing of red ginseng (*Panax ginseng*, C.A. Meyer), has been confirmed to exert a hepatoprotective effect in alcohol-induced oxidative damage in mice. However, its beneficial effects on acetaminophen (APAP)-induced hepatotoxicity and the related molecular mechanisms remain unclear. The purpose of this article was to investigate the protective effect and elucidate the mechanisms of action of maltol on APAP-induced liver injury in vivo. Maltol was administered orally at 50 and 100 mg/kg daily for seven consecutive days, then a single intraperitoneal injection of APAP (250 mg/kg) was performed after the final maltol administration. Liver function, oxidative indices, inflammatory factors—including serum alanine and aspartate aminotransferases (ALT and AST), tumor necrosis factor α (TNF-α), interleukin-1β (IL-1β), liver glutathione (GSH), superoxide dismutase (SOD), malondialdehyde (MDA), cytochrome P450 E1 (CYP2E1) and 4-hydroxynonenal (4-HNE) were measured. Results demonstrated that maltol possessed a protective effect on APAP-induced liver injury. Liver histological changes and Hoechst 33258 staining also provided strong evidence for the protective effect of maltol. Furthermore, a maltol supplement mitigated APAP-induced inflammatory responses by increasing phosphorylated nuclear factor-kappa B (NF-κB), inhibitor kappa B kinase α/β (IKKα/β), and NF-kappa-B inhibitor alpha (IκBα) in NF-κB signal pathways. Immunoblotting results showed that maltol pretreatment downregulated the protein expression levels of the B-cell-lymphoma-2 (Bcl-2) family and caspase and altered the phosphorylation of phosphatidylinositol 3-kinase/protein kinase B (PI3K/Akt) in a dose-dependent manner. In conclusion, our findings clearly demonstrate that maltol exerts a significant liver protection effect, which may partly be ascribed to its anti-inflammatory and anti-apoptotic action via regulation of the PI3K/Akt signaling pathway.

Keywords: maltol; acetaminophen; liver injury; oxidative stress; apoptosis; inflammation response

1. Introduction

It is well known that drug-induced liver injury (DILI) is a common problem that leads to acute liver failure (ALF) in clinical application and severely affects human health [1]. Acetaminophen (APAP), an antipyretic-analgesic agent, is used in the clinic at therapeutic doses [2]. However, non-intentional misuse may result in hepatic toxicity and high mortality [3]. Initially, cytochrome P450 (CYP) bio-transforms APAP into *N*-acetyl-P-aminophenol (NAPQI), a toxic reactive intermediate, which consumes glutathione (GSH) and leads to mitochondrial dysfunction, oxidative stress, cellular necrosis, and apoptosis and eventually exerts toxic effects on the organism [4,5]. Therefore, seeking novel drugs or supplementary alternatives to prevent APAP-induced liver damage effectively is urgent for researchers.

APAP-mediated hepatotoxicity is closely related to oxidative stress, inflammatory response, and apoptosis. Excessive APAP exposure can cause mitochondrial dysfunction, severe energy debt, and oxidative stress, which induces reactive oxygen species (ROS) and further damage to hepatocytes [6]. Nuclear factor-kappa B (NF-κb) is an important nuclear transport factor and is related to immunoregulation, inflammatory response, and embryo development. Moreover, NF-κB also upregulates death receptors, including tumor necrosis factor-α (TNF-α), FAS (fatty acid synthetase), and the apoptotic proteins of the B-cell-lymphoma-2 (Bcl-2) family [7]. Meanwhile, apoptosis is also regarded as an important research subject in the development of liver diseases. At present, some studies have reported that the phosphatidylinositol 3-kinase/ protein kinase B (PI3K/AKT) signaling pathway is associated with the development of APAP-induced liver injury and early liver regeneration [8]. Based on the above, most researchers have speculated that the inhibition of ROS, apoptosis, and inflammation was a potential target for hepatoprotection.

Maltol (3-hydroxy-2-methyl-4-pyrrolidone) is a flavor enhancer, natural antioxidant, and one of the Maillard reaction products of heated-processed ginseng [9]. Maltol was also found in the roasted Korean ginseng root [10]. In recent years, maltol has been widely used in the fields of catalysis, pharmaceutical preparations, and food chemistry [11,12]. Previous studies have shown that maltol inhibited hexanal oxidation in a dose-dependent manner, and neuroprotective effects of maltol against oxidative stress were also proposed by Kim and Wei et al. [13,14]. Importantly, our previous study confirmed its potent antioxidant properties in TAA (thioacetamide) and alcohol-induced hepatic injury in vivo, which might be attributed to the prevention of lipid peroxidation and alleviation of the inflammation response [15–17].

Although maltol was demonstrated to contribute greatly to hepatic organ protection in an alcohol liver injury model, its protective effect on APAP-induced hepatotoxicity has not been further evaluated. Therefore, our aim was to prove the protective effects of maltol on APAP-induced liver injury and explore potential mechanisms of action to develop a reasonable prevention plan for APAP-induced hepatotoxicity.

2. Materials and Methods

2.1. Chemicals and Reagents

Maltol and APAP were purchased from Sigma-Aldrich (St. Louis, MO, USA). Alanine aminotransferase (ALT), aspartate aminotransferase (AST), GSH, superoxide dismutase (SOD), malondialdehyde (MDA) commercial kits, and haematoxylin and eosin (H&E) dye kits were provided by Nanjing Jiancheng Bioengineering Research Institute (Nanjing, China). Two-site sandwich enzyme-linked immunosorbent assay (ELISA) kits for the detection of mouse TNF-α and IL-1β were purchased from R&D systems (Minneapolis, MN, USA). Antibodies against the rabbit proteins CYP2E1, 4-HNE, Bax, Bcl-2, Bcl-XL, caspase 3, 8, 9, PI3K and p-PI3K, Akt and p-Akt, NF-κB and p-NF-κB, IKKα/β and p-IKKα/β, IκBα and p-IκBα, and β-actin were provided by Cell Signaling Technology (Danvers, MA, USA) and Proteintech (Rosemont, IL, USA). Hoechst 33258 and DyLight 488-SABC immunofluorescence staining kits were provided by Beyotime Institute of Biotechnology (Shanghai, China) and BOSTER Biological Technology (Wuhan, China) dividedly. Other reagents were obtained from Beijing Chemical Factory (Beijing, China).

2.2. Animal Experiments

Male ICR mice of 22–25 g body weight were provided by YISI Experimental Animals Co., Ltd. (Changchun, China). Food and water were freely available, and the animals were housed at

23.0 ± 2.0 °C, 50–70% humidity, and light/dark cycle of 12 h. The procedures for all laboratory animals were carried out in strict accordance with the ethical principles adopted in the Laboratory Animal Care and Use Guide (Ministry of Science and Technology of China, 2016). All animal experiments were approved by the Experimental Animal Ethics Committee of Jilin Agricultural University (Ethical Code: ECLA-JLAU-18062). To measure the anti-APAP-induced-hepatotoxicity effect of maltol in mice, the mice were randomly assigned to four groups after one week of adaptation to the environment: a normal group, an APAP (250 mg/kg i.p.) group, and two groups receiving different doses of maltol (50 and 100 mg/kg) ($n = 8$). Maltol was suspended in 0.9% saline. Maltol was administered to mice in two treatment groups for 7 days at doses of 50 mg/kg and 100 mg/kg, respectively. The normal and APAP groups received only 0.9% saline, in the same way. Then, all treatment groups, except the normal group, received a single injection of APAP (250 mg/kg, i.p.) after the final administration to induce acute liver injury. Mice from all groups were euthanized 24 h later. Tissues and blood samples were collected instantly. The serum was stored at −80 °C to determine transaminase after centrifugation (3500 rpm, 10 min, and 4 °C). The liver sections of each group were fixed in formalin for further use.

2.3. Analysis of ALT and AST Biochemical Markers

The liver biochemical indicators of serum ALT and AST were measured using commercial detection kits. The samples were transferred to a 96-well plate containing the substrate or a buffer solution and incubated at 37 °C, and the absorbance at 510 nm was measured after adding the developer. All data were expressed as U/L.

2.4. Analysis of GSH, SOD, and MDA Oxidative Markers

GSH, SOD, and MDA levels in liver tissues were determined according to commercial reagent methods. The lipid peroxides contained in the sample reacted with thiobarbituric acid (TBA) to form a red mixture. Absorbance at 532 nm was measured. The supernatant of liver tissues was centrifuged at 3500 rpm for 5 min, and then analyzed to determine SOD activity and GSH content.

2.5. Analysis of TNF-α and IL-1β Inflammation Markers

After serum samples were obtained, the concentrations of TNF-α and IL-1β were determined using ELISA kits according to the protocols provided by the manufacturer. In brief, prepared reagents, sample standards, and antibodies labeled with enzymes were added, then the reaction was carried out at 37 °C for 1 h. After adding the stopping solution, the absorbance at 450 nm was measured via an ELISA reader (Bio-Rad, Hercules, CA, USA).

2.6. Histopathological Examination

For histopathological examination, the liver samples were fixed over 24 h with 10% buffered formaldehyde before paraffin embedding and sectioning into 5 μm thickness. The liver tissues were routinely stained with H&E dye kits (Nanjing Jiancheng Bioengineering Research Institute, Nanjing, China) for conventional morphological evaluation using a light microscope (Olympus BX-60, Olympus Corporation, Tokyo, Japan).

2.7. Hoechst 33258 Staining

To observe the nuclear changes of hepatocytes, Hoechst 33258 staining was performed as described previously [18]. The sections were stained with Hoechst 33258 solution (10 μg/mL). UV excitation in a fluorescence microscope allowed us to observe the stained nuclei (Leica TCS SP8, Leica Microsystems, Wetzlar, Germany). The fluorescent intensity was quantified using Image-Pro plus 6.0 software (Media Cybernetics, Rockville, MD, USA).

2.8. Immunohistochemistry and Immunofluorescence Staining

As previously described, paraffin sections were deparaffinized and rehydrated prior to dyeing. After antigen retrieval, the slides were incubated with 1% BSA (bovine serum albumin) for 1 h and then with B-associated X (Bax) and Bcl-2 primary antibodies at 4 °C overnight, followed by secondary antibodies for half an hour at room temperature. Positive cells showing a brownish-yellow color in the cytoplasm or nucleus after DAB (diaminobenzidine) and hematoxylin staining were observed [19]. Fluorescence microscopy (Olympus BX-60, Olympus Corporation, Tokyo, Japan) was used for photographing, and positive cells were analyzed by Image-Pro Plus 6.0 software.

Immunofluorescence staining was used to measure CYP2E1 and 4-HNE proteins [20]. Briefly, the sections were incubated with primary antibodies at 4 °C for 12 h, then marked with a secondary antibody for 30 min at room temperature after washing the slides. Finally, the slides were exposed to DyLight 488-SABC. 4, 6 diamidino-2-phenylindole (DAPI) staining used for visualizing the cell nucleui and fluorescence intensities were analyzed by a Leica TCS SP8 microscope.

2.9. Western Blot Analysis

Total protein extracts from liver tissues were prepared with RIPA buffer (1:10, *g/v*), and a BCA Protein Assay Kit (Beyotime Biotechnology, Shanghai, China) was used for the determination of total protein content. An equal amount of proteins were isolated from the liver tissues and separated by 12% sodium dodecyl sulfate-polyacrylamide gel electrophoresis (SDS-PAGE), then transferred to a polyvinylidene fluoride membrane (PVDF). The membrane was blocked with 5% (*w/v*) defatted milk for 2 h and then probed with different primary antibodies at 4 °C for 12 h. After being washed by Tris-buffered saline-Tween (TBS-T), the proteins were incubated for 1.5 h in the presence of secondary antibodies. Protein bands were visualized using Quantity One software (Bio-Rad Laboratories, Hercules, CA, USA).

2.10. Statistical Analysis

Results are presented as mean ± standard deviation (S.D.) All samples were tested in triplicate. GraphPad Prism 6.0 (ISI, GraphPad Software, San Diego, CA, USA) was used to analyze the data. ANOVA, followed by the Bonferroni post-test, was used for comparing the differences among groups: $p < 0.05$ or $p < 0.01$ were considered statistically significant.

3. Results

3.1. Maltol Ameliorated APAP-Induced Hepatic Dysfunction

The liver levels of ALT and AST were elevated after APAP (250 mg/kg) injection ($p < 0.01$, $p < 0.05$) compared to those of the normal group, which indicated that hepatocellular damage induced by APAP was successfully established. Supplementation with maltol (50 and 100 mg/kg) for 1 week inhibited the increase in ALT and AST levels after exposure to APAP treatment ($p < 0.01$, $p < 0.05$) (Figure 1A,B).

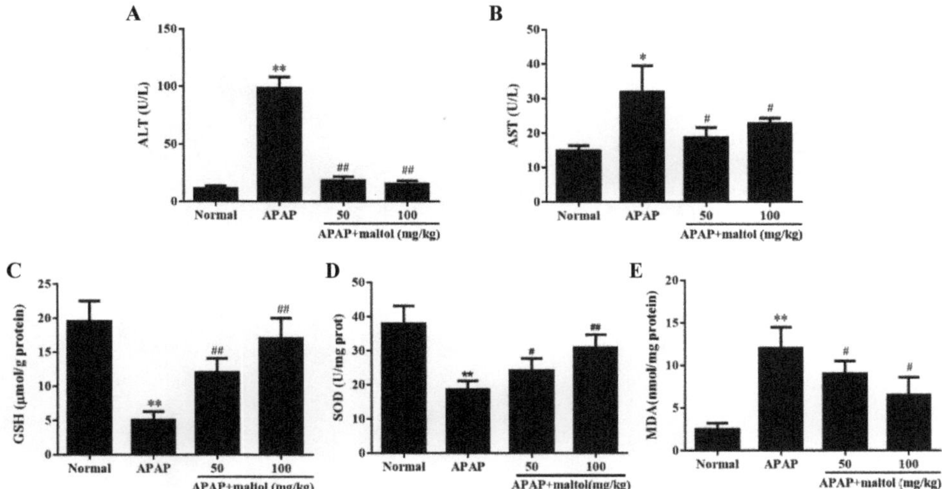

Figure 1. Effects of maltol pretreatment on hepatic dysfunction and histopathological changes caused by an overdose of acetaminophen (APAP). (**A**) serum alanine aminotransferase (ALT) and (**B**) aspartate aminotransferase (AST) activities; (**C**) liver glutathione (GSH) and (**D**) superoxide dismutase (SOD) amount; (**E**) liver malondialdehyde (MDA) content. All data were expressed as mean ± S.D; $n = 8$, * $p < 0.05$, ** $p < 0.01$, vs. normal group; # $p < 0.05$, ## $p < 0.01$ vs. APAP group.

3.2. Maltol Mitigated APAP-Induced Oxidative Stress Injury

APAP-induced oxidative stress injury is associated with the antioxidant defense system. Compared to the normal group, GSH and SOD contents significantly decreased in the APAP group ($p < 0.01$). However, maltol inhibited the depletion of GSH and restored hepatic SOD activity caused by APAP (Figure 1C,D) ($p < 0.01$, $p < 0.05$). In addition, maltol could also block the APAP-induced increase of MDA level in the liver (Figure 1E) ($p < 0.05$). These results clearly demonstrated that maltol reduced the oxidative stress injury caused by APAP.

For further evaluation of the hepatoprotective activity of maltol on APAP-induced oxidative stress during the progress of acute liver injury, immunofluorescence staining was used to determine CYP2E1 and 4-HNE expression levels in liver tissues. The results showed that maltol treatment significantly reversed the over-expression of CYP2E1 and 4-HNE caused by APAP compared to the normal group ($p < 0.01$) (Figure 2). These findings further confirmed that maltol ameliorated the oxidative stress injury induced by an overdose of APAP.

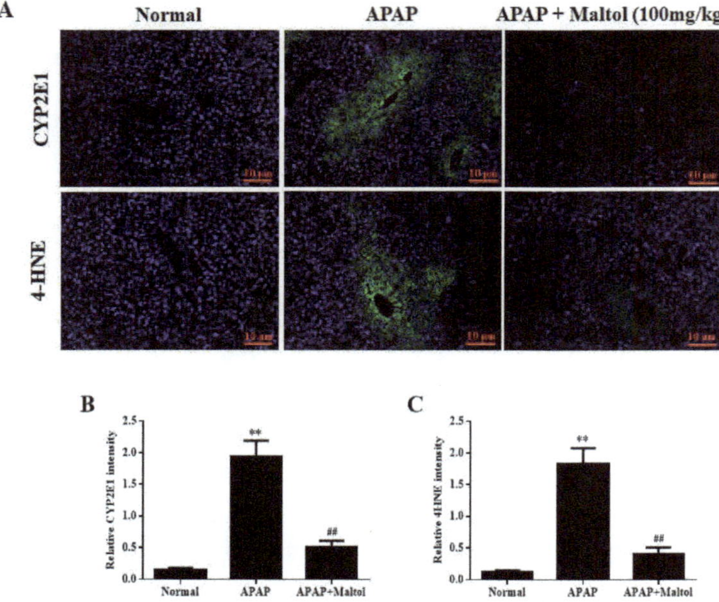

Figure 2. (**A**) Effects of maltol pretreatment on the expression of cytochrome P450 E1 (CYP2E1) and 4-hydroxynonenal (4-HNE) in liver tissues. The expression levels of 4-HNE and CYP2E1 (green) were determined by immunofluorescence (magnification ×200), and nuclear counterstaining (blue) was performed by 4, 6-Diamidino-2-phenylindole (DAPI). Quantitative fluorescence intensities of CYP2E1-positive cells (**B**) and 4-HNE-positive cells (**C**). All data are expressed as mean ± S.D.; $n = 8$. ** $p < 0.01$ vs. normal group; ## $p < 0.01$ vs. APAP group.

3.3. Maltol Mitigated APAP-Induced Liver Histopathological Changes

The results of H&E staining showed that the liver tissue in the normal group was normal and intact, presenting normal cell nuclei and the hepatic central vein. However, severe liver injury characterized by liver structural damage, intrahepatic hemorrhage, and inflammatory infiltration was observed in the APAP group. After maltol treatment for 1 week, inflammation and apoptosis were significantly attenuated (Figure 3).

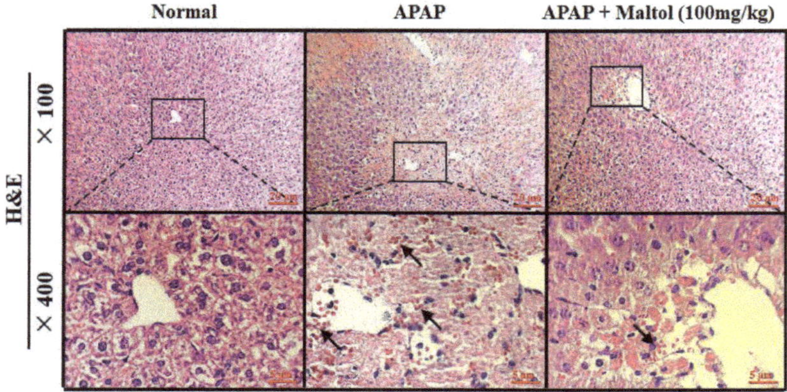

Figure 3. Liver tissue sections were stained with haematoxylin and eosin (H&E) for evaluation of pathological changes.

3.4. Maltol Mitigated APAP-Induced Apoptosis

Immunohistochemistry staining, Hoechst 33258 staining, and western bolt analysis were performed to investigate the molecular mechanism of the maltol-mediated beneficial effect against APAP-induced apoptosis by detecting the expression of the apoptotic proteins Bax, Bcl-XL, Bcl-2, caspase 3, 8, 9, and cleaved caspase 3, 8, 9 in liver tissues. Immunohistochemistry staining results demonstrated that APAP exposure caused hepatic cell apoptosis, as indicated by the higher Bax and lower Bcl-2 levels. Nevertheless, maltol could significantly mitigate these changes (Figure 4A). Hoechst 33258 staining supported the above results (Figure 4B). The western bolt analysis results showed that APAP injection markedly increased hepatic Bax and cleaved caspase 3, 8, 9 and decreased Bcl-XL and Bcl-2 levels ($p < 0.01$). Oppositely, apoptosis could be attenuated by maltol (Figure 5) ($p < 0.01$, $p < 0.05$). All results showed that maltol pretreatment dramatically prevented hepatic caspase-mediated apoptosis.

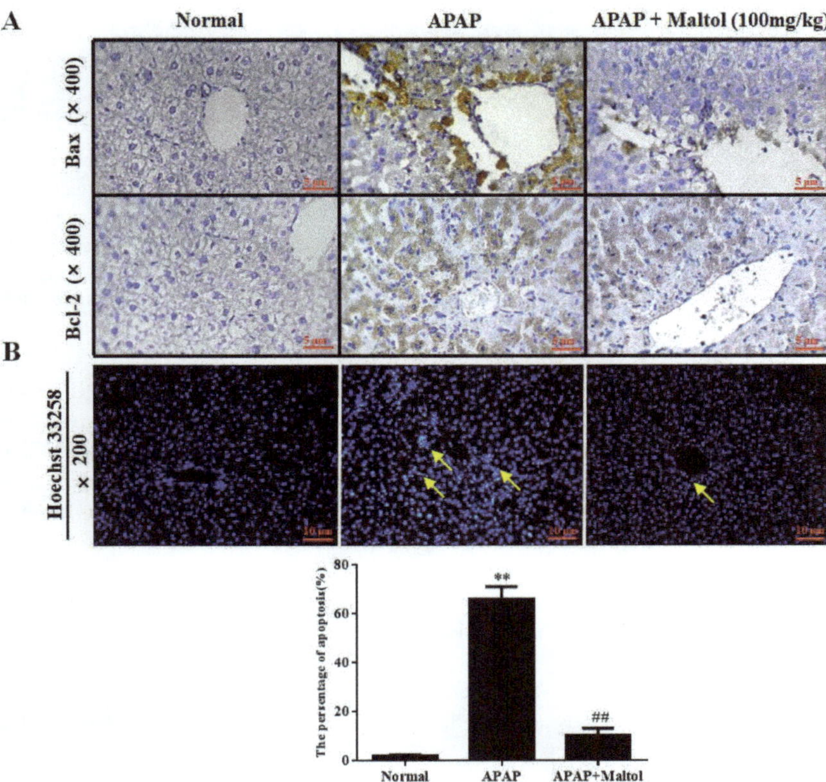

Figure 4. Effects of maltol pretreatment on the expression of B-associated X (Bax) and B-cell-lymphoma-2 (Bcl-2) and Hoechst 33258 staining in liver tissues. (**A**) The protein expression levels of Bax and Bcl-2 were examined by immunohistochemistry in liver tissues (magnification, ×400). (**B**) Hoechst 33258 staining (magnification, ×200). Arrows show necrotic and injured cells. All data are expressed as mean ± S.D., $n = 8$; ** $p < 0.01$ vs. normal group; ## $p < 0.01$ vs. APAP group.

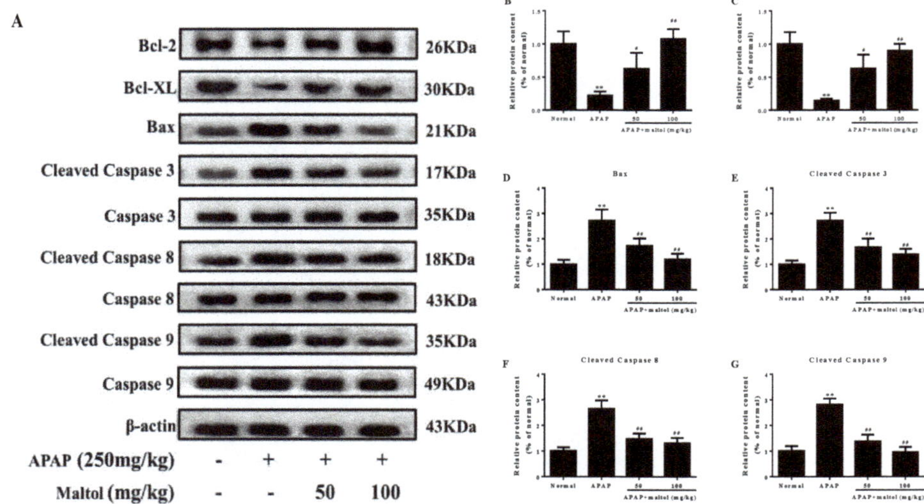

Figure 5. Effects of maltol pretreatment on apoptosis signaling pathways in APAP-triggered acute liver injury (ALI) mice. (**A**) Protein expression levels of Bax, Bcl-2, Bcl-XL, caspase 3, cleaved caspase 3, caspase 8, cleaved caspase 8, caspase 9, and cleaved caspase 9 were measured by western blotting analysis. (**B–G**) Quantification of relative protein expression levels was performed by densitometric analysis. All data are expressed as mean ± S.D., $n = 8$; ** $p < 0.01$ vs. normal group; ## $p < 0.01$, # $p < 0.05$ vs. APAP group.

3.5. Maltol Mitigated APAP-Induced Inflammatory Responses

APAP induced a series of inflammatory changes that mediated liver injury. Therefore, a western blot was also used to analyze the anti-inflammatory effects of maltol on APAP-activated NF-κB signal pathway. As shown in Figure 6A–E, APAP resulted in evidently higher levels of phosphorylated NF-κB. The upstream regulators, IKKα/β and IκBα, were also upregulated ($p < 0.01$). However, maltol treatment (50 and 100 mg/kg) prominently suppressed the release of NF-κB phosphorylation and blocked IKKα/β and IκBα phosphorylation ($p < 0.01$, $p < 0.05$), indicating that maltol prevented APAP-triggered inflammatory reaction partly via inhibiting the NF-κB pathway.

In addition, TNF-α and IL-1β are two key proinflammatory cytokines involved in the above progression. In our present study, as shown in Figure 6F,G, APAP injection caused a dramatic increase in the serum levels of TNF-α and IL-1β compared to those in the normal group ($p < 0.01$). However, pretreatment with maltol significantly inhibited the overproduction of TNF-α and IL-1β ($p < 0.01$, $p < 0.05$).

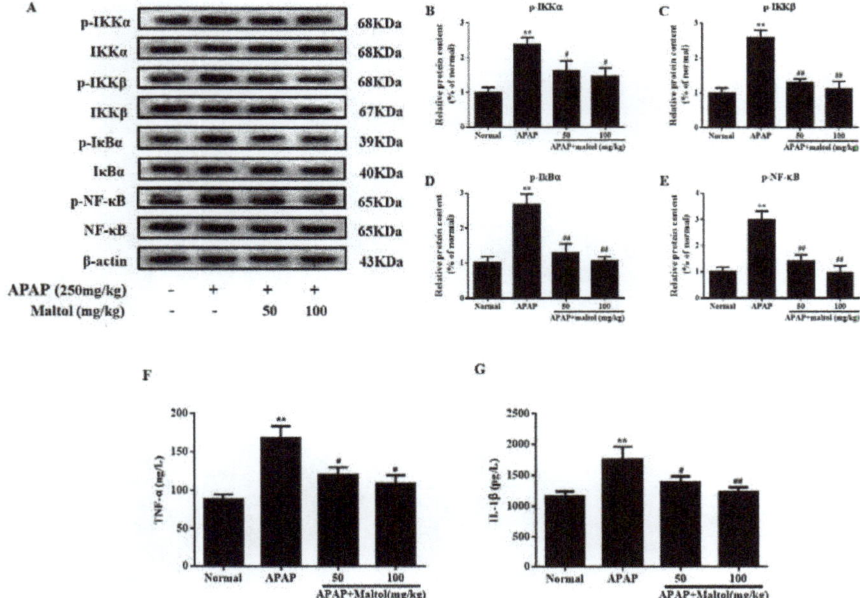

Figure 6. Effects of maltol pretreatment on APAP-induced activation of the nuclear factor-kappa B (NF-κB) signaling pathway in ALI mice. (**A**) Protein expression of phosphorylated and total inhibitor kappa B kinase α/β (IKKα/β), NF-kappa-B inhibitor alpha (IκBα) and NF-κB were measured by western blotting, and β-actin protein was used as a loading control. (**B–E**) The relative protein expression levels were quantified by densitometric analysis. The levels of (**F**) tumor necrosis factor α (TNF-α) and (**G**) interleukin-1β (IL-1β) in the serum of mice. Data are expressed as mean ± S.D., $n = 8$; ** $p < 0.01$ vs. normal group; ## $p < 0.01$, # $p < 0.05$ vs. APAP group.

3.6. Maltol Regulated the PI3K/Akt Signaling Pathway

In order to explore the protective role of the PI3K/Akt signaling pathway, we investigated the effects of maltol on the protein molecules in this signal pathway. From the results of the western bolt analysis, we found clearly that a single exposure to APAP decreased PI3K level and Akt phosphorylation ($p < 0.01$), which is consistent with our previous study [21]. However, maltol could reverse these changes in a dose-independent manner (Figure 7) ($p < 0.01$, $p < 0.05$). The above results showed that maltol exerted a potential protective effect by preventing APAP-induced acute liver toxicity at least partially through modulation of the PI3K/Akt signaling pathway.

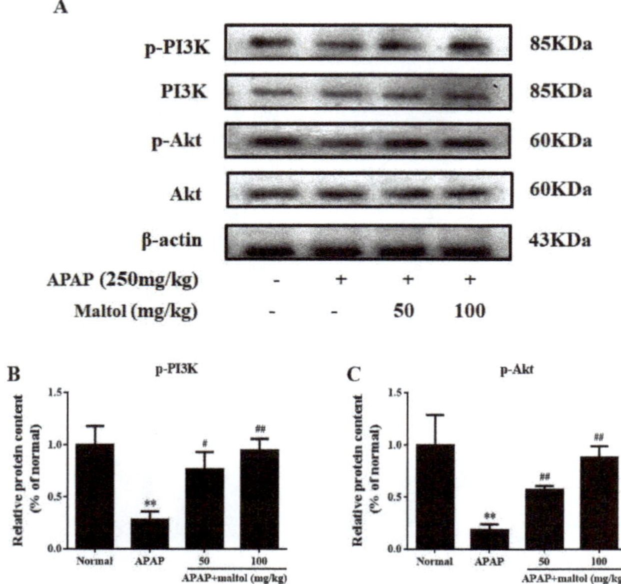

Figure 7. Effects of maltol pretreatment on the phosphatidylinositol 3-kinase/protein kinase B (PI3K/Akt) signaling pathway against APAP-induced liver injury. (**A**) The protein expression levels of phosphorylated and total PI3K and Akt were measured by western blotting with specific primary antibodies, and β-actin protein was used as a loading control. (**B**,**C**) Quantification of relative protein expression levels was performed by densitometric analysis. All data are expressed as mean ± S.D., $n = 8$. ** $p < 0.01$ vs normal group; ## $p < 0.01$, # $p < 0.05$ vs. APAP group.

4. Discussion

According to previous reports, APAP is a common harmful agent when misused or ingested in an excess dose [22]. Hepatotoxicity induced by overdose of APAP has become the most common cause of acute liver failure, replacing viral hepatitis in many developed countries [23]. However, the therapeutic options for this kind of liver injury disease are rather limited. A previous study has shown that maltol exerted beneficial anti-oxidative stress and anti-inflammatory actions in vitro [24]. Given that maltol was confirmed to have various medicinal activities, we evaluated whether maltol has a protective effect for APAP hepatotoxicity. Our former work indicated that maltol pretreatment exerted an important potential and beneficial effect on APAP-triggered acute liver injury and found that its molecular mechanisms of action were related to the alteration of oxidative stress-mediated inflammation and apoptosis, partly via regulation of the PI3K/Akt pathway.

Due to the conjugation with APAP metabolic product NAPQI, the GSH antioxidant system is key to decreasing the toxicity caused by APAP, which causes a sharp depletion of GSH content and then results in the necrosis of hepatocytes [25]. In the present study, it was found that excessive APAP could result in hepatic oxidative stress and cellular necrosis through reducing GSH and SOD levels and increasing MDA production, which were significantly reversed by maltol pretreatment for seven days. These results suggest a potential antioxidant capacity of maltol, in agreement with our previous study [17].

Previous studies have confirmed that oxidative stress is the central mediator of APAP-induced hepatotoxicity [26]. APAP-induced liver injury activates biochemical signaling pathways that originate mainly from CYP2E1-mediated formation of the reactive metabolite NAPQI [27]. It is well known that CYP2E1 has the greatest effect on acute hepatotoxicity caused by APAP, which is a potent inducer of 4-HNE lipid peroxide [28]. Therefore, we evaluated the oxidative stress injury caused by an overdose of APAP by analyzing CYP2E1 and 4-HNE expression. Our results showed that maltol pretreatment could effectively suppress APAP-induced CYP2E1 and 4-HNE overexpression.

Apoptosis plays a critical role in the pathology of tissues and presents with morphological and biochemical features such as DNA fragmentation, cell contraction, and Bcl-2 family protein activation [29]. A key step in apoptotic signaling is the mitochondrial release of cytochrome c, which promotes the formation of apoptotic bodies and the activation of caspase 9, followed by caspase 3 [30]. Activated caspase 3 can promote the cleavage of caspase 8 and amplify the pro-apoptotic signaling pathway through mitochondria. Moreover, cleaved downstream targets perpetuate the apoptotic pathway [31]. In addition, the regulatory factors Bcl-2 and Bcl-XL are two anti-apoptotic molecules of the pro-apoptotic protein Bax heterodimer in the mitochondrial outer membrane, which can prevent the permeability of the outer membrane [32]. A recent study related to mitochondria-dependent apoptosis proved that expression/stability of Bcl-2 could result in the release of the cytochrome and in turn, establish caspase-dependent pathways [33]. As expected, our results clearly indicated that the protein expression of cleaved caspase3, 8, 9 and Bax was remarkably inhibited, while that of Bcl-2 and Bcl-XL was enhanced, indicating that maltol exerted a certain anti-apoptosis effect in APAP-caused hepatotoxicity. Additionally, the result obtained from the Hoechst 33258 staining provided further support that APAP exposure led to a high density of apoptosis, whereas pretreatment with maltol significantly reversed the apoptosis in the liver, corroborating that maltol could dramatically inhibit hepatocyte apoptosis.

Oxidative stress can upregulate pro-inflammatory gene expression [34], and then inflammatory cells can trigger ROS overproduction, which would form a vicious circle and trigger the development of liver damage [35]. NF-κB is a major transcription factor, participating in immunity and inflammation processes, regulating apoptotic genes expression, and then causing apoptosis [36]. Previous literature reported that extracellular stimuli induced the rapid phosphorylation of I-κB and lead to the dissociation of NF-κB from I-κB [37]. Subsequently, activated NF-κB caused transcription of some inflammatory genes, including TNF-α and IL-1β [38]. The pro-inflammatory cytokine TNF-α can activate IKK. I-κB is phosphorylated by activated IKK, and subsequently, the inflammatory signal can also further lead to free NF-κB [39]. In our study, maltol was found to inhibit NF-κB activation by restraining the phosphorylation of IKKα, IKKβ, and I-κBα in a dose-dependent manner. Based on a preceding report describing the anti-inflammatory action of maltol [24], this study suggests that maltol could potentially exert a protective mechanism against APAP liver toxicity that might be partly attributed to the blockade of NF-κB signal activation.

PI3K is an intracellular phosphatidylinositol kinase. Akt is a key downstream effector of PI3K, and its anti-apoptotic effect is mainly achieved by phosphorylation of multiple target proteins in downstream pathways [40]. Our previous study proved that inhibited phosphorylation of Akt contributes to APAP-induced liver injury in mice [41]. In this study, maltol was shown to prevent APAP-induced liver injury by activating the PI3K/Akt signaling pathways. A more complete mechanism of maltol anti-APAP hepatotoxicity could include several key signaling pathways (Figure 8).

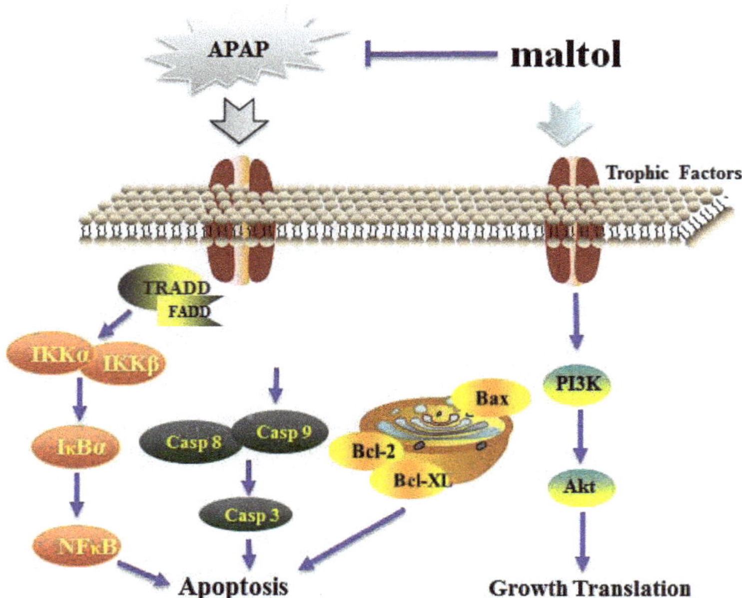

Figure 8. The possible mechanism of action underlying the protective effects of maltol against APAP-induced hepatic injury through inhibition of oxidative stress-mediated activation of the NF-κB pathway and apoptosis and regulation of the PI3K/Akt pathway. Tumor necrosis factor receptor-associated death domain protein (TRADD): Fas-Associated protein with Death Domain (FADD); inhibitor kappa B kinase α (IKKα); inhibitor kappa B kinase β (IKKβ); Caspase 3; Caspase 8; Caspase 9; B-cell-lymphoma-2 (Bcl-2); B-cell-lymphoma-XL (Bcl-XL); B-associated X (Bax); Protein kinase B (Akt); Phosphatidylinositol 3-kinase (PI3K).

5. Conclusions

In conclusion, the present study proved that maltol exerted a potential therapeutic effect against APAP-induced acute liver injury, which is attributed to its anti-apoptosis, anti-inflammatory, and anti-oxidation activities. The molecular mechanisms of action of maltol involved the suppression of the NF-κB signaling pathway and caspase-dependent cascade and the activation of the PI3K/Akt signaling pathway.

Author Contributions: W.L. and Y.W. conceived and designed the experiments; W.H. and S.J. performed the experiments; J.H. and Y.H. analyzed the data; X.M. and S.R. contributed reagents/materials/analysis tools; Z.W. and X.L. wrote the paper.

Funding: This work was supported by the grants of Scientific Research Foundation for the Returned Overseas Chinese Scholars (Jilin Province, 2016), Jilin Science & Technology Development Plan (No. 20180201083YY, 20190103092JH, and 20190304003YY), 13th Five-year Plan Science & Technology Project from the Education Department of Jilin Province (JJKH20190946kj), and the Program for the Young Top-Notch and Innovative Talents of Jilin Agricultural University (2016).

Conflicts of Interest: The authors declare no conflict of interest.

Abbreviations

The following abbreviations are used in this manuscript:

APAP	Acetaminophen
ALF	Acute Liver Failure
ALT	Alanine aminotransferase

AST	Aspartate aminotransferase
GSH	Glutathione
MDA	Malondialdehyde
ROS	Reactive Oxygen Species
CYP2E1	Cytochrome P450 E1
4-HNE	4-hydroxynonenal
NAPQI	N-acetyl-P-aminophenol
H&E	Hematoxylin and Eosin
NF-κB	Nuclear factor-kappa B
Akt	Protein kinase B
PI3K	Phosphatidylinositol 3-kinase
IKKα	Inhibitor kappa B kinase α
IKKβ	Inhibitor kappa B kinase β

References

1. Jaeschke, H. Acetaminophen: Dose-dependent drug hepatotoxicity and acute liver failure in patients. *Dig. Dis.* **2015**, *33*, 464–471. [CrossRef] [PubMed]
2. Gandillet, A.; Vidal, I.; Alexandre, E.; Audet, M.; Chenard-Neu, M.P.; Stutzmann, J.; Heyd, B.; Jaeck, D.; Richert, L. Experimental models of acute and chronic liver failure in nude mice to study hepatocyte transplantation. *Cell Transpl.* **2005**, *14*, 277–290. [CrossRef] [PubMed]
3. Yoon, E.; Babar, A.; Choudhary, M.; Kutner, M.; Pyrsopoulos, N. Acetaminophen-induced hepatotoxicity: A comprehensive update. *J. Clin. Transl. Hepatol.* **2016**, *4*, 131–142. [PubMed]
4. Hinson, J.A.; Reid, A.B.; McCullough, S.S.; James, L.P. Acetaminophen-induced hepatotoxicity: Role of metabolic activation, reactive oxygen/nitrogen species, and mitochondrial permeability transition. *Drug Metab. Rev.* **2004**, *36*, 805–822. [CrossRef] [PubMed]
5. Zhao, X.; Cong, X.; Zheng, L.; Xu, L.; Yin, L.; Peng, J. Dioscin, a natural steroid saponin, shows remarkable protective effect against acetaminophen-induced liver damage in vitro and in vivo. *Toxicol. Lett.* **2012**, *214*, 69–80. [CrossRef]
6. Brown, J.M.; Kuhlman, C.; Terneus, M.V.; Labenski, M.T.; Lamyaithong, A.B.; Ball, J.G.; Lau, S.S.; Valentovic, M.A. S-adenosyl-l-methionine protection of acetaminophen mediated oxidative stress and identification of hepatic 4-hydroxynonenal protein adducts by mass spectrometry. *Toxicol. Appl. Pharmacol.* **2014**, *281*, 174–184. [CrossRef]
7. Jayasooriya, R.G.; Moon, D.O.; Yun, S.G.; Choi, Y.H.; Asami, Y.; Kim, M.O.; Jang, J.H.; Kim, B.Y.; Ahn, J.S.; Kim, G.Y. Verrucarin A enhances TRAIL-induced apoptosis via NF-kappaB-mediated Fas overexpression. *Food Chem. Toxicol.* **2013**, *55*, 1–7. [CrossRef]
8. Leng, J.; Wang, Z.; Fu, C.L. NF-κB and AMPK/PI3K/Akt signaling pathways are involved in the protective effects of *Platycodon grandiflorum saponins* against acetaminophen-induced acute hepatotoxicity in mice. *Phytother. Res.* **2018**, *32*, 1–12. [CrossRef]
9. Li, W.; Su, X.; Han, Y.; Xu, Q.; Zhang, J.; Wang, Z.; Wang, Y. Maltol, a Maillard reaction product, exerts anti-tumor efficacy in H22 tumor-bearing mice via improving immune function and inducing apoptosis. *Rsc. Adv.* **2015**, *5*, 101850–101859. [CrossRef]
10. Sha, J.Y.; Zhou, Y.Y.; Yang, J.Y.; Leng, J.; Li, J.H.; Hu, J.N.; Liu, W.; Jiang, S.; Wang, Y.P.; Chen, C.; et al. Maltol (3-hydroxy-2-methyl-4-pyrone) slows D-galactose-induced brain aging process by damping the Nrf2/HO-1-mediated oxidative stress in mice. *J. Agric. Food Chem.* **2019**. [CrossRef]
11. Anwar-Mohamed, A.; El-Kadi, A.O. Induction of cytochrome P450 1a1 by the food flavoring agent, maltol. *Toxicol In Vitro* **2007**, *21*, 685–690. [CrossRef]
12. Krishnakumar, V.; Barathi, D.; Mathammal, R.; Balamani, J.; Jayamani, N. Spectroscopic properties, NLO, HOMO-LUMO and NBO of maltol. *Spectrochim. Acta A Mol. Biomol. Spectrosc.* **2014**, *121*, 245–253. [CrossRef]
13. Kim, Y.B.; Oh, S.H.; Sok, D.E.; Kim, M.R. Neuroprotective effect of maltol against oxidative stress in brain of mice challenged with kainic acid. *Nutr. Neurosci.* **2004**, *7*, 33–39.
14. Wei, A.; Mura, K.; Shibamoto, T. Antioxidative activity of volatile chemicals extracted from beer. *J. Agric. Food Chem.* **2001**, *49*, 4097–4101. [CrossRef]

15. Mi, X.J.; Hou, J.G.; Jiang, S.; Liu, Z.; Tang, S.; Liu, X.X.; Wang, Y.P.; Chen, C.; Wang, Z.; Li, W. Maltol mitigates thioacetamide-induced liver fibrosis through TGF-beta1-mediated activation of PI3K/Akt signaling pathway. *J. Agric. Food Chem.* **2019**, *67*, 1392–1401. [CrossRef]
16. Liu, W.; Wang, Z.; Hou, J.G.; Zhou, Y.D.; He, Y.F.; Jiang, S.; Wang, Y.P.; Ren, S.; Li, W. The liver protection effects of maltol, a flavoring agent, on carbon tetrachloride-induced acute liver injury in mice via inhibiting apoptosis and inflammatory response. *Molecules* **2018**, *23*, 2120. [CrossRef]
17. Han, Y.; Xu, Q.; Hu, J.N.; Han, X.Y.; Li, W.; Zhao, L.C. Maltol, a food flavoring agent, attenuates acute alcohol-induced oxidative damage in mice. *Nutrients* **2015**, *7*, 682–696. [CrossRef]
18. Li, W.; Yan, M.H.; Liu, Y.; Liu, Z.; Wang, Z.; Chen, C.; Zhang, J.; Sun, Y.S. Ginsenoside Rg5 ameliorates cisplatin-induced nephrotoxicity in mice through inhibition of inflammation, oxidative stress, and apoptosis. *Nutrients* **2016**, *8*, 566. [CrossRef]
19. Li, R.Y.; Zhang, W.Z.; Yan, X.T.; Hou, J.G.; Wang, Z.; Ding, C.B.; Liu, W.C.; Zheng, Y.N.; Chen, C.; Li, Y.R.; et al. Arginyl-fructosyl-glucose, a major maillard reaction product of red ginseng, attenuates cisplatin-induced acute kidney injury by regulating nuclear factor kappaB and phosphatidylinositol 3-kinase/protein kinase B signaling pathways. *J. Agric. Food Chem.* **2019**, *67*, 5754–5763. [CrossRef]
20. Xu, X.Y.; Hu, J.N.; Liu, Z.; Zhang, R.; He, Y.F.; Hou, W.; Wang, Z.Q.; Yang, G.; Li, W. Saponins (Ginsenosides) from the leaves of panax quinquefolius ameliorated acetaminophen-induced hepatotoxicity in mice. *J. Agric. Food Chem.* **2017**, *65*, 3684–3692. [CrossRef]
21. Yan, X.T.; Sun, Y.S.; Ren, S.; Zhao, L.C.; Liu, W.C.; Chen, C.; Wang, Z.; Li, W. Dietary alpha-mangostin provides protective effects against acetaminophen-induced hepatotoxicity in mice via Akt/mTOR-mediated inhibition of autophagy and apoptosis. *Int. J. Mol. Sci.* **2018**, *19*, 1335. [CrossRef]
22. Michael Brown, J.; Ball, J.G.; Wright, M.S.; Van Meter, S.; Valentovic, M.A. Novel protective mechanisms for S-adenosyl-L-methionine against acetaminophen hepatotoxicity: Improvement of key antioxidant enzymatic function. *Toxicol. Lett.* **2012**, *212*, 320–328. [CrossRef]
23. Lin, Z.; Wu, F.; Lin, S.; Pan, X.; Jin, L.; Lu, T.; Shi, L.; Wang, Y.; Xu, A.; Li, X. Adiponectin protects against acetaminophen-induced mitochondrial dysfunction and acute liver injury by promoting autophagy in mice. *J. Hepatol.* **2014**, *61*, 825–831. [CrossRef]
24. Song, Y.; Hong, S.; Iizuka, Y.; Kim, C.Y.; Seong, G.J. The neuroprotective effect of maltol against oxidative stress on rat retinal neuronal cells. *Korean J. Ophthalmol.* **2015**, *29*, 58–65. [CrossRef]
25. Wang, Z.; Hu, J.N.; Yan, M.H.; Xing, J.J.; Liu, W.C.; Li, W. Caspase-mediated anti-apoptotic effect of ginsenoside Rg5, a main rare ginsenoside, on acetaminophen-induced hepatotoxicity in mice. *J. Agric. Food Chem.* **2017**, *65*, 9226–9236. [CrossRef]
26. Hu, J.N.; Xu, X.Y.; Li, W.; Wang, Y.M.; Liu, Y.; Wang, Z.; Wang, Y.P. Ginsenoside Rk1 ameliorates paracetamol-induced hepatotoxicity in mice through inhibition of inflammation, oxidative stress, nitrative stress and apoptosis. *J. Ginseng Res.* **2019**, *43*, 10–19. [CrossRef]
27. McGill, M.R.; Jaeschke, H. Metabolism and disposition of acetaminophen: Recent advances in relation to hepatotoxicity and diagnosis. *Pharm. Res.* **2013**, *30*, 2174–2187. [CrossRef]
28. Hau, D.K.; Gambari, R.; Wong, R.S.; Yuen, M.C.; Cheng, G.Y.; Tong, C.S.; Zhu, G.Y.; Leung, A.K.; Lai, P.B.; Lau, F.Y.; et al. Phyllanthus urinaria extract attenuates acetaminophen induced hepatotoxicity: Involvement of cytochrome P450 CYP2E1. *Phytomedicine* **2009**, *16*, 751–760. [CrossRef]
29. Shou, Y.; Li, N.; Li, L.; Borowitz, J.L.; Isom, G.E. NF-kappaB-mediated up-regulation of Bcl-X(S) and Bax contributes to cytochrome c release in cyanide-induced apoptosis. *J. Neurochem.* **2002**, *81*, 842–852. [CrossRef]
30. McGill, M.R.; Sharpe, M.R.; Williams, C.D.; Taha, M.; Curry, S.C.; Jaeschke, H. The mechanism underlying acetaminophen-induced hepatotoxicity in humans and mice involves mitochondrial damage and nuclear DNA fragmentation. *J. Clin. Invest.* **2012**, *122*, 1574–1583. [CrossRef]
31. Jaeschke, H.; Duan, L.; Akakpo, J.Y.; Farhood, A.; Ramachandran, A. The role of apoptosis in acetaminophen hepatotoxicity. *Food Chem. Toxicol.* **2018**, *118*, 709–718. [CrossRef] [PubMed]
32. Hetz, C. BCL-2 protein family. Essential regulators of cell death. Preface. *Adv. Exp. Med. Biol.* **2010**, *687*, vii–viii. [PubMed]
33. Moshari, S.; Nejati, V.; Najafi, G.; Razi, M. Nanomicelle curcumin-induced DNA fragmentation in testicular tissue; Correlation between mitochondria dependent apoptosis and failed PCNA-related hemostasis. *Acta Histochem.* **2017**, *119*, 372–381. [CrossRef] [PubMed]

34. Arthur, M.J.; Bentley, I.S.; Tanner, A.R.; Saunders, P.K.; Millward-Sadler, G.H.; Wright, R. Oxygen-derived free radicals promote hepatic injury in the rat. *Gastroenterology* **1985**, *89*, 1114–1122. [CrossRef]
35. Jaeschke, H. Reactive oxygen and mechanisms of inflammatory liver injury: Present concepts. *J. Gastroenterol. Hepatol.* **2011**, *26*, 173–179. [CrossRef] [PubMed]
36. Luo, J.L.; Kamata, H.; Karin, M. IKK/NF-kappaB signaling: Balancing life and death—A new approach to cancer therapy. *J. Clin. Invest.* **2005**, *115*, 2625–2632. [CrossRef]
37. Robinson, S.M.; Mann, D.A. Role of nuclear factor kappaB in liver health and disease. *Clin. Sci.* **2010**, *118*, 691–705. [CrossRef]
38. Bieghs, V.; Trautwein, C. The innate immune response during liver inflammation and metabolic disease. *Trends Immunol.* **2013**, *34*, 446–452. [CrossRef]
39. Yu, L.; She, T.; Li, M.; Shi, C.; Han, L.; Cheng, M. Tetramethylpyrazine inhibits angiotensin II-induced cardiomyocyte hypertrophy and tumor necrosis factor-alpha secretion through an NF-kappaB-dependent mechanism. *Int. J. Mol. Med.* **2013**, *32*, 717–722. [CrossRef]
40. Rasul, A.; Ding, C.; Li, X.; Khan, M.; Yi, F.; Ali, M.; Ma, T. Dracorhodin perchlorate inhibits PI3K/Akt and NF-kappaB activation, up-regulates the expression of p53, and enhances apoptosis. *Apoptosis* **2012**, *17*, 1104–1119. [CrossRef]
41. Wang, L.; Zhang, S.; Cheng, H.; Lv, H.; Cheng, G.; Ci, X. Nrf2-mediated liver protection by esculentoside A against acetaminophen toxicity through the AMPK/Akt/GSK3β pathway. *Free Radic. Biol. Med.* **2016**, *101*, 401–412. [CrossRef] [PubMed]

Article

Sulfuric Odor Precursor *S*-Allyl-L-Cysteine Sulfoxide in Garlic Induces Detoxifying Enzymes and Prevents Hepatic Injury

Yusuke Yamaguchi [1], Ryosuke Honma [1], Tomoaki Yazaki [1], Takeshi Shibuya [1], Tomoya Sakaguchi [1], Harumi Uto-Kondo [2] and Hitomi Kumagai [1,*]

1 Department of Chemistry and Life Science, Nihon University, 1866 Kameino, Fujisawa-shi 252-0880, Japan
2 Department of Bioscience in Daily Life, Nihon University, 1866 Kameino, Fujisawa-shi 252-0880, Japan
* Correspondence: kumagai.hitomi@nihon-u.ac.jp; Tel.: +81-466-84-3946

Received: 9 August 2019; Accepted: 8 September 2019; Published: 10 September 2019

Abstract: *S*-Allyl-L-cysteine sulfoxide (ACSO) is a precursor of garlic-odor compounds like diallyl disulfide (DADS) and diallyl trisulfide (DATS) known as bioactive components. ACSO has suitable properties as a food material because it is water-soluble, odorless, tasteless and rich in bulbs of fresh garlic. The present study was conducted to examine the preventive effect of ACSO on hepatic injury induced by CCl_4 in rats. ACSO, its analogs and garlic-odor compounds were each orally administered via gavage for five consecutive days before inducing hepatic injury. Then, biomarkers for hepatic injury and antioxidative state were measured. Furthermore, we evaluated the absorption and metabolism of ACSO in the small intestine of rats and NF-E2-related factor 2 (Nrf2) nuclear translocation by ACSO using HepG2 cells. As a result, ACSO, DADS and DATS significantly suppressed the increases in biomarkers for hepatic injury such as the activities of aspartate transaminase (AST), alanine transaminase (ALT) and lactate dehydrogenase (LDH), and decreases in antioxidative potency such as glutathione (GSH) level and the activities of glutathione *S*-transferase (GST) and glutathione peroxidase (GPx). We also found ACSO was absorbed into the portal vein from the small intestine but partially metabolized to DADS probably in the small intestine. In in vitro study, ACSO induced Nrf2 nuclear translocation in HepG2 cells, which is recognized as an initial trigger to induce antioxidative and detoxifying enzymes. Taken together, orally administered ACSO probably reached the liver and induced antioxidative and detoxifying enzymes by Nrf2 nuclear translocation, resulting in prevention of hepatic injury. DADS produced by the metabolism of ACSO in the small intestine might also have contributed to the prevention of hepatic injury. These results suggest potential use of ACSO in functional foods that prevent hepatic injury and other diseases caused by reactive oxygen species (ROS).

Keywords: organosulfur compound; odor precursor; garlic; hepatic injury; Nrf2

---

1. Introduction

Sulfur is one of the key elements involved in the regulation of biological functions in the human body. Pivotal roles of organosulfur compounds are to maintain redox balance and to detoxify toxic agents. Reduced glutathione (GSH) is ubiquitously expressed in cells and reduces oxidative agents, such as hydroxyl radicals, oxide anion radicals, and hydrogen peroxide [1,2], playing a central role in detoxification [3,4]. Although GSH is important for such defense, oral intake of GSH does not necessarily increase these antioxidative and detoxifying activities in the human body [5,6]. Orally administered GSH can be delivered to various organs of the human body [7]; however, the effects of GSH are cancelled by the metabolite L-cysteine-L-glycine (Cys-Gly), which serves as a pro-oxidant to readily produce thiyl radicals and reactive oxygen species (ROS), resulting in oxidative stress [8].

Therefore, in order to reduce oxidative stress via oral administration of a natural compound, it is not sufficient for a functional compound to merely reach the target organ. The compound needs to increase the antioxidative and detoxifying activities by regulating transcription of biomolecules involved in these activities as electrophiles in food [9,10]. In addition, for maximum effectiveness, the metabolites of the compound are also required to increase the antioxidative and detoxifying activities.

Garlic has been used as a medicinal food since ancient times. This plant is reported to exert anti-cancer [11–13], anti-atherosclerotic [14], anti-diabetic [15], anti-tumoral [16], anti-bacterial [17], antioxidative [18], and detoxifying [19] activities in animal studies. Most of these effects have been attributed to organosulfur compounds, such as diallyl disulfide (DADS) and diallyl trisulfide (DATS), which have distinctive garlic odors produced when the garlic bulb is crushed or sliced [20]. During these processes, a garlic odor precursor, *S*-allyl-L-cysteine sulfoxide (ACSO; also known as alliin) in the cytoplasm collides with cysteine *S*-conjugate beta-lyase (C-S lyase) that leaked from the vacuole, producing garlic odor compounds along with pyruvic acid (Figure 1). Although studies have demonstrated that the physiological effects of garlic are due to the odor components DADS and DATS, odor precursors may also contribute to these effects. Indeed, some previous studies have reported that ACSO exerts anti-diabetic [21], anti-myocardial ischemia [22], hypoglycemic [23], and hypolipidemic effects [23] in animal studies. We have also reported that ACSO inhibits platelet aggregation [24] and suppresses increases in blood ethanol concentration [25] in animal studies. Considering that raw garlic contains up to 14 mg ACSO/g fresh weight [26] and ACSO can be retained just by heating bulbs before cutting to inactivate enzymes, garlic extract or powder rich in ACSO can be easily prepared from preheated bulbs. In addition, as ACSO is odorless and works as a taste enhancer [27], ACSO or garlic with high ACSO level can be added to various foods. Therefore, if oral administration of ACSO and/or its metabolites effectively increases the antioxidative and detoxifying activities of garlic, ACSO or such ACSO-rich garlic material would be a promising functional-food additive to prevent diseases caused by oxidative stress.

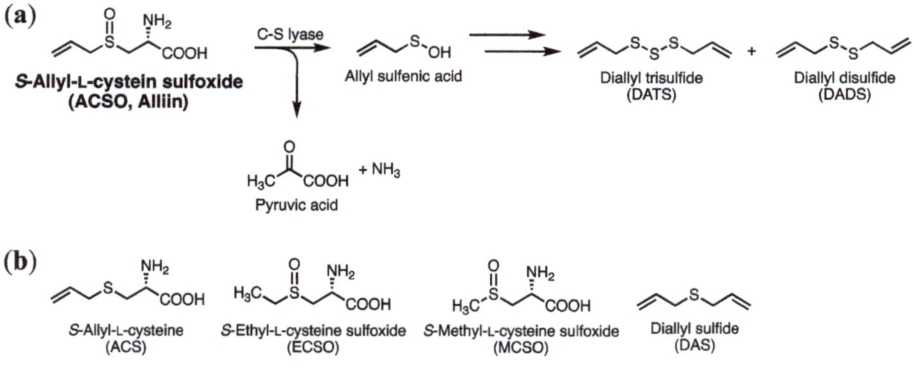

Figure 1. Production of garlic odor molecules from *S*-allyl-L-cysteine sulfoxide (ACSO) (**a**) and its related compounds tested in this study (**b**).

Antioxidative and detoxifying effects of food components can be evaluated by examining its preventive effect on acute hepatic injury induced by carbon tetrachloride (CCl_4) [28–31]. Intraperitoneally administered CCl_4 is transported to the liver and reduced by phase I enzymes, such as CYP450, to yield trichloromethyl radical (CCl_3 radical) [32,33], which forms covalent bonds with proteins, lipids, and nucleic acids to impair their functions [34,35]. In addition, CCl_3 radical oxidizes lipids to produce lipid peroxide, which destroys the lipid bilayer of cell membranes, resulting in leakage of the liver cell contents, such as aspartate transaminase (AST), alanine transaminase (ALT), and lactate dehydrogenase (LDH), into the blood [36]. Therefore, CCl_4-induced hepatic injury causes increases in the amounts of lipid peroxide in the liver and AST, ALT, and LDH activities in the blood.

In order to prevent the symptoms of CCl_4-induced hepatic injury, detoxification of the peroxide in vivo by phase II enzymes, including glutathione-*S*-transferase (GST) and quinone reductase (QR) and antioxidative enzymes such as glutathione reductase (GR) and glutathione peroxidase (GPx), should occur immediately [28–31]. GSH is consumed by detoxification with GST as well as the reduction of peroxide with GPx and the oxidation of GSH, while GSH is recovered by the reduction of oxidized GSH with GR [28–31]. Therefore, if intake of ACSO can prevent CCl_4-induced hepatic injury, increases in AST, ALT, LDH, and lipid peroxide as well as decreases in the amount of GSH will be suppressed by the concomitant increases in GST, QR, GR and GPx activities. Phase II enzymes and antioxidative enzymes are induced by the activation of the NF-E2-related factor 2 antioxidant response element (Nrf2-ARE) pathway [37,38]. This activation is triggered by Nrf2 nuclear translocation. To activate the pathway in the liver, orally administered ACSO and/or its bioactive metabolites are required to be transported to the liver, inducing translocation of Nrf2 to the nucleus.

The present study was conducted to examine the effect of oral administration of ACSO on the prevention of hepatic injury induced by CCl_4. Furthermore, we investigated the mechanism underlying the preventive effect on CCl_4-induced hepatic injury. We also evaluated Nrf2 nuclear translocation in liver cells following ACSO administration and the absorption of ACSO from the small intestine into the portal vein, which leads to the liver. Moreover, in order to examine whether intact ACSO is absorbed from the small intestine or metabolized during absorption, we also measured the ACSO and the ACSO metabolite pyruvic acid in the blood after injection of ACSO in the ligated loop of the small intestine in rats.

2. Materials and Methods

2.1. Materials

S-Allyl-L-cysteine sulfoxide (ACSO), *S*-methyl-L-cysteine sulfoxide (MCSO), and *S*-ethyl-L-cysteine sulfoxide (ECSO) were synthesized from the corresponding alk(en)yl bromide and L-cysteine followed by the addition of hydrogen peroxide [25,39]. *S*-Allyl-L-cysteine (ACS), diallyl sulfide (DAS), diallyl disulfide (DADS), and diallyl trisulfide (DATS) were purchased from Tokyo Chemical Industry Co., Ltd. (Tokyo, Japan). Chemical reagents for the experiments were purchased from Wako Pure Chemical Industries, Ltd. (Osaka, Japan), Oriental Yeast Co., Ltd. (Tokyo, Japan), Cosmobio Co., Ltd. (Tokyo, Japan), and Roche Diagnostics GmbH Co., Ltd. (Mannheim, Germany).

2.2. In Vivo and Ex Vivo Experiments

2.2.1. Animals

Seven-week-old male Sprague Dawley (SD) rats were purchased from Japan SLC, Inc. (Tokyo, Japan). All animal experiments were performed in accordance with the Guidelines for Animal Experiments of the College of Bioresource Sciences, Nihon University (approval code: AP13B010 and AP16B139). The feeding facility was maintained at an ambient temperature of 21–22 °C with 12-h light-dark cycling. Rats were housed in individual stainless-steel wire cages with free access to food (CE-2, Clea Japan, Tokyo, Japan) and water during a 1-week-acclimation period prior to the experiments.

2.2.2. Effect of ACSO and Its Related Compounds on Suppression of Hepatic Injury Induced by CCl_4

Rats were divided into 11 groups of six rats (Figure 2). ACSO, MCSO, ECSO, ACS, DAS, DADS, and DATS at a dosage of 50 µmol/mL/day were orally administered via gavage to rats of the corresponding group for seven consecutive days. ACSO, MCSO, ECSO, and ACS were dissolved in distilled water (DW), while DAS, DADS, and DATS were dissolved in olive oil at the time of use. Control rats received only either DW or olive oil. CCl_4 was intraperitoneally administered at a dosage of 1 mL/kg body weight after oral administration of the sulfoxides and sulfides on the seventh day. Then, the rats were subjected to fasting for 24 h and thereafter sacrificed. The liver was excised, and the microsomal and cytosolic fractions were prepared by centrifugation as described [40]. Briefly, serum was separated from sodium citrate-treated whole blood by centrifugation at 1500× g for 15 min at 4 °C.

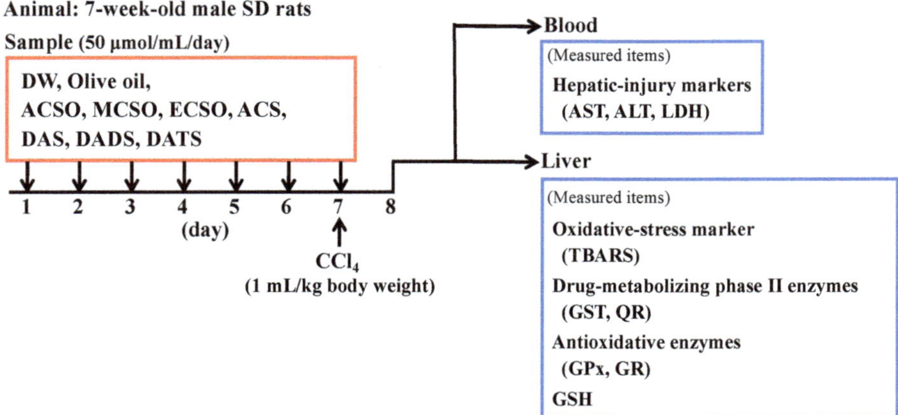

Figure 2. Animal experimental design for evaluation of the suppressive effects of S-allyl-L-cysteine sulfoxide (ACSO) and its related compounds on hepatic injury induced by CCl_4.

(a) Determination of AST, ALT and LDH Levels

The obtained serum in 2.2.2. was used to examine acute hepatic-injury enzyme markers (i.e., AST, ALT, and LDH). These levels in the serum were determined by the enzymatic method using an automatic analyzer, Spotchem EZ SP-4430 (Liver-2, Arkray, Inc., Kyoto, Japan).

(b) Measurement of Lipid Peroxide

The content of lipid peroxide in the liver was determined by thiobarbituric acid (TBA) reactive substances (TBARS) assay, which detects aldehydes produced from the decomposition of lipid hydroperoxide [41]. First, 0.5 mL of the liver homogenate was mixed with 0.3 mL of 1% phosphoric acid and 1 mL of 0.67% TBA aqueous solution. The mixture was incubated at 95 °C for 45 min to produce aldehyde-TBA adduct possessing absorbance at 535 nm. After cooling the reaction mixture to room temperature, 4 mL of *n*-butanol was added to dissolve the adduct. The mixture was centrifuged at 1500× g for 10 min, and the supernatant containing aldehyde-TBA adduct was obtained. The absorbance of the supernatant at 535 nm was measured. The difference in absorbance at 535 nm between the mixture with and without the adduct was used as the TBARS value. Malondialdehyde (MDA) was used as a standard and the TBARS value was expressed as MDA equivalent.

(c) Measurement of GST Activity

The activity of glutathione S-transferase (GST) in the cytosol of the liver was spectrophotometrically assayed [42]. A 1.48 mL solution containing 0.1 M potassium phosphate buffer at pH 7.4, 30 mM reduced glutathione (GSH) was preincubated at 25 °C for 5 min. Then, 60 µL of 30 mM 1-chloro-2,4-dinitrobenzene (CDNB) and 300 µL cytosol fraction were added to the solution to initiate the reaction of CDNB to produce S-2,4-dinitrophenylglutathione, and the increase in absorbance at 340 nm was recorded. The amount of S-2,4-dinitrophenylglutathione was calculated by its extinction coefficient ($\varepsilon = 9.6\ mM^{-1}\ cm^{-1}$). The activity of GST was expressed as the amount of S-2,4-dinitrophenylglutathione produced per minute per milligram of cytosol protein. The amount of cytosol protein was determined by the bicinchoninic acid (BCA) method using BCA Protein Assay Kit (Thermo Fisher Scientific, Waltham, MA, USA).

(d) Measurement of QR Activity

The activity of quinone reductase (QR) in the cytosol of liver was assayed spectrophotometrically [43]. An 800 μL solution containing 50 mM Tris-HCl at pH 8.0, 0.2% Tween-20, 40 μM oxidized 2,6-dichlorophenolindophenol (DCIP), and 0.3 mM NADPH was preincubated at 25 °C for 5 min. Then, 150 μL of cytosol fraction was added to the solution to produce reduced DCIP from oxidized DCIP. A decrease in the absorbance at 600 nm was recorded, and the activity of QR was calculated from the difference between the absorbance with and without the cytosol fraction. The amount of oxidized DCIP was calculated by its extinction coefficient (ε = 21 mM^{-1} cm^{-1}). The activity of QR was expressed as the amount of oxidized DCIP consumed per minute per milligram of cytosol protein. The amount of cytosol protein was determined by the BCA method.

(e) Measurement of GR Activity

The activity of glutathione reductase (GR) in the liver was measured as described by Carlberg and Mannervik [44]. A 600 μL of solution containing 0.1 M potassium phosphate buffer at pH 7.6 containing 1 mM ethylenediaminetetraacetic acid (EDTA), 0.1 mM reduced nicotinamide-adenine dinucleotide phosphate (NADPH), 1 mM oxidized glutathione (GSSG), and 0.1% bovine serum albumin (BSA) was mixed with 100 μL of the cytosol fraction to produce GSH from GSSG. The decrease in absorbance of NADPH at 340 nm was monitored at 25 °C, and the amount of NADPH consumed was calculated by using its molar extinction coefficient (ε = 6.22 mM^{-1} cm^{-1}). The activity of GR was expressed as the amount of NADPH consumed per minute per milligram of cytosol protein. The amount of cytosol protein was determined by the BCA method.

(f) Measurement of GPx Activity

The activity of glutathione peroxidase (GPx) in the liver was determined spectrophotometrically using GSH and hydrogen peroxide (H_2O_2) as substrates [45]. First, 20 μL of 0.1 M GSH, 100 μL of 10 unit/mL GR, and 100 μL of 2 mM NADPH were mixed with 100 μL of 0.1 M sodium phosphate buffer and 2 mM NaN_3 at pH 7.0 in a sample cuvette. Then, 10 μL of the cytosol fraction was added to the mixture, while 10 μL of buffer was added to the reference cuvette. The total volume of the solution was adjusted to 1 mL by adding 660 μL of DW into each cuvette. After preincubation at 37 °C for 2 min, the reaction was started by adding 10 μL of 1.5 mM H_2O_2. The oxidation of NADPH to oxidized nicotinamide-adenine dinucleotide phosphate ($NADP^+$) along with the conversion of GSSG to GSH by GR was followed by the absorbance of NADPH at 340 nm, and the amount of NADPH consumed was calculated by using its molar extinction coefficient (ε = 6.22 mM^{-1} cm^{-1}). The activity of GPx was expressed as the amount of NADPH consumed per minute per milligram of cytosol protein. The amount of cytosol protein was determined by the BCA method.

(g) Measurement of total GSH Level

The level of total glutathione was measured as described by Habig et al. [42]. First, a mixture of 100 μL of the cytosol fraction, 50 μL of 4 mM NADPH, 100 μL of 6 unit/mL GR in 500 μL of 10 mM sodium phosphate buffer at pH 7.5 was preincubated at 37 °C for 5 min to convert GSSG to GSH. Then, 50 μL of 10 mM 5,5′-dithiobis-2-nitrobenzoic acid (DTNB) was added to the mixture. The absorbance of 5-mercapto-2-nitrobenzoic acid at 412 nm produced by the reaction of DTNB and GSH was measured.

2.2.3. Effect of ACSO and Sulfides on Liver Function of Normal Rats

Rats were divided into four groups of six rats. ACSO, DADS, and DAS at a dosage of 50 μmol/mL/day were orally administered to normal healthy rats of the corresponding group for seven consecutive days. ACSO was dissolved in DW, while DADS and DAS were dissolved in olive oil for use. Control rats received only DW or olive oil. The rats were subjected to fasting for 24 h following oral administration of ACSO, DADS, and DAS on the seventh day and then sacrificed. The activities of GST and GPx were measured as described in 2.2.2.

2.2.4. ACSO Absorption and Metabolism in the Small Intestine

(a) ACSO Absorption from the Small Intestine

Rats were divided into two groups (n = 3) designated as control and ACSO. After acclimation, rats were subjected to fasting overnight. The abdomen was opened on a thermal heat table under anesthesia, and the small intestine was ligated to be closed. The blood was collected from the portal vein through a cannula and designated as the sample at 0 min. Then, the ligated small intestine was injected with phosphate buffered salts (PBS) in the control group and with 100 mM ACSO in PBS for the ACSO group. Blood was collected from the portal vein through the cannula 10, 20 and 30 min after this injection. The small intestine was then excised, and the luminal solution was collected and heated at 80 °C for 30 min for measurement of ACSO and pyruvic-acid content. The excised small intestine from the control group was used for analysis of ACSO metabolism.

(b) Measurement of ACSO and Pyruvic-Acid Content

For the measurement of ACSO content, the obtained blood or luminal solution was mixed with 1 M hydrogen peroxide and placed on ice for 10 min. The solution was centrifuged at 10,000× g at 4 °C for 10 min, and the supernatant was mixed with 0.7 M potassium carbonate and placed on ice for an additional 10 min. Then, the solution was centrifuged at 2300× g at 4 °C for 10 min, and the resultant supernatant was kept cool on ice until use. For measurement of ACSO, the supernatant was mixed with 9-fluorenylmethyl chloroformate (Fmoc-Cl), and Fmoc-derivatized compounds were analysed using HPLC (Alliance e2695, Waters, Milford, MA, USA) equipped with an ODS column (Inertsil ODS-4, GL Sciences, Tokyo, Japan) and fluorescent detector (Waters 2475, Waters, Ex. 263 nm, Em. 313 nm).

For the measurement of pyruvic-acid content, the obtained blood or luminal solution was mixed with 1 M hydrogen peroxide and placed on ice for 10 min. The solution was centrifuged at 10,000× g at 4 °C for 10 min, and the supernatant was mixed with 6 N HCl. The solution was incubated with 1,2-diamino-4,5-methylenedioxybenzene (DMB), 9 N HCl, 2-mercaptoethanol, and $Na_2S_2O_4$ at 100 °C for 45 min. After cooling, the supernatant was collected and filtered. The obtained solution containing DMB-derivatized compounds was analyzed with HPLC (Alliance e2695) equipped with an ODS column (Separar C18G, Rikaken, Japan) and fluorescent detector (Waters 2475, Ex.367 nm, Em. 446 nm).

(c) Analysis of Volatile Metabolites from ACSO in the Small Intestine

The excised small intestines were rinsed with 0.02 M K_2HPO_4-KH_2PO_4 buffer at pH 7.5 and homogenized with the same solution before centrifugation at 15,000× g at 4 °C for 30 min. The supernatant was centrifuged at 105,000× g at 4 °C for 60 min. Ammonium sulfate was added to the obtained supernatant to yield precipitated protein. The concentration of ammonium sulfate was increased stepwise, and the precipitate was recovered at each stage. The protein precipitated by 60–80% ammonium sulfate was collected and dialyzed against 0.02 M K_2HPO_4-KH_2PO_4 buffer. The dialyzed solution was purified with cation-exchange chromatography (CM Sepharose FF, GE Healthcare, Chicago, IL, USA), and the obtained fraction was mixed with 50 mM ACSO in 0.02 M K_2HPO_4-KH_2PO_4 buffer at 30 °C for 30 min in a sealed vial. Volatiles in the headspace were collected by solid-phase microextraction using divinylbenzene/carboxen/polydimethylsiloxane fiber (57328-U, Sigma-Aldrich, St. Louis, MO, USA) for 30 min while the solution was continuously stirred. The absorbed volatiles were analyzed by GC-Atomic Emission Detector (HP 6890GC, HP G2350A AED, Agilent technology, Santa Clara, CA, USA) equipped with analytical column (DB-1, J&W Scientific, Folsom, CA, USA). Sulfuric compounds were detected at 181 nm, and carbon compounds were detected at 193 nm.

2.3. *In vitro Experiments*

2.3.1. Cell Culture

Human Caucasian hepatocyte carcinoma HepG2 cells (HB-8065, Lot No: 16K046, passages 10–20, ATCC, Manassas, VA, USA) were grown in Eagle's Minimum Essential Medium (EMEM) supplemented with 10% (*v*/*v*) fetal bovine serum, 100 U/mL penicillin, and 100 g/mL streptomycin. Cultures were grown at 37 °C under 5% CO_2. Stock cultures were grown in 75-cm^2 flasks (Corning, Tokyo, Japan), and media was replaced every 2 days. Cells were routinely subcultured by trypsinization upon reaching 80–90% confluency.

2.3.2. Nrf2 Content in the Nucleus, Cytoplasm, and Whole Cells

A suspension of HepG2 cells at a density of 7×10^5 cells/mL in EMEM medium containing 10% FBS was incubated at 37 °C under 5% CO_2 for 48 h in a 100-mm dish. The medium was then removed, and EMEM medium containing 1% BSA and 0.25–1.0 mM ACSO was added. The cells were further incubated at 37 °C under 5% CO_2 for 6 h. After removal of the medium, cells were detached in 5 mL PBS using a cell scraper. The obtained suspension was centrifuged at 300× *g* for 5 min at room temperature to pellet the cells. To obtain whole cells, the precipitate was mixed with lysis buffer and sonicated. The resulting suspension was centrifuged at 1000× *g* at room temperature for 10 min, and then the supernatant was designated as the whole-cell fraction. Nuclei and cytoplasm of HepG2 cells were extracted from the obtained precipitate using the Nuclear Extraction Kit (RayBiotech, Peachtree Corners, GA, USA), according to the manufacturer's instructions. These samples were used for SDS-PAGE analysis with a 12.5% acrylamide gel. After transfer of protein from the gel, the PVDF membrane was blocked for 30 min in 150 mM NaCl; 10 mM Tris/HCl, pH 7.4; 0.05% (*v*/*v*) Tween 20, and 0.5% (w/v) skim milk powder. PVDF membranes were incubated overnight at 4 °C with primary antibodies raised against Nrf2 (E-AB-32280, Lot No: DK7634, Elabscience, Houston, TX, USA; 1:2000) and β-actin (sc-47778; Lot No: K1418, Santa Cruz Biotechnology, Dallas, TX, USA; 1:1000). After this incubation, membranes were incubated HRP-linked secondary anti-rabbit IgG (7074S, Lot No: 27, Cell Signaling Technology, Danvers, MA, USA; 1:10,000) or anti-mouse IgG (7076S, Lot No: 33, Cell Signaling Technology, 1:5000). Protein was visualized and quantified using ECL Western blot detection system (RPN2235; GE Healthcare, Tokyo, Japan) and Image Lab software (ChemiDoc XRS Plus; Bio-Rad Laboratories, Hercules, CA, USA).

2.4. *Statistical Analysis*

All data were expressed as the mean ± SEM, and the significance of the differences (*p*-values) between groups was evaluated using a one-way ANOVA followed by Duncan's test (for the animal experiments shown in Figures 3–5) or the Tukey-Kramer test (for the cell experiments shown in Figure 6).

3. Results

3.1. *Effect of ACSO, Its Analogs, and Garlic-Odor Compounds on Suppression of Hepatic Injury Induced by CCl_4*

The activities of AST, ALT, and LDH were measured as enzyme markers for acute hepatic injury (Figure 3a–c). AST, ALT, and LDH activities were increased following injection of CCl_4 in the control groups, and these increases were significantly suppressed by oral administration of ACSO, ACS, DADS, and DATS ($p < 0.01$). In addition, TBARS was measured to evaluate the peroxidation of lipids produced by acute hepatic injury (Figure 3d). TBARS was increased by the injection of CCl_4 in the control groups, while oral administration of ACSO, ACS, DADS and DATS significantly suppressed the increase in TBARS after the injection of CCl_4 ($p < 0.01$).

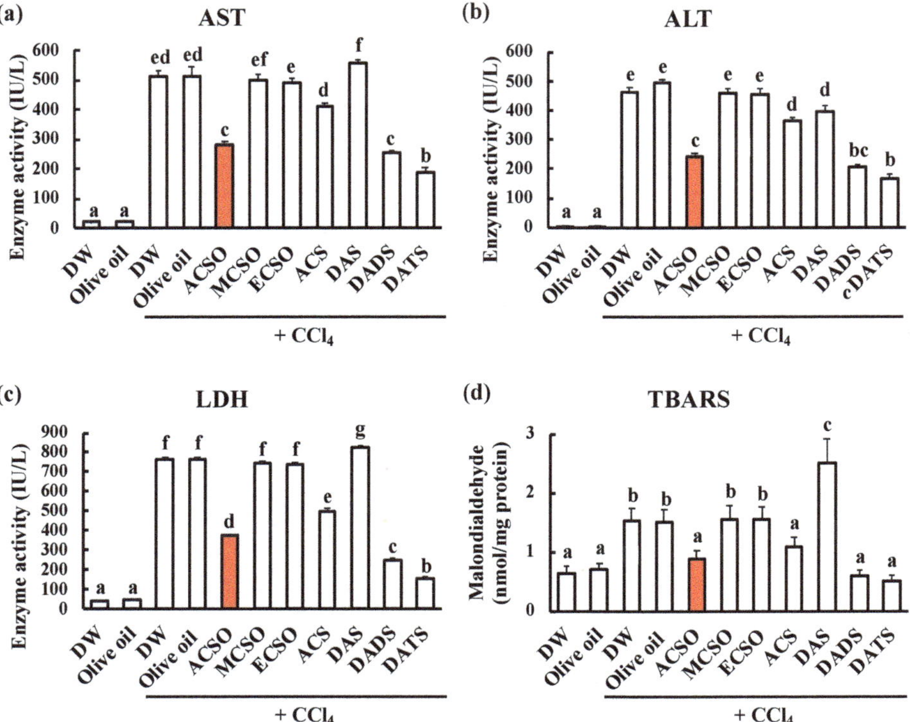

Figure 3. Effect of ACSO, its analogs, and garlic-odor compounds on markers for hepatic injury and oxidative stress. Aspartate transaminase (AST) activity (**a**), alanine transaminase (ALT) activity (**b**), lactate dehydrogenase (LDH) activity (**c**), and the amount of thiobarbituric acid reactive substances (TBARS) (**d**) in the blood of rats with CCl_4-induced hepatic injury. Each value represents the mean of six rats ± S.E. The different letters in the figures indicate a significant difference between the groups ($p < 0.01$).

We next measured the effect of oral administration of sulfoxides and sulfides on the enzymatic activities of GST, QR, GR and GPx, and GSH content in the liver after the injection of CCl_4 (Figure 4). All measured enzyme activities were lower in the control with acute hepatic injury resulting from CCl_4 than in the control groups without injection of CCl_4 ($p < 0.01$). GST activities in the ACSO, ACS, DADS, and DATS groups were significantly higher than those in the control, MCSO, ECSO, and DAS groups following injection of CCl_4 (Figure 4a, $p < 0.01$). QR and GR activities in the ACSO, DADS, and DATS groups were significantly higher than those in the control, MCSO, ECSO, ACS, and DAS groups following injection of CCl_4 (Figure 4b and c, $p < 0.01$). GPx activities in the ACSO, DADS, and DADS groups were significantly higher than those in the control, MCSO, ECSO, and DAS groups (Figure 4d, $p < 0.01$). GSH content in the ACSO, ACS, DADS, and DATS groups was significantly higher than that in the control, MCSO, ECSO, and DAS groups following injection of CCl_4 (Figure 4e, $p < 0.01$) and lower than that in the control groups without acute hepatitis. Thus, the activities of the phase II and antioxidative enzymes and the GST content in the ACSO as well as the DADS and DATS groups were higher than those in the control groups even after the induction of hepatic injury.

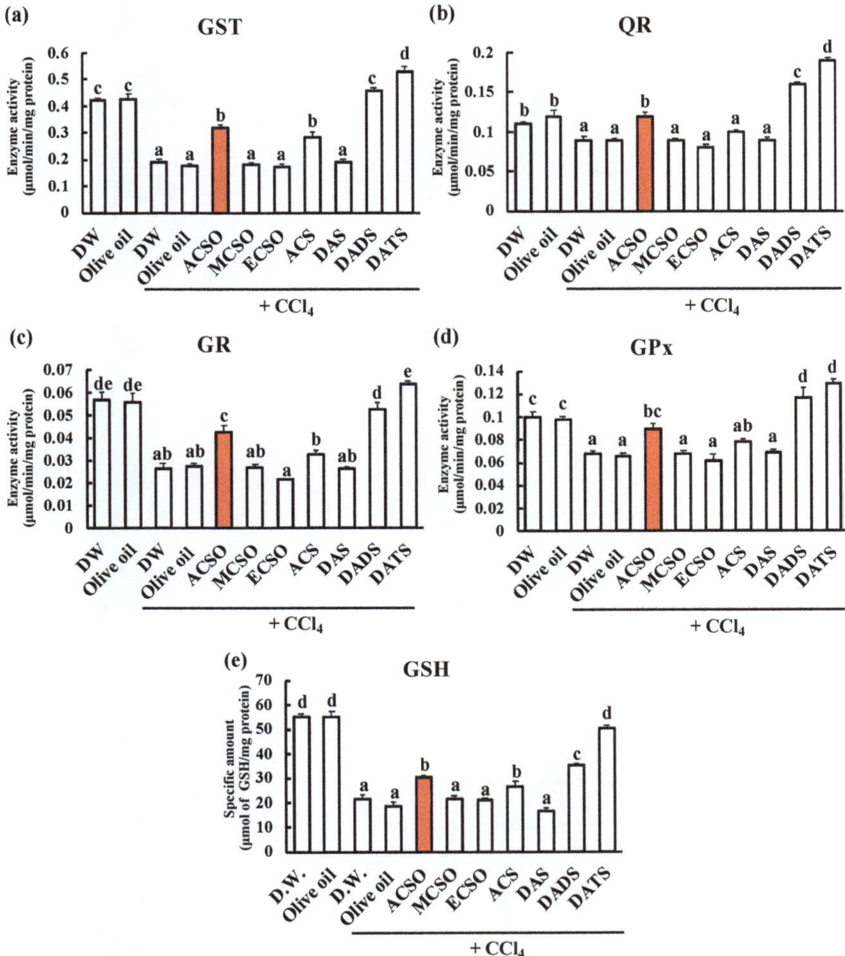

Figure 4. Effect of ACSO, its analogs, and garlic-odor compounds on Phase II and antioxidative enzyme activities and glutathione (GSH) content. Enzymatic activities of (**a**) glutathione-*S*-transferase (GST), (**b**) quinone reductase (QR), (**c**) glutathione reductase (GR), (**d**) glutathione peroxidase (GPx), and (**e**) GSH, in the livers of rats with CCl_4-induced hepatic injury. Each value represents the mean of six rats ± S.E. The different letters in the figures indicate a significant difference between the groups ($p < 0.01$).

3.2. Effect of ACSO and Sulfides on Liver Function

Figure 5 shows the effect of oral administration of ACSO, DADS, and DATS on GST and GPx activities in the livers of rats without the injection of CCl_4. GST and GPx activities in the ACSO and DADS groups were significantly higher than those in the control and DAS groups ($p < 0.01$).

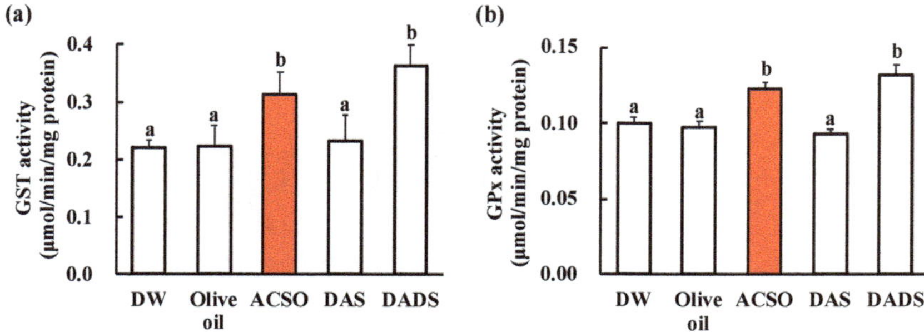

Figure 5. Effect of ACSO and sulfides on liver function. GST activity (**a**) and GPx activity (**b**) in the livers of normal rats after administration of CCl_4. Each value represents the mean of six rats ± S.E. The different letters in the figures indicate a significant difference between the groups ($p < 0.01$).

3.3. Effect of ACSO on Nrf2 Nuclear Translocation in HepG2 Cells

The Nrf2/β-actin ratio in HepG2 cells was determined by Western blot analysis (Figure 6). Nrf2/β-actin ratio in whole HepG2 cells was increased with the addition of ACSO, and the ratio was significantly higher than that in the control group following addition of 1 mM ACSO (Figure 6a, $p < 0.05$). Nrf2/β-actin ratios in the cytoplasm were not significantly different between the groups; however (Figure 6b), the Nrf2/β-actin ratio in the nucleus increased as the concentration of added ACSO increased (Figure 6c). The Nrf2/β-actin ratio in the nuclei of HepG2 cells was significantly higher following the addition of 0.5 and 1 mM ACSO than that in the nuclei of HepG2 cells without the addition of ACSO ($p < 0.05$).

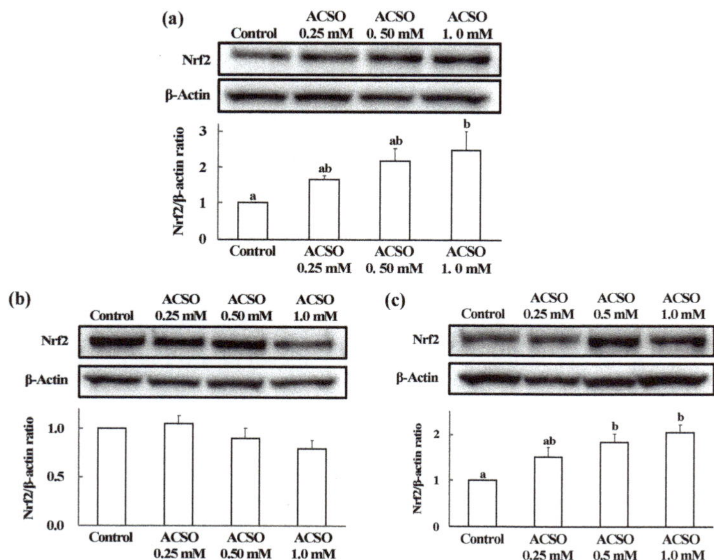

Figure 6. Effect of ACSO on Nrf2 nuclear translocation in HepG2 cells. The expression levels of Nrf2 were analyzed by Western blot analysis in whole cell (**a**), cytoplasm (**b**), and nuclei (**c**) preparations of HepG2 cells. Expression was quantified as the ratio of Nrf2 expression to β-actin expression. Each value represents the mean of three experiments ± S.E. The different letters indicate a significant difference between the groups ($p < 0.05$).

3.4. Absorption and Metabolism of ACSO in the Small Intestine

The changes in ACSO and pyruvic-acid concentrations in the portal vein after the injection of ACSO into the ligated loop of the small intestine were measured to evaluate the absorption and metabolism of ACSO. Concentrations of both ACSO and pyruvic acid increased as time proceeded (Figure 7). The concentration of ACSO in the portal vein reached 4 mM 30 min after injection of ACSO into the small intestine, while that of pyruvic acid reached 0.2 mM at this time point. ACSO and pyruvic-acid concentrations in the luminal liquid of the small intestine 30 min after injection of PBS into the small intestine were 0 mM and 0.04 mM, respectively. While ACSO and pyruvic-acid concentrations in the luminal liquid of the small intestine 30 min after injection of 100 mM ACSO into the small intestine were 8.02 mM and 0.81 mM, respectively (Table 1).

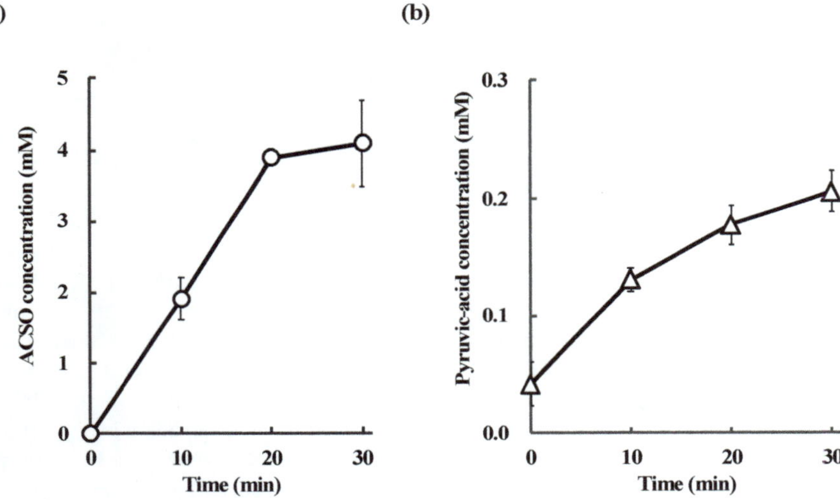

Figure 7. ACSO (**a**) and pyruvic-acid (**b**) concentrations in the portal vein were measured after injection of ACSO into the small intestine. Each value represents the mean of three excited small intestines ± S.E.

Table 1. ACSO and pyruvic-acid concentrations in rat small intestines 30 min after injection of ACSO or PBS into the small intestine.

Sample	ACSO Concentration (mM)	Pyruvic-Acid Concentration (mM)
PBS	Not detected	0.04 ± 0.01
ACSO	8.02 ± 1.84	0.81 ± 0.07

Each value is the mean of three excited small intestines ± S.E.

Volatile components produced from the mixture of ACSO and the crude protein extracted from the small intestine were analyzed using solid-phase microextraction and GC-AED in order to further evaluate the metabolism of ACSO in the small intestine. DAS and DADS were detected by GC-AED at retention times of 14 and 29 min, respectively. The relative intensity of DADS was higher than that of DAS (Figure 8).

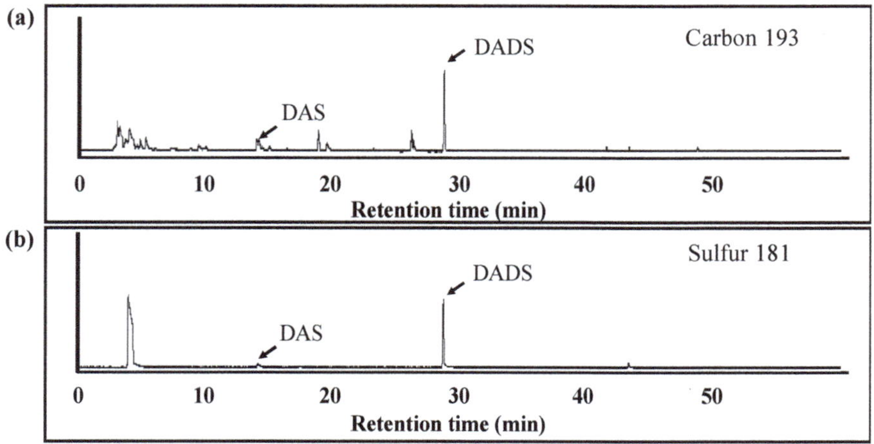

Figure 8. Element chromatograms of volatile compounds obtained from the mixture of ACSO and proteins of the small intestine to assess ACSO metabolism. Carbon was detected at 193 nm (**a**) and sulfur was detected at 181 nm (**b**).

4. Discussion

We prepared ACSO and its analogs MCSO, ECSO, and ACS to compare their effects on the prevention of hepatic injury. MCSO is a minor sulfuric component in garlic and the content is up to 2 mg/g fresh weight [26]. ACS is also a minor sulfuric component: The content is less than 30 μg/g fresh weight, but rich in aged garlic [46]. ECSO is not a naturally occurring compound in garlic. We also examined the garlic odor components DAS, DADS, and DATS in these experiments. ACS [47,48], DADS, and DATS [49,50] have previously been reported to prevent hepatic injury when they are intraperitoneally injected, while DAS did not show such an effect. In this study, we found that oral administration of ACSO suppressed acute hepatic injury induced by CCl_4 in addition to that of DADS and DATS (Figure 3). The enzymatic activities of hepatic injury markers, such as AST, ALT, and LDH, increased following intraperitoneal injection of CCl_4; however, these increases were suppressed by oral administration of ACSO as well as that of DADS and DATS. Analysis of the reactive aldehydes produced from lipid hydroperoxides as a malondialdehyde equivalent using the TBARS method revealed that the increase in TBARS after injection of CCl_4 was remarkably suppressed by oral administration of ACSO, DADS, and DATS to levels approximately the same as that of the control group without CCl_4 injection. These results suggest that oral administration of ACSO prevented acute hepatic injury induced by CCl_4 in rats. As MCSO, ECSO, and ACS were not effective, the allyl and sulfoxide groups in ACSO are essential for this preventive effect.

The preventive effect of ACSO on hepatic injury is probably attributable to the induction of detoxifying and antioxidative enzymes, as scavenging of free radicals is crucial for these effects on hepatic injury induced by CCl_4 and ACSO itself would not function as a free radical scavenger. The activities of GST, QR, GPx, and GR as well as the level of GSH were reduced by the injection of CCl_4 in the vehicle-treated control group. These reductions in enzymatic activities likely resulted from protein denaturation by radicals, such as CCl_3, that are produced by CCl_4. This decrease was suppressed by oral administration of ACSO, DADS, and DATS (Figure 4). The suppressive effect of ACSO stems from its ability to induce GST and GPx activities, as shown in normal rats (Figure 5). The enhanced induction of GST and GPx relieves oxidative stress, preventing protein denaturation and enzyme inactivation and leading to the reduction of oxidized GSH to reduced GSH. Therefore, consecutive oral administration of ACSO induced both phase II and oxidative enzymes, resulting in attenuation of the symptoms of hepatic injury.

Some phase II and antioxidative enzymes, such as GST and GR, are regulated by the Nrf2-Kelch-like ECH-associated protein 1 (Keap1) system. Increases in such enzymes often prevent or attenuate diseases, such as Alzheimer's disease [51], vascular diseases [52], and cancers [53], that are considered to be caused by oxidative stress. Activation of this pathway is triggered by the release of the Nrf2 transcriptional factor from Keap1, which is a marker of ubiquitination [37]. The release occurs in response to modification of the Cys residues of Keap1 with electrophiles [54]. Translocation of Nrf2 into the nucleus activates transcription of antioxidative and detoxifying enzymes. DADS and DATS have been reported to induce antioxidative and detoxifying enzymes via the activation of the Nrf2-ARE pathway [50]. Since disulfide and trisulfide bonds are readily cleaved by nucleophiles to form covalent linkages with biomolecules [55], DADS and DATS form covalent bonds with Keap1, thus activating the Nrf2-ARE pathway to induce antioxidative and detoxifying enzymes [56]. As the precise mechanism underlying the induction of phase II and antioxidative enzymes by ACSO has not yet been thoroughly investigated, we examined the ability of ACSO to cause translocation of Nrf2 into the nucleus of HepG2 liver cells and demonstrated that the addition of ACSO to these cells induced translocation of Nrf2 into nucleus (Figure 6). Nrf2-ARE pathway controls the expression of a variety of antioxidative enzymes including superoxide dismutase (SOD) and catalase [38], so that ACSO could also increase such enzymes activities not tested in the current study.

Although ACSO was shown to promote Nrf2 nuclear translocation in vitro, the functional molecules to induce phase II and antioxidative enzymes in vivo were not known. One plausible mechanism for the prevention of hepatic injury by oral administration of ACSO was that ACSO itself was absorbed and delivered to the liver and/or metabolized to garlic odor components, which induced phase II and antioxidative enzymes. In vivo experiments showed that the concentrations of both ACSO and pyruvic acid, a metabolite of ACSO, increased in the portal vein after injection of ACSO in the ligated loop of the small intestine (Figure 7). In addition, the pyruvic-acid concentration increased after injection of ACSO in the small intestine (Table 1). These results indicate that ACSO was not only transported to the portal vein but also metabolized to allyl sulfenic acid and pyruvic acid, probably by some enzyme or bacterium present in the small intestine. As allyl sulfenic acid is quite reactive, sulfides such as DADS would be produced in the small intestine. Therefore, ACSO was mixed with crude proteins extracted from the small intestine to examine the production of volatile bioactive sulfides. In these experiments, DAS and DADS were detected as volatile components (Figure 8), and the amount of DADS was greater than that of DAS. These results suggest that orally administered ACSO is partially metabolized to afford DADS, a known inducer of antioxidative and detoxifying enzymes in the liver [49,50]. Taken together, the suppressive effect of oral administration of ACSO on hepatic injury induced by CCl_4 may stem from the activities of both ACSO itself and its metabolites, including DADS, that induce phase II and antioxidative enzymes in the liver by promoting Nrf2 nuclear translocation.

In this study, 50 μmol/mL/day of ACSO was orally administered to a rat of approximately 200 g body weight, which simply corresponds to about 45 mg/kg bodyweight/day (molecular weight of ACSO: 177.22). If a person of 50 kg body weight takes the proportional amount, it would become about 2.2 g of ACSO. As fresh garlic contains 14 mg/g weight of ACSO, the amount of garlic corresponding to 2.2 g of ACSO would be approximately 150 g that might be difficult to take daily. However, ACSO is water-soluble and odorless, 2.2 g of ACSO can be easily added to variety of foods. As ACSO is known to enhance richness of taste [27], foods fortified with ACSO may provide both enhanced palatability and health benefit. In the present study, side effects of ACSO were not observed without showing any distinctive changes in the liver weight and the appearance compared to the vehicle group. Although the safety of ACSO to human body should be more precisely investigated, our findings suggest that ACSO has potential to be used as a functional-food additive to prevent diseases caused by oxidative stress.

5. Conclusions

Oral administration of ACSO induced phase II and antioxidative enzymes to suppress acute hepatic injury induced by CCl_4. ACSO was absorbed from the small intestine to the portal vein but was also metabolized to a certain extent in vivo to yield garlic odor components, such as DADS. Because ACSO induced nuclear translocation of Nrf2, ACSO, in addition to DADS, may be an important molecular factor involved in the induction of phase II and antioxidative enzymes and suppression of acute hepatic injury. Oral administration of ACSO may therefore be effective for increasing antioxidative potency and preventing other diseases caused by ROS.

Author Contributions: Y.Y. and H.K. designed the study. Y.Y., R.H., T.Y., T.S. (Takeshi Shibuya), T.S. (Tomoya Sakaguchi), and H.U.-K. conducted the research. Y.Y., R.H., T.Y., T.S. (Takeshi Shibuya), H.U-K., and H.K. analyzed the data. Y.Y. and H.K. wrote the manuscript. H.K. had primary responsibility for final content.

Acknowledgments: Y.Y. and H.K. are grateful to Takeshi Saito of ACERA Co., Ltd. for his useful comments on the experiment of Nrf2 nuclear translocation.

Conflicts of Interest: The authors declare that they have no conflict of interest.

References

1. Hayes, J.D.; McLellan, L.I. Glutathione and glutathione-dependent enzymes represent a co-ordinately regulated defence against oxidative stress. *Free Radic. Res.* **1999**, *31*, 273–300. [CrossRef] [PubMed]
2. Schafer, F.Q.; Buettner, G.R. Redox environment of the cell as viewed through the redox state of the glutathione disulfide/glutathione couple. *Free Radic. Biol. Med.* **2001**, *30*, 1191–1212. [CrossRef]
3. Ketterer, B.; Coles, B.; Meyer, D.J. The role of glutathione in detoxication. *Environ. Health Perspect.* **1983**, *49*, 59–69. [CrossRef] [PubMed]
4. Coles, B.; Ketterer, B. The role of glutathione and glutathione transferases in chemical carcinogenesis. *Crit. Rev. Biochem. Mol. Biol.* **1990**, *25*, 47–70. [CrossRef] [PubMed]
5. Allen, J.; Bradley, R.D. Effects of oral glutathione supplementation on systemic oxidative stress biomarkers in human volunteers. *J. Altern. Complement. Med.* **2011**, *17*, 827–833. [CrossRef] [PubMed]
6. Schmitt, B.; Vicenzi, M.; Garrel, C.; Denis, F.M. Effects of *N*-acetylcysteine, oral glutathione (GSH) and a novel sublingual form of GSH on oxidative stress markers: A comparative crossover study. *Redox Biol.* **2015**, *6*, 198–205. [CrossRef] [PubMed]
7. Favilli, F.; Marraccini, P.; Iantomasi, T.; Vincenzini, M.T. Effect of orally administered glutathione on glutathione levels in some organs of rats: Role of specific transporters. *Br. J. Nutr.* **1997**, *78*, 293–300. [CrossRef] [PubMed]
8. Stark, A.-A.; Zeiger, E.; Pagano, D.A. Glutathione metabolism by γ-glutamyltranspeptidase leads to lipid peroxidation: Characterization of the system and relevance to hepatocarcinogenesis. *Carcinogenesis* **1993**, *14*, 183–189. [CrossRef]
9. Nakamura, Y.; Miyoshi, N. Electrophiles in Foods: The current status of isothiocyanates and their chemical biology. *Biosci. Biotechnol. Biochem.* **2010**, *74*, 242–255. [CrossRef] [PubMed]
10. Parvez, S.; Long, M.J.C.; Poganik, J.R.; Aye, Y. Redox signaling by reactive electrophiles and oxidants. *Chem. Rev.* **2018**, *118*, 8798–8888. [CrossRef]
11. Thomson, M.; Ali, M. A review of its potential use as an anti-cancer agent. *Curr. Cancer Drug Tar.* **2003**, *3*, 67–81. [CrossRef]
12. Shukla, Y.; Kalra, N. Cancer chemoprevention with garlic and its constituents. *Cancer Lett.* **2007**, *247*, 167–181. [CrossRef] [PubMed]
13. Powolny, A.A.; Singh, S.V. Multitargeted prevention and therapy of cancer by diallyl trisulfide and related allium vegetable-derived organosulfur compounds. *Cancer Lett.* **2008**, *269*, 305–314. [CrossRef] [PubMed]
14. Campbell, J.H.; Efendy, J.L.; Smith, N.J.; Campbell, G.R. Molecular basis by which garlic suppresses atherosclerosis. *J. Nutr.* **2001**, *31*, 1006S–1009S. [CrossRef] [PubMed]
15. Ohaeri, O.C. Effect of garlic oil on the levels of various enzymes in the serum and tissue of streptozotocin diabetic rats. *Biosci. Rep.* **2001**, *21*, 19–24. [CrossRef] [PubMed]

16. Agarwal, M.K.; Iqbal, M.; Athar, M. Garlic oil ameliorates ferric nitrilotriacetate (Fe-NTA)-induced damage and tumor promotion: Implications for cancer prevention. *Food Chem. Toxicol.* **2007**, *45*, 1634–1640. [CrossRef] [PubMed]
17. Fujisawa, H.; Watanabe, K.; Suma, K.; Origuchi, K.; Matsufuji, H.; Seki, T.; Ariga, T. Antibacterial potential of garlic-derived allicin and its cancellation by sulfhydryl compounds. *Biosci. Biotechnol. Biochem.* **2009**, *73*, 1948–1955. [CrossRef] [PubMed]
18. Banerjee, S.K.; Maulik, M.; Mancahanda, S.C.; Dinda, A.K.; Gupta, S.K.; Maulik, S.K. Dose-dependent induction of endogenous antioxidants in rat heart by chronic administration of garlic. *Life Sci.* **2002**, *70*, 1509–1518. [CrossRef]
19. Wu, C.-C.; Sheen, L.-Y.; Chen, H.-W.; Kuo, W.-W.; Tsai, S.-J.; Lii, C.-K. Differential effects of garlic oil and its three major organosulfur components on the hepatic detoxification system in rats. *J. Agric. Food Chem.* **2002**, *50*, 378–383. [CrossRef] [PubMed]
20. Jones, M.G.; Hughes, J.; Tregova, A.; Milne, J.; Tomsett, A.B.; Collin, H.A. Biosynthesis of the flavour precursors of onion and garlic. *J. Exp. Bot.* **2004**, *55*, 1903–1918. [CrossRef]
21. Augusti, K.T.; Sheela, C.G. Antiperoxide Effect of S-allyl cysteine sulfoxide, an insulin secretagogue, in diabetic rats. *Experientia* **1996**, *52*, 115–119. [CrossRef] [PubMed]
22. Sangeetha, T.; Quine, S.D. Preventive Effect of *S*-allyl cysteine sulfoxide (Alliin) on cardiac marker enzymes and lipids in isoproterenol-induced myocardial injury. *J. Pharm. Pharmacol.* **2006**, *58*, 617–623. [CrossRef] [PubMed]
23. Zhai, B.; Zhang, C.; Sheng, Y.; Zhao, C.; He, X.; Xu, W.; Huang, K.; Luo, Y. Hypoglycemic and hypolipidemic effect of *S*-allyl-cysteine sulfoxide (alliin) in DIO mice. *Sci. Rep.* **2018**, *8*. [CrossRef] [PubMed]
24. Akao, M.; Shibuya, T.; Shimada, S.; Sakurai, H.; Kumagai, H. In vivo production of bioactive compounds from *S*-allyl-L-cysteine sulfoxide, garlic odor precursor, that inhibit platelet aggregation. *J. Clin. Biochem. Nutr. Supple.* **2008**, *43*, 1–3.
25. Uto-Kondo, H.; Hase, A.; Yamaguchi, Y.; Sakurai, A.; Akao, M.; Saito, T.; Kumagai, H. *S*-allyl-L-cysteine sulfoxide, a garlic odor precursor, suppresses elevation in blood ethanol concentration by accelerating ethanol metabolism and preventing ethanol absorption from gut. *Biosci. Biotech. Biochem.* **2018**, *82*, 724–731. [CrossRef] [PubMed]
26. Lawson, L.D. Garlic: A review of its medicinal effects and indicated active compounds. In *ACS Symposium Series*; American Chemical Society: Washington, DC, USA, 1998; pp. 176–209. [CrossRef]
27. Ueda, Y.; Sakaguchi, M.; Hirayama, K.; Miyajima, R.; Kimizuka, A. Characteristic flavor constituents in water extract of garlic. *Agric. Biol. Chem.* **1990**, *54*, 163–169.
28. Jayakumar, T.; Ramesh, E.; Geraldine, P. Antioxidant activity of the oyster mushroom, pleurotus ostreatus, on CCl_4-induced liver injury in rats. *Food. Chem. Toxicol.* **2006**, *44*, 1989–1996. [CrossRef] [PubMed]
29. Huo, H.Z.; Wang, B.; Liang, Y.K.; Bao, Y.Y.; Gu, Y. Hepatoprotective and antioxidant effects of licorice extract against CCl_4-induced oxidative damage in rats. *Int. J. Mol. Sci.* **2011**, *12*, 6529–6543. [CrossRef] [PubMed]
30. Cheng, N.; Ren, N.; Gao, H.; Lei, X.; Zheng, J.; Cao, W. Antioxidant and hepatoprotective effects of schisandra chinensis pollen extract on CCl_4-induced acute liver damage in mice. *Food. Chem. Toxicol.* **2013**, *55*, 234–240. [CrossRef]
31. Pan, Y.; Long, X.; Yi, R.; Zhao, X. Polyphenols in liubao tea can prevent CCl_4-induced hepatic damage in mice through its antioxidant capacities. *Nutrients* **2018**, *10*, 1280. [CrossRef]
32. Raucy, J.L.; Kraner, J.C.; Lasker, J.M. Bioactivation of halogenated hydrocarbons by cytochrome P4502E1. *Crit. Rev. Toxicol.* **1993**, *23*, 1–20. [CrossRef] [PubMed]
33. Weber, L.W.D.; Boll, M.; Stampfl, A. Hepatotoxicity and mechanism of action of haloalkanes: Carbon tetrachloride as a toxicological model. *Crit. Rev. Toxicol.* **2003**, *33*, 105–136. [CrossRef] [PubMed]
34. Recknagel, R.O. A New direction in the study of carbon tetrachloride hepatotoxicity. *Life Sci.* **1983**, *33*, 401–408. [CrossRef]
35. Slater, T.F. Free-radical mechanisms in tissue injury. *Biochem. J.* **1984**, *222*, 1–15. [CrossRef]
36. Ramaiah, S.K. A Toxicologist guide to the diagnostic interpretation of hepatic biochemical parameters. *Food Chem. Toxicol.* **2007**, *45*, 1551–1557. [CrossRef]
37. Kensler, T.W.; Wakabayashi, N.; Biswal, S. Cell survival responses to environmental stresses via the Keap1-Nrf2-ARE pathway. *Annu. Rev. Pharmacol. Toxicol.* **2007**, *47*, 89–116. [CrossRef]

38. Bataille, A.M.; Manautou, J.E. Nrf2: A potential target for new therapeutics in liver disease. *Clin. Pharmacol. Ther.* **2012**, *92*, 340–348. [CrossRef]
39. Hakamata, W.; Koyama, R.; Tanida, M.; Haga, T.; Hirano, T.; Akao, M.; Kumagai, H.; Nishio, T. A simple synthesis of alliin and allo-alliin: X-ray diffraction analysis and determination of their absolute configurations. *J. Agric. Food Chem.* **2015**, *63*, 10778–10784. [CrossRef]
40. Haber, D.; Siess, M.-H.; De Waziers, I.; Beaune, P.; Suschetet, M. Modification of hepatic drug-metabolizing enzymes in rat fed naturally occurring allyl sulphides. *Xenobiotica* **1994**, *24*, 169–182. [CrossRef]
41. Uchiyama, M.; Mihara, M. Determination of malonaldehyde precursor in tissues by thiobarbituric acid test. *Anal. Biochem.* **1978**, *86*, 271–278. [CrossRef]
42. Habig, W.H.; Pabst, M.J.; Jakoby, W.B. Glutathione *S*-transferases. *J. Biol. Chem.* **1974**, *249*, 7130–7139. [PubMed]
43. Ernster, L.; Danielson, L.; Ljunggren, M. DT diaphorase. I. Purification from the soluble fraction of rat-liver cytoplasm, and properties. *Biochim. Biophys. Acta* **1962**, *58*, 171–188. [CrossRef]
44. Carlberg, I.; Mannervik, B. Glutathione reductase. *Method Enzymol.* **1985**, *113*, 484–490.
45. Yamamoto, Y.; Takahashi, K. Glutathione peroxidase isolated from plasma reduces phospholipid hydroperoxides. *Arch. Biochem. Biophys.* **1993**, *305*, 541–545. [CrossRef] [PubMed]
46. Kodera, Y.; Suzuki, A.; Imada, O.; Kasuga, S.; Sumioka, I.; Kanezawa, A.; Taru, N.; Fujikawa, M.; Nagae, S.; Masamoto, K.; et al. Physical, chemical, and biological properties of S-allylcysteine, an amino acid derived from garlic. *J. Agric. Food Chem.* **2002**, *50*, 622–632. [CrossRef]
47. Mizuguchi, S.; Takemura, S.; Minamiyama, Y.; Kodai, S.; Tsukioka, T.; Inoue, K.; Okada, S.; Suehiro, S. *S*-allyl cysteine attenuated CCl_4-induced oxidative stress and pulmonary fibrosis in rats. *Biofactors* **2006**, *26*, 81–92. [CrossRef] [PubMed]
48. Kodai, S.; Takemura, S.; Minamiyama, Y.; Hai, S.; Yamamoto, S.; Kubo, S.; Yoshida, Y.; Niki, E.; Okada, S.; Hirohashi, K.; et al. *S*-allyl cysteine prevents CCl_4-induced acute liver injury in rats. *Free Radic. Res.* **2007**, *41*, 489–497. [CrossRef]
49. Fukao, T.; Hosono, T.; Misawa, S.; Seki, T.; Ariga, T. The effects of allyl sulfides on the induction of phase II detoxification enzymes and liver injury by carbon tetrachloride. *Food Chem. Toxicol.* **2004**, *42*, 743–749. [CrossRef]
50. Lee, I.-C.; Kim, S.-H.; Baek, H.-S.; Moon, C.; Kang, S.-S.; Kim, S.-H.; Kim, Y.-B.; Shin, I.-S.; Kim, J.-C. The involvement of Nrf2 in the protective effects of diallyl disulfide on carbon tetrachloride-induced hepatic oxidative damage and inflammatory response in rats. *Food. Chem. Toxicol.* **2014**, *63*, 174–185. [CrossRef]
51. Markesbery, W.R. Oxidative stress hypothesis in alzheimer's disease. *Free Radic. Biol. Med.* **1997**, *23*, 134–147. [CrossRef]
52. Madamanchi, N.R.; Vendrov, A.; Runge, M.S. Oxidative stress and vascular disease. *Arterioscler. Thromb. Vasc. Biol.* **2005**, *25*, 29–38. [CrossRef] [PubMed]
53. Reuter, S.; Gupta, S.C.; Chaturvedi, M.M.; Aggarwal, B.B. Oxidative stress, inflammation, and cancer: How are they linked? *Free Radic Biol. Med.* **2010**, *49*, 1603–1616. [CrossRef] [PubMed]
54. Hu, C.; Eggler, A.L.; Mesecar, A.D.; van Breemen, R.B. Modification of keap1 cysteine residues by sulforaphane. *Chem. Res. Toxicol.* **2011**, *24*, 515–521. [CrossRef] [PubMed]
55. Hosono, T.; Fukao, T.; Ogihara, J.; Ito, Y.; Shiba, H.; Seki, T.; Ariga, T. Diallyl trisulfide suppresses the proliferation and induces apoptosis of human colon cancer cells through oxidative modification of beta-tubulin. *J. Biol. Chem.* **2005**, *280*, 41487–41493. [CrossRef] [PubMed]
56. Kim, S.; Lee, H.G.; Park, S.A.; Kundu, J.K.; Keum, Y.S.; Cha, Y.N.; Na, H.K.; Surh, Y.J. Keap1 cysteine 288 as a potential target for diallyl trisulfide-induced Nrf2 activation. *PLoS ONE* **2014**, *9*, e85984. [CrossRef] [PubMed]

Article

Polydatin Encapsulated Poly [Lactic-co-glycolic acid] Nanoformulation Counteract the 7,12-Dimethylbenz[a] Anthracene Mediated Experimental Carcinogenesis through the Inhibition of Cell Proliferation

Sankaran Vijayalakshmi [1], Arokia Vijaya Anand Mariadoss [1], Vinayagam Ramachandran [1], Vijayakumar Shalini [1], Balupillai Agilan [1], Casimeer C. Sangeetha [2], Periyasamy Balu [3], Venkata Subbaih Kotakadi [4], Venkatachalam Karthikkumar [5,*] and David Ernest [1,*]

1 Department of Biotechnology, Thiruvalluvar University, Serkadu, Vellore 632 115, Tamilnadu, India
2 Department of Physics, Sri Padmavati Mahila Visvavidyalayam, Tirupati 517502, Andra Pradesh, India
3 Department of Chemistry, Thiruvalluvar University, Serkadu, Vellore 632 115, Tamilnadu, India
4 DST-PURSE Center, Sri Venkateswara University, Tirupathi 517 502, Andhra Pradesh, India
5 Department of Pharmacology and Therapeutics, College of Medicine and Health Sciences, UAE University, Al Ain 17666, UAE
* Correspondence: karthikjega@gmail.com (V.K.); ernestdavid2009@gmail.com (D.E.)

Received: 30 June 2019; Accepted: 23 August 2019; Published: 5 September 2019

Abstract: In the present study, the authors have attempted to fabricate Polydatin encapsulated Poly [lactic-co-glycolic acid] (POL-PLGA-NPs) to counteract 7,12-dimethyl benzyl anthracene (DMBA) promoted buccal pouch carcinogenesis in experimental animals. The bio-formulated POL-PLGA-NPs were characterized by dynamic light scattering (DLS), Fourier transform infrared (FTIR) spectroscopy, X-ray powder diffraction (XRD) pattern analysis, and transmission electron microscope (TEM). In addition, the nano-chemopreventive potential of POL-PLGA-NPs was assessed by scrutinizing the neoplastic incidence and analyzing the status of lipid peroxidation, antioxidants, phase I, phase II detoxification status, and histopathological changes and in DMBA-treated animals. In golden Syrian hamsters, oral squamous cell carcinoma (OSCC) was generated by painting with 0.5% DMBA in liquid paraffin three times a week for 14 weeks. After 100% tumor formation was observed, high tumor volume, tumor burden, and altered levels of biochemical status were observed in the DMBA-painted hamsters. Intra-gastric administration of varying concentration of POL-PLGA-NPs (7.5, 15, and 30 mg/kg b.wt) to DMBA-treated hamsters assumedly prevents oncological incidences and restores the status of the biochemical markers. It also significantly enhances the apoptotic associated and inhibits the cancer cell proliferative markers expression (p53, Bax, Bcl-2, cleaved caspase 3, cyclin-D1). The present study reveals that POL-PLGA-NPs is a penitential candidate for nano-chemopreventive, anti-lipid peroxidative, and antioxidant potential, and also has a modulating effect on the phase I and Phase II detoxification system, which is associated with reduced cell proliferation and induced apoptosis in experimental oral carcinogenesis.

Keywords: polydatin; PLGA; nanoformulation; antioxidant; cell proliferation; apoptosis

1. Introduction

Cancer has a high mortality rate, and around 18.1 million people are diagnosed with cancer each year. According to the World Health Organization (WHO) statistics, by 2030, this number will be almost double [1], and a recent report from the Indian council of Medical Research Council states that by 2020, 1.73 million new cancer cases will be detected, and over 8.8 lakh deaths will occur due to cancer [2]. The drastic incidence and mortality rate of cancer is associated with age, sex, and race. Risk factors for cancer incidence include (i) tobacco smoking, which causes lung, head, and neck cancer; (ii) drinking alcohol, which causes liver, esophageal, breast, oral and other cancers; (iii) physical inactivity; and (iv) a diet low in fruit and vegetables, which can increase the risk of colon, breast, and possibly other types of cancers [3,4].

The conventional therapeutic management of cancer, e.g., surgery, chemotherapy, radiation, and hormonal therapy, are still ineffective for the management of cancer progression. Hence, less toxic and more effective anti-cancer agents for the management of cancer are urgently needed. Polydatin ($C_{20}H_{22}O_8$) is a monocrystalline glycisidic phyto-compound found in Sitka spruce, grape, peanut, hop cones, red wines, hop pellets, and cocoa [5]. Pharmacological and clinical studies have revealed that polydatin has anti-arteriosclerosis, anti-tumor, anti-oxidative, anti-inflammatory, anti-proliferative, anti-angiogenic, hepatoprotective, and immunoregulatory effects. In recent times, the cancer preventive potential of polydatin has also been examined. It act as repressor candidate of tumorogenesis, through the hindrance of cell proliferation, invasion, migration, and induced cell apoptosis [6]. In addition, Chen et al. (2017) have reported that polydatin suppressed the cell cycle progression and enhanced the apoptotic associated gene expression in human cancer cell lines [7]. It also suppresses the breast carcinogenesis in MCF-7 cells and gradually down-regulates the expression of phosphor-NF-κB p65 and activation of NF-κB pathway in non-small cell lung cancer [8]. In a recent study, Hu et al. (2018) suggested that polydatin modulated the VEGF-induced angiogenesis by suppressing the phosporylatin of Akt, eNOS, and Erk [9]. In addition, it induces autophagy and apoptosis in multiple myeloma cells through the inhibition of mTOR/p70s6k pathway [10].

Biodegradable polymeric agents have been extensively used to improve the bioavailability of plant based chemotherapeutic agents. Considering this, we utilized the poly-lactic-co-glycolic acid (PLGA) for the synthesis of polydatin nanoparticles. PLGA is one of the most extensively used biodegradable polymers because its hydrolysis leads to endogenous and easily metabolized monomers of lactic acid and glycolic acid [11]. Recent publications suggest that PLGA-NPs functionalized with (i) ß-Sitosterol, a natural phytosterol, (ii) resveratrol, a natural polyphenol, and (iii) tea polyphenols of theaflavin and epigallocatechin-3-gallate [12–14] might be potential candidates for cancer treatment. Many studies have revealed that encapsulation of PLGA nanoparticles improves the biocompatibility, tunable mechanical property, and controllable degradation of several chemotherapeutic drugs including paclitaxel, tamoxifen, and anthracyclines. Wang et al. (2014) used a soy-phospholipid based liposome system to improve the solubility and bioavailability of polydatin [15]. Yallapu et al. (2010) documented the nano-formulation of PLGA improve the therapeutical efficacy of curcumin in human ovarian and metastatic breast cancer cell lines [16]. Since, there are no reports are available the combinational physiochemical features of polydatin loaded PLGA nanoparticles. To the best of our knowledge, this is the first report on the biosynthesis of Polydatin-loaded PLGA nanoparticles (POL-PLGA-NPs). The findings of this study validate that the high negatively charged synthesized nanoparticles have the ability to penetrate into inside the tumor cells via sustainable drug releasing profile. It could be promising. The feasible outcome of this study, hopefully provide new insights of nanochemopreventive potential POL-PLGA-NPs.

Hence, the aim of the study is to employ a simple method for the bio-fabrication of POL-PLGA-NPs. The efficiency of the structural modification of POL-PLGA-NPs was evaluated by FTIR and FRD analysis. The physical-chemical characteristics, namely, average size, morphological features, zeta potential, drug loading efficiency, and encapsulation efficacy, were determined. Further, the apoptotic activating efficacy of POL-PLGA-NPs in DMBA induced buccal pouch carcinogenesis was investigated.

2. Materials and Methods

2.1. Chemicals and Reagents

DMBA, PLGA (lactide: glycolide 75:25, Mw 76,000–115,000), polydatin and bovine serum albumin (BSA) were obtained from Sigma-Aldrich Chemical (St. Louis, MO, USA). Primary antibodies, such as mutant p53, Bax, Bcl-2, cleaved caspase 3, cyclin-DI and ß-actin were purchased from Santa Cruz Biotechnology (Santa Cruz, CA, USA). All other chemicals and solvents were supplied from Himedia laboratories, Mumbai, India and Fisher Inorganic and aromatic Limited (Chennai, India).

2.2. Synthesis of Polydatin Encapsulated PLGA Nanoparticles [POL-PLGA-NPs]

Polydatin encapsulated PLGA nanoparticles [POL-PLGA-NPs] were fabricated by an oil/water emulsion method with minor modification [17]. 50 mg of PLGA was dissolved in 5 mL of dichloromethane and acetone (prepared as 3:2) to form well-proportioned PLGA solution in a round-bottomed flask. Then, 10 mg of polydatin was added to the solution and sonicated at 200 W for 10 min to make a primary emulsion (organic phase) and the resultant primary emulsion was added dropwise to BSA solution (1% *w/v*) (aqueous phase) and the mixture was sonicated at 200 W for 15 min to make an oil/water (O/W) emulsion. To diffuse the O/W emulsion, 15 mL of deionized water was added and stirred vigorously to eliminate the residual organic solvent. After continuous stirring for 2–3 h, the solution was centrifuged at 14,000 rpm for 30 min, the supernatant was discharged and the pellet was washed repeatedly with deionized water. After the centrifugation of 10,000 rpm for 20 min, the POL-PLGA-NPs which settled down was collected and lyophilized by freeze drying and stored at 4 °C.

2.3. Characterization of Nanoparticles

After the successful synthesis of Polydatin-encapsulated PLGA nanoparticles were processed for physicochemical characterization techniques. Particle size, polydispersity index and zeta potential of PLGA-NPS/POL-PLGA-NPs was investigated by dynamic light scattering (DLS) using Horiba Scientific-SZ-100 (Horiba, Kyoto, Japan). X-ray diffraction pattern (XRD) of the crystalline phase was recorded using an Ultima IV X-ray diffractometer (X'pert-pro MPD-PANalytical, Netherland) at the angle range of 2θ (10–80°). Surface chemistry of the nanoparticles and functional group analysis was done by Fourier transfer infrared spectroscopy (FTIR) (FTIR PerkinElmer Paragon 500, USA). Particle size and topological features of the nanoparticles were recorded by Transmission electron microscope using Philips CM120 M (80 kV; Philips, Eindhoven, Netherlands) and the three-dimensional features of the individual and the groups of particles are investigated by atomic force microscope (AFM) using AFM-Solver Next (NT-MDT, Moscow, Russia).

2.4. Determination of Encapsulation and Loading Efficiency of POL-PLGA-NPs

The encapsulation efficiency (EE) and drug loading efficiency (LE) of POL-PLGA-NPs were determined by spectrophotometric method. Briefly, 3 mg of POL-PLGA-NPs was dissolved in 6 mL of PBS and centrifuged at 12,000 rpm for 30 min. The content of free polydatin in the supernatant was measured by UV-Vis spectrophotometer (Elico SL 196, Hyderabad, India) at 230 nm. The percentage of EE and LE was calculated from this equation:

$$\text{EE } (\%) = W_0/W_1 \times 100; \text{ LE } (\%) = W_0/W \times 100 \quad (1)$$

Here, W_0 is the amount of polydatin enveloped in the PLGA nanoparticles, W is the amount of polydatin encapsulated nanoparticles, and W_1 is the amount of polydatin added in the system.

2.5. In Vitro Releasing Profile of Polydatin

The amounts of polydatin released from the polydatin-encapsulated PLGA nanoparticles were determined by the spectroscopic method using dialysis bag. In this study, we were chosen for two different pH of 4.8 and 7.4 to simulate the extracellular and lysosomal environment, respectively. In briefly, 10 mg of sample were immersed in a dialysis bag and flooded into 50 mL of phosphate buffer saline (PBS) at different pH (4.8 and 7.4) at under constant and continuous shaking (100 rpm at 37 °C). At the scheduled time intervals (2, 4, 8, 12, 24, 48 h), samples were taken from the solution and the volume was replaced with fresh PBS and the released content of polydatin was measured by UV-visible spectrophotometer (Elico SL 196, Hyderabad, India).

2.6. Animals

Eight- to 10-week-old Syrian hamsters weighing 90–120 g were obtained from Indian Council of Medical Research (ICMR)-National Animal Resource Facility for Bio-Medical Research (NARFBR), Hyderabad, India. The animals were housed in ventilated cages under the constant conditions (22 °C, 12 h light/dark cycle). The animals were fed with a normal pellet diet (Hindustan Lever Ltd., India) and water ad libitum. Animals' care, experimental procedure, and euthanasia procedure was performed by the guidelines of the committee for the purpose of control and supervision on experiments on animals (CPCSEA) and the protocol was approved by the institutional ethical committee (1282/PO/Re/S/09/CPCSEA).

2.7. Treatment Protocol

After allowing the animals one week of acclimation to their new environmental conditions, they were randomized into control and experimental groups and separated into six groups ($n = 6$ animals). Group 1 animals served as control. The animals in the groups (2–5) were painted with 0.5% solution of DMBA in mineral oil using number 4 hair brushes to induce oral carcinogenesis. Every application giving 0.4 mg DMBA load. Carcinogenic control animals had not received any other treatment (Group-2). Groups of 3–5 animals (Nanoparticles treated group) were orally treated with (intra-gastric mode-infant feeding tube No: 5) different concentrations of POL-PLGA-NPs (7.5, 15, and 30 mg/kg b.wt; dissolved in 0.2% DMSO) by intragastric intubation thrice a week on alternate days of the DMBA application. Groups of 6 animals were orally administrated with 30 mg kg/b.wt of POL-PLGA-NPS to check its adverse effects. Vehicle control animals were painted with liquid paraffin throughout the study (Group-1). After the treatment schedule, the animals were sacrificed; blood, liver, and buccal pouches were used to biochemical, histopathological, and molecular studies. The body weights of all hamsters were recorded until the end of the experiment. Tumor incidence, tumor weight and tumor volume were accessed by the method of Geren et al. [18].

2.8. Histological Studies

Part of the buccal tissue was surgically removed and immersed in 10% formalin for 24 h for fixation. Then the tissue was processed and embedded in paraffin wax, 4–5 µm sections were sliced and stained with hematoxylin and eosin. The sections were examined under a light microscope and photo-micrograph was documented.

2.9. Biochemical Estimations

The tissue lipid peroxidative byproducts known as thiobarbituric acid reactive substances (TBARS) level was measured as described by Ohkawa et al., and the formation of the pink-colored chromogen was measured at 532 nm [19]. Lipid hydroperoxides (LOOH) content was estimated by the method of Jiang et al. [20] and the Conjugated dienes (CD) levels was measured by the method of Rao and Recknagel [21]. Superoxide dismutase (SOD, EC.1.15.1.1) activity was estimated by the method of Kakkar et al., and the percentage of inhibition of formazan development was calculated. The amount of enzyme required for 50% inhibition of NBT reduction/min/mg protein defined as one unit of the enzyme [22]. Catalase (CAT, EC.1.11.16) activity was assayed by the method of Sinha. The reaction of tissue homogenate with H_2O_2, the presence of buffer was arrested by the addition of a dichromate acetic acid reagent, and the formation of chromic acetate was measured at 590 nm [23]. The values are expressed as μmoles of H_2O_2 utilized/min/mg protein. The levels of reduced glutathione (GSH) was assessed by the method of Ellman, which is based on the reduction of 5, 5′ dithiobis 2-nitrobenzoic acid (DTNB) and the glutathione concentration was expressed as μmoles of -SH content/g tissue [24]. Vitamin E level was measured by the method of Palan et al., This method incorporates the reduction of Fe^{3+} to Fe^{2+} by a-tocopherol and the formation of a colored complex was measured at 520 nm [25]. Vitamin C level was estimated by the method of Omaye et al., This method involves the oxidation of ascorbic acid to form dehydro-ascorbic acid and diketogluconic acid and the development of the yellow-orange colored composite was measured at 520 nm and defined as μg/mg protein [26]. Glutathione peroxidase (GPx, EC.1.11.1.9) activity was assessed using the method of Rotruck et al., which is based on the reduction of hydrogen peroxide by GSH for 5 min and the values are expressed as μmoles of GSH utilized/min/mg protein [27]. Glutathione-S-transferase (GST) activity was measured by using the method of Habig et al., which is based on the conjugation of the thiol group of glutathione with the 1-chloro, 2-4dinitrobenzene (CDNB) and the values are expressed as μmol of CDNB-GSH conjugate formed min/mg protein [28]. Glutathione reductase (GR) activity was measured by the method of Carlberg and Mannervick: Based on the reduction of glutathione disulfide to reduced glutathione, one unit of enzyme activity is defined as the nmoles of NADPH consumed/min/mg protein [29]. Cytochrome p450 and cytochrome b5 activity were measured according to the method proposed by Omura and Sato, the formation of carbon monoxide (CO) adduct reduced cytochrome p450 with CO, and the spectral difference between reduced and oxidized cytochrome b5 measured respectively [30].

2.10. Western Blot Analysis

According to the manufacturer's instruction of protein isolation kit, total proteins were extracted from the buccal tissues of control and experimental groups. Each protein (50 mg) samples were separated through SDS-PAGE and then transferred to PVDF membranes by electrophoretically. The membrane was blocked with 5% nonfat dry milk for 2 h to block unspecific binding sites. The membrane was kept overnight incubation with 1:1000 dilutions of primary monoclonal antibodies Mutant p53 (catalogue No: ab32049; Abcam, UK), Bax, Bcl-2, cleaved caspase 3, cyclin-D1 and β-actin at 4 °C and detected with horseradish peroxidase-conjugated secondary antibody for 1 h. Finally, the transferred protein bands were visualized using enhanced chemiluminescence reagents and quantitated by ImageJ, a public domain Java image processing software (Wayne Rasband, NIH, Bethesda, MD, USA).

2.11. Statistical Analysis

Data were expressed as mean ± standard deviation. Statistical differences compared between treated groups and the untreated group were analyzed by one-way analysis of variance (ANOVA) and followed by Turkey HSD with IBM SPSS version 23.0 (SPSS Inc., NY, USA).

3. Results

3.1. Physiochemical Analysis of Polydatin Loaded Nanoparticles for the Determination of Size, Potential, and Morphological Features

Size and shape of the nanoparticles are a key factor in designing of drug delivery systems. The high surface–volume ratios of smaller sized nanoparticles efficiently interact with active compounds and polymers which ultimately enhance the therapeutically efficacy of the drug. In this study, DLS study was undertaken to ascertain the particle size, distribution, polydispersity index, and the potential of the fabricated nanoparticles (Figure 1). It was seen that the average size of biosynthesized PLGA-NPs was 119.6 nm (PDI index: 1.412) and POL-PLGA-NPs was 187.3 nm (PDI index: 0.256). Surface charge of PLGA-NPs was found to be −35.2 mV and POL-PLGA-NPs was −23.8 mV.

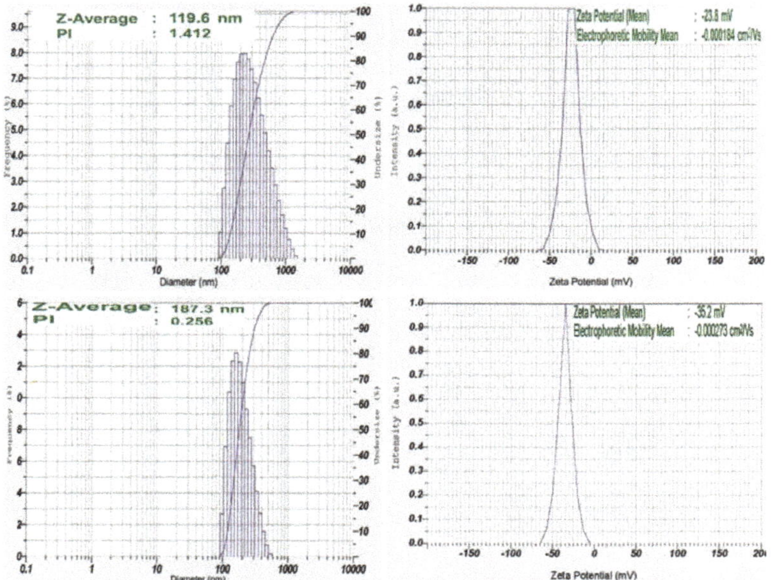

Figure 1. Dynamic light scattering (DLS) analysis (mean particle size, polydispersity index and Zeta potential) of poly-lactic-co-glycolic acid nanoparticles (PLGA-NPs) and POL-PLGA-NPs. POL-PLGA-NPs has 187.3 (average size), 0.256 (PDI index), and −23.8 mV of Zeta potential.

TEM and SEM analysis were investigated to find out the surface morphology of synthesized nanoparticles. TEM images revealed that the smooth surface without agglomeration and fabricated nanoparticles appeared spherical in shape, with the average size of the particles ranging from 105 to 200 nm (Figure 2). Figure 3 showed the analysis of surface morphology and size distribution of synthesized POL-PLGA-NPs using atomic force microscopy. The results indicate that the NPs are spherical in shape and the size distribution of nanoparticles is between 120 to 210 nm (Figure 3A–D). The results are similar that of Particle size analysis and TEM analysis.

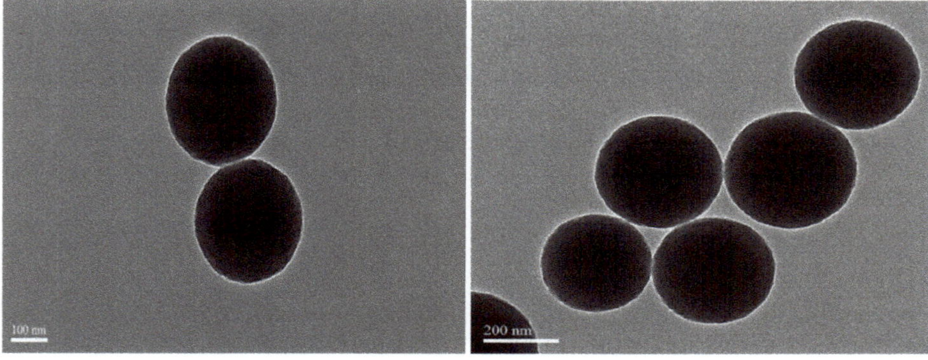

Figure 2. Transmission electron microphotograph of POL-PLGA-NPs. It shows the smooth surface without agglomeration and NPs appeared spherical in shape, with an average size of 105 to 200 nm.

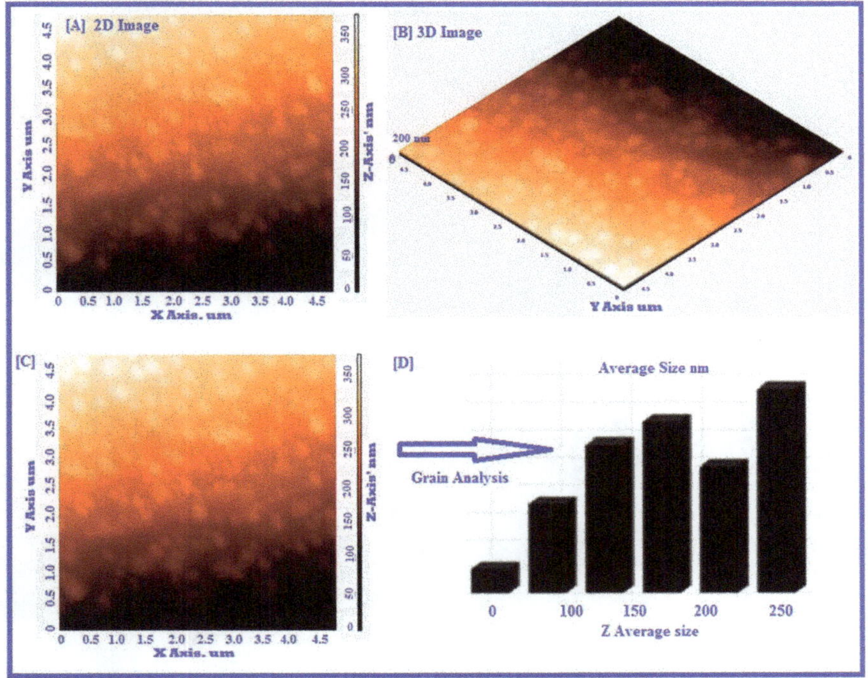

Figure 3. Atomic force microscopy analysis of green synthesized POL-PLGA-NPs. The results disclose that biosynthesized NPs appears to be spherical in shape by the seeing the nanoparticle's topology and morphology of 2D image (**A**). Nova-Px 3.2.0.rev soft ware provided by NT-MDT was used to detect the grain size of the AFM image. An analysis of the results reveals that the NPs are varied in size that is 150 nm ± 10 nm to 200 nm ± 10 nm, we have also carried out grain analysis of the AFM 3D image using Nova-Px 3.2.0.rev software (**B**). Whereas the average size of the grains was found to be 160 ± 10 nm nm by using grain analysis (**C&D**).

3.2. Elemental Analysis of FTIR and XRD Analysis

Fourier-transform infrared spectroscopy (FTIR) analysis was conducted to identify the functional group analysis of Polydatin, PLGA and POL-PLGA-NPs (Figure 4). FTIR spectrum of polydatin was observed at 3485 cm^{-1} (O–H stretching), 2945 and 2888 cm^{-1} (C–H stretching), 1596 cm^{-1} (C=C stretching), and 11797 1081 cm^{-1} (C–O stretching), and intense peaks at 1506 cm^{-1} and 1449 cm^{-1} due to C-H bending, and 1327 cm^{-1} due to O-H bending for alcohol. Moreover, FTIR spectrum of PLGA showed distinct peaks at 3503 cm^{-1} (O–H stretching for acid group), 3000 and 2954 cm^{-1} (C-H stretching), 1747 cm^{-1} due to C=O stretching for carbonyl group), 1626 cm^{-1} (alkyl C=C stretching), 1386 cm^{-1} (O-H bending), 1122 cm^{-1} (C–OH stretching), 868 and 750 cm^{-1}(C–H bending). On the other hand, the FTIR spectrum of POL-PLGA-NPs showed 3503 and 3485 cm^{-1} shifted to a lower frequency at 3395 cm^{-1} due to the encapsulation of polymer. The sharp peaks at 1747 and 1596 cm^{-1} were reduced to 1752 and 1588 cm^{-1} due to carbonyl groups. It was seen that all the characteristic peaks of polydatin and PLGA are visible in polydatin loaded PLGA nanoparticles.

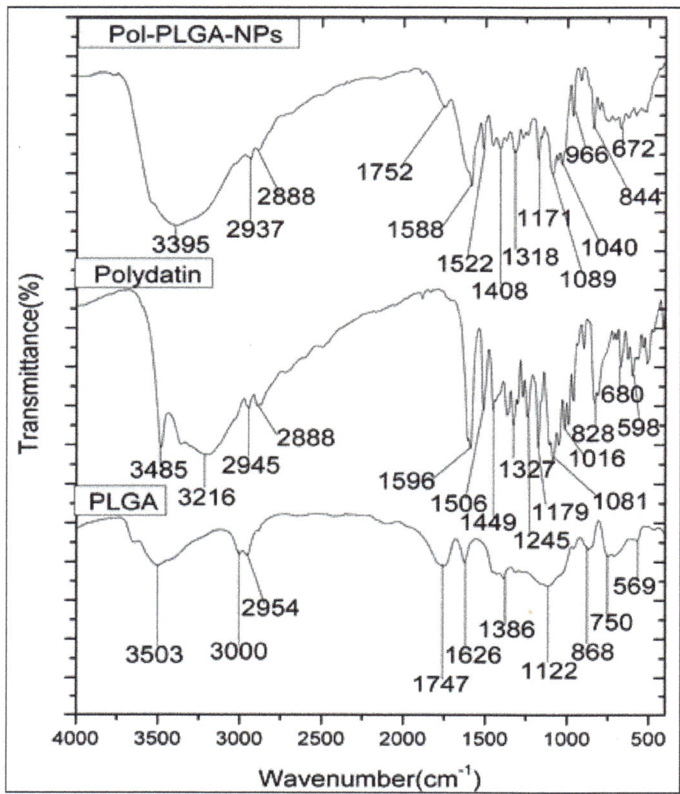

Figure 4. FTIR spectrum of polydatin, PLGA and POL-PLGA-NPs. FTIR spectrum of POL-PLGA-NPs showing the peaks of 3503, 3485 cm^{-1} shift to a lower frequency at 3395 cm^{-1} due to the encapsulation of PLGA.

The XRD pattern of polydatin clearly showed many intense and sharp peaks at 2θ at 12.71°, 14.07°, 16,98°, 17.69°, 19.87°, 21.41°, 23.23°, 26.77°, 28.23°, 29.86°, and 32.04°, which suggested its crystalline nature [29] as it has strong crystalline peaks. PLGA exhibits amorphous nature due to the presence of hump peaks. Besides, POL-PLGA-NPs showed peaks at 14.18°, 21.98°, and 35.62°, clearly indicating that the drug entrapped in nanoparticles and has amorphous nature (Figure 5).

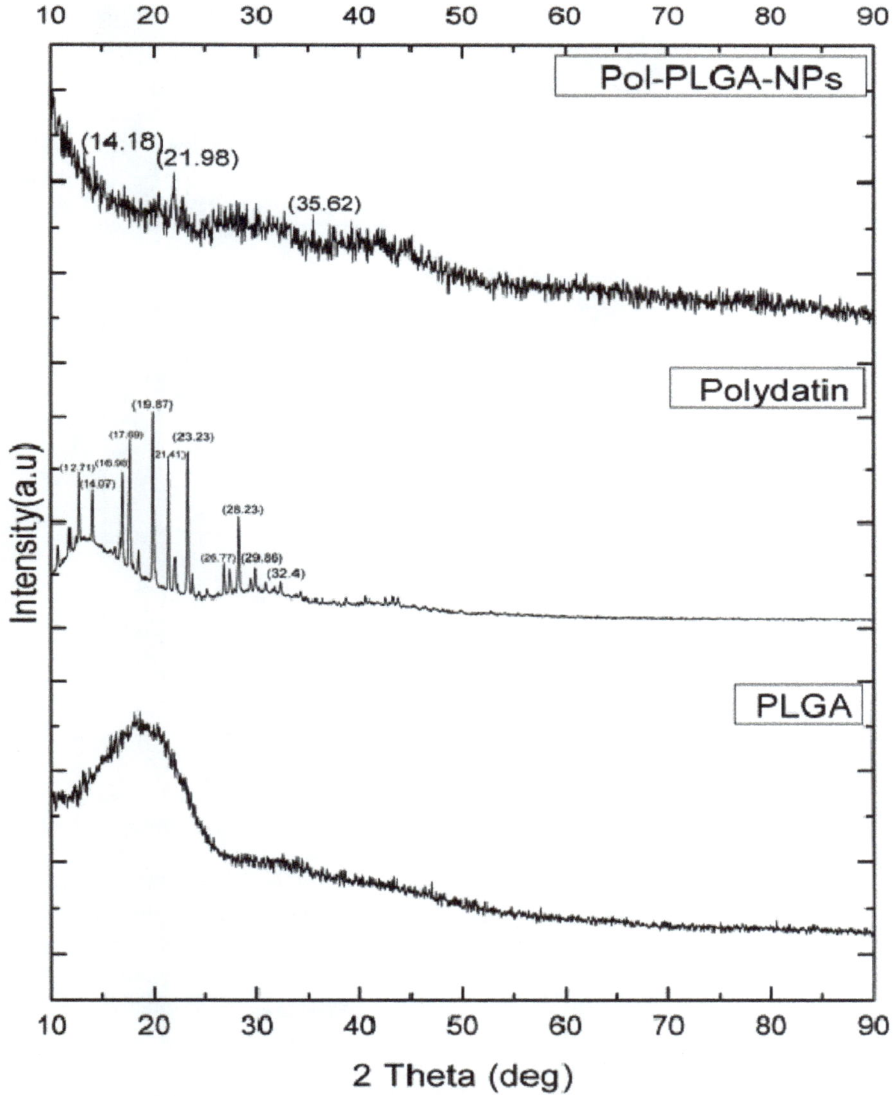

Figure 5. XRD pattern of polydatin, PLGA, and POL-PLGA-NPs. POL-PLGA-NPs showings peaks in 14.18°, 21.98°, and 35.62° indicates the entrapment of polydatin in synthesized PLGA nanoparticles.

3.3. Encapsulation Efficiency, Drug Loading, and Drug Releasing Profile of POL-PLGA-NPs

Table 1 shows the drug loading and encapsulation efficiency of POL-PLGA-NPs with different concentrations of POL., i.e., at 1, 3, and 5 mg/mL. The nanoparticles with 5 mg/mL of POL showed remarkable drug loading and encapsulation efficiency of 8.71 ± 0.74% and 94.52 ± 9.23%, respectively. As shown in Figure 6, the releasing patterns of POL-PLGA-NPs reveals that the fabricated nanoparticle has the pH-independent drug releasing profile. Burst and fast releasing patterns were recorded at pH 5.5. Nearly 50% of polydatin was released in the initial 2 h, and later the release was very slow. A maximum of 68% polydatin was released from the nanoformulation of POL-PLGA-NPs at 48 h. No more release was recorded after that. Also, the sustained drug releasing profile was recorded at pH

7.4. About 16% of the drug was released in an initial period of 2 h, and only 23% drug was released in 48 h.

Table 1. Loading and encapsulation efficiency of polydatin loaded nanoparticles.

Concentration of Polydatin	1 mg/mL	3 mg/mL	5 mg/mL
Loading efficiency (%)	3.81 ± 0.25	7.29 ± 0.63	10.71 ± 0.74
Encapsulation efficiency (%)	22.78 ± 1.37	83.15 ± 6.22	96.54 ± 8.03

Values are expressed as the mean ± SD (n = 3).

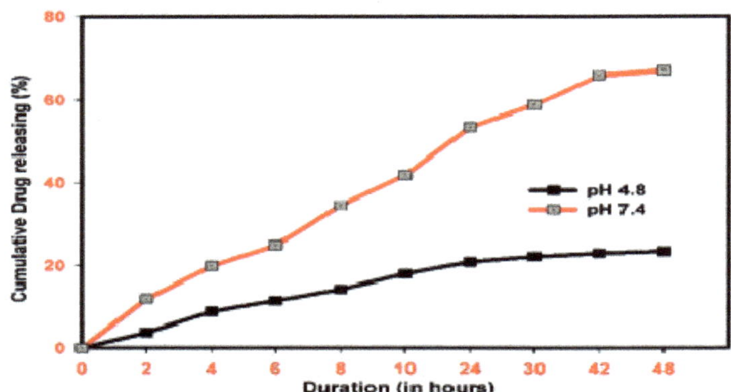

Figure 6. In vitro release pattern of polydatin loaded PLGA nanoparticles. Burst and fast releasing pattern were recorded in pH 4.8, Almost 68% polydatin was released from the nanoformulation.

3.4. POL-PLGA-NPs Suppress the DMBA Induced Neoplastic Changes

Body weight changes, tumor formation, and multiplicity POL-PLGA-NPs treated groups showed a significant gradual increase in body weight (149.97 ± 8.61, 151.16 ± 10.81, and 168.33 ± 14.17). The decreased body weight was evident in carcinogen-alone treated animals (112.73 ± 4.21), whereas the mean body weight of control animals was 192.45 ± 7.17 (Table 2). The administration of POL-PLGA-NPs did not show any clinical sign of toxicity, thus confirming the non-toxic effects of biosynthesized POL-PLGA-NPs and their dosage levels. The site-specific carcinogen DMBA caused 100% of tumor incidence in all carcinogens-alone painted animals, which shows the potential of the carcinogen (Table 2). The total number of tumors and number of tumors and tumor-bearing animals was significantly high in group 2 among all DMBA treated animals ($p < 0.05$). In addition, the high tumor volume indicates the aggressiveness of the disease. Administration of POL-PLGA-NPs to DMBA treated animals (groups 3–5) showed a remarkable decrease in tumor volume and percentage incidence. There were no tumors found in the control (group 1) and drug control (group 6) animals. DMBA-alone painted tumor-bearing animals (group 2) showed the histological characterization of tumor, such as severity in keratosis, hyperplasia, dysplasia, and moderate levels of squamous cell carcinoma, whereas the treatment with POL-PLGA-NPs reduced the tumor histological characteristics from severe to moderate, and inhibited the formation of squamous cell carcinoma. No histological abnormalities were found in the control and drug control animals (groups 1 and 6).

Table 2. Body weight, incidence of oral neoplasm and histological features in POL-PLGA-NPs treated control and experimental animals.

Parameters	Control	DMBA	DMBA+ POL-PLGA-NPs (7.5 mg/kg b.wt.)	DMBA+ POL-PLGA-NPs (15 mg/kg b.wt.)	DMBA+ POL-PLGA-NPs (30 mg/kg b.wt.)	POL-PLGA-NP alone (30 mg/kg b.wt.)
Initial Bodyweight (g)	125.24 ± 6.47^{a}	120.15 ± 3.03^{b}	131.15 ± 9.03^{b}	130.24 ± 12.16^{c}	126.82 ± 8.04^{e}	130.47 ± 7.95^{a}
Final Bodyweight (g)	193.45 ± 7.17^{a}	136.73 ± 9.01^{b}	149.97 ± 8.61^{c}	151.16 ± 9.81^{d}	168.33 ± 9.17^{e}	183.12 ± 8.07^{a}
Weight Gain(g)	68.21 ± 6.72^{a}	16.54 ± 1.72^{b}	18.82 ± 5.72^{c}	20.92 ± 6.74^{d}	43.51 ± 6.38^{e}	52.65 ± 5.21^{a}
Tumor Incidence	-	100	80	68	20	-
Total number of tumors/animals	-	12/(6)	10/(6)	7/(6)	2/(6)	-
Tumor Burden	-	2024.76 ± 82.6 *	1586.2 ± 62.45	942.61.54.83	105.73 ± 7.11 ***	
Tumor Volume	-	168.73 ± 6.43 *	158.6 ± 5.84	134.3 ± 4.81	52.86 ± 1.33 ***	-
Keratosis	Not observed	Severe	Moderate	Moderate	Mild	Not observed
Hyperplasia	Not observed	Severe	Moderate	Moderate	Mild	Not observed
Dysplasia	Not observed	Severe	Moderate	Moderate	Mild	Not observed
Squamous cell carcinoma	Not observed	Well differentiated	Severe	Moderate	Mild	Not observed

Values are expressed as the mean ± SD for 6 hamsters in each group. * Significantly differ from control group ($p < 0.05$), *** Significantly differ from DMBA group ($p < 0.05$) (Oneway ANOVA). Groups not sharing a common superscript letter (a–e) differ significantly at $p < 0.05$ (Oneway ANOVA)

3.5. POL-PLGA-NPs Enhances the Lipid Peroxidative Byproducts

The levels of TBARS, LOOH, and CD in circulation and buccal mucosa of DMBA-treated and control groups are shown in Figure 7. Exposure to carcinogen exhibited significant increase ($p < 0.05$) in the levels of TBARS, LOOH, and CD at the end of 16 weeks. On supplementation of POL-PLGA-NPs to DMBA exposed animals revealed significantly reduced levels ($p < 0.05$) of lipid peroxidative byproducts levels.

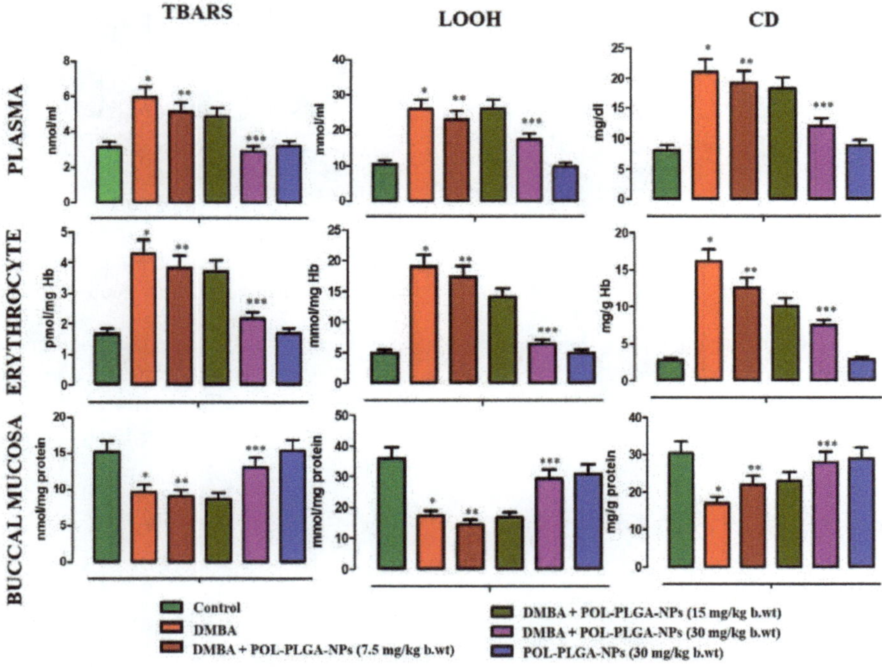

Figure 7. Effects of POL-PLGA-NPs on 7, 12-dimethyl benzyl anthracene (DMBA) -induced changes in the activity of thiobarbituric acid reactive substances (TBARS), Lipid hydroperoxides (LOOH), and CD. Values are expressed as mean ± SD ($n = 6$). * significantly differ from control group ($p < 0.05$), ** Significantly differ from DMBA group ($p < 0.001$), *** significantly differ from DMBA group ($p < 0.05$) (One-way ANOVA).

3.6. Enzymic and Non Enzymic Antioxidant Status

Figure 8 presents the activities of enzymic antioxidant levels in circulation and buccal mucosa of control and DMBA-exposed animals. DMBA-alone painted animals show the reduced levels of enzymic antioxidants such as SOD, CAT, and GPx levels, whereas oral supplementation of POL-PLGA-NPs to DMBA-painted animals significantly improved ($p < 0.05$) the levels of above said antioxidants. There were no significant differences between control and drug control animals.

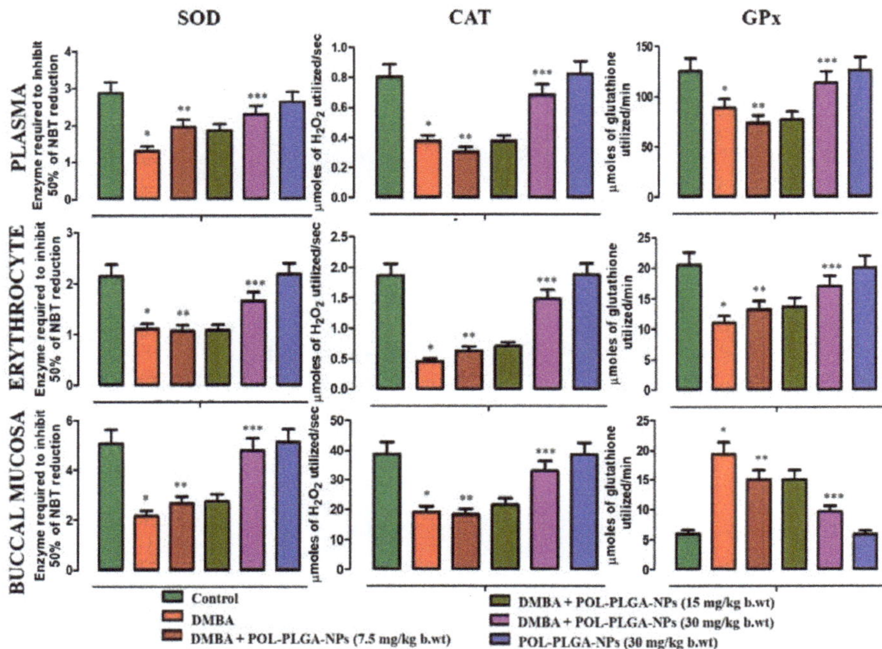

Figure 8. Effects of POL-PLGA-NPs on DMBA induced changes in the activity of enzymatic antioxidants. Values are expressed as mean ± SD ($n = 6$). * Significantly differ from control group ($p < 0.05$), ** Significantly differ from DMBA group ($p < 0.001$), *** significantly differ from DMBA group ($p < 0.05$).

The levels of non-enzymic antioxidants in the circulation and buccal mucosa of experimental animals were shown in Figure 9. A significant reduction was observed in the levels of non-enzymic antioxidants such as Vitamins E, C, and reduced glutathione in carcinogen-treated unsupplemented animals. Upon treatment with POL-PLGA-NPs significantly ($p < 0.05$) increases the levels of those non-enzymic antioxidants to bring back near control values.

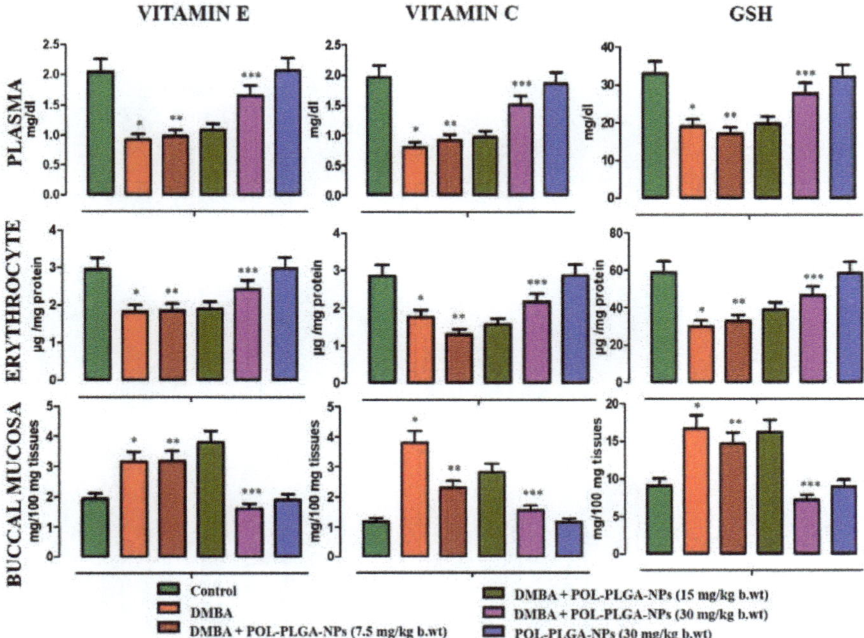

Figure 9. Effects of POL-PLGA-NPs on DMBA induced changes in the level of non-enzymatic antioxidants. Values are expressed as mean ± SD ($n = 6$). * Significantly differ from control group ($p < 0.05$), ** Significantly differ from DMBA group ($p < 0.001$), *** Significantly differ from DMBA group ($p < 0.05$) (Oneway ANOVA).

3.7. Xenobiotic Metabolizing Enzymes

The xenobiotic metabolizing enzymes levels of control and DMBA painted animals were shown in Figure 10. Carcinogen alone expose animal showed a considerable increase in Phase I and Phase II metabolizing enzymes such as Cyt p450, Cyt b5, GST, GGT, and GR activities. POL-PLGA-NPs supplementation to DMBA painted animals reduces the levels of those phases I and II enzymes on a dose-dependent basis, which was more pronounced in POL-PLGA-NPs (30 mg/kg b.wt).

3.8. Effect of POL-PLGA-NPs on the Histopathological Features of the DMBA Induced Buccal Pouch Carcinogenesis

The histopathological evaluation of the buccal tissues of control, carcinogen-alone and POL-PLGA-NPs-treated animals are presented in Figure 11. At the end of 16 weeks, squamous cell carcinoma was evident in the carcinogen-alone exposed group initiated with DMBA. Hyperkeratosis, along with hyperplasia and dysplasia was also observed in DMBA-alone exposed animals. POL-PLGA-NPs-treated DMBA-painted animals displayed mild keratosis as well as mild hyperplasia and dysplasia.

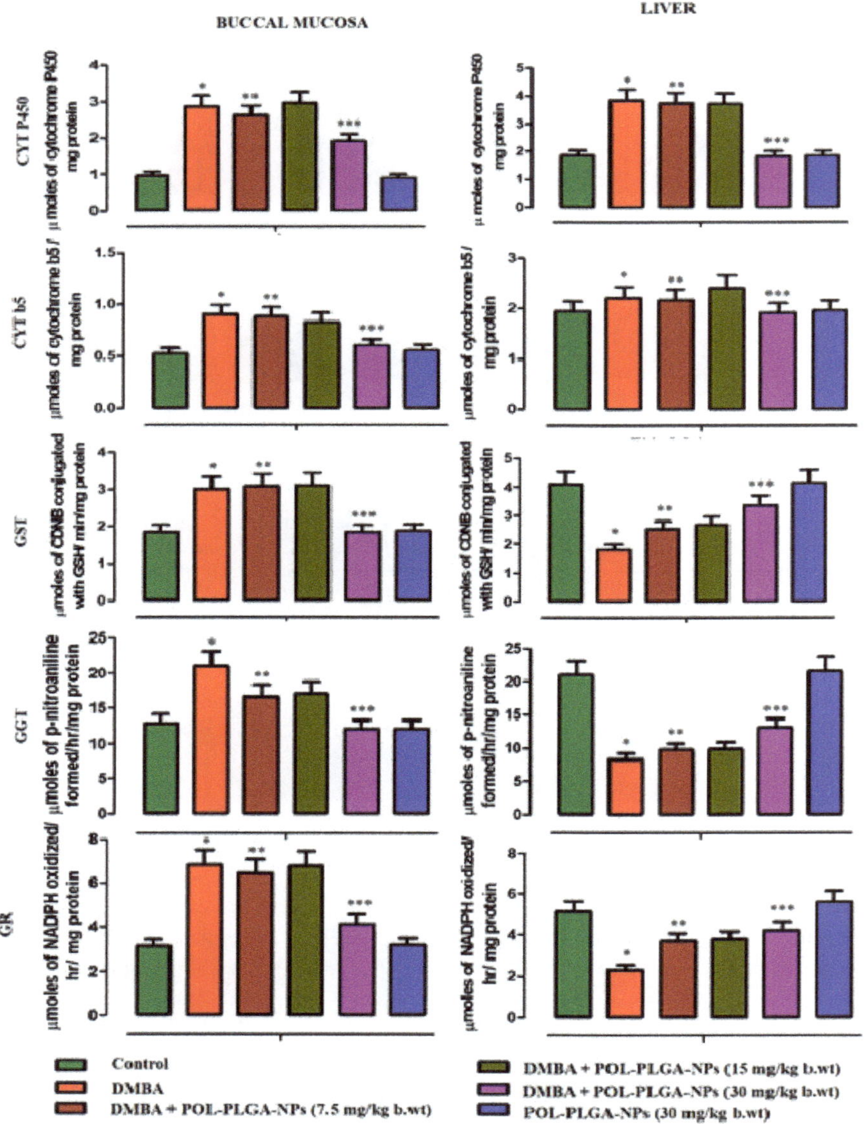

Figure 10. Modulating effects of POL-PLGA-NPs on DMBA induced Xenobiotic metabolizing enzymes. Values are expressed as mean ± SD (n = 6). * significantly differ from control group ($p < 0.05$), ** significantly differ from DMBA group ($p < 0.001$), *** significantly differ from DMBA group ($p < 0.05$) (One-way ANOVA).

3.9. Effect of POL-PLGA-NPs on Apoptotic and Proliferative Marker Expressions in DMBA Induced Buccal Pouch Carcinogenesis

The role of POL-PLGA-NPs and/or DMBA-mediated protein expression of apoptotic and proliferative markers were studied by western blotting analysis (Figure 12). The proapoptotic marker Bax, cleaved caspase-3 was highly down regulated and proliferative marker mutant p53, Bcl-2, and cyclin-D1 were extensively over expressed during the exposure of DMBA in rat buccal pouch. On the other hand, the delivery of POL-PLGA-NPs induces apoptotic mediators such as Bax, cleaved caspase-3 and inhibits DMBA induced mutant p53, Bcl-2, and cyclin-D1 expressions in a dose-dependent manner. We noticed that POL-PLGA-NPs (30 mg/kg.b.wt) treatment was a more efficiently remarkable activity in DMBA-exposed hamsters, whereas POL-PLGA-NPs (30 mg/kg.b.wt) alone produced no toxicity.

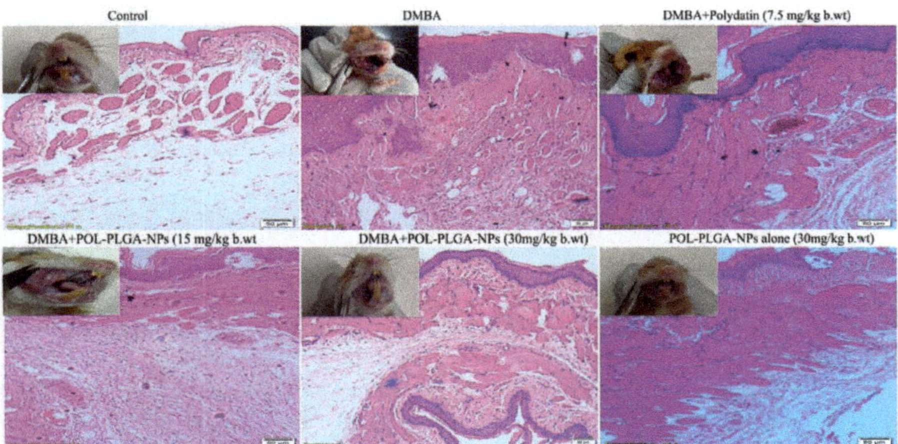

Figure 11. Histopathological analysis of buccal tissue of control and experimental animals (10x). Control and POL-PLGA-NPs showing normal architecture. DMBA alone treated sections showing a well-defined squamous cell carcinoma with hyper chromatic nuclei containing epithelial and keratin pearls. DMBA+POL-PLGA-NPs showing a mild to moderate dysplasia and hyperplasia.

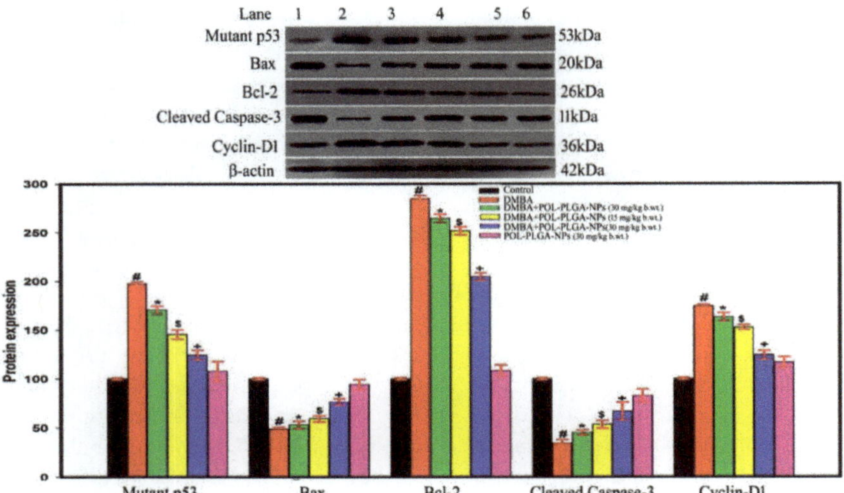

Figure 12. Immunoblot analysis of mutant p53, bax, Bcl-2, cleaved caspase-3, and cyclin D1 in control an experimental animal. 1-Control; 2-DMB; 3-DMBA+POL-PLGA-NPs (7.5 mg/kg b.wt.); 4-DMBA +POL-PLGA-NPs (15 mg/kg b.wt.); 5-DMBA +POL-PLGA-NPs (30 mg/kg b.wt.); 6-POL-PLGA-NP alone (30 mg/kg b.wt.). POL-PLGA-NPs enhance the apoptotic mediators such as Bax, cleaved caspase-3 and inhibit p53, Bcl-2 and cyclin-D1 expressions in a dose-dependent manner. Bar diagram represents the proteins expression; each bar represents the mean ± SD of three independent analysis.

4. Discussion

Ploydatin has strong antioxidant activities. Owing to conjugated double bonds in its molecular structure, it shows many beneficial pharmacological activities such as improving learning and memory, lipid lowering, and extending lifespan. Conjugated compounds can absorb electrons and form adducts with oxygen species (e.g., epoxides, diols and other structures), and thus behave like antioxidants. Although they are antioxidants as well, they form adducts with protein SH groups, thereby activating Nrf-2 antioxidant signaling pathways [31]. Through hydrophobic stacking and hydrogen bonds, polydatin can interact with neurotensin (NT). The polyphenol–protein complexes seem to affect NT metabolism and diminish the NT-induced metabolic activation of colon carcinoma cells [32]. Mikulski and Molski (2010) reported that the presence of 4′-OH group is primarily responsible for the antioxidant capacity [33]. However, to improve the meditative potential of phytocompounds, polymer-based nanoparticles can be attainable feasible approach to improve the biocompatibility and shield against digestive enzymes and pH changes.

Polymer based nanoparticles have attracted the attention of the modern scientific community due to their fascinating applications in biomedical sciences. Scientific evidences have demonstrated that PLGA-based nanoparticles are capable of inducing apoptosis and arresting the cell proliferation in cancerous conditions acts as a carrier molecule to enhance the stability and pharmacological activity of polydatin [34]. Based on this information, the oil/water emulsion method was commonly used in the preparation of nanocarrier with therapeutic agent embedded with hydrophobic or polymeric lattice. This method allows for rapid access of nanospheres or nanocapsules in large quantity and scale up pharmaceuticals industries. Based on this literature, the oil/water emulsion method was used for fabrication of polydatin loaded PLGA nanoparticles (POL-PLGA-NPs). There was a strong ionic interaction between the polydatin and PLGA facilitate the formation of nano-sized particles with the help of stirring and sonication. In the study, average diameter of the fabricated POL-PLGA-NPs was found to be 187.3 nm. In the study, the average diameter of the fabricated POL-PLGA-NPs was found to be 187.3 nm. Due to the presence of terminal carboxyl groups in the PLGA ensures the negative potential, which again ensures abiding stability and avoids particle aggregation. This result correlates with the previous findings that nanoparticles with the average size of 400–600 nm are able to penetrate the endothelial gap of the tumor tissue [35]. The findings of the TEM analysis and 3D analysis of AFM studies were confirmed that the synthesized nanoparticles were typically uniform and spherical shaped nanoparticles with an average size range of 144 nm to 200 nm. Similarly, Lozano et al. reported that the nanoencapsulated quercetin was found to be spherical in shape with an average from 90 nm to 165 nm [36]. The particle size recognized from TEM and AFM analysis strongly supports the findings of DLS analysis.

X-ray diffraction (XRD) analysis is a non-destructive technique generally used to scrutinize the crystallinity and physical nature of the nanoparticles. XRD patterns of polydatin, PLGA, and POL-PLGA-NPs were acquired and compared the significant differences in the molecular state of the nanoformulation. A hump peaks at 20° (2θ), which is pinpointing of the amorphous nature of PLGA. Whereas a sharp peak observed at 19.87°, 23.23°, and 28.23° indicating the crystalline nature of the polydatin. Upon the integration of polydatin into PLGA nanoformulation showed the less intensity of peaks at 21.98° clearly indicates the amorphization nature. Earlier studies also documented that XRD pattern of the encapsulation NPs were exhibited less intensity of peaks when compared with plant-based phytochemicals, which clearly indicate the reduction in the crystallity form the nanoparticles. Similar observation was made our study. In the present study, FTIR patterns of POL-PLGA-NPs strongly suggest that PLGA nanoparticles were successfully encapsulated with a bioactive molecule of polydatin by oil/water emulsion method. The major peaks at 3395 and 1588 cm^{-1} became wider and flatter; indicating that hydrogen bond was enhanced, that there were no loss of functional groups in nanoformulation, and that the crystalline structure was imported to the PLGA nanoparticles.

Encapsulation and loading efficiency of nanoparticles is considered to be one of the critical qualities that improve bioavailability of the drug. The particles having higher loading ability form inefficient delivery systems. The nanoparticles with 5 mg/mL of POL showed remarkable drug loading and encapsulation efficiency of 10.71 ± 0.74% and 96.54 ± 8.03%, respectively. It was revealed that the drug releasing profile of POL-PLGA-NPs was directly equivalent to the concentration of POL. The loaded formulation of POL significantly enhanced the drug loading efficiency of the nanoparticles due to the strong hydrophobic interaction of PLGA. This speculation is in line with numerous studies which testified that PLGA-mediated nanoparticles enhance the drug loading efficiency [37,38]. The findings of the in vitro drug releasing pattern of POL-PLGA-NPs revealed that it is minimally released in normal healthy cells and tissue (pH 7.4), whereas at pH 4.8, the formulated polydatin may attack the tumor tissue and ultimately enter inside the cancer cells to selectively kill the cancer cells due to its nano size and high negative potential.

DMBA is classified as a polycyclic aromatic hydrocarbon. It is an indirect carcinogen that needs metabolic activation to yield an ultimate carcinogen. Initially, the oxidation reaction converts DMBA to DMBA-3,4-epoxide by phase I xenobiotic metabolizing enzymes, especially cytochrome p450 [39]. The epoxide hydratase, another phase I enzyme, converts the epoxide to DMBA-3, 4- diol, theproximate carcinogen. Following a series of oxidation steps by cytochrome leads to synthesis of DMBA-3,4 dioll,2-epoxide, the ultimate carcinogen, which reacts with purine molecules to form DNA adducts [40].

Weight loss is a common characteristic in tumor-bearing animals. DMBA-alone painted animals show a drastic weight reduction along with reduced growth rate, showing the alteration in body metabolism which breaks down the proteins and lipids. In particular, glucose metabolism in cancerous-stage whole-body glucose turnover rate may increase, which increases hepatic glucose synthesis, or gluconeogenesis, from substrates derived from proteolysis and lipolysis [41]. The site-specific carcinogen DMBA induces multiple tumors in buccal tissue with malignant features. The tumor incidence was 100% in carcinogen-alone animals, validating the potency of the carcinogen, and its characteristics revealing the aggressiveness of the disease. Intragastric administrations of POL-PLGA-NPs at the different dosages inhibit the formation of tumors and prevent the tumor growth.

The number and percentage of tumors also reduced in treatment with POL-PLGA-NPs. In particular, at 30 mg/kg b.w., the chemopreventive potential is realized either by preventing or inhibiting the formation of tumor. Martano et al. (2018) supported the use of polydatin in oral cancer prevention and/or as alimentary support associated with anti-tumoral therapy, which is evident from the present study [42]. The PLGA coated nanoparticles may penetrate epithelial cells to enter into the circulation, and accumulate inside the tumor to prevent further progression. Oral cancer was histopathologically confirmed as well differentiated squamous cell carcinoma. DMBA requires metabolic activation by cytochrome p450 to form diolepoxide and other ROS that are known to increase intracellular oxidation, causing severe damage to DNA, lipids and proteins, and thereby contribute to carcinogenesis [43].

ROS mediated oxidative stress has been implicated in the membrane lipid peroxidation, which include increased membrane fragility, decreased red cell fluidity, altered cell function, and structural integrity [44]. Several studies reported the relationship between ROS-mediated lipid peroxidation and several diseases, including oral cancer. The byproducts of lipid peroxidation, reactive aldehydes often form bioactive adducts with macromolecules that are important for the pathophysiology of living cells, thus simulating the impacts of reactive oxygen species (ROS) even in the lack of serious oxidative stress [45,46]. Blood can reflect the liability of the entire animal to oxidative circumstances and it is also a major target of oxy radical assault [47]. Free radicals released into circulation eventually cause hemolysis [48]. When there is an imbalance between prooxidants and antioxidants, it results in increased free radical production and excessive antioxidant consumption, which are the causative factors for oxidative damage [49]. The enhanced lipid peroxidation in the circulation of tumor-bearing animals reflect excessive free radical generation exacerbated by a decreased efficiency of the host antioxidant defense mechanisms. Tumor cells generate and release peroxides into the circulation which can subsequently oxidize GSH. Tumor cells also sequester antioxidants from circulation to promote tumor growth. This may be one of the reasons for the declined antioxidant status with enhanced lipid peroxidation in the circulation of the DMBA treated animals. Increased plasma TBARS observed in tumor-bearing animals are probably due to the overproduction and diffusion of lipid peroxidation byproducts from the damaged tissues with consequent leakage to the plasma. On supplementation with POL-PLGA-NPs, the levels of plasma and erythrocytes TBARS in DMBA-treated animals significantly decreased. This suggests that POL-PLGA-NPs have anti-lipid peroxidative potential during oral carcinogenesis. Numerous studies have demonstrated the increased lipid peroxidation and declining antioxidant status in experimental oral carcinogenesis [50]. On the other hand, tumor tissue has the ability to prevent lipid peroxidation through the highly evolved protective mechanisms so that rapid cell proliferation can occur [51]. Cancer cells are known to acquire certain characteristics that benefit proliferation [52] and they tend to proliferate faster when lipid peroxidation is low. Moreover, malignant tissues are less susceptible and more resistant to free radical attack and hence lipid peroxidation is less intense [53]. Thus, we observed decreased lipid peroxidative that rapid progression of tumor. However, the administration of POL-PLGA-NPs to tumor-bearing animals brings back the lipid peroxidative byproduct levels to near control.

The enzymic antioxidants such as SOD, CAT, and GPx function as the front line of defense against oxidative stress by virtue of their ability to catalyze the disproportionation reactions of their substrate free radicals that are spontaneously generated by in vivo oxidative phosphorylation, cytochrome p450 metabolism, and inflammatory processes [54]. Catalase is a catalyst that changes H_2O_2 to nonpartisan items, O_2, and H_2O. GPx is an initiated protein that acts against oxidative damage, and this requires glutathione as a cofactor. It catalyzes the oxidation of GSH to GSSG to the detriment of H_2O [55]. In the carcinogen-treated animals, the activities of SOD, CAT, and GPx were reduced, which shows the high utilization of endogenous antioxidants and need of more to scavenge the radicals induced by DMBA.

Administration of POL-PLGA-NPs to DMBA-exposed animals reduces the scavenging activities of antioxidants, thereby reducing the oxidative stress. In addition to the above, DMBA-treated animals lessened the levels of Vitamins E and C in the blood and buccal mucosa, and the exercises of these catalysts were impeded because of rehashed exacerbation by the carcinogen. On supplementation of POL-PLGA-NPs, the levels of Vitamins E and C were recovered to near control. In contrast, GSH levels were upheld in the carcinogen-exposed animals, due to uncontrolled proliferation of tumor cells; however, the POL-PLGA-NPs minimize the utilization of glutathione and inhibit the tumor growth process. The cytochrome p450 (oxidizing phase I metabolizing enzymes) is a group of enzymes playing a central role in oxidative metabolic activity [56]. The metabolic activation of DMBA produces diol epoxides, and various ROS, RNS are known to cause damages to lipids, protein, and nucleic acids [57]. Supplementation of POL-PLGA-NPs on DMBA treated animals bring back the phase I and II enzymes levels to near control in buccal mucosa and liver tissues. This finding suggests that POL-PLGA-NPs play a crucial role in the detoxification of DMBA.

Inductions of oral carcinogenesis have been associated with the failure of apoptosis and subsequent activation of proliferation. Bax, Bcl-2, and caspases are involved in the proapoptosis process. Cyclin-D1 and mutant-p53 are deeply involved in proliferation. These apoptotic and proliferative markers are substantially analyzed by western blotting. The in active form of caspase-3 (pro enzyme), is cleaved at an aspartate residue to yield a p12 and p17 subunit to form the active caspase-3 enzyme (cleaved caspase-3) that is responsible for morphological and biochemical changes in apoptosis and is useful in scoring the apoptotic index. Aberrant caspase-3 protein expression has been extensively studied. On this basis, we investigated the expressions of cleaved caspase-3 in tumorigenesis. The inhibition of cell proliferation was measured by evaluating the protein expression levels of Bcl-2. Bcl-2 is an integral membrane protein located mainly on the outer membrane of mitochondria. Overexpression of Bcl-2 prevents cells from undergoing apoptosis in response to a variety of stimuli. Cytosolic cytochrome c is necessary for the initiation of the apoptotic program, suggesting a possible connection between Bcl-2 and cytochrome c, which is normally located in the mitochondrial intermembrane space. Cells undergoing apoptosis were found to have an elevation of cytochrome c in the cytosol and a corresponding decrease in the mitochondria. Overexpression of Bcl-2 prevented the efflux of cytochrome c from the mitochondria and the initiation of apoptosis. Thus, one possible role of Bcl-2 in prevention of apoptosis is to block cytochrome c release from mitochondria. Moreover, dysregulation of cell death genes leading to overexpression of Bcl-2 or reduction in Bax expression, for example, would alter the Bcl-2: Bax ratio which is considered to be anticarcinogenic, and vice versa [58]. As expected, the levels of Bcl-2 in carcinogen treated animals were elevated and treatment with POL-PLGA-NPs to tumor-bearing animals reduces the Bcl-2 protein level. On the other hand, Bax protein levels were increased in tumor-bearing POL-PLGA-NPs-treated an animal, which shows the anticarcinogenic potential of POL-PLGA-NPs. Moreover, POL-PLGA-NPs treatment induces apoptotis through the over-expression of cleaved caspase-3 and inhibits DMBA-induced mutant p53 and cyclin-D1 expressions in a dose-dependent manner. We noticed that POL-PLGA-NPs (30 mg/kg.b.wt) treatment more efficient and remarkable in DMBA-exposed rats. These results are closely correlated with the activity of detoxification enzymes. Previously, metformin-encapsulated PLGA-PEG nanoparticles induced apoptosis by the expression of p53, Bax and caspase-3 in ovarian cancer [59]. It was concluded that the nanoformulation of polydatin may enhance the mitochondrial-mediated apoptotic mechanism in DMBA-treated hamsters.

5. Conclusions

In conclusion, overall findings proposed that the green based POL-PLGA-NPs formulation inhibited the progression of tumor and its growth during DMBA initiated carcinogenesis in golden Syrian hamsters. In addition, the synthesized POL-PLGA-NPs shows strong antioxidant activities and reduces the tissue lipid peroxidation and spares the function of xenobiotic metabolizing enzymes, thereby shows potent chemopreventive efficacy evidenced by pathological reports (Scheme 1). These findings hopefully provide new insights of nanochemopreventive potential of POL-PLGA-NPs. This might pave the way for next generation of nano drug, which might be less expensive and with minimum side effects. Further studies on the bioavailability of the synthesized POL-PLGA-NPs are warranted in experimental animals. In the future, POL-PLGA-NPs may be useful for cancer therapies as an individual drug or in combination with other drugs.

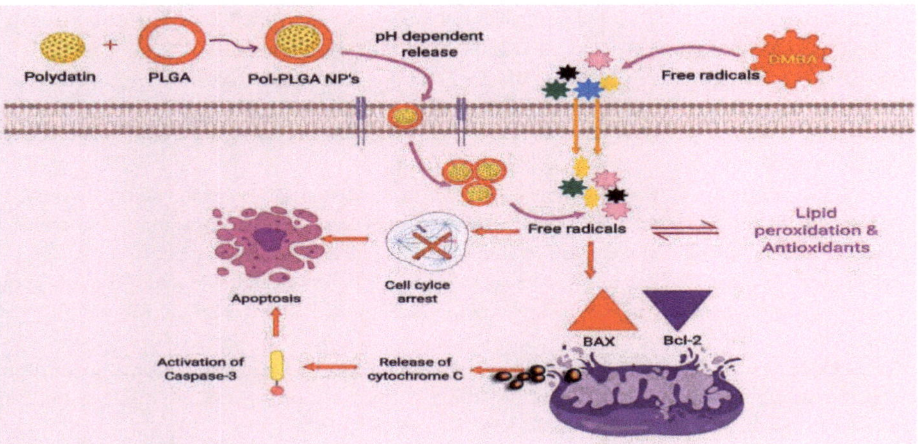

Scheme 1. Proposed mechanism involved in the polydatin loaded PLGA nanoparticles enhance the apoptosis and inhibit the cell proliferation in DMBA induced experimental carcinogenesis.

Author Contributions: S.V.: Investigation and writing-original draft preparation; A.V.A.M.: project administration; V.R. and B.A.: formal analysis; V.S.: data curation; C.C.S.: validation; P.B. and V.S.K.: Data curation; V.K. and D.E.: conceptualization and writing—review and editing.

Funding: This research received no external funding.

Acknowledgments: The authors are grateful to Thiruvalluvar University, Centre for Stem Cell Research, Christian Medical College and Vellore Institute of Technology, Vellore, Tamilnadu, for providing necessary infrastructural facilities. Special thanks DST-PURSE Center, SV University, Tirupathi, Andhra Pradesh, India for the instrumentation facility. Authors also thanks to Jeevakarunyam Sathiya Jeeva, Department of Oral Pathology and Microbiology, Annapoorna Medical College and Hospital, Salem—636 308, Tamil Nadu, India for their valuable assistance in histological studies.

Conflicts of Interest: The authors declare no conflict of interest.

References

1. Bray, F.; Ferlay, J.; Soerjomataram, I.; Siegel, R.L.; Torre, L.A.; Jemal, A. Global cancer statistics 2018: GLOBOCAN estimates of incidence and mortality worldwide for 36 cancers in 185 countries. *CA Cancer J. Clin.* **2018**, *68*, 394–424. [CrossRef] [PubMed]
2. Gunasheela, D.; Menon, J.; Ashwin, N. Strategies for fertility preservation in young patients with cancer. *Onco Fertil. J.* **2018**, *1*, 86. [CrossRef]
3. Arem, H.; Loftfield, E. Cancer Epidemiology: A Survey of modifiable risk factors for prevention and survivorship. *Am. J. Lifestyle Med.* **2017**, *12*, 200–210. [CrossRef] [PubMed]
4. Jethwa, A.R.; Khariwala, S.S. Tobacco-related carcinogenesis in head and neck cancer. *Cancer Metastasis Rev.* **2017**, *36*, 411–423. [CrossRef] [PubMed]
5. Romero-Perez, A.I.; Ibern-Gomez, M.; Lamuela-Raventos, R.M.; de La Torre-Boronat, M.C. Piceid, the major resveratrol derivative in grape juices. *J. Agric. Food Chem.* **1999**, *47*, 1533–1536. [CrossRef] [PubMed]
6. Du, Q.H.; Peng, C.; Zhang, H. Polydatin: A review of pharmacology and pharmacokinetics. *Pharm. Biol.* **2013**, *51*, 1347–1354. [CrossRef]
7. Chen, S.; Tao, J.; Zhong, F.; Jiao, Y.; Xu, J.; Shen, Q.; Wang, H.; Fan, S.; Zhang, Y. Polydatin down-regulates the phosphorylation level of Creb and induces apoptosis in human breast cancer cell. *PLoS ONE* **2017**, *12*, e0176501. [CrossRef]
8. Zhang, Y.; Zhuang, Z.; Meng, Q.; Jiao, Y.; Xu, J.; Fan, S. Polydatin inhibits growth of lung cancer cells by inducing apoptosis and causing cell cycle arrest. *Oncol. Lett.* **2014**, *7*, 295–301. [CrossRef]
9. Hu, W.H.; Wang, H.Y.; Kong, X.P.; Xiong, Q.P.; Poon, K.K.; Xu, L.; Duan, R.; Chan, G.K.; Dong, T.T.; Tsim, K.W. Polydatin suppresses VEGF-induced angiogenesis through binding with VEGF and inhibiting its receptor signaling. *FASEB J.* **2018**, *33*, 532–544. [CrossRef]
10. Yang, B.; Zhao, S. Polydatin regulates proliferation, apoptosis and autophagy in multiple myeloma cells through mTOR/p70s6k pathway. *Onco Targets Ther.* **2017**, *10*, 935–944. [CrossRef]
11. Makadia, H.K.; Siegel, S.J. Poly Lactic-co-Glycolic Acid (PLGA) as biodegradable controlled drug delivery carrier. *Polymers* **2011**, *3*, 1377–1397. [CrossRef] [PubMed]
12. Andima, M.; Costabile, G.; Isert, L.; Ndakala, A.; Derese, S.; Merkel, O. Evaluation of β-Sitosterol loaded PLGA and PEG-PLA nanoparticles for effective treatment of breast cancer: Preparation, physicochemical characterization, and antitumor activity. *Pharmaceutics* **2018**, *10*, 232. [CrossRef] [PubMed]
13. Nassir, A.M.; Shahzad, N.; Ibrahim, I.A.; Ahmad, I.; Md, S.; Ain, M.R. Resveratrol-loaded PLGA nanoparticles mediated programmed cell death in prostate cancer cells. *Saudi Pharm. J.* **2018**, *26*, 876–885. [CrossRef] [PubMed]
14. Singh, M.; Bhatnagar, P.; Mishra, S.; Kumar, P.; Shukla, Y.; Gupta, K.C. PLGA-encapsulated tea polyphenols enhance the chemotherapeutic efficacy of cisplatin against human cancer cells and mice bearing Ehrlich ascites carcinoma. *Int. J. Nanomed.* **2015**, *10*, 6789. [CrossRef] [PubMed]
15. Wang, X.; Luo, Z.; Xiao, Z. Preparation, characterization and thermal stability of β-cyclodextrin/soybean lecithin inclusion complex. *Carbohydr. Polym.* **2014**, *101*, 1027–1032. [CrossRef] [PubMed]

16. Yallapu, M.M.; Gupta, B.K.; Jaggi, M.; Chauhan, S.C. Fabrication of curcumin encapsulated PLGA nanoparticles for improved therapeutic effects in metastatic cancer cells. *J. Colloid Interface Sci.* **2010**, *351*, 19–29. [CrossRef] [PubMed]
17. Manchanda, R.; Fernandez-Fernandez, A.; Nagesetti, A.; McGoron, A.J. Preparation and characterization of a polymeric (PLGA) nanoparticulate drug delivery system with simultaneous incorporation of chemotherapeutic and thermo-optical agents. *Colloids Surf. B Biointerfaces* **2010**, *75*, 260–267. [CrossRef] [PubMed]
18. Green, D.M.; Breslow, N.E.; Beckwith, J.B.; Takashima, J.; Kelalis, P.; D'Angio, G.J. Treatment outcomes in patients less than 2 years of age with small, stage I, favorable-histology Wilms' tumors: A report from the National Wilms' Tumor Study. *J. Clin. Oncol.* **1993**, *11*, 91–95. [CrossRef] [PubMed]
19. Ohkawa, H.; Ohishi, N.; Yagi, K. Assay for lipid peroxides in animal tissues by thiobarbituric acid reaction. *Anal. Biochem.* **1979**, *95*, 351–358. [CrossRef]
20. Jiang, Z.Y.; Hunt, J.V.; Wolff, S.P. Ferrous ion oxidation in the presence of xylenol orange for detection of lipid hydroperoxides in low density lipoprotein. *Anal. Biochem.* **1992**, *202*, 384–389. [CrossRef]
21. Rao, K.S.; Recknagel, R.O. Early onset of lipoperoxidation in rat liver after carbon tetrachloride administration. *Exp. Mol. Pathol.* **1968**, *9*, 271–278. [PubMed]
22. Kakkar, P.; Das, B.; Viswanathan, P.N. A Modified spectrophotometric assay of superoxide dismutase. *Indian J. Biochem. Biophys.* **1984**, *21*, 131–132.
23. Sinha, A.K. Colorimetric assay of catalase. *Anal. Biochem.* **1972**, *47*, 389–394. [CrossRef]
24. Ellman, G.L. Tissue sulphydryl groups. *Arch. Biochem. Biophys.* **1959**, *82*, 70–77. [CrossRef]
25. Palan, P.R.; Mikhail, B.S.; Basu, J.; Romney, S.L. Plasma levels of antioxidant betacarotene and alpha-tocopherol in uterine cervix dysplasias and cancer. *Nutr. Cancer* **1973**, *15*, 13–20. [CrossRef] [PubMed]
26. Omaye, S.T.; Turbull, T.P.; Sauberchich, H.C. Selected methods for the determination of ascorbic acid in animal cells, tissues, and fluids. *Methods Enzymol.* **1979**, *62*, 3–11. [PubMed]
27. Rotruck, J.T.; Pope, A.L.; Ganther, H.E.; Swanson, A.B.; Hafeman, D.G.; Hoekstra, W. Selenium: Biochemical role as a component of glutathione peroxidase. *Science* **1973**, *179*, 588–590. [CrossRef]
28. Habig, W.H.; Pabst, M.J.; Jakoby, W.B. Glutathione S-transferases the first enzymatic step in mercapturic acid formation. *J. Biol. Chem.* **1974**, *249*, 7130–7139.
29. Carlberg, I.; Mannervik, B. Glutathione reductase. *Methods Enzymol.* **1985**, *113*, 484–490.
30. Omura, T.; Sato, R. The carbon monoxide-binding pigment of liver microsomes I. Evidence for its hemoprotein nature. *J. Biol. Chem.* **1964**, *239*, 2370–2378.
31. Liu, B.; Li, Y.; Xiao, H.; Liu, Y.; Mo, H.; Ma, H.; Liang, G. Characterization of the super molecular structure of polydatin/6-O-α-maltosyl-β-cyclodextrin inclusion complex. *J. Food Sci.* **2015**, *80*, C1156–C1161. [CrossRef] [PubMed]
32. Briviba, K.; Abrahamse, S.L.; Pool-Zobel, B.L.; Rechkemmer, G. Neurotensin-and EGF-induced metabolic activation of colon carcinoma cells is diminished by dietary flavonoid cyanidin but not by its glycosides. *Nutr. Cancer* **2001**, *41*, 172–179. [CrossRef] [PubMed]
33. Mikulski, D.; Molski, M. Quantitative structure-antioxidant activity relationship of trans-resveratrol oligomers, trans-4,4′-dihydroxystilbene dimer, trans-resveratrol-3-O-glucuronide, glucosides: Trans-piceid, cis-piceid, trans-astringin and trans-resveratrol-4′-O-beta-D-glucopyranoside. *Eur. J. Med. Chem.* **2010**, *45*, 2366–2380. [PubMed]
34. Mukerjee, A.; Vishwanatha, J.K. Formulation, characterization and evaluation of curcumin-loaded PLGA nanospheres for cancer therapy. *Anticancer Res.* **2009**, *29*, 3867–3875.
35. Yuan, F.; Dellian, M.; Fukumura, D.; Leunig, M.; Berk, D.A.; Torchilin, V.P.; Jain, R.K. Vascular permeability in a human tumor xenograft: Molecular size dependence and cutoff size. *Cancer Res.* **1995**, *55*, 3752–3756.
36. Lozano, O.; Lazaro-Alfaro, A.; Silva-Platas, C.; Oropeza-Almazan, Y.; Torres-Quintanilla, A.; Bernal-Ramirez, J.; Alves-Figueiredo, H.; Garcia-Rivas, G. Nanoencapsulated quercetin improves cardioprotection during hypoxia-reoxygenation injury through preservation of mitochondrial function. *Oxidative Med. Cell. Longev.* **2019**, *2019*, 7683051. [CrossRef]
37. Rezvantalab, S.; Drude, N.I.; Moraveji, M.K.; Güvener, N.; Koons, E.K.; Shi, Y.; Lammers, T.; Kiessling, F. PLGA-based nanoparticles in cancer treatment. *Front. Pharmacol.* **2018**, *9*, 1260. [CrossRef]

38. Arya, G.; Das, M.; Sahoo, S.K. Evaluation of curcumin loaded chitosan/PEG blended PLGA nanoparticles for effective treatment of pancreatic cancer. *Biomed. Pharmacother.* **2018**, *102*, 555–566. [CrossRef]
39. Wilson, N.M.; Christou, M.; Turner, C.R.; Wrighton, S.A.; Jefcoate, C.R. Binding and metabolism of benzo[a]pyrene and 7,12-dimethylbenz[a]anthracene by seven purified forms of cytochrome P-450. *Carcinogenesis* **1984**, *5*, 1475–1483. [CrossRef]
40. Anand, M.A.; Suresh, K. Biochemical profiling and chemopreventive activity of phloretin on 7, 12-Dimethylbenz (a) anthracene induced oral carcinogenesis in male golden Syrian hamsters. *Toxicol. Int.* **2014**, *21*, 179–185.
41. Younes, R.N.; Noguchi, Y. Pathophysiology of cancer cachexia. *Rev. Hosp. Clín.* **2000**, *55*, 181–193. [CrossRef]
42. Martano, M.; Stiuso, P.; Facchiano, A.; De Maria, S.; Vanacore, D.; Restucci, B.; Lo Muzio, L. Aryl hydrocarbon receptor, a tumor grade associated marker of oral cancer, is directly downregulated by polydatin: A pilot study. *Oncol. Rep.* **2018**, *40*, 1435–1442. [CrossRef] [PubMed]
43. Suresh, K.; Manoharan, S.; Vijayaanand, M.A.; Sugunadevi, G. Chemopreventive and antioxidant efficacy of (6)-paradol in 7,12-dimethylbenz (a) anthracene induced hamster buccal pouch carcinogenesis. *Pharmacol. Rep.* **2010**, *62*, 1178–1185. [CrossRef]
44. Halliwell, B. Free radicals, antioxidants, and human disease: Curiosity, cause, or consequence? *Lancet* **1994**, *344*, 721–724. [CrossRef]
45. Mariadoss, A.V.; Vinayagam, R.; Xu, B.; Venkatachalam, K.; Sankaran, V.; Vijayakumar, S.; Bakthavatsalam, S.R.; Mohamed, S.A.; David, E. Phloretin loaded chitosan nanoparticles enhance the antioxidants and apoptotic mechanisms in DMBA induced experimental carcinogenesis. *Chem. Biol. Interact.* **2019**, *308*, 11–19. [CrossRef]
46. Zhang, Z.; Li, M.; Wang, H.; Agrawal, S.; Zhang, R. Antisense therapy targeting MDM2 oncogene in prostate cancer: Effects on proliferation, apoptosis, multiple gene expression, and chemotherapy. *Proc. Natl. Acad. Sci. USA* **2003**, *100*, 11636–11641. [CrossRef] [PubMed]
47. Zarkovic, K.; Uchida, K.; Kolenc, D.; Hlupic, L.J.; Zarkovic, N. Tissue distribution of lipid peroxidation product acrolein in human colon carcinogenesis. *Free Radic. Res.* **2006**, *40*, 543–552. [CrossRef] [PubMed]
48. Zarkovic, K.; Jakovcevic, A.; Zarkovic, N. Contribution of the HNE-immunohistochemistry to modern pathological concepts of major human diseases. *Free Radic. Biol. Med.* **2017**, *111*, 110–125. [CrossRef]
49. Miki, M.; Ramai, H.; Mino, M.; Yamamoto, Y.; Niki, E. Free radical chain oxidation of rat red blood cells by molecular oxygen and its inhibition by α-toxopherol. *Arch. Biochem. Biophys.* **1987**, *258*, 373–380. [CrossRef]
50. Della Rovere, F.; Granata, A.; Saija, A.; Broccio, M.; Tomaino, A.; Zirilli, A.; De Caridi, G.; Broccio, G. SH groups and glutathione in cancer patients. *Anticancer Res.* **2000**, *20*, 1595–1598.
51. Rajalingam, K.; Sugunadevi, G.; Arokia Vijayaanand, M.; Kalaimathi, J.; Suresh, K. Anti-tumour and anti-oxidative potential of diosgenin against 7,12-dimethylbenz(a)anthracene induced experimental oral carcinogenesis. *Pathol. Oncol. Res.* **2012**, *18*, 405–412. [CrossRef] [PubMed]
52. Kenneth, A.C. Dietary antioxidants during cancer chemotherapy: Impact on chemotherapeutic effectiveness and development of side effects. *Nutr. Cancer* **2000**, *37*, 1–18.
53. Nakagami, K.; Uchida, T.; Okwada, S. Increased choline kinase activity in 1,2 dimethylhydrazine induced rat colon cancer. *Jpn. J. Cancer Res.* **1990**, *90*, 1212–1217. [CrossRef] [PubMed]
54. Masotti, L.; Casali, E.; Gesmundo, N.; Sartor, G.; Galeotti, T.; Borrello, S.; Piretti, M.V.; Pagliuca, G. Lipid peroxidation in cancer cells: Chemical and physical studies. *Ann. N.Y. Acad. Sci.* **1988**, *551*, 47–57. [CrossRef] [PubMed]
55. Ray, G.; Husain, S.A. Oxidants, antioxidants and carcinogenesis. *Indian J. Exp. Biol.* **2002**, *40*, 1213–1232. [PubMed]
56. Soujanya, J.; Silambujanaki, P.; Krishna, V.L. Anticancer efficacy of Holoptelea integrifolia, Planch. against 7,12-dimethyl benz (a) anthracene induced breast carcinoma in experimental rats. *Int. J. Pharm. Pharm. Sci.* **2011**, *3*, 103–106.
57. Balamurugan, M.; Sivakumar, K.; Mariadoss, A.V.; Suresh, K. Modulating effect of hypneamusciformis (red seaweed) on lipid peroxidation, antioxidants and biotransforming enzymes in 7, 12-Dimethylbenz(a)anthracene induced mammary carcinogenesis in experimental animals. *Pharmacogn. Res.* **2017**, *9*, 108–115.

58. Mariadoss, A.V.; Kathiresan, S.; Muthusamy, R.; Kathiresan, S. Protective effects of [6]-paradol on histological lesions and immunohistochemical gene expression in DMBA induced hamster buccal pouch carcinogenesis. *Asian Pac. J. Cancer Prev.* **2013**, *14*, 3123–3129. [CrossRef]
59. Faramarzi, L.; Dadashpour, M.; Sadeghzadeh, H.; Mahdavi, M.; Zarghami, N. Enhanced anti-proliferative and pro-apoptotic effects of metformin encapsulated PLGA-PEG nanoparticles on SKOV3 human ovarian carcinoma cells. *Artif. Cells Nanomed. Biotechnol.* **2019**, *47*, 737–746. [CrossRef]

Article

Supplementation of Saponins from Leaves of *Panax quinquefolius* Mitigates Cisplatin-Evoked Cardiotoxicity via Inhibiting Oxidative Stress-Associated Inflammation and Apoptosis in Mice

Jing-Jing Xing [1,2], Jin-Gang Hou [1,3], Ying Liu [1,4], Ruo-Bing Zhang [1,2], Shuang Jiang [1,2], Shen Ren [1,2], Ying-Ping Wang [1,2], Qiong Shen [1,2], Wei Li [1,2], Xin-Dian Li [1,*] and Zi Wang [1,2,*]

1 College of Chinese Medicinal Materials, Jilin Agricultural University, Changchun 130118, China
2 National & Local Joint Engineering Research Center for Ginseng Breeding and Development, Changchun 130118, China
3 Intelligent Synthetic Biology Center, Daejeon 34141, Korea
4 College of Life Science, Kyung Hee University, Seoul 446-701, Korea
* Correspondence: xdli2018@126.com (X.-D.L.); wangzi8020@126.com (Z.W.); Tel./Fax: +86-4318-453-3304 (ext. 8011) (Z.W.)

Received: 27 June 2019; Accepted: 24 July 2019; Published: 1 September 2019

Abstract: Background: Although kidney injury caused by cisplatin has attracted much attention, cisplatin-induced cardiotoxicity is elusive. Our previous studies have confirmed that saponins (ginsenosides) from *Panax quinquefolius* can effectively reduce acute renal injuries. Our current study aimed to identify the potential effects of saponins from leaves of *P. quinquefolius* (PQS) on cisplatin-evoked cardiotoxicity. Methods: Mice were intragastrically with PQS at the doses of 125 and 250 mg/kg daily for 15 days. The mice in cisplatin group and PQS + cisplatin groups received four times intraperitoneal injections of cisplatin (3 mg/kg) two days at a time from the 7th day, respectively. All mice were killed at 48 h following final cisplatin injection. Body weights, blood and organic samples were collected immediately. Results: Our results showed that cisplatin-challenged mice experienced a remarkable cardiac damage with obvious histopathological changes and elevation of lactate dehydrogenase (LDH), creatine kinase (CK), creatine kinase isoenzyme MB (CK-MB) and cardiac troponin T (cTnT) concentrations and viabilities in serum. Cisplatin also impaired antioxidative defense system in heart tissues manifested by a remarkable reduction in reduced glutathione (GSH) content and superoxide dismutase (SOD) activity, demonstrating the overproduction of reactive oxygen species (ROS) and oxidative stress. Interestingly, PQS (125 and 250 mg/kg) can attenuate cisplatin-evoked changes in the above-mentioned parameters. Additionally, PQS administration significantly alleviated the oxidation resulted from inflammatory responses and apoptosis in cardiac tissues via inhibition of overexpressions of TNF-α, IL-1β, Bax, and Bad as well as the caspase family members like caspase-3, and 8, respectively. Conclusion: Findings from our present research clearly indicated that PQS exerted significant effects on cisplatin-induced cardiotoxicity in part by inhibition of the NF-κB activity and regulation of PI3K/Akt/apoptosis mediated signaling pathways.

Keywords: ginsenosides; cisplatin; cardiotoxicity; PI3K/Akt/GSK-3β; oxidative stress; inflammation; apoptosis

1. Introduction

Cisplatin is considered as one of the most potent chemotherapeutic agent against a verity of tumors [1]. However, its effectiveness can be often limited by tissues toxicity such as nephrotoxicity and ototoxicity, which are reported previously [2]. Currently, other factors like acute and cumulative cardiovascular complications are also been reported, which can impair the quality of patient's life [3]. Electrocardiographic changes, myocarditis as well as cardiomyopathy are considered as its major clinical symptoms [4]. These cardiac changes leading to the max-dose reduction of cisplatin, moreover, it will also necessitate the discontinuation of chemotherapy employment [5]. Although we have not formed a comprehensive understanding on cardiotoxicity induced by cisplatin, recent researches have demonstrated that oxidative stress, apoptosis, and inflammation are commonly involved in the occurrence of cisplatin-induced injury [6]. Generally, cisplatin induces mitochondrial dysfunction [7] and decrease of antioxidants in tissues of cancer patients during cisplatin therapy [8], which lead to the overproduction of ROS and subsequent oxidative stress. Importantly, overwhelmed oxidative stress causes changes in the heart after several injections of cisplatin, like fibrosis and edema [9]. Furthermore, excessive ROS production can generate inflammation reactions through NF-κB signal pathway activation resulting in increased expression and secretion of proinflammatory cytokines in cisplatin-induced pathologies [10]. Bcl-2 plays an indispensable part in the process of cardiomyocytes apoptosis, while Bax is a main regulator of Bcl-2 activity [11]. When cisplatin induces generation of ROS, Bax is activated and transported to the mitochondrial outer membrane and changes its permeability, resulting the opening of the mitochondrial permeability transition pores (MPTPs) and the release of cytochrome C into the cytosol, and therefore causing activation of caspase 9 and its downstream caspases-dependent manner [12].

Overproduction of proinflammatory factors, injuries of immune cells, as well as turbulence of the PI3K/Akt signaling pathway, activate apoptosis altogether. Previous studies have confirmed that cisplatin induced irreversible renal dysfunctions owing to excessive cell death, which can be reduced through regulating of PI3K/Akt/GSK-3β signaling pathways. Cisplatin has been shown to modulate PI3K/Akt signal pathway to induce apoptosis in a variety of tissues [13], but its mechanism of action on cardiomyocytes remains unclear. In our study, we also testified that PI3K/Akt signaling pathway is closely related to the effect of cardioprotective effects. The activity of apoptosis-related protein kinase like caspase family members, and Bax can be stimulated by GSK-3β, finally causes apoptosis [14]. Moreover, GSK-3β can be mediated by PI3K/Akt signal pathway in a mouse model and it can also be considered as an indispensable part in the occurrence of Akt [15]. Previously, researches have confirmed that PI3K/Akt plays a vital role in the evolution of myocardial infarction as well as diabetic cardiac injuries [16]. Moreover, the up-regulation of PI3K/Akt pathway attenuates myocardial damages induced by doxorubicin [17]. GSK3β has been shown to play a defensive role against oxidation and toxicological stress through elevation of antioxidant and detoxifying enzyme levels [18].

NF-κB, which is considered response factor in an early stage, exerts significant effects in stimulating generations of various proinflammatory factors [19]. In the meantime, NF-κB combines with inhibiting NF-κB proteins (IκBs) to form a trimmer that is retained in the cytoplasm. Once IκBs are phosphorylated and degraded, NF-κB moves from the cytosol to the nucleus to regulate its target genes [20,21].

The roots of *Panax quinquefolius*, named American ginseng, has been recognized widely herb of genus *Panax* in the US and Canada, its roots and rhizomes have been employed extensively for more than 300 years in China [19]. Like the roots, the leaves of *P. quinquefolius* was rich in saponins including ginsenosides Rb1, Rb2, Rc, Rb3, Rd, Rg1, and Re. Previous studies have focused more on pharmacological activities of several saponins, which are extracted from leaves of *P. quinquefolius* (PQS), including kidney protection [22], anti-inflammation [23], anti-oxidation [24], hypoglycemic effect, etc.

A recent report from our group has confirmed that PQS exerted significant reno-protective effects on cisplatin-evoked renal damages in mice through suppression of oxidative stress, inflammation and apoptosis [19]. Considering PQS's better activity on cisplatin-resulted nephrotoxicity, it will be of great significance to study the protective potential of PQS on cisplatin-caused cardiac toxicities. According to the above works, from our present investigations, we supposed that PQS may have protecting potential on cardiotoxicity in a mouse model. Interestingly, we have confirmed the cardioprotective effect of PQS in cisplatin-treated mice.

2. Materials and Methods

2.1. Chemicals and Reagents

All standards were at least 95% pure, as confirmed by HPLC. HPLC-grade acetonitrile and methanol were purchased from Merck (Darmstadt, Germany). Cisplatin (purity ≥ 99%), was supplied from Sigma Chemicals (St. Louis, MO, USA). Hematoxylin and eosin (H&E), malondialdehyde (MDA), glutathione (GSH), superoxide dismutase (SOD), lactic dehydrogenase (LDH), and myeloperoxidase (MPO) commercial assay kits were obtained from Nanjing Jiancheng Bioengineering Research Institute (Nanjing, China). The primary rabbit monoclonal antibodies including caspase-3, cleaved caspase-3, caspase-8, cleaved caspase-8, caspase-9, cleaved caspase-9, Bax, Bcl-2, β-actin, and secondary rabbit antibodies were purchased from Cell Signaling Technology (Danvers, MA, USA) or DBOSTER Bio-Engineer Co., Ltd. (Wuhan, China). TUNEL apoptosis detection kits were provided with Roche Applied Science (No. 11684817910). Hoechst 33258 dye kits were obtained from Shanghai Beyotime Co, Ltd. (Shanghai, China). DyLight 488-labeled and SABC-Cy3 secondary antibodies were provided by BOSTER Bio-Engineer Co., Ltd. (Wuhan, China). TNF-α, IL-1β, CK, CK-MB, and cTnT commercial ELISA kits were all provided by R&D systems (Minneapolis, MN, USA).

2.2. Animal and Experiments Design

ICR mice (Eight-week-old, male), weighting 25~30 g, provided by Changchun YISI Experimental Animals Co., Ltd. (Changchun, China). The mice were given a standard laboratory diet and water *ad libitum* and maintained at 12 h light/dark cycle at constant temperature (23 ± 2 °C). All experimental animals' processing project were strictly performed according to the Guide for the Care and Use of Laboratory Animals (2016). Animal experiments conducted in line with experimental protocols, and have been acknowledged and confirmed by Jilin Agricultural University Ethical Committee (Permit No.: ECLA-JLAU-18090). The selected 10 mice were randomly took in a group, 5 groups in total, and raised for two weeks before the start of formal experiment, Group 1: normal group, Group 2: cisplatin group (3 mg/kg), Group 3: PQS groups (250 mg/kg), Group 4 and Group 5: cisplatin + 125 or 250 mg/kg PQS groups, respectively. PQS was dissolved in 0.05% carboxymethylcellulose sodium in advance. Mice in group 2, 4 and 5 received four times intraperitoneal injection of cisplatin with 3 mg/kg (body weight) on the 7th, 9th, 11th, and 13th day, and mice in group 4 and 5 were administered with PQS at different doses (125 and 250 mg/kg) for 15 days. Mice in group 3 were administrated with PQS (250 mg/kg) only. Then, all mice were killed at 48 h after final injection of cisplatin. Body weights, blood and tissue samples collections were handled immediately for different purpose. Five hearts in each groups were swiftly and carefully been put into liquid nitrogen, while other hearts were fixed in formalin. Heart serum sample collections were also been promptly segregated by refrigerated centrifuge for the following analysis.

2.3. Biochemical Parameters Determination

2.3.1. Cardiac Biomarkers

Activities of serum cardiac enzymes CK (Cat. No. MM-58997), CK-MB (Cat. No. MM-0839M1), CK-MB (Cat. No. MM-43703M1), and cTnT (Cat. No. MM-0945M1) were measured by using ELISA kits according to the commercial protocols.

2.3.2. Assessment of Cardiac Oxidative Stress

Heart homogenates were used to estimate different oxidative stress parameters. The heart tissues were homogenated in 50 mM phosphate buffer (pH 7.4). The resulting suspension was then centrifuged at 3000× *g* for 10 min twice at 4 °C, and the supernatant was used for the detection of GSH, MDA and SOD. In brief, the levels of oxidative indexes in heart homogenates were detected by commercial kits.

2.3.3. Assessment of Proinflammatory Cytokine

MPO was determined via tissues homogenate. In order to measure the MPO activity, we also measured the rate of oxidation of odianisidine and the absorbance was 460 nm, the MPO activity was calculated and expressed by U/mg protein. Moreover, serum TNF-α and IL-1β were assayed by mouse TNF-α and IL-1β reading ELISA plates at a wavelength of 450 nm.

2.4. H&E Staining

The hearts sections from the normal as well as cisplatin groups were disposed with paraffin 10% buffered formalin. Sections were cut into approximately 5 μm thickness and were stained with H&E staining observe and identify sections histology by light microscope (Leica TCS SP8, Leica Microsystems, Mannheim, Germany) [20].

2.5. Immunohistochemistry

Briefly, the sections were deparaffinized and rehydrated with xylene and various concentrations of alcohol solutions [21]. TBS was used to wash all sections, then they were incubated with 1% BSA for 2 h. Then, they were washed, and were incubated at 4 °C for 12 h with primary antibodies including mouse polyclonal anti-Bax (1:200) and anti-Bcl-2 (1:200), followed by mouse and rabbit secondary antibodies for 1 h. Substratum was added to the tissues for 1 h after DAB staining. The positive staining was detected majorly by brown color in the cytoplasm or nucleus of the positive cells. A light microscopy (Leica, DN750, Berlin Germany) was also used to observe and record the changes.

2.6. Immunofluorescence and Hoechst 33258 Staining

We used primary antibodies like COX-2 (1:200) and iNOS (1:200) in 4 °C overnight, and then all selected sections were exposed to Dylight448-labeled secondary antibody. DAPI staining was used to visualize nucleus followed by PBS washing. Light microscope (LEICA DM 2500, Berlin, Germany) was used to observe their changes. Hoechst 33258 was conducted as mentioned earlier with slight modifications. Briefly, the heart tissues were removed out and sealed in 10% formalin solution. After randomly chose three tissues from every group. Three other samples were chopped into 5 μm sections and dyed by specific stains (10 μg/mL). And then, we used PBS to wash all the sections for three times, fluorescence microscope was used to observe stained nuclei. We also used Image-Pro plus 6.0 to quantify the staining results.

2.7. Western Blotting

Radio Immunoprecipitation Assay (RIPA) buffer was used to split proteins. We prepared 12% SDS polyacrylamide gels and transferred the proteins (50 μg/lane) to a polyvinylidene difluoride (PVDF) membrane. 5% non-fat milk insulted with Tris-buffered saline (TBS) which was made up of 0.1% Tween-20 for more than 2 h at room temperature, then PBS was used to wash the PVDF membrane three times before incubating in primary antibodies at 4 °C for 12 h. Thereafter, the membrane was shacked for half an hour at room temperature before being washed three times by TBST, and 8 min for each time. Latterly, secondary mouse and rabbit antibodies was separately incubated the membrane for 2 h. Eventually, Emitter Coupled Logic (ECL) substrate (Pierce Chemical Co., Rockford, IL, USA), which preserved in 4 °C was 1:1 mixed to detect the expressions of all proteins. We also used Image plus 6.0 software (Media Cybernetics, Rockville, MD, USA) to analyze date.

2.8. Statistical Analysis

All data referenced were expressed as the mean ± S.D. and analyzed with SPSS 19.0 (SPSS, Chicago, IL, USA). Differences among experimental groups were conducted by one-way of variance (ANOVA). Statistical significance was defined as $p < 0.05$ or $p < 0.01$.

3. Results

3.1. Typical HPLC Chromatogram of PQS

High performance liquid chromatography (HPLC) was used to determine and identify the components of PQS. We authenticate all compositions like Rg1, Re, Rf, Rb1, Rc, Rb2, Rb3; Rd, Rg6, F4, Rk3, Rh4, (S)-Rg3, and (R)-Rg3 via comparing the retention times in mixed saponins standard. The chromatograms and structures are concluded in Figure 1.

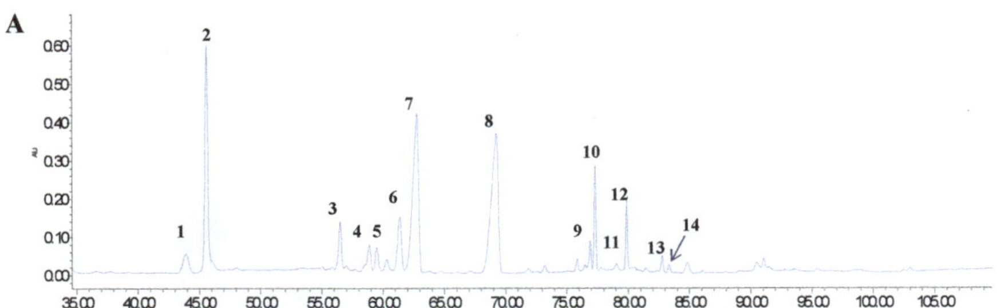

1.Rg1, 2.Re, 3.Rf, 4.Rb1, 5.Rc, 6.Rb2, 7.Rb3, 8.Rd, 9,Rg6, 10.F4, 11.Rk3, 12.Rh4, 13.(S)-Rg3,14. (R)-Rg3

B

20(S)-Protopanaxadiol

	R1	R2
Rb1	-OGlc(2-1)Glc	-OGlc(6-1)Glc
Rc	-OGlc(2-1)Glc	-OGlc(6-1)Ara(p)
Rb2	-OGlc(2-1)Glc	-OGlc(6-1)Ara(f)
Rb3	-OGlc(2-1)Glc	-OGlc(6-1)Xyl
Rd	-OGlc(2-1)Glc	-OGlc
(S)-Rg3 (R)-Rg3	-OGlc(2-1)Glc	-OH

20(S)-Protopanaxatriol

	R1	R2	
Rg1	-OGlc	-OGlc	
Re	-OGlc(2-1)Rha	-OGlc	
Rf	-OGlc(2-1)Glc	-OH	
Rg6 F4	-OGlc(2-1)Rha	=	Dehydration isomerization
Rk3 Rh4	-OGlc	=	

Figure 1. (**A**) Fourteen saponins from leaves of *P. quinquefolius* are confirmed by HPLC analysis, chromatograms and chemical structures of these saponins mainly includes panaxadiol-type ginsenosides Rb1, Rc, Rb2, Rb3; Rd, 20(S)-ginsenoside Rg3, 20(R)-ginsenoside Rg3 and panaxatriol-type ginsenosides Rg1, Re, Rf, Rg6, F4, Rk3, and Rh4. (**B**) The structures of these fourteen saponins.

3.2. PQS Protects Against Cisplatin-Induced Cardiotoxicity

Figure 2 showed that administration of cisplatin injection (5 mg/kg) for 4 times leaded to elevation of CK, CK-MB competence as well as cTnT level comparing with normal group. These changes indicated that cardiac injury can be induced by cisplatin *in vivo*. However, PQS resulted in reduction ($p < 0.05$) in the above-mentioned indicators. Furthermore, histologic sections from the normal group shows that cardiac muscle fibers is regular, however, in the cisplatin-injected group, abundant degeneration in cardiac muscle fibers can be noticed (Figure 2B, E). Mice pretreated with PQS indicated similar forms, which showed that PQS (250 mg/kg body weight) was exerting no impairments on heart

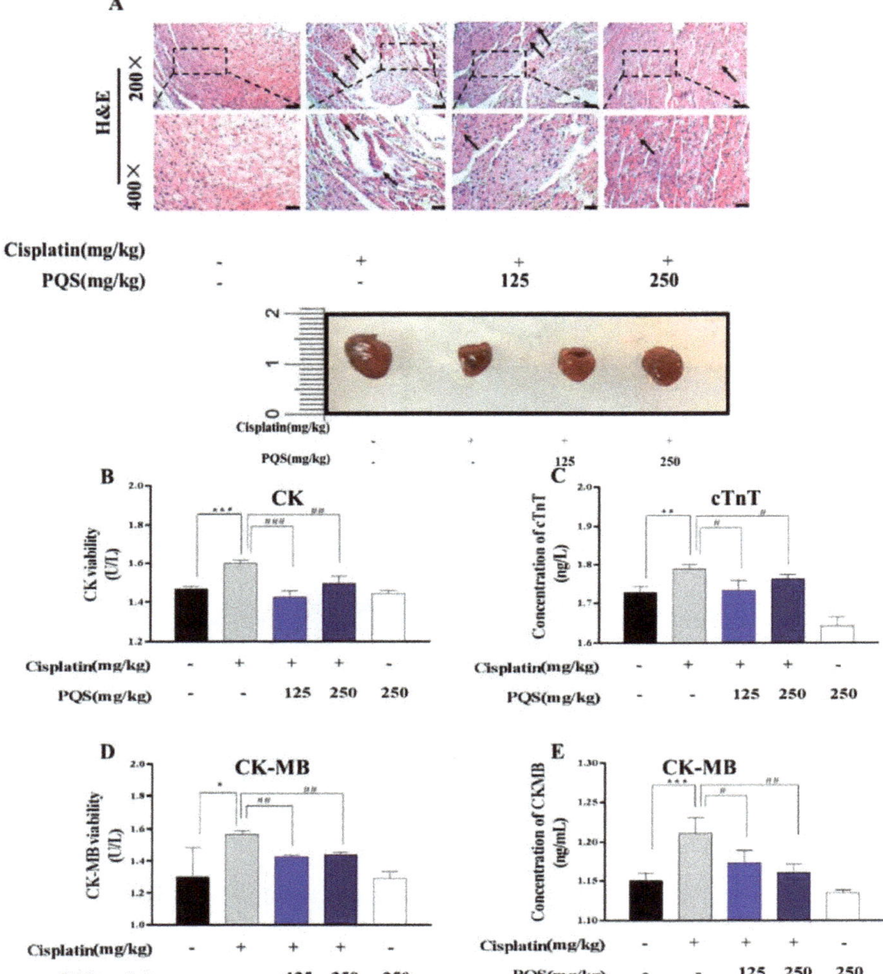

Figure 2. Effect of PQS on cisplatin-induced changes in heart tissues of mice (**A**). Cisplatin + PQS (125 mg/kg), Cisplatin + PQS (250 mg/kg) groups (H&E × 200). Effects of cisplatin and PQS on the serum levels of related markers CK (**B**), cTnT (**C**), viability of CK-MB (**D**), and concentration of CK-MB (**E**). All data were expressed as mean ± S.D. * $p < 0.05$ or ** $p < 0.01$ or *** $p < 0.01$ comparing with normal group. # $p < 0.05$ or ## $p < 0.01$ or ### $p < 0.01$ comparing with cisplatin group.

3.3. PQS Inhibits Oxidative Stress Induced by Cisplatin Treatment

To assess the cardiac markers of oxidative stress injury, GSH and SOD level in heart tissues were tested. As shown in Figure 3, PQS attenuated significantly the decline of myocardial SOD induced by cisplatin ($p < 0.05$) compared to the normal group. GSH content was significantly reduced by cisplatin, compared with the normal group ($p < 0.05$), which were ameliorated by PQS administration evidently ($p < 0.05$). MDA is an important parameter reflecting the potential antioxidant capacity of the body, which can reflect the lipid peroxidation rate and intensity of the body, and can also indirectly reflect the degree of tissue peroxidation damage. The level of MDA was increased after injections of cisplatin, however, a significant decrease of MDA level was observed after treatment with PQS. These data clearly demonstrated that PQS alleviated cisplatin-caused cardiac oxidative stress injuries (Figure 3A–C).

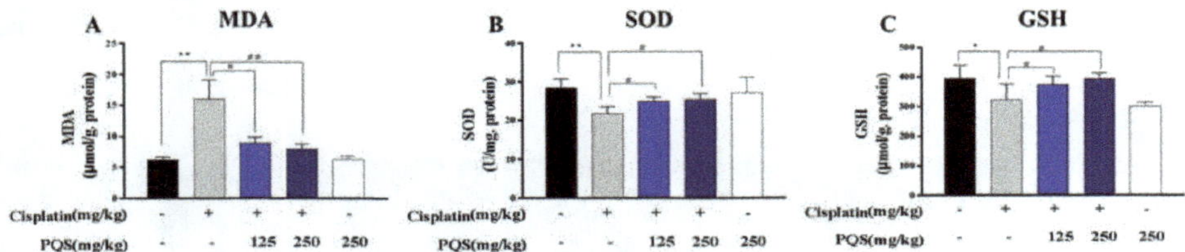

Figure 3. Effects of PQS on the levels of (**A**) MDA, (**B**) superoxide dismutase (SOD) and (**C**) glutathione (GSH). All data were expressed as mean ± S.D. * $p < 0.05$ or ** $p < 0.01$ comparing with normal group. # $p < 0.05$ or ## $p < 0.01$ comparing with cisplatin group.

3.4. Effect of PQS on Cardiac Inflammation

In order to better understand the anti-inflammatory effects of PQS, levels of TNF-α and IL-1β in serum, and activities of LDH and MPO as markers for reflecting neutrophil infiltration were detected. Serum levels of TNF-α and IL-1β showed remarkable elevation for nearly more than 2-folds in mice treated with cisplatin only, and near 1-fold on mice treated with PQS comparing to normal group ($p < 0.001$). Likewise, MPO activity in heart tissues were higher in cisplatin group than that in normal group. Co-administration of PQS significantly abolished the MPO activity. A significant rise in LDH activity illustrated the impairment of heart induced by cisplatin challenging, and PQS significantly decreased these serum-marker enzymes (Figure 4A,C,D,E) ($p < 0.01$). Moreover, pro-inflammatory COX-2 and iNOS levels in the heart were assessed through immunofluorescence. As evidence from immunofluorescence staining, COX-2 and iNOS levels were elevated in cisplatin group. However, mice receiving PQS lowered the expressions than the mice treated with cisplatin alone, such alterations were significantly inhibited (Figure 4B,F,G).

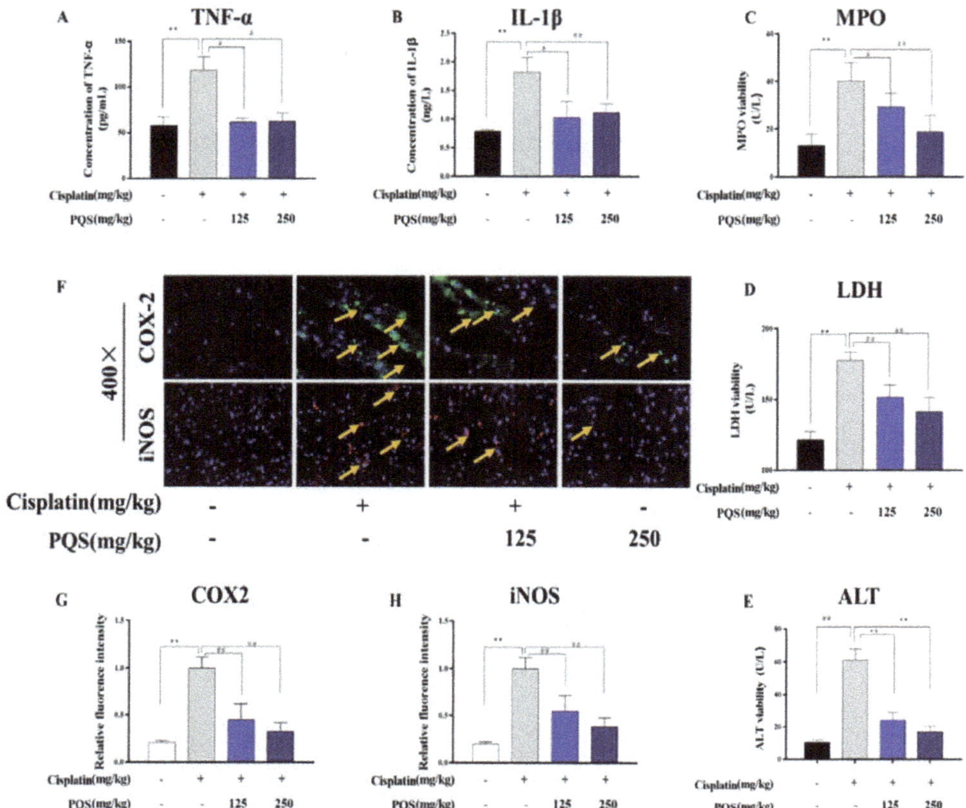

Figure 4. Effect of PQS on cisplatin-induced changes in inflammatory markers in heart tissues of mice. (**A**) Tumor necrosis factor-a (TNF-α); (**B**) Interleukin-1β (IL-1β); (**C**) Myeloperoxidase (MPO) activity; (**D**) Lactate dehydrogenase (LDH); (**E**) ALT activity; (**F**) PQS exerted great changes on expression of COX-2 and iNOS in heart tissues, the expression levels of COX2 (Green) and iNOS (Red) in tissue section isolated from different groups were assessed by immunofluorescence. (**G**) Quantitative analysis of scanning densitometry for cleaved COX-2 (**H**) Quantitative analysis of scanning densitometry for cleaved iNOS. All data were expressed as mean ± S.D. ** $p < 0.01$ comparing with normal group. # $p < 0.05$ or ## $p < 0.01$ comparing with cisplatin group.

3.5. Effects of PQS Treatment on Cisplatin-Induced Inflammatory Markers

Additionally, reduction of overproduction of iNOS and COX-2 in the heart tissues by the PQS pretreatment were confirmed by western blotting analysis (Figure 5) ($p < 0.05$, $p < 0.01$). The expression levels of TNF-α and IL-1β in the heart tissues were also determined by western blotting. As expected, expression levels of proinflammatory factors like TNF-α and IL-1β induced by cisplatin induced by 2.37- and 2.44-fold respectively, when compared to the normal group (Figure 5) ($p < 0.05$ or $p < 0.01$) comparing with normal group.

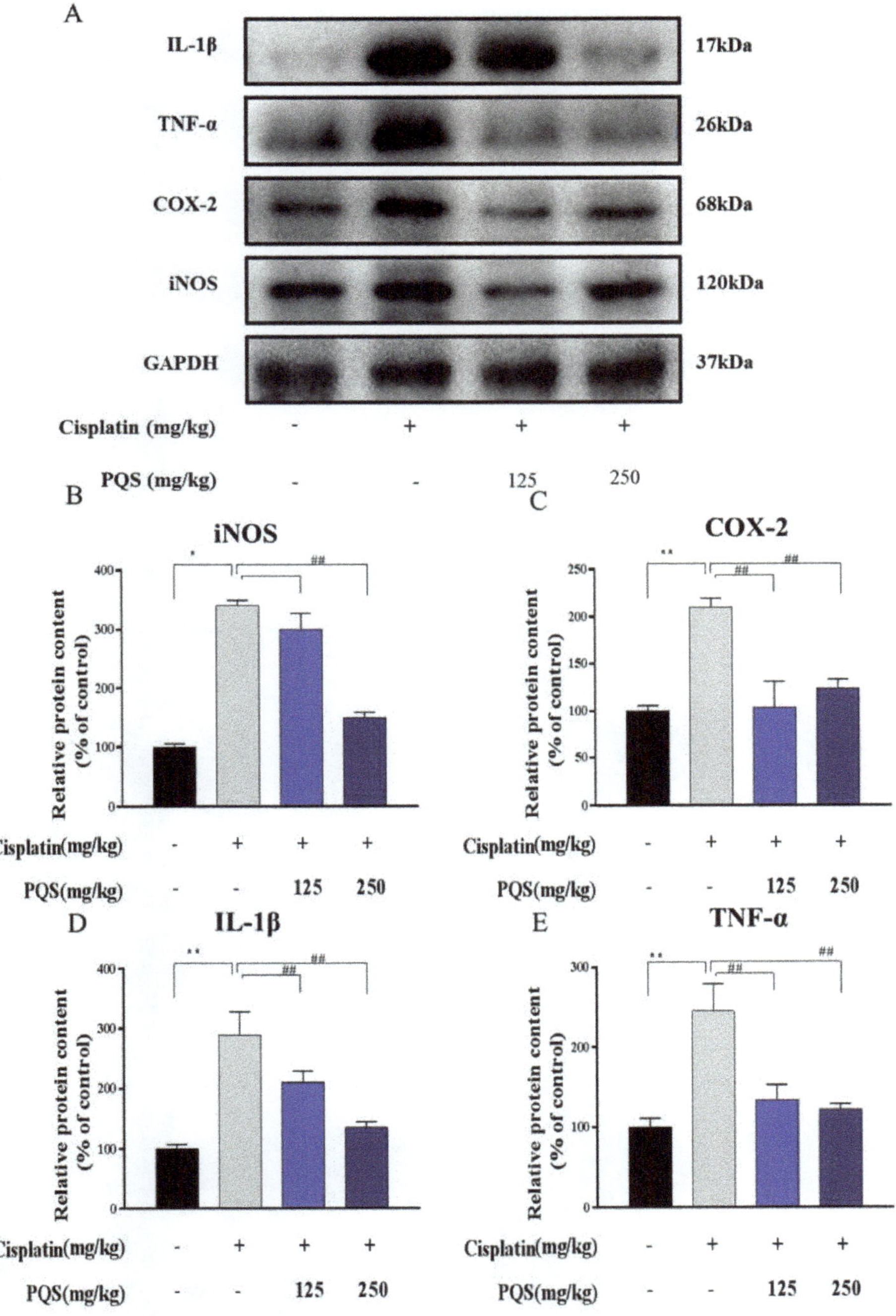

Figure 5. The expression of iNOS, COX-2, TNF-α and IL-1β were measured by western blotting analysis (**A**). Quantitative analysis of scanning densitometry for iNOS (**B**); COX-2 (**C**); IL-1β (**D**); TNF-α (**E**). All data were expressed as mean ± S.D. * $p < 0.05$ or ** $p < 0.01$ comparing with normal group. ## $p < 0.01$ comparing with cisplatin group.

3.6. Effects of PQS on the NF-κB Signaling Pathway

To determine whether PQS can improve cisplatin-induced cardiotoxicity by reducing cisplatin caused inflammation. The effects of PQS pretreatment on cisplatin-activated NF-κB signal pathway were tested by western blotting analysis in this study. Treatment with PQS (125 and 250 mg/kg) decreased the levels of p-IKKα/β, p-IκBs and p-NF-κB, which demonstrated that PQS can effectively recede inflammation evoked by cisplatin (Figure 6A–E).

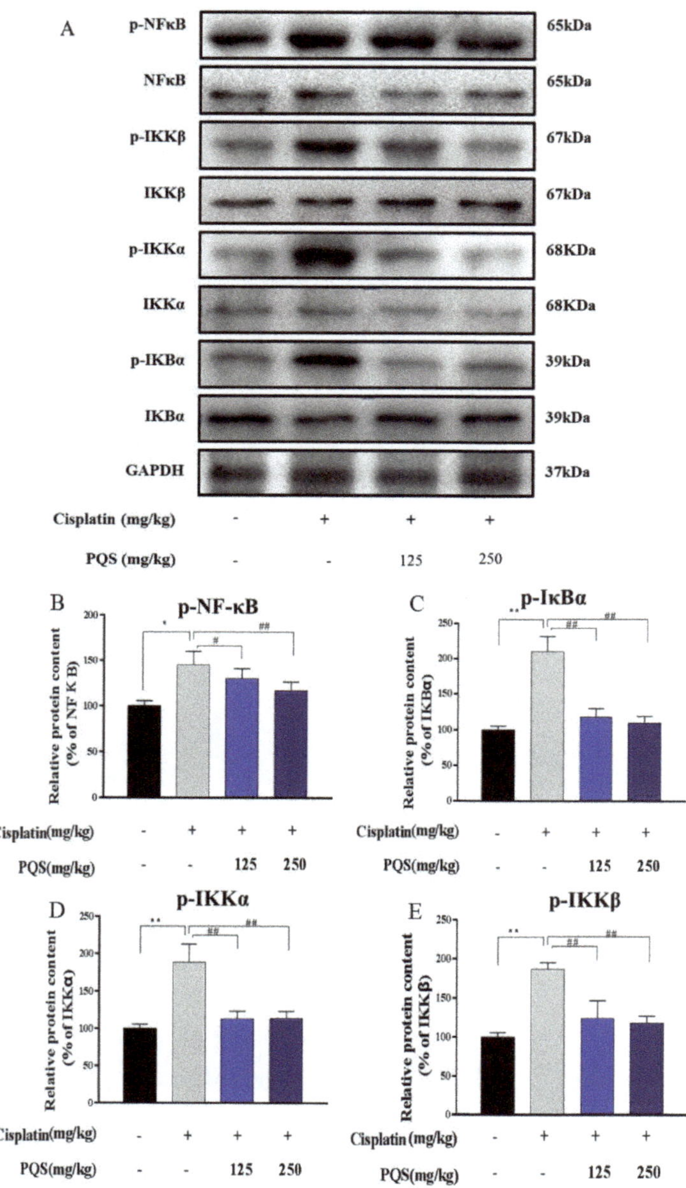

Figure 6. (**A**). Quantitative analysis of scanning densitometry for p-NF-κB (**B**); p-IκBα (**C**); p-IKKα (**D**); p-IKKβ (**E**). All data were expressed as mean ± S.D. * $p < 0.05$ or ** $p < 0.01$ comparing with normal group. # $p < 0.05$ or ## $p < 0.01$ comparing with cisplatin group.

3.7. PQS Attenuates the Intrinsic Mitochondrial Apoptotic Pathway In Vivo

In order to explore the underlying mechanism of the reduction in apoptosis in PQS treatment, we evaluated Bcl-2 and Bax by western blotting (Figure 7B). Mice subjected to cisplatin, the ratio of Bcl-2 and Bax was reduced in comparison to the normal group ($p < 0.05$). Activation of Bax and decrease of Bcl-2 (Figure 7E) were confirmed by immunohistochemical analysis. Treatment with PQS (125 and 250 mg/kg) significantly reversed Bax, Bad, caspase-3, caspase-8, and caspase-9, however, elevated the level of Bcl-2. All data supported that PQS can inhibit apoptotic pathway activation. Interestingly, treatment with PQS led to an increase in this ratio to 1.3 for Bcl-2, and decreased this ratio by 42.7% for Bax, these findings indicated that PQS might block apoptosis through a mitochondrial pathway mediated by the relative ratio of expression of Bcl-2 and Bax in the mitochondria. In this study, apoptosis was further verified by Hoechst 33258 staining in heart tissues. As shown in Figure 7E, mice injected with cisplatin alone showed significantly increased positive staining cells with condensed nuclei. However, comparing with cisplatin group, less apoptotic cells were observed in the cardiac sections in PQS pretreatment group.

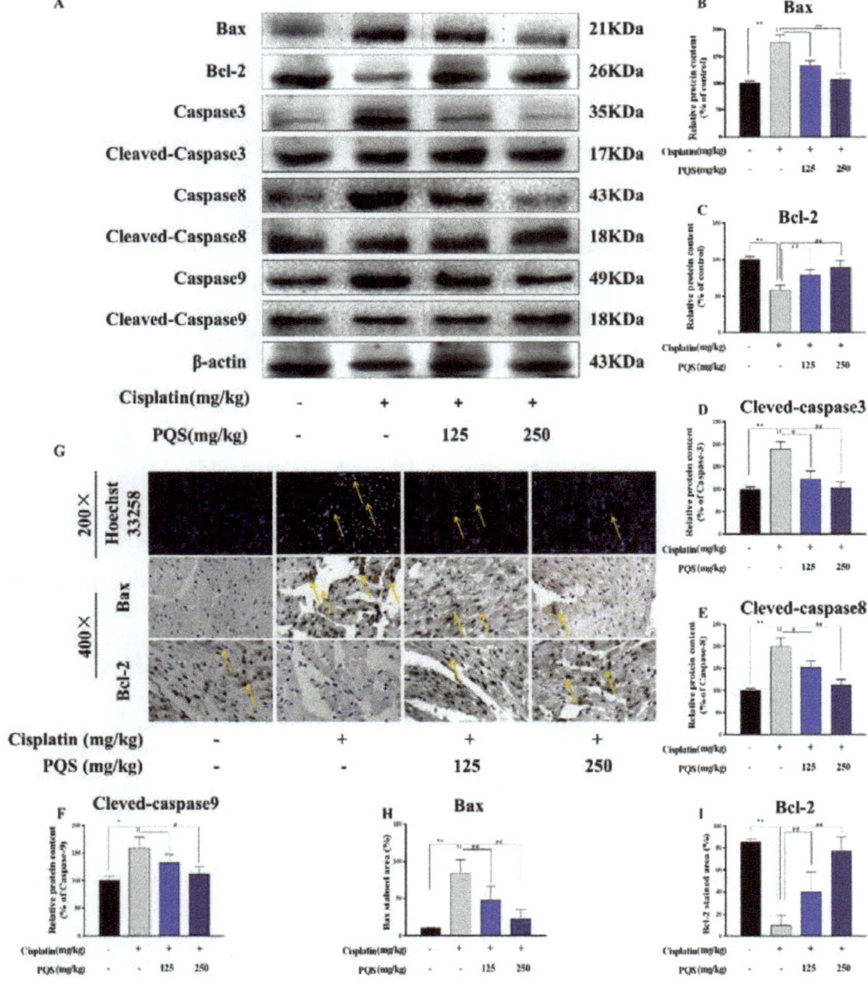

Figure 7. (**A**) The blots of Bax, Bcl-2, Bad, caspase 3, cleaved caspase 3, caspase 8, cleaved caspase 8 and caspase 9, cleaved caspase 9 were standardized to that of β-actin; Quantitative analysis of scanning densitometry for Bax (**B**), Bcl-2 (**C**), Caspase-3 (**D**), Caspase-8 (**E**), Caspase-9 (**F**). (**G**) Representative photomicrographs of cardiac immunohistochemically staining for Hoechst 33258, (**H**) Bax staining area. The percentage of apoptosis (**I**) and Bcl-2 staining area in indicated groups. Scale bars data were expressed as mean ± S.D. * $p < 0.05$ or ** $p < 0.01$ comparing with normal group. # $p < 0.05$ or ## $p < 0.01$ comparing with cisplatin group.

3.8. PQS Attenuates Cisplatin-Induced through PI3K/Akt/GSK-3β Signal Pathway

Mice pretreated with PQS or saline for 7 days, given with or without cisplatin at day 7th, 9th, 11th and 13th. (Figure 8A) Representative western blots depicting total and phosphorylated PI3K, Akt, GSK-3β. Quantitative analyses of the p-PI3K/PI3K, p-Akt/Akt, and p-GSK-3β/GSK-3β expression ratios are shown (Figure 8B–D).

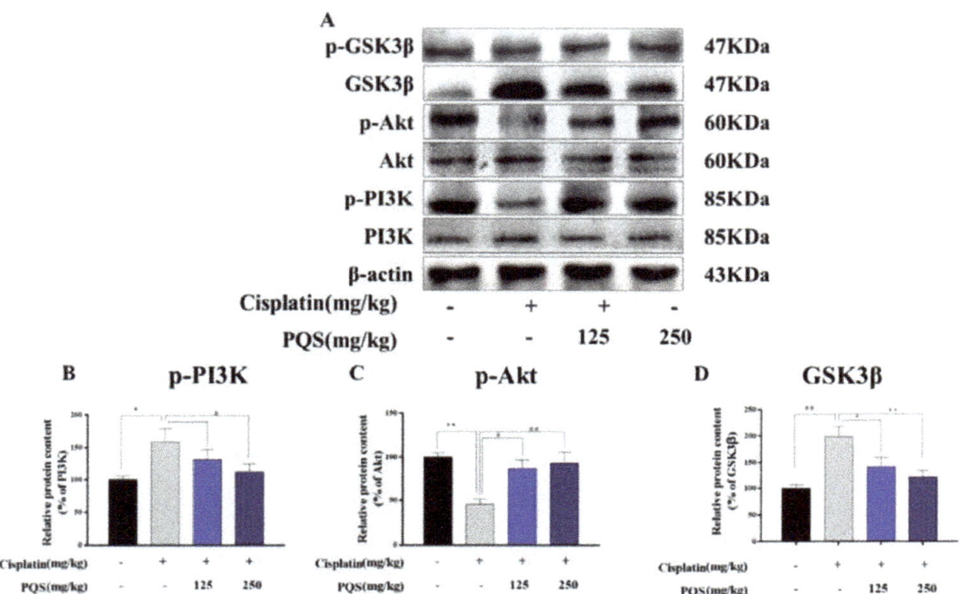

Figure 8. Mice pretreated with PQS or saline for 7days, given with or without cisplatin at day 7th, 9th, 11th and 13th. (**A**) Representative western blots depicting total and phosphorylated PI3K, Akt, GSK-3β. Quantitative analyses of the p-PI3K/PI3K, p-Akt/Akt, and p-GSK-3β/GSK-3β expression ratios are shown (**B–D**). All data were expressed as mean ± S.D. * $p < 0.05$, ** $p < 0.01$ comparing with normal group. # $p < 0.05$, ## $p < 0.01$ comparing with cisplatin group.

3.9. Mechanism of PQS Improving Cardiac Toxicity Induced by Cisplatin

When cisplatin enters the body, it promotes the expression of Bax and inhibits the increase of Bcl-2, which leads to mitochondrial dysfunction and disorder of ATP synthesis, leading to cell apoptosis and necrosis, and finally cardiac toxicity. However, PQS improve cisplatin-induced cardiac toxicity through PI3K/Akt /GSK-3β pathway and caspase family protein expressions as described in Figure 9.

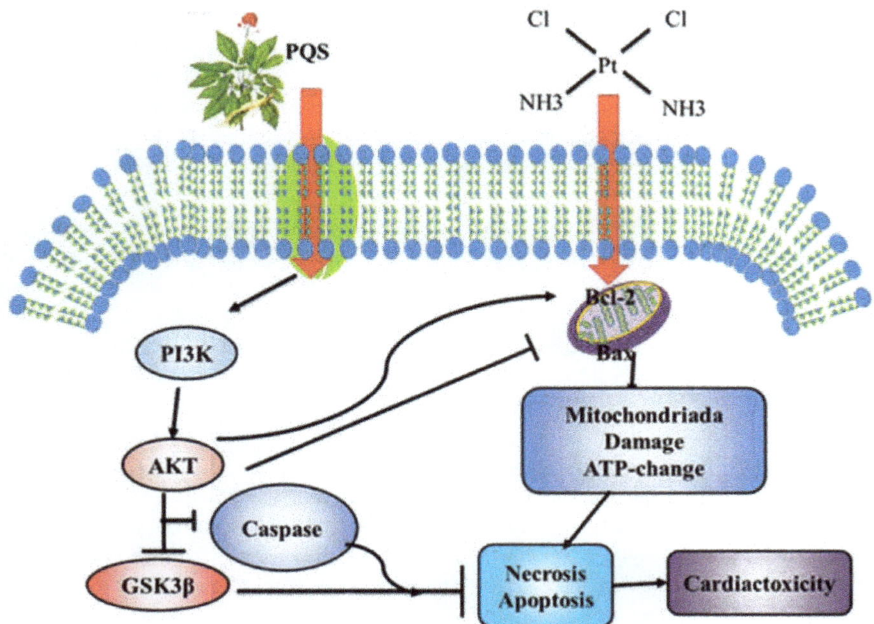

Figure 9. Scheme Summarizing the Inhibition of Cisplatin-Induced Cardiotoxicity by PQS via the Upregulation of PI3K/Akt/GSK-3β Mediated Inhibition of Oxidative Stress Inflammation and Apoptosis.

3.10. PQS Attenuates Cisplatin-Induced through PI3K/Akt/GSK-3β Signal Pathway

Mice pretreated with PQS or saline for 7 days, given with or without cisplatin at day 7th, 9th, 11th and 13th. Mice were dissected at 15th day and serum and tissue samples were collected for future detection as described in Figure 10.

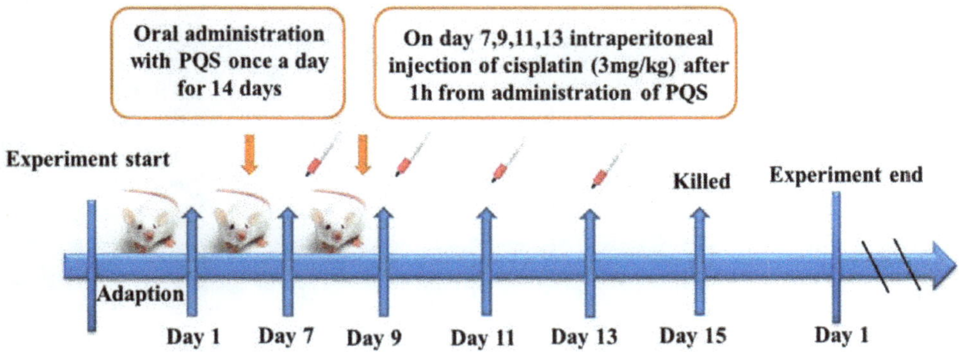

Figure 10. Experimental Flow Chart.

4. Discussion

Cisplatin is acknowledged as one of the widely employed chemotherapy agents to treat solid tumors in clinic. Nevertheless, its toxicity can also limit its usages, moreover higher doses of cisplatin also induced serious cardiotoxicity. However, its major and concrete mechanism of cardiotoxicity resulted by cisplatin is poorly understood. Some evidence has previously confirmed that oxidative stress, inflammation and apoptosis plays an important part in cardiotoxicity evoked by cisplatin [25].

The total saponins (ginsenosides, PQS) from leaves are accessible to extract and isolate [26]. Previous studies have confirmed that it had beneficially effective effects on treating coronary heart disease. Moreover, researches have confirmed that it has been extensively used in acute myocardial infarction clinically. From the results of our current works, we have investigated its effective effects on cardiac protection. Moreover, our current researches have also provided its medicinal value and exploited more market for PQS to reduce patients' side effects induced by cisplatin on cardiac injury in clinical research.

Cisplatin can result in over-generation of ROS, which will alter the cells' structures, functions and integrity. CTnT, which is acknowledged as a specific marker of cardiac dysfunction, because it can be released into the body after chemotherapeutic treatments. Moreover, myocardial damages can be resulted by overproduction of cTnT [27]. Chemotherapy diminishes the normal homeostasis of the body, which is particularly applicable for cisplatin treatment. Previous studies have reported that cisplatin exposure usually results in cardiotoxicity, which could be a secondary event following increased lipid peroxidation of cardiac membranes that results in irreversible modification of membrane structures and functions with the consequent leakage of cardiac enzymes as well as cTnT [20]. In line with previous study, our results showed that the cTnT concentration and some cardiotoxicity-associated markers, for example, LDH, CK and CK-MB were increased after cisplatin challenge when compared to the normal group. Furthermore, we also found that protective potentials of PQS in myocardial tissue were shown evidently in histopathological examinations. Mice administered with cisplatin alone showed histopathological changes in myocardial tissues. However, PQS plus cisplatin can alleviate the above-mentioned changes. We also found that normal myocardial morphology structure can be observed in mice treated with PQS (250 mg/kg) alone.

The decreasing levels in GSH levels and decreased viabilities of SOD enzyme can be also marks of oxidative stress caused by cisplatin. All radicals can result massive damages in various tissues, Oxidative stress and inflammation are closely interrelated in the biological systems and this interaction facilitates the progression of cardiotoxicity induced by cisplatin [21]. ROS has not only direct deleterious effects on bodies, but induces the activation of the redox sensitive NF-κB when the concentration of ROS is high, then causing the high levels of the proinflammatory cytokines, such as TNF-α and IL-1β which play a vital role in some inflammation-associated signaling molecules [22]. Therefore, it is also reasonable to conclude that cardiac toxicities resulted by challenging of cisplatin maybe partly due to the activation of inflammation response. Some studies have also confirmed that inflammation can also promote the lipolysis [23], which can be a persuasive explanation on weight loss caused by cisplatin. Moreover, as one of the leading genes for NF-κB, COX-2 is placed at the center of a myriad of mechanisms on tissue injuries during the produce of vasoactive and pro-inflammatory responses [24]. Augmented oxidative stress can be resulted by TNF-α and IL-1β through activating COX-2 in inflammatory reactions [25]. Admittedly, the activation of neutrophils on inflammation responses can induce physical damages via the production of many pro-inflammatory and pro-oxidative enzymes, such as MPO, GSH, SOD and MDA [26]. Our results were consistent with the reported findings previously, significant increase of TNF-α and IL-1β accompanied by elevation of COX-2 protein and MPO can be observed in the cisplatin-injected groups. When tissue cells are damaged by stimulations; ALT in the cells will be released, resulting in increased ALT value [27]. We also found that cisplatin could increase the level of ALT. Interestingly, PQS can remarkably reduce these alterations indicating its potentials on anti-inflammatory. All results are in accordance with other reports on anti-inflammatory effect of PQS, which may be attributed to the scavenging of free radicals [28,29].

During the pathogenesis cardiotoxicity, inflammation, which has been extensively acknowledged as a vital contributor [30]. The transcription of inflammatory indicators, for example, TNF-α, IL-1β, COX-2 and iNOS can be also specifically triggered via activating the NF-κB [31] pathway. Moreover, IL-1β is also play a vital role in the process of inflammation [32]. Our current researches showed that, PQS can diminish inflammation responses resulted by TNF-α and IL-1β. These results were also confirmed by reducing levels of iNOS and COX-2 levels, suggesting suppression of inducible enzymatic pathways. In conclusion, these findings provide a potential for PQS on improving cardiotoxicity associated inflammatory response. In this study, PQS suppressed NF-κB activation in cisplatin induced cardiotoxicities, evidenced by decreased expressions of p-IKKα, p-IKKβ, p-IκBα, and p-NF-κB.

Recent evidence suggested cisplatin-induced cardiotoxicity is related to apoptosis. Various substrates apoptosis-associated proteins like Bax, Bad and caspase family members can trigger the apoptotic responses [33,34]. Other proteins like Bcl-2, Bcl-XL can act as antiapoptotic markers, which may cause inhibitions in some apoptotic reactions. Previous investigations have demonstrated that PI3K/Akt is considered as a prosurvival function in cardiac tissues challenged by oxidations and apoptosis induced by various stimulations [35,36]. Once Akt signal pathway is activated, it can confer cell exist by activating its cytoplasmic targets, like its downstream proteins - GSK-3*β* as well as other apoptotic indicators like Bcl-2, Bax, and caspase-9 [37]. In this study, we showed that phosphorylated PI3K/Akt levels on cisplatin-treated mice were lower than those mice untreated with cisplatin. From our results, we demonstrated GSK3β can be significantly promoted in the cisplatin-treated mice. These results can be served to decrease cardiomyocyte apoptosis, and preserve heart functions. Moreover, our western blotting analysis indicated that elevation of PI3K/Akt signaling pathway as well as the increased expression of GSK3β can serve a cardioprotective effect in this mouse model. Our data are in line with previous studies. Furthermore, we also showed that levels of caspase family members' proteins in the myocardium of PQS-pretreated mice were less than those of the untreated mouse. Importantly, we also found that the expression level of Bcl-2 can be reversed by pretreatment with PQS.

5. Conclusions

In conclusion, our researches revealed the protective potentials of PQS against cisplatin-evoked cardiotoxicity, and its mechanism may be partly attributed to the inhibition of oxidative stress, inflammation and apoptosis via PI3K/Akt/GSK-3β signaling pathway.

Author Contributions: Z.W. and X.-D.L. conceived and designed the experiments; R.-B.Z., J.-J.X. and Q.S. performed the experiments; W.L. Q.S. and Y.L. analyzed the data; S.J., Y.-P.W. and S.R. contributed reagents/materials/analysis tools; J.-J.X. and J.G.-H wrote the paper.

Funding: This work was supported by the grants of Scientific Research Foundation for the Returned Overseas Chinese Scholars (Jilin Province, 2016), Jilin Science & Technology Development Plan (No. 20180201083YY, 20190103092JH, and 20190304003YY), 13th Five-year Plan Science & Technology Project from the Education Department of Jilin Province (JJKH20190946kj), and the Program for the Young Top-notch and Innovative Talents of Jilin Agricultural University (2016).

Conflicts of Interest: The authors declare no conflict of interest

Abbreviations

The following abbreviations are used in this manuscript:

CK	Creatine kinase
MPO	Myeloperoxidase
CK-MB	Creatine kinase isoenzyme MB
CTnT	Plasma cardiac troponin T Glutathione
PQS	Panax quinquefolius saponins
IL-1β	Interleukin-1β
SOD	Superoxide dismutase
LDH	Lactate dehydrogenase
TNF-α	Tumor necrosis factor-α
H&E	Hematoxylin and Eosin
NF-κB	Nuclear factor-kappa B
Akt	Protein kinase B
PI3K	Phosphatidylinositol 3-kinase

References

1. Dasari, S.; Tchounwou, P.B. Cisplatin in cancer therapy: Molecular mechanisms of action. *Eur. J. Pharmacol.* **2014**, *740*, 364–378. [CrossRef] [PubMed]
2. Karasawa, T.; Steyger, P.S. An integrated view of cisplatin-induced nephrotoxicity and ototoxicity. *Toxicol. Lett.* **2015**, *237*, 219–227. [CrossRef] [PubMed]
3. Topal, I.; Özbek Bilgin, A.; Çimen, F.K.; Kurt, N.; Süleyman, Z.; Bilgin, Y.; Özçiçek, A.; Altuner, D. The effect of rutin on cisplatin-induced oxidative cardiac damage in rats. *Anatol. J. Cardiol.* **2018**, *20*, 136–142. [CrossRef] [PubMed]
4. Choi, Y.-M.; Kim, H.-K.; Shim, W.; Anwar, M.A.; Kwon, J.-W.; Kwon, H.-K.; Kim, H.J.; Jeong, H.; Kim, H.M.; Hwang, D.; et al. Mechanism of Cisplatin-Induced Cytotoxicity Is Correlated to Impaired Metabolism Due to Mitochondrial ROS Generation. *PLoS ONE* **2015**, *10*, e0135083. [CrossRef] [PubMed]
5. Varga, Z.V.; Ferdinandy, P.; Liaudet, L.; Pacher, P. Drug-induced mitochondrial dysfunction and cardiotoxicity. *Am. J. Physiol. Heart Circ. Physiol.* **2015**, *309*, H1453–H1467. [CrossRef] [PubMed]
6. Dugbartey, G.J.; Peppone, L.J.; De Graaf, I.A. An integrative view of cisplatin-induced renal and cardiac toxicities: Molecular mechanisms, current treatment challenges and potential protective measures. *Toxicology* **2016**, *371*, 58–66. [CrossRef] [PubMed]
7. Zhao, C.; Chen, Z.; Qi, J.; Duan, S.; Huang, Z.; Zhang, C.; Wu, L.; Zeng, M.; Zhang, B.; Wang, N.; et al. Drp1-dependent mitophagy protects against cisplatin-induced apoptosis of renal tubular epithelial cells by improving mitochondrial function. *Oncotarget* **2017**, *8*, 20988–21000. [CrossRef]
8. Rubera, I.; Duranton, C.; Melis, N.; Cougnon, M.; Mograbi, B.; Tauc, M. Role of CFTR in oxidative stress and suicidal death of renal cells during cisplatin-induced nephrotoxicity. *Cell Death Dis.* **2013**, *4*, e817. [CrossRef]
9. Alhoshani, A.R.; Hafez, M.M.; Husain, S.; Al-Sheikh, A.M.; Alotaibi, M.R.; Al Rejaie, S.S.; Alshammari, M.A.; Almutairi, M.M.; Al-Shabanah, O.A. Protective effect of rutin supplementation against cisplatin-induced Nephrotoxicity in rats. *BMC Nephrol.* **2017**, *18*, 194. [CrossRef]
10. Mukhopadhyay, P.; Horváth, B.; Kechrid, M.; Tanchian, G.; Rajesh, M.; Naura, A.S.; Boulares, A.H.; Pacher, P. Poly(ADP-ribose) polymerase-1 is a key mediator of cisplatin-induced kidney inflammation and injury. *Free Radic. Biol. Med.* **2011**, *51*, 1774–1788. [CrossRef]
11. Zhu, X.; Jiang, X.; Li, A.; Zhao, Z.; Li, S. S-Allylmercaptocysteine Attenuates Cisplatin-Induced Nephrotoxicity through Suppression of Apoptosis, Oxidative Stress, and Inflammation. *Nutrients* **2017**, *9*, 166. [CrossRef] [PubMed]
12. Marullo, R.; Werner, E.; Degtyareva, N.; Moore, B.; Altavilla, G.; Ramalingam, S.S.; Doetsch, P.W. Cisplatin Induces a Mitochondrial-ROS Response That Contributes to Cytotoxicity Depending on Mitochondrial Redox Status and Bioenergetic Functions. *PLoS ONE* **2013**, *8*, e81162. [CrossRef] [PubMed]
13. Jing, D.; Bai, H.; Yin, S. Renoprotective effects of emodin against diabetic nephropathy in rat models are mediated via PI3K/Akt/GSK-3β and Bax/caspase-3 signaling pathways. *Exp. Ther. Med.* **2017**, *14*, 5163–5169. [CrossRef] [PubMed]
14. Wang, Y.; Hao, Y.; Alway, S.E. Suppression of GSK-3β activation by M-cadherin protects myoblasts against mitochondria-associated apoptosis during myogenic differentiation. *J. Cell Sci.* **2011**, *124*, 3835–3847. [CrossRef]
15. Xu, L.; Jiang, X.; Wei, F.; Zhu, H. Leonurine protects cardiac function following acute myocardial infarction through anti-apoptosis by the PI3K/AKT/GSK3β signaling pathway. *Mol. Med. Rep.* **2018**, *18*, 1582–1590. [CrossRef]
16. He, H.; Qiao, X.; Wu, S. Carbamylated erythropoietin attenuates cardiomyopathy via PI3K/Akt activation in rats with diabetic cardiomyopathy. *Exp. Ther. Med.* **2013**, *6*, 567–573. [CrossRef] [PubMed]
17. Hang, P.; Zhao, J.; Sun, L.; Li, M.; Han, Y.; Du, Z.; Li, Y. Brain-derived neurotrophic factor attenuates doxorubicin-induced cardiac dysfunction through activating Akt signalling in rats. *J. Cell. Mol. Med.* **2017**, *21*, 685–696. [CrossRef]
18. Mobasher, M.A.; González-Rodriguez, A.; Santamaría, B.; Ramos, S.; Martín, M.Á.; Goya, L.; Rada, P.; Letzig, L.; James, L.P.; Cuadrado, A.; et al. Protein tyrosine phosphatase 1B modulates GSK3β/Nrf2 and IGFIR signaling pathways in acetaminophen-induced hepatotoxicity. *Cell Death Dis.* **2013**, *4*, e626. [CrossRef]

19. Ma, Z.-N.; Li, Y.-Z.; Li, W.; Yan, X.-T.; Yang, G.; Zhang, J.; Zhao, L.-C.; Yang, L.-M. Nephroprotective Effects of Saponins from Leaves of Panax quinquefolius against Cisplatin-Induced Acute Kidney Injury. *Int. J. Mol. Sci.* **2017**, *18*, 1407. [CrossRef]
20. Li, R.Y.; Zhang, W.Z.; Yan, X.T.; Hou, J.G.; Wang, Z.; Ding, C.B.; Liu, W.C.; Zheng, Y.N.; Chen, C.; Li, Y.R.; et al. Arginyl-fructosyl-glucose, a Major Maillard Reaction Product of Red Ginseng, Attenuates Cisplatin-Induced Acute Kidney Injury by Regulating Nuclear Factor kappaB and Phosphatidylinositol 3-Kinase/Protein Kinase B Signaling Pathways. *J Agric Food Chem* **2019**, *67*, 5754–5763. [CrossRef]
21. Mi, X.J.; Hou, J.G.; Jiang, S.; Liu, Z.; Tang, S.; Liu, X.X.; Wang, Y.P.; Chen, C.; Wang, Z.; Li, W. Maltol Mitigates Thioacetamide-induced Liver Fibrosis through TGF-beta1-mediated Activation of PI3K/Akt Signaling Pathway. *J. Agric. Food Chem.* **2019**, *67*, 1392–1401. [CrossRef] [PubMed]
22. Xing, J.J.; Hou, J.G.; Ma, Z.N.; Wang, Z.; Ren, S.; Wang, Y.P.; Liu, W.C.; Chen, C.; Li, W. Ginsenoside Rb3 provides protective effects against cisplatin-induced nephrotoxicity via regulation of AMPK-/mTOR-mediated autophagy and inhibition of apoptosis in vitro and in vivo. *Cell Prolif.* **2019**, *52*, e12627. [CrossRef] [PubMed]
23. Bahadır, A.; Ceyhan, A.; Öz Gergin, Ö.; Yalçın, B.; Ülger, M.; Özyazgan, T.M.; Yay, A.; Bahadir, A. Protective effects of curcumin and beta-carotene on cisplatin-induced cardiotoxicity: An experimental rat model. *Anatol. J. Cardiol.* **2018**, *19*, 213–221. [CrossRef] [PubMed]
24. Li, S.; Hong, M.; Tan, H.-Y.; Wang, N.; Feng, Y. Insights into the Role and Interdependence of Oxidative Stress and Inflammation in Liver Diseases. *Oxidative Med. Cell. Longev.* **2016**, *2016*, 4234061. [CrossRef] [PubMed]
25. Xu, X.; Grijalva, A.; Skowronski, A.; Van Eijk, M.; Serlie, M.J.; Ferrante, A.W. Obesity Activates a Program of Lysosomal-Dependent Lipid Metabolism in Adipose Tissue Macrophages Independently of Classic Activation. *Cell Metab.* **2013**, *18*, 816–830. [CrossRef] [PubMed]
26. Xu, X.Y.; Hu, J.N.; Liu, Z.; Zhang, R.; He, Y.F.; Hou, W.; Wang, Z.Q.; Yang, G.; Li, W. Saponins (Ginsenosides) from the Leaves of Panax quinquefolius Ameliorated Acetaminophen-Induced Hepatotoxicity in Mice. *J. Agric. Food Chem.* **2017**, *65*, 3684–3692. [CrossRef]
27. Perez, C.E.R.; Nie, W.; Sinnett-Smith, J.; Rozengurt, E.; Yoo, J. TNF-α potentiates lysophosphatidic acid-induced COX-2 expression via PKD in human colonic myofibroblasts. *Am. J. Physiol. Liver Physiol.* **2011**, *300*, G637–G646.
28. Decean, H.; Fischer-Fodor, E.; Tatomir, C.; Perde-Schrepler, M.; Somfelean, L.; Burz, C.; Hodor, T.; Orasan, R.; Virag, P. Vitis vinifera seeds extract for the modulation of cytosolic factors BAX-α and NF-kB involved in UVB-induced oxidative stress and apoptosis of human skin cells. *Clujul Med.* **2016**, *89*, 72–81. [CrossRef]
29. Ben Saad, A.; Rjeibi, I.; Ncib, S.; Zouari, N.; Zourgui, L. Ameliorative Effect of Cactus (Opuntia ficus indica) Extract on Lithium-Induced Nephrocardiotoxicity: A Biochemical and Histopathological Study. *BioMed Res. Int.* **2017**, *2017*, 8215392.
30. Kou, N.; Xue, M.; Yang, L.; Zang, M.-X.; Qu, H.; Wang, M.-M.; Miao, Y.; Yang, B.; Shi, D.-Z. Panax quinquefolius saponins combined with dual antiplatelet drug therapy alleviate gastric mucosal injury and thrombogenesis through the COX/PG pathway in a rat model of acute myocardial infarction. *PLoS ONE* **2018**, *13*, e0194082. [CrossRef]
31. Hu, J.-N.; Xu, X.-Y.; Li, W.; Wang, Y.-M.; Liu, Y.; Wang, Z.; Wang, Y.-P. Ginsenoside Rk1 ameliorates paracetamol-induced hepatotoxicity in mice through inhibition of inflammation, oxidative stress, nitrative stress and apoptosis. *J. Ginseng Res.* **2019**, *43*, 10–19. [CrossRef]
32. Sun, S.-C. The non-canonical NF-κB pathway in immunity and inflammation. *Nat. Rev. Immunol.* **2017**, *17*, 545–558. [CrossRef]
33. Ishida, Y.; Goto, Y.; Kondo, T.; Kurata, M.; Nishio, K.; Kawai, S.; Osafune, T.; Naito, M.; Hamajima, N. Eradication rate of Helicobacter pylori according to genotypes of CYP2C19, IL-1B, and TNF-A. *Int. J. Med. Sci.* **2006**, *3*, 135–140. [CrossRef]
34. Almutairi, M.M.; Alanazi, W.A.; Alshammari, M.A.; Alotaibi, M.R.; Alhoshani, A.R.; Al-Rejaie, S.S.; Hafez, M.M.; Al-Shabanah, O.A. Neuro-protective effect of rutin against Cisplatin-induced neurotoxic rat model. *BMC Complement. Altern. Med.* **2017**, *17*, 472. [CrossRef]
35. Mi, X.-J.; Hou, J.-G.; Wang, Z.; Han, Y.; Ren, S.; Hu, J.-N.; Chen, C.; Li, W. The protective effects of maltol on cisplatin-induced nephrotoxicity through the AMPK-mediated PI3K/Akt and p53 signaling pathways. *Sci. Rep.* **2018**, *8*, 15922. [CrossRef]

36. Thakur, B.; Ray, P. Cisplatin triggers cancer stem cell enrichment in platinum-resistant cells through NF-κB-TNFα-PIK3CA loop. *J. Exp. Clin. Cancer Res.* **2017**, *36*, 164. [CrossRef]
37. Shu, X.-R.; Wu, J.; Sun, H.; Chi, L.-Q.; Wang, J.-H. PAK4 confers the malignance of cervical cancers and contributes to the cisplatin-resistance in cervical cancer cells via PI3K/AKT pathway. *Diagn. Pathol.* **2015**, *10*, 177. [CrossRef]

Article

Qualitative Chemical Characterization and Multidirectional Biological Investigation of Leaves and Bark Extracts of *Anogeissus leiocarpus* (DC.) Guill. & Perr. (Combretaceae)

Giustino Orlando [1], Claudio Ferrante [1,*,†], Gokhan Zengin [2,*,†], Kouadio Ibrahime Sinan [2], Kouadio Bene [3], Alina Diuzheva [4], József Jekő [5], Zoltán Cziáky [5], Simonetta Di Simone [1], Lucia Recinella [1], Annalisa Chiavaroli [1], Sheila Leone [1], Luigi Brunetti [1], Carene Marie Nancy Picot-Allain [6], Mohamad Fawzi Mahomoodally [6,*] and Luigi Menghini [1]

1 Department of Pharmacy, University "G. d'Annunzio" of Chieti-Pescara, 66100 Chieti, Italy
2 Department of Biology, Faculty of Science, Selcuk University, Konya 42130, Turkey
3 Laboratoire de Botanique et Phytothérapie, Unité de Formation et de Recherche Sciences de la Nature, Université Nangui Abrogoua, 02 BP 801 Abidjan 02, Ivory Coast
4 Department of Forest Protection and Entomology, Faculty of Forestry and Wood Sciences, Czech University of Life Sciences, 16500 Prague, Czech Republic
5 Agricultural and Molecular Research and Service Institute, University of Nyíregyháza, 4400 Nyíregyháza, Hungary
6 Department of Health Sciences, Faculty of Science, University of Mauritius, 230 Réduit, Mauritius
* Correspondence: claudio.ferrante@unich.it (C.F.); gokhanzengin@selcuk.edu.tr (G.Z.); f.mahomoodally@uom.ac.mu (M.F.M.)
† These Authors contributed equally to this work.

Received: 8 August 2019; Accepted: 22 August 2019; Published: 1 September 2019

Abstract: *Anogeissus leiocarpus* (DC.) Guill. & Perr. (Combretaceae) has a long history of use by folk populations for the management of multiple human ailments. Based on the published literature, there has been no attempt to conduct a comparative assessment of the biological activity and the phytochemical profiles of the leaves and stem bark of *A. leiocarpus* extracted using methanol, ethyl acetate, and water. By high-performance liquid chromatography with electrospray ionization mass spectrometric detection (HPLC-ESI-MS^n) analysis, quinic, shikimic, gallic, and protocatechuic acids were tentatively identified from all the extracts, while chlorogenic, caffeic, ferulic, and dodecanedioic acids were only characterised from the leaves extracts. Additionally, a pharmacological study was carried out to evaluate potential protective effects that are induced by the extracts in rat colon and colon cancer HCT116 cell line. In general, the methanol and water extracts of *A. leiocarpus* leaves and stem bark showed potent radical scavenging and reducing properties. It was noted that the stem bark extracts were more potent antioxidants as compared to the leaves extracts. The methanol extract of *A. leiocarpus* leaves showed the highest acetyl (4.68 mg galantamine equivalent/g) and butyryl (4.0 mg galantamine equivalent/g) cholinesterase inhibition. Among ethyl acetate extracts, the pharmacological investigation suggested stem bark ethyl acetate extracts to be the most promising. This extract revealed ability to protect rat colon from lipopolysaccharide-induced oxidative stress, without exerting promoting effects on HCT116 cell line viability and migration. As a conclusion, *A. leiocarpus* represents a potential source of bioactive compounds in the development of novel therapeutic agents.

Keywords: *Anogeissus*; bioactive compounds; antioxidant; enzyme inhibition; ulcerative colitis

1. Introduction

Anogeissus leiocarpus (DC.) Guill. & Perr. (Combretaceae), also known as chewing stick or axlewood tree, has a long history of traditional use for the management of multiple human ailments. The leaves of *A. leiocarpus* are used in the treatment of skin diseases, fever, diarrhoea, malaria, and stomach infections [1]. *A. leiocarpus* is used by the Yoruba people in Nigeria to treat bacterial infections and the roots and twigs of the plant are used as chewing sticks for dental hygiene. Various parts of the plant (roots, leaves, stem bark, and twigs) are used in the management of gonorrhoea, cough, wounds, acute respiratory tract infections, stomach infections, fever, tuberculosis, dysentery, giardiasis, malaria, trypanosomiasis, yellow fever, jaundice, and pathogenic microbial infections [2]. The water extract of *A. leiocarpus* stem bark was recently found to combat erectile dysfunction in paroxetine-induced sexually impaired male Wistar rats [3]. A spontaneous decrease in serum glucose level in alloxan-induced diabetic rats administered with the aqueous extract of *A. leiocarpus* leaves [4] was linked to the α-amylase and α-glucosidase inhibitory action of the extract [5]. The aqueous extract of *A. leiocarpus* trunk bark was reported to exert significant antihypertensive effects in NG-nitro-L-arginine methyl ester (L-NAME)-induced hypertensive rats [6]. The methanolic and ethyl acetate extracts of *A. leiocarpus* leaves exhibited antioxidant and antibacterial properties [7]. The stem bark methanolic extract of *A. leiocarpus* demonstrated antitrypanosomal activity against four *Trypanosoma* strains [8] and leishmanicidal activity [9]. The methylene chloride extract of *A. leiocarpus* (IC_{50} value of 3.8 μg/mL) showed in vitro antiplasmodial activity against *Plasmodium falciparum*, the protozoan parasite that is responsible for malaria in human [10]. Lately, a group of researchers investigated the effect of *A. leiocarpus* methanolic extract on the liver function in mice that were infected with *Plasmodium berghei* [11].

From the literature, several studies attempted to investigate the biological activity, mainly, the antibacterial properties, of different extracts of *A. leiocarpus*. However, as far as our literature search could ascertain, no study was focused on the comparative evaluation of the phytochemical profiles of the methanol, ethyl acetate, and water extracts of the leaves and stem bark of *A. leiocarpus*. Additionally, in the present study, the authors present the antioxidant and inhibitory action of *A. leiocarpus* extracts on key enzymes that are related to diabetes type II, Alzheimer's disease, and skin hyperpigmentation. Finally, while considering the potential antiproliferative effects that are exerted by *A. latofolia* on colon cancer cells [12], the antiproliferative effects of *A. leiocarpus* extracts were tested on human colon cancer HCT116 cell line. Additionally, the same extracts were tested for their putative antioxidant/anti-inflammatory effects on isolated rat colon specimens that were exposed to *E. coli* lipopolysaccharide (LPS), in order to reproduce the burden of oxidative stress and inflammation occurring in ulcerative colitis [13]. To this regard, selected biomarkers of oxidative stress/inflammation, including prostaglandin (PG)E_2, 8-iso-$PGF_{2\alpha}$, and serotonin (5-HT) were selected. It is expected that detailed phytochemical profiles of the different extracts will enable tentative identification of phytochemical/s, which might be responsible for the observed biological activity.

2. Materials and Methods

2.1. Plant Material and Preparation of Extracts

The sampling of the plant species was done in Gontougo region (Sandegue) of Ivory Coast in the year 2018. Botanical authentication of the plant was done by the botanist Dr. Kouadio Bene (Laboratoire de Botanique et Phytothérapie, Université Nangui Abrogoua, Abidjan, Ivory Coast). The leaves and stem barks were dried at room temperature (in shade, about 10 days). These materials were then powdered by using a laboratory mill.

Methanol and ethyl acetate extracts were prepared through maceration techniques (five grams of plant samples were mixed with one hundred ml of each solvents for 24 h). After maceration, the extracts were subjected to filtration and evaporation in vacuo at 40 °C. Traditional infusion was selected to prepare the water extract (five grams of plant samples were infused with one hundred mL of boiling water for 20 min.). After preparation, the water extract was subjected to filtration and freeze drying. Finally, the extracts were stored at 4 °C until phytochemical and pharmacological analysis.

2.2. Profile of Bioactive Compounds

Total phenols, flavonoids, phenolic acids, and flavonols were assayed through spectrophotometric assays [14,15], The extract concentrations of phenolics, flavonoids, phenolic acids, flavonols and tannins, and saponins were determined through spectrophotometric assays [14,15], and were expressed as equivalents of gallic acid (mg GAE/g dry extract), rutin (mg RE/g dry extract), caffeic acid (mg CAE/g dry extract), catechin (mg CE/g dry extract), and quillaja (mg QE/g dry extract), respectively.

The qualitative analysis of *A. santonicum* extracts (5 mg/mL) was carried out according the protocol that was described by Zengin et al. [16].

An high performance liquid chromatography (HPLC)-fluorimetric analysis was carried out in order to quantify the selected phenolic compounds, in *A. santonicum* extracts (5 μg/mL). To this regard, an HPLC apparatus (MOD. 1525, Waters Corporation, Milford, MA, USA) coupled to fluorimetric detector (MOD. 2475, Waters Corporation, Milford, MA, USA) and a C18 reversed-phase column (Phenomenex Kinetex, Torrance, CA, USA, 150 mm × 4.6 mm i.d., 2.6 μm) were used. The HPLC gradient conditions were selected, as previously mentioned by Rodrıguez-Delgado and coworkers [17]. In agreement with the same authors, $\lambda ex = 278$ nm and $\lambda em = 360$ nm were selected in order to analyze the following phenolic compounds: gallic acid, catechin, and epicatechin.

2.3. Determination of Antioxidant and Enzyme Inhibitory Effects

The evaluation of anti-α-amylase, anti-α-glucosidase, anti-cholinesterases, and anti-tyrosinase activities was carried out as previously described by Uysal and coworkers [18]. The enzyme inhibitory results were evaluated in terms of standard equivalents; galatamine for cholinesterase (mg GALAE/g dry extract); kojic acid for tyrosinase (mg KAE/g dry extract), acarbose for amylase, and glucosidase (mmol ACAE/g dry extract). According to the same paper [18], the antiradical activity of the extracts was measured through the use of ferric reducing antioxidant power (FRAP), 2,2′-azino-bis(3-ethylbenzothiazoline-6-sulphonic acid) (ABTS) cupric reducing antioxidant capacity (CUPRAC), 2,2-diphenyl-1-picrylhydrazyl (DPPH), phosphomolydenum, and metal chelating tests. The antioxidant results were explained as equivalents of trolox (mg TE/g dry extract) and ethylenediaminetetraacetic acid (EDTA) (in metal chelating assay) (mg EDTAE/g dry extract). One-way ANOVA, followed by Tukey's post hoc test, were applied for comparing the samples in terms of bioactive compounds content and biological activities. The MCA (multiple correspondence analysis) and Clustering Image Map were performed for the discrimination between the samples based on their chemical compositions and Venn graph was built to identify the chemical profile differences among those samples. Before, the data of the chemical composition were attributed to classes with two modalities (e.g., + for presence and – for the absence of compounds in extracts). Afterwards, Multiple datasets supervised analysis, namely DIABLO, was achieved to find out the key factor (parts and solvents) that is responsible for variation in datasets. Subsequently, the correlation between the bioactive compounds and biological activities were estimated. All of the statistical tests were conducted by using R 3.5.1 software environment.

2.4. Pharmacological Assays

2.4.1. Allelopathy Assay

As previously described [19], the allelopathy bioassay was carried out in 90 mm diameter Petri dishes, which represented the substrate for the germination of seeds, whereas *A. leiocarpus* extracts (0.1–10 mg/mL) were dissolved in imbibition water. During the incubation period (three days at 4 °C), seeds were monitored in order to evaluate their uniform size and integrity. To this regard, lettuce could be considered as one of the most suitable dicotyledon for allelopathy assay. This is due to both fast germination rate and high sensitivity. After the third day of treatment, a root length ≥ 1 mm was the condition to consider positive the germination of seeds [19]. The experiments were carried out in triplicate and means ± SEM were determined through the use of GraphPad Prism software (version 5.01).

2.4.2. Brine Shrimp Lethality Assay

Artemia salina lethality bioassay was performed, as previously reported [13]. The larvae of brine shrimps were exposed to the extracts (0.01–10 mg/mL) at 25–28 °C for 24 h. At the end of the incubation period, brine shrimp lethality was evaluated with the equation $((T - S)/T) \times 100$, being T and S the total number of larvae that were exposed to extracts and living nauplii, respectively. The experiments were carried out in triplicate.

2.5. *In Vitro Studies*

HCT116 cell line (ATCC® CCL-247™) culture and differentiation were carried out as previously described in our published paper [13]. To evaluate the biocompatibility of *A. leiocarpus* extracts (0.1 mg/mL), the 3-(4,5-dimethylthiazol-2-yl)-2,5-diphenyltetrazolium bromide (MTT) viability test was carried out, as recently described [13]. The effects of extracts (0.1 mg/mL) on HCT116 cell viability was evaluated in comparison to the untreated control group 24 h after treatment. Finally, the effects of extracts on HCT116 cell spontaneous migration, through the use of wound healing test, as recently reported. Briefly, the cells were challenged with *A. leiocarpus* extracts (0.1 mg/mL), and spontaneous migration was monitored at different time points (0 and 24 h). The Image-J software (NIH) was used to quantify the scratch area, whereas GraphPad software was employed to calculate mean data at 0 and 24 h and express them as percentage variation with reference to relative 100% of that at 0 h.

2.6. *Ex Vivo Studies*

Male adult Sprague-Dawley rats (200–250 g) were sacrificed by CO_2 inhalation, and colon specimens were immediately stimulated with *Escherichia coli* lipopolysaccharide (LPS) 10 μg/mL for 4 h (incubation period), as recently described [13]. Italian Health Minstry (authorization N. F4738.N.XTQ, delivered on 11th Novembre 2018) approved the experimental procedures.

During the incubation period, the colon specimens were also treated with *A. leiocarpus* extracts (0.1 mg/mL). Subsequently, extraction and chromatographic quantification of 5-HT (ng/mg wet tissue) was carried out in colon homogenate, as previously reported [20,21]. Additionally, colon homogenate was assayed for measuring PGE_2 and 8-iso-$PGF_{2\alpha}$ via radioimmunoassay [22,23].

2.7. *Statistical Analysis*

Data were means ± SEM and analyzed by one-way analysis of variance (ANOVA), followed by Newman-Keuls post hoc test (GraphPad Prism version 5.01 for Windows, GraphPad Software, San Diego, CA, USA). The data were considered to be significant for p values less than 0.05. With the aim to apply 3Rs (Reduction, Refinement and Reduction) approach to the ex vivo procedures, the number of animals was determined through the "Resource Equation" $N = (E + T)/T$ $(10 \leq E \leq 20)$ [24]), where N, T, and E represent the number of animals, pharmacological treatments, and degree freedom in ANOVA, respectively.

3. Results

3.1. Phytochemical Profile

Table 1 presents phytochemical evaluations of the different extracts of *A. leiocarpus* leaves and stem bark. Quantitative determination showed that the stem barks extracts of *A. leiocarpus* (water extract = methanol extract > ethyl acetate extract) possessed significant amounts of phenolics when compared to their respective leaves extracts. The water extract of *A. leiocarpus* leaves showed the highest flavonoid (89.0 mg RE/g) and phenolic acid (14 mg CAE/g) contents. Highest tannin (77.0 mg CE/g), flavanol (79 mg CE/g), and saponin (438 mg QE/g) contents were recorded from the methanol extract of *A. leiocarpus* leaves. Phenolic acids, such as, protocatechuic acid, chlorogenic acid, caffeic acid, and ferulic acid, were previously reported to be soluble in polar protic solvents, like methanol, except gallic acid, which was readily soluble in water [25]. From Table 1, it is noted that no phenolic acid was recorded in the ethyl acetate extracts of *A. leiocarpus* leaves and stem bark. Caffeic acid, a phenolic acid, was found to be minimally soluble in ethyl acetate [26].

HPLC-fluorimeter analysis that was performed on selected phenolic compounds (i.e., gallic acid, catechin and epicatechin), revealed that leaf extract could have, overall, a higher content of these metabolites than stem bark extract (Table 2). This is consistent with the observed content of the total phenolic acids and flavanols.

Detailed analysis of the chemical composition of the ethyl acetate, methanol, and water extracts of *A. leiocarpus* leaves and stem bark by HPLC-ESI-MS^n, while using both positive and negative ionisation modes, was also conducted. The detailed results are given as supplemental materials (Tables S1–S6). The results are also summarized in Table 3.

Table 1. Quantitative phytochemical determinations of *A. leiocarpus* leaves and stem bark extracts.

Samples	Total Phenolic Content (mg GAE/g)	Total Flavonoid Content (mg RE/g)	Total Flavonol (mg CE/g)	Total Phenolic acid (mg CAE/g)	Total Tannin Content (mg CE/g)	Total Saponin Content (mgQE/g)
Leaves-EA	49 ± 1^{e}	35.0 ± 0.6^{c}	6.0 ± 0.1^{d}	nd	6.0 ± 0.4^{e}	190 ± 17^{bc}
Leaves-MeOH	223 ± 2^{c}	54.0 ± 0.6^{b}	79 ± 3^{a}	8 ± 1^{b}	77.0 ± 0.7^{a}	438 ± 54^{a}
Leaves-Water	257 ± 3^{b}	89.0 ± 0.2^{a}	3.46 ± 0.02^{e}	14 ± 1^{a}	18 ± 4^{c}	200 ± 30^{bc}
Stem barks-EA	207 ± 2^{d}	16 ± 0.3^{f}	14.0 ± 0.3^{c}	nd	17 ± 0.5^{c}	171 ± 24^{bc}
Stem barks-MeOH	271 ± 1^{a}	27.0 ± 0.3^{e}	28.0 ± 0.4^{b}	nd	33 ± 1^{b}	230 ± 34^{b}
Stem barks-Water	274 ± 2^{a}	33.0 ± 0.2^{d}	2.09 ± 0.01^{e}	7.0 ± 0.6^{c}	10.0 ± 0.1^{d}	163 ± 28^{c}

Values expressed are means ± SD of three parallel measurements. GAE: Gallic acid equivalent; RE: Rutin equivalent. CE: Catechin equivalent; CAE: Caffeic acid equivalent; QE: Quillaja equivalent. nd: not detected. Superscripts in the same column indicate significant difference in the tested extracts ($p < 0.05$).

Table 2. Phenol content of *A. leiocarpus* leaves and stem bark ethyl acetate (EA), methanol (MeOH) and water extracts.

Phenolic Compound	Leaves-EA mg/g	Leaves-MeOH mg/g	Leaves-Water Extract mg/g	Stem Barks-EA mg/g	Stem Barks MeOH mg/g	Stem Barks-Water mg/g
Gallic acid	226 ± 21	89 ± 28	30 ± 3	65 ± 6	38 ± 2	37 ± 1
Catechin	7.0 ± 0.8	3.0 ± 0.2	9 ± 1	1.0 ± 0.1	0.29 ± 0.02	0.28 ± 0.02
Epicatechin	0.29 ± 0.02	2.0 ± 0.1	0.27 ± 0.02	0.27 ± 0.02	0.18 ± 0.01	1.0 ± 0.1

Table 3. Chemical composition of *A. leiocarpus* extracts.

No.	Name	Formula	[M + H]+	[M – H]−	Leaves-EA	Leaves-MeOH	Leaves-Water	Stem Bark-EA	Stem Bark-MeOH	Stem Bark-Water
1	Quinic acid	$C_7H_{12}O_6$		19,105,557	+	+	+	+	+	+
2	Hexahydroxydiphenoylhexose	$C_{20}H_{18}O_{14}$		48,106,184	–	–	–	+	+	+
3	Shikimic acid	$C_7H_{10}O_5$		17,304,500	+	+	+	+	+	+
4	Galloylquinic acid isomer 1	$C_{14}H_{16}O_{10}$		34,306,653	–	–	–	–	+	+
5	Galloylhexose isomer 1	$C_{13}H_{16}O_{10}$		33,106,653	–	–	–	+	+	+
6	Galloylhexose isomer 2	$C_{13}H_{16}O_{10}$		33,106,653	–	–	–	+	+	+
7[1]	Gallic acid (3,4,5-Trihydroxybenzoic acid)	$C_7H_6O_5$		16,901,370	+	+	+	+	+	+
8	Galloylhexose isomer 3	$C_{13}H_{16}O_{10}$		33,106,653	–	–	–	+	+	+
9	Galloylquinic acid isomer 2	$C_{14}H_{16}O_{10}$		34,306,653	–	–	–	–	+	+
10	Galloylquinic acid isomer 3	$C_{14}H_{16}O_{10}$		34,306,653	–	–	–	–	+	+
11	Gallocatechin	$C_{15}H1_4O_7$		30,506,613	–	+	+	+	+	+
12	Protocatechuic acid (3,4-Dihydroxybenzoic acid)	$C_7H_6O_4$		15,301,879	+	+	+	+	+	+
13	3-Hydroxybenzaldehyde	$C_7H_6O_2$	12,304,461		+	+	+	–	–	–
14	Procyanidin B isomer 1	$C_{30}H_{26}O_{12}$		57,713,460	–	–	–	+	+	+
15	Punicalagin	$C_{48}H_{28}O_{30}$		108,305,872	+	+	+	+	+	+
16	Kynurenic acid	$C_{10}H_7NO_3$	19,005,042		–	+	+	–	–	–
17	Procyanidin B isomer 2	$C_{30}H_{26}O_{12}$		57,713,460	–	–	–	+	+	+
18[1]	Catechin	$C_{15}H_{14}O_6$		28,907,121	+	+	+	+	+	+
19[1]	Epigallocatechin	$C_{15}H_{14}O_7$		30,506,613	–	+	+	+	+	+
20	Casuarinin	$C_{41}H_{28}O_{26}$		93,507,906	+	+	+	+	+	+
21	Chlorogenic acid (3-O-Caffeoylquinic acid)	$C_{16}H_{18}O_9$	35,510,291		+	+	+	–	–	–
22	Caffeic acid	$C_9H_8O_4$		17,903,444	+	+	+	+	+	+
23	Cornusiin B or isomer	$C_{48}H_{30}O_{30}$		108,507,437	+	+	+	+	+	–
24	Ampelopsin (Dihydromyricetin)	$C_{15}H_{12}O_8$		31,904,540	+	+	+	+	+	+
25	Tellimagrandin I or isomer	$C_{34}H_{26}O_{22}$		78,508,375	+	+	+	+	+	+
26	Coumaroylquinic acid	$C_{16}H_{18}O_8$		33,709,235	+	+	+	–	–	–
27	Corilagin or isomer	$C_{27}H_{22}O_{18}$		63,307,279	+	+	+	–	–	–
28	Caffeoylshikimic acid	$C_{16}H_{16}O_8$		33,507,670	–	+	+	–	–	–
29	Procyanidin B isomer 3	$C_{30}H_{26}O_{12}$		57,713,460	–	–	–	+	+	–
30	Digalloylhexose	$C_{20}H_{20}O_{14}$		48,307,749	–	–	–	+	+	+
31[1]	Epigallocatechin-3-O-gallate (Teatannin II)	$C_{22}H_{18}O_{11}$		45,707,709	–	+	+	+	+	+
32[1]	Epicatechin	$C_{15}H_{14}O_6$		28,907,121	–	+	+	+	+	+
33	Punicacortein C or D	$C_{48}H_{28}O_{30}$		108,305,872	–	–	–	+	+	+

Table 3. *Cont.*

No.	Name	Formula	[M + H]+	[M – H]–	Leaves-EA	Leaves-MeOH	Leaves-Water	Stem Bark-EA	Stem Bark-MeOH	Stem Bark-Water
34	Trigalloylhexose isomer 1	$C_{27}H_{24}O_{18}$		63,508,844	–	–	–	+	+	+
35	Trigalloylhexose isomer 2	$C_{27}H_{24}O_{18}$		63,508,844	–	–	–	+	+	+
36	Di-O-methylcoruleoellagic acid	$C_{16}H_{10}O_{10}$		36,101,958	–	–	–	+	–	–
37	Mangiferin	$C_{19}H_{18}O_{11}$		42,107,709	–	–	–	+	+	+
38	Trigalloylhexose isomer 3	$C_{27}H_{24}O_{18}$		63,508,844	–	–	–	+	+	+
39[1]	Taxifolin (Dihydroquercetin)	$C_{15}H_{12}O_7$		30,305,048	+	+	+	+	+	+
40	Ferulic acid	$C_{10}H_{10}O_4$		19,305,009	+	+	+	–	–	–
41	Tetragalloylhexose	$C_{34}H_{28}O_{22}$		78,709,940	–	–	–	+	+	–
42[1]	Epicatechin-3-O-gallate	$C_{22}H_{18}O_{10}$		44,108,218	+	+	+	+	+	+
43	Chebulagic acid	$C_{41}H_{30}O_{27}$		95,308,963	+	+	+	–	–	–
44	Ellagic acid O-glucuronide	$C_{20}H_{14}O_{14}$		47,703,054	–	–	–	+	+	+
45	Ellagic acid O-hexoside isomer 1	$C_{20}H_{16}O_{13}$		46,305,127	+	+	+	+	+	+
46	Ellagic acid O-hexoside isomer 2	$C_{20}H_{16}O_{13}$		46,305,127	+	+	+	+	+	+
47	Coumaroylshikimic acid	$C_{16}H_{16}O_7$		31,908,178	+	+	+	–	–	–
48	O–Methylellagic acid O-hexoside isomer 1	$C_{21}H_{18}O_{13}$		47,706,692	–	–	–	+	+	+
49	Ellagic acid C-hexoside isomer 1	$C_{20}H_{16}O_{13}$	46,506,692		–	–	–	+	+	+
50	Myricetin-O-hexoside	$C_{21}H_{20}O_{13}$		47,908,257	+	+	+	+	+	+
51	Quercetin-O-galloylhexoside	$C_{28}H_{24}O_{16}$		61,509,862	+	+	+	–	–	–
52	Vitexin (Apigenin-8-C-glucoside)	$C_{21}H_{20}O_{10}$	43,311,348		–	–	–	+	+	+
53	Pentagalloylhexose	$C_{41}H_{32}O_{26}$		93,911,036	–	–	–	+	+	–
54	Di-O-methylflavellagic acid O-hexoside	$C_{22}H_{20}O_{14}$		50,707,749	–	–	–	+	+	+
55	Theaflavin or isomer	$C_{29}H_{24}O_{12}$	56,513,461		–	+	–	–	+	–
56	O-Methylellagic acid O-hexoside isomer 2	$C_{21}H_{18}O_{13}$		47,706,692	–	–	–	+	+	+
57	Aromadendrin (Dihydrokaempferol)	$C_{15}H_{12}O_6$		28,705,557	+	+	+	+	+	–
58	3,3′-Di-O-methylellagic acid-4-O-glucoside	$C_{22}H_{20}O_{13}$		49,108,257	+	+	+	+	+	+
59	Quercetin-3-O-glucuronide	$C_{21}H_{18}O_{13}$		47,706,692	+	+	+	–	–	–
60	Isoquercitrin (Hirsutrin, Quercetin-3-O-glucoside)	$C_{21}H_{20}O_{12}$		46,308,765	+	+	+	–	–	–
61	Rutin (Quercetin-3-O-rutinoside)	$C_{27}H_{30}O_{16}$	61,116,122		+	+	+	–	–	–
62	Luteolin-7-O-glucoside (Cynaroside)	$C_{21}H_{20}O_{11}$		44,709,274	–	–	–	–	–	+

Table 3. *Cont.*

No.	Name	Formula	$[M + H]^+$	$[M - H]^-$	Leaves-EA	Leaves-MeOH	Leaves-Water	Stem Bark-EA	Stem Bark-MeOH	Stem Bark-Water
63	Luteolin-O-deoxyhexosylhexoside	$C_{27}H_{30}O_{15}$		59,315,065	–	–	–	–	–	+
64	Isovitexin (Apigenin-6-C-glucoside)	$C_{21}H_{20}O_{10}$	43,311,348		–	–	–	+	+	+
65	Coatline A isomer	$C_{21}H_{24}O_{10}$		43,512,913	–	+	+	+	+	–
66	Ellagic acid O-pentoside	$C_{19}H_{14}O_{12}$		43,304,071	–	–	–	+	+	+
67	Ellagic acid C-hexoside isomer 2	$C_{20}H_{16}O_{13}$	46,506,692		–	–	–	+	+	+
68	Eschweilenol C (Ellagic acid-4-O-rhamnoside)	$C_{20}H_{16}O_{12}$		44,705,636	+	+	+	+	+	+
69	Reinutrin (Quercetin-3-O-xyloside)	$C_{20}H_{18}O_{11}$		43,307,709	+	+	+	–	–	–
70	Ellagic acid	$C_{14}H_6O_8$		30,099,845	+	+	+	+	+	+
71	Avicularin (Quercetin-3-O-arabinoside)	$C_{20}H_{18}O_{11}$		43,307,709	+	+	+	–	–	–
72 [1]	Myricetin (3,3′,4′,5,5′,7-Hexahydroxyflavone)	$C_{15}H_{10}O_8$		31,702,974	+	+	+	+	+	+
73	Guaijaverin (Quercetin-3-O-arabinoside)	$C_{20}H_{18}O_{11}$		43,307,709	+	+	+	–	–	–
74	Isorhamnetin-O-glucuronide isomer 1	$C_{22}H_{20}O_{13}$		49,108,257	–	+	–	–	–	–
75	Quercitrin (Quercetin-3-O-rhamnoside)	$C_{21}H_{20}O_{11}$		44,709,274	+	+	+	–	–	–
76	Di-O-methylflavellagic acid O-pentoside	$C_{21}H_{18}O_{13}$		47,706,692	–	–	–	+	–	–
77	Ducheside A (3-O-Methylellagic acid-4′-O-xyloside)	$C_{20}H_{16}O_{12}$		44,705,636	–	–	–	+	+	+
78	Eriodictyol	$C_{15}H_{12}O_6$		28,705,557	+	+	–	+	+	+
79	Isorhamnetin-3-O-glucoside	$C_{22}H_{22}O_{12}$		47,710,330	+	+	+	–	–	–
80	Dimethoxy-tetrahydroxy(iso) flavone-O-hexoside	$C_{23}H_{24}O_{13}$		50,711,387	–	+	+	–	–	–
81	Isorhamnetin-O-glucuronide isomer 2	$C_{22}H_{20}O_{13}$		49,108,257	–	+	–	–	–	–
82	Isorhamnetin-O-glucuronide	$C_{22}H_{20}O_{13}$		49,108,257	–	–	+	–	–	–
83	Isorhamnetin-3-O-rutinoside (Narcissin)	$C_{28}H_{32}O_{16}$		62,316,122	+	+	+	–	–	–
84	4-Methoxycinnamic acid	$C_{10}H_{10}O_3$		17,907,082	–	–	–	–	+	–
85	Di-O-methylellagic acid-O-pentoside	$C_{21}H_{18}O_{12}$		46,107,200	+	+	+	+	+	+

Table 3. *Cont.*

No.	Name	Formula	$[M + H]^+$	$[M - H]^-$	Leaves-EA	Leaves-MeOH	Leaves-Water	Stem Bark-EA	Stem Bark-MeOH	Stem Bark-Water
86	3,3′,4-Tri-O-methylflavellagic acid-4-O-glucoside	$C_{23}H_{22}O_{14}$		52,109,314	+	+	+	+	+	+
87	Quercetin	$C_{15}H_{10}O_7$		30,103,483	+	+	+	–	–	–
88	3-O-Methylellagic acid	$C_{15}H_8O_8$		31,501,410	–	+	+	+	+	+
89	Di-O-methylellagic acid-O-deoxyhexoside	$C_{22}H_{20}O_{12}$		47,508,766	–	–	–	+	+	+
90[1]	Naringenin	$C_{15}H_{12}O_5$		27,106,065	+	+	+	+	+	+
91	Di-O-methylflavellagic acid isomer 1	$C_{16}H_{10}O_9$		34,502,466	–	–	–	+	+	+
92[1]	Luteolin (3′,4′,5,7-Tetrahydroxyflavone)	$C_{15}H_{10}O_6$		28,503,991	+	+	+	–	+	–
93	3,3′-Di-O-methylellagic acid	$C_{16}H_{10}O_8$		32,902,975	+	+	+	+	+	+
94	Kaempferol (3,4′,5,7-Tetrahydroxyflavone)	$C_{15}H_{10}O_6$	28,705,556		+	+	+	–	–	–
95	Isorhamnetin (3′-Methoxy-3,4′,5,7 -tetrahydroxyflavone)	$C_{16}H_{12}O_7$		31,505,048	+	+	+	–	–	–
96	Dimethoxy-trihydroxy(iso)flavone	$C_{17}H_{14}O_7$		32,906,613	+	+	–	–	–	–
97	Di-O-methylflavellagic acid isomer 2	$C_{16}H_{10}O_9$		34,502,466	–	–	–	+	+	+
98	Tetra-O-methylflavellagic acid isomer 1	$C_{18}H_{14}O_9$		37,305,596	–	–	–	+	+	+
99	Apigenin (4′,5,7-Trihydroxyflavone)	$C_{15}H_{10}O_5$		26,904,500	–	+	–	–	–	+
100	3,3′,4-Tri-O-methylellagic acid	$C_{17}H_{12}O_8$		34,304,540	+	+	+	+	+	+
101	Undecanedioic acid	$C_{11}H_{20}O_4$		21,512,834	+	+	+	–	–	–
102	Tetra-O-methylflavellagic acid isomer 2	$C_{18}H_{14}O_9$		37,305,596	–	–	–	+	+	+
103	3,3′,4-Tri-O-methylflavellagic acid	$C_{17}H_{12}O_9$		35,904,031	+	+	+	+	+	+
104	Dihydroxy-dimethoxy(iso)flavone	$C_{17}H_{14}O_6$		31,307,122	+	+	–	–	–	–
105	Pinocembrin (5,7-Dihydroxyflavanone)	$C_{15}H_{12}O_4$		25,506,573	+	+	–	–	+	–
106	Dihydroxy-trimethoxy(iso)flavone	$C_{18}H_{16}O_7$		34,308,178	+	+	+	–	–	–
107	Dodecanedioic acid	$C_{12}H_{22}O_4$		22,914,399	+	+	+	–	–	–
108	Hexadecanedioic acid	$C_{16}H_{30}O_4$		28,520,659	+	+	–	–	–	–

[1] Confirmed by standard.

3.2. Phenolic Acids

Quinic ($[M - H]^-$ at *m/z* 191), shikimic ($[M - H]^-$ at *m/z* 173), gallic ($[M - H]^-$ at *m/z* 169), and protocatechuic ($[M - H]^-$ at *m/z* 153) acids were tentatively identified from all the extracts of *A. leiocarpus* leaves and stem bark. Chlorogenic ($[M - H]^+$ at *m/z* 355), caffeic ($[M - H]^-$ at *m/z* 179), ferulic ($[M - H]^-$ at *m/z* 193), and dodecanedioic ($[M - H]^-$ at *m/z* 229) acids were tentatively identified from the leaves extracts of *A. leiocarpus* only.

3.3. Flavonoids

Several compounds belonging to the flavonoid family were identified from *A. leiocarpus* extracts. The compound suffering deprotonation at *m/z* 269 $[M - H]^-$ and fragment ions at *m/z* 225, 151, 149, and 117 was characterised as apigenin and was present in the methanol extract of *A. leiocarpus* leaves and water extract of *A. leiocarpus* stem bark. C-glucosides of apigenin ($[M - H]^+$ at *m/z* 433), namely, vitexin and isovitexin, only occurred in the stem bark extracts of *A. leiocarpus*. Luteolin, suffering deprotonation at *m/z* 285 $[M - H]^-$ and fragment ions at *m/z* 217, 199, 175, 151, and 133 was tentatively characterised. Other flavonoids, such as, catechin ($[M - H]^-$ at *m/z* 289), naringenin ($[M - H]^-$ at *m/z* 271), myricetin ($[M - H]^-$ at *m/z* 317), taxifolin ($[M - H]^-$ at *m/z* 303), and pinocembrin ($[M - H]^-$ at *m/z* 255) were also identified.

3.4. Tannins Derivatives

Some tannin derivatives were tentatively characterised. Casuarinin with deprotonation at *m/z* 935 $[M - H]^-$ and fragment ions at *m/z* 917, 783, 633, 300, and 275, was tentatively characterized in all the extracts. Hydrolysable tannins, namely chebulagic acid and punicalagin [27], were characterized at $[M - H]^-$ *m/z* 953 and 1083, respectively. While punicalagin was identified in all of the extracts, chebulagic acid was only characterised in the leaves extracts.

3.5. Antioxidant Activities

Phytochemicals, which are ubiquitously present in plants, have been identified to possess antioxidant activity and they are capable of managing oxidative stress related diseases [28]. In this study, three types of antioxidant mechanisms were used, namely, radical scavenging, reducing power, and metal chelating. Table 4 presents the ability of *A. leiocarpus* extracts to scavenge DPPH and ABTS radicals. DPPH, which is a stable radical, is widely used to assess the free radical scavenging abilities of plant extract. By proton transfer, there is the DPPH change in the non-radical form, characterized by a yellow chromophore [29]. On the other side, the ABTS method is based on the monitoring of electron or hydrogen transfer-induced ABTS radical-cation decay, which is characterized by the disappearance of the corresponding blue-green radical [30]. From Table 4, it is observed that, in general, the methanol and water extracts of *A. leiocarpus* leaves and stem bark showed potent radical scavenging properties compared to the ethyl acetate extracts of *A. leiocarpus* leaves and stem bark. Besides, it was noted that the stem bark extracts were more potent radical scavengers when compared to the leaves extracts. Furthermore, the strong radical scavenging activities of the extracts was related to their high phenolic contents. Reducing power, as described by electron transfer ability, is considered to be one of the key indicators of antioxidant capacity of plant extracts [28]. In this study, the reducing power of *A. leiocarpus* extracts was assessed using FRAP and CUPRAC assays, which are characterized by the reduction of Fe^{3+} to Fe^{2+} and Cu^{2+} to Cu^{+}, respectively [30,31]. Comparable to the free radical scavenging assessment, the methanol and water extracts of *A. leiocarpus* showed potent reducing properties as compared to the ethyl acetate extracts. The total antioxidant capacity that was determined by the phosphomolybdenum assay demonstrated that methanol and water extracts of *A. leiocarpus* leaves and stem bark were more potent antioxidants as compared to the ethyl acetate extracts. This finding is in line with radical scavenging and reducing power evaluations. The ability of *A. leiocarpus* extracts to chelate metal was also evaluated and is presented in Table 4. Given the recognised role of iron in oxidative stress, which is understood as an increase in oxygen radical intermediates concentration, eventually leading to dysregulation, the development of metal chelators having the ability to restore metal homeostasis and oxidative status appears to be a valuable challenge, particularly if the chelators possess other important biological activities that might mitigate other diseases [32]. Data that were gathered from this study revealed that the water extract of *A. leiocarpus* leaves, possessing the highest flavonoid content, was the most activity metal chelator. It has previously been proposed that flavonoids were potent chelators of iron [33]. Naringenin, quercetin, luteolin, and catechin compounds, belonging to the flavonoid family, were identified in the water extract of *A. leiocarpus* leaves and they were reported to possess metal chelating abilities [33,34]. For instance, quercetin was reported to form different complexes with Fe^{2+} through its 5-OH and 4-carbonyl groups [35].

Table 4. Antioxidant properties of *A. leiocarpus* extracts.

Samples	DPPH (mmol TE/g)	ABTS (mmol TE/g)	CUPRAC (mmol TE/g)	FRAP (mmol TE/g)	Metal Chelating (mg EDTAE/g)	Phosphomolybdenum (mmol TE/g)
Leaves-EA	30 ± 0.01^{f}	0.26 ± 0.02^{e}	0.50 ± 0.04^{e}	0.26 ± 0.03^{d}	10.0 ± 0.8^{f}	2.0 ± 0.1^{e}
Leaves-MeOH	5.0 ± 0.1^{d}	3.0 ± 0.4^{c}	7.0 ± 0.2^{c}	4.0 ± 0.1^{c}	47.0 ± 0.8^{c}	4.0 ± 0.1^{d}
Leaves-Water	5.0 ± 0.1^{c}	4.0 ± 0.3^{b}	7.0 ± 0.2^{b}	6.0 ± 0.4^{b}	79.0 ± 0.9^{a}	4.0 ± 0.1^{d}
Stem barks-EA	3.0 ± 0.1^{e}	2.0 ± 0.1^{d}	5.0 ± 0.2^{d}	4.0 ± 0.1^{c}	30.0 ± 0.5^{e}	4.0 ± 0.1^{c}
Stem barks-MeOH	6.0 ± 0.1^{a}	5.0 ± 0.1^{a}	8.0 ± 0.2^{a}	6.0 ± 0.3^{a}	45.0 ± 0.6^{d}	6.0 ± 0.2^{a}
Stem barks-Water	5.0 ± 0.1^{b}	4.0 ± 0.4^{b}	8.0 ± 0.1^{b}	6.0 ± 0.2^{a}	61.0 ± 0.4^{b}	5.0 ± 0.1^{a}

Values expressed are means ± S.D. of three parallel measurements. DPPH: 2-diphenyl-1-picrylhydrazyl; ABTS: 2,2′-azino-bis(3-ethylbenzothiazoline-6-sulphonic acid; CUPRAC: cupric reducing antioxidant capacity; FRAP: ferric reducing antioxidant power; TE: Trolox equivalent; EDTAE: ethylenediaminetetraacetic acid equivalent. Superscripts in the same column indicate significant difference in the tested extracts ($p < 0.05$).

3.6. Enzyme Inhibitory Activities

While the antioxidant activity of plant extract is often linked to the phenolic content, the enzyme inhibitory properties of extracts mainly involves the interaction of phytochemicals with the enzyme or enzyme-substrate complex. To the best of our knowledge, this is the first report on the assessment of the inhibitory potential of *A. leiocarpus* leaves and stem bark on enzymes related to Alzheimer's disease and skin hyperpigmentation. A previous study has appraised the amylase and glucosidase inhibitory action of *A. leiocarpus* leaves [5]. However, no comparison has been made with the stem bark extract of the plant and the possible effect of different extraction solvents. Among the five food drug administration (FDA)-approved Alzheimer's disease treatments, four are acetyl cholinesterase inhibitors [36]. Cholinesterase inhibitors designed for the management of Alzheimer's disease stem from the cholinergic hypothesis, which is the leading theory proposed to explain the pathogenesis of Alzheimer's disease [37]. It has been recognised that cholinergic neurons loss in brain area that is responsible for cognition and behaviour was the hallmark of Alzheimer's disease. While the role of acetyl cholinesterase has been clearly claimed, the exact mechanism that involves butyryl cholinesterase remains elusive. Butyryl cholinesterase, previously underestimated in the pathogenesis of Alzheimer's disease, was found to be up-regulated in advanced stages of the condition and plays a key role in the disease maintenance and progression [38]. From this perspective, it can be stated that cholinesterase inhibitors targeting both acetyl and butyryl cholinesterases are in need. Table 5 reports the acetyl cholinesterase inhibitory activity of the different extracts of *A. leiocarpus* leaves and stem bark ranging from 3.51 to 4.68 mg GALAE/g. With regards to the butyryl cholinesterase inhibitory action, the values ranged from 0.45 to 4.0 mg GALAE/g. Interestingly, the water extract of *A. leiocarpus* leaves only inhibited acetyl cholinesterase. It can be suggested that this extract might be targeted at the initial stage of the disease, when butyryl cholinesterase activity is not pronounced. The methanol extract of *A. leiocarpus* leaves (4.68 and 4.0 mg GALAE/g) showed potent inhibition against both cholinesterases. Over the past decades, there has been an emerging trend of naturally derived cosmetic products. This shift has encouraged researchers to find new cosmeceuticals and the focus has geared towards plants. Plant extracts have witnessed increased global demand for de-pigmenting agents due to their safety and compatibility with all skin types [39]. The inhibitory action of *A. leiocarpus* extracts on tyrosinase, a copper-containing enzyme responsible for the biosynthesis of melanin [40], was investigated. The data collected showed potent tyrosinase inhibition with values ranging from 113.0 to 155.26 mg KAE/g, the highest values was recorded for methanol extract of *A. leiocarpus* stem bark. Pinocembrin, shikimic acid, and vitexin, tentatively identified in the methanol extract of *A. leiocarpus* stem bark, were previously reported to inhibit tyrosinase [41–43]. A group of researchers [5] have reported the amylase (IC_{50} value of 242.17 μg/mL) and glucosidase (IC_{50} value of 196.35 μg/mL) inhibitory activity of *A. leiocarpus* leaves water extract. In the present investigation and, as opposed to the previous study, low inhibition was recorded against amylase, while no inhibitory action was observed against glucosidase in the presence of the water extract of *A. leiocarpus* leaves. The different activity that was recorded in our study might be related to the geographical location along with the environmental conditions of the studied *A. leiocarpus* plants. In this study, it was observed that the different *A. leiocarpus* extracts were poor inhibitors of amylase, with values that ranged from 0.19 to 1.13 mmol ACAE/g. Only ethyl acetate extracts inhibited glucosidase and the values were higher as compared to amylase. The inhibition of glucosidase is considered as strategic in the management of diabetes type II. Indeed, it has been advocated that the inhibition of glucosidase reduced post-prandial glucose rise and it was associated to less side effects.

Table 5. Enzyme inhibitory properties of *A. leiocarpus* extracts.

Samples	AChE Inhibition (mg GALAE/g)	BChE Inhibition (mg GALAE/g)	Tyrosinase Inhibition (mg KAE/g)	Amylase Inhibition (mmol ACAE/g)	Glucosidase Inhibition (mmol ACAE/g)
Leaves-EA	4.0 ± 0.2[d]	3.0 ± 0.2[b]	131.0 ± 0.2[d]	0.79 ± 0.04[b]	15.0±0.1[b]
Leaves-MeOH	4.68 ± 0.02[a]	4.0 ± 0.1[a]	154.0 ± 0.2[b]	1.0 ± 0.1[a]	nd
Leaves-Water	4.19 ± 0.04[c]	nd	113 ± 1[e]	0.53 ± 0.04[c]	nd
Stem barks-EA	4.0 ± 0.1[bc]	2.0 ± 0.2[c]	152.0 ± 0.6[c]	1.0 ± 0.1[b]	15.0 ± 0.1[a]
Stem barks-MeOH	4.0 ± 0.1[b]	1.0 ± 0.3[c]	155.26 ± 0.04[a]	0.85 ± 0.03[b]	nd
Stem barks-Water	4.0 ± 0.1[e]	0.5 ± 0.1[d]	113.0 ± 0.6[e]	0.19 ± 0.01[d]	nd

Values expressed are means ± S.D. of three parallel measurements. AChE: Acetylcholinesterase; BChE: Butyrylcholinesterase; GALAE: Galatamine equivalent; KAE: Kojic acid equivalent; ACAE: Acarbose equivalent. nd: not detected. Superscripts in the same column indicate significant difference in the tested extracts (p <0.05).

3.7. Multivariate Analysis

An unsupervised multiple correspondence analysis (MCA), a Heatmap clustering approach, and Venn graph were applied on the chemical composition of *A. leiocarpus* samples to obtain a typology of the samples and to characterize the chemical profile differences among those samples. MCA is commonly used qualitative variables to examine a set of observations described by a set of nominal variables. Figure 1A displays the proportion of explained inertia per component and the projection of the samples and chemical compounds on the first two dimensions. Two components were required to summarize approximately 97.7% of the variance. The first component explained 96,1% of variance, while the second accounted for 1, 6%. As we could notice, the samples were predominantly separated by the first component of MCA, with the extracts of stem bark being grouped on the negative side of the factors and the leaves extract on the positive side (Figure 1B). In agreement with MCA, Heatmap separated and categorized the samples into two groups, with each group being divided into two sub-clusters (Figure 1C). On the other hand, we observed that the stem bark extracts were more homogeneous than the leaves extracts. Indeed, ethyl acetate extract of leaves was clearly separated from the other two extracts (methanol and water), which were close (Figure 1B). This indicates that solvent used have a large influence on the leaves secondary metabolite extraction than that of stem bark. Moreover, when cross-checking a list of 108 compounds that were identified in all the leaves extracts against those were recognized in all stem bark extracts, we observed that a total of 38 were commonly detected in the two organs, whereas there were 38 and 32 unique compounds found in stem bark and leaves, respectively (Figure 2).

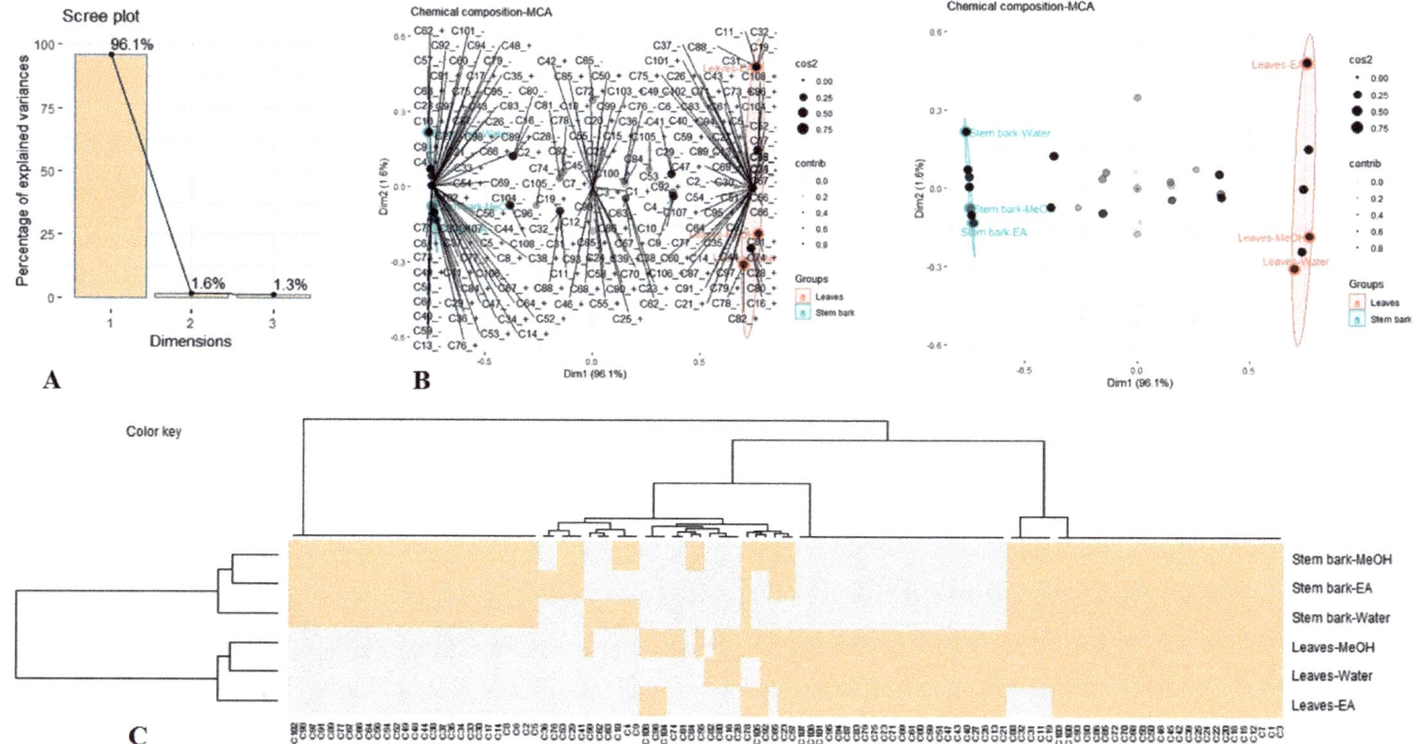

Figure 1. Multivariate analysis using multiple correspondence analysis (MCA) and Heatmap clustering analysis of chemical composition in *A. leiocarpus* extracts. (**A**): Percentage of explained variance per component. (**B**): projection of extracts and chemical compounds into the subspace spanned by the first two components of MCA. (**C**): Clustered Image Map (Euclidean Distance, Ward linkage). Gray colour: Absence, Wheat colour: Presence.

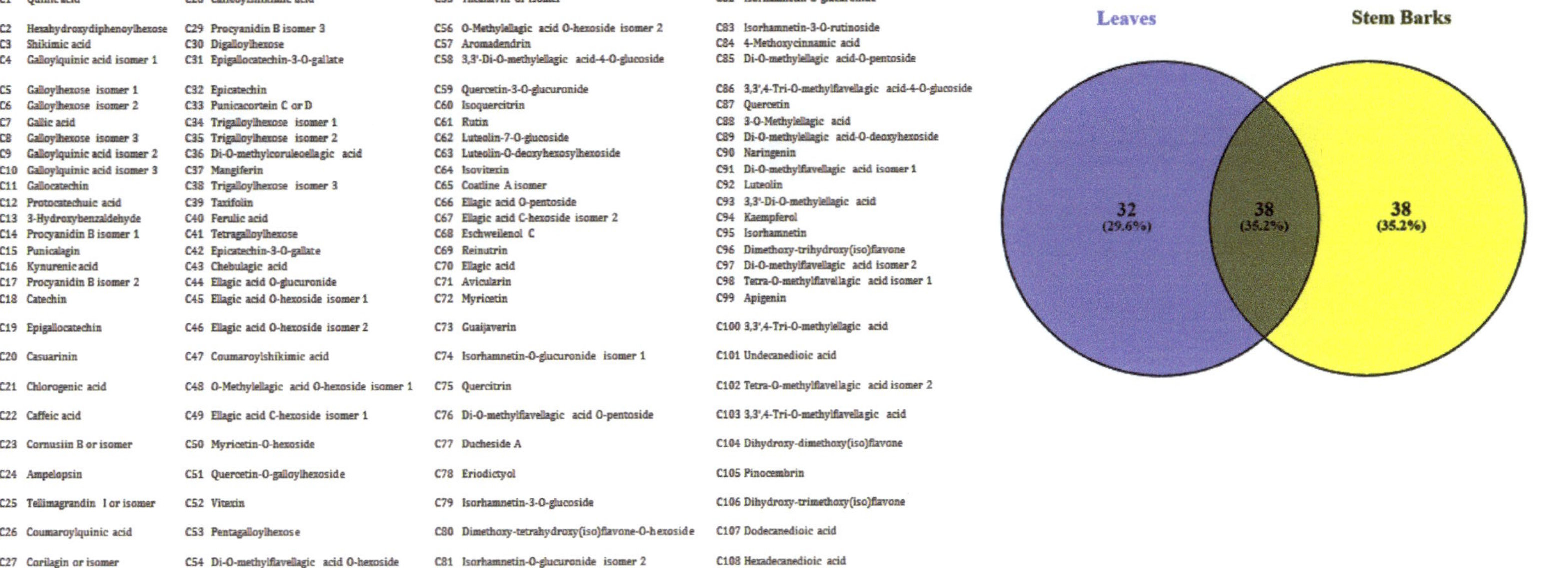

Figure 2. Venn diagram representing the overlap of compounds on the two organs.

The analysis showed the variation of chemical composition of *A. leiocarpus* depending to part used, as well as the influence of solvent on the extraction of compounds, with a more pronounced effect on the leaves. In agreement with this observation, we decided to ascertain whether the plant parts and solvent types used had any statistically significant effect on both total bioactive compounds content and biological activities of *A. leiocarpus*. Thus, multivariate methods for integrative large biological data sets, namely DIABLO, was applied to total bioactive compounds content and biological activities data-sets. DIABLO is a highly flexible supervised multivariate method that enables to classify in an optimal and reliable manner the studied samples and to construct a predictive multi-omics model that can be used to classify new samples. Figure 3 shows the multivariate analysis results. From Figure 3A,B, it is clear that there were parts and solvents effects on the total bioactive compounds content and biological activities of *A. leiocarpus*. In fact, as we could observe in Figure 3A, the factorial plan discriminated leaves parts with the stem bark parts effectively in both bioactive compounds and biological activities data sets. A similar outcome was provided with the second studied factor, in short, a clear segregation between the solvents was achieved (Figure 3B). Furthermore, a better separation of solvents was found while using biological activities data than when using the bioactive compounds data. The different extracts of the stem bark were relatively close as we have seen on the plots. As well by observing the samples plot using bioactive compounds content data (Figure 3A,B, Block: Bioactive compounds), this view was echoed with a consolidation of stem bark extracts contrast to a high variability between the leaves extracts. Accordingly, the extraction solvent, by extension the change of polarity, greatly influenced the bioactive compounds content of leaves than those of stem bark. The present result indicated that *A. leiocarpus* leaves, unlike stem bark, contain chemical molecules with varying polarity and solubility that are sensitive to the variation of solvent.

Figure 3D,G shows that the first three and two components, respectively, of bioactive compounds dataset were positively correlated to biological activities dataset, which allowed for us to say that DIABLO analysis was able to model a good agreement between our datasets. Subsequently, to compare DIABLO models that include/exclude the repeated measures experimental design, we examined the ROC assay (Receiver Operating Characteristic Curve). As we could observe in Figure 3E, the AUC (area under the curve) for the first three component for bioactive compounds and biological activities were 0.89 and 0.77, respectively. As for the second model, the AUC for the first two component were 1 for both bioactive compounds and biological activities (Figure 3H). Finally, the performance of each model was evaluated by estimating the classification error rate. Centroids distance was used as prediction distance and 10×5-fold CV as repeated stratified cross-validation. Thus, by observing Figure 3C,F, the best performance was obtained for 3 and 2 component, respectively, which suggests a satisfactory result on our model.

Circos plot and network were carried out to analyze the correlative relationships between total bioactive compound contents of each extract with their biological activities (Figure 4). The analysis revealed that AChE had a positive correlation with total tannin content (TTC) ($r = 0.79$), total saponin content (TSC) ($r = 0.78$) and total flavonol content (TFvLC) ($r = 0.86$), whereas tyrosinase was correlated with TFvLC ($r = 0.75$). Likewise total phenolic content (TPC) and total phenolic acid (TPaC) were positively related to Radical Scavenging Activity ABTS ($r = 0.74$; $r = 0.77$) and DPPH ($r = 0.74$; $r = 0.77$), Reducing Power ability FRAP ($r = 0.77$; $r = 0.81$) and CUPRAC ($r = 0.75$; $r = 0.78$) and Ferrous ion Chelation (MCA) ($r = 0.73$; $r = 0.81$). Accordingly, it was obvious that the phenolic compounds especially phenolic acid compounds were mostly responsible for the antioxidant activities and metal chelating ability of *A. leiocarpus* extracts.

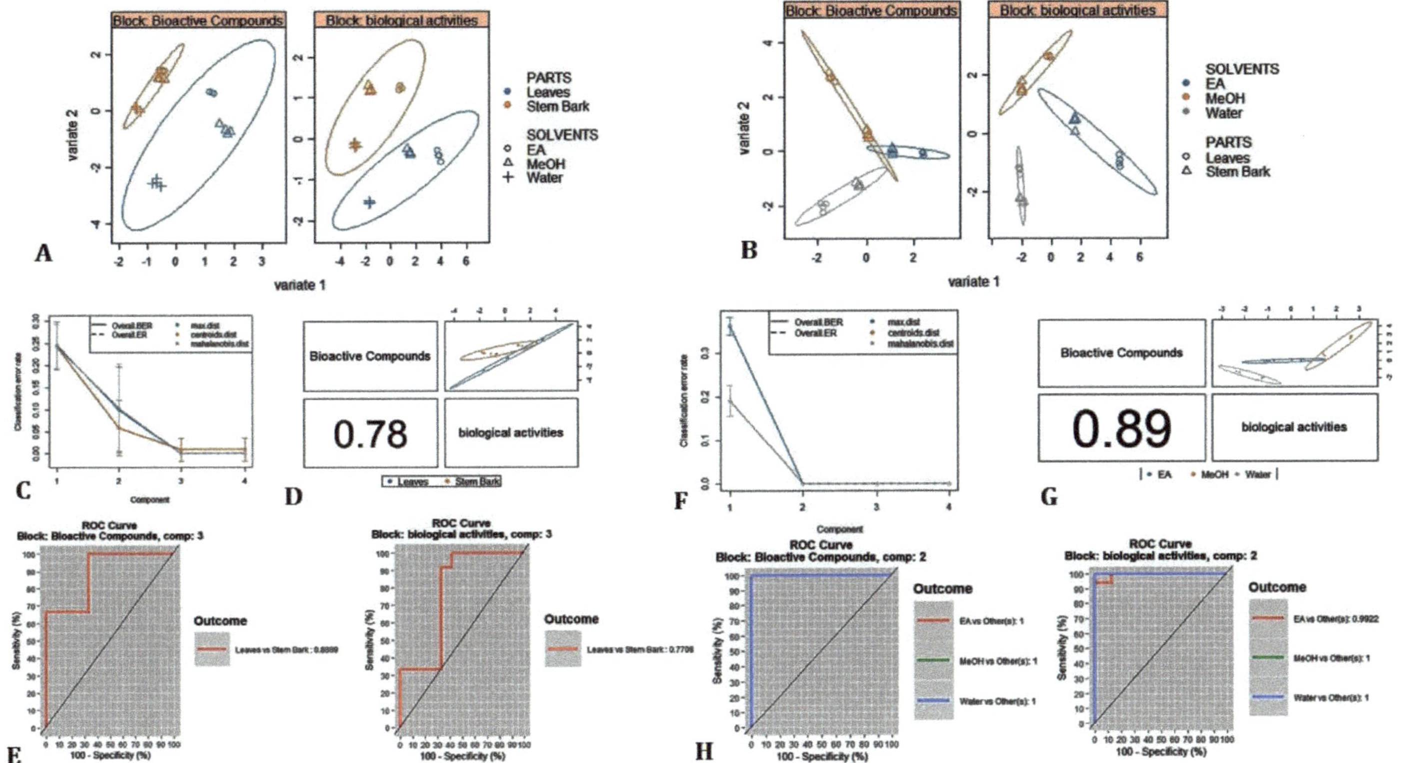

Figure 3. N-integration across multiple datasets analysis on *Anogeissus leiocarpus* bioactive compounds content and biological activities according to two factors (solvents and parts). (**A**,**B**): Sample plot with confidence ellipse according to the parts of plant and the extracting solvent as factor, respectively. (**C**,**F**): The model performance per component for Centroids Distance using 5-fold CV repeated 10 times. (**D**,**G**): the global overview of the relationship between the two datasets at the two first component level. (**E**,**H**): AUC (area under the curve) average and ROC (Receiver Operating Characteristic Curve) curve using one-vs-all comparisons.

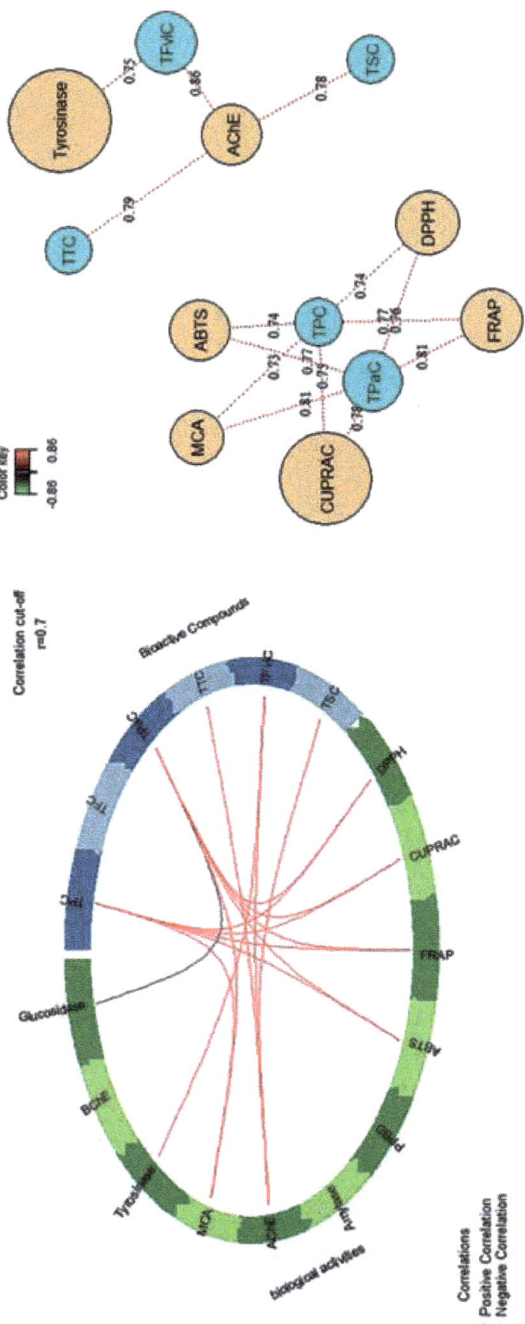

Figure 4. Circos-plot and network showing the relationship between total bioactive compounds and evaluated biological activities (cut-off: $r = 0.7$).

3.8. Pharmacological Studies

With the aim of investigating extract biological activity, EA, MeOH, and water extracts of *A. leiocarpus* leaves and stem barks were assayed through the allelopathy test. To this regard, the seeds of the commercial *Lollo bionda* lettuce cultivar were exposed to scalar extract concentrations (0.1–10 mg/mL), and the seedling germination and growth were monitored. After incubation of seeds with extracts, we observed a null effect on the seedling germination (Figure 5), thus obtaining a preliminary index of the extract biocompatibility.

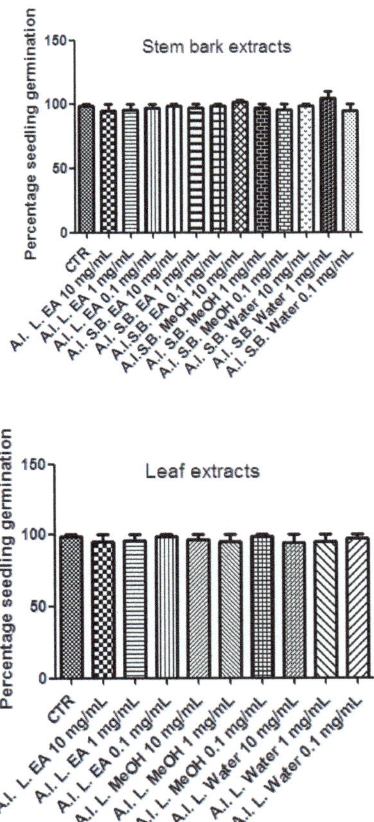

Figure 5. Effects of *A. leiocarpus* leaf (L) and stem bark (SB) extracts (0.1–10 mg/mL) on *Lollo bionda* lettuce root elongation rate. Data, expressed as mean length distribution of germinated seeds, are means ± SD of three experiments performed in triplicate. After exposing lettuce roots to the extracts, a null effect on seedling germination was observed. EA: Ethyl acetate; MeOH: Methanol.

As a further approach to evaluate potential toxicity, *A. leiocarpus* extracts, in the concentration range 0.01–10 mg/mL, were tested on brine shrimp lethality assay, performed on the brine shrimp *Artemia salina* Leach, which is recognized as a valuable tool to predict potential cytotoxicity related to plant extracts [44]. The experimental procedure was conducted in agreement with a previous published paradigm [45]. The results of this assay indicated LC_{50} values in the range 0.26–2.04 mg/mL (Figure 6), which were indicatory to choose the extract concentration for the in vitro and ex vivo investigations in order to elucidate putative protective effects, in the colon.

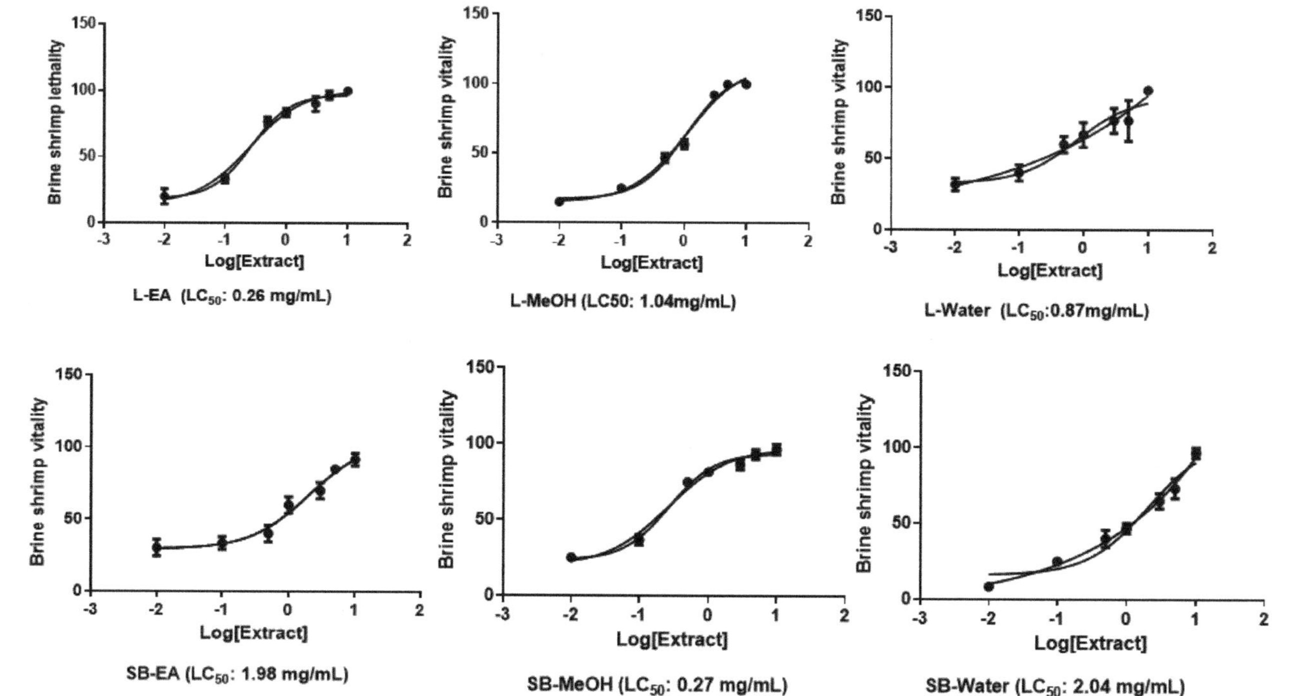

Figure 6. Effects of *A. leiocarpus* leaf (L) and stem bark (SB) extracts (0.01–10 mg/mL) on *Artemia salina* Leach lethality (Brine shrimp lethality test). Data are means ± SD of three experiments performed in triplicate. After exposing brine shrimps to the extracts, LC_{50} values in the range 0.26–2.04 mg/mL were recorded. EA: Ethyl acetate; MeOH: Methanol.

Particularly, we selected the concentration 0.1 mg/mL that was at least two-fold lower than LC_{50} and in agreement with previous investigations that demonstrated the antioxidant effects on isolated porcine tissue [46]. While considering these findings, we assayed extract effects on rat colon stimulated with LPS, ex vivo, in order reproduce the burden of oxidative stress and inflammation that characterize ulcerative colitis [20,47,48]. All extracts, with the only exception of stembark methanol extract, revealed effective in reducing LPS-induced 8-iso-$PGF_{2\alpha}$ level (Figure 7). On the other hand, all of the extracts blunted LPS-stimulated PGE_2 colon level (Figure 8), whereas leaf water and stem bark methanol extracts failed to reduce 5-HT concentration (Figure 9). Finally, when the extracts were tested on colon cancer HCT116 cell line, only stem bark ethyl acetate extract revealed biocompatibility, exerting a null effect on cell proliferation (Figures 10 and 11). Conversely, the other extracts displayed stimulatory effects on either viability or spontaneous migration of HCT116 cells.

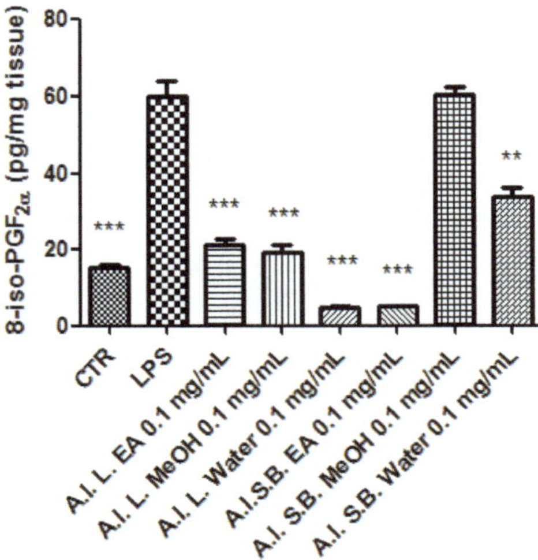

Figure 7. Effect *A. leiocarpus* leaf (L) and stem bark (SB) extracts (0.01 mg/mL) on lipopolysaccharide (LPS)-induced 8-iso-prostaglandin(PG)$F_{2\alpha}$ level in isolated rat colon. EA: Ethyl acetate; MeOH: Methanol. ANOVA, $p < 0.0001$; post hoc, $^{**}p < 0.01$, $^{***}p < 0.001$ vs. LPS.

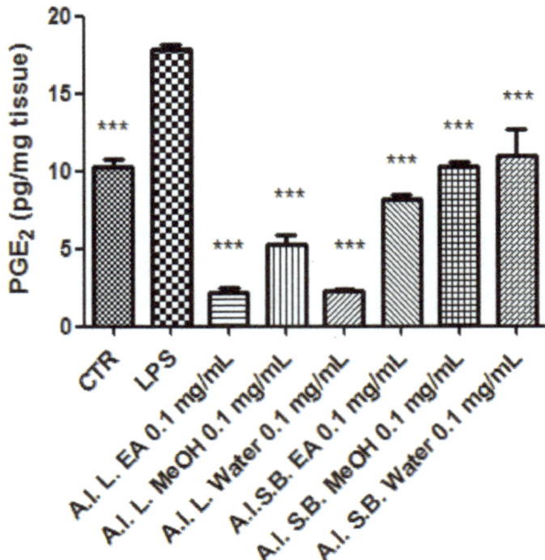

Figure 8. Effect *A. leiocarpus* leaf (L) and stem bark (SB) extracts (0.01 mg/mL) on lipopolysaccharide (LPS)-induced prostaglandin(PG)E_2 level in isolated rat colon. EA: Ethyl acetate; MeOH: Methanol. ANOVA, $p < 0.0001$; post hoc, $^{***}p < 0.001$ vs. LPS.

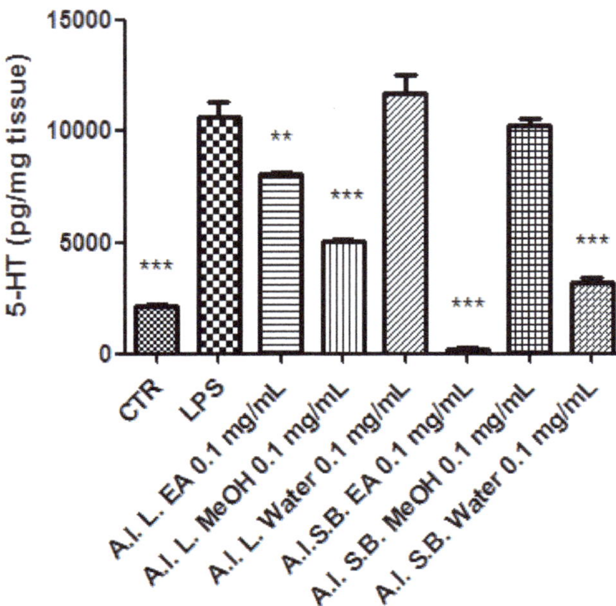

Figure 9. Effect *A. leiocarpus* leaf (L) and stem bark (SB) extracts (0.01 mg/mL) on lipopolysaccharide (LPS)-induced serotonin (5-HT) level in isolated rat colon. EA: Ethyl acetate; MeOH: Methanol. ANOVA, $p < 0.0001$; post hoc, $^{**}p < 0.01$, $^{***}p < 0.001$ vs. LPS.

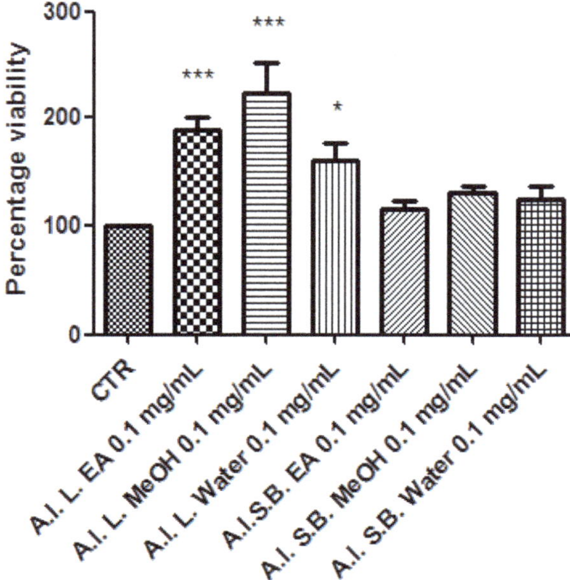

Figure 10. Effect *A. leiocarpus* leaf (L) and stem bark (SB) extracts (0.01 mg/mL) on human colon cancer HCT116 cell line viability (MTT assay). EA: Ethyl acetate; MeOH: Methanol. ANOVA, $p < 0.0001$; post hoc, $^{*}p < 0.05$, $^{***}p < 0.001$ vs. CTR (Control group).

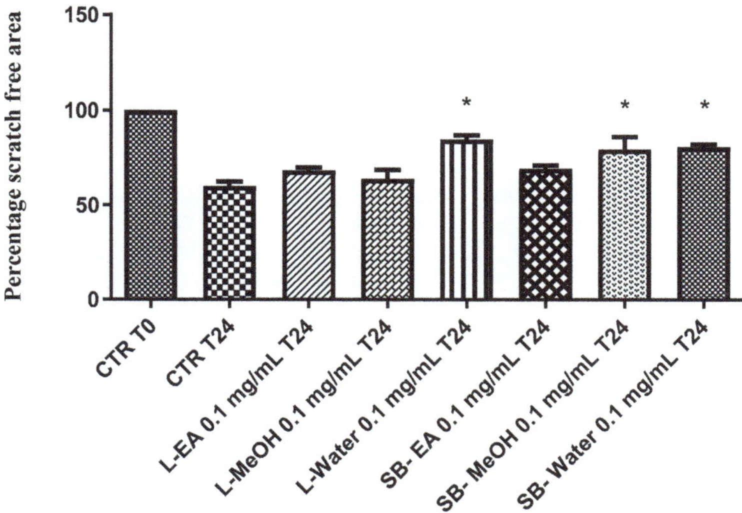

Figure 11. Effect *A. leiocarpus* leaf (L) and stem bark (SB) extracts (0.01 mg/mL) on human colon cancer HCT116 cell line spontaneous migration (wound healing assay). EA: Ethyl acetate; MeOH: Methanol. ANOVA, $p < 0.01$; post hoc, $*p < 0.05$ vs. CTR T24 (Control group).

4. Discussion

Oxidative stress is characterized by the overproduction of reactive oxygen/nitrogen (ROS/RNS) species that could drive to lipid peroxidation [49], which displays a key role in the pathogenesis of ulcerative colitis [50]. 8-iso-PGF$_{2\alpha}$, deriving from ROS/RNS peroxidation of membrane arachidonic acid, represents a stable marker of lipid peroxidation and tissue damage, in vivo [51], whereas the blunting effects on 8-iso-PGF$_{2\alpha}$ production that are induced by herbal extracts were related to protective effects [48,52]. Consistently with the reported antioxidant effects and the findings by Belemnaba et al. [46], the tested extracts blunted LPS-induced 8-iso-PGF$_{2\alpha}$ production, with the only exception being represented by stem bark MeOH extract. On the other hand, stem bark MeOH extract showed the ability to reduce the level of malondialdehyde (MDA), which is another key marker of lipid peroxidation, in vivo [53]. This discrepancy could depend on more than one speculation. On one side, the differences in the employed experimental ex vivo and in vivo paradigms that were chosen by us and Akanbi and colleagues [53], respectively, could be crucial. On the other side, we should also consider that isoprostanes derive only by arachidonic acid peroxidation, whereas MDA could originate from various polyunsaturated acids [54], and this could represent a limit in the evaluation of lipid peroxidation.

The effects of *A. leiocarpus* extracts on LPS-induced levels of colon PGE$_2$ were investigated as well. PGE$_2$ is a cyclooxygenase (COX)-2-derived pro-inflammatory mediator, whose upregulation has been long involved in colon inflammation and damage, whereas the antioxidants were revealed to be effective in blunting the colon levels of this prostaglandin [47,55]. All of the tested extracts proved able in down-regulating LPS-induced PGE$_2$ level (Figure 8), consistently with the reported antioxidant effects.

A blunting effect on LPS-induced 5-HT level was observed after treating the colon specimens with *A. leiocarpus* extract, as well. 5-HT pro-inflammatory role in ulcerative colitis was previously suggested [56], which possibly involved the activation of 5-HT$_3$ receptors [57]. With the only exception of leaf water and stem bark MeOH extracts, the *A. leiocarpus* extracts displayed a significant inhibition of 5-HT steady state level, in the colon. This could be, albeit partially, related to decreased neurotransmitter synthesis and release, in the colon tissue [58,59].

Collectively, all of the tested *A. leiocarpus* extracts could play a noteworthy anti-inflammatory role, as indicated by their blunting effects on LPS-stimulated PGE_2 level. On the other hand, the lack of efficacy that was exerted by the leaf water and stem bark MeOH extracts on LPS-induced levels of 5-HT and 8-iso-$PGF_{2\alpha}$ suggests that the lower quantitative profile of gallic acid, catechin, and epicatechin could limit the antioxidant potency, as compared to the other tested extracts.

Finally, the extracts were tested for their putative anti-proliferative role against the human colon HCT116 cell line, which was previously found to be sensitive to different polarity extracts from *A. latifolia* [12]. In the present study, the anti-proliferative effects were investigated through validated in vitro tests, including MTT and wound healing assays. A different pattern of effects on cell viability was observed after exposing HCT116 cells to *A. leiocarpus* extracts. On one side, stem bark extracts displayed a null effect on cell viability (Figure 10), which resulted in the range of biocompatibility (>70% and <130% as compared to control group). On the other side, leaf extract increased significantly HCT116 cell viability (>140% as compared to control group). Actually, the stimulating effect on HCT116 cell viability induced by leaf extracts could be related to the their higher qualitative content of metabolites that are related to quercetin, which was found to exert protective effects on this cell line [60]. Additionally, water leaf and water and MeOH stem bark extracts induced spontaneous HCT116 cell migration in wound healing assay (Figure 11), thus also indicating a potential stimulating-effect on invasion capacity. Conversely, stem bark EA and leaf EA and MeOH extracts did not exert any influence on spontaneous HCT116 cell migration, in the 24 h following experimental lesion induced on cell monolayer. Actually, the different effects that were showed by the tested extracts in the wound healing paradigm could be related, at least in part, to the different content of gallic acid, which was found to inhibit spontaneous cell migration [61]. As a conclusive note of the pharmacological investigation, it resulted of particular interest the antioxidant/anti-inflammatory profile that was exerted by the stem bark EA extract, together with its null effect on colon cancer cell proliferation. The highest inhibitory effect that is exerted by this extract on colon 5-HT level could be one of the main causes leading to the null effect on HCT116 viability and spontaneous migration [62].

5. Conclusions

Data that are presented in this study highlighted the key role of solvent choice in the quest for novel bioactive compounds from plants. It was demonstrated that water and methanol were good solvents for the extraction of phytochemicals having antioxidant properties. The methanol extract of *A. leiocarpus* leaves was also an active cholinesterase inhibitor, while the methanol extract of the stem bark inhibited tyrosinase. On the other hand, the ethyl acetate extract of *A. leiocarpus* leaves and stem bark showed potent inhibition against α-glucosidase. The pharmacological study that was carried out on isolated colon and HCT116 cell line further deepened the spectrum of potential application of the present extracts. Noteworthy interest derives from the stem bark EA extract that, besides exerting the best antioxidant/anti-inflammatory profile, was the only one that was unable to stimulate the proliferation of human colon cancer HCT116 cell line, thus supporting potential application in the prevention of the oxidative stress-induced tissue damage occurring in ulcerative colitis. As a conclusion, *A. leiocarpus* stem bark EA extract represents a potential source of bioactive compounds for the development of novel therapeutic agents.

Supplementary Materials: The following are available online at http://www.mdpi.com/2076-3921/8/9/343/s1. Table S1: Chemical composition of *A. leiocarpus* leaves ethyl acetate extract. Table S2: Chemical composition of *A. leiocarpus* leaves methanol extract. Table S3: Chemical composition of *A. leiocarpus* leaves water extract. Table S4: Chemical composition of *A. leiocarpus* stem bark ethyl acetate extract. Table S5: Chemical composition of *A. leiocarpus* stem bark methanol extract. Table S6: Chemical composition of *A. leiocarpus* stem bark water extract.

Author Contributions: Conceptualization, C.F. and G.Z.; methodology, G.O.; L.M.; software, L.M.; validation, C.F.; G.Z.; L.M.; G.O.; formal analysis, C.F.; G.Z.; investigation, K.I.S.; K.B.; A.D.; J.J.; L.R.; A.C.; S.L.; Z.C.; S.D.S.; C.M.N.P.-A.; resources, C.F.; G.O.; L.M.; data curation, C.F.; G.Z.; writing—original draft preparation, M.F.M.; writing—review and editing, C.F., G.Z.; G.O.; L.M.; visualization, L.B.; supervision, L.B.; project administration, C.F.; L.M.; G.O.; G.Z.; funding acquisition, C.F.; G.O.; L.M.

Acknowledgments: This work was supported by grants from the Italian Ministry of University (FAR 2016 granted to Claudio Ferrante; FAR 2018 granted to Giustino Orlando; FAR 2016 granted to Luigi Menghini).

Conflicts of Interest: The authors declare no conflict of interest.

References

1. Chaabi, M.; Benayache, S.; Benayache, F.; N'Gom, S.; Koné, M.; Anton, R.; Weniger, B.; Lobstein, A. Triterpenes and polyphenols from *Anogeissus leiocarpus* (Combretaceae). *Biochem. Syst. Ecol.* **2008**, *36*, 59–62. [CrossRef]
2. Salih, E.Y.A.; Kanninen, M.; Sipi, M.; Luukkanen, O.; Hiltunen, R.; Vuorela, H.; Julkunen-Tiitto, R.; Fyhrquist, P. Tannins, flavonoids and stilbenes in extracts of African savanna woodland trees *Terminalia brownii, Terminalia laxiflora* and *Anogeissus leiocarpus* showing promising antibacterial potential. *S. Afr. J. Bot.* **2017**, *108*, 370–386. [CrossRef]
3. Ademosun, A.O.; Adebayo, A.A.; Oboh, G. *Anogeissus leiocarpus* attenuates paroxetine-induced erectile dysfunction in male rats via enhanced sexual behavior, nitric oxide level and antioxidant status. *Biomed. Pharmacother.* **2019**, *111*, 1029–1035. [CrossRef] [PubMed]
4. Etuk, E.; Mohammed, B. Informant consensus selection method: A reliability assessment on medicinal plants used in north western Nigeria for the treatment of diabetes mellitus. *Afr. J. Pharm. Pharmacol.* **2009**, *3*, 496–500.
5. Adefegha, S.A.; Oboh, G.; Omojokun, O.S.; Jimoh, T.O.; Oyeleye, S.I. In vitro antioxidant activities of African birch (*Anogeissus leiocarpus*) leaf and its effect on the α-amylase and α-glucosidase inhibitory properties of acarbose. *J. Taibah Univ. Med. Sci.* **2016**, *11*, 236–242. [CrossRef]
6. Belemnaba, L.; Nitiema, M.; Ilboudo, S.; Ouedraogo, N.; Ouedraogo, G.G.; Belemlilga, M.B.; Ouedraogo, S.; Guissou, I.P. O8 Study on antihypertensive activity of an aqueous extract of *Anogeissus leiocarpus* (AEAL) DC Guill et Perr bark of trunk in L-NAME-induced hypertensive rats. *Biochem. Pharmacol.* **2017**, *139*, 112. [CrossRef]
7. Victor, B.Y.; Grace, A. Phytochemical studies, in-vitro antibacterial activities and antioxidant properties of the methanolic and ethyl acetate extracts of the leaves of *Anogeissus leiocarpus*. *Int. J. Biochem. Res. Rev.* **2013**, *3*, 137. [CrossRef]
8. Shuaibu, M.N.; Wuyep, P.T.; Yanagi, T.; Hirayama, K.; Ichinose, A.; Tanaka, T.; Kouno, I. Trypanocidal activity of extracts and compounds from the stem bark of *Anogeissus leiocarpus* and *Terminalia avicennoides*. *Parasitol. Res.* **2008**, *102*, 697–703. [CrossRef]
9. Shuaibu, M.N.; Pandey, K.; Wuyep, P.A.; Yanagi, T.; Hirayama, K.; Ichinose, A.; Tanaka, T.; Kouno, I. Castalagin from *Anogeissus leiocarpus* mediates the killing of Leishmania in vitro. *Parasitol. Res.* **2008**, *103*, 1333–1338. [CrossRef]
10. Vonthron-Sénécheau, C.; Weniger, B.; Ouattara, M.; Bi, F.T.; Kamenan, A.; Lobstein, A.; Brun, R.; Anton, R. In vitro antiplasmodial activity and cytotoxicity of ethnobotanically selected Ivorian plants. *J. Ethnopharmacol.* **2003**, *87*, 221–225. [CrossRef]
11. Akanbi, O.M.; Omonkhua, A.A.; Cyril-Olutayo, C.M. Effect of crude methanolic extract of *Anogeissus leiocarpus* on the liver function of *P. berghei* infected mice. *Int. J. Infect. Dis.* **2014**, *21*, 196. [CrossRef]
12. Diab, K.A.; Guru, S.K.; Bhushan, S.; Saxena, A.K. In vitro anticancer activities of *Anogeissus latifolia, Terminalia bellerica, Acacia catechu* and *Moringa oleiferna* Indian plants. *Asian Pac. J. Cancer Prev.* **2015**, *16*, 6423–6428. [CrossRef] [PubMed]
13. Ferrante, C.; Recinella, L.; Ronci, M.; Menghini, L.; Brunetti, L.; Chiavaroli, A.; Leone, S.; Di Iorio, L.; Carradori, S.; Tirillini, B.; et al. Multiple pharmacognostic characterization on hemp commercial cultivars: Focus on inflorescence water extract activity. *Food Chem. Toxicol.* **2019**, *125*, 452–461. [CrossRef] [PubMed]
14. Zengin, G.; Aktumsek, A. Investigation of antioxidant potentials of solvent extracts from different anatomical parts of *Asphodeline anatolica* E. Tuzlaci: An endemic plant to Turkey. *Afr. J. Tradit. Complement. Altern. Med.* **2014**, *11*, 481–488. [CrossRef] [PubMed]
15. Zengin, G.; Llorent-Martínez, E.J.; Fernández-de Córdova, M.L.; Bahadori, M.B.; Mocan, A.; Locatelli, M.; Aktumsek, A. Chemical composition and biological activities of extracts from three *Salvia* species: *S. blepharochlaena, S. euphratica* var. *leiocalycina*, and *S. verticillata* subsp. *amasiaca*. *Ind. Crops Prod.* **2018**, *111*, 11–21. [CrossRef]

16. Zengin, G.; Uysal, A.; Diuzheva, A.; Gunes, E.; Jekő, J.; Cziáky, Z.; Picot-Allain, C.M.N.; Mahomoodally, M.F. Characterization of phytochemical components of *Ferula halophila* extracts using HPLC-MS/MS and their pharmacological potentials: A multi-functional insight. *J. Pharm. Biomed. Anal.* **2018**, *160*, 374–382. [CrossRef] [PubMed]
17. Rodrıguez-Delgado, M.; Malovana, S.; Perez, J.; Borges, T.; Montelongo, F.G. Separation of phenolic compounds by high-performance liquid chromatography with absorbance and fluorimetric detection. *J. Chromatogr. A* **2001**, *912*, 249–257. [CrossRef]
18. Uysal, S.; Zengin, G.; Locatelli, M.; Bahadori, M.B.; Mocan, A.; Bellagamba, G.; De Luca, E.; Mollica, A.; Aktumsek, A. Cytotoxic and enzyme inhibitory potential of two Potentilla species (*P. speciosa* L. and *P. reptans* Willd.) and their chemical composition. *Front. Pharmacol.* **2017**, *8*, 290. [CrossRef]
19. Mahmoodzadeh, H.; GhasemI, M.; Zanganeh, H. Allelopathic effect of medicinal plant *Cannabis sativa* L. on *Lactuca sativa* L. seed germination. *Acta Agric. Slov.* **2015**, *105*, 233–239. [CrossRef]
20. Brunetti, L.; Orlando, G.; Ferrante, C.; Recinella, L.; Leone, S.; Chiavaroli, A.; Di Nisio, C.; Shohreh, R.; Manippa, F.; Ricciuti, A.; et al. Peripheral chemerin administration modulates hypothalamic control of feeding. *Peptides* **2014**, *51*, 115–121. [CrossRef]
21. Ferrante, C.; Orlando, G.; Recinella, L.; Leone, S.; Chiavaroli, A.; Di Nisio, C.; Shohreh, R.; Manippa, F.; Ricciuti, A.; Vacca, M.; et al. Central inhibitory effects on feeding induced by the adipo-myokine irisin. *Eur. J. Pharmacol.* **2016**, *791*, 389–394. [CrossRef] [PubMed]
22. Chiavaroli, A.; Brunetti, L.; Orlando, G.; Recinella, L.; Ferrante, C.; Leone, S.; Di Michele, P.; Di Nisio, C.; Vacca, M. Resveratrol inhibits isoprostane production in young and aged rat brain. *J. Biol. Regul. Homeost. Agents* **2010**, *24*, 441. [PubMed]
23. Locatelli, M.; Macchione, N.; Ferrante, C.; Chiavaroli, A.; Recinella, L.; Carradori, S.; Zengin, G.; Cesa, S.; Leporini, L.; Leone, S. Graminex pollen: Phenolic pattern, colorimetric analysis and protective effects in immortalized prostate cells (PC3) and rat prostate challenged with LPS. *Molecules* **2018**, *23*, 1145. [CrossRef] [PubMed]
24. National Centre for Replacement Refinement& Reduction of Animals in Research. Available online: https://www.nc3rs.org.uk/experimental-designstatistics (accessed on 16 August 2019).
25. Galanakis, C.M.; Goulas, V.; Tsakona, S.; Manganaris, G.A.; Gekas, V. A Knowledge Base for the Recovery of Natural Phenols with Different Solvents. *Int. J. Food Prop.* **2013**, *16*, 382–396. [CrossRef]
26. Ji, W.; Meng, Q.; Ding, L.; Wang, F.; Dong, J.; Zhou, G.; Wang, B. Measurement and correlation of the solubility of caffeic acid in eight mono and water+ethanol mixed solvents at temperatures from (293.15 to 333.15) K. *J. Mol. Liq.* **2016**, *224*, 1275–1281. [CrossRef]
27. Lin, L.-T.; Chen, T.-Y.; Chung, C.-Y.; Noyce, R.S.; Grindley, T.B.; McCormick, C.; Lin, T.-C.; Wang, G.-H.; Lin, C.-C.; Richardson, C.D. Hydrolyzable Tannins (Chebulagic Acid and Punicalagin) Target Viral Glycoprotein-Glycosaminoglycan Interactions To Inhibit Herpes Simplex Virus 1 Entry and Cell-to-Cell Spread. *J. Virol.* **2011**, *85*, 4386–4398. [CrossRef] [PubMed]
28. Perera, H.D.S.M.; Samarasekera, J.K.R.R.; Handunnetti, S.M.; Weerasena, O.V.D.S.J. In vitro anti-inflammatory and anti-oxidant activities of Sri Lankan medicinal plants. *Ind. Crops Prod.* **2016**, *94*, 610–620. [CrossRef]
29. Holtz, R.W. Chapter 13—In Vitro Methods to Screen Materials for Anti-aging Effects. In *Skin Aging Handbook*; Dayan, N., Ed.; William Andrew Publishing: Norwich, NY, USA, 2009; pp. 329–362.
30. Cerretani, L.; Bendini, A. Chapter 67—Rapid Assays to Evaluate the Antioxidant Capacity of Phenols in Virgin Olive Oil. In *Olives and Olive Oil in Health and Disease Prevention*; Preedy, V.R., Watson, R.R., Eds.; Academic Press: San Diego, CA, USA, 2010; pp. 625–635.
31. Özyürek, M.; Güçlü, K.; Tütem, E.; Başkan, K.S.; Erçağ, E.; Celik, S.E.; Baki, S.; Yıldız, L.; Karaman, Ş.; Apak, R. A comprehensive review of CUPRAC methodology. *Anal. Methods* **2011**, *3*, 2439–2453. [CrossRef]
32. Puntarulo, S. Iron, oxidative stress and human health. *Mol. Asp. Med.* **2005**, *26*, 299–312. [CrossRef]
33. Mira, L.; Fernandez, M.T.; Santos, M.; Rocha, R.; Florencio, M.H.; Jennings, K.R. Interactions of flavonoids with iron and copper ions: A mechanism for their antioxidant activity. *Free Radic. Res.* **2002**, *36*, 1199–1208. [CrossRef]
34. Mandel, S.; Amit, T.; Reznichenko, L.; Weinreb, O.; Youdim, M.B. Green tea catechins as brain-permeable, natural iron chelators-antioxidants for the treatment of neurodegenerative disorders. *Mol. Nutr. Food Res.* **2006**, *50*, 229–234. [CrossRef] [PubMed]

35. Liu, Y.; Guo, M. Studies on Transition Metal-Quercetin Complexes Using Electrospray Ionization Tandem Mass Spectrometry. *Molecules* **2015**, *20*, 8583–8594. [CrossRef] [PubMed]
36. Santos, M.A.; Chand, K.; Chaves, S. Recent progress in multifunctional metal chelators as potential drugs for Alzheimer's disease. *Coord. Chem. Rev.* **2016**, *327*, 287–303. [CrossRef]
37. Ragab, H.M.; Teleb, M.; Haidar, H.R.; Gouda, N. Chlorinated tacrine analogs: Design, synthesis and biological evaluation of their anti-cholinesterase activity as potential treatment for Alzheimer's disease. *Bioorg. Chem.* **2019**, *86*, 557–568. [CrossRef] [PubMed]
38. de Andrade, P.; Mantoani, S.P.; Gonçalves Nunes, P.S.; Magadán, C.R.; Pérez, C.; Xavier, D.J.; Hojo, E.T.S.; Campillo, N.E.; Martínez, A.; Carvalho, I. Highly potent and selective aryl-1,2,3-triazolyl benzylpiperidine inhibitors toward butyrylcholinesterase in Alzheimer's disease. *Bioorg. Med. Chem.* **2019**, *27*, 931–943. [CrossRef] [PubMed]
39. Liyanaarachchi, G.D.; Samarasekera, J.K.R.R.; Mahanama, K.R.R.; Hemalal, K.D.P. Tyrosinase, elastase, hyaluronidase, inhibitory and antioxidant activity of Sri Lankan medicinal plants for novel cosmeceuticals. *Ind. Crops Prod.* **2018**, *111*, 597–605. [CrossRef]
40. Wang, R.; Wang, G.; Xia, Y.; Sui, W.; Si, C. Functionality study of lignin as a tyrosinase inhibitor: Influence of lignin heterogeneity on anti-tyrosinase activity. *Int. J. Biol. Macromol.* **2019**, *128*, 107–113. [CrossRef]
41. Chen, Y.H.; Huang, L.; Wen, Z.H.; Zhang, C.; Liang, C.H.; Lai, S.T.; Luo, L.Z.; Wang, Y.Y.; Wang, G.H. Skin whitening capability of shikimic acid pathway compound. *Eur. Rev. Med. Pharmacol. Sci.* **2016**, *20*, 1214–1220.
42. Mapunya, M.B.; Hussein, A.A.; Rodriguez, B.; Lall, N. Tyrosinase activity of *Greyia flanaganii* (Bolus) constituents. *Phytomedicine* **2011**, *18*, 1006–1012. [CrossRef]
43. Yao, Y.; Cheng, X.; Wang, L.; Wang, S.; Ren, G. Mushroom tyrosinase inhibitors from mung bean (*Vigna radiatae* L.) extracts. *Int. J. Food Sci. Nutr.* **2012**, *63*, 358–361. [CrossRef]
44. Ohikhena, F.U.; Wintola, O.A.; Afolayan, A.J. Toxicity Assessment of Different Solvent Extracts of the Medicinal Plant, *Phragmanthera capitata* (Sprengel) Balle on Brine Shrimp (*Artemia salina*). *Int. J. Pharmacol.* **2016**, *12*, 701–710.
45. Taviano, M.F.; Marino, A.; Trovato, A.; Bellinghieri, V.; Melchini, A.; Dugo, P.; Cacciola, F.; Donato, P.; Mondello, L.; Güvenç, A. *Juniperus oxycedrus* L. subsp. *oxycedrus* and *Juniperus oxycedrus* L. subsp. macrocarpa (Sibth. & Sm.) Ball. "berries" from Turkey: Comparative evaluation of phenolic profile, antioxidant, cytotoxic and antimicrobial activities. *Food Chem. Toxicol.* **2013**, *58*, 22–29. [PubMed]
46. Belemnaba, L.; Ouédraogo, S.; Nitiéma, M.; Chataigneau, T.; Guissou, I.P.; Schini-Kerth, V.B.; Bucher, B.; Auger, C. An aqueous extract of the *Anogeissus leiocarpus* bark (AEAL) induces the endothelium-dependent relaxation of porcine coronary artery rings involving predominantly nitric oxide. *J. Basic Clin. Physiol. Pharmacol.* **2018**, *29*, 599–608. [CrossRef] [PubMed]
47. Locatelli, M.; Ferrante, C.; Carradori, S.; Secci, D.; Leporini, L.; Chiavaroli, A.; Leone, S.; Recinella, L.; Orlando, G.; Martinotti, S. Optimization of aqueous extraction and biological activity of *Harpagophytum procumbens* root on ex vivo rat colon inflammatory model. *Phytother. Res.* **2017**, *31*, 937–944. [CrossRef] [PubMed]
48. Menghini, L.; Ferrante, C.; Leporini, L.; Recinella, L.; Chiavaroli, A.; Leone, S.; Pintore, G.; Vacca, M.; Orlando, G.; Brunetti, L. An hydroalcoholic chamomile extract modulates inflammatory and immune response in HT29 cells and isolated rat colon. *Phytother. Res.* **2016**, *30*, 1513–1518. [CrossRef]
49. Uttara, B.; Singh, A.V.; Zamboni, P.; Mahajan, R. Oxidative stress and neurodegenerative diseases: A review of upstream and downstream antioxidant therapeutic options. *Curr. Neuropharmacol.* **2009**, *7*, 65–74. [CrossRef] [PubMed]
50. Achitei, D.; Ciobica, A.; Balan, G.; Gologan, E.; Stanciu, C.; Stefanescu, G. Different profile of peripheral antioxidant enzymes and lipid peroxidation in active and non-active inflammatory bowel disease patients. *Dig. Dis. Sci.* **2013**, *58*, 1244–1249. [CrossRef] [PubMed]
51. Praticò, D. Alzheimer's disease and oxygen radicals: New insights. *Biochem. Pharmacol.* **2002**, *63*, 563–567. [CrossRef]
52. Brunetti, L.; Menghini, L.; Orlando, G.; Recinella, L.; Leone, S.; Epifano, F.; Lazzarin, F.; Chiavaroli, A.; Ferrante, C.; Vacca, M. Antioxidant effects of garlic in young and aged rat brain in vitro. *J. Med. Food* **2009**, *12*, 1166–1169. [CrossRef]

53. Akanbi, O.M.; Omonkhua, A.A.; Cyril-Olutayo, C.M.; Fasimoye, R.Y. The antiplasmodial activity of *Anogeissus leiocarpus* and its effect on oxidative stress and lipid profile in mice infected with *Plasmodium bergheii*. *Parasitol. Res.* **2012**, *110*, 219–226. [CrossRef]
54. Tsikas, D. Assessment of lipid peroxidation by measuring malondialdehyde (MDA) and relatives in biological samples: Analytical and biological challenges. *Anal. Biochem.* **2017**, *524*, 13–30. [CrossRef] [PubMed]
55. Yu, L.; Yan, J.; Sun, Z. D-limonene exhibits anti-inflammatory and antioxidant properties in an ulcerative colitis rat model via regulation of iNOS, COX-2, PGE2 and ERK signaling pathways. *Mol. Med. Rep.* **2017**, *15*, 2339–2346. [CrossRef] [PubMed]
56. Regmi, S.C.; Park, S.-Y.; Ku, S.K.; Kim, J.-A. Serotonin regulates innate immune responses of colon epithelial cells through Nox2-derived reactive oxygen species. *Free Radic. Biol. Med.* **2014**, *69*, 377–389. [CrossRef] [PubMed]
57. Mousavizadeh, K.; Rahimian, R.; Fakhfouri, G.; Aslani, F.; Ghafourifar, P. Anti-inflammatory effects of 5-HT3 receptor antagonist, tropisetron on experimental colitis in rats. *Eur. J. Clin. Investig.* **2009**, *39*, 375–383. [CrossRef] [PubMed]
58. Bungo, T.; Shiraishi, J.-i.; Yanagita, K.; Ohta, Y.; Fujita, M. Effect of nociceptin/orphanin FQ on feeding behavior and hypothalamic neuropeptide expression in layer-type chicks. *Gen. Comp. Endocrinol.* **2009**, *163*, 47–51. [CrossRef]
59. Clark, K.A.; MohanKumar, S.M.; Kasturi, B.S.; MohanKumar, P.S. Effects of central and systemic administration of leptin on neurotransmitter concentrations in specific areas of the hypothalamus. *Am. J. Physiol. Regul. Integr. Comp. Physiol.* **2006**, *290*, R306–R312. [CrossRef] [PubMed]
60. Salem, I.B.; Boussabbeh, M.; Graiet, I.; Rhouma, A.; Bacha, H.; Essefi, S.A. Quercetin protects HCT116 cells from Dichlorvos-induced oxidative stress and apoptosis. *Cell Stress Chaperones* **2016**, *21*, 179–186. [CrossRef] [PubMed]
61. Wang, X.; Liu, K.; Ruan, M.; Yang, J.; Gao, Z. Gallic acid inhibits fibroblast growth and migration in keloids through the AKT/ERK signaling pathway. *Acta Biochim. Biophys. Sin.* **2018**, *50*, 1114–1120. [CrossRef]
62. Sui, X.; Zhu, J.; Tang, H.; Wang, C.; Zhou, J.; Han, W.; Wang, X.; Fang, Y.; Xu, Y.; Li, D.; et al. p53 controls colorectal cancer cell invasion by inhibiting the NF-κB-mediated activation of Fascin. *Oncotarget* **2015**, *6*, 22869–22879. [CrossRef]

Article

Increase of 4-Hydroxybenzoic, a Bioactive Phenolic Compound, after an Organic Intervention Diet

Sara Hurtado-Barroso [1,2,3], Paola Quifer-Rada [4], María Marhuenda-Muñoz [1,2,3], Jose Fernando Rinaldi de Alvarenga [1,3], Anna Tresserra-Rimbau [2,5] and Rosa M. Lamuela-Raventós [1,2,3,*]

1 Department of Nutrition, Food Sciences and Gastronomy, School of Pharmacy and Food Sciences, University of Barcelona, 08028 Barcelona, Spain
2 Consorcio CIBER, M.P. Fisiopatología de la Obesidad y Nutrición (CIBERObn), Instituto de Salud Carlos III (ISCIII), 28029 Madrid, Spain
3 INSA-UB, Nutrition and Food Safety Research Institute, University of Barcelona, 08028 Barcelona, Spain
4 Department of Endocrinology & Nutrition, CIBER of Diabetes and Associated Metabolic Diseases, Biomedical Research Institute Sant Pau, Hospital de la Santa Creu i Sant Pau, 08041 Barcelona, Spain
5 Unitat de Nutrició Humana, Hospital Universitari Sant Joan de Reus, Departament de Bioquímica i Biotecnologia, Institut d'Investigació Pere Virgili (IISPV), Universitat Rovira i Virgili, 43007 Reus, Spain
* Correspondence: lamuela@ub.edu; Tel.: +34-934-034-843

Received: 24 July 2019; Accepted: 22 August 2019; Published: 24 August 2019

Abstract: Consumption of organic products is increasing yearly due to perceived health-promoting qualities. Several studies have shown higher amounts of phytochemicals such as polyphenols and carotenoids in foods produced by this type of agriculture than in conventional foods, but whether this increase has an impact on humans still needs to be assessed. A randomized, controlled and crossover study was carried out in nineteen healthy subjects aged 18–40 years, who all followed an organic and conventional healthy diet, both for a 4-week period. Analysis of biological samples revealed a significant increase on the excretion of 4-hydroxybenzoic acid (4-HBA), a phenolic metabolite with biological activity, after the organic intervention. However, no changes were observed in the other variables analyzed.

Keywords: healthy diet; phenolic acid; 4-HBA; crossover study; carotenes; microbiota metabolites; intervention; humans; metals

1. Introduction

Organic food consumption has been increasing yearly over the last decade due to growing public awareness of its environmental benefits and alleged healthy properties [1,2]. The general belief that organic produce is healthier due to a lower use of chemical agents, such as pesticides, fertilizers and antibiotics, [3] is supported by studies reporting lower concentrations of pesticide residues in individuals consuming organic food [4–8]. Differences in nutritional composition associated with the cropping system have also been found, but more studies are needed to draw conclusions [9]. Factors known to influence the nutritional composition of food include crop variety, geographical location, climatic conditions, soil type, season and state of maturity from harvest to storage. Organic food seems to have higher amounts of bioactive compounds such as polyphenols and carotenoids than conventionally produced food [10–16]. When exposed to a stressful environment, plants activate defense mechanisms. Accordingly, a lack of synthetic protectors (pesticides, chemical fertilizers, etc.) induces organic crops to produce phytochemicals. Phenolic acids represent one third of the phenol group in a diet, but also many of them are produced from dietary polyphenols through microbiota metabolism. Approximately 90% of polyphenols are not absorbed in the small intestine reaching the colon, where they are transformed to other compounds such as the phenolic acids [17,18]. In addition, lower concentrations of cadmium have been observed in organic versus conventional cereals [19], as well as differences in the content of fatty acids and proteins [20–23]. Among foods of animal origin, total polyunsaturated fatty acid (PUFA) and n-3 PUFA concentrations are higher in organically rather than conventionally produced milk [21]. A similar profile has been observed in meat, although the evidence is weak [23].

The few studies to evaluate the effect of organic foods on human biochemical parameters and health have employed methodologies with some limitations and provide inconclusive results [9], so further intervention studies are needed to corroborate their possible beneficial effects. Consumers of organic produce are associated with having a higher quality diet, lower body mass index (BMI), greater physical activity [24,25] and a generally healthier and more holistic lifestyle [26,27]. Thus, the question is the following: Are the consumers of organic food healthier due to their lifestyle or also because their diet has a superior nutritional value?

The aim of this study was to evaluate the effect of an intervention with organic diet versus a conventional one on biological parameters, inorganic elements, bioactive compounds, and phenolic acids and carotenes in healthy subjects.

2. Materials and Methods

2.1. Study Subjects

Twenty-one healthy volunteers aged 18–40 years were included in the intervention, 19 of whom completed the study and two dropped out alleging personal reasons. Participants had previous interest in healthy diets and organic food, and they were recruited from the Food and Nutrition Torribera Campus of the University of Barcelona and surroundings. Exclusion criteria were history of cancer, cardiovascular diseases, hypertension and dyslipidemia, chronic illness or homeostatic disorder, as well as toxic habits such as tobacco and other drugs and an excessive alcohol intake.

After approval of the protocol by the Ethics Committee of Clinical Investigation of the University of Barcelona (Barcelona, Spain), the study was registered (ISRCTN29145931). Each participant signed an informed consent prior to the start, which was conducted according to the principles of the Declaration of Helsinki.

2.2. Study Design

An open, crossover, randomized and controlled study was carried out (Figure 1). Each volunteer consumed an organic diet (OD) and a conventional diet (CD), both for 4 weeks, and received dietary advice to support adherence. Organic products represented at least 80% of the OD and no organic foods were allowed in the CD. In both diets, subjects were encouraged to follow a healthy Mediterranean diet with a similar food pattern. Additionally, during the OD intervention participants were given weekly vouchers from Ecoveritas S.A., as well as products (oil, wine, snacks and canned vegetables) from other organic food companies to facilitate dietary compliance. At the end of each intervention, the absence of differential dietary patterns was checked. Interventions were separated by a washout period of two months. The study was run in the Department of Nutrition, Food Science and Gastronomy of the Food Science and Nutrition Torribera Campus of the University of Barcelona (Spain).

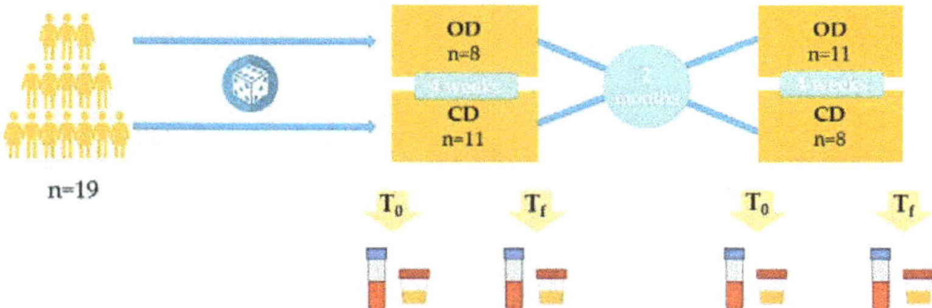

Figure 1. Study design. OD: Organic diet; CD: Conventional diet; T_0: Initial time point (before interventions); T_f: Final time point (after interventions).

2.3. Assessment of Diet and Physical Activity

Before the study, adherence to the Mediterranean diet and physical activity were measured through a 14-item questionnaire [28] and the validated Spanish version of the Minnesota Leisure-Time Physical Activity Questionnaire [29], respectively. Also, at baseline, participants were asked about the frequency of organic food and beverage intake. After each intervention, a 137-item semi-quantitative food frequency questionnaire was filled in with the help of the study staff to assess nutrient and food intake [30].

2.4. Anthropometric and Clinical Data Measurements

Body weight was measured using an electronic scale and height with a stadiometer. The BMI was calculated from body weight and height. Waist and hip circumferences were measured with a measuring tape accurate to 0.1 cm. The waist-hip ratio (WHR) was calculated from these parameters.

Diastolic and systolic blood pressure (DBP and SBP) and heart rate were measured in fasting conditions with an OMRON M6 monitor in triplicate at each visit.

2.5. Sample Collection

Fasting blood was collected before and after each intervention. Blood samples were collected from the arm via venipuncture using tubes containing ethylenediaminetetraacetic acid (EDTA). After centrifugation of blood samples at 1902× g for 15 min at 4 °C, plasma was obtained. In addition, 24 h urine was collected at each visit. Plasma and urine were stored at −80 °C.

2.6. Laboratory Evaluations

Biochemical analyses were performed by an external accredited laboratory (mdb.lad Durán Bellido) as follows. C-reactive protein (CRP) was assayed by an immunoturbidimetry method. The lipid parameters (high density lipoprotein (HDL), low density lipoprotein (LDL) and total cholesterol and triglycerides) were tested by an enzymatic method. Urea and uric acid were measured by enzymatic and enzymatic/chromogen methods, respectively, and creatinine by the Jaffe method (as modified by Larsen) [31]. The concentration of total proteins was quantified by a Biuret reaction to the final point and amount of albumin by a bromocresol green method.

2.7. Analysis of Inorganic Elements in Plasma

Plasma samples were digested with nitric acid (HNO_3) (Instra, J.T. Baker) in Teflon reactors. After incubation at 90 °C overnight, Milli-Q water was added to the reactors. An aliquot was transferred into assay tubes and stored at 4 °C for the chromatographic analyses. The inorganic compounds (Inorganic ventures, Christiansburg, VA, USA) used as standards were the following: Iron (Fe), arsenic (As), copper (Cu), cadmium (Cd), uranium (U), lead (Pb), zinc (Zn), calcium (Ca), magnesium (Mg), potassium (K) and sodium (Na). Fe, As, Cu, Cd, U, Pb and Zn were analyzed by ICP-MS (NexIon

350D. Perkin Elmer, Waltham, MA, USA) and Ca, Mg, K, P and Na, by ICP-OES (Optima8300. Perkin, Waltham, MA, USA). The analyses were performed in the facilities of the CCIT (Centres Científics i Tecnològics) of the University of Barcelona.

2.8. Extraction and Quantification of Phenolic Acids from Urine

Urinary phenolic compounds were extracted by solid phase extraction using a Waters Oasis HLB 96-well plate 30 μm (30 mg; Waters Oasis, Milford, MA, USA) [32]. Chromatographic analysis of phenolic compounds was performed by ultra-high performance liquid chromatography tandem mass spectrometry (UHPLC-MS/MS), using an API 3000 triple-quadrupole mass spectrometer (Sciex, Framingham, MA, USA). The separation was carried out with Milli-Q water and acetonitrile (Panreac Quimica S.A., Barcelona, Spain) with 0.025% formic acid in both solvents (Scharlau Chemie S.A., Barcelona, Spain), according to a method validated by our group [32]. A Waters BEH C18 column 1.7 μm (50 mm × 2.1 mm) and an Acquity UPLC BEH C18 VanGuard pre-column 1.7 μm (2.1 mm × 2.0 mm) were used.

The pool of standards was prepared in synthetic urine and included 3-(4-hydroxyphenyl) propionic acid (3,4-HPPA), 4-hydroxybenzoic acid (4-HBA), 3,4-dihydroxyphenylacetic acid (3,4-DHPAA), 3-hydroxyphenylacetic acid (3-HPAA), dihydrocaffeic acid (DHCA), hippuric acid, homovanillic acid, caffeic acid (CA), m-coumaric acid (m-Cou), p-coumaric (p-Cou) and gallic acid (GA) (Sigma-Aldrich, St. Louis, MO, USA) and 4-hydroxyhippuric acid (4-HH) (Bachem Americas Inc, Torrance, CA, USA). Ethylgallate (Extrasynthese, Genay, France) was the internal standard.

2.9. Extraction and Quantification of Carotenoids from Plasma

Carotenoids were extracted from plasma samples by liquid–liquid extraction [33]. Chromatographic analysis of carotenoids was performed by high performance liquid chromatography with ultraviolet diode-array detector (HPLC-UV-DAD), using an HP 1100 HPLC system (Hewlett-105 Packard, Waldbronn, Germany) containing a quaternary pump, coupled to a DAD G1315B. The separation was carried out with Milli-Q water, methanol and methyl-tert-butyl ether (Panreac Quimica S.A., Barcelona, Spain), according to a procedure previously validated by our group [33]. A Waters reversed-phase column YMC Carotenoid S-5 μm (250 mm × 4.6 mm) and a precolumn YMC Guard Cartridge Carotenoid S-5 μm (20 mm × 4.0 mm) were used.

The standards used were α-carotene, β-carotene, and all-E-lycopene (Sigma-Aldrich, St. Louis, MO, USA) and 5-Z-licopene (CaroteNature GmbH, Ostermundigen, Switzerland). These were pooled and prepared in synthetic human plasma (Sigma-Aldrich, St. Louis, MO, USA).

2.10. Statistical Analysis

Normality of distribution was assessed by a Shapiro-Wilk test. A non-parametric Wilcoxon signed-rank test was used for all statistical analysis due to the small sample size and the non-normality distribution. First, baseline measures were compared to corroborate similar pre-intervention conditions. As no significant differences between interventions at baseline were observed, the final analysis was performed with post-intervention measures (n = 19). Baseline values of variables were calculated from the mean of 38 observations (2 measurements for each subject). Differences were considered statistically significant when $p < 0.05$. Statistical analysis was performed using SPSS Version 23.0 for Windows (SPSS Inc, Chicago, IL, USA).

3. Results

3.1. Participant Characteristics

Table 1 shows the baseline characteristics of participants. Nineteen healthy subjects (9 males and 10 females) completed the study. Approximately three out of every four individuals were occasional consumers of organic products (foods and beverages). The mean age was 30 years and subjects were physically active. The baseline adherence to the Mediterranean diet was high in 7 individuals (≥ 10 points); moderate in 11 (6–9 points) and low in 1 (≤ 5 points).

Table 1. Baseline characteristics of the participants (*n* = 38).

Characteristics	
Males, *n* (%)	9 (47)
Occasional intake of organic products, *n* (%)	14 (74)
Age (years)	30 ± 1
Physical activity in leisure time (METS-min/week)	3814 ± 489
14-item MedDiet score (points)	9 ± 0.3
Weight (kg)	63 ± 2
BMI (kg/m^2)	22.1 ± 0.4
Waist (cm)	76 ± 1
WHR	0.79 ± 0.01
DBP (mmHg)	75 ± 2
SBP (mmHg)	116 ± 2
Heart rate (bpm)	68 ± 2
CRP (mg/dL)	0.14 ± 0.03
HDL (mg/dL)	62 ± 3
LDL (mg/dL)	93 ± 5
Total cholesterol (mg/dL)	169 ± 6
Triglycerides (mg/dL)	69 ± 4
Urea (mg/dL)	29 ± 1
Creatinine (mg/dL)	0.80 ± 0.02
Uric acid (mg/dL)	4.60 ± 0.18
Total proteins (g/L)	72 ± 1
Albumin (g/L)	44 ± 0

Data are mean ± SEM unless otherwise specified. BMI: body mass index, WHR: waist–hip ratio, DBP: diastolic blood pressure, SBP: systolic blood pressure, CRP: C-reactive protein, HDL: high density lipoprotein, LDL: low density lipoprotein.

Baseline anthropometric (weight, BMI, waist and WHR), clinical (DBP, SBP and heart rate) and biochemical (CRP, HDL, LDL, total cholesterol, triglycerides, urea, creatinine, uric acid, total protein and albumin) measurements are also given in Table 1. The baseline concentrations of inorganic elements and bioactive compounds (phenolic acids and carotenes) are available as Supplementary Material Tables S1, S2 and S3, respectively.

3.2. Mean Dietary Composition of Participants During the Interventions

Participants followed a similar dietary pattern in both interventions (Table 2), although the OD was lower in protein ($p = 0.036$) and fish/seafood ($p = 0.042$). The mean proportion of macronutrients was the same in both diets (57% carbohydrates, 24% fats and 19% proteins). A borderline p was obtained comparing dairy products and vegetables ($p = 0.051$ and 0.055). However, the differences between both diets considering individual food were not significant (data not shown). In addition, a significantly lower amount of calcium and phosphorus was ingested in the OD.

Table 2. Nutrient and food intake of participants in both diets ($n = 19$).

	OD	CD	p
	Nutrient intake		
Energy (kcal/d)	1965 ± 203	2062 ± 204	0.070
Carbohydrates (g/d)	211 ± 21	220 ± 22	0.260
Total fat (g/d)	88 ± 9	92 ± 9	0.091
SFA (g/d)	22 ± 3	23 ± 3	0.064
MUFA (g/d)	45 ± 4	46 ± 4	0.136
PUFA (g/d)	12 ± 2	12 ± 1	0.136
Protein (g/d)	68 ± 9	72 ± 9	**0.036** *
Ca (mg/d)	780 ± 111	847 ± 110	**0.024** *
Mg (mg/d)	344 ± 37	353 ± 39	0.376
P (mg/d)	1352 ± 171	1433 ± 169	**0.018** *
Fe (mg/d)	16 ± 1	16 ± 2	0.376
	Food intake		
Dairy products (g/d)	192 ± 52	207 ± 50	0.051
Meat (g/d)	98 ± 20	102 ± 19	0.202
Eggs (g/d)	28 ± 3	31 ± 3	0.180
Fish and seafood (g/d)	56 ± 16	66 ± 16	**0.042** *
Vegetables (g/d)	296 ± 32	366 ± 39	0.055
Fruits (g/d)	360 ± 68	377 ± 70	0.650
Nuts (g/d)	13 ± 5	12 ± 5	0.950
Legumes (g/d)	26 ± 5	26 ± 5	0.528
Cereals (g/d)	98 ± 11	98 ± 10	0.717
Oils (g/d)	40 ± 4	40 ± 4	0.317
Cocoa (g/d)	18 ± 5	21 ± 8	0.812
Coffee (g/d)	62 ± 16	59 ± 16	0.600
Tea (g/d)	22 ± 7	17 ± 7	0.106
Wine (g/d)	54 ± 23	62 ± 28	0.634

Data are mean ± SEM. **p*-value < 0.05. SFA: Saturated fatty acid, MUFA: monounsaturated fatty acid, PUFA: polyunsaturated fatty acid.

3.3. Physiological Parameters of Participants After the Interventions

Table 3 shows anthropometric, clinical and biochemical data of the participants after following the OD and CD.

Table 3. Anthropometric, clinical and biochemical measurements after the interventions ($n = 19$).

	OD	CD	p
	Anthropometric measurements		
Weight (kg)	64 ± 2	63 ± 2	0.365
BMI	22.1 ± 0.6	22.2 ± 0.6	0.352
Waist (cm)	76 ± 1	76 ± 1	0.549
WHR	0.80 ± 0.01	0.79 ± 0.01	0.822
	Clinical measurements		
DBP (mmHg)	79 ± 2	73 ± 2	0.074
SBP (mmHg)	119 ± 4	118 ± 3	0.979
Heart rate (bpm)	70 ± 3	66 ± 2	0.326
	Biochemical measurements		
CRP (mg/dL)	0.17 ± 0.07	0.26 ± 0.11	0.438
HDL (mg/dL)	62 ± 4	60 ± 4	0.301
LDL (mg/dL)	92 ± 9	90 ± 7	0.653
Total cholesterol (mg/dL)	168 ± 9	164 ± 7	0.494
Triglycerides (mg/dL)	66 ± 4	68 ± 4	0.421
Urea (mg/dL)	29 ± 2	29 ± 2	0.913
Creatinine (mg/dL)	0.80 ± 0.03	0.79 ± 0.02	0.763
Uric acid (mg/dL)	4.55 ± 0.29	4.68 ± 0.26	0.456
Total proteins (g/L)	73 ± 1	71 ± 1	0.145
Albumin (g/L)	44 ± 1	43 ± 1	0.136

Data are mean ± SEM. BMI: body mass index, WHR: waist–hip ratio, DBP: diastolic blood pressure, SBP: systolic blood pressure, CRP: C-reactive protein, HDL: high density lipoprotein, LDL: low density lipoprotein.

3.4. Inorganic Elements in Plasma

No significant differences were observed in the plasmatic concentration of minerals and heavy metals between the two diets (Table 4).

Table 4. Inorganic elements in plasma after the interventions (n = 19).

	OD	CD	p
Na (ppm)	2991 ± 20	2992 ± 19	0.445
K (ppm)	839 ± 14	844 ± 11	0.778
Ca (ppm)	88 ± 1	88 ± 1	0.717
Mg (ppm)	18 ± 0	18 ± 0	0.778
P (ppm)	104 ± 3	100 ± 3	0.136
Fe (ppb)	1252 ± 127	1339 ± 107	0.601
Zn (ppb)	778 ± 98	785 ± 46	0.376
Cu (ppb)	858 ± 64	856 ± 71	0.904
As (ppb)	4.35 ± 2.27	3.99 ± 0.95	0.221
Pb (ppb)	BLD	BLD	-
Cd (ppb)	BLD	BLD	-
U (ppb)	BLD	BLD	-

Data are mean ± SEM. BLD: Below limit of detection.

3.5. Phenolic Acids in Urine

Several polyphenols, mainly phenolic acids generated by microbiota metabolism and their derivatives, were evaluated in urine after the interventions (Table 5). A significant increase was observed in 4-HBA (p = 0.028) after the OD compared to the CD, but no changes were detected in the rest of the phenols.

Table 5. Urinary phenolic acids excretion after the interventions (n = 19).

	OD	CD	p
	Phenylacetic acids		
3,4-DHPAA (nmol)	90 ± 35	35 ± 9	0.42
3-HPAA (nmol)	943 ± 594	941 ± 440	0.717
Homovanillic (nmol)	154 ± 60	108 ± 27	0.868
	Phenylpropionic acids		
3,4-HPPA (nmol)	10 ± 3	27 ± 12	0.407
DHCA (nmol)	1.2 ± 0.4	1.2 ± 0.4	0.955
	Hydroxybenzoic and derivatives		
4-HBA (nmol)	205 ± 123	70 ± 35	**0.028** *
4-HH (nmol)	471 ± 225	212 ± 85	0.306
Hippuric (nmol)	1281 ± 235	1463 ± 211	0.231
	Hydroxycinnamic and derivatives		
CA (nmol)	7 ± 2	10 ± 2	0.349
m-Cou (nmol)	0.5 ± 0.3	0.26 ± 0.07	0.501
p-Cou (nmol)	0.3 ± 0.7	0.54 ± 0.19	0.554
GA (nmol)	0.48 ± 0.45	0.07 ± 0.03	0.878

Data are mean ± SEM. * p-value < 0.05. 3,4-DHPAA: 3,4-dihydroxyphenylacetic acid, 3-HPAA: 3-hydroxyphenylacetic acid, 3,4-HPPA: 3-(4-hydroxyphenyl) propionic acid, DHCA: dihydrocaffeic acid, 4-HBA: 4-hydroxybenzoic acid, 4-HH: 4-hydroxyhippuric, CA: caffeic acid, *m*-Cou: *m*-coumaric acid, *p*-Cou: *p*-coumaric acid, GA: gallic acid.

3.6. Carotenoids in Plasma

No significant differences were observed in plasmatic concentrations of carotenes (Table 6).

Table 6. Plasmatic carotenoids after the interventions ($n = 19$).

	OD	CD	p
α-carotene (nmol/mL)	0.39 ± 0.09	0.27 ± 0.06	0.552
β-carotene (nmol/mL)	1.03 ± 0.24	0.95 ± 0.22	0.744
E-lycopene (nmol/mL)	0.7 ± 0.17	0.78 ± 0.18	0.913
Z-lycopene (nmol/mL)	0.15 ± 0.04	0.20 ± 0.05	0.379

Data are mean ± SEM.

4. Discussion

A randomized, controlled and crossover pilot study with nineteen healthy subjects was carried out to assess whether following an OD for 4 weeks changes health parameters and biomarkers compared to a CD.

In this study, the phenol 4-HBA increased approximately three times at the end of the OD compared to the CD ($p = 0.028$). 4-HBA can come from a diet, nevertheless, the intake of food rich in this phenol, such as berries, beer, etc., did not change significantly between both interventions (data not shown). However, this compound is produced from anthocyanins catabolism, as a metabolite of pelargonidin [34–36], and it can be formed by the colonic microbiota [36,37]. The metabolite 4-HBA has shown anticancer and neuroprotective effects [37–40]. Moreover, this compound is a precursor of the coenzyme Q10, showing cardioprotective properties [41,42].

No significant differences in the urinary concentration of the rest of the phenols were observed between the two diets, although vegetable intake was borderline lower in the OD. Stracke et al. carried out a study in which healthy men consumed 500 g of organic or conventional apples for four weeks. Twenty-four hours after the last intake, polyphenol concentrations in plasma and urine were not higher in the organic consumers [43].

Studies on the carotenoid content in organically grown fruits have provided inconclusive results [13,19,44]. In the present work, no effects of the OD on carotenoid levels were detected. In contrast to our results, a previous observational study reported significant differences in both carotenes and other fat-soluble micronutrients after consumption of organic food [45].

No changes in the concentration of inorganic elements in plasma were observed after either intervention. According to other authors, organic agriculture does not affect dietary copper [45,46] or zinc absorption [46]. However, a cohort from the NutriNet-Santé study presented a higher level of magnesium after following an OD, whereas no differences were found in iron absorption [45]. Higher magnesium, iron and phosphorus levels have been described in organic versus conventional plant-derived foods [47]. In contrast, concentrations of cadmium have been reported to be lower in organic food, due to the type of plant fertilizer used, but lower levels in consumers of organic produce were not observed [19,45,48]. In the present study, cadmium was not detected in plasma, nor was lead or uranium. Marchioni et al. showed that the content of cadmium and lead in coffee was influenced by temperature and mass, respectively, but not by the type of crop [49]. Although uranium is used more in conventional than in organic agriculture [50,51], a higher uranium content was not evidenced in conventional produce [52]. We found calcium and phosphate intake was lower in the OD, likely due to a lower consumption of dairy products. Although previous studies have described a higher concentration of phosphate in conventional foods due to crop fertilizers [53], here no differences were detected in the plasma levels between the two diets. Previous findings from the Environmental Defense Fund indicate that organic foods are as likely as conventional foods to contain heavy metals, because the organic standard is focused on pesticides and not these contaminants [54].

ODs are generally believed to be healthier and to provide more bioactive compounds. Some authors have observed a higher concentration of some phytochemicals in organic food, but without considering their bioavailability. In addition, when assessing the nutritional value of food, other influential factors need to be considered, including crop variety, maturity, soil and climate. On the other hand, consumers of organic products tend to be more concerned with health-related issues than the general population, which can bias the results of observational studies.

Organic foods are appreciated for the limited use of synthetic compounds (fertilizers, pesticides and antibiotics) in their production. Nevertheless, conventional crops are also regulated in this respect, and long-term studies are required to corroborate the effect of these compounds on health. To date, evidence suggesting that organic products are more nutritive or healthier is still lacking. Therefore, further carefully designed research is needed to evaluate the effect of an OD on bioactive compounds in biological fluids and health-related biomarkers.

The strongest point of the current study is its crossover design and the evaluation of a dietary pattern instead of only one or a few foods. Also, few such clinical assays have been conducted to date, with most studies being observational. Limitations of the work include a small sample size, the short duration of interventions and some differences in dietary patterns between the two interventions. However, this may be considered a pilot study to assess the short-term effects of organic food consumption. The increase of the phenolic compound arising from microbiota metabolism (4-HBA) in consumers following the OD need to be corroborated by further research with a higher number of subjects, which may shed light on a potential mechanism and possible health beneficial effects. In addition, a better control of factors as crop variety, maturity, soil and climate would provide more reliable results.

5. Conclusions

This intervention study for only one month found a significant difference in the concentration of a phenolic acid, the 4-HBA, after the OD. No changes were observed in the rest of the bioactive compounds analyzed nor in the other health-related biomarkers considered, neither in the results of minerals and heavy metals. The relation between the organic or conventional foods consumed and the concentration of bioactive compounds in the organism should be further researched. Longer studies and with larger sample sizes could reach significant values in other biochemical and heathy variables, demonstrating the health benefits of an OD.

Supplementary Materials: The following are available online at http://www.mdpi.com/2076-3921/8/9/340/s1, Table S1. Baseline concentrations of inorganic elements in plasma; Table S2. Baseline concentrations of urinary excretion of phenolic acids; Table S3. Baseline concentrations of carotenes in plasma.

Author Contributions: Conceptualization, S.H.-B., P.Q.-R., J.F.R.d.A., A.T.-R. and R.M.L.-R.; data curation, S.H.-B. and P.Q.-R.; formal analysis, S.H.-B., P.Q.-R. and A.T.-R.; investigation, S.H.-B.; methodology, S.H.-B., M.M.-M. and J.F.R.d.A.; supervision, R.M.L.-R.; writing—original draft, S.H.-B.; all participants reviewed and approved the paper; and R.M.L.-R. was the main person responsible for the project and the final content.

Funding: This study was supported by CYCIT from the Ministerio de Ciencia, Innovación y Universidades (grant number AGL2016-75329-R), the Instituto de Salud Carlos III—CIBEROBN (C03-01) and Generalitat de Catalunya (SGR 2017)—Departament d'Agricultura, Ramaderia, Pesca i Alimentació—Direcció General d'Agricultura i Ramaderia under grant 53 05012 2016.

Acknowledgments: First: we thank all the participants of the study and sponsors (Ecoveritas S.A., Conservas José Salcedo Soria S.L., Paul & Pippa Gourmet Food S.L., Artfood S.L., Can Feixes, Grupo Codorniu Aceites Borges Pont S.A. and Molí dels Torms, S.L.) who kindly provided organic wine, olive oil, snacks and canned vegetables, and weekly vouchers to shop in organic supermarkets. S.H.B. and M.M.M. received support from the Ministerio de Educación, Cultura y deporte (MECD) through predoctoral scholarship FPU (FPU14/01715 and FPU17/00513, respectively). P.Q.-R. is thankful for the Sara Borrell postdoctoral program from the Instituto de Salud Carlos III (ISCIII). J.F.R.d.A. is grateful to the Science without Borders program for the predoctoral scholarship from Conselho Nacional de Desenvolvimento Científico e Tecnológico (CNPq)—Brazil (233576/2014-2). A.T.-R. thanks the Juan de la Cierva postdoctoral program (FJCI-2016-28694) from the Ministerio de Economía, Industria y Competitividad.

Conflicts of Interest: Dra. R.M.L.-R. reports receiving lecture fees from Cerveceros de España and receiving lecture fees and travel support from Adventia. Moreover, weekly vouchers and other organic products have been provided by Ecoveritas S.A. and sponsors previously named. Nevertheless, these foundations and sponsors were not involved in the study design, the collection, analysis and interpretation of data, the writing of the manuscript, or the decision to submit the manuscript for publication.

References

1. Magnusson, M.K.; Arvola, A.; Koivisto Hursti, U.-K.; Åberg, L.; Sjödén, P.-O. Choice of organic foods is related to perceived consequences for human health and to environmentally friendly behaviour. *Appetite* **2003**, *40*, 109–117. [CrossRef]
2. Dean, M.; Lampila, P.; Shepherd, R.; Arvola, A.; Saba, A.; Vassallo, M.; Claupein, E.; Winkelmann, M.; Lähteenmäki, L. Perceived relevance and foods with health-related claims. *Food Qual. Prefer.* **2012**, *24*, 129–135. [CrossRef]
3. Hoefkens, C.; Verbeke, W.; Aertsens, J.; Mondelaers, K.; Van Camp, J. The nutritional and toxicological value of organic vegetables. *Br. Food J.* **2009**, *111*, 1062–1077. [CrossRef]
4. Curl, C.L.; Fenske, R.A.; Elgethun, K. Organophosphorus pesticide exposure of urban and suburban preschool children with organic and conventional diets. *Environ. Health Perspect.* **2003**, *111*, 377–382. [CrossRef]
5. Lu, C.; Toepel, K.; Irish, R.; Fenske, R.A.; Barr, D.B.; Bravo, R. Organic diets significantly lower children's dietary exposure to organophosphorus pesticides. *Environ. Health Perspect.* **2006**, *114*, 260–263. [CrossRef]
6. Bradman, A.; Quirós-Alcalá, L.; Castorina, R.; Schall, R.A.; Camacho, J.; Holland, N.T.; Barr, D.B.; Eskenazi, B. Effect of Organic Diet Intervention on Pesticide Exposures in Young Children Living in Low-Income Urban and Agricultural Communities. *Environ. Health Perspect.* **2015**, *123*, 1086–1093. [CrossRef]
7. Oates, L.; Cohen, M.; Braun, L.; Schembri, A.; Taskova, R. Reduction in urinary organophosphate pesticide metabolites in adults after a week-long organic diet. *Environ. Res.* **2014**, *132*, 105–111. [CrossRef]
8. Baudry, J.; Debrauwer, L.; Durand, G.; Limon, G.; Delcambre, A.; Vidal, R.; Taupier-Letage, B.; Druesne-Pecollo, N.; Galan, P.; Hercberg, S.; et al. Urinary pesticide concentrations in French adults with low and high organic food consumption: Results from the general population-based NutriNet-Santé. *J. Expo. Sci. Environ. Epidemiol.* **2019**, *29*, 366–378. [CrossRef]
9. Hurtado-Barroso, S.; Tresserra-Rimbau, A.; Vallverdú-Queralt, A.; Lamuela-Raventós, R.M. Organic food and the impact on human health. *Crit. Rev. Food Sci. Nutr.* **2019**, *59*, 704–714. [CrossRef]
10. Vallverdú-Queralt, A.; Lamuela-Raventós, R.M. Foodomics: A new tool to differentiate between organic and conventional foods. *Electrophoresis* **2016**, *37*, 1784–1794. [CrossRef]
11. Vallverdú-Queralt, A.; Medina-Remón, A.; Casals-Ribes, I.; Amat, M.; Lamuela-Raventós, R.M. A metabolomic approach differentiates between conventional and organic ketchups. *J. Agric. Food Chem.* **2011**, *59*, 11703–11710. [CrossRef]
12. Vallverdú-Queralt, A.; Medina-Remón, A.; Casals-Ribes, I.; Lamuela-Raventos, R.M. Is there any difference between the phenolic content of organic and conventional tomato juices? *Food Chem.* **2012**, *130*, 222–227. [CrossRef]
13. Vallverdú-Queralt, A.; Martínez-Huélamo, M.; Casals-Ribes, I.; Lamuela-Raventós, R.M. Differences in the carotenoid profile of commercially available organic and conventional tomato-based products. *J. Berry Res.* **2014**, *4*, 69–77. [CrossRef]
14. Borguini, R.G. *Avaliação do Potencial Antioxidante e de Algumas Características Físico-Químicas do Tomate (Lycopersicon Esculentum) Orgânico em Comparação ao Convencional*; Biblioteca Digital de Teses e Dissertações, Universidade de São Paulo: São Paulo, Brazil, 2006.
15. Györe-Kis, G.; Deák, K.; Lugasi, A.; Csúr-Vargaa, A.; Helyes, L. Comparison of conventional and organic tomato yield from a three-year-term experiment. *Acta Aliment.* **2012**, *41*, 486–493. [CrossRef]
16. Roussos, P.A.; Gasparatos, D. Apple tree growth and overall fruit quality under organic and conventional orchard management. *Sci. Hortic.* **2009**, *123*, 247–252. [CrossRef]
17. Williamson, G.; Manach, C. Bioavailability and bioefficacy of polyphenols in humans. II. Review of 93 intervention studies. *Am. J. Clin. Nutr.* **2005**, *81*, 243S–255S. [CrossRef]
18. Thilakarathna, S.; Rupasinghe, H. Flavonoid Bioavailability and Attempts for Bioavailability Enhancement. *Nutrients* **2013**, *5*, 3367–3387. [CrossRef]
19. Barański, M.; Średnicka-Tober, D.; Volakakis, N.; Seal, C.; Sanderson, R.; Stewart, G.B.; Benbrook, C.; Biavati, B.; Markellou, E.; Giotis, C.; et al. Higher antioxidant and lower cadmium concentrations and lower incidence of pesticide residues in organically grown crops: A systematic literature review and meta-analyses. *Br. J. Nutr.* **2014**, *112*, 794–811. [CrossRef]
20. Palupi, E.; Jayanegara, A.; Ploeger, A.; Kahl, J. Comparison of nutritional quality between conventional and organic dairy products: A meta-analysis. *J. Sci. Food Agric.* **2012**, *92*, 2774–2781. [CrossRef]

21. Średnicka-Tober, D.; Barański, M.; Seal, C.J.; Sanderson, R.; Benbrook, C.; Steinshamn, H.; Gromadzka-Ostrowska, J.; Rembiałkowska, E.; Skwarło-Sona, K.; Eyre, M.; et al. Higher PUFA and n-3 PUFA, conjugated linoleic acid, α-tocopherol and iron, but lower iodine and selenium concentrations in organic milk: A systematic literature review and meta-and redundancy analyses. *Br. J. Nutr.* **2016**, *115*, 1043–1060. [CrossRef]
22. Magkos, F.; Arvaniti, F.; Zampelas, A. Organic food: Nutritious food or food for thought? A review of the evidence. *Int. J. Food Sci. Nutr.* **2003**, *54*, 357–371. [CrossRef]
23. Średnicka-Tober, D.; Barański, M.; Seal, C.; Sanderson, R.; Benbrook, C.; Steinshamn, H.; Gromadzka-Ostrowska, J.; Rembiałkowska, E.; Skwarło-Sońta, K.; Eyre, M.; et al. Composition differences between organic and conventional meat: A systematic literature review and meta-analysis. *Br. J. Nutr.* **2016**, *23*, 1–18. [CrossRef]
24. Baudry, J.; Allès, B.; Péneau, S.; Touvier, M.; Méjean, C.; Hercberg, S.; Galan, P.; Lairon, D.; Kesse-Guyot, E. Dietary intakes and diet quality according to levels of organic food consumption by French adults: Cross-sectional findings from the NutriNet-Santé Cohort Study. *Public Health Nutr.* **2016**, *20*, 638–648. [CrossRef]
25. Eisinger-Watzl, M.; Wittig, F.; Heuer, T.; Hoffmann, I.; Verhagen, H.; Scientific Advisor, S. Customers Purchasing Organic Food—Do They Live Healthier? Results of the German National Nutrition Survey II. *Eur. J. Nutr. Food Saf.* **2015**, *5*, 59–71. [CrossRef]
26. Baudry, J.; Méjean, C.; Allès, B.; Péneau, S.; Touvier, M.; Hercberg, S.; Lairon, D.; Galan, P.; Kesse-Guyot, E. Contribution of organic food to the diet in a large sample of French adults (The NutriNet-Santé cohort study). *Nutrients* **2015**, *7*, 8615–8632. [CrossRef]
27. Goetzke, B.; Nitzko, S.; Spiller, A. Consumption of organic and functional food. A matter of well-being and health? *Appetite* **2014**, *77*, 96–105. [CrossRef]
28. Schröder, H.; Fitó, M.; Estruch, R.; Martínez-González, M.A.; Corella, D.; Salas-Salvadó, J.; Lamuela-Raventós, R.; Ros, E.; Salaverría, I.; Fiol, M.; et al. A Short Screener Is Valid for Assessing Mediterranean Diet Adherence among Older Spanish Men and Women. *J. Nutr.* **2011**, *141*, 1140–1145. [CrossRef]
29. Elosua, R.; Marrugat, J.; Molina, L.; Pons, S.; Pujol, E. Validation of the Minnesota Leisure Time Physical Activity Questionnaire in Spanish men. The MARATHOM Investigators. *Am. J. Epidemiol.* **1994**, *139*, 1197–1209. [CrossRef]
30. Fernández-Ballart, J.D.; Piñol, J.L.; Zazpe, I.; Corella, D.; Carrasco, P.; Toledo, E.; Perez-Bauer, M.; Martínez-González, M.Á.; Salas-Salvadó, J.; Martín-Moreno, J.M. Relative validity of a semi-quantitative food-frequency questionnaire in an elderly Mediterranean population of Spain. *Br. J. Nutr.* **2010**, *103*, 1808–1816. [CrossRef]
31. Larsen, K. Creatinine assay by a reaction-kinetic principle. *Clin. Chim. Acta* **1972**, *41*, 209–217. [CrossRef]
32. Martínez-Huélamo, M.; Tulipani, S.; Jáuregui, O.; Valderas-Martinez, P.; Vallverdú-Queralt, A.; Estruch, R.; Torrado, X.; Lamuela-Raventós, R. Sensitive and Rapid UHPLC-MS/MS for the Analysis of Tomato Phenolics in Human Biological Samples. *Molecules* **2015**, *20*, 20409–20425. [CrossRef]
33. Colmán-Martínez, M.; Martínez-Huélamo, M.; Miralles, E.; Estruch, R.; Lamuela-Raventós, R.M. A New Method to Simultaneously Quantify the Antioxidants: Carotenes, Xanthophylls, and Vitamin A in Human Plasma. *Oxid. Med. Cell Longev.* **2016**, *2016*, 1–10. [CrossRef]
34. de Ferrars, R.M.; Czank, C.; Zhang, Q.; Botting, N.P.; Kroon, P.A.; Cassidy, A.; Kay, C.D. The pharmacokinetics of anthocyanins and their metabolites in humans. *Br. J. Pharmacol.* **2014**, *171*, 3268–3282. [CrossRef]
35. Amini, A.M.; Muzs, K.; Spencer, J.P.; Yaqoob, P. Pelargonidin-3-O-glucoside and its metabolites have modest anti-inflammatory effects in human whole blood cultures. *Nutr. Res.* **2017**, *46*, 88–95. [CrossRef]
36. El Mohsen, M.A.; Marks, J.; Kuhnle, G.; Moore, K.; Debnam, E.; Kaila Srai, S.; Rice-Evans, C.; Spencer, J.P.E. Absorption, tissue distribution and excretion of pelargonidin and its metabolites following oral administration to rats. *Br. J. Nutr.* **2006**, *95*, 51–58. [CrossRef]
37. Sannino, F.; Sansone, C.; Galasso, C.; Kildgaard, S.; Tedesco, P.; Fani, R.; Marino, G.; de Pascale, D.; Ianora, A.; Parrilli, E.; et al. Pseudoalteromonas haloplanktis TAC125 produces 4-hydroxybenzoic acid that induces pyroptosis in human A459 lung adenocarcinoma cells. *Sci. Rep.* **2018**, *8*, 1190. [CrossRef]
38. Wang, X.-N.; Wang, K.-Y.; Zhang, X.-S.; Yang, C.; Li, X.-Y. 4-Hydroxybenzoic acid (4-HBA) enhances the sensitivity of human breast cancer cells to adriamycin as a specific HDAC6 inhibitor by promoting HIPK2/p53 pathway. *Biochem. Biophys. Res. Commun.* **2018**, *504*, 812–819. [CrossRef]

39. Winter, A.N.; Brenner, M.C.; Punessen, N.; Snodgrass, M.; Byars, C.; Arora, Y.; Linseman, D.A. Comparison of the Neuroprotective and Anti-Inflammatory Effects of the Anthocyanin Metabolites, Protocatechuic Acid and 4-Hydroxybenzoic Acid. *Oxid. Med. Cell Longev.* **2017**, *2017*, 1–13. [CrossRef]
40. Distelmaier, F. 4-Hydroxybenzoic acid for multiple system atrophy? *Parkinsonism Relat. Disord.* **2018**, *50*, 119–120. [CrossRef]
41. Timoshchuk, S.V.; Vavilova, H.L.; Strutyns'ka, N.A.; Talanov, S.A.; Petukhov, D.M.; Kuchmenko, O.B.; Donchenko, H.V.; Sahach, V.F. Cardioprotective action of coenzyme Q in conditions of its endogenous synthesis activation in cardiac ischemia-reperfusion in old rats. *BMC Pharmacol. Texicol.* **2009**, *55*, 58–63.
42. Kumchenko, E.B.; Petukhov, D.N.; Donchenko, G.V.; Mkhitarian, L.S.; Timoshchuk, S.V.; Strutinskaia, N.A.; Vavilova, G.L.; Sagach, V.F. Effect of precursors and modulators of coenzyme Q biosynthesis on the heart mitochondria function in aged rats. *Biomed. Khim.* **2010**, *56*, 244–256.
43. Stracke, B.A.; Rüfer, C.E.; Bub, A.; Seifert, S.; Weibel, F.P.; Kunz, C.; Watzl, B. No effect of the farming system (organic/conventional) on the bioavailability of apple (Malus domestica Bork., cultivar Golden Delicious) polyphenols in healthy men: A comparative study. *Eur. J. Nutr.* **2010**, *49*, 301–310. [CrossRef]
44. Cardoso, P.C.; Tomazini, A.P.B.; Stringheta, P.C.; Ribeiro, S.M.R.; Pinheiro-Sant'Ana, H.M. Vitamin C and carotenoids in organic and conventional fruits grown in Brazil. *Food Chem.* **2011**, *126*, 411–416. [CrossRef]
45. Baudry, J.; Ducros, V.; Druesne-Pecollo, N.; Galan, P.; Hercberg, S.; Debrauwer, L.; Amiot, M.J.; Lairon, D.; Kesse-Guyot, E. Some Differences in Nutritional Biomarkers are Detected Between Consumers and Nonconsumers of Organic Foods: Findings from the BioNutriNet Project. *Curr. Dev. Nutr.* **2019**, *3*, nzy090. [CrossRef]
46. Mark, A.B.; Kápolna, E.; Laursen, K.H.; Halekoh, U.; Rasmussen, S.K.; Husted, S.; Larsen, E.H.; Bügel, S. Consumption of organic diets does not affect intake and absorption of zinc and copper in men-evidence from two cross-over trials. *Food Funct.* **2013**, *4*, 409–419. [CrossRef]
47. Rembiałkowska, E. Review Quality of plant products from organic agriculture. *J. Sci. Food Agric.* **2007**, *87*, 2757–2762. [CrossRef]
48. Kratz, S.; Schick, J.; Schnug, E. Trace elements in rock phosphates and P containing mineral and organo-mineral fertilizers sold in Germany. *Sci. Total Environ.* **2016**, *542*, 1013–1019. [CrossRef]
49. Marchioni, C.; de Oliveira, F.M.; de Magalhães, C.S.; Luccas, P.O. Assessment of Cadmium and Lead Adsorption in Organic and Conventional Coffee. *Anal. Sci.* **2015**, *31*, 165–172. [CrossRef]
50. Schnug, E.; Haneklaus, N. *Uranium in Phosphate Fertilizers—Review and Outlook. Uranium—Past and Future Challenges*; Springer International Publishing: Cham, Switzerland, 2015; pp. 123–130.
51. De Kok, L.J.; Luit, J.; Schnug, E. *Loads and Fate of Fertilizer-Derived Uranium*; Backhuys Publishers, 2008; p. 229.
52. Mie, A.; Andersen, H.R.; Gunnarsson, S.; Kahl, J.; Kesse-Guyot, E.; Rembiałkowska, E.; Quaglio, G.; Grandjean, P. Human health implications of organic food and organic agriculture: A comprehensive review. *Environ. Health* **2017**, *16*, 111. [CrossRef]
53. Syers, J.K.; Johnston, A.E.; Curtin, D. Efficiency of Soil and Fertilizer Phosphorus Use Reconciling Changing Concepts of Soil Phosphorus Behaviour with Agronomic Information. 2008. Available online: http://www.fao.org/3/a-a1595e.pdf (accessed on 10 June 2019).
54. Environmental Defense Fund (EDF). Consumer Reports Study Finds "Concerning Levels" of Heavy Metals in Baby and Toddler Foods. Available online: https://www.edf.org/media/consumer-reports-study-finds-concerning-levels-heavy-metals-baby-and-toddler-foods (accessed on 10 June 2019).

Article

Application of Natural Flavonoids to Impart Antioxidant and Antibacterial Activities to Polyamide Fiber for Health Care Applications

Ya-Dong Li [1], Jin-Ping Guan [1,2], Ren-Cheng Tang [1,2,*] and Yi-Fan Qiao [1]

1 College of Textile and Clothing Engineering, Soochow University, 178 East Ganjiang Road, Suzhou 215021, China

2 National Engineering Laboratory for Modern Silk, Soochow University, 199 Renai Road, Suzhou 215123, China

* Correspondence: tangrencheng@suda.edu.cn; Tel.: +86-512-6716-4993

Received: 22 June 2019; Accepted: 8 August 2019; Published: 12 August 2019

Abstract: Polyamide fiber has the requirements for antioxidant and antibacterial properties when applied to produce functional textiles for heath care purposes. In this work, three natural flavonoids (baicalin, quercetin, and rutin) were used to simultaneously impart antioxidant and antibacterial functions to polyamide fiber using an adsorption technology. The relations of the chemical structures of flavonoids with their adsorption capability, adsorption mechanisms, and antioxidant and antibacterial activities were discussed. The Langmuir–Nernst adsorption model fitted the adsorption isotherms of the three flavonoids well. The adsorption kinetics of the three flavonoids conformed to the pseudo second-order kinetic model. Quercetin exhibited the highest affinity and adsorption capability, and imparted the highest antioxidant and antibacterial activities to polyamide fiber; and moreover, its antioxidant and antibacterial functions had good washing durability. This study demonstrates that the treatment using natural flavonoids is an effective way to exhance the health care functions of polyamide fiber.

Keywords: antioxidant activity; antibacterial activity; flavonoids; polyamide fiber; adsorption

1. Introduction

Health care textiles have attracted increasing attention in recent years. Textiles or apparel possessing antibacterial and antioxidant functions can offer health care effects. However, most textiles lack these two functions. Textile materials are easily infested by microbes. Microbial growth and proliferation on textiles can give rise to dermal infections, cross-infections, mildew formation, disease spread, allergic reactions, and foul odors [1]. As a consequence of the great importance of antibacterial properties, the antibacterial functionalization of textiles has attracted more and more interest. Up to now, the antioxidant properties of textiles have been less studied. Textiles containing antioxidants can function as a reservoir system steadily delivering antioxidants to the skin. When in contact with skin, such textiles have the ability to scavenge free radicals caused by skin degeneration, and protect skin tissues from oxidative stress and damage [2,3]. Antioxidant and antibacterial textiles can be utilized to prepare facial masks, patient clothes, and daily clothes for people who have skin diseases. Nowadays consumers are pursuing healthy and comfortable fiber materials which can provide and maintain an optimal microenvironment for healing some disorders or avoiding disease [4]. This promotes the development of novel healthcare textiles possessing antibacterial and antioxidant functions.

Polyamide fiber is one of the three major synthetic fibers, and its consumption is the second after polyester fiber. Polyamide 6 {poly[imino(1-oxohexane-1,6-diyl)]} and polyamide 6,6 {poly[N,N'-(hexane-1,6-diyl)adipamide]} are most extensively used in fiber and textile industries.

Polyamide fiber possesses excellent performance such as good abrasion resistance, elastic resilience, mechanical properties, chemical resistance, temperature resistance, and processability [5], and has found wide applications in underwear, sports/leisure wear, and outerwear. In particular, polyamide fiber is often employed to manufacture socks, leggings, bras, knickers, tight sportswear, restrictive clothing, etc. As a result, these polyamide products have frequent contact with the skin when in use. In order to protect the skin and promote existential health, biological and cosmetic functions such as antioxidant and antibacterial activities as well as pleasant feeling, slimming, refreshing, skin glowing, anti-ageing, body care, fitness and health, etc. [6] are expected to be imparted to polyamide textiles.

Antibacterial polyamide fiber can be produced using two ways: the addition of inorganic antibacterial agents (e.g., silver and zinc oxide nanoparticles) into polyamide during fiber spinning [7,8], and the treatment of polyamide fiber using antibacterial agents in wet processing [9–13]. Because of the processing convenience, the latter is most often adopted. The antibacterial agents used in wet processing include cationic non-surfactant and surfactant agents (e.g., chlorhexidine, cetylpyridinium chloride) [9,10], chitosan [11], silver nanoparticles [12], and metal salts [13]. However, the aforementioned antibacterial agents cannot impart antioxidant properties to polyamide fiber. Plant extracts seem to be more preferred for imparting health care functions to textiles because of their non-toxicity, eco-friendliness, low irritation, and potential multi-functional properties [1,4]. In previous researches, the use of some natural dyes such as berberine, turmeric, madder, safflower yellow, and colors from walnut shells was found to confer good antibacterial functions to polyamide fiber [14–18]. Additionally, resveratrol as well as carotenoids from tomato processing wastes were used to treat polyamide fiber for enhanced antioxidant activity [2,19]. The action of the resveratrol treated polyamide fiber on the skin was assessed, and an improved antioxidant capacity of the skin was revealed [2].

In the light of the requirements for the functional properties of polyamide fiber for health care purposes, this study aims to use natural flavonoids as functional agents to treat polyamide fiber to simultaneously enhance its antioxidant and antibacterial activities. In this work, polyamide fiber was treated by means of an adsorption technique with three flavonoids (baicalin, quercetin, and rutin, whose chemical structures are depicted in Figure 1). The pH dependence of flavonoid adsorption was discussed, and the adsorption kinetics and isotherms as well as the adsorption mechanisms of flavonoids were studied. Furthermore, the antioxidant and antibacterial activities of treated polyamide fiber as well as their washing durability were evaluated.

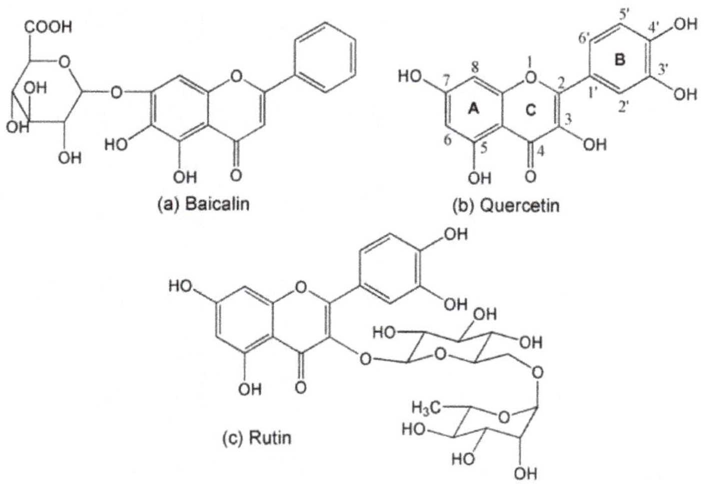

Figure 1. Chemical structures of the studied flavonoids.

2. Materials and Methods

2.1. Materials

A knitted polyamide fabric with a weight of 168.3 g/m^2 was provided by Kunshan Teng Fei Underwear Technology Co. Ltd. (Kunshan, China). Baicalin with a purity of 85%, quercetin with a purity of 95%, and rutin with a purity of 95% were all purchased from Xi'an Qing Yue Biotechnology Co. Ltd. (Xi'an, China). 2,2′-Azino-bis(3-ethylbenzothiazoline-6-sulphonic acid) diammonium salt (ABTS) was purchased from Sigma–Aldrich Trading Co. Ltd. (Shanghai, China). Citric acid, hydrochloric acid, sodium hydroxide, disodium hydrogen phosphate, monopotassium phosphate, potassium persulfate, and potassium chloride were of analytical reagent grade. A commercial detergent was obtained from Shanghai Zhengzhang Laundering and Dyeing Co. Ltd. (Shanghai, China); the pH of 2 g/L detergent solution was about 6.5. Nutrient agar and nutrient broth were bought from Sinopharm Chemical Reagent Co. Ltd. (Shanghai, China) and Shanghai Sincere Biotech Co. Ltd. (Shanghai, China), respectively.

2.2. Flavonoids Adsorption and Polyamide Fabric Treatment

All the adsorption and treatment of flavonoids were carried out in sealed conical flasks placed in a XW-ZDR low-noise oscillated dyeing machine (Jingjiang Xinwang Dying and Finishing Machinery Factory, Jingjiang, China). The liquor ratio (the ratio of liquor volume to fabric weight) was 50:1, and the fabric weight was 2 g. A McIlvaine buffer consisting of citric acid and disodium hydrogen phosphate was added to adjust pH. At the end of the treatment process, the fabrics were rinsed in distilled water and then dried in the open air. In order to investigate the pH dependence of flavonoid adsorption, polyamide fabric was treated with 2% owf (on the weight of fabric) flavonoids whose pH values were adjusted within the range of 2.7 to 7.2; the temperature was started at 25 °C, and elevated to 70 °C at a heating rate of 2 °C/min, and subsequently the treatment was continued at 70 °C for 60 min. In order to study the adsorption rates of flavonoids, polyamide fabric was treated with 2% owf flavonoids at pH 2.79 and 70 °C for different times. The equilibrium adsorption isotherms for flavonoids on polyamide fabric were measured in a series of flavonoid solutions of various concentrations (1–12% owf) at pH 2.79 at 70 °C for 150 min; the isotherms were determined on the basis of the adsorption for 150 min as this time was enough for the equilibrium adsorption to be achieved. In order to determine the effect of the initial concentration of flavonoids on their uptake by polyamide fiber, polyamide fabric was treated with different concentration (2–10% owf) of flavonoids at pH 2.83; the temperature was started at 25 °C, and raised at a heating rate of 2 °C/min up to 70 °C with a holding time of 60 min. In adsorption researches, three parallel experiments were performed and their average results were used.

2.3. Measurements

2.3.1. Uptake of Flavonoids

The absorption spectra and absorbance of flavonoid solutions were measured using a Shimadzu UV-1800 UV–vis spectrophotometer (Shimadzu Co., Kyoto, Japan). The percentage of flavonoid exhaustion was determined using a previously established absorbance/concentration relationship at the maximum absorption wavelength of flavonoid solutions using Equation (1):

$$\text{Exhaustion } (\%) = \frac{m_0 - m_1}{m_0} \times 100 \quad (1)$$

where m_0 and m_1 are the quantities of flavonoids before and after adsorption, respectively. The quantity of flavonoids on polyamide fabric was calculated by the difference in the initial and final concentrations of flavonoids in solution as well as the weight of the dried fabric.

2.3.2. Zeta Potential

The zeta potential and isoelectric point of polyamide fabric were determined using the streaming potential method applied in a SurPASS electrokinetic analyzer (Anton Paar GmbH, Graz, Austria). A pair of fabric samples (ca. 10 × 20 mm^2) was equilibrated in a 1 mM KCl supporting electrolyte solution at 20 °C. During the measurement, the electrolyte solution was forced through the packed fabric samples between two perforated Ag/AgCl electrodes in a measuring cell. The pH of solution was adjusted to the range of 3.2–8.5 with 0.1 M HCl and 0.1 M NaOH.

2.3.3. Antioxidant Activity

The samples treated with 2–10% owf flavonoids in the section of building-up ability were used to evaluate the antioxidant activity. The antioxidant activity of polyamide fabric was measured using a previously reported ABTS radical cation decolorization assay [20]. ABTS was first dissolved in water to a 7 mM concentration, and then the ABTS stock solution was employed to react with 2.45 mM potassium persulfate (final concentration) so as to produce the ABTS radical cation ($ABTS^{\cdot+}$). The mixture was allowed to stand in the dark at room temperature for 12–16 h. Before use, the $ABTS^{\cdot+}$ solution was diluted with a phosphate buffer (0.1 M, pH 7.4) to reach an absorbance of 0.700 ± 0.025 at 734 nm, and afterwards 10 mg of fabric sample was immersed into 10 mL of $ABTS^{\cdot+}$ solution. After 30 min, the scavenging capability of $ABTS^{\cdot+}$ at 734 nm was calculated according to Equation (2):

$$\text{Antioxidant activity } (\%) = \frac{A_{ctrl} - A_{spl}}{A_{ctrl}} \times 100 \tag{2}$$

where A_{ctrl} is the initial absorbance of the $ABTS^{\cdot+}$, and A_{spl} is the absorbance of the remaining $ABTS^{\cdot+}$ in the presence of fabric sample. The average of three tests for antioxidant activity was reported.

2.3.4. Antibacterial Activity

The samples treated with 2% and 10% owf flavonoids in the section of building-up ability were used to evaluate the antibacterial activity. The test on the antibacterial activity of polyamide fabric against *Staphylococcus aureus* (*S. aureus*, ATCC 6538) and *Escherichia coli* (*E. coli*, ATCC 8099) was performed according to GB/T 20944.3-2008 [21]. In the test, standard cotton fabric and tested polyamide fabric were used. The fabric fragments (0.75 g) were dipped into the conical flasks with bacteria, which were oscillated in a shaker at a required temperature (24 °C for *S. aureus* and 30 °C for *E. coli*) for 24 h. After completion of vibration, the bacteria suspension was diluted 1000 times. Subsequently, the diluted bacteria solution was inoculated onto the agar plates which were stored at 37 °C for a desired time (48 h for *S. aureus* and 24 h for *E. coli*). In the end, the quantity of visual bacterial colonies was counted, and the antibacterial activity was calculated using Equation (3):

$$\text{Antibacterial activity } (\%) = \frac{N_{ctrl} - N_{spl}}{N_{ctrl}} \times 100 \tag{3}$$

where N_{ctrl} and N_{spl} are the quantities of the visual bacterial colonies for standard cotton fabric and tested polyamide fabric, respectively. The average of three tests for antibacterial activity was reported.

2.3.5. Durability of Antioxidant and Antibacterial Effects

The samples treated with 2% owf and 10% owf flavonoids were subjected to repeated laundrying. After repeated laundrying, the antioxidant and antibacterial activities were measured. The washing test was briefly described in the following: the samples were immersed into the washing solution containing 2 g/L commercial detergent using a liquor ratio of 50:1. Afterwards, the samples were stirred and left for 10 min at 40 ± 2 °C in a Wash Tec-P fastness tester (Roaches International, West Yorkshire, UK). After washing, the fabrics were gently squeezed and rinsed with distilled water. This treatment was repeated 1, 5, 10, 20 and 30 times.

3. Results and Discussion

3.1. Adsorption Characteristics of Flavonoids

3.1.1. pH Dependence of Flavonoids Adsorption

Because pH can affect the surface potential of polyamide fiber, the ionization of functional groups in flavonoids, and the stability of flavonoids, it would have an impact on the adsorption of flavonoids. Thus, pH was considered as a significant factor having impact on the adsorption of flavonoids. The impact of pH on the uptake of flavonoids is depicted in Figure 2. From Figure 2, it can be interestingly observed that pH had different effects on the adsorption of three flavonoids. The uptake of baicalin by polyamide fiber was significantly dependent on pH. The uptake of baicalin was 77.4% at pH 2.79, and it dropped with an increase in pH. It implies that the adsorption of baicalin is primarily caused by the electrostatic attractions between anionic baicalin and protonated polyamide fiber. In order to explain the electrostatic attractions between baicalin and polyamide fiber having amphoteric character at different pHs, the surface electric potential of polyamide fiber was determined. Figure 2 shows that as pH increased, the surface electric potential had a shift from positive to negative. The zero point of net charge (isoelectric point) was at pH 5.63 or so. At a low pH, the protonation of amino groups in polyamide fiber is increased. Thus the electrostatic interactions between baicalin and polyamide fiber is enhanced, thereby resulting in the high adsorption of baicalin. The similar pH dependence of adsorption was also observed in our earlier study about the treatment of silk by baicalin [22].

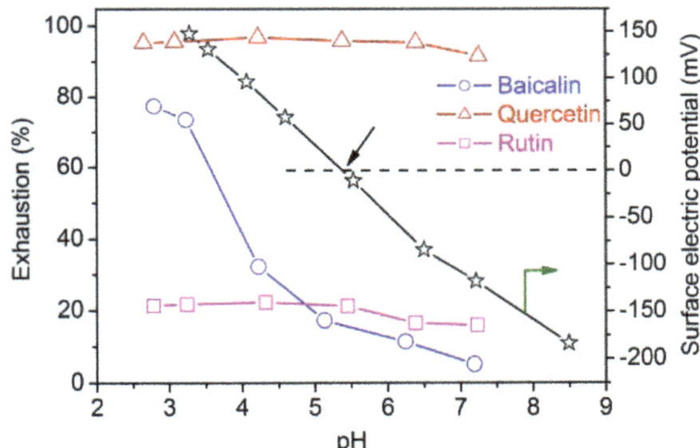

Figure 2. pH Dependence of flavonoids adsorption and surface electric potential of polyamide fiber.

Figure 2 shows that pH had a very small impact on the uptake of quercetin and rutin. The uptake of quercetin and rutin remained almost unchanged within pH 2.65–6.39 and 2.78–5.45, respectively. Quercetin and rutin had a slightly decreased uptake when pH exceeded 6.39 and 5.45, respectively, most likely due to their reduced stability at increasing pH. In addition, it is clear from Figure 2 that on the whole, quercetin showed the maximum adsorption on polyamide fiber, rutin displayed very poor adsorption ability, and baicalin had an adsorption extent between the adsorption of quercetin and rutin. The three flavonoids exhibited rather different pH dependence and capability of adsorption. These results are likely indicative of different adsorption mechanisms of three flavonoids.

3.1.2. Adsorption Kinetics of Flavonoids

The adsorption kinetics is vital for controlling the efficiency and uniformity of adsorption process. The adsorption kinetics of flavonoids was studied through the relationships between their adsorption amount (C_t) and time (t). As depicted in Figure 3a, the adsorption rates of three flavonoids on polyamide fiber were not fast because flavonoids were applied at a moderate temperature (70 °C).

The adsorption quantity of three flavonoids increased gradually with prolonged time. After about 100 min, the adsorption quantity of flavonoids remained almost unchanged. It indicates that the adsorption equilibrium was achieved. It seemed that the adsorption rate of quercetin was the fastest while that of rutin was the slowest.

To compare the adsorption behaviors of three flavonoids, the pseudo second-order kinetic model was utilized to simulate the experimental data. This model can be expressed using Equation (4) [23]:

$$\frac{t}{C_t} = \frac{1}{kC_e^2} + \frac{1}{C_e}t \tag{4}$$

where k denotes the rate constant of adsorption, and C_t and C_e denote the adsorption quantity of flavonoids at time t and at equilibrium.

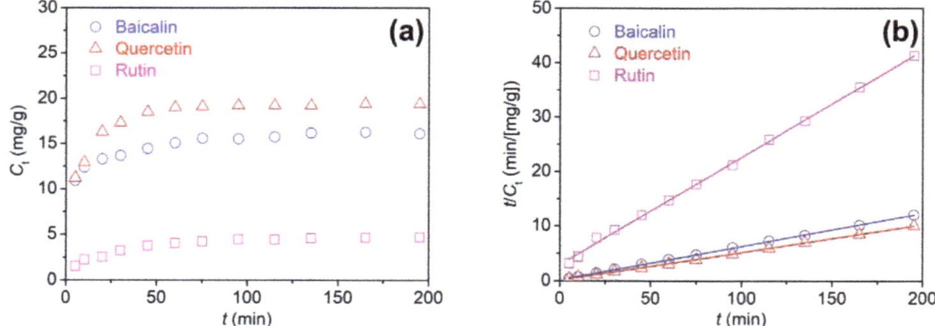

Figure 3. Adsorption rates of three flavonoids for polyamide fiber: (**a**) C_t~t and (**b**) t/C_t~t.

If the adsorption conforms to the aforementioned model, t/C_t and t have a linear relation. Figure 3b displays the linear relation of the t/C_t~t plot. By regression analysis, the fitted straight line was got. Its slope and intercept were utilized to calculate k and C_e. Additionally, based on this model, the half adsorption time ($t_{1/2}$) and initial adsorption rate (r_i) were estimated according to Equations (5) and (6), respectively:

$$t_{1/2} = \frac{1}{kC_e} \tag{5}$$

$$r_i = kC_e^2 \tag{6}$$

Table 1 shows the correlation coefficients (R^2) of this model. The R^2 values were higher than 0.998 for three flavonoids. It indicates that this model is applicable to describe the adsorption rates of flavonoids for polyamide fiber. Table 1 shows the great differences in the kinetic parameters between three flavonoids. Quercetin exhibited the highest initial adsorption rate and the shortest half adsorption time with the moderate adsorption rate constant. Baicalin also had a short half adsorption time and a high initial adsorption rate. Rutin displayed a very low initial adsorption rate and a very long half adsorption time. Moreover, the equilibrium adsorption of flavonoids increased in the following order: rutin < baicalin < quercetin. These observations indicate that quercetin would have the highest affinity to polyamide fiber whereas the affinity of rutin is the lowest.

3.1.3. Adsorption Isotherms of Flavonoids

The research on the equilibrium adsorption isotherms is used to explore the adsorption mechanisms of flavonoids on polyamide fiber, and help to discuss the interactions between flavonoids and polyamide fiber. The adsorption isotherms of three flavonoids on polyamide fiber are depicted in Figure 4. Three equilibrium adsorption equations (Langmuir, Freundlich and Langmuir–Nernst) were utilized to simulate the experimental data.

Table 1. Kinetic parameters for flavonoid adsorption.

Flavonoid	r_i (mg/[g·min])	$t_{1/2}$ (min)	k ($\times 10^{-2}$g/[mg·min])	C_e (mg/g)	R^2
Baicalin	3.43	4.826	1.251	16.56	0.9995
Quercetin	5.03	3.954	1.272	19.88	0.9998
Rutin	0.34	15.100	1.303	5.08	0.9987

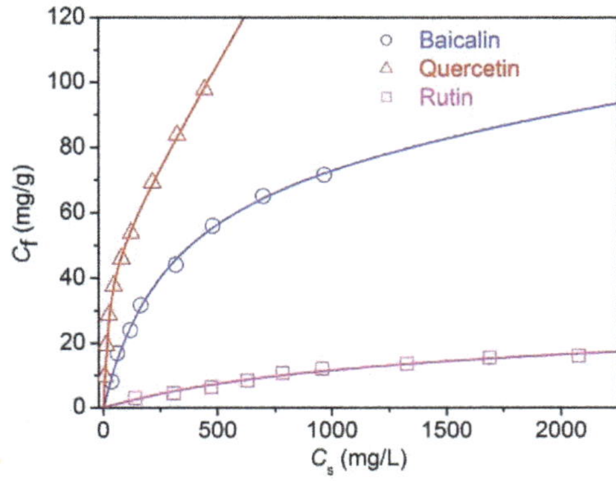

Figure 4. Adsorption isotherms of flavonoids for polyamide fiber and Langmuir–Nernst plots.

The Langmuir isotherm is expressed using the following equation:

$$C_f = \frac{SK_L C_s}{1 + K_L C_s} \quad (7)$$

where C_f (mg/g) and C_s (mg/L) denote the concentration of flavonoids on polyamide fiber and in solution at equilibrium, respectively; S denotes the adsorption saturation of flavonoids on polyamide fiber; K_L denotes the Langmuir affinity constant.

The Freundlich isotherm is expressed using the following equation:

$$C_f = K_F C_s^n \quad (8)$$

where K_F denotes the Freundlich affinity constant, and n reflects adsorption intensity or surface heterogeneity.

The dual Langmuir–Nernst isotherm is expressed using the following equation:

$$C_f = C_P + C_L = K_P C_S + \frac{SK_L C_s}{1 + K_L C_s} \quad (9)$$

where C_P and C_L denote the concentration of flavonoids on polyamide fiber following Nernst type partitioning and Langmuir adsorption mechanisms, respectively; S denotes the saturation adsorption of flavonoids on polyamide fiber by Langmuir mechanism; K_P and K_L denote the partition coefficient and the Langmuir affinity constant, respectively.

The experimental data in Figure 4 were fitted using a nonlinear least-squares fitting approach. To assess the fitting degree, the normalized deviation (ND) of the experimental data was determined using the following equation:

$$ND(\%) = 100 \times \frac{1}{N} \sum_{i=1}^{N} \left(\frac{C_{f,\exp,i} - C_{f,calc,i}}{C_{f,\exp,i}} \right) \quad (10)$$

where $C_{f,exp,i}$ and $C_{f,calc,i}$ denote the experimental and calculated amount of flavonoids adsorption by polyamide fiber, respectively; the index "*i*" denotes to the serial number of data points; *N* denotes the sum of data points.

Table 2 lists the *ND* values of the experimental data. Overall, the Langmuir–Nernst model had the lowest *ND*, and gave the best fitting to the experimental points. The Langmuir–Nernst curves of Figure 4 almost passed all the experimental points precisely. These findings suggest that the Langmuir–Nernst model is the most suitable to characterize the adsorption performance of flavonoids on polyamide fiber. According to this model, the electrostatic interaction operating between flavonoids and polyamide fiber is responsible for Langmuir adsorption, whereas the non-electrostatic interactions are responsible for Nernst adsorption. In this work, the adsorption isotherm measurement was carried out at pH 2.79 which was below the isoelectric point (5.63) of polyamide fiber. At this pH, the protonation extent of amino groups in polyamide fiber is great. The protonated amino groups in polyamide fiber can adsorb the negatively charged flavonoids through electrostatic interaction, contributing to Langmuir adsorption.

Table 2. Fitting degree of three equilibrium adsorption models.

Flavonoid	*ND* (%)		
	Langmuir	**Freundlich**	**Langmuir–Nernst**
Baicalin	4.59	12.81	5.04
Quercetin	15.76	7.55	3.13
Rutin	7.13	9.07	6.64

Additionally, in some previous investigations, macroporous polymer containing amino groups exhibited high adsorption ability towards phenolic compounds because of hydrogen bonding or acid-base interactions between them [24,25]. In our previous research, we also found that the electrostatic and hydrogen bond interactions between tea polyphenols and polyamide fiber contributed to Langmuir adsorption [26]. These previous findings suggest that in addition to electrostatic interaction, the hydrogen bond between flavonoids and polyamide fiber also can contribute to Langmuir adsorption. Flavonoids have abundant hydroxyl groups. Polyamide fiber contains abundant amide groups as well as a small amount of amino and carboxyl groups. Thus, flavonoids interact with polyamide fiber through hydrogen bond.

For the adsorption of baicalin, the electrostatic interaction between baicalin and polyamide fiber is readily explained. The dissociation constant (pK_a) of 7-glucuronic acid in A ring is 5.05 [27]. The pK_a values of 6-OH and 5-OH in A ring are 7.6 and 10.1, respectively [28]. At pH 5.05, the ionization degree of carboxyl groups is 50%. At pH 2.79 set in the present study, the partially ionized carboxyl groups in baicalin interact with the protonated amino groups in polyamide fiber through electrostatic attraction. After these ionized baicalin molecules are adsorbed by polyamide fiber, the ionization balance of 7-glucuronic acid is broken, which accelerates the further dissociation of 7-glucuronic acid. Thus, baicalin can continue to be adsorbed by polyamide fiber. Carboxyl groups have higher dissociation degree than hydroxyl groups. Therefore, it is reasonable to conclude that the electrostatic interaction between the ionized carboxyl groups in baicalin and the protonated amino groups in polyamide fiber is responsible for Langmuir adsorption.

Quercetin and rutin have no carboxyl groups in their structures. During the adsorption process their ion-ion interactions with polyamide fiber are associated with the deprotonated phenolic hydroxyl groups in their structures. The dissociation constants of quercetin obtained from different literatures showed significant variation. The list of the reported values of pK_{a1} gives pK_{a1} = 5.7, 6.6, 6.7, 7.03, 7.4, 7.7, 8.2, and 9.0 [29,30]. In general, the acidity of OH groups in different substitution sites decreases in the order: 7-OH > 4′-OH > 3-OH. It is supposed that after a small amount of hydroxyl groups in quercetin are deprotonated, they can be adsorbed by polyamide fiber through electrostatic interaction. Thus, the dynamic ionization equilibrium of quercetin is broken which facilitates the further disassociation of quercetin and the subsequent adsorption of quercetin on polyamide fiber. The dissociation constant (pK_{a1}) of rutin is 7.1 [31]. The ion-ion interactions of rutin with polyamide fiber are similar to those of quercetin.

In addition to ion-ion interaction and hydrogen bonding, there exist van der Waals attraction and hydrophobic interactions between flavonoids and polyamide molecules. These interactions can be responsible for Nernst adsorption. Polyamide fiber contains considerable methylene groups, which have interactions with the aromatic moieties of flavonoids through hydrophobic and non-polar der Waals forces.

Table 3 lists the adsorption parameters for the Langmuir–Nernst isotherms. Baicalin exhibited the highest saturation adsorption, and much lower K_L and K_P values than quercetin, due to the fact that carboxyl groups in baicalin have higher dissociation degree than hydroxyl groups in quercetin. Quercetin displayed relatively high saturation adsorption, and the highest K_L and K_P values. This indicates that quercetin has the highest affinity to polyamide fiber as compared with baicalin and rutin. Moreover, the Nernst adsorption caused by hydrophobic interaction and non-polar van der Waals force has an important contribution to total adsorption. Therefore, it is not difficult to understand why quercetin exhibits the fastest adsorption and high adsorption quantity as aforementioned.

Table 3. Parameters for the Langmuir–Nernst isotherm of flavonoids adsorption.

Flavonoid	S (mg/g)	K_L (10^{-3} L/mg)	K_P (10^{-3} L/mg)
Baicalin	82.43	3.39	9.23
Quercetin	46.48	46.53	121.67
Rutin	22.10	0.91	1.13

It is worth noting that the K_L and K_P values as well as the adsorption quantity of rutin are remarkably lower than those of quercetin. Rutin and quercetin have similar chemical structures. Their only difference is that rutin has a glycosidic linkage at position 3. The presence of a glycosidic moiety not only increases molecular weight and size, but also causes a steric hindrance. Both these factors exert a negative impact on the diffusion of rutin into the interior of polyamide fiber whose physical structure is compact, and accordingly decrease the adsorption quantity of rutin.

3.1.4. Initial Concentration Dependence of Flavonoids Adsorption

The function relation of the adsorption quantity and initial concentration of flavonoids reflects the building-up property which is very important for industrial application [32]. The flavonoids possessing excellent building-up performance can draw attention of manufacturers due to the advantage of eco-friendliness [33], high utilization, and sufficient functionalities. Hence, taking the practical application conditions in consideration, the test on the building-up properties of three flavonoids onto polyamide fiber was performed using a heating and holding approach.

The adsorption quantity (C_f) and exhaustion of flavonoids onto polyamide fiber are shown in Figure 5. In the case of baicalin and quercetin, the extent of adsorption almost increased linearly with increasing flavonoid dosage from 2% to 10% owf, and what is more, both of them kept high uptake at high dosages, indicating their great building-up ability towards polyamide fiber. On the contrary, rutin displayed very low exhaustion and adsorption quantity, indicating its poor building-up property and low utilization rate which would give rise to a high application cost.

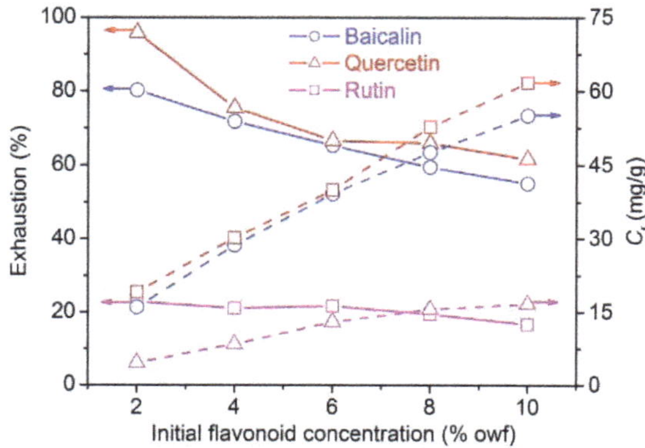

Figure 5. Uptake of flavonoids by polyamide fiber at different initial concentrations.

3.2. *Antioxidant and Antibacterial Properties of Flavonoids Treated Polyamide Fiber*

3.2.1. Antioxidant Property

The antioxidant properties of flavonoids have already been well discussed [30,31,34,35]. They are mainly associated with the substitution positions and total number of hydroxyl groups [35]. In general, the relationships between the antioxidant activity and structure of flavonoids are as follows [35,36]: the B-ring hydroxyl configuration of flavonoids is the most important factor in deciding radical scavenging ability due to its donation of hydrogen and an electron to radicals; a 3′,4′-catechol in the B-ring has great relevance to increased radical scavenging ability; flavonoid heterocycle (C-ring) is responsible for radical scavenging ability due to the conjugate effect of aromatic ring and 3-OH, whereas the substitution of 3-OH by glycosyl moiety inhibits radical scavenging; in comparison with the hydroxylation of B-ring, the influence of substituent groups in A-ring on radical scavenging is small. According to these above rules and by comparison of the chemical structures of the three flavonoids (Figure 1), it seems that the antioxidant activity of three flavonoids decreases in the following order: quercetin > rutin > baicalin. The aforementioned antioxidant activity of flavonoids mainly refers to foods and medical care. However, the antioxidant activity of flavonoids on polymeric fiber has been less studied and reported.

Figure 6a shows the antioxidant property of polyamide fibers treated using three flavonoids at different concentrations (2–10% owf). As can be seen in Figure 6a, pristine polyamide fiber had a low antioxidant activity value of about 10%. After the treatment using three flavonoids, polyamide fiber displayed significantly improved antioxidant property. Moreover, the antioxidant property of polyamide fiber increased with increasing initial dosages of baicalin and rutin, whereas it always kept a very high level at different dosages of quercetin. As the dosage of three flavonoids reached 10% owf, the antioxidant activity of all the samples exceeded 80%.

At a dosage of 2% owf flavonoids, the antioxidant activity of polyamide fiber was 37.5% for baicalin, 97.2% for quercetin, and 39.9% for rutin, respectively. At this flavonoid dosage, the adsorption quantity of baicalin, quercetin, rutin on polyamide fiber was 16.06, 19.18, and 4.58 mg/g, respectively (Figures 5 and 6b). Figure 6b reveals the relation between the antioxidant property and adsorption amount of three flavonoids. In the case that the adsorption amount was virtually at the same level, quercetin imparted the highest antioxidant activity to polyamide fiber, rutin was the second efficient although its adsorption was low, and baicalin had the lowest antioxidant activity. The great antioxidant ability of quercetin originates from its catechol structure of B-ring and multiple hydroxyl groups. The poor antioxidant ability of baicalin is due to the lack of catechol structure and 3-OH. Although rutin has a low adsorption extent, it bears the catechol structure and four phenolic hydroxyl groups and accordingly has better antioxidant property than baicalin.

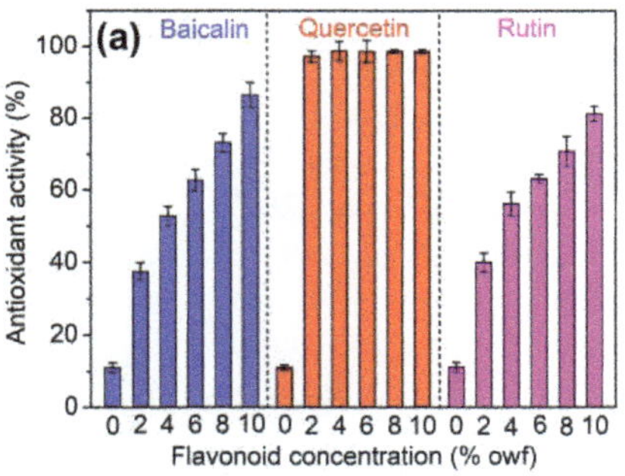

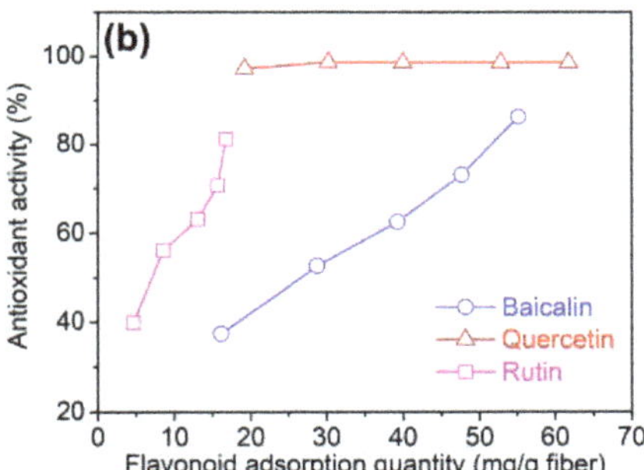

Figure 6. Antioxidant property of polyamide fibers treated using flavonoids at different dosages (**a**) and adsorption quantities (**b**).

3.2.2. Antibacterial Property

It has been reported that baicalin, quercetin and rutin are potential antibacterial agents [37–39]. Some natural dyes containing baicalin, quercetin and rutin were able to improve the antibacterial properties of silk and wool fibers [40–42]. The antibacterial activity of natural dyes is usually considered to be mainly related to phenolic hydroxyl groups in their structures. In this study, three flavonoids were also expected to impart antibacterial activity to polyamide fiber.

Figure 7a displays the antibacterial property of polyamide fibers treated using baicalin, quercetin, and rutin at different concentrations. Pristine polyamide fiber showed poor antibacterial properties. Its antibacterial rate was 38.2% for *S. aureus* and 32.7% for *E. coli*. However, the treated fibers, especially the ones treated with quercetin, exhibited excellent antibacterial property. Moreover, the antibacterial activity increased with increasing initial dosages of flavonoids. At 2% owf flavonoids, the fibers treated using baicalin, quercetin and rutin exhibited good antibacterial property with an inhibition rate of 79.3%, 86.8%, and 79.1% against *S. aureus*, respectively, and 88.0%, 96.7%, and 89.4% against *E. coli*, respectively. Obviously, *S. aureus* is more tolerant to the treated polyamide fiber than *E. coli*. Figure 7b reveals the relation between the antibacterial property and adsorption amount of three flavonoids. At almost the same flavonoid adsorption amount, quercetin and rutin provided better antibacterial function than baicalin. This order of antibacterial activity is similar to that of antioxidant activity discussed above. But the difference of antibacterial activity among three flavonoids is evidently smaller than that of antioxidant activity.

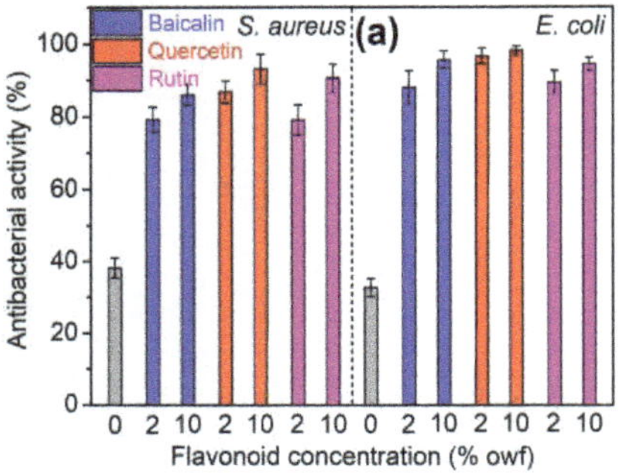

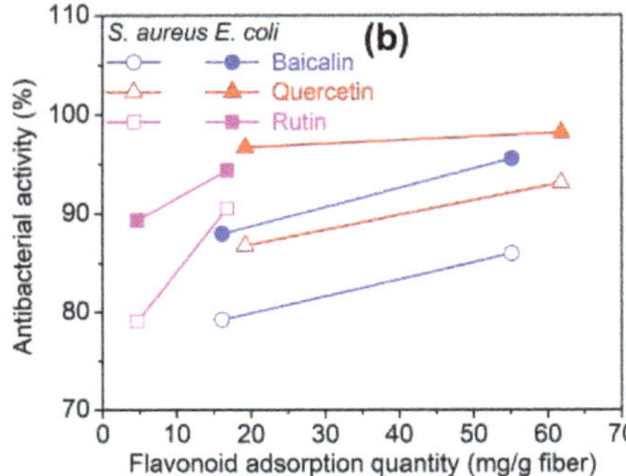

Figure 7. Antibacterial property of polyamide fibers treated using flavonoids at different dosages (**a**) and adsorption quantities (**b**).

3.2.3. Laundering Durability of Antioxidant and Antibacterial Properties

Polyamide fiber is frequently subjected to repeated laundering when in use. Therefore, the laundering durability of antioxidant and antibacterial activities imparted to polyamide fiber is very important. Figure 8 displays the antioxidant activity of polyamide fibers treated with 2% and 10% owf flavonoids after 1, 5 and 10 cycles of washing. At two dosages of flavonoids, the antioxidant property of all the fibers declined gradually as the washing cycle increased. The reduction in antioxidant activity was the lowest for quercetin, whereas that was the highest for baicalin. Figure 9 displays the antibacterial property of polyamide fibers treated using 10% owf flavonoids after 1, 5 and 10 cycles of washing. As compared with antioxidant activity, antibacterial activity displayed the similar decrease tendency as the washing cycle increased. As pointed out in the adsorption isotherm section, quercetin has the highest affinity to polyamide fiber, and hence its desorption degree should be the lowest in the washing process of the treated polyamide fiber. Thus, quercetin can still provide very high antioxidant and antibacterial properties after repeated washings. For baicalin, its high decrease in antioxidant and antibacterial properties might be associated with its higher water solubility than that of quertin and rutin.

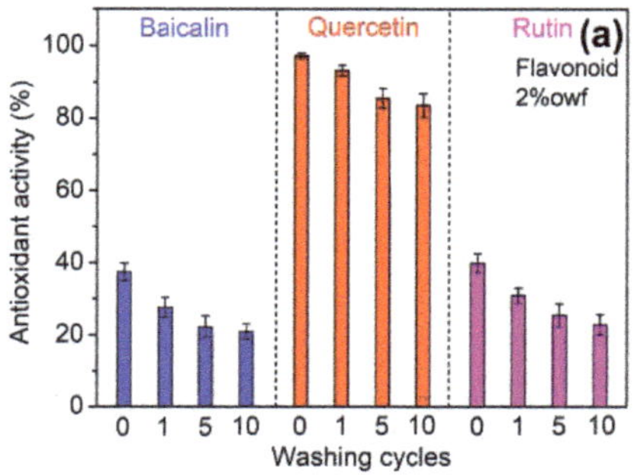

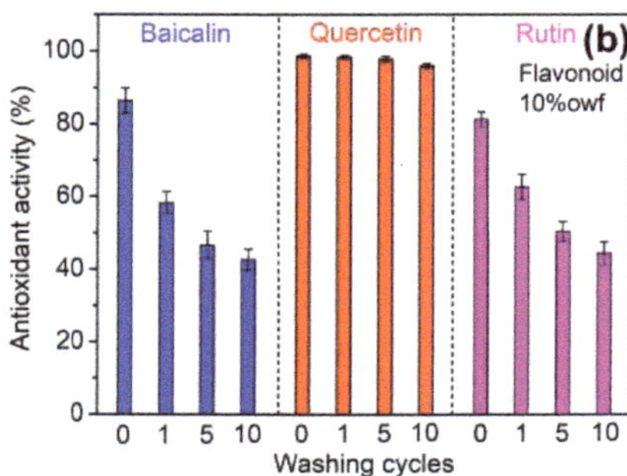

Figure 8. Antioxidant properties of polyamide fibers treated using 2% owf (**a**) and 10% owf (**b**) flavonoids after repeated laundering.

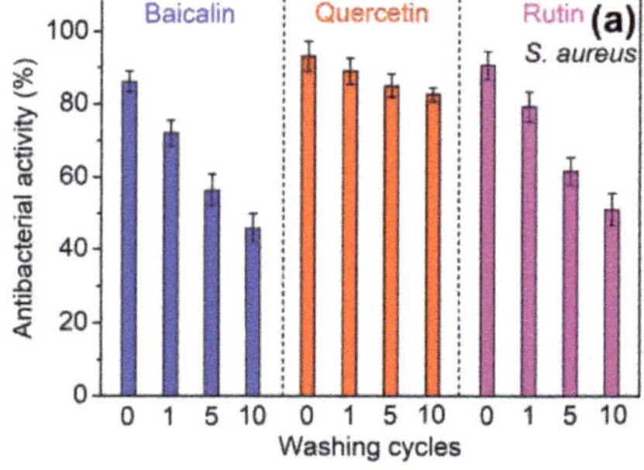

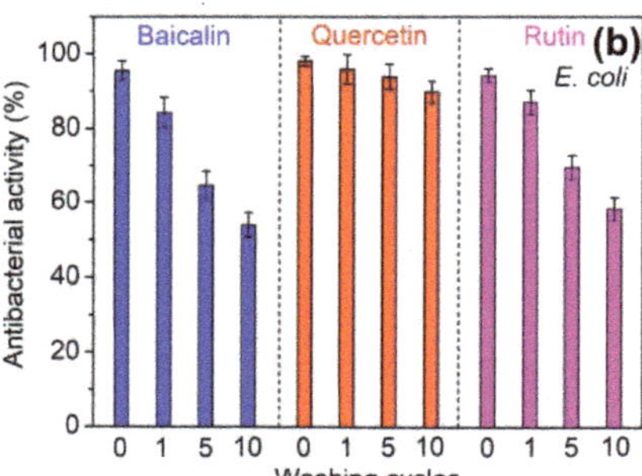

Figure 9. Antibacterial properties against *S. aureus* (**a**) and *E. coli* (**b**) of polyamide fibers treated using 10% owf flavonoids after repeated laundering.

Because quercetin exhibited high adsorption, excellent functions and good durability, the polyamide fabric treated using 10% quercetin was subjected to more repeated washings. Figure 10 displays that after 30 washings, the treated sample had an antioxidant activity of above 65%. After 20 washings, the antibacterial rate for both *S. aureus* and *E. coli* was higher than 70%. After 30 washings, the antibacterial rate for *S. aureus* was lower than 70%. According to GB/T 20944.3–2008 [21], the antibacterial textiles are required to have an antibacterial rate of above 70% for *S. aureus*, and above 60% for *E. coli*. Therefore the polyamide fabric treated using 10% quercetin can be resistant to 20 cycles of washing.

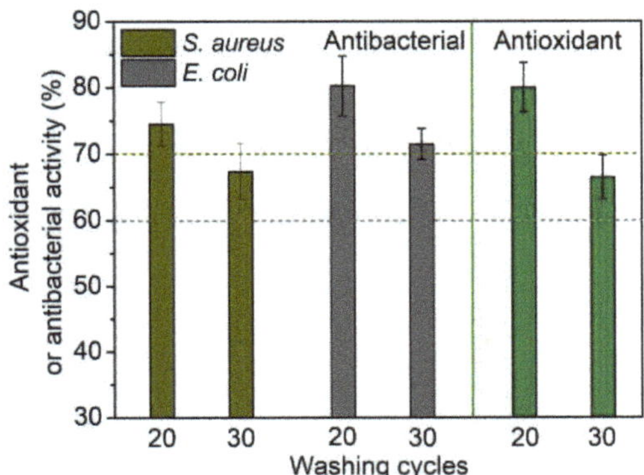

Figure 10. Antioxidant and antibacterial properties of polyamide fiber treated using 10% owf quercetin after 20 and 30 cycles of laundering.

4. Conclusions

In the present study, three natural flavonoids were employed to simultaneously impart antioxidant and antibacterial activities to polyamide fiber using an adsorption technology. The adsorption of baicalin on polyamide fiber greatly depended on the pH of its solutions. The adsorption of the three flavonoids conformed to the pseudo second-order kinetic model. The adsorption isotherms of three flavonoids fitted the Langmuir–Nernst model. Quercetin showed the highest affinity to polyamide fiber and adsorption quantity, followed by baicalin, whereas rutin displayed poor adsorption capability due to the presence of a glycosidic moiety. Quercetin imparted very high antioxidant and antibacterial activities to polyamide fiber as compared with baicalin and rutin, and these functions displayed good resistance to washing. Although rutin provided slightly higher antioxidant and antibacterial activities than baicalin, and its low adsorption quantity would increase the processing cost of polyamide fabric, hence limiting its application. The present study suggests that the simultaneous antioxidant and antibacterial functionalization of polyamide fiber can be realized by treatments using natural flavonoids.

Author Contributions: Conceptualization, Y.-D.L. and R.-C.T.; methodology, Y.-D.L., J.-P.G., R.-C.T. and Y.-F.Q.; formal analysis and investigation, Y.-D.L., J.-P.G. and R.-C.T.; data curation, Y.-D.L. and Y.-F.Q.; writing—original draft preparation, Y.-D.L.; writing—review and editing, Y.-D.L., J.-P.G. and R.-C.T.; project administration and funding acquisition, R.-C.T.; supervision, R.-C.T.

Funding: This research was funded by Science and Technology Support Program of Jiangsu Province (BE2015066).

Conflicts of Interest: The authors declare no conflicts of interest.

References

1. Shahid, M.; Mohammad, F. Green chemistry approaches to develop antimicrobial textiles based on sustainable biopolymers—A review. *Ind. Eng. Chem. Res.* **2013**, *52*, 5245–5260. [CrossRef]
2. Alonso, C.; Martí, M.; Martínez, V.; Rubio, L.; Parra, J.L.; Coderch, L. Antioxidant cosmeto-textiles: Skin assessment. *Eur. J. Pharm. Biopharm.* **2013**, *84*, 192–199. [CrossRef] [PubMed]
3. Mocanu, G.; Nichifor, M.; Mihai, D.; Oproiu, L.C. Bioactive cotton fabrics containing chitosan and biologically active substances extracted from plants. *Mater. Sci. Eng. C* **2013**, *33*, 72–77. [CrossRef] [PubMed]
4. Yi, E.; Hong, J.Y.; Yoo, E.S. A novel bioactive fabric dyed with unripe *Citrus grandis Osbeck* extract part 2: Effects of the Citrus extract and dyed fabric on skin irritancy and atopic dermatitis. *Text. Res. J.* **2010**, *80*, 2124–2131. [CrossRef]
5. Eltahir, Y.A.; Saeed, H.A.M.; Xia, Y.; Yong, H.; Wang, Y. Mechanical properties, moisture absorption, and dyeability of polyamide 5,6 fibers. *J. Text. Inst.* **2016**, *107*, 208–214. [CrossRef]
6. Singh, M.K.; Varun, V.K.; Behera, B.K. Cosmetotextiles: State of art. *Fibers Text. East. Eur.* **2011**, *19*, 27–33. [CrossRef]

7. Erem, A.D.; Ozcan, G.; Skrifvars, M.; Cakmak, M. In vitro assesment of antimicrobial activity and characteristics of polyamide 6/silver nanocomposite fibers. *Fibers Polym.* **2013**, *14*, 1415–1421. [CrossRef]
8. Tang, L.; Wang, D.; Xu, Q.; Wang, C.; Wang, H.; Huang, Q. Preparation and characterization of antibacterial nylon 6 fiber. *Mater. Sci. Forum* **2017**, *898*, 2254–2262. [CrossRef]
9. Giménez-Martín, E.; López-Andrade, M.; Moleón-Baca, J.A.; López, M.A.; Ontiveros-Ortega, A. Polyamide fibers covered with chlorhexidine: Thermodynamic aspects. *J. Surf. Eng. Mater. Adv. Technol.* **2015**, *5*, 190–206. [CrossRef]
10. Kim, H.-S.; Hwang, J.-Y.; Lim, S.-H.; Lim, J.-N.; Son, Y.-A. Preparation, physical characteristics and antibacterial finishing of PCM/nylon fibers having sheath/core structure. *Text. Color. Finish.* **2014**, *26*, 311–321. [CrossRef]
11. Shi, Z. Grafting chitosan oxidized by potassium persulfate onto nylon 6 fiber, and characterizing the antibacterial property of the graft. *J. Polym. Res.* **2014**, *21*, 534. [CrossRef]
12. Montazer, M.; Shamei, A.; Alimohammadi, F. Stabilized nanosilver loaded nylon knitted fabric using BTCA without yellowing. *Prog. Org. Coat.* **2012**, *74*, 270–276. [CrossRef]
13. Zhang, M.; Gao, Q.; Yang, C.; Pang, L.; Wang, H.; Li, R.; Xing, Z.; Hu, J.; Wu, G. Preparation of antimicrobial MnO_4^- doped nylon-66 fibers with excellent laundering durability. *Appl. Surf. Sci.* **2017**, *422*, 1067–1074. [CrossRef]
14. Haji, A.; Shoushtari, A.M.; Mirafshar, M. Natural dyeing and antibacterial activity of atmospheric-plasma-treated nylon 6 fabric. *Color. Technol.* **2014**, *130*, 37–42. [CrossRef]
15. Mirjalili, M.; Karimi, L. Antibacterial dyeing of polyamide using turmeric as a natural dye. *Autex Res. J.* **2013**, *13*, 51–56. [CrossRef]
16. Mirjalili, M.; Karimi, L. Extraction and characterization of natural dye from green walnut shells and its use in dyeing polyamide: Focus on antibacterial properties. *J. Chem.* **2013**, *2013*, 375352. [CrossRef]
17. Coman, D.; Vrînceanu, N.; Oancea, S.; Rîmbu, C. Coloristic and antimicrobial behavior of polymeric substrates using bioactive substances. *IOP Conf. Ser. Mater. Sci. Eng.* **2016**, *145*, 032003. [CrossRef]
18. Ibrahim, N.A.; El-Zairy, W.M.; El-Zairy, M.R.; Ghazal, H.A. Enhancing the UV-protection and antibacterial properties of polyamide-6 fabric by natural dyeing. *Text. Light Ind. Sci. Technol.* **2013**, *2*, 36–41.
19. Baaka, N.; Ksibi, I.E.; Mhenni, M.F. Optimization of the recovery of carotenoids from tomato processing wastes: Application on textile dyeing and assessment of its antioxidant activity. *Nat. Prod. Res.* **2017**, *31*, 196–203. [CrossRef]
20. Re, R.; Pellegrini, N.; Proteggente, A.; Pannala, A.; Yang, M.; Rice-Evans, C. Antioxidant activity applying an improved ABTS radical cation decolorization assay. *Free Radic. Biol. Med.* **1999**, *26*, 1231–1237. [CrossRef]
21. China's General Administration of Quality Supervision; Inspection and Quarantine and Standardization Administration of China. *Textiles—Evaluation for Antibacterial Activity—Part 3: Shake Flask Method*; GB/T 20944.3-2008; Beijing, China, 2008.
22. Zhou, Y.; Yang, Z.-Y.; Tang, R.-C. Bioactive and UV protective silk materials containing baicalin—The multifunctional plant extract from *Scutellaria baicalensis* Georgi. *Mater. Sci. Eng. C* **2016**, *67*, 336–344. [CrossRef] [PubMed]
23. Ho, Y.S.; McKay, G. Pseudo-second order model for sorption processes. *Process Biochem.* **1999**, *34*, 451–465. [CrossRef]
24. Pan, B.C.; Xiong, Y.; Su, Q.; Li, A.M.; Chen, J.L.; Zhang, Q.X. Role of amination of a polymeric adsorbent on phenol adsorption from aqueous solution. *Chemosphere* **2003**, *51*, 953–962. [CrossRef]
25. Pan, B.C.; Zhang, X.; Zhang, W.M.; Zheng, J.Z.; Pan, B.J.; Chen, J.L.; Zhang, Q.X. Adsorption of phenolic compounds from aqueous solution onto a macroporous polymer and its aminated derivative: Isotherm analysis. *J. Hazard. Mater.* **2005**, *121*, 233–241. [CrossRef] [PubMed]
26. Tang, R.-C.; Tang, H.; Yang, C. Adsorption isotherms and mordant dyeing properties of tea polyphenols on wool, silk, and nylon. *Ind. Eng. Chem. Res.* **2010**, *49*, 8894–8901. [CrossRef]
27. Wang, H.; Chen, J.M.; Zhang, Q.M. Determination of the physical chemistry constants of baicalin. *J. Shenyang Pharm. Univ.* **2000**, *17*, 105–106. (In Chinese)
28. Liang, R.; Han, R.-M.; Fu, L.-M.; Ai, X.-C.; Zhang, J.-P.; Skibsted, L.H. Baicalin in radical scavenging and its synergistic effect with β-carotene in antilipoxidation. *J. Agric. Food Chem.* **2009**, *57*, 7118–7124. [CrossRef]

29. Ramešová, Š.; Sokolová, R.; Degano, I.; Bulíčková, J.; Žabka, J.; Gál, M. On the stability of the bioactive flavonoids quercetin and luteolin under oxygen-free conditions. *Anal. Bioanal. Chem.* **2012**, *402*, 975–982. [CrossRef]
30. Musialik, M.; Kuzmicz, R.; Pawłowski, T.S.; Litwinienko, G. Acidity of hydroxyl groups: An overlooked influence on antiradical properties of flavonoids. *J. Org. Chem.* **2009**, *74*, 2699–2709. [CrossRef]
31. Jovanovic, S.V.; Steenken, S.; Tosic, M.; Marjanovic, B.; Simic, M.G. Flavonoids as antioxidants. *J. Am. Chem. Soc.* **1994**, *116*, 4846–4851. [CrossRef]
32. Sun, S.-S.; Tang, R.-C. Adsorption and UV protection properties of the extract from honeysuckle onto wool. *Ind. Eng. Chem. Res.* **2011**, *50*, 4217–4224. [CrossRef]
33. Samanta, A.K.; Agarwal, P. Application of natural dyes on textiles. *Indian J. Fiber Text. Res.* **2009**, *34*, 384–399.
34. Pietta, P.-G. Flavonoids as antioxidants. *J. Nat. Prod.* **2000**, *63*, 1035–1042. [CrossRef] [PubMed]
35. Heim, K.E.; Tagliaferro, A.R.; Bobilya, D.J. Flavonoid antioxidants: Chemistry, metabolism and structure-activity relationships. *J. Nutr. Biochem.* **2002**, *13*, 572–584. [CrossRef]
36. Kumar, S.; Pandey, A.K. Chemistry and biological activities of flavonoids: An overview. *Sci. World J.* **2013**, *2013*, 162750. [CrossRef]
37. Liu, I.X.; Durham, D.G.; Richards, R.M.E. Baicalin synergy with β-lactam antibiotics against methicillin-resistant *Staphylococcus aureus* and other β-lactam-resistant strains of *S. Aureus*. *J. Pharm. Pharmacol.* **2000**, *52*, 361–366. [CrossRef]
38. Li, M.; Xu, Z. Quercetin in a lotus leaves extract may be responsible for antibacterial activity. *Arch. Pharm. Res.* **2008**, *31*, 640–644. [CrossRef]
39. Vaquero, M.J.R.; Alberto, M.R.; de Nadra, M.C.M. Antibacterial effect of phenolic compounds from different wines. *Food Control* **2007**, *18*, 93–101. [CrossRef]
40. Baliarsingh, S.; Panda, A.K.; Jena, J.; Das, T.; Das, N.B. Exploring sustainable technique on natural dye extraction from native plants for textile: Identification of colorants, colorimetric analysis of dyed yarns and their antimicrobial evaluation. *J. Clean. Prod.* **2012**, *37*, 257–264. [CrossRef]
41. Ghaheh, F.S.; Mortazavi, S.M.; Alihosseini, F.; Fassihi, A.; Nateri, A.S.; Abedi, D. Assessment of antibacterial activity of wool fabrics dyed with natural dyes. *J. Clean. Prod.* **2014**, *72*, 139–145. [CrossRef]
42. Khan, M.I.; Ahmad, A.; Khan, S.A.; Yusuf, M.; Shahid, M.; Manzoor, N.; Mohammad, F. Assessment of antimicrobial activity of catechu and its dyed substrate. *J. Clean. Prod.* **2011**, *19*, 1385–1394. [CrossRef]

Article

Antioxidant Properties of a Traditional Vine Tea, *Ampelopsis grossedentata*

Kun Xie [1], Xi He [2], Keyu Chen [1], Jihua Chen [3], Kozue Sakao [1,4] and De-Xing Hou [1,4,*]

[1] Biological Science and Technology, United Graduate School of Agricultural Sciences, Kagoshima University, Kagoshima 890-0065, Japan

[2] College of Animal Science and Technology, Hunan Agricultural University, Changsha 410128, China

[3] Xiangya School of Public Health, Central South University, Changsha 410128, China

[4] Department of Food Science and Biotechnology, Faculty of Agriculture, Kagoshima University, Kagoshima 890-0065, Japan

* Correspondence: hou@chem.agri.kagoshima-u.ac.jp; Tel.: +81-99-285-8649

Received: 28 June 2019; Accepted: 7 August 2019; Published: 9 August 2019

Abstract: *Ampelopsis grossedentata*, also called vine tea, has been used as a traditional beverage in China for centuries. Vine tea contains rich polyphenols and shows benefit to human health, but the chemical and antioxidant properties of vine tea polyphenols from different locations remain unclear. This study aims to investigate the chemical and antioxidant properties of vine tea from three major production areas in China including Guizhou, Hunan, and Guangxi Provinces. The highest amount of polyphenol from vine tea was extracted by 70% ethanol at 70 °C for 40 min with ultrasonic treatment. The major compound in vine tea polyphenols (VTP) was determined as dihydromyricetin (DMY) by high-performance liquid chromatography (HPLC) and the content was estimated as 21.42%, 20.17%, and 16.47% of dry weight basis from Hunan, Guizhou, and Guangxi products, respectively. The antioxidant activities were investigated *in vitro* and in culture hepatic cells. VTP and DMY showed strong 1,1-Diphenyl-2-picrylhydrazyl free radical (DPPH) scavenging ability and high oxygen radical absorption capacity (ORAC) value *in vitro*. VTP and DMY also increased the level of nicotinamide adenine dinucleotide phosphate (NADPH):quinone oxidoreductase (NQO1) in HepG2 cells. Moreover, VTP and DMY enhanced the level of nuclear factor erythroid 2-related factor 2 (Nrf2) and reduced the level of Kelch-like ECH-associated protein 1 (Keap1). Taken together, our data demonstrated that the extraction of vine tea by 70% ethanol with ultrasonic treatment is a novel method to efficiently obtain components possessing stronger antioxidant activity. Furthermore, the results from the culture cells suggest that the bioactive component of vine tea might exert the antioxidant activity by activating the cellular Nrf2/Keap1 pathway.

Keywords: *Ampelopsis grossedentata*; dihydromyricetin; antioxidant ability; Nrf2/Keap1

1. Introduction

Polyphenols are natural substances occurring in fruits, vegetables, beverages, and essential oils. These compounds can protect plants from oxidative stress and insects, and maintain the bioactivity in plant-derived food for humans. Plant polyphenols as antioxidant agents are now used to keep the properties of food in the aspects of both preservation and nutrition. Furthermore, dietary polyphenols intake has been linked to a lowered risk of the most common chronic diseases that are known to be caused by oxidative stress [1,2].

Oxidative stress is an imbalance status in oxidants and antioxidants, and is considered as a major factor in the pathogenesis of chronic disease [3,4]. The proper level of reactive oxygen species (ROS) in our body in a low-moderate concentration has positive effects such as involvement in energy production, regulation of cell growth, and intercellular signaling [5]. On the other hand, excess ROS can attack lipids in cell membranes, proteins in tissues or enzymes, and DNA to cause oxidation, which leads to lipid peroxidation and DNA damage [6]. This oxidation damage is considered to be an important factor of aging and aging-associated disease such as heart disease, cognitive dysfunction, and cancer [7]. A number of polyphenolic compounds have been reported to possess antioxidant properties in human study, and enhance the expression of cellular antioxidant enzymes through the nuclear factor erythroid 2-related factor 2 (Nrf2)-mediated pathway [8,9]. These activities account for the disease-preventing effects of polyphenol diets. Epidemiological studies have revealed an inverse correlation between the intake of fruits, vegetables, wine, tea, and the incidence of certain cancers and cardiovascular disease [10]. It has been reported that dietary polyphenols enhanced the function of antioxidant vitamins and enzymes to defend against the oxidative stress caused by excess ROS [11]. Thus, it is now well recognized that a daily intake of polyphenols in the diet is important for preventing some chronic diseases.

Green tea and black tea are the most consumed beverages worldwide, and their antioxidant properties are well investigated [12–14]. On the other hand, some traditional or folk teas from various edible plant leaves are also popular in Asia [15–17]. *Ampelopsis grossedentata*, also called vine tea, is a traditional herb widely used in medicine and health supplements in the southwest of China. The traditional manufacturing process of vine tea is similar to green tea, and people also usually drink vine tea by soaking it in boiling water as a health beverage. However, the growing environment of vine tea is different, and there is no standard for manufacturing process. Recently, it has been reported that the antioxidant capacity and major polyphenol composition of teas are affected by geographical location, plantation elevation, and leaf grades [18]. To clarify whether the antioxidant capacity and major polyphenol composition of vine tea are affected by the geographical locations and plantation elevation, we chose vine tea samples from three principle producing areas in China including Guizhou, Hunan, and Guangxi Provinces in this study. First, we used different solvents to optimize an efficient extraction method to obtain the highest content of polyphenol, and the major compounds in vine tea were then determined by high-performance liquid chromatography (HPLC). Second, we estimated the antioxidant capacity of the vine tea polyphenol (VTP) extract and its major compound, dihydromyricetin (DMY), from the above three locations. Finally, we used a culture cell line, HepG2, to investigate their antioxidant mechanisms in culture cells, focusing on their effect on the expression of Nrf2/Kelch-like ECH-associated protein 1 (Keap1)-mediated antioxidant enzymes.

2. Material and Methods

2.1. Samples and Chemical Reagents Preparation

The dried leaves and stems of vine tea were purchased from Hunan, Guizhou, and Guangxi Provinces, P.R. China. The gallic acid standard (3,4,5-trihydroxybenzoic acid, CAS Number 5995-86-8, purity ≥ 98%) and 6-hydroxy-2,5,7,8 tetramethyl chroman-2-carboxylic acid (Trolox) were purchased from Sigma-Aldrich (St. Louis, MO, USA). Dihydromyricetin (DMY) (CAS Number 27200-12-0, purity ≥ 98%) standard was purchased from Yuanye Bio-Technology Co. Ltd (Shanghai, China). Ethanol, methanol, acetone, and ethyl acetate were purchased from Sinopharm Chemical Reagent Co. Ltd (Shanghai, China). Phosphoric acid (HPLC purity) and methanol (HPLC purity) were obtained from Sigma (St. Louis, MO, USA). Fetal bovine serum (FBS) was obtained from Equitech-Bio (Kerrville, TX, USA). The antibodies against Nrf2, Keap1, Heme oxygenase 1 (HO-1), nicotinamide adenine dinucleotide phosphate (NADPH): quinone oxidoreductase (NQO1), and β-actin were from Santa Cruz Biotechnology (Santa Cruz, CA, USA).

2.2. Extraction and Total Polyphenol Content Analysis

Vine tea was comminuted by a grinder to pass a 0.60 mm sifter and stored in a −20 °C freezer for further analysis. One gram of vine tea dry powder extracted with different solvents was placed in glass tubes and a condenser pipe was connected to the glass tubes to prevent solvent evaporation. The tubes were in a thermostatic water bath set at 15 °C in a fume hood, and the ratio of vine tea to solvent was 1:5. It has been reported that the extracting solvents significantly affected the total polyphenol content and antioxidant activity of the green tea extracts [19]. Thus, we used distilled water, ethanol, methanol, acetone, and ethyl acetate as the extracting solvents, and further investigated the effect of solvent concentration, extraction time, and temperature on the extraction efficiency, respectively. The extracts were filtrated with a 0.45 μm organic filter and concentrated by a rotary evaporator. The concentrated extracts were further purified by nonionic polystyrene-divinylbenzene resin, and freeze-dried for three days. The powder obtained was used as VTP (Figure 1).

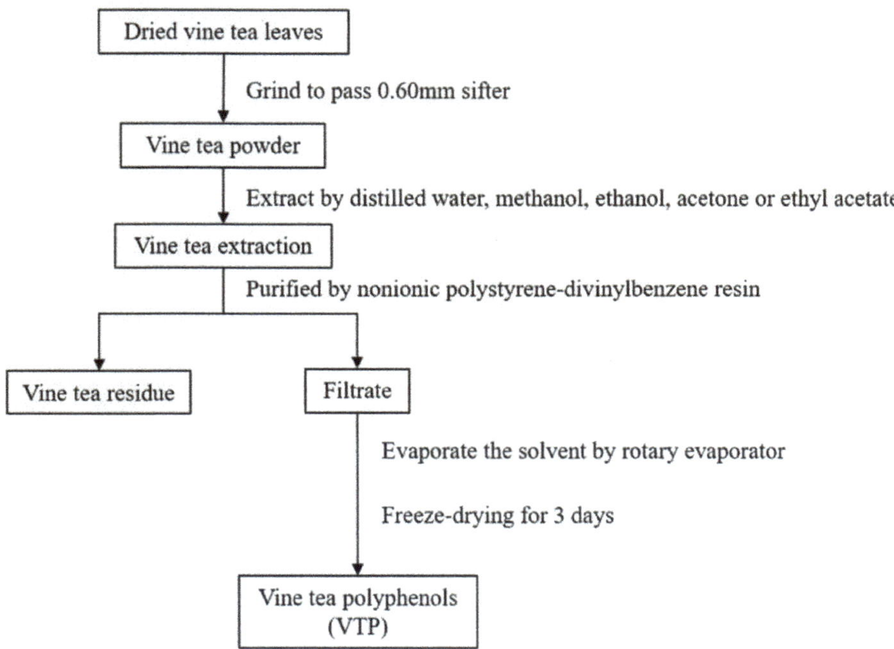

Figure 1. Diagram of vine tea polyphenol (VTP) extraction. All vine tea leaves from different locations were freeze-dried to ensure the same moisture content.

The total polyphenol content was determined by the Folin–Ciocalteu method [20]. In brief, gallic acid (3,4,5-trihydroxybenzoic acid) standards were set at 1, 0.5, 0.25, 0.125, and 0.06125 mg/mL. Vine tea extracts were prepared at 0.5 mg/mL. One hundred microliters of vine tea extracts and standards were diluted with 900 μL distilled water in a 10 mL tube, 4 mL 0.2 N Folin and Ciocalteu's phenol reagent (Sigma-Aldrich, Shanghai, China.), and 4 mL of 10% sodium carbonate aqueous solution were added into the tube. The tube was placed in a thermostat water bath set at 25 °C, the absorbance of each reaction was measured at 760 nm by a spectrophotometer (Thermo Fisher scientific, Oulu, Finland) after 2 h. Results are expressed as mg gallic acid equivalents (GAE) per g dry-matter of vine tea.

2.3. HPLC Analysis

HPLC analysis of VTP was performed in a Jasco MD-2015Plus HPLC (JASCO International Co. Ltd., Tokyo, Japan) equipped with a Cosmocore 2.6 C18 Packed Column (Nacalai Tesque Inc., Kyoto. Japan). Mobile phase A was 0.1% (*v/v*) phosphoric acid (aqueous), and mobile phase B was acetonitrile. A mobile phase consisting of 85% A and 15% B was delivered to the column at a flow rate of 1.00 mL/min at 27 °C. DMY was prepared at 2, 1, 0.5, 0.25, 0.125, and 0.06125 mg/mL in methanol. VTP extracted from different locations were prepared at 1 mg/mL in methanol. The injection volume was set at 10 μL. UV absorption spectra were obtained from 200 nm to 400 nm, and in particular, the absorbance at 290 nm was recorded.

2.4. Assay of 2,2-diphenyl-1-picrylhydrazy (DPPH) Free Radical Scavenging Activity

The radical scavenging activity of different locations of VTP and dihydromyricetin were measured by the DPPH method [21]. All of the VTP and DMY samples were prepared at 12.5, 25, 50, 100, 200, and 400 μg/mL in 70% ethanol. Trolox standards were prepared at 12.5, 25, 50, 100, 200, and 400 μM in 70% ethanol. Briefly, ten microliters of each sample were mixed with 190 μL of 0.2 mM DPPH in 96-well plates, and the final concentrations of VTP samples and DMY were 0.625, 1.25, 2.5, 5, 10, and 20 μg/mL. The absorbance was then measured at 492 nm with a microplate reader (Thermo scientific Multiscan FC, Tokyo, Japan) after the plate covered with aluminum foil was left for 30 min at 25 °C. The percentage scavenging rate of DPPH was calculated according to the formula:

$$\text{DPPH Scavenging rate} = (A0 - As)/A0 \times 100\%$$

where A0 represents the absorption of the blank sample, and as represents the absorption of VTP or other standards.

2.5. Assay of Oxygen Radical Absorbance Capacity (ORAC)

ORAC was measured according to the method as described previously [22]. In brief, one hundred microliters of fluorescein (7.5 nM), 10 μL of Trolox or VTP or DMY were added in the 96-well plate and incubated at 37 °C for 15 min. After the incubation, 40 μL of 2,2'-Azobis(2-amidinopropane) dihydrochloride (AAPH) (100 mM) was added rapidly to start the reaction, and the microplate was automatically shaken prior to each reading. The fluorescence was recorded every 2 min by a multilabel counter (PerkinElmer Co. Ltd., Tokyo, Japan). The ORAC values were calculated based on the area under curve (AUC) of the sample standardized by blank and Trolox standards. The data were expressed as Trolox equivalents (μmol TE/g).

2.6. Cell Culture and Western Blot Analysis

Human hepatoblastoma HepG2 cells obtained from the Cancer Cell Repository (Tohoku University, Sendai, Japan) were cultured in Dulbecco's modified Eagle's medium (DMEM) containing 10% FBS at 37 °C in a 5% CO_2 atmosphere. HepG2 cells (5×10^5 cells/dish) were precultured in 10 cm culture dishes for 24 h and then treated with various concentrations of VTP and DMY in 0.1% dimethyl sulfoxide (DMSO), and the control group was 0.1% DMSO only. The cells were harvested with modified RIPA buffer (50 mM Tris-HCl (pH 8.0), 150 mM NaCl, 1 mM EDTA, 1% Nonidet P-40, 0.25% Na-deoxycholate, 1 mM sodium fluoride, 1 mM sodium orthovanadate, 1 mM phenylmethylsulfonyl fluoride) plus proteinase inhibitor cocktail (Nacalai Tesque, Inc., Kyoto, Japan). Equal amounts of lysate protein were separated on sodium dodecyl sulfate-polyacrylamide gel electrophoresis (SDS-PAGE) and transferred to a polyvinylidene difluoride (PVDF) membrane electrophoretically (GE Healthcare UK Ltd., Amersham, England). After the membrane was blocked with TBST buffer (500 mM NaCl, 20 mM Tris-HCl (pH 7.4), and 0.1% Tween 20) containing 5% non-fat dry milk, the membrane was incubated overnight with the primary antibodies (β-actin, Nrf2, HO-1, and NQO1) at 4 °C and further incubated with HRP-conjugated secondary antibodies for another 1 h. The target proteins were detected using the enhanced chemiluminescence (ECL) system. The relative amounts of proteins bound with a specific antibody were quantified with Lumi Vision Imager software (TAITEC Co. Ltd., Saitama, Japan).

2.7. Statistical Analysis

The experiment results were presented as mean ± SD. The statistical differences between groups were performed by one-way analysis of variance tests, followed by Fisher's least significant difference

(LSD) and Duncan's multiple range tests with the SPSS statistical program (version 19.0, IBM Corp., NY, USA.). A probability of $p < 0.05$ was considered as significant.

3. Results

3.1. Extraction Conditions for Vine Tea Polyphenol

As shown in Figure 2A, the extracts obtained by organic solvents including ethanol, methanol, acetone, and ethyl acetate showed significant higher polyphenol content than that by water ($p < 0.05$). Due to the use of ethanol being recognized in the food industry, we chose ethanol as the solvent to extract polyphenol from vine tea in this study. To optimize the efficiency conditions, the ethanol concentration, extraction time, and temperature were further investigated. The polyphenol yield was increased in a concentration-dependent manner from 10–70% ethanol, and the highest yield of polyphenol was obtained by 70% ethanol extraction (Figure 2B). The polyphenol yield was also observed in the extraction time and temperature-dependent manner from 20–60 min at 40–100 °C, respectively. Extraction with 70% ethanol at 70 °C for 40 min yielded the highest VTP (Figure 2C,D).

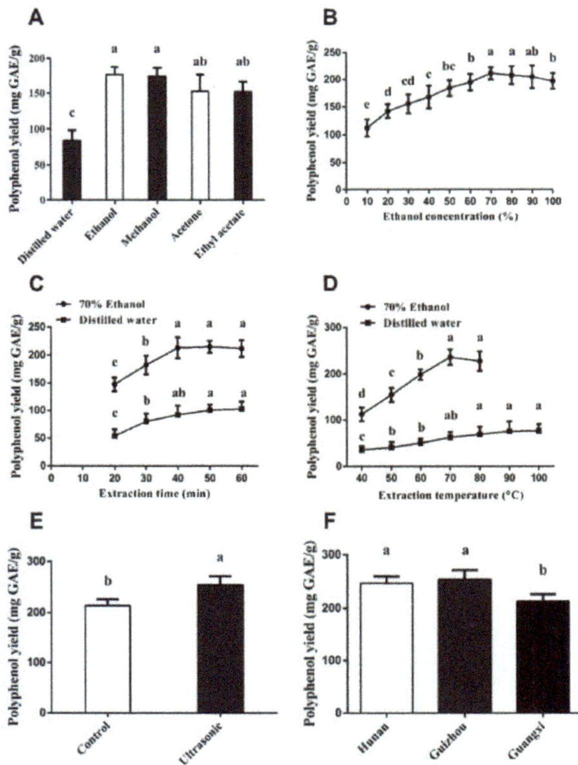

Figure 2. Conditions for extracting the vine tea polyphenols. (**A**) Polyphenol yield extracted by different solvents. (**B**) Polyphenol yield extracted by different concentrations of ethanol. (**C**) Polyphenol yield extracted by 70% ethanol with different times. (**D**) Polyphenol yield extracted by 70% ethanol at different temperatures. (**E**) Polyphenol yield extracted by 70% ethanol alone (control) and 70% ethanol plus ultrasonic treatment (ultrasonic). (**F**) Polyphenol yield of vine tea from different locations. The data represent mean ± SD with three repeats, and different letters in the same column indicate significant differences ($p < 0.05$).

An ultrasonic extraction technique was reported to increase the polyphenol contents extracted from tea [23]. Our data also revealed that extraction plus ultrasonic treatment in the above conditions could significantly increase the yield of VTP than that by ethanol alone ($p < 0.05$, Figure 2E). Finally, we used these optimized extraction conditions to extract the polyphenols from three locations. As shown in Figure 2F, the total polyphenol contents of vine tea from Hunan and Guizhou Provinces were significantly higher than that from Guangxi Province ($p < 0.05$).

3.2. DMY Determination in Vine Tea Polyphenol by HPLC

It has been reported that the main component of vine tea is DMY [24], therefore, we determined DMY in VTP by HPLC with a standard DMY. Figure 3A–C show the HPLC profile with a main peak from Hunan, Guizhou, and Guangxi Provinces, respectively. Figure 3D shows the HPLC profile of a mixture containing 1 mg VTP and 1 mg standard DMY, where the main peak increased to about twice as high than that in VTP alone. These data indicate that the main peak in VTP is DMY. Furthermore, we estimated that the DMY content in vine tea came from three different locations, according to the dihydromyricetin standard curve. The DMY content was estimated as 21.67%, 20.79%, and 16.42% in the dry powder of vine tea (white bar, Figure 3E), and as 64.44%, 62.36%, and 56.22% in the vine tea polyphenol (VTP) extract (black bar, Figure 3E) from Hunan, Guizhou, and Guangxi Provinces respectively (Figure 3E). The DMY content from Guizhou and Hunan Provinces was significantly higher than that from Guangxi Province ($p < 0.05$).

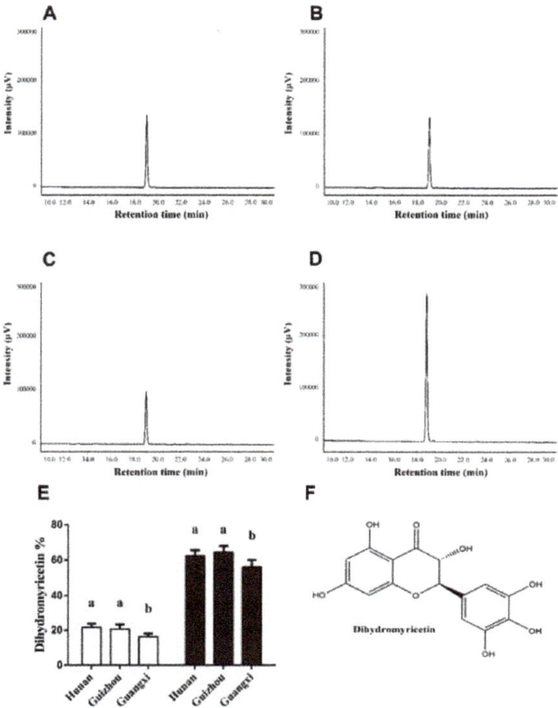

Figure 3. Dihydromyricetin determination in vine tea polyphenols by high-performance liquid chromatography (HPLC). The HPLC profiles of vine tea polyphenols (VTP) from Hunan Province (**A**), Guizhou Province (**B**), Guangxi Province (**C**) as well as the VTP plus standard dihydromyricetin (DMY) (**D**). (**E**) DMY content in the vine tea dry powder (white bar) and in VTP (black bar) from the above three locations. (**F**) Chemical structure of DMY. The data represent mean ± SD from three repeats, different letters in the same column indicate significant differences ($p < 0.05$).

3.3. DPPH Radical Scavenging Activity and ORAC Values

The DPPH radical is one of the few stable organic nitrogen radicals and can be simple and accurately measured [25]. Thus, we first used the DPPH assay to screen the radical scavenging activity of vine tea extract from three different locations. As shown in Figure 4A, a concentration-dependent manner was observed in the range of 0.6125 to 10 µg/mL. The concentration for scavenging 50% DPPH radicals (IC_{50}) by VTP from Hunan, Guizhou, and Guangxi products were estimated as 4.51 µg/mL, 4.06 µg/mL, and 4.31 µg/mL, respectively. Moreover, the IC_{50} of pure DMY was estimated as 3.24 µg/mL, which was 0.7-fold of the IC_{50} than that of the VTP and significantly lower (Figure 4B). As we measured above, the DMY content was as high as 64.44%, 62.36%, and 56.22% in VTP from three locations. These data indicated that DMY is a major DPPH radical scavenger in VTP. The ORAC assay utilizes a controllable source of peroxyl radicals that can stimulate the antioxidant reactions with lipids in both food and physiological systems, which cannot be assayed by the DPPH assay [25]. Therefore, we further used the ORAC assay to evaluate the oxygen radical absorbance capacity of vine tea from three different locations. As shown in Figure 4C,D, the ORAC value of vine tea from Hunan, Guizhou, and Guangxi Provinces were estimated as 3116.97, 2941.61, and 2791.32 µmol of Trolox equivalent (TE)/g, respectively. No significant difference in ORAC value was observed between three products VTP ($p > 0.05$). The ORAC value of DMY was also significantly higher than that of VTP from three locations ($p < 0.05$), indicating DMY is also a major compound for ORAC in VTP.

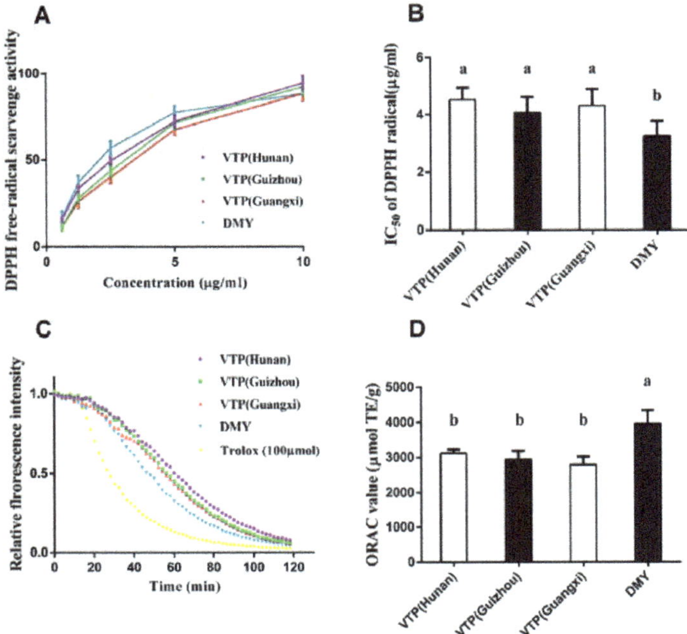

Figure 4. Antioxidant ability of VTP. (**A**) 2,2-diphenyl-1-picrylhydrazy (DPPH) free-radical scavenging rate of VTP from three locations and DMY. (**B**) IC_{50} value of the DPPH free-radical scavenging of VTP from three locations and DMY. (**C**) Relative florescence intensity of oxygen radical absorbance capacity (ORAC) of VTP from three locations and DMY in 2 h. (**D**) ORAC value (Trolox equivalent/g sample) of VTP from three locations and DMY. The data represent mean ± SD with three repeats, different letters in the same bar indicate significant differences ($p < 0.05$).

3.4. Effect of VTP and DMY on Expression of Antioxidant Enzymes in HepG2 Cells

The results from the *in vitro* data indicated that VTP and its main component DMY possessed antioxidant activity. To clarify whether the antioxidant activity was also observed in the cells, we further investigated the effect of VTP and DMY on the expression of antioxidant enzymes such as NQO1, which is typical antioxidant enzyme in liver and is regulated by the Nrf2/Keap1 pathway. In a time-course experiment, HepG2 cells were treated with 40 μM DMY (Figure 5A) and VTP (equivalent to 40 μM DMY) (Figure 5B) from 0–12 h. Both VTP and DMY enhanced the NQO1 level from 3–12 h, Nrf2 level from 3–6 h, and reduced Keap1 level from 12 h. In a dose-experiment, HepG2 cells were treated with 0–120 μM DMY (Figure 5C) and VTP (equivalent to 0–120 μM DMY) (Figure 5D) for 9 h. Both DMY and VTP enhanced the NQO1 and Nrf2 level from 20–120 μM, and also reduced the Keap1 level in this dose range.

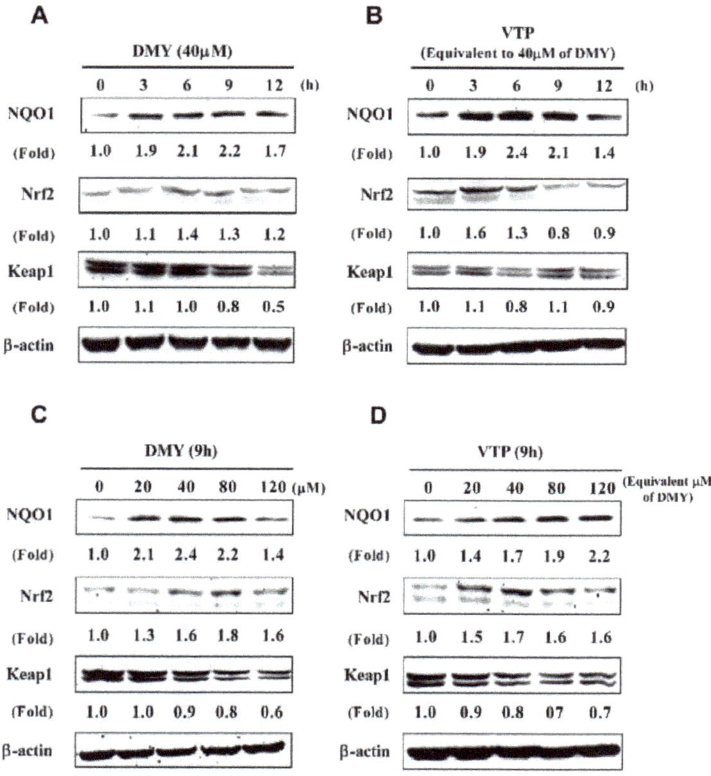

Figure 5. Effects of DMY and VTP on the level of nicotinamide adenine dinucleotide phosphate (NADPH):quinone oxidoreductase (NQO1), nuclear factor erythroid 2-related factor 2 (Nrf2) and Kelch-like ECH-associated protein 1 (Keap1) proteins. A time-effect of DMY (**A**) and VTP (**B**) on the level of NQO1, Nrf2 ,and Keap1 proteins. HepG2 cells were treated with DMY (40 μM) or VTP (equivalent to 40 μM DMY) for 0–12 h. A dose-effect of DMY (**C**) and VTP (**D**) on the level of NQO1, Nrf2, and Keap1 proteins. HepG2 cells were treated with DMY (20–120 μM) and VTP (equivalent to 20–120 μM DMY) for 9 h. The fold was normalized with the control protein, β-actin, and obtained from triplicate blot data.

4. Discussion

4.1. Extraction and Determination of Vine Tea Polyphenols

Vine tea as a traditional herb is widely used in medicine and health supplements in southwest China. A previous study reported that the extraction of tea compounds by ethanol was more efficient than that by water [26]. In order to precisely estimate the chemical and antioxidant properties in vine tea, we optimized the extraction condition by investigating the extraction solvents, solvent concentration, extraction time, and temperature in this study. The extraction condition with 70% ethanol at 70 °C for 40 min with ultrasonic treatment obtained the highest VTP, significantly higher than that by water. Although tea is usually consumed by soaking it into boiling water, there has been an increasing utilization of tea extracts in a variety of foods such as bread, biscuits, and meat products [14,27], and especially in health supplements. Moreover, ethanol is approved as a generally recognized as safe (GRAS) substance by the Food and Drug Administration(FDA) [28]. Furthermore, the extracts of vine tea by 70% ethanol contained a higher total polyphenol content, which seemed to be 2-fold higher than that in green tea, and was 7-fold higher than black tea through a comparison with the results of the polyphenol content in aqueous or ethanol extracts of green tea, black tea, and another 11 leafy herb teas [29]. Thus, the extracts of vine tea by 70% ethanol with ultrasonic treatment will have great potential in efficiently utilizing vine tea.

Generally, tea is a complex mixture containing a range of polyphenols and other components, many of which have well-recognized antioxidant properties [29]. In this study, we found that there was only one major component, DMY, which is as high as 60% in VTP. DMY belongs to flavonoids, which is a special class of phenolic compounds with a structure based on the diphenylpropane carbon skeleton. It is known that DMY is an antioxidant agent in food preservation [30], and also increases antioxidant ability in animal model experiments [31].

4.2. Antioxidant Capacity of Vine Tea Polyphenols

Vine tea possesses high polyphenol contents. Accumulated data have indicated that polyphenol content has a significant positive correlation to antioxidant ability [32]. The DPPH free radical assay is an electron transfer reaction, and this assay is rapid and widespread in antioxidant screening. However, the DPPH assay is not a competitive reaction because the small molecules tend to find it easier to bind with the radical site and have a higher value. Meanwhile, the ORAC assay can simulate a human physiological antioxidant situation based on the transfer reaction mechanism. Thus, we used both assays to investigate the antioxidant ability of VTP and DMY. The high antioxidant activity of VTP and DMY were observed in both assays, suggesting that the extracted VTP possessed strong antioxidant capacity and could be developed as an antioxidant agent used in human biology. Moreover, vine tea seems to have higher ORAC value when compared with most consumed green tea ethanol extracts in China [17].

The contents of VTP and DMY showed some differences between the three locations. The order of both contents was Guizhou = Hunan > Guangxi, in particular, the DMY content of Guangxi was significantly lower than that from Guizhou and Hunan. We further researched the plantation elevation of the three samples, and found that the plantation elevation of vine tea in Guizhou and Hunan was 800–1300 m, and was 800 m Guangxi. A recent study reported that the antioxidant ability and polyphenol composition were affected by geographic location, growing environment, and leaf grades [18], and black tea contained about 20% more polyphenols when plants were at low elevation. However, our study showed the opposite trend where vine tea from higher plantation elevations contained 30% more polyphenols, and 15% more DMY than those from lower elevations. This result may be due to the different varieties, but this is still the first report and in further study will be required to find the relationship between the growing environment and polyphenolic active compound contents.

Previous studies have reported that DMY could increase antioxidant defense through activation of the ERK and Akt signaling pathways, which induces heme oxygenase-1 expression and thereby protects PC12 cells from H_2O_2 induced apoptosis [33,34], and DMY could protect endothelial cells from oxidative stress, and increase the production of nitric oxide [35]. Based on this information, we investigated the effect of VTP and DMY on the Nrf2/Keap1 pathway, which is a master cellular antioxidant defense system against oxidative stress. Our data revealed that VTP and its major component DMY enhanced the level of Nrf2, a positive factor for the Nrf2/Keap1 pathway, and reduced the level of Keap1, a negative factor for the Nrf2/Keap1 pathway. Sequentially, the level of downstream antioxidant enzyme, NQO1, was increased by VTP and DMY in a dose- and time-dependent manner. These data demonstrated that vine tea and its major compound DMY might exert an antioxidant activity in culture cells by activating the Nrf2/Keap1 pathway.

5. Conclusions

In conclusion, the extraction of vine tea by 70% ethanol with ultrasonic treatment is a novel method to efficiently obtain the bioactive components that possess stronger DPPH scavenging ability and ORAC *in vitro*. Moreover, they enhanced the level of the antioxidant enzyme, NQO1, in culture cells by activating the Nrf2/Keap1 pathway. These findings will help us understand the mechanism of the health function of a traditional vine tea.

Author Contributions: Conceptualization, D.-X.H. and K.X., Methodology, K.X., K.C., and J.C.; Software, K.X.; Validation, K.S., X.H. ,and D.-X.H.; Formal Analysis, K.X. and D.-X.H.; Investigation, K.X. and K.C.; Resources, X.H. and D.-X.H.; Data Curation, K.X. and D.-X.H.; Writing-Original Draft Preparation, K.X.; Writing-Review & Editing, D.-X.H; Project Administration, K.X. and D.-X.H.

Funding: This research was supported by the Scholar Research of Kagoshima University Fund to D.-X.H. (grant number 70030117), and the National Key R&D Program of Intergovernmental Key Projects of China (grant number 2018YFE0101700) to X.H.

Conflicts of Interest: The authors declare no conflicts of interest.

References

1. Zhang, H.; Tsao, R. Dietary polyphenols, oxidative stress and antioxidant and anti-inflammatory effects. *Curr. Opin. Food Sci.* **2016**, *8*, 33–42. [CrossRef]
2. Amiot, M.J.; Riva, C.; Vinet, A. Effects of dietary polyphenols on metabolic syndrome features in humans: A systematic review. *Obes. Rev.* **2016**, *17*, 573–586. [CrossRef] [PubMed]
3. Moylan, J.S.; Reid, M.B. Oxidative stress, chronic disease, and muscle wasting. *Muscle Nerve* **2007**, *35*, 411–429. [CrossRef] [PubMed]
4. Dhalla, N.S.; Elmoselhi, A.B.; Hata, T.; Makino, N. Status of myocardial antioxidants in ischemia-reperfusion injury. *Cardiovasc. Res.* **2000**, *47*, 446–456. [CrossRef]
5. Shadel, G.S.; Horvath, T.L. Mitochondrial ROS Signaling in Organismal Homeostasis. *Cell* **2015**, *163*, 560–569. [CrossRef]
6. Birben, E.; Sahiner, U.M.; Sackesen, C.; Erzurum, S.; Kalayci, O. Oxidative stress and antioxidant defense. *World Allergy Organ. J.* **2012**, *5*, 9–19. [CrossRef]
7. Pietta, P.G. Flavonoids as antioxidants. *J. Nat. Prod.* **2000**, *63*, 1035–1042. [CrossRef]
8. Tanigawa, S.; Fujii, M.; Hou, D.X. Action of Nrf2 and Keap1 in ARE-mediated NQO1 expression by quercetin. *Free Radic. Biol. Med.* **2007**, *11*, 1690–1703. [CrossRef]
9. Li, J.; Sapper, T.N.; Mah, E.; Rudraiah, S.; Schill, K.E.; Chitchumroonchokchai, C.; Moller, M.V.; Mcdonald, J.D.; Rohrer, P.R.; Manautou, J.E.; et al. Green tea extract provides extensive Nrf2-independent protection against lipid accumulation and NFκB pro-inflammatory responses during nonalcoholic steatohepatitis in mice fed a high-fat diet. *Mol. Nutr. Food Res.* **2016**, *60*, 858–870. [CrossRef]
10. Rice-Evans, C.A.; Miller, N.J.; Paganga, G. Antioxidant properties of phenolic compounds. *Trends Plant Sci.* **1997**, *2*, 152–159. [CrossRef]
11. Tsao, R. Chemistry and biochemistry of dietary polyphenols. *Nutrients* **2010**, *2*, 1231–1246. [CrossRef] [PubMed]
12. Almajano, M.P.; Carbó, R.; Jiménez, J.A.L.; Gordon, M.H. Antioxidant and antimicrobial activities of tea infusions. *Food Chem.* **2008**, *108*, 55–63. [CrossRef]

13. El-Shahawi, M.S.; Hamza, A.; Bahaffi, S.O.; Al-Sibaai, A.A.; Abduljabbar, T.N. Analysis of some selected catechins and caffeine in green tea by high performance liquid chromatography. *Food Chem.* **2012**, *134*, 2268–2275. [CrossRef] [PubMed]
14. Namal Senanayake, S.P.J. Green tea extract: Chemistry, antioxidant properties and food applications—A review. *J. Funct. Foods* **2013**, *5*, 1529–1541. [CrossRef]
15. Hayat, K.; Iqbal, H.; Malik, U.; Bilal, U.; Mushtaq, S. Tea and Its Consumption: Benefits and Risks. *Crit. Rev. Food Sci. Nutr.* **2015**, *55*, 939–954. [CrossRef] [PubMed]
16. Frei, B.; Higdon, J.V. Antioxidant Activity of Tea Polyphenols In Vivo: Evidence from Animal Studies. *J. Nutr.* **2018**, *133*, 3275S–3284S. [CrossRef] [PubMed]
17. Jin, L.; Li, X.B.; Tian, D.Q.; Fang, X.P.; Yu, Y.M.; Zhu, H.Q.; Ge, Y.Y.; Ma, G.Y.; Wang, W.Y.; Xiao, W.F.; et al. Antioxidant properties and color parameters of herbal teas in China. *Ind. Crops Prod.* **2016**, *87*, 198–209. [CrossRef]
18. Zhang, C.; Suen, C.L.C.; Yang, C.; Quek, S.Y. Antioxidant capacity and major polyphenol composition of teas as affected by geographical location, plantation elevation and leaf grade. *Food Chem.* **2018**, *244*, 109–119. [CrossRef]
19. Turkmen, N.; Sari, F.; Velioglu, Y.S. Effects of extraction solvents on concentration and antioxidant activity of black and black mate tea polyphenols determined by ferrous tartrate and Folin–Ciocalteu methods. *Food Chem.* **2006**, *99*, 835–841. [CrossRef]
20. Ainsworth, E.A.; Gillespie, K.M. Estimation of total phenolic content and other oxidation substrates in plant tissues using Folin–Ciocalteu reagent. *Nat. Protoc.* **2007**, *2*, 875. [CrossRef]
21. Zar, P.P.K.; Morishita, A.; Hashimoto, F.; Sakao, K.; Fujii, M.; Wada, K.; Hou, D.X. Anti-inflammatory effects and molecular mechanisms of loquat (*Eriobotrya japonica*) tea. *J. Funct. Foods* **2014**, *6*, 523–533. [CrossRef]
22. Dávalos, A.; Gómez-Cordovés, C.; Bartolomé, B. Extending Applicability of the Oxygen Radical Absorbance Capacity (ORAC-Fluorescein) Assay. *J. Agric. Food Chem.* **2004**, *52*, 48–54. [CrossRef] [PubMed]
23. Albu, S.; Joyce, E.; Paniwnyk, L.; Lorimer, J.P.; Mason, T.J. Potential for the use of ultrasound in the extraction of antioxidants from Rosmarinus officinalis for the food and pharmaceutical industry. *Ultrason. Sonochem.* **2004**, *11*, 261–265. [CrossRef] [PubMed]
24. Du, Q.; Cai, W.; Xia, M.; Ito, Y. Purification of (+)-dihydromyricetin from leaves extract of *Ampelopsis grossedentata* using high-speed countercurrent chromatograph with scale-up triple columns. *J. Chromatogr. A* **2002**, *973*, 217–220. [CrossRef]
25. Prior, R.L.; Wu, X.; Schaich, K. Standardized methods for the determination of antioxidant capacity and phenolics in foods and dietary supplements. *J. Agric. Food Chem.* **2005**, *53*, 4290–4302. [CrossRef] [PubMed]
26. Friedman, M.; Levin, C.E.; Choi, S.H.; Kozukue, E.; Kozukue, N. HPLC analysis of catechins, theaflavins, and alkaloids in commercial teas and green tea dietary supplements: Comparison of water and 80% ethanol/water extracts. *J. Food Sci.* **2006**, *71*, C328–C337. [CrossRef]
27. Ye, L.; Wang, H.; Duncan, S.E.; Eigel, W.N.; O'Keefe, S.F. Antioxidant activities of Vine Tea (*Ampelopsis grossedentata*) extract and its major component dihydromyricetin in soybean oil and cooked ground beef. *Food Chem.* **2015**, *172*, 416–422. [CrossRef] [PubMed]
28. Gad, S.E.; Sullivan, D.W. Generally Recognized as Safe (GRAS). In *Encyclopedia of Toxicology*, 3rd ed.; Academic Press: Cambridge, MA, USA, 2014; ISBN 9780123864543.
29. Del Rio, D.; Stewart, A.J.; Mullen, W.; Burns, J.; Lean, M.E.J.; Brighenti, F.; Crozier, A. HPLC-MSn Analysis of Phenolic Compounds and Purine Alkaloids in Green and Black Tea. *J. Agric. Food Chem.* **2004**, *52*, 2807–2815. [CrossRef] [PubMed]
30. Yang, J.G.; Liu, B.G.; Liang, G.Z.; Ning, Z.X. Structure-activity relationship of flavonoids active against lard oil oxidation based on quantum chemical analysis. *Molecules* **2009**, *14*, 46–52. [CrossRef]
31. Zheng, X.J.; Xiao, H.; Zeng, Z.; Sun, Z.W.; Lei, C.; Dong, J.Z.; Wang, Y. Composition and serum antioxidation of the main flavonoids from fermented vine tea (*Ampelopsis grossedentata*). *J. Funct. Foods* **2014**, *9*, 290–294. [CrossRef]
32. Dudonné, S.; Vitrac, X.; Coutiére, P.; Woillez, M.; Mérillon, J.M. Comparative study of antioxidant properties and total phenolic content of 30 plant extracts of industrial interest using DPPH, ABTS, FRAP, SOD, and ORAC assays. *J. Agric. Food Chem.* **2009**, *57*, 1768–1774. [CrossRef] [PubMed]

33. Kou, X.; Shen, K.; An, Y.; Qi, S.; Dai, W.X.; Yin, Z. Ampelopsin inhibits H2O2-induced apoptosis by ERK and Akt signaling pathways and up-regulation of heme oxygenase-1. *Phyther. Res.* **2012**, *26*, 988–994. [CrossRef] [PubMed]
34. Kou, X.; Chen, N. Pharmacological potential of ampelopsin in Rattan tea. *Food Sci. Hum. Wellness* **2012**, *1*, 14–18. [CrossRef]
35. Hou, X.; Tong, Q.; Wang, W.; Xiong, W.; Shi, C.; Fang, J. Dihydromyricetin protects endothelial cells from hydrogen peroxide-induced oxidative stress damage by regulating mitochondrial pathways. *Life Sci.* **2015**, *130*, 38–46. [CrossRef] [PubMed]

Article

Inhibition of LPS-Induced Oxidative Damages and Potential Anti-Inflammatory Effects of *Phyllanthus emblica* Extract via Down-Regulating NF-κB, COX-2, and iNOS in RAW 264.7 Cells

Hui Min-David Wang [1,2,3,4,†], Ling Fu [2,†], Chia Chi Cheng [5], Rong Gao [6], Meng Yi Lin [7], Hong Lin Su [5], Nathania Earlene Belinda [8], Thi Hiep Nguyen [9], Wen-Hung Lin [10], Po Chun Lee [11,*] and Liang Po Hsieh [12,*]

1 College of Oceanology and Food Science, Quanzhou Normal University, Quanzhou 362000, China
2 Graduate Institute of Biomedical Engineering, National Chung Hsing University, Taichung City 402; Taiwan
3 Graduate Institute of Medicine, College of Medicine, Kaohsiung Medical University, Kaohsiung 807, Taiwan
4 Department of Medical Laboratory Science and Biotechnology, China Medical University, Taichung City 404, Taiwan
5 Department of Life Science, National Chung Hsing University, Taichung City 402, Taiwan
6 Deloitte Institute of Biology, Yangtze River Delta Research Institute, Tsinghua University, Beijing 100084, China
7 Department of Chemical and Materials Engineering, Tunghai University Taichung City 407, Taiwan
8 Undergraduate Study Program of Biomedical Engineering, Department of Physics, Faculty of Science & Technology, Airlangga University, Surabaya 60115, Indonesia
9 Tissue Engineering and Regenerative Medicine Laboratory, Department of Biomedical Engineering, International University, Vietnam National University, Ho Chi Minh City 700000, Vietnam
10 Department of Biomedical Informatics, Postdoctoral researcher for Biomedical Informatics, National Defense Medical Center, Taipei 114, Taiwan
11 Cardiovascular Clinic, Kaohsiung Armed Forces General Hospital, Kaohsiung City 802, Taiwan
12 Neurology, Internal Medicine, Cheng Ching Hospital, Taichung City 407, Taiwan
* Correspondence: chyun0124@gmail.com (P.C.L.); lphsieh624@yahoo.com.tw (L.P.H.); Tel.: +886-8-7225671 (P.C.L.); +886-4-24632000 (L.P.H.); Fax: +886-7-7491320 (P.C.L.); +886-4-24635962 (L.P.H.)
† These authors contributed equally to this work.

Received: 22 July 2019; Accepted: 31 July 2019; Published: 2 August 2019

Abstract: *Phyllanthus emblica* is an edible nutraceutical and functional food in the Asia area with medicinal and nutritive importance. The fruit extract of *P. emblica* is currently considered to be one of the effective functional foods for flesh maintenance and disease treatments because of its antioxidative and immunomodulatory properties. We examined the antioxidant abilities of the fruit extract powder by carrying out 2,2-diphenyl-1-picrylhydrazyl (DPPH) free radical scavenging, iron reducing power, and metal chelating activity analysis and showed excellent antioxidative results. In 3-(4,5-dimethylthiazol-2-yl)-2,5-diphenyltetrazolium bromide (MTT) assay, the result showed that the samples had no cytotoxic effect on RAW 264.7 cells even at a high concentration of 2 mg/mL. To investigate its immunomodulatory function, our estimation was to treat it with lipopolysaccharide (LPS) in RAW 264.7 cells to present anti-inflammatory capacities. The extract decreased reactive oxygen species (ROS) production levels in a dose-dependent manner measured by flow cytometry. We also examined various inflammatory mRNAs and proteins, including nuclear factor-κB (NF-κB), inducible nitric oxide synthases (iNOS), and cyclooxygenase-2 (COX-2). In quantitative reverse transcription polymerase chain reaction (qRT-PCR) and western blotting assay, all three targets were decreased by the extract, also in a dose-dependent manner. In conclusion, *P. emblica* fruit extract powder not only lessened antioxidative stress damages, but also inhibited inflammatory reactions.

Keywords: *Phyllanthus emblica*; antioxidant; COX-2; iNOS; NF-κB

1. Introduction

Innate immune response, also called nonspecific immune response, is the first barrier to stop detrimental materials invading our bodies and granulocytes, macrophages, and inflammatory biomolecules are involved. Inflammation, a common but complex reaction after the immune system recognizes external pathogens or damaged cells, occurs in all types of human tissues and usually presents a protective effect. Thus, a normal inflammatory response has been regarded as a guard to protect the human body from extrinsic pathogens and intrinsic injury [1]. Vital physiological symptoms, for example, increased blood flow, vasodilation, elevated cellular metabolism, a release of proinflammatory mediators, cellular influx, and an accumulation of fluid are hallmarks of inflammatory responses. Generally, an inflammatory reaction is good for humans. However, abnormal inflammation has been reported to be related to several human chronic diseases, including rheumatoid arthritis, atherosclerosis, and diabetes [2,3]. To heal immoderate inflammation, proinflammatory mediators are aimed as targets because inflammatory cells recruit these materials to the scene site.

Proinflammatory mediators such as nuclear factor-κB (NF-κB), cyclooxygenase-2 (COX-2) and inducible nitric oxide synthase (iNOS) are pivotal to the evaluation of inflammation levels. Incorrect regulation of NF-κB has been reported to be linked to cancers, inflammatory and autoimmune diseases, viral and bacterial infections, and improper immune responses [4]. Because there is a variety of proinflammatory gene expressions induced by NF-κB and the regulation of inflammation, down-regulating of NF-κB activation contributes to various inflammatory diseases [5]. NF-κB also participates in the transcription of another inflammatory association enzyme, iNOS. Dependent activation of the iNOS promoter supports an inflammation-mediated stimulation of this transcript. Nitric oxide (NO) is a critical signaling molecule as a retrograde neurotransmitter which is associated with neural development, immune response, angiogenesis, and one vital feature of inflammation, i.e., vasodilation [6]. NO is mediated in humans by three major types of nitric oxide synthases (NOS) (i.e., endothelial NOS (eNOS), neuronal NOS (nNOS), and iNOS) [7]. When iNOS is activated by cytokines, NO is released. NO is an activating factor of cyclooxygenase (COX), which forms a five coordination with the COX structure, causing a conformational change in COX. COX is officially called prostaglandin endoperoxide synthase, and it is responsible for the biosynthesis of prostanoid, such as thromboxane and prostaglandins, from arachidonic acid. In humans, one of two cyclooxygenases, COX-2, responds by mediating inflammatory reactions [8]. Therefore, COX-2 inhibitors are often used as anti-inflammatory drugs.

Phyllanthus emblica fruit, an Indian traditional medicine and an effective functional food, has been used to test its anti-inflammatory activity for centuries, and provides potential therapeutics for a variety of maladies [9]. *P. emblica* fruit contains high levels of vitamin C, tannins, polyphenols (gallic acid and ellagic acid), minerals, fibers, and so on [10]. Recently, several hydrolysable tannins, flavonoids, and alkaloids have been identified in *P. emblica* fruit. Not surprisingly, vitamin C, gallic acid, and ellagic acid, which are present in *P. emblica* fruit, are known to be potent antioxidants, flavonoids, and other biofunctional constitutes that assist inflammation reduction [11]. Although some materials have been proven to improve the symptoms of the inflammation, the mechanism of *P. emblica* fruit on its anti-inflammation activity is still not well known. As an edible food or food additive, *P. emblica* fruit extract powder can be used as an antioxidant and anti-inflammatory diet, and its fruit may help us to deal with these related diseases.

2. Materials and Methods

2.1. Materials

The testing sample, *P. emblica* fruit extract powder, was obtained from SHENG GUO Biotech Co., Ltd, Miaoli, Taiwan. Dimethyl sulfoxide (DMSO); lipopolysaccharide (LPS) (*Escherichia coli*

055: B5); vitamin C; 2,2-diphenyl-1-picrylhydrazyl (DPPH); ethylenediaminetetraacetic acid (EDTA); 3-tert-butyl-4-hydroxyanisole (BHA); potassium ferricyanide [$K_3Fe(CN)_6$] trichloroacetic acid, $FeCl_3$, $FeCl_2{\cdot}4H_2O$, and 3-(4,5-dimethylthiazol-2-yl)-2,5-diphenyl tetrazolium bromide (MTT); 2,7-dichlorofluorescein diacetate (DCFDA, D6883); and bicinchoninic acid (BCA) were purchased from Sigma-Aldrich Corp., USA. Dulbecco's Modified Eagle's Medium (DMEM), fetal bovine serum (FBS), penicillin, streptomycin, and amphotericin B (PSA) were purchased from GIBCO BRL (Gaithersburg, MD, USA).

2.2. *P. emblica* Fruit Powder Extracts Preparation

The extraction of *P. emblica* fruit was carried out using a custom freeze-drying procedure using a freeze dryer (FD-1, CHIANG DING Technology co., ltd., Taiwan) to make the *P. emblica* fruit at −35 °C for 10–12 h, and then dried at 60 °C ± 5% for 35 h. In order to freeze and dry the water in the fruit of *P. emblica* and make it into a powder, after the fruit was freeze dried, the moisture inside the fruit had to be less than 5%, and this was detected using a moisture analyzer (ML-50, A&D Technology, Inc., Japan). The dried fruits were extracted with 85–95 °C hot water at 5 liters per kilogram of fruit to make a liquid extract with 5% soluble content. The extract was filtered through a 10 microns polypropylene filter bag to remove insoluble materials. After vacuum concentration, the concentration was increased to 10% w/v. Maltodextrin was used as the carrier, which was added at a 1:1 ratio (10% *P. emblica* soluble content, 10% maltodextrin, w/v). The concentrate was frozen at −35 °C followed by freeze drying for 72 h (0–50 h at 0 °C, and then by a temperature ramp for 50–72 h to 45 °C) and pulverized to gain testing samples using a 1HPTable Type Pulverizing Machine (Product ID: RT-34).

2.3. Free Radical Scavenging Activity

The DPPH reagent which accepts an electron or hydrogen radical becomes a stable molecule to detect oxidative activities. When DPPH reacts with antioxidant agents, hydrogen is supplied, reducing the amount of DPPH and decreasing its absorbance, the optical density (OD) values at 517 nm [12,13]. Compared to other antioxidants, vitamin C (100 μM) is a great positive control because of its prominent antioxidant capacity. We added 1 μL at different concentrations of *P. emblica* fruit extracts and primary-filtered water to 99 μL DPPH (0.1 mg/mL). The absorbance was measured using the spectrophotometer and the remaining DPPH amount was plotted to determine the initial concentration of DPPH reduced by the antioxidant. Various sample amounts were dissolved in methanol for each well, and the final working volume was 100 μL. The clearance capacity (%) is calculated as follows:

$$\text{Clearance capacity } (\%) = \frac{\left(A_{\text{blank}} - A_{\text{sample}}\right)}{A_{\text{blank}}} \times 100\%$$

2.4. Ferric Reducing Antioxidant Power (FRAP) Assay

We carried out the reducing power assay analysis to examine the reductive ability of *P. emblica* fruit extract samples. The samples were dissolved in DMSO at a suitable concentration mix of 85 μL, phosphate buffer (0.2 M, pH 4.4) and 20% potassium ferricyanide (2.5 μL). The mixture was kept at 50 °C for 20 min, and then 160 μL of 10% trichloroacetic acid (TCA) was added. Subsequently, the solution was centrifuged at 3000× g for 10 min collecting supernatant (75 μL), and 25 μL $FeCl_3$ (2%) was added to the supernatant. After a 10 min reaction, the absorbance of the solution was measured at OD_{700} nm [14,15]. BHA was used as a positive control at 100 μM. A higher absorbance value means a better reduction activity.

2.5. Metal Chelating Ability Test

The chelation of ferrous ions in our sample was estimated by our previously published method [13]. Briefly, 10 μL of 2 mM $FeCl_2{\cdot}4H_2O$ was added to 1 μL of various concentrations (0.5–50 mg/mL) of samples, and the reaction was initiated by the addition of 20 μL of 5 mM ferrozine. This assay is based on the complexes of ferrous ions and ferrozine that change color at 562 nm, and the lower absorbance means the better metal chelating activity. EDTA at 100 μM acted as a positive control, and the chelating power activity is calculated by:

$$\text{Metal chelating activity } (\%) = \frac{\left(A_{\text{control}} - A_{\text{sample}}\right)}{A_{\text{control}}} \times 100\%$$

2.6. Cell Culture and Treatment

Mouse macrophage cell lines, RAW 264.7, were purchased from Bioresource Collection and Research Center (BCRC number: 60001) and cultured in DMEM containing 10% FBS and 1% PSA. Cells were incubated at 37 °C in a humidified incubator with 5% CO_2 atmosphere [16]. Samples were dissolved in DMSO and then diluted by using DMEM medium (0.25, 0.5, 1, and 2 mg/mL). The cells were pretreated with testing samples for 1 h and stimulated with LPS (5 μg/mL) for 6 h, and untreated cells served as a blank control [17,18].

2.7. Cell Viability Assay

RAW 264.7 cell viability was evaluated using MTT colorimetric assay [19,20]. The cells were cultured in DMEM containing 10% FBS and 1% PSA at 37 °C in 5% CO_2. All of the cells were seeded in 96-well microplates. After seeding the cells for 24 h, samples with concentration ranges from 0.005 to 10 mg/mL were added. After another 24 h, cells were treated with MTT solution (0.5 mg/mL) for 2 h, followed by incubation at 37 °C. After 2 h of MTT treatment, the medium was removed, and 100 μL of DMSO was added in each well to dissolve the purple formazan crystals. The dishes were gently shaken for 20 min in the dark to ensure maximal dissolutions of formazan crystals, and OD values of the supernatant were measured at 595 nm. The cell viability was presented as the percentage of live cells in each well, and was calculated according to the following formula:

$$\text{Cell viability } (\%) = \frac{\left(A_{\text{sample}} - A_{\text{blank}}\right)}{\left(A_{\text{control}} - A_{\text{blank}}\right)} \times 100\%$$

2.8. Measurement of Intracellular ROS Level

The ROS-sensitive fluorescent dye, DCFDA, was used to determine LPS-upregulated intracellular ROS level in RAW 264.7 cells. DCFDA is generally non-fluorescent, but in the presence of ROS (when this reagent is oxidized) it turns into green fluorescent. For an observation of intracellular ROS product through the oxidation of DCFDA, cells were pretreated with or without *P. emblica* samples (0.5–2.0 mg/mL) for 1 h, and stimulated with LPS (5 μg/mL) for 6 h. Afterward, we rinsed them with warm phosphate-buffered saline (PBS) buffer, and incubated them in PBS containing 20 μM DCFDA at 37 °C, 5% CO_2 for 30 min. PBS containing DCFDA was removed and replaced with fresh cell medium. The cells were washed at least 3 times with PBS and detached with trypsin/ EDTA. The fluorescence intensity of the cells was analyzed using a Guava®easyCyte Flow Cytometers (Merck KGaA, Darmstadt, Germany) at 485 nm excitation and 530 nm emission for 2,7-dichlorofluorescein (DCF) [21].

2.9. Quantitative Reverse Transcription Polymerase Chain Reaction (qRT-PCR)

For the qRT-PCR, a 10 μL reaction contained a 3 mixture of two reverse transcriptases: 10 μL of 2 × AceQ qPCR SYBR Green Master Mix (Vazyme Biotech Co.,Ltd, Nanjing, China) with the hot start Taq polymerase, 0.5 μL of primers, and 0.5 μL (20 ng/mL) of template. The primer sequences are listed in Table 1. The StepOnePlus™ System (Version 2.3) was used for all real-time PCR assays [22]. The reaction activated the AceTaq®DNA polymerase at 95 °C for 5 min. This was then amplified for 40 cycles at 95 °C for 3 s for denaturation, annealing, and acquisition at 60 °C for 40 s. It was finally elongated at 95 °C for 15 s. Fluorescence was measured after the annealing phase. With an Applied Biosystems™ MicroAmp™ (N8010560) Fast Optical 96-Well Reaction Plate, 10 μL of the reaction mix was added, as well as the 96-SYBR-Green assays on the StepOnePlus™ Real Time System [23]. To prepare the assay, all of the reagents were kept either on a cooling block or on ice. The ΔΔCt method was used in calculations. For each sample, three independent qPCR experiments were performed. Each experiment involved three replicates for each gene. An expression of GAPDH was used as an internal control. Duplicate SGPERT reactions were performed on each lysate sample. Using qPCR software and instruments, an ABI 7300 with its threshold determined manually and a LightCycler®480 with its maximum second derivative method generated the cycle of quantification (Cq) values. Using the same software for both instruments, the melting peaks were also automatically calculated.

Table 1. Primers used for quantitative reverse transcription polymerase chain reaction for the analysis of inflammatory gene expressions.

NF-κB
Forward: 5′-TATGTGTGTGAAGGCCCATCA-3′
Reverse: 5′-ACCAACTGAACGATAACCTTTGC-3′
iNOS
Forward: 5′-CGAGACGGATAGGCAGAGATTG-3′
Reverse: 5′-CTCTTCAAGCACCTCCAGGAA-3′
COX-2
Forward: 5′-CCAGCACTTCACCCATCAGTTT-3′
Reverse: 5′-TCTGTCCAGAGTTTCACCATAAATG-3′

2.10. Western Blotting

A total of 10^5 cells were treated with sample groups or the blank vehicle control for one day, respectively. The RAW 264.7 cells were harvested and lysed with the lysis buffer (Thermo Scientific Pierce RIPA Buffer, 1 mM EDTA, 10% glycerol, 1% Nonidet P-40, 2 μM leupeptin, 50 mM Tris-HCl, 137 mM sodium chloride, 50 mM sodium fluoride, 10 mM sodium pyrophosphate, 20 mM β-glycerophosphate, 1 mM phenylmethylsulfonyl fluoride, 0.1 mM sodium orthovanadate, and 2 μg/mL aprotinin; pH 7.5). Afterwards, the lysate was cleaved on ice for 30 min, then centrifuged at 12,000 × g for 30 min, and then placed in an incubator for 30 min. The protein quantitation in the supernatant was measured by a BCA protein assay kit. The amounts of protein were taken in equal quantities and separated by sodium dodecyl sulfate-polyacrylamide gel electrophoresis (SDS–PAGE) on 10% gel, and electrotransferred to a polyvinylidene difluoride (PVDF) nitrocellulose membrane (PALL Life Science, Ann Arbor, MI, USA). The transfer film was gently removed from the wet transfer tank, then closed the PVDF membrane with 5% skim milk for 1 h. After this, a mild rinse of 1 × TBS-T was carried out to eliminate any traces of skim milk. In each case, the membrane was incubated with a corresponding anti-mouse primary antibody. In each case, the membrane was incubated with a corresponding anti-mouse and anti-rabbit primary antibody. We used antibodies, including anti-β-actin (St John's Laboratory, STJ97040), anti-NF-κB (Cell signaling, C22B4), anti-COX-2 (Elabscience, E-AB-27666), and anti-iNOS (Thermo Fisher Corp., PA1-036). We added descriptions of the multiple dilutions of the various primary antibodies as follows: anti-β-actin 1:5000, anti-NF-κB 1:1000, anti-COX-2 1:1000, and anti-iNOS 1:500. Washed at least three times with TBST buffer (TBS containing 0.1% Tween 20) and dipped in horseradish peroxidase-conjugated secondary antibodies against the corresponding primary antibody. Then treated with enhanced chemiluminescence (ECL) detection reagents (PerkinElmer, ECL1:ECL2 = 1:1) and exposed to a Mini Size Chemiluminescent Imaging System from Life Science to specify the time intervals for detecting the protein bands and visualizing the stained blots [24].

2.11. Statistical Analysis

All the experiments in each platform were carried out in triplicate and presented as mean ± standard error. For statistical analysis, all data were analyzed by Student's *t*-test for multiple comparisons. A significant difference (*) was defined as $p < 0.05$.

3. Results

3.1. Antioxidant Activity of P. emblica Fruit Extracts Powder

As a functional food, antioxidant properties of *P. emblica* samples were assessed using various biochemical assays with different objectives, namely, DPPH, power reducing, and metal chelating activity. The first oxidation inhibitory assay was the DPPH radical scavenging test. This is a simple and economical experimental platform, in which antioxidants act to prevent oxidation products. Antioxidants change the color of the stable radical DPPH reagent from purple to the light yellow of diphenyl-picrylhydrazine. As shown in Table 2, *P. emblica* exhibited excellent radical scavenging ability and scavenged 88.7 ± 0.3% of the DPPH free radical, and vitamin C scavenged 89.9 ± 0.17%. In the power reducing assay, the color of the testing solutions changed from yellow to different shades of green and blue depending upon the reducing power of these antioxidants. The presence of antioxidant substance induces the reduction of the Fe^{3+}/ferricyanide complex to the ferrous form. As shown in Table 2, BHA at 100 μM has a reducing power value of 0.6 ± 0.002%, and *P. emblica* at 50 mg/mL has a reducing power value of 2.31 ± 0.05% as compared with BHA. The ferrous ion-chelating activities of *P. emblica* samples are shown in Table 2, and ferrozine could form complexes with Fe^{2+} quantitatively. With the presence of chelating agents, the complex construction was disrupted, resulting in a lightening of the red color of the complex. Compared with EDTA, although the testing samples showed a lower level of Fe^{2+} scavenging ability, its antioxidant activity still showed an increasing trend. *P. emblica* at the concentration of 50 mg/mL presented 16.9% ± 0.11% inhibition. The positive control, EDTA, had approximately 94.4 ± 0.21% ion-chelating capacities at 100 μM.

Table 2. The effect of antioxidative activity assays on *Phyllanthus emblica* at different concentrations.

Concentration (mg/mL)	DPPH Free Radical Scavenging Activity (%)	Reducing Power (OD_{700})	Metal Chelating Activity (%)
0	0 ± 0	0.105 ± 0.001	0 ± 0
0.5	3.92 ± 0.07	0.151 ± 0.001	5.66 ± 0.20
1	7.04 ± 0.10	0.180 ± 0.002	7.72 ± 0.69
2	16.43 ± 0.25	0.205 ± 0.002	11.63 ± 0.66
5	42.43 ± 0.51	0.436 ± 0.006	15.06 ± 0.13
10	67.03 ± 0.07	0.796 ± 0.023	16.12 ± 0.25
50	88.71 ± 0.30	2.311 ± 0.054	16.92 ± 0.11
Vitamin C [a]	89.97 ± 0.17	-	-
BHA [b]	-	0.604 ± 0.002	-
EDTA [c]	-	-	94.43 ± 0.21

[a] Vitamin C is the positive control of DPPH radical scavenging capacity assay with the concentration of 100 μM; [b] BHA is the positive control of reducing power assay with the concentration of 100 μM; [c] EDTA is the positive control of metal chelating activity assay with the concentration of 100 μM.

3.2. Cell Viability Effect of P. emblica Fruit Extract Treatment

As a potent food additive, the component should be harmless, without undesirable cytotoxic side effects. To evaluate the optimal dose of *P. emblica* fruit extract samples, the cytotoxicity of its varying concentrations (0.005–10 mg/mL) were applied to RAW 264.7 cells for 24 h. It was initially determined using MTT assay (Figure 1). The results showed that the low concentrations of the testing samples contributed to proliferations on the RAW 264.7 cells, and had cellular survival rates of 66.7 ± 0.9% and 52.7 ± 2.6% even at high concentrations of 5 and 10 mg/mL, respectively. It proved that the extract of *P. emblica* fruit did not affect the cell viability in RAW 264.7 cells. At 2 mg/mL, the samples had no severe cytotoxic effect on the RAW 264.7 cells, and thus the dosage was optimally deliberated in all the following experiments.

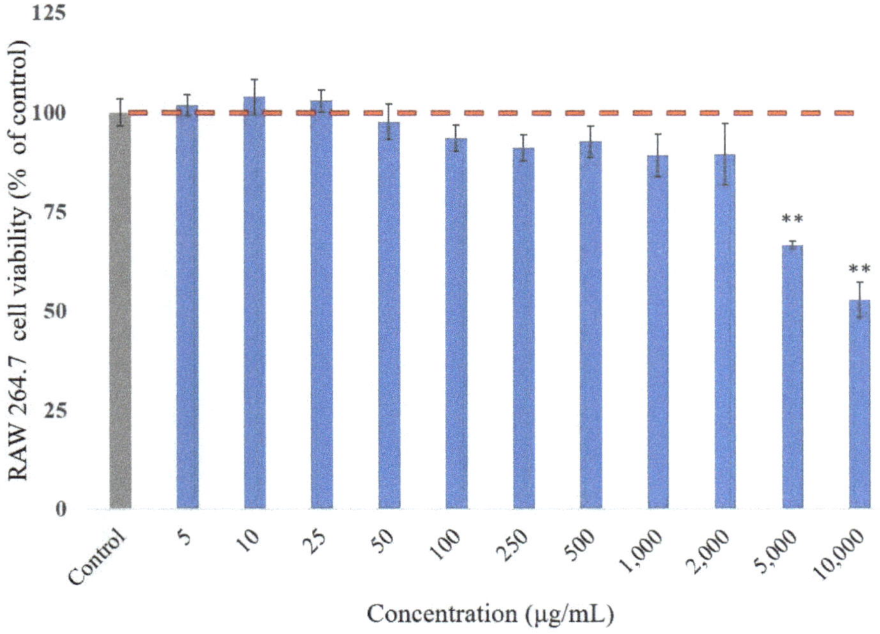

Figure 1. The cell viability of RAW 264.7 cells was measured by MTT assay after 24 h treatment of *P. emblica* samples. The data represented mean ± S.D. of three independent experiments performed. The red dash line is the trend line for cell survival rate of 100%. * $p < 0.05$, ** $p < 0.01$.

3.3. ROS Scavenge by P. emblica Fruit Extract Powder

To determine whether *P. emblica* fruit extract powder treatment induces cellular oxidative statuses, we investigated ROS generation in RAW 264.7 cells. The intracellular H_2O_2 results of the DCFDA staining, which is often quantified to measure the oxidative stress, can be defined as the presence of oxidation. Typically, DCFDA is introduced into target cells through a small amount of aqueous solution, and then rapidly diffuses through the cell membrane as a colorless probe. Once the two acetate groups are cleaved by esterases within the cell, the DCFDA fluorescence is detectable. A valuable property of DCFDA is that it cannot be exited within the cell once it has been cleaved in the cell. This increases the period of time, and DCFDA can be used as a cellular indicator. As shown in Figure 2, increases of *P. emblica* sample concentrations gradually decreased oxidative stresses. DCF fluorescent intensity was reduced to 69.8 ± 0.5% at 0.25 mg/mL, indicating that the treatment of the samples reduced the production of cellular ROS.

3.4. Quantitative Reverse Transcription Polymerase Chain Reaction Analysis for NF-κB, iNOS, and COX-2

To observe the effect of *P. emblica* fruit extract powder on cytokine expression in RAW 264.7 cells, the cells were pretreated with proper concentrations (0.25–2 mg/mL) for 1 h and then stimulated with LPS (5 μg/mL) for 6 h. When the cells are traumatized or infected by gram-negative bacteria, the bacterial cell wall component, LPS, induces the activation of *NF-κB* triggering inflammatory cytokines. During an inflammation, LPS primarily actuates the reaction of proinflammatory genes, including *iNOS* and *COX-2*, producing significant amounts of NO. The inflammatory mediator gene, *NF-κB*, also plays an important role in inflammation-related diseases, which is related to the above gene modulation expressions. The expressions of *iNOS* and *COX-2* lead to an increased production of proinflammatory bio-molecules, which eventually lead to the progression of inflammatory cytokines. Transcriptional changes in *NF-κB*, *COX-2*, and *iNOS* were confirmed by qRT-PCR, as shown in Figure 3. When cells were stimulated with LPS for 6 h, gene expressions of *NF-κB*, *COX-2*, and *iNOS* were increased. After different concentrations of the extract were incubated with LPS, we observed that the levels of *NF-κB*, *COX-*,*2* and *iNOS* were reduced to 14.8 ± 0.6%, 25.6 ± 0.4%, and 44.1 ± 0.1%, respectively.

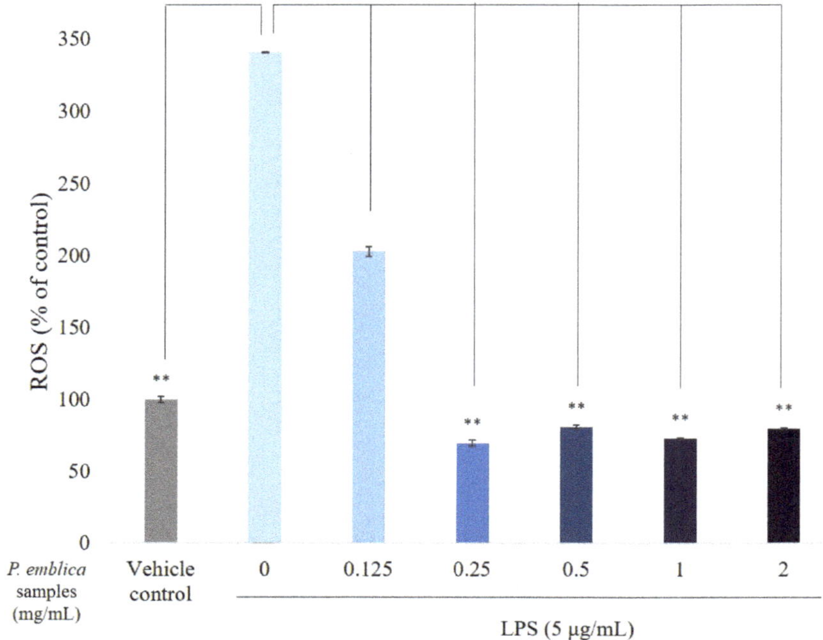

Figure 2. The reactive oxygen species (ROS) percentage measured by a flow cytometry. RAW 264.7 cells were pretreated with *P. emblica* samples (0.125–2 mg/mL) for 1 h, and then stimulated with LPS (5 μg/mL) for 6 h. The data represented mean ± S.D of three independent experiments performed. * $p < 0.05$, ** $p < 0.01$.

3.5. Western Blotting Analyses for NF-κB, iNOS, and COX-2

We carried out western blotting to analyze the inhibitory effects of *P. emblica* fruit extract powder on expressions of NF-κB, iNOS, and COX-2. The RAW 264.7 cells were treated at fitting sample concentrations, and then stimulated with LPS (5 μg/mL) for 6 h. The inflammatory mediators, NF-κB, iNOS, and COX-2, reflect the states of inflammations and are often used to estimate the severities of the inflammation. The stimulations with LPS led the expressions of three proteins upregulating, as shown in Figure 4A. As we predicted, their levels were down-regulated by *P. emblica* fruit extract to 1.16 ± 0.2%, 1.74 ± 0.06%, and 1.51 ± 0.03%, respectively. Quantifications of the western blotting are shown in Figure 4B1–B3. These results suggest that the extract plays an anti-inflammatory role in LPS-stimulated macrophage RAW 264.7 cells.

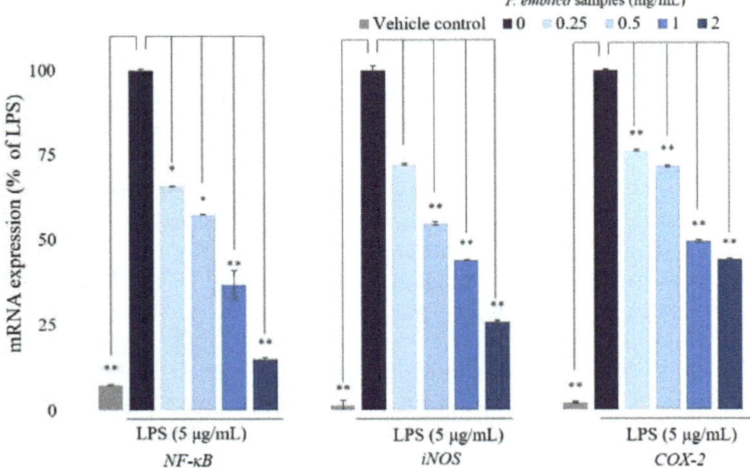

Figure 3. The inflammation-related mRNA expressions in RAW 264.7 cells. RNA expression levels of *NF-κB, iNOS, COX-2* in RAW 264.7 cells treated with different concentrations of *P. emblica* samples (0.25–2 mg/mL) were evaluated by qRT-PCR and normalized to the *GAPDH* gene. The data represented mean ± S.D of three independent experiments performed. * $p < 0.05$, ** $p < 0.01$.

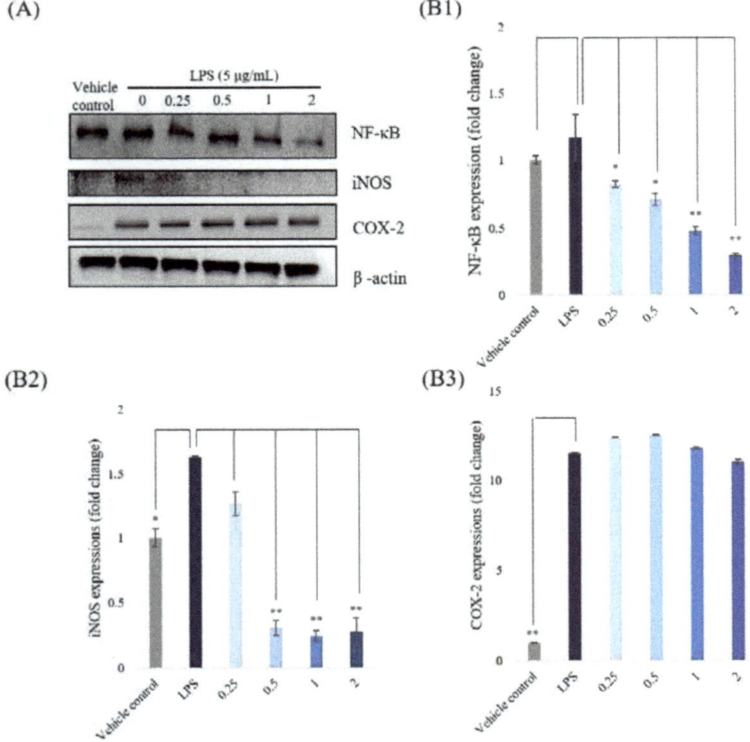

Figure 4. The inflammation-related protein expressions. (**A**) NF-κB, iNOS, and COX-2 expressions in RAW 264.7 cells were pretreated with *P. emblica* samples (0.25–2 mg/mL) for 1 h, and then were stimulated with LPS (5 µg/mL) for 6 h. (**B1**) Protein quantification of NF-κB (**B2**) iNOS (**B3**) COX-2 in western blotting. β-Actin was viewed as an internal control. * $p < 0.05$, ** $p < 0.01$.

4. Discussion

Flavonoids naturally have excellent antioxidant capacity, and tannins are known for their anti-inflammation and antioxidant activities. According to one study, several tannins are considered to be potential cytotoxic and anti-inflammatory agents [25]. *P. emblica* fruit in nature is an edible that contains flavonoids, tannins, and other compounds which have excellent antioxidative capacities. Therefore, it can be used in general food for ingestion and as a supplementary food to enhance human health [8]. In view of this, we are interested in the role of *P. emblica* fruit due to the existence of all the above-mentioned important classes of bioactive properties. The main components of the fruit of *P. emblica* include phenolic constituents, flavonoids, polysaccharides, sterols, fatty acids, vitamins, proteins, amino acids, trace elements, anthraquinone, and alkaloids, etc [10]. These main ingredients all have antioxidative potentials. It can be explained that *P. emblica* extract repairs the LPS-induced oxidative damage and inflammatory symptom of RAW 264.7 cells [13].

Free radicals are substances produced by the metabolism of oxygen in the body. They are extremely active and can react strongly with any substance. In physiological conditions, when bacteria, mold, viruses, etc., invade the body, the defense system will notify the immune cells to prepare for the battle in the body [19]. Thus, the immune cells are catalyzed by the enzymes to produce superoxide anion radicals to remove bacteria and infected cells. In other words, the body needs some free radicals as a weapon to prevent disease. Once the free radicals in the body exceeds the normal range, a free radical chain reaction will occur, which will promote the oxidation of proteins, carbohydrates, lipids and other basic constituent substances into new free radicals. In the continuous circulation, the functions of the human body will be corrupted. Antioxidants are chemicals that do not only reduce the rate of oxidation of cells and biomolecules, but also protect the body from free radicals. Adding antioxidant-rich foods can prevent free radical damage. In order to maintain a healthy body we should not only eat a variety of fruits and vegetables in a balanced manner, but also supplementary antioxidants-rich foods [19]. Antioxidant studies showed that *P. emblica* fruit extract has the capacity to either inhibiting free radical ability or to be a free radical scavenger. In this study, we analyzed the DPPH, metal chelating activity, reducing power, and cellular ROS to estimate the free radical scavenging ability of the extract in various concentrations and also carried out qRT-PCR experiments on different inflammatory genes. We confirmed that *P. emblica* fruit extract powder is an effective antioxidant which also has the ability to regulate inflammatory genes.

Once a human gets damaged by foreign objects, the body produces a protective response which is the inflammation. A controlled inflammatory response is beneficial to the body and provides protection against the site of infection. However, once the inflammatory response is dysregulated, it may become harmful. Therefore, the inflammation may evolve into an adaptive response to restore homeostasis. In order to resolve the inflammatory response in the body, the main site of infection promotes the aggregation and mediated of macrophages and T cells which repairs the inflamed parts. The inflammatory response consists of a many media that form a complex regulatory network [26]. Chronic inflammation is also associated with many death-related diseases. Various interconnecting signaling pathways are related to the development of inflammation.

NF-κB is an extremely important molecule in the inflammatory reaction. When the cells receive stimulation from the outside of the cell, NF-κB in the cytoplasm is released and activated by the original IκB inhibition. NF-κB is also involved in the transcription of iNOS. When iNOS is activated by cytokines, NO will be released, and NO is an activator of COX [7,8]. Almost all mammal cell types have NF-κB, which consists of a family of transcription factors and is associated with inflammatory cytokine production, cell survival, activation, and differentiation of innate immune cells and inflammatory T helper cells. It regulates a large array of genes which takes part in the immune and inflammatory responses [27]. *NF-κB* is involved in several cellular responses to stimuli such as stress, free radicals, heavy metals, ultraviolet irradiation, oxidized LDL, and pathogens like bacterial or viral antigens. iNOS produces multitudinous amounts of NO that can activate immune cells in inflamed tissue, and thus speed up pathological changes [28]. The proinflammatory cytokines, prostaglandins, and NO are produced by activated macrophages which play decisive roles in inflammatory diseases such as Parkinson's disease and Alzheimer's disease. Compared to the critical calcium-calmodulin dependent regulation isoenzymes (nNOS and eNOS), iNOS has been reported as calcium insensitive, maybe due to its tight noncovalent interaction with the calcium-calmodulin complex. iNOS produces larger quantities of NO than eNOS and nNOS upon stimulation, such as by proinflammatory cytokines. iNOS binds calmodulin at physiologically relevant concentrations to synthesize a free radical with an unpaired electron to present an immune defense mechanism. The high iNOS activity typically occurs in an oxidative environmental stimulation, and the overexpressive levels of NO by proinflammatory cytokines have the opportunity to interact with superoxide leading to cell toxicity and peroxynitrite production [27]. These properties may define the roles of iNOS in human immune response, especially the stimulation of inflammation caused by macrophages [8]. Related to the generation of prostaglandin, the major effect which COX-2 causes in inflammation is the generation of pain. Prostaglandin controls the role of vasodilation and inhibits the aggregation of blood plates. In inflammation, these roles have an influence on the accession of blood flow, such as regulating the contraction of smooth muscle tissue and preventing needless clot formation. Thus, COX-2 indirectly increases the sensitization of peripheral nociceptors and generation of hypersensitivity pain. In pathology, several pharmaceutical inhibitions of COX have been used so that they can provide relief caused by the symptoms of inflammation and pain, such as aspirin and ibuprofen [29]. Therefore, the inhibition of proinflammatory cytokines or iNOS and COX-2 expressions in inflammatory cells provides a novel therapeutic method for treatment of inflammation. We used LPS to irritate macrophages as an in vitro model of inflammation. The *P. emblica* sample treatments extenuated LPS-induced inflammation. This study illustrated that iNOS, COX-2, and NF-κB levels increase significantly in LPS-induced cells, whereas, they were evidently decreased by treatment with *P. emblica* fruit extract powder. It means that *P. emblica* fruit samples protect the cell and prevent inflammation symptoms via decreasing the expressions of iNOS, COX-2, and NF-κB at the transcriptional levels and protein expressions, as shown in Figure 5.

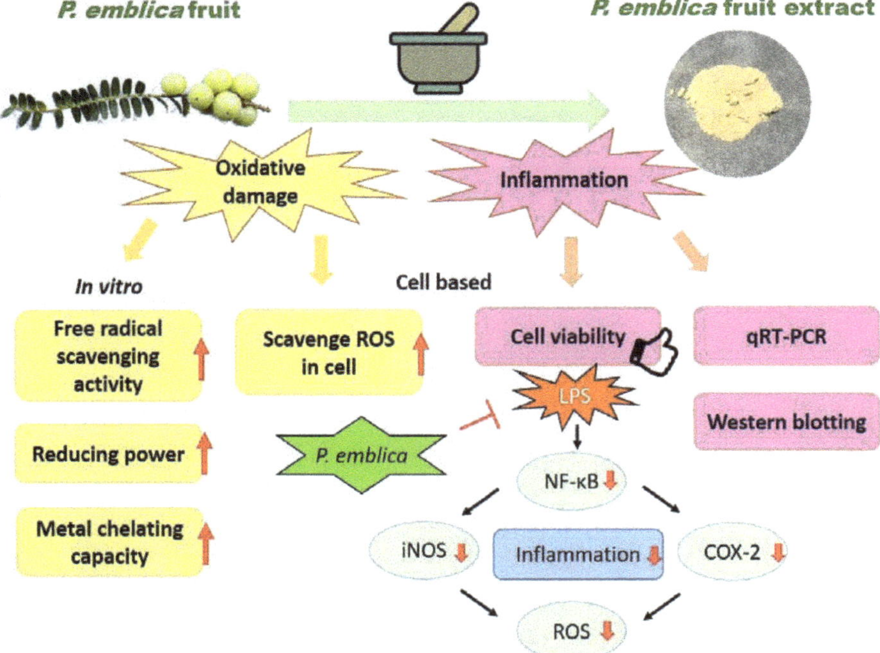

Figure 5. The cartoon illustration of antioxidation and anti-inflammation mechanisms of *P. emblica* fruit extract powder.

5. Conclusions

This study is about antioxidative properties and anti-inflammatory effects from *P. emblica* fruit extracts induced by LPS and provides evidence of the possible beneficial health advantages of this native Taiwan fruit. On the basis of the results from the antioxidant experiments, we found that *P. emblica* fruit extracts showed excellent antioxidative activity, that immune cells could be regulated via *P. emblica* substances, and that, at low concentration, the fruit extract powder increased RAW 264.7 cell proliferations. LPS stimulation in RAW 264.7 cells enhanced the immunological activity on the accumulation of intracellular ROS and upregulations of inflammatory related genes (*NF-κB, iNOS,* and *COX-2*). *P. emblica* samples reduced the cellular ROS productions in a dose dependent manner from 0.125 to 2 mg/mL and decreased the above genes and proteins. *P. emblica* samples showed good antioxidant activities and anti-inflammation properties to be useful as a functional food additive.

Author Contributions: H.M.-D.W., L.F., W.-H.L., and P.C.L. conceived and designed the experiments; L.F., C.C.C., and M.Y.L. performed the experiments; H.M.-D.W., L.F., C.C.C., R.G., H.L.S., T.H.N., and L.P.H. analyzed the data; H.M.-D.W., L.F., C.C.C., and N.E.B. wrote the paper.

Acknowledgments: This work was supported by grants from the Ministry of Science and Technology (MOST107-2221-E-005-063); we also thank the Research Center for Sustainable Energy and Nanotechnology, NCHU 107S0203B, and Cheng Ching Hospital, CH10700222B.

Conflicts of Interest: The authors declare that they have no conflict of interest.

References

1. Huang, S.-Y.; Feng, C.-W.; Hung, H.-C.; Chakraborty, C.; Chen, C.-H.; Chen, W.-F.; Jean, Y.-H.; Wang, H.-M.D.; Sung, C.-S.; Sun, Y.-M.; et al. A novel zebrafish model to provide mechanistic insights into the inflammatory events in carrageenan-induced abdominal edema. *PLoS ONE* **2014**, *9*, e104414. [CrossRef]
2. Du, C.; Bhatia, M.; Tang, S.C.W.; Zhang, M.; Steiner, T. Mediators of inflammation: Inflammation in cancer, chronic diseases, and wound healing. *Mediat. Inflamm.* **2015**, *2015*, 1–2. [CrossRef] [PubMed]
3. Tamura, Y.; Harada, Y.; Nishikawa, S.-I.; Yamano, K.; Kamiya, M.; Shiota, T.; Kuroda, T.; Kuge, O.; Sesaki, H.; Imai, K.; et al. Tam41 is a CDP-diacylglycerol synthase required for cardiolipin biosynthesis in mitochondria. *Cell Metab.* **2013**, *17*, 709–718. [CrossRef]
4. Liu, T.; Zhang, L.; Joo, D.; Sun, S.C. NF-κB signaling in inflammation. *Signal Transduct. Target Ther.* **2017**, *2*, 17023. [CrossRef] [PubMed]
5. Liu, P.-L.; Chong, I.-W.; Lee, Y.-C.; Tsai, J.-R.; Wang, H.-M.; Hsieh, C.-C.; Kuo, H.-F.; Liu, W.-L.; Chen, Y.-H.; Chen, H.-L. Anti-inflammatory effects of resveratrol on hypoxia/reoxygenation-induced alveolar epithelial cell dysfunction. *J. Agric. Food Chem.* **2015**, *63*, 9480–9487. [CrossRef] [PubMed]
6. Frattaruolo, L.; Carullo, G.; Brindisi, M.; Mazzotta, S.; Bellissimo, L.; Rago, V.; Curcio, R.; Dolce, V.; Aiello, F.; Cappello, A.R. Antioxidant and anti-inflammatory activities of flavanones from Glycyrrhiza glabra L. (licorice) leaf phytocomplexes: Identification of Licoflavanone as a modulator of NF-κB/MAPK pathway. *Antioxidants* **2019**, *8*, 186. [CrossRef]
7. Rao, T.P.; Okamoto, T.; Akita, N.; Hayashi, T.; Kato-Yasuda, N.; Suzuki, K. Amla (Emblica officinalis Gaertn.) extract inhibits lipopolysaccharide-induced procoagulant and pro-inflammatory factors in cultured vascular endothelial cells. *Br. J. Nutr.* **2013**, *110*, 2201–2206.
8. Shih, C.-C.; Hwang, H.-R.; Chang, C.-I.; Su, H.-M.; Chen, P.-C.; Kuo, H.-M.; Li, P.-J.; Wang, H.-M.D.; Tsui, K.-H.; Lin, Y.-C.; et al. Anti-inflammatory and antinociceptive effects of ethyl acetate fraction of an edible red macroalgae Sarcodia ceylanica. *Int. J. Mol. Sci.* **2017**, *18*, 2437. [CrossRef]
9. Gaire, B.P.; Subedi, L. Phytochemistry, pharmacology and medicinal properties of Phyllanthus emblica Linn. *Chin. J. Integr. Med.* **2014**, 1–8. [CrossRef]
10. Khanna, S.; Das, A.; Spieldenner, J.; Rink, C.; Roy, S. Supplementation of a standardized Extract from Phyllanthus emblica improves cardiovascular risk factors and platelet aggregation in overweight/class-1 obese adults. *J. Med. Food* **2015**, *18*, 415–420. [CrossRef]
11. Wang, F.; Pan, T.; Yuan, R.; Li, C.; Li, K. Optimization of extraction process of flavonoids in *Phyllanthus emblica* L. by response surface methodology and content determination. *Indian J. Tradit. knowl.* **2015**, *14*, 213–219.

12. Wu, P.-F.; Wang, H.-M.D.; Chen, C.-Y. Isophilippinolide A arrests cell cycle progression and induces apoptosis for anticancer inhibitory agents in human melanoma cells. *J. Agric. Food Chem.* **2014**, *62*, 1057–1065.
13. Li, P.-H.; Chiu, Y.-P.; Shih, C.-C.; Wen, Z.-H.; Ibeto, L.K.; Huang, S.-H.; Chiu, C.-C.; Ma, D.-L.; Leung, C.-H.; Chang, Y.-N.; et al. Biofunctional activities of Equisetum ramosissimum extract: Protective effects against oxidation, melanoma, and melanogenesis. *Oxidative Med. Cell. Longev.* **2016**, *2016*, 1–9.
14. Chen, Y.; Huang, J.-Y.; Lin, Y.; Lin, I.-F.; Lu, Y.-R.; Liu, L.-H.; Wang, H.-M.D. Antioxidative and antimelanoma effects of various tea extracts via a green extraction method. *J. Food Qual.* **2018**, *2018*, 1–6. [CrossRef]
15. Zhao, C.-N.; Tang, G.-Y.; Cao, S.-Y.; Xu, X.-Y.; Gan, R.-Y.; Liu, Q.; Mao, Q.-Q.; Shang, A.; Li, H.-B. Phenolic profiles and antioxidant activities of 30 tea infusions from green, black, oolong, white, yellow and dark Teas. *Antioxidants* **2019**, *8*, 215. [CrossRef]
16. Rossin, D.; Barbosa-Pereira, L.; Iaia, N.; Testa, G.; Sottero, B.; Poli, G.; Zeppa, G.; Biasi, F. A dietary mixture of oxysterols induces in vitro intestinal inflammation through TLR2/4 activation: The protective effect of cocoa bean shells. *Antioxidants* **2019**, *8*, 151. [CrossRef]
17. Wang, H.-M.; Chiu, C.-C.; Wu, P.-F.; Chen, C.-Y. Subamolide E from Cinnamomum subavenium induces sub-G1 cell-cycle arrest and caspase-dependent apoptosis and reduces the migration ability of human melanoma cells. *J. Agric. Food Chem.* **2011**, *59*, 8187–8192. [CrossRef]
18. Wang, S.; Suh, J.H.; Hung, W.-L.; Zheng, X.; Wang, Y.; Ho, C.-T. Use of UHPLC-TripleQ with synthetic standards to profile anti-inflammatory hydroxycinnamic acid amides in root barks and leaves of Lycium barbarum. *J. Food Drug Anal.* **2018**, *26*, 572–582. [CrossRef]
19. Lin, C.Y.; Lee, C.H.; Chang, Y.W.; Wang, H.M.; Chen, C.Y.; Chen, Y.H. Pheophytin a inhibits inflammation via suppression of LPS-induced nitric oxide synthase-2, prostaglandin E2, and interleukin-1beta of macrophages. *Int. J. Mol. Sci.* **2014**, *15*, 22819–22834. [CrossRef]
20. Wang, C.C.; Huang, S.Y.; Huang, S.H.; Wen, Z.H.; Huang, J.Y.; Liu, W.S.; Wang, H.M.D. A synthetic biological secondary metabolite, LycogenTM, produced and extracted from Rhodobacter sphaeroides WL-APD911 in an optimizatioal scale-up strategy. *Food Sci. Hum. Wellness* **2017**, *6*, 195–201. [CrossRef]
21. Li, P.-H.; Liu, L.-H.; Chang, C.-C.; Gao, R.; Leung, C.-H.; Ma, D.-L.; Wang, H.-M.D. Silencing stem cell factor gene in fibroblasts to regulate paracrine factor productions and enhance c-Kit expression in melanocytes on melanogenesis. *Int. J. Mol. Sci.* **2018**, *19*, 1475. [CrossRef] [PubMed]
22. Chen, Y.-T.; Kao, C.-J.; Huang, H.-Y.; Huang, S.-Y.; Chen, C.-Y.; Lin, Y.-S.; Wen, Z.-H.; Wang, H.-M.D. Astaxanthin reduces MMP expressions, suppresses cancer cell migrations, and triggers apoptotic caspases of in vitro and in vivo models in melanoma. *J. Funct. Foods* **2017**, *31*, 20–31. [CrossRef]
23. Galletti, E.; Bonilauri, P.; Bardasi, L.; Fontana, M.C.; Ramini, M.; Renzi, M.; Dosa, G.; Merialdi, G. Development of a minor groove binding probe based real-time PCR for the diagnosis and quantification of *Leishmania infantum* in dog specimens. *Res. Vet. Sci.* **2011**, *91*, 243–245. [CrossRef] [PubMed]
24. Lin, L.-C.; Chen, C.-Y.; Kuo, C.-H.; Lin, Y.-S.; Hwang, B.H.; Wang, T.K.; Kuo, Y.-H.; Wang, H.-M.D. 36H: A novel potent inhibitor for antimelanogenesis. *Oxidative Med. Cell. Longev.* **2018**, *2018*, 1–12. [CrossRef] [PubMed]
25. Parveen, R.; Shamsi, T.N.; Singh, G.; Athar, T.; Fatima, S. Phytochemical analysis and in-vitro biochemical characterization of aqueous and methanolic extract of Triphala, a conventional herbal remedy. *Biotechnol. Rep.* **2018**, *17*, 126–136. [CrossRef] [PubMed]
26. Medzhitov, R. Origin and physiological roles of inflammation. *Nat.* **2008**, *454*, 428–435. [CrossRef] [PubMed]
27. Huang, S.-H.; Wu, S.-H.; Lee, S.-S.; Chang, K.-P.; Chai, C.-Y.; Yeh, J.-L.; Lin, S.-D.; Kwan, A.-L.; Wang, H.-M.D.; Lai, C.-S. Fat grafting in burn scar alleviates neuropathic pain via anti-inflammation effect in scar and spinal cord. *PLoS ONE* **2015**, *10*, e0137563. [CrossRef] [PubMed]
28. Wang, R.; Yang, Z.; Zhang, J.; Mu, J.; Zhou, X.; Zhao, X. Liver injury induced by carbon tetrachloride in mice is prevented by the antioxidant capacity of Anji white tea polyphenols. *Antioxidants* **2019**, *8*, 64. [CrossRef] [PubMed]
29. Ning, C.; Wang, H.-M.D.; Gao, R.; Chang, Y.-C.; Hu, F.; Meng, X.; Huang, S.-Y. Marine-derived protein kinase inhibitors for neuroinflammatory diseases. *Biomed. Eng. Online* **2018**, *17*, 46. [CrossRef]

Article

Antioxidant Activity of *Graptopetalum paraguayense* E. Walther Leaf Extract Counteracts Oxidative Stress Induced by Ethanol and Carbon Tetrachloride Co-Induced Hepatotoxicity in Rats

Wen-Wan Chao [1], Shu-Ju Chen [2], Hui-Chen Peng [2], Jiunn-Wang Liao [3] and Su-Tze Chou [2,*]

[1] Department of Nutrition and Health Sciences, Kainan University, Taoyuan 33857, Taiwan
[2] Department of Food and Nutrition, Providence University, Taichung 43301, Taiwan
[3] Graduate Institute of Veterinary Pathobiology, National Chung Hsing University, Taichung 40227, Taiwan
[*] Correspondence: stchou@pu.edu.tw; Tel.: +88-64-2632-8001 (ext. 15327)

Received: 26 June 2019; Accepted: 25 July 2019; Published: 28 July 2019

Abstract: (1) Background: *Graptopetalum paraguayense* E. Walther is a traditional Chinese herbal medicine. In our previous study, 50% ethanolic *G. paraguayense* extracts (GE50) demonstrated good antioxidant activity. (2) Methods: To investigate the hepatoprotective effects of GE50 on ethanol and carbon tetrachloride (CCl_4) co-induced hepatic damage in rats, Sprague–Dawley rats were randomly divided into five groups (Control group; GE50 group, 0.25 g/100 g BW; EC group: Ethanol + CCl_4, 1.25 mL 50% ethanol and 0.1 mL 20% CCl_4/100 g BW; EC + GE50 group: Ethanol + CCl_4 + GE50; EC + silymarin group: ethanol + CCl_4 + silymarin, 20 mg/100 g BW) for six consecutive weeks. (3) Results: Compared with the control group, EC group significantly elevated the serum aspartate aminotransferase (AST), alanine aminitransferase (ALT), and lactate dehydrogenase (LDH). However, GE50 or silymarin treatment effectively reversed these changes. GE50 had a significant protective effect against ethanol + CCl_4 induced lipid peroxidation and increased the levels of glutathione (GSH), vitamin C, E, total antioxidant status (TAS), and the activities of superoxide dismutase (SOD), glutathione peroxidase (GPx), catalase (CAT), and glutathione S-transferases (GST). Furthermore, in EC focal group, slight fat droplet infiltration was observed in the livers, while in the GE50 or silymarin treatment groups, decreased fat droplet infiltration. HPLC phytochemical profile of GE50 revealed the presence of gallic acid, flavone, genistin, daidzin, and quercetin. (4) Conclusions: The hepatoprotective activity of GE50 is proposed to occur through the synergic effects of its chemical component, namely, gallic acid, flavone, genistin, daidzin, and quercetin. Hence, *G. paraguayense* can be used as a complementary and alternative therapy in the prevention of alcohol + CCl_4-induced liver injury.

Keywords: *Graptopetalum paraguayense* E. Walther; ethanol; carbon tetrachloride; antioxidant activity; hepatoprotective

1. Introduction

The liver, being a dynamic and vital organ, actively participates in multi-metabolic functions of foods, drugs, chemicals, biologicals, and xenobiotics, as well as detoxification of viral and bacterial products. These models, induced by toxins such as carbon tetrachloride (CCl_4), dimethylnitrosamine (DMN), acetaminophen, or thioacetamide, can represent chronic or acute/fulminant hepatitis. Experimentally induced cirrhotic response in rat by CCl_4 is shown to be similar to liver cirrhosis in the humans. Hepatotoxicity of CCl_4 is largely due to its degraded metabolites trichloromethyl (CCl_3) and trichloromethyl peroxyl (CCl_3O_2) formed by hepatic microsomal enzyme [1,2]. The hepatotoxicity of CCl_4 is considered to be mediated by highly reactive trichloromethyl free radical (CCl_3•) and/or peroxyl radical ($CCl_3OO\bullet$) activated forms of CCl_4 formed by the action of the cytochrome P450 system, including CYP2B1, CYP2B2, CYP2E1, and CYP3A [3]. Liver injury induced by CCl_4 has been widely used as an experimental model to screen medicine drugs [4–6].

Alcoholic liver disease (ALD), one of the most common causes of chronic liver disease in the world, is mainly caused by the excessive intake of alcohol. Its misuse represents a major risk factor for many organs, including the heart, brain, pancreas, and particularly, the liver. Oxidative stress associated with alcohol toxicity is mainly caused by reactive oxygen species (ROS) generated by the mitochondrial respiratory chain [7]. Oxidative stress is considered to be a key risk factor in the development of hepatic diseases. Ethanol is metabolized into acetaldehyde by alcohol metabolizing enzymes, including alcohol dehydrogenase (present in cytosol), CYP2E1 (present in microsomes), and catalase (CAT) (present in peroxisomes). The acetaldehyde is further oxidized into acetic acid by aldehyde dehydrogenase in the mitochondria [8]. Chronic alcohol consumption induces an increase in cellular nicotinamide adenine dinucleotide hydrate concentration and acetaldehyde dehydrogenase activity, which leads to severe free fatty acid overload, triglyceride accumulation, and subsequent steatosis in hepatic tissue [9]. Alcohol consumption is a common cause of death in adults, and CCl_4 is a commonly used model for the hepatoprotective activity of drugs.

Plant phenols, such as flavonoids and anthocyanins, are widely distributed in the human diet through vegetables, fruits, cereals, beans, coffee, tea, natural herbs, and extracts, and they have been found to possess significant antioxidant activities. Silymarin is the extract of *Silybum marianum* and consists of seven flavonoglignans (silibinin, isosilibinin, silychristin, isosilychristin, and silydianin) and a flavonoid (taxifolin). *Silybum marianum* is one of the oldest and thoroughly researched plants in the treatment of liver diseases. Silymarin, found in milk thistle, also inhibits CYP2E1. Silymarin demonstrated potent antioxidative, anti-inflammatory, and immunomodulatory activities against liver diseases in various animal models [10,11]. Silymarin, a clinical antifibrotic agent, is widely accepted and used for treating liver diseases.

Graptopetalum paraguayense E. Walther is a traditional Chinese herbal medicine belonging to the *Crassulaceae* family. In Taiwan, *G. paraguayense* is a medicinal plant that is regarded as a vegetable with health benefits. The general composition of *G. paraguayense* contains 95–96% water, 0.82% dietary fiber, 0.54% protein, 0.52% fat, and the total phenolic compounds and anthocyanins levels are in the range of 11—34 mg gallic acid equivalent/g and 0.03–1.29 μmol/g [12]. It has also demonstrated dose-dependent ACE inhibitory activity, and the kinetics of ACE inhibition reveal that the *G. paraguayense* extracts are mixed-type inhibitors [13]. Our previous study also demonstrated that the leaf extracts of *G. paraguayense* are safe and have a potential antioxidative activity [14–16]. The previous studies have shown that *G. paraguayense* plays neuroprotective effects of brain injury in ischemic rats [17] and reduces oxidative stress in hypercholesterolemia patients [18]. Furthermore, according to the Chinese prescription, *G. paraguayense* is able to alleviate hepatic disorders. In our previous studies, we have shown that *G. paraguayense* have hepatoprotective effects in human hepatoma cell line-HepG2 by induced G1 phase arrest and apoptosis [14]. In vivo, *G. paraguayense* can attenuate toxin-induced hepatic damage and fibrosis, while in vitro, it can inhibit HSC and Kupffer cell activation. Duh et al. suggested that *G. paraguayense* exerts hepatoprotection through antioxidative and anti-inflammatory properties and can protect against CCl_4-induced oxidative stress and liver injury [19]. The treatment of a rat model with diethylnitrosamine (DEN)-induced liver cancer with *G. paraguayense* extracts decreased hepatic collagen contents and inhibited tumor growth [20]. The abovementioned suggest that *G. paraguayense* could be considered as a complementary therapy agent for liver disease [21]. In this study, we investigated the protective effects of *G. paraguayense* extracted with 50% ethanol (GE50) against ethanol + CCl_4 (EC)-induced hepatic damage in rats. Silymarin, an antioxidant flavonoid, was used as the control preventive agent in our experiment.

2. Materials and Methods

2.1. Preparation of GE50 Extract of G. Paraguayense

The 50% ethanolic extract of *G. paraguayense* E. Walther (GE50) was prepared according to the previously described procedures [12]. The *G. paraguayense* E. Walther was grown in a pot, and the leaves were cleaned, washed, cut into small pieces, and then freeze-dried by a vacuum freeze-dryer. Each 20 g of freeze-dried leaves was extracted with 700 mL of 50% ethanol at 85 °C for 3 h. The decoction was filtered and dried using a vacuum freeze-dryer. The GE50 extracts were sealed in plastic bottles and stored at −70 °C until use.

2.2. Animal Treatment

Fifty male weanling Sprague–Dawley (SD) rats were obtained and fed commercial chow diets (Fwusow Industry Co., LTD, Taiwan). They were randomly divided into five groups, each containing ten animals. The control group was gavaged with 1.25 mL of normal saline daily for six weeks. The GE50 group was gavaged with GE50, dissolved in 1.25 mL normal saline, at a dose of 0.25 g/100 g BW for six weeks. The ethanol + CCl_4 (EC) group was gavaged with 50% ethanol 1.25 mL 50% ethanol/100 g BW (equal to 0.5 g ethanol/100 g BW) and 0.1 mL of 20% CCl_4 in olive oil twice a week and administered 1.25 mL of normal saline daily for six weeks. The ethanol + CCl_4 + GE50 (EC + GE50) group and ethanol + CCl_4 + silymarin (EC + silymarin) group were gavaged with GE50 (0.25 g/100 g BW) and silymarin (20 mg/100 g BW), respectively, daily for six weeks and received ethanol/CCl_4 in the same manner as EC group. The control and GE50 groups, which are not administered CCl_4, received 0.1 mL of olive oil/100 g BW at the same time points. The animals were kept under standard laboratory conditions of light/dark cycle, a temperature of 22 ± 2 °C and humidity of 50 ± 10%. This animal research and all the procedures were reviewed and approved by the Animal Research Ethics Committee at Providence University, Taichung, Taiwan (Approval No: 20071210-A05).

2.3. Serum and Liver Tissue Preparation

After six weeks of feeding, the blood was collected. The serum was analyzed for aspartate aminotransferase (AST), alanine aminotransferase (ALT), lactate dehydrogenase (LDH), and total antioxidant status (TAS). The livers were homogenized in an ice-cold phosphate buffer (0.05 M, pH 7.4) using a Potter-Elvehjem-type homogenizer with a Teflon pestle. One portion of this tissue homogenate (0.3 g/mL) was used for assaying the levels of malondialdehyde (MDA), vitamin C, vitamin E, and reduced glutathione (GSH). After centrifuged at 12,000× *g* and 4 °C for 10 min. The resulting supernatant was used to determine the activities of superoxide dismutase (SOD), glutathione peroxidase (GPx), CAT, and glutathione S-transferase (GST).

2.4. Determination of AST, ALT, LDH, and TAS Serum Levels in Rats

The levels of AST and ALT in the serum samples were determined by enzymatic methods using an automatic analyzer at a commercial analytical service center (Lian-Ming Co., Taiwan, ROC). The levels of LDH and TAS were determined using commercial kits from Randox Laboratories Ltd. (Antrim, UK).

2.5. Measurement of MDA and GSH Levels, and GPx, SOD, and CAT, and GST Activities

To measure activities in the liver, MDA, GSH, GPx, SOD, and CAT were performed in accordance with our previously reported procedures. Tissue MDA levels were used to spectrophotometrically estimate thiobarbituric acid-reactive substances (TBARS) at 535 nm. GSH contents were measured by HPLC. SOD activity was determined spectrophotometrically at 325 nm. One unit of SOD activity was defined as half the rate of reduction of pyrogallol autoxidation over a 1-min period at 15-s intervals. GPx activity was determined by an enzyme coupled method with glutathione reductase (GR) using cumene hydroperoxide as the substrate at 30 °C. The rate of decrease in the NADPH concentration was observed at 340 nm over a 3-min period at 30-s intervals. One unit of GPx activity was defined as the amount of enzyme that catalyzed the oxidation of 1 µmol of NADPH/min/mL. CAT activity was determined using H_2O_2 as the substrate. The rate of H_2O_2 dismutated to H_2O and O_2 was proportional to the CAT activity. The decrease in the amount of H_2O_2 was observed at 240 nm over a 1-min period at 15-s intervals. One unit of CAT activity was defined as 1 mmole H_2O_2 remaining per minute [22]. GST activity was determined at 340 nm by an enzyme-coupled method with glutathione-1-chloro-2,4-dinitrobenzene (CDNB) as the substrate at 25 °C [23]. One unit of GST activity was defined as the amount of enzyme that catalyzed the formation of glutathione-CDNB/min/ml. The protein content of the tissue cytosols was determined based on the Biuret reaction using a BCA kit. The specific activity of the enzyme was expressed as unit/mg protein.

2.6. Measurement of Antioxidants

The liver vitamin C content was stabilized by MPA and cysteine solution and determined by HPLC [24]. The vitamin E standard and the tissue cytosols were diluted in methanol solution containing 0.25% BHT and 0.2% ascorbate before HPLC analysis. The tissue vitamin E and GSH contents were measured by HPLC [25,26]. The tissue GSH was reduced by dithiothrietol and the monobromobimane derivative was produced before the HPLC analysis. As the eluent, 30 mM TBA in methanol solution was used at a flow rate of 1 mL/min.

2.7. Histopathology

The liver tissue of rats was immobilized in 10% formalin solution for 24 h. Then, the liver tissue was dehydrated, made transparent, waxed, embedded, and sectioned. Liver tissues were trimmed to 2-mm thickness. Then, the liver tissue was stained with hematoxylin and eosin (H&E).

2.8. Characterization of Phenolic Compounds in GE50 Extracts

Determine the polyphenolic compounds in GE50 extracts, HPLC analysis was performed in accordance with our previously described method with modifications [16,27]. GE50 was dissolved in 50% ethanol, filtered, and analyzed by HPLC. Peak areas and concentrations were determined. The identification of each compound was based on a combination of retention time and spectral matching by comparison with those of known standards.

2.9. Statistical Analysis

Data are expressed as mean ± standard deviation (SD). Statistical analyses were conducted using SPSS (v.16.0; SPSS, USA). One-way ANOVA and Scheffe's method were used to analyze the differences between the means. Differences with $p < 0.05$ were considered statistically significant.

3. Results

3.1. Effects of GE50 on Body Weight Gain, Feed Efficiency, and Liver Index in Ethanol + CCl_4 -Treated Group

Table 1 shows the daily body weight gain, feed efficiency, and liver index of the rats in each group. The effects of GE50 on the body weight were evaluated. Silymarin, a polyphenolic flavonoid, was used in this study as a reference drug. Significant weight loss was observed in ethanol + CCl_4 (EC)-treated group. The change in body weight was highest in the EC-treated group (3.56 ± 0.66 g/day/rat) compared with the control (5.58 ± 0.66 g/day/rat), followed by the GE50 (5.92 ± 0.81 g/day/rat), EC + GE50 (4.68 ± 0.59 g/day/rat), and EC + silymarin (3.56 ± 0.74 g/day/rat) groups. The change in feed efficiency was the highest in EC-treated group compared with the other groups. The relative liver weight in the EC-treated group was the highest compared with the other groups. The administration of GE50 or silymarin over six weeks significantly reversed the ethanol + CCl_4 effects, inducing body weight gain and improving feed efficiency.

Table 1. Effects of oral administration of 50% ethanolic *Graptopetalum paraguayense* E. Walther (GE50) over six consecutive weeks on body weight gain, feed efficiency, and relative liver weight in rats treated with ethanol + carbon tetrachloride (CCl_4).

Groups	Daily Body Weight Gain (g/Day/Rat)	Feed Efficiency (g Gain/g Feed)	Relative Liver Weight (g/100 g Body Weight)
Control	5.58 ± 0.66 ^{a}	0.21 ± 0.02 ^{a}	2.96 ± 0.14 ^{c}
GE50	5.92 ± 0.81 ^{a}	0.21 ± 0.02 ^{ab}	3.05 ± 0.19 ^{bc}
Ethanol + CCl_4 (EC)	3.56 ± 0.66 ^{c}	0.15 ± 0.01 ^{c}	3.30 ± 0.11 ^{a}
EC + GE50	4.68 ± 0.59 ^{b}	0.20 ± 0.01 ^{b}	3.10 ± 0.12 ^{b}
EC + silymarin	4.32 ± 0.74 ^{b}	0.20 ± 0.02 ^{ab}	3.17 ± 0.11 ^{b}

The data are presented as mean ± S.D. of 10 rats. One-way ANOVA and Scheffe's method were used to analyze the differences between the means. $^{a-c}$ Mean values with different letters in the same row are significantly different ($p < 0.05$) according to Duncan's multiple-range test. Control group; GE50 group, 0.25 g/100 g BW; EC group: Ethanol + CCl_4, 1.25 mL 50% ethanol and 0.1 mL 20% CCl_4/100 g BW; EC + GE50 group: Ethanol + CCl_4 + GE50; EC + silymarin group: ethanol + CCl_4 + silymarin, 20 mg/100 g BW.

3.2. *Effects of GE50 on Ethanol + CCl_4 -Induced Hepatotoxicity*

To evaluate possible hepatocellular damage caused by ethanol and CCl_4, the activities of AST and ALT were assessed. Ethanol and CCl_4 co-treatment significantly increased the activity of these enzymes in plasma, indicating intense hepatic damage. As shown in Table 2, the serum AST, ALT, LDH, and TAS levels were measured. A significantly higher serum levels of AST, ALT, and LDH were observed in the EC group than in the control group (AST: 184.40 ± 25.5 vs. 103.98 ± 14.0 U/L; ALT: 66.89 ± 4.9 vs. 44.10 ± 5.4 U/L; LDH: 563.04 ± 103.7 vs. 273.82 ± 94.9 U/L). However, EC + GE50 treatment efficiently decreased the AST level to 110.95 ± 14.0 U/L, ALT level to 50.01 ± 4.2, and LDH level to 403.20 ± 79.4 U/L. Similar results were obtained when the rats were treated with silymarin, which is a known hepatoprotective chemical.

Table 2. Effects of oral administration of GE50 over six consecutive weeks on serum aspartate aminotransferase (AST), alanine aminotransferase (ALT), and lactate dehydrogenase (LDH) levels, and total antioxidant status (TAS) in rats treated with ethanol + CCl_4.

Groups	AST (U/L)	ALT (U/L)	LDH (U/L)	TAS (nmole/L)
Control	103.98 ± 14.0 ^{c}	44.10 ± 5.4 ^{c}	273.82 ± 94.9 ^{c}	0.39 ± 0.1 ^{a}
GE50	97.25 ± 19.8 ^{c}	44.28 ± 3.8 ^{c}	218.39 ± 35.2 ^{c}	0.37 ± 0.1 ^{a}
Ethanol + CCl_4 (EC)	184.40 ± 25.5 ^{a}	66.89 ± 4.9 ^{a}	563.04 ± 103.7 ^{a}	0.15 ± 0.1 ^{c}
EC + GE50	110.95 ± 14.0 ^{bc}	50.01 ± 4.2 ^{b}	403.20 ± 79.4 ^{b}	0.28 ± 0.1 ^{b}
EC + silymarin	122.28 ± 24.2 ^{b}	42.05 ± 6.6 ^{c}	347.13 ± 43.9 ^{b}	0.26 ± 0.1 ^{b}

The data are presented as mean ± S.D. of 10 rats. One-way ANOVA and Scheffe's method were used to analyze the differences between the means. $^{a-c}$ Mean values with different letters in the same row are significantly different ($p < 0.05$) according to Duncan's multiple-range test. Control group; GE50 group; EC group: Ethanol + CCl_4; EC + GE50 group: Ethanol + CCl_4 + GE50; EC + silymarin group: ethanol + CCl_4 + silymarin.

The serum TAS level in rats showed a significantly decreased in the EC group compared to the control group (0.15 ± 0.1 vs. 0.39 ± 0.1 nmole/L). The EC + GE50 treatment reversed the TAS level to 0.28 ± 0.1 nmole/L. GE50 treatment alone did not affect the serum AST and ALT levels.

3.3. *Histological Analyses*

The hepatoprotective effects were confirmed by histological examinations. Hepatic steatosis represents an excess accumulation of fat in hepatocytes. To assess the degree of fatty liver, we examined the accumulation of hepatic triglyceride by H&E staining of the rat liver. Ethanol administration caused degenerative morphological changes, which were exhibited by fat droplets in the liver sections.

The results of H&E staining indicated that ethanol and CCl_4 co-administration presented extensive changes in the liver morphology, including marked enlarged areas of hepatocellular degeneration and infiltration inflammatory cells. No histological abnormality was observed in the control group. As illustrated in Figure 1, ethanol + CCl_4 (EC)-induced injury included increased vacuole formation, neutrophil infiltration, and necrosis. Liver section from the EC-treated group showed highly deformed liver architecture with fatty lesion due to intensive fatty infiltration (FI) and signs of necrosis (N). EC + GE50 group and EC + silymarin group demonstrated improved hepatocellular architecture with intact nucleus (IN) and normal sinusoids (NS).

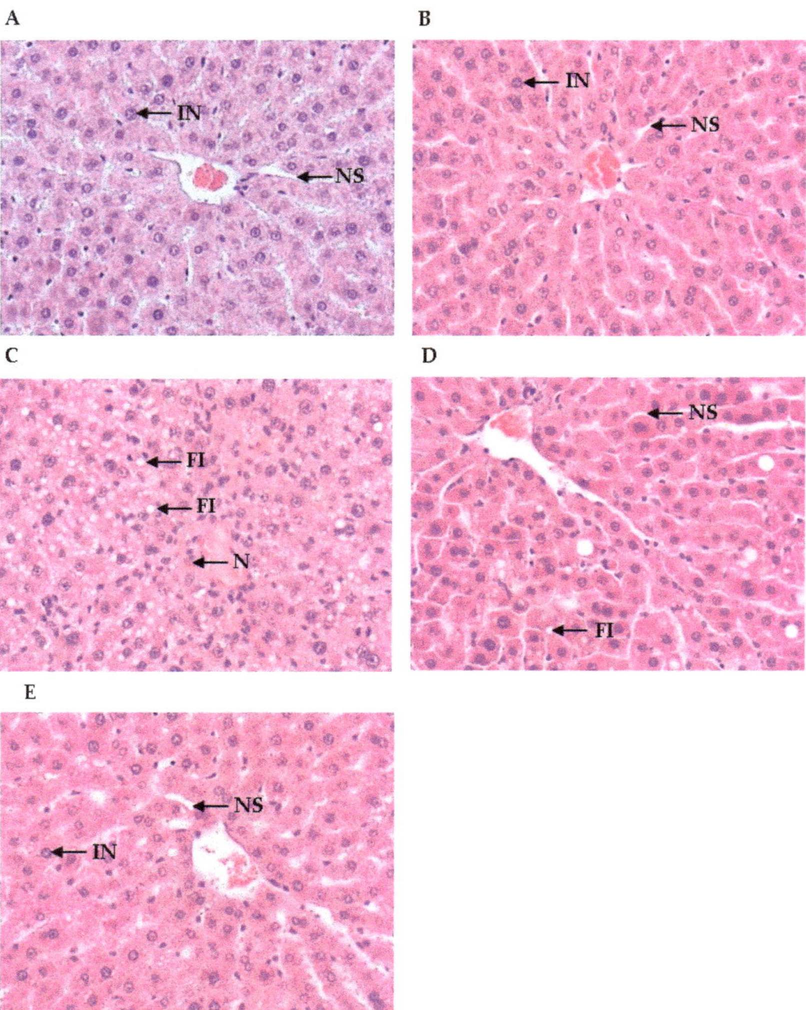

Figure 1. Effects of oral administration of GE50 over six consecutive weeks on histopathological changes in the livers of rats treated with ethanol + CCl_4. Hematoxylin/Eosin staining (H&E), 400×. (**A**) Liver section from the control group demonstrating normal hepatic architecture with intact nucleus (IN) and normal sinusoids (NS); (**B**) GE50; (**C**) Section of ethanol + CCl_4 (EC)-induced damaged liver showing highly deformed hepatic architecture with fatty lesion due to fatty infiltration (FI) and necrosis (N); (**D**) EC+ GE50; (**E**) Liver section from EC + silymarin-treated group showing improved hepatocellular architecture with IN and NS.

3.4. *Effects of GE50 on Hepatic MDA, Vitamin C, Vitamin E, and GSH Levels in Ethanol + CCl_4-Treated Group*

In our study, the hepatic MDA levels were significantly elevated in the ethanol + CCl_4 (EC) group (2.40 ± 0.13 nmol/mg protein) compared with that in the control (1.28 ± 0.21 nmol/mg protein) and GE50 (1.19 ± 0.21 nmol/mg protein) groups ($p < 0.05$). In contrast, in the EC + GE50 group and EC + silymarin group, hepatic MDA levels were markedly decreased compared with the EC group (Figure 2). This observation indicated that the plant extracts may provide phytochemicals that inhibit lipid peroxidation in the rat liver.

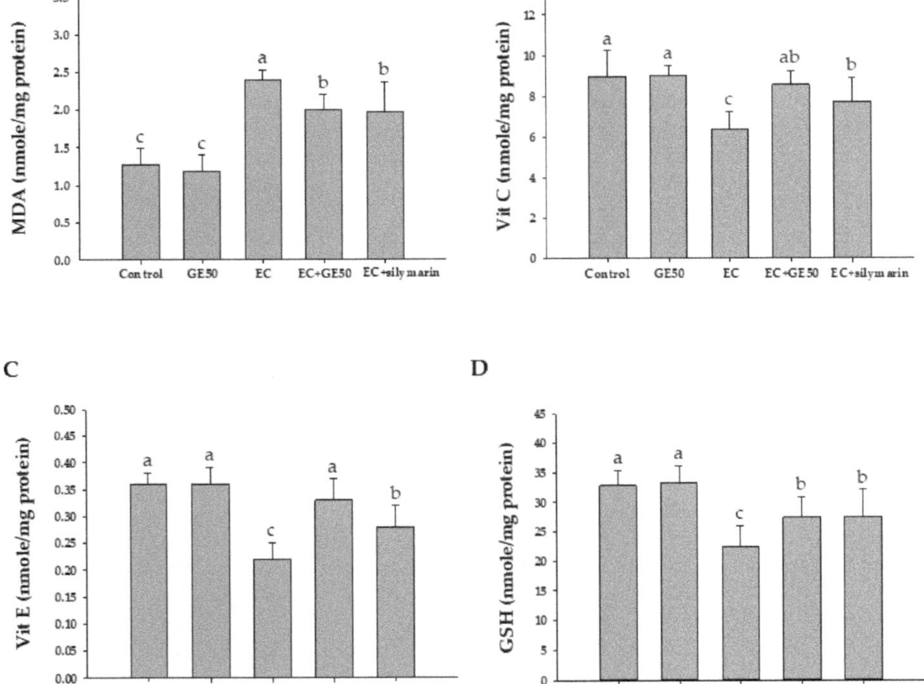

Figure 2. Effects of oral administration of GE50 over six consecutive weeks on malondialdehyde (MDA) (**A**), vitamin C (**B**), vitamin E (**C**), glutathione (GSH) (**D**) levels in the livers of rats treated with ethanol + CCl_4. The data are presented as mean ± S.D. of 10 rats. One-way ANOVA and Scheffe's method were used to analyze the differences between the means. [a–c] Mean values with different letters in the same row are significantly different ($p < 0.05$) according to Duncan's multiple-range test. Control group; GE50 group; EC group: Ethanol + CCl_4; EC + GE50 group: Ethanol + CCl_4 + GE50; EC + silymarin group: ethanol + CCl_4 + silymarin.

Vitamin C level was significantly decreased in the EC group (6.37 ± 0.87 nmol/mg protein) compared with the control group (8.97 ± 1.26 nmol/mg protein). Vitamin E level was also significantly decreased in the EC group (0.22 ± 0.03 nmol/mg protein) compared with the control group (0.36 ± 0.02 nmol/mg protein). GSH level was significantly decreased in the EC group (22.30 ± 3.62 nmol/mg protein) compared with the control group (32.78 ± 2.55 nmol/mg protein). However, EC + GE50 treatment reversed the vitamin C level to 8.55 ± 0.68 nmol/mg protein, vitamin E level to 0.33 ± 0.04 nmol nmol/mg protein, and GSH level to 27.41 ± 3.45 nmol/mg protein. The decline of GSH level in the EC group might be due to its excessively generated quantity of free radicals leading to hepatic injury.

3.5. Effects of GE50 on Antioxidant Enzymatic Activities in Ethanol + CCl_4-Treated Group

The activities of SOD, GPx, CAT, and GST were measured to evaluate the antioxidant effects of GE50 (Figure 3). The SOD activity significantly decreased in the ethanol + CCl_4 (EC) group (3.27 ± 0.11 unit/mg protein) compared with the control (3.95 ± 0.11 unit/mg protein) and GE50 (3.91 ± 0.19 unit/mg protein) groups ($p < 0.05$). The GPx activity significantly decreased in the EC group (627.25 ± 79.43 unit/mg protein) compared with the control (711.73 ± 37.97 unit/mg protein) and GE50 (702.96 ± 37.71 unit/mg protein) groups. The CAT activity significantly decreased in the EC group (19.61 ± 1.11 unit/mg protein) compared with the control (24.11 ± 1.44 unit/mg protein) and GE50 (23.87 ± 0.91 unit/mg protein) groups. The GST activity significantly decreased in the EC group (908.03 ± 88.92 unit/mg protein) compared with the control (1079.68 ± 41.13 unit/mg protein) and GE50 (1045.72 ± 52.27 unit/mg protein) groups ($p < 0.05$). GE50 treatment successfully recovered these enzymes levels to almost normal levels. The effect of GE50 was similar to silymarin, which has been previously shown to have a significant protective effect in rats.

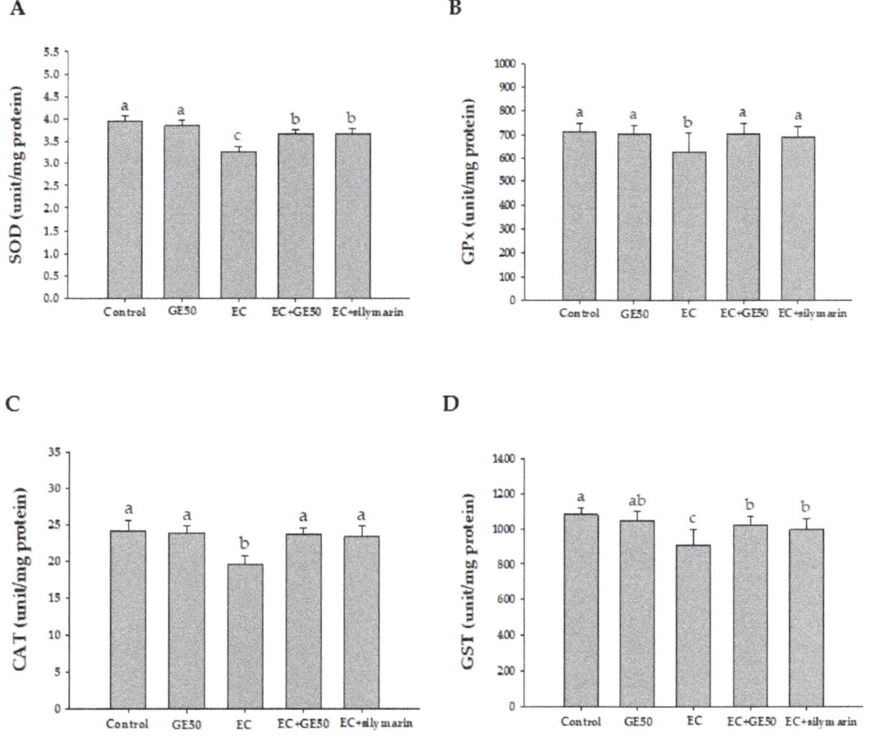

Figure 3. Effects of oral administration of GE50 over six consecutive weeks on superoxide dismutase (SOD) (**A**), glutathione peroxidase (GPx), (**B**), catalase (CAT), (**C**), and glutathione S-transferases (GST) (**D**) activities in the livers of rats treated with ethanol + CCl_4. The data are presented as mean ± S.D. of 10 rats. One-way ANOVA and Scheffe's method were used to analyze the differences between the means. $^{a-c}$ Mean values with different letters in the same row are significantly different ($p < 0.05$) according to Duncan's multiple-range test. Control group; GE50 group; EC group: Ethanol + CCl_4; EC + GE50 group: Ethanol + CCl_4 + GE50; EC + silymarin group: ethanol + CCl_4 + silymarin.

3.6. Polyphenolic Profile in GE50

The HPLC chromatogram showed that gallic acid, quercetin, genistin, and daidzin were the major components among organic molecules found in GE50, which had maximum absorbance at 270 nm. *G. paraguayense* E. Walther mainly contained flavonoids that were identified in our study (Figure 4). It has also been reported that *G. paraguayense* E. Walther contains various antioxidants, such as gallic acid, quercetin, genistin, and daidzin [16,17]. The protective effects of GE50 may be attributed to the presence of polyphenolic compounds such as gallic acid, flavone, genistin, daidzin, and quercetin. The pharmacological fundamental constituents of the plant are flavonoids.

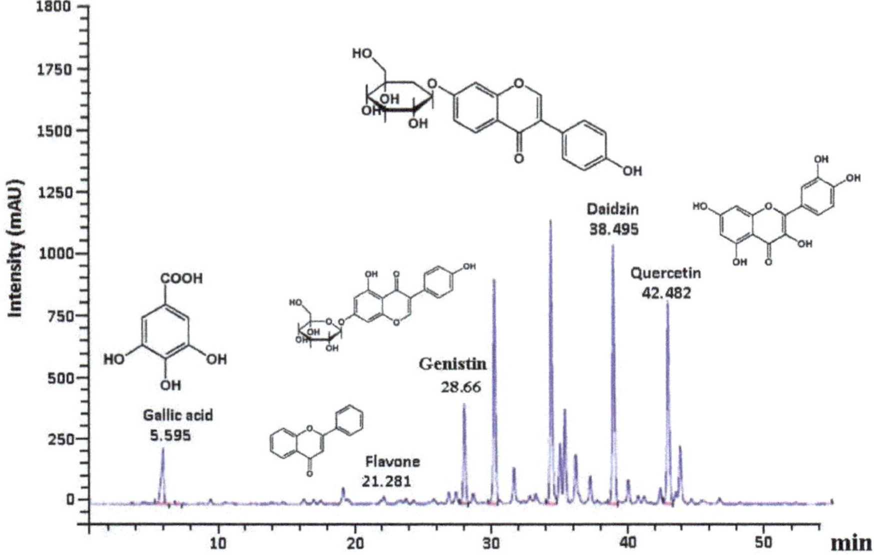

Figure 4. HPLC phytochemical profile of GE50 extract detected at 270 nm. GE50 chemical profile was identified by relative retention times using authentic standards. Key to peak identities: Gallic acid, flavone, genistin, daidzin, and quercetin.

4. Discussion

CCl_4 is a typical hepatotoxic substance, and its mechanism of action is complex. CCl_4-mediated hepatotoxicity was chosen as the experimental model. Due to ethanol and CCl_4 toxicity, relative liver weight in the ethanol + CCl_4 (EC)-treated group was higher than that in the control group (Table 1). Liver weight is known to increase due to hepatic damage inflicted by trichloromethyl radical, as well as consequent liver fibrosis and hypertrophy. Changes in body and liver weight after ethanol and CCl_4 intoxication provides direct evidence of the overall hepatic damage [28,29]. The liver, the largest and the most metabolically complex organ in the body, is involved in the storage and biosynthesis metabolism. It is also responsible for detoxification and metabolic homeostasis. Ethanol and CCl_4 resulted in loss of body weight, which is considered a putative indicator of health. We demonstrated that GE50 markedly ameliorated ethanol + CCl_4 (EC)-induced chronic hepatitis in test rats, accompanied by reduced relative liver weight. Similar results were obtained when the rats were treated with silymarin, a commercial hepatoprotective agent. Silymarin was used as the positive control in this study.

AST and ALT are aminotransferases linked with liver parenchymal cells. If the hepatocellular plasma membrane is damaged, these will leak out from the cytosol into the bloodstream. Serum AST and ALT levels markedly increased in the ethanol + CCl_4 (EC)-treated group, indicating altered permeability of membranes and hepatotoxicity. The results revealed that the serum AST and ALT levels significantly decreased after treatment with GE50. These results demonstrate the hepatoprotective effect of GE50 against ethanol + CCl_4-induced liver injury in rats. Damage to the liver cells results in elevations of the both ALT and AST, which have been widely accepted as major biomarkers to assess the hepatic injury [30].

Alcoholic hepatitis should at least include inflammation, steatosis, fibrosis, and cell damage. The activity of alcohol dehydrogenase and aldehyde dehydrogenase causes a reduction in NAD^+/NADH ratio, which is the process that causes a reduced mitochondrial capacity to metabolize lipids [11,31]. Otherwise, CCl_4-induced liver injury is characterized by oxidative stress and activated hepatic macrophage, leading to hepatocyte damage and death [32]. Chronic alcohol consumption increases cellular nicotinamide adenine dinucleotide hydrate concentration and acetaldehyde dehydrogenase activity, which leads to severe free fatty acid overload, triglyceride accumulation, and subsequent hepatic steatosis [9]. Liver injury can lead to the transfer of fatty acids to the liver, resulting in an increase in the TG content in the liver. These results demonstrate that GE50 treatment significantly alleviated the degree of liver injury.

Lipid peroxidation is one of the major characteristics of CCl_4-induced hepatotoxicity [33]. MDA, the end product of lipid peroxidation, is widely used as a marker of lipid peroxidation injury. Antioxidants, such as N-acetyl-cysteine, α-tocopherol, phenols, selenium, and vitamin C and E, have been proposed and used as therapeutic agents in liver damage [34]. Vitamin E is believed to be the most important lipophilic antioxidant in biological tissues. Cheeseman et al. demonstrated that an increased vitamin E level in liver protects from acute CCl_4-induced damage by preventing lipid peroxidation [35].

Astrocytes contain one of the highest cytosolic concentrations of GSH amongst mammalian cells. GSH is the major non-protein thiol that plays a vital role in maintaining the antioxidant defense mechanism in the body. GSH levels in the liver dropped in CCl_4-treated mice. The depletion of GSH may also be a consequence of liver damage. Ethanol inhibits the synthesis of reduced GSH. Moreover, acetaldehyde promotes GSH depletion, free radical-mediated toxicity, and lipid peroxidation [36]. The hepatoprotective effects of some compounds, such as silymarin, may be attributable to its ability to increase cellular GSH [37]. The increase in hepatic GSH by GE50 may be one of the ways in which *G. paraguayense* protects the liver against ethanol and CCl_4 co-induced hepatotoxicity in rats.

CCl_4 initiates lipid peroxidation, as well as reduces tissue GPx, GR, CAT, and SOD activities. In experimental animals, the induction of an SOD–CAT-insensitive free radical species by diet and alcohol facilitates liver damage [38]. The GST family represents one of the main detoxification systems in the hepatocytes. GST regulates apoptosis by influencing the lipid peroxidation pathway [39]. GPx and GSH are well-known reductants that metabolize toxic chemicals, drugs, and xenobiotics. In general, the liver can combat the free radical damage by biotransformation of these toxic agents in less reactive compounds through cytochrome P-450 and GPx [40]. CAT plays a role in the metabolism of ethanol. In addition to alcohol dehydrogenase and CYP2E1, CAT is involved in the metabolism of ethanol in the body.

The treatment of SD rats with alcohol and CCl_4 caused the development of significant hepatocellular damage, as was evident from a marked increase in the serum activities of AST, ALT, and LDH compared with the control rats (Table 2). We observed a large number of inflammatory cell infiltration in the ethanol + CCl_4 (EC) co-treatment group (Figure 1). Rats treated with the GE50 had lower concentration of MDA in the liver cells. Alcohol + CCl_4 (EC) co-treatment also caused a considerable increase ($p < 0.05$) in hepatic MDA formation (Figure 2) and simultaneously induced a marked depletion ($p < 0.05$) in vitamin C, vitamin E, GSH, and SOD, GPx, CAT, and GST levels in the liver (Figure 3) compared with the control rats. Our studies showed a decrease in food intake and increase in oxidative stress in rats co-treated ethanol and CCl_4. Our results demonstrated that GE50 significantly enhanced the GSH, SOD, CAT, and GR levels, and may contribute to the important mechanisms underlying the liver preventive effects.

Plant flavonoid compounds are a gifted class of so-called "nutraceuticals" that include the ability to protect hepatic damage [41–43]. Phytochemicals are naturally occurring chemicals in plant that promote the prevention and treatment of various diseases. Plants are a good source of useful hepatoprotective agents that can modulate the activities of free radicals [41]. A significant decrease in lipid peroxides in liver tissues following co-treatment with CCl_4 and antioxidants indicated the protective effects of polysaccharide from *Angelica* and *Astragalus* [44] and *Fagonia schweinfurthii* [45] through the scavenging of free radicals produced by CCl_4. Several reports have shown that *Antrodia camphorate*, also known as *Antrodia cinnamomea* (*Niuchangchih*), is a precious fungus grown in Taiwan. It has been reported to prevent ethanol-, CCl_4-, and cytokine-induced liver injury, ameliorate fatty liver and liver fibrosis, and inhibit hepatoma cells [46–48]. In our study, GE50 and *Niuchangchih* were identical to the positive drug Silymarin.

The structural characteristics of plant polyphenols provide them with strong antioxidant and free radical scavenging abilities. Hepatoprotection using edible phytochemicals is a novel strategy for the treatment of various hepatic dysfunctions. Gallic acid (3,4,5-trihydroxybenzoic acid), a phenolic acid with strong antioxidant effect, is abundant in tea leaves, as well as white, red, and black mulberry, blackberry, raspberry, strawberry, dragon fruit, guava, mangosteen, papaya, and other plants. Gallic acid downregulated CYP2E1 expression in liver tissues while increasing SOD activities. These results support its ability to regulate the antioxidant enzymes activities and inhibit lipid peroxidation. Many studies have demonstrated its hepatoprotective effects [49–51].

Isoflavones are dietary phytoestrogens occurring naturally in legumes such as soybeans. Two major isoflavones found in soybean are daidzin and genistin, respectively. In soy and soy products, 95–99% of genistein exists in the conjugated form genistin (glycoside). Many studies demonstrated that daidzein and genistein alleviate hepatic damage [52,53].

Quercetin (3,5,7,30,40-pentahydroxyflavone) is a major flavonoid found in fruits, vegetables, and red wine and displays several healthy properties, including antioxidative, anti-inflammatory, anti-apoptotic, and hepatoprotective activities against various hepatic ailments. Several studies examined the protective effects of quercetin on chronic ethanol-induced liver injury [43,54]. Quercetin ameliorated lipid metabolism and ethanol-induced liver damage by inducing antioxidant enzymes, increasing GSH levels, and reducing CYP2E1 activity [55]. Quercetin, a well-known flavonoid, ameliorates CCl_4 and ethanol-induced liver injury in vivo by alleviating oxidative stress [56–58]. The antioxidant and antifibrotic effect of GE50 may be due to presence of quercetin. In our previous study, we also demonstrated that GE50 contains high levels of quercetin. The major components isolated were quercetin 3-O-[6′′-(3-hydroxyl-3-methylglutaroyl)]-β-d- glucopyranoside and quercetin 3-O-[6′′-(3-hydroxyl-3-methylglutaroyl)-2′′-acetyl]- β-d-glucopyranoside [16]. Quercetin also improves the antioxidant defense mechanism by increasing the levels of enzymatic and nonenzymatic antioxidants in cells, thus reducing oxidative stress.

It is, therefore, reasonable to assume that the hepatoprotective activities of GE50 is attributed to its gallic acid, flavone, genistin, daidzin, and quercetin components that most possibly act synergistically. *G. paraguayense* is an edible vegetable that has also been used in traditional Taiwanese folk medicine for protection against liver injury.

5. Conclusions

In conclusion, the results of our study indicate that GE50 enhances hepatic antioxidant enzyme activities, as well as inhibits lipid peroxidation in ethanol and CCl_4 (EC) co-induced liver injury, and its effect is similar to that of silymarin. The hepatoprotective activity of GE50 is proposed to occur through the synergic effects of its chemical component, namely gallic acid, flavone, genistin, daidzin, and quercetin. These results confirm that the in vivo hepatoprotective activity of GE50 may be associated with the phenolic phytochemicals present in the extract, which are known for their antioxidant potential. GE50 can be used as a functional food or even a pharmacological agent for the prevention of liver diseases.

Author Contributions: Conceived and designed the experiments: S.-T.C.; Performed the experiments: H.-C.P., J.-W.L., S.-J.C.; Analyzed the data: H.-C.P., S.-J.C., W.-W.C.; Wrote the manuscript: W.-W.C.; Reviewed the manuscript and coordinated the submission processes: S.-T.C. All authors had reviewed and approved the final version of manuscript for submission.

Funding: This research received no external funding. The authors are grateful for financial support from the Ministry of Science and Technology of the Republic of China (NSC98-2313-B-235-002-MY3, NSC102-2313-B-126-004-MY3).

Conflicts of Interest: The authors declare no conflict of interest.

References

1. Zhang, H.Y.; Wang, H.L.; Zhong, G.Y.; Zhu, J.X. Molecular mechanism and research progress on pharmacology of traditional Chinese medicine in liver injury. *Pharm. Biol.* **2018**, *56*, 594–611. [CrossRef] [PubMed]
2. Ingawale, D.K.; Mandlik, S.K.; Naik, S.R. Models of hepatotoxicity and the underlying cellular, biochemical and immunological mechanism(s): A critical discussion. *Environ. Toxicol. Pharmacol.* **2014**, *37*, 118–133. [CrossRef] [PubMed]
3. Stoyanovsky, D.A.; Cederbaum, A.I. Thiol oxidation and cytochrome P450-dependent metabolism of CCl_4 triggers Ca^{2+} release from liver microsomes. *Biochemistry* **1996**, *35*, 15839–15845. [CrossRef] [PubMed]
4. Chao, W.W.; Lin, B.F. Hepatoprotective diterpenoids isolated from *Andrographis paniculata*. *Chin. Med.* **2012**, *3*, 136–143. [CrossRef]
5. Wang, M.Y.; Srinivasan, M.; Dasari, S.; Narvekar, P.; Samy, A.L.P.; Dontaraju, V.S.; Peng, L.; Anderson, G.L.; Munirathinam, G. Antioxidant activity of Yichun Blue Honeysuckle (YBHS) berry counteracts CCl_4-induced toxicity in liver injury model of mice. *Antioxidants* **2017**, *6*, 50. [CrossRef] [PubMed]
6. Wang, R.; Yang, Z.; Zhang, J.; Mu, J.; Zhou, X.; Zhao, X. Liver injury induced by carbon tetrachloride in mice is prevented by the antioxidant capacity of Anji white tea polyphenols. *Antioxidants* **2019**, *8*, 64. [CrossRef] [PubMed]
7. Albano, E. Oxidative mechanisms in the pathogenesis of alcoholic liver disease. *Mol. Asp. Med.* **2008**, *29*, 9–16. [CrossRef]
8. Das, M.; Basu, S.; Banerjee, B.; Sen, A.; Jana, K.; Datta, G. Hepatoprotective effects of green *Capsicum annum* against ethanol induced oxidative stress, inflammation and apoptosis in rats. *J. Ethnopharmacol.* **2018**, *227*, 69–81. [CrossRef]
9. Choi, Y.; Abdelmegeed, M.A.; Song, B.J. Preventive effects of indole-3-carbinol against alcohol-induced liver injury in mice via antioxidant, anti-inflammatory, and anti-apoptotic mechanisms: Role of gut-liver-adipose tissue axis. *J. Nutr. Biochem.* **2018**, *55*, 12–25. [CrossRef]
10. Hellerbrand, C.; Schattenberg, J.M.; Peterburs, P.; Lechner, A.; Brignoli, R. The potential of silymarin for the treatment of hepatic disorders. *Clin. Phytosc.* **2016**, *2*, 7–20. [CrossRef]
11. Federico, A.; Dallio, M.; Loguercio, C. Silymarin/silybin and chronic liver disease: A marriage of many years. *Molecules* **2017**, *22*, 191. [CrossRef] [PubMed]
12. Chung, Y.C.; Chen, S.J.; Hsu, C.K.; Chang, C.T.; Chou, S.T. Studies on the antioxidative activity of *Graptopetalum paraguayense* E. Walther. *Food Chem.* **2005**, *91*, 419–424. [CrossRef]
13. Chen, S.J.; Chang, C.T.; Chung, Y.C.; Chou, S.T. Studies on the inhibitory effect of *Graptopetalum paraguayense* E. Walther extracts on the angiotensin converting enzyme. *Food Chem.* **2007**, *100*, 1032–1036. [CrossRef]
14. Chen, S.J.; Chung, J.G.; Chung, Y.C.; Chou, S.T. In vitro antioxidant and antiproliferative activity of the stem extracts from *Graptopetalum paraguayense*. *Am. J. Chin. Med.* **2008**, *36*, 369–383. [CrossRef] [PubMed]
15. Chung, Y.C.; Chou, S.T.; Jhan, J.K.; Liao, J.W.; Chen, S.J. In vitro and in vivo safety of aqueous extracts of *Graptopetalum paraguayense* E. Walther. *J. Ethnopharmacol.* **2012**, *140*, 91–97. [CrossRef]
16. Liu, H.Y.; Peng, H.Y.; Hsu, S.L.; Jong, T.T.; Chou, S.T. Chemical characterization and antioxidative activity of four 3-hydroxyl-3-methylglutaroyl (HMG)-substituted flavonoid glycosides from *Graptopetalum paraguayense* E. Walther. *Bot. Stud.* **2015**, *56*, 8–16. [CrossRef]
17. Kao, T.K.; Ou, Y.C.; Raung, S.L.; Chen, W.Y.; Yen, Y.J.; Lai, C.Y.; Chou, S.T.; Chen, C.J. *Graptopetalum paraguayense* E. Walther leaf extracts protect against brain injury in ischemic rats. *Am. J. Chin. Med.* **2010**, *38*, 495–516. [CrossRef]
18. Lin, Y.L.; Peng, H.Y.; Hsieh, H.M.; Lin, C.H.; Chou, S.T. Effects of *Graptopetalum paraguayense* consumption on serum lipid profiles and antioxidative status in hypercholesteremic subjects. *J. Sci. Food Agric.* **2011**, *91*, 1230–1235.

19. Duh, P.D.; Lin, S.L.; Wu, S.C. Hepatoprotection of *Graptopetalum paraguayense* E. Walther on CCl4-induced liver damage and inflammation. *J. Ethnopharmacol.* **2011**, *134*, 379–385. [CrossRef]
20. Hsu, W.H.; Chang, C.C.; Huang, K.W.; Chen, Y.C.; Hsu, S.L.; Wu, L.C.; Tsou, A.P.; Lai, J.M.; Huang, C.Y. Evaluation of the medicinal herb *Graptopetalum paraguayense* as a treatment for liver cancer. *PLoS ONE* **2015**, *10*, e0121298. [CrossRef]
21. Su, L.J.; Chang, C.C.; Yang, C.H.; Hsieh, S.J.; Wu, Y.C.; Lai, J.M.; Tseng, T.L.; Huang, C.Y.F.; Hsu, S.L. *Graptopetalum paraguayense* ameliorates chemical-induced rat hepatic fibrosis in vivo and inactivates stellate cells and kupffer cells in vitro. *PLoS ONE* **2013**, *8*, e53988. [CrossRef] [PubMed]
22. Chou, S.T.; Lai, C.C.; Lai, C.P.; Chao, W.W. Chemical composition, antioxidant, anti-melanogenic and anti-inflammatory activities of *Glechoma hederacea* (Lamiaceae) essential oil. *Ind. Crops Prod.* **2018**, *122*, 675–685. [CrossRef]
23. Habig, W.H.; Pabst, M.J.; Jakoby, W.B. Glutathione *S*-transferase: The first enzymatic step in mercapturic acid formation. *J. Biol. Chem.* **1974**, *249*, 7130–7139. [PubMed]
24. Mitton, K.P.; Trevithick, J.R. High-performance liquid chromatography-electrochemical detection of antioxidants in vertebrate lens: Glutathione, α-tocopherol and ascorbate. *Meth. Enzymol.* **1994**, *233*, 523–539. [PubMed]
25. Chou, S.T.; Ko, L.E.; Yang, C.S. Determination of tissue α- tocopherol in senescence-accelerated mice by high-performance liquid chromatography with fluorimetric detection. *Anal. Chim. Acta* **2000**, *419*, 81–86. [CrossRef]
26. Yang, C.S.; Chou, S.T.; Lin, L.; Tsai, P.J.; Kuo, J.S. Effect of ageing on human plasma glutathione concentrations as determined by high-performance liquid chromatography with fluorimetric detection. *J. Chromatogr. B* **1995**, *674*, 23–30. [CrossRef]
27. Li, Q.; Jia, Y.; Xu, L.; Wang, X.; Shen, Z.; Liu, Y.; Bi, K. Simultaneous determination of protocatechuic acid, syringin, chlorogenic acid, caffeic acid, liriodendrin and isofraxidin in *Acanthopanax senticosus* Harms by HPLC-DAD. *Biol. Pharm. Bull.* **2006**, *29*, 532–534. [CrossRef]
28. Chang, Y.Y.; Liu, Y.C.; Kuo, Y.H.; Lin, Y.L.; Wu, Y.H.S.; Chen, J.W.; Chen, Y.C. Effects of antrosterol from *Antrodia camphorata* submerged whole broth on lipid homeostasis, antioxidation, alcohol clearance, and antiinflammation in livers of chronic-alcohol fed mice. *J. Ethnopharmacol.* **2017**, *202*, 200–207. [CrossRef]
29. Tiwary, B.K.; Dutta, S.; Dey, P.; Hossain, M.; Kumar, A.; Bihani, S.; Nanda, A.K.; Chaudhuri, T.K. Radical scavenging activities of *Lagerstroemia speciosa* (L.) Pers. Petal extracts and its hepatoprotection in CCl_4-intoxicated mice. *BMC Complement. Altern. Med.* **2017**, *17*, 55–67. [CrossRef]
30. Kim, T.W.; Lee, D.R.; Choi, B.K.; Kang, H.K.; Jung, J.Y.; Lim, S.W.; Yang, S.H.; Suh, J.W. Hepatoprotective effects of polymethoxyflavones against acute and chronic carbon tetrachloride intoxication. *Food Chem. Toxicol.* **2016**, *91*, 91–99. [CrossRef]
31. Bishayee, A.; Mandal, A.; Atterjee, M. Prevention of alcohol-carbon tetrachloride-induced signs of early hepatotoxicity in mice by *Trianthema portulacastrum* L. *Phytornedicine* **1996**, *3*, 155–161. [CrossRef]
32. McCay, P.B.; Lai, E.K.; Poyer, J.L.; DuBose, C.M.; Janzen, E.G. Oxygen- and carboncentered free radical formation during carbon tetrachloride metabolism. Observation of lipid radicals in vivo and in vitro. *J. Biol. Chem.* **1984**, *259*, 2135–2143.
33. Basu, S. Carbon tetrachloride-induced lipid peroxidation: Eicosanoid formation and their regulation by antioxidant nutrients. *Toxicology* **2003**, *189*, 113–127. [CrossRef]
34. Turkodgan, M.K.; Agaoglu, Z.; Yener, Z.; Sekeroglu, R.; Akkan, H.A.; Avci, M.E. The role of antioxidant vitamins (C and E), selenium and nigella sativa in the prevention of liver fibrosis and cirrhosis in rabbits: New hopes. *Dtsch. Tierarztl. Wochenschr.* **2001**, *108*, 71–73.
35. Cheeseman, K.H.; Davies, M.J.; Emery, S.; Maddix, S.P.; Slater, T.F. Effects of alpha-tocopherol on carbon tetrachloride metabolism in rat liver microsomes. *Free Radic. Res. Commun.* **1987**, *3*, 325–330. [CrossRef]
36. McBean, G. Cysteine, glutathione, and thiol redox balance in astrocytes. *Antioxidants* **2017**, *6*, 62. [CrossRef]
37. Fraschini, F.; Demartini, G.; Esposti, D. Pharmacology of silymarin. *Clin. Drug Investig.* **2002**, *22*, 51–65. [CrossRef]
38. Koch, O.; Farre, S.; De Leo, M.E.; Palozza, P.; Palazzotti, B.; Borrelo, S.; Palombini, G.; Cravero, A.; Galeotti, T. Regulation of manganese superoxide dismutase (MnSOD) in chronic experimental alcoholism: Effects of vitamin E supplemented and -deficient diets. *Alcohol Alcohol.* **2000**, *35*, 159–163. [CrossRef]

39. Hayes, P.C.; Bouchier, J.A.D.; Beckett, G.J. Glutathione S-transferase in humans in heath and disease. *Gut* **1991**, *32*, 813–818. [CrossRef]
40. Hiraganahalli, D.; Chandrasekaran, C.; Dethe, S.; Mundkinajeddu, D.; Pandre, M.; Balachandran, J. Hepatoprotective and antioxidant activity of standardized herbal extracts. *Pharm. Mag.* **2012**, *8*, 116–123.
41. Kerimi, A.; Williamson, G. At the interface of antioxidant signaling and cellular function: Key polyphenol effects. *Mol. Nutr. Food Res.* **2016**, *60*, 1770–1788. [CrossRef]
42. Sun, X.; Yamasaki, M.; Katsube, T.; Shiwaku, K. Effects of quercetin derivatives from mulberry leaves: Improved gene expression related hepatic lipid and glucose metabolism in short-term high-fat fed mice. *Nutr. Res. Pract.* **2015**, *9*, 137–143. [CrossRef]
43. Miltonprabu, S.; Tomczyk, M.; Skalicka-Woaniak, K.; Rastrelli, L.; Daglia, M.; Nabavi, S.F.; Alavian, S.M.; Nabavi, S.M. Hepatoprotective effect of quercetin: From chemistry to medicine. *Food Chem. Toxicol.* **2017**, *108*, 365–374. [CrossRef]
44. Pu, X.; Fan, W.; Yu, S.; Li, Y.; Ma, X.; Liu, L.; Ren, J.; Zhang, W. Polysaccharides from *Angelica* and *Astragalus* exert hepatoprotective effects against carbon-tetrachloride-induced intoxication in mice. *Can. J. Physiol. Pharmacol.* **2015**, *93*, 39–43. [CrossRef]
45. Pareek, A.; Godavarthi, A.; Issarani, R.; Nagori, B.P. Antioxidant and hepatoprotective activity of *Fagonia schweinfurthii* (Hadidi) Hadidi extract in carbon tetrachloride induced hepatotoxicity in HepG2 cell line and rats. *J. Ethnopharmacol.* **2013**, *150*, 973–981. [CrossRef]
46. Ao, Z.H.; Xu, Z.H.; Lu, Z.M.; Xu, H.Y.; Zhang, X.M.; Dou, W.F. *Niuchangchih* (*Antrodia camphorata*) and its potential in treating liver diseases. *J. Ethnopharmacol.* **2009**, *121*, 194–212. [CrossRef]
47. Li, Z.W.; Kuang, Y.; Tang, S.N.; Li, K.; Huang, Y.; Qiao, X.; Yu, S.W.; Tzeng, Y.M.; Lo, J.Y.; Ye, M. Hepatoprotective activities of *Antrodia camphorata* and its triterpenoid compounds against CCl_4-induced liver injury in mice. *J. Ethnopharmacol.* **2017**, *206*, 31–39. [CrossRef]
48. Wu, Y.; Tian, W.J.; Gao, S.; Liao, Z.J.; Wang, G.H.; Lo, J.M.; Lin, P.H.; Zeng, D.Q.; Qiu, D.R.; Liu, X.Z.; et al. Secondary metabolites of petri-dish cultured *Antrodia camphorata* and their hepatoprotective activities against alcohol-induced liver injury in mice. *Chin. J. Nat. Med.* **2019**, *17*, 33–42. [CrossRef]
49. Wang, J.; Tang, L.; White, J.; Fang, J. Inhibitory effect of gallic acid on CCl_4-mediated liver fibrosis in mice. *Cell Biochem. Biophys.* **2014**, *69*, 21–26. [CrossRef]
50. Chen, Y.; Zhou, Z.; Mo, Q.; Zhou, G.; Wang, Y. Gallic acid attenuates dimethylnitrosamine-induced liver fibrosis by alteration of *Smad phosphoisoform* signaling in rats. *BioMed Res. Int.* **2018**, *2018*, 1682743. [CrossRef]
51. Hsu, W.H.; Liao, S.C.; Chyan, Y.J.; Huang, K.W.; Hsu, S.L.; Chen, Y.C.; Siu, M.L.; Chang, C.C.; Chung, Y.S.; Huang, C.Y.F. *Graptopetalum paraguayense* inhibits liver fibrosis by blocking TGF-β signaling in vivo and in vitro. *Int. J. Mol. Sci.* **2019**, *20*, 2592. [CrossRef]
52. Chiang, C.M.; Wang, D.S.; Chang, T.S. Improving free radical scavenging activity of soy isoflavone glycosides daidzin and genistin by 30-hydroxylation using recombinant *Escherichia coli*. *Molecules* **2016**, *21*, 1723. [CrossRef]
53. Zhao, L.; Wang, Y.; Liu, J.; Wang, K.; Guo, X.X.; Ji, B.; Wu, W.; Zhou, F. Protective effects of genistein and puerarin against chronic alcohol-induced liver injury in mice via antioxidant, anti-inflammatory, and anti-apoptotic mechanisms. *J. Agric. Food Chem.* **2016**, *64*, 7291–7297. [CrossRef]
54. Qiu, P.; Dong, Y.; Zhu, T.; Luo, Y.Y.; Kang, X.J.; Pang, M.X.; Li, H.Z.; Xu, H.; Gu, C.; Pan, S.H.; et al. Semen hoveniae extract ameliorates alcohol-induced chronic liver damage in rats via modulation of the abnormalities of gut-liver axis. *Phytomedicine* **2019**, *52*, 40–50. [CrossRef]
55. Surapaneni, K.M.; Jainu, M. Comparative effect of pioglitazone, quercetin and hydroxy citric acid on the status of lipid peroxidation and antioxidants in experimental non-alcoholic steatohepatitis. *J. Physiol. Pharmacol.* **2014**, *65*, 67–74.
56. Kemelo, M.K.; Pierzynova, A.; Canova, N.K.; Kucera, T.; Farghali, H. The involvement of sirtuin 1 and heme oxygenase 1 in the hepatoprotective effects of quercetin against carbon tetrachloride-induced sub-chronic liver toxicity in rats. *Chem. Biol. Interact.* **2017**, *269*, 1–8. [CrossRef]

57. Ma, J.Q.; Li, Z.; Xie, W.R.; Liu, C.M.; Liu, S.S. Quercetin protectsmouse liver against CCl_4-induced inflammation by the TLR2/4 and MAPK/NF-κB pathway. *Int. Immunopharmacol.* **2015**, *28*, 531–539. [CrossRef]
58. Zakarua, Z.A.; Mahmood, N.D.; Omar, M.H.; Taher, M.; Basir, R. Methanol extract of *Muntingia calabura* leaves attenuates CCl_4-induced liver injury: Possible synergistic action of flavonoids and volatile bioactive compounds on endogenous defence system. *Pharm. Biol.* **2019**, *57*, 335–344. [CrossRef]

Article

Nutritional Composition, Total Phenolic Content, Antioxidant and α-Amylase Inhibitory Activities of Different Fractions of Selected Wild Edible Plants

Ziaul Hasan Rana [1,*], Mohammad Khairul Alam [2] and Mohammad Akhtaruzzaman [2]

[1] Department of Nutritional Sciences, Texas Tech University, Lubbock, TX 79409, USA
[2] Institute of Nutrition and Food Science, University of Dhaka, Dhaka-1000, Bangladesh
* Correspondence: ziaul.h.rana@ttu.edu; Tel.: +1-(806)-500-9335

Received: 16 June 2019; Accepted: 29 June 2019; Published: 1 July 2019

Abstract: Wild plants are considered the richest source of essential nutrients and other beneficial phytochemicals. Hence, the objective of this study was to evaluate the nutritional composition, antioxidant- and α-amylase inhibition activities of leaves and roots of selected Bangladeshi wild plants. These wild plants were found to have high fiber (13.78–22.26 g/100 g), protein (7.08–21.56 g/100 g) and ash (8.21–21.43 g/100 g) contents. The total phenolic and total flavonoid contents were significantly higher in the leaves than the roots. Additionally, antioxidant activity was evaluated using ferric-reducing antioxidant power, 2, 2-diphenyl-1-picrylhydrazyl radical (DPPH) and trolox equivalent antioxidant capacity assays and was strongly correlated with phenolic compounds. The leaf extracts of the selected plants also exhibited potent α-amylase inhibition (~71%) and were significantly higher than their root counterparts. Thus, the study findings concluded that the investigated plants were good sources of fiber, protein, mineral, natural antioxidant compounds and α-amylase inhibitors, and their increased intake could provide health benefits. The principal component analysis (PCA) of analyzed variables divided the samples into three clear groups, and the first two principal components accounted for 86.05% of the total data set variance.

Keywords: antioxidants; α-amylase; Bangladesh; nutritional profile; total phenolic content; wild plants

---

1. Introduction

Vegetables are an integral part of the daily human diet and provide essential nutrients (vitamins and minerals) required for active and healthy life. They are a locally available and cheap source of nutrient-dense foods and considerably contribute to human health, nutrition and food security. The edible parts of vegetable plants include the leaves, roots, stems, fruits or seeds, and they can be consumed cooked as well as in raw forms. Evidence from epidemiological studies indicate that daily consumption of fruits and vegetables is correlated with a lower prevalence of many chronic diseases, including diabetes, infections, cardiovascular and neurological disorders and cancers [1–3]. Wild vegetables are considered to be a potential source of essential nutrients such as vitamin C, minerals, vitamins, proteins, fibers [4–6] and are also good dietary sources of antioxidants such as flavonoids and other polyphenolic constituents [4,7].

A diet low in antioxidants and high in processed foods (e.g., red meat) can augment the production of endogenous reactive oxygen species (ROS) [8–10] which can lead to many of the chronic diseases stated above [11]. This demonstrates the need for natural antioxidant compounds which can prevent the overproduction of ROS. It has been shown that there is an inverse relationship between morbidity and mortality from degenerative disorders and the ingestion of natural antioxidants [12]. Currently, wild or traditional plants, as a source of natural antioxidants, have received increased attention because of their ability to scavenge ROS and also as sources of trace elements, and their additional health properties, such as antidiabetic, antibacterial and anticancer activity, make them valuable for incorporation into the daily diet [13–16].

Phenolic compound-rich food consumption has also been shown to be inversely associated with type-2 diabetes [17]. By binding to the non-specific site of the enzyme, phenolic compounds can inactivate the starch-digesting enzyme α-amylase [18]. Thus, wild edible plants, as a source of high phenolic compounds, can inhibit α-amylase activity which subsequently decreases postprandial rises in blood glucose by suppressing the rate of glucose release and absorption in the small intestine [19].

Several wild edible plants are traditionally consumed along with staple foods, especially in rural areas and a few urban communities, in Bangladesh. These plants play a vital role in fulfilling the demand for nutritional, minerals and antioxidant compounds in the diet of indigenous communities [4,20]; besides these factors, they are also used in treating certain medical conditions, for example, diabetes, in these local tribes [21]. However, there is lack of information about the nutritional composition of wild edible plants in Bangladesh and their ability to inhibit digestive enzymes. Therefore, the aim of this study was to assess the proximate and mineral composition, antioxidant potential, and α-amylase inhibition activity of the leaves and roots of three indigenous wild edible plants (*Achyranthes aspera* L., *Eclipta alba* L., and *Vitex negundo* L.) consumed by different local communities in Bangladesh. The findings of the present study will provide the preliminary data on the nutritional and neutraceutical potential of wild edible plants in Bangladesh and thus could be incorporated into food composition databases and used for further utilization as dietary supplements and/or functional foods.

2. Materials and Methods

2.1. Reagents

Analytical-grade acetone, petroleum ether, n-hexane, dichloromethane, sodium carbonate, Folin–Ciocalteu reagent and acetic acid were purchased from Merck (Darmstadt, Germany). Gallic acid (Pub Chem CID:370) was purchased from Tokyo Chemical Industry Co., (Tokyo, Japan) and 2,20-azinobis (3-ethylbenothiazoline-6-sulfonic acid) diammonium salt (ABTS) was purchased from Wako Pure Chemical Industries, Ltd. (Osaka, Japan). α-amylase, 2, 2-diphenyl-1-picrylhydrazyl radical (DPPH), Tri(2-pyridyl)-s-triazine (TPTZ), trolox, potassium persulfate, and mineral standards were obtained from Sigma Aldrich (Steinheim, Germany). All chemicals used for the analysis were of analytical grade.

2.2. Sample Collection and Preparation

To determine the proximate and mineral composition, total phenolic contents (TPC), total flavonoid contents (TFC), antioxidant capacities (DPPH, ferric-reducing antioxidant power (FRAP) and trolox equivalent antioxidant capacity (TEAC)), and α-amylase inhibition activity, three wild plant samples were collected from different locations in Bangladesh. Two to three samples (300–600 g) were collected for each of the wild plants from every growing location. These were then mixed to make three analytes or composite test samples. The study samples were *Achyranthes aspera* L. (Upat Lengra), *Eclipta alba* L. (Kalokeshi), and *Vitex negundo* L. (Nirgundi). The samples were selected based on their traditional use, by interviewing local people, in treating diabetes. The identification of the samples was confirmed by a taxonomist of the Department of Botany, University of Dhaka, who accompanied the collection team, after examining the morphological characteristics. Photographs of these samples are shown in Figure 1. After collection, the leaves and roots of the samples were separated and gently washed with tap water immediately to remove sand and other extraneous material before being washed with distilled water and then air-dried. Then, the samples were cut into small pieces and freeze-dried (il Shin lab.Co. Ltd., Korea). The freeze-dried samples were ground and homogenized into a fine powder using a grinder. The homogenized samples were sieved to obtain an even particle size, then placed in an air-tight zipper bag and stored at −20 °C until further analysis.

Figure 1. Photograph of selected samples.

2.3. Determination of Proximate Composition

The proximate composition (moisture, total protein, total fat, total dietary fiber including soluble and insoluble, ash and total available carbohydrate content) of the selected samples was estimated according to the method described previously [22]. Moisture and ash contents of the sample were calculated by the weight difference method, whereas the total fat content of the samples was estimated by the Association of Official Analytical Chemists (AOAC) method using petroleum ether as solvent. The total protein content was determined by using the micro-Kjeldhal method (nitrogen content of the samples × 6.25). The gravimetric method was utilized for the estimation of total dietary fiber (soluble and insoluble). Total available carbohydrate contents were calculated by difference using the formula below:

$$\text{Carbohydrate content (\%)} = 100 - [\text{total protein (\%)} + \text{ash content (\%)} + \text{total fat (\%)} + \text{total fiber (\%)}]. \quad (1)$$

2.4. Determination of Mineral Composition

Mineral concentrations in the plants sample were calculated by using an atomic absorption spectrophotometric method described previously [23]. Briefly, approximately 500 mg of plant samples after drying were subjected to wet digestion with nitric acid and perchloric acid (2:1 ratio) in an auto-digestor at 325 °C to accelerate the discharge of mineral in the plant matrix. After digestion and appropriate dilution, the digested sample was aspirated into an air–acetylene flame to burn the elements into atomic components, which were then detected in a spectrophotometer at their relevant wavelengths. Proportions of calcium, magnesium, sodium, zinc, copper and iron were evaluated by atomic absorption spectrophotometry (Model-AA-7000S, Shimadzu, Tokyo, Japan). The amount of potassium was determined by flame photometry (Jenway flame photometer model PFP7, Origin UK). A standard calibration curve was plotted for each of the minerals using the respective mineral standard obtained from Sigma Chemical Co., USA.

2.5. Plant Extraction

The extraction of plants was carried out according to the previously described procedure using methanol and 1N HCl [24], and the extract was stored at 4 °C for the determination of TPC, TFC, antioxidant activity and α-amylase activity.

2.6. Determination of Total Phenolic Content

The TPC in plant extracts was estimated by the Folin–Ciocalteu colorimetric method as described previously [22,24]. Briefly, for each sample, 150 μL of plant extracts were taken in test tubes. To this, 900 μL distilled water was added. 225 μL of diluted Folin–Ciocalteu reagent (2-fold) was added to the solution and allowed to stand for 5 min at room temperature. Then, 1.125 mL of 2% Na_2CO_3 solution was added, mixed well and left for 15 min at room temperature. Finally, the absorbance was measured at 750 nm by a UV-VIS spectrophotometer (UV-1800, Shimadzu, Kyoto, Japan). The TPC was calculated using a standard curve based on gallic acid. Results were expressed as milligrams of gallic acid equivalent (GAE) per gram dry weight (DW) (mg GAE/g DW).

2.7. Determination of Total Flavonoid Content

The TFC was estimated by means of the colorimetric method according to Miao et al. [25] with slight modification. Briefly, 250 μL of the extract was mixed with 1.125 ml of distilled water in a test tube. To these, 75 μL of 5% $NaNO_2$ solution was added. After 6 min, 150 μL of 10% $AlCl_3{\cdot}6H_2O$ solution was added. The solution was left to stand for another 5 min, and 500 μL of 1 M NaOH was added. Finally, the mixture was vortexed, and the absorbance was measured immediately at 510 nm by a UV-VIS spectrophotometer (UV-1800, Shimadzu, Kyoto, Japan). The TFC in the plant extract was calculated using a standard curve based on quercetin and results were expressed as milligrams quercetin equivalent (QE) per gram of dry weight (mg QE/g DW).

2.8. Evaluation of Antioxidant Capacities

2.8.1. DPPH Free Radical Scavenging Assay

The antioxidant activity of the plant extracts was evaluated by utilizing 2,2-diphenyl-1 -pycrylhydrazyl (DPPH) free radical according to Alam et al. [24]. The DPPH free radical inhibition capacity was calculated according to the following equation:

$$\%\ \text{DPPH inhibition} = ((1 - ((\text{Abs}_{\text{sample}} - \text{Abs}_{\text{blank}})/(\text{Abs}_{\text{control}} - \text{Abs}_{\text{blank}}))) \times 100 \quad (2)$$

where $\text{Abs}_{\text{blank}}$ is the absorbance of the blank (containing only methanol), $\text{Abs}_{\text{control}}$ is the absorbance of the control reaction (containing all reagents minus plant extracts), and $\text{Abs}_{\text{sample}}$ is the absorbance of the plant extracts. The plant extract concentration required for the 50% inhibition of DPPH free radical (IC_{50}) was estimated from the dose–response graph plotted with percentage inhibition and concentrations of plant extract.

2.8.2. Ferric Reducing Antioxidant Power (FRAP) Assay

This assay was carried out according to Miao et al. [25] with little modification. Briefly, the FRAP reagent was made from by combining 10 mmol/L 2,4,6-tripyridyls-triazine (TPTZ) solution, 300 mmol/L acetate buffer (pH 3.6), and 20 mmol/L $FeCl_3$ solution in a ratio of 1:10:1 (*v/v*), respectively. The FRAP reagent was freshly prepared and was incubated at 37 °C in a water bath before using. 100 μl of plant extracts were added to 3 mL of the FRAP reagent. The mixture was vortexed, and absorbance of the solution was then measured at 593 nm (UV-1800, Shimadzu, Kyoto, Japan) after incubating at 37 °C for 30 min. Various concentrations (50–600 μmol/L) of Fe^{2+} solution was used to prepare the standard curve. The results were expressed as μmol Fe^{2+} per gram of dry weight (μmol Fe^{2+}/g DW).

2.8.3. Trolox Equivalent Antioxidant Capacity (TEAC) Assay

This assay was performed by the advanced $ABTS^{\bullet+}$ method as described by Miao et al. [25] with little modification. $ABTS^{\bullet+}$ radical cation was produced by dissolving ABTS and potassium persulfate in distilled water to give a final concentration of 7 mmol/L and 2.45 mmol/L, respectively. The solutions were mixed, and the reaction mixture was left in the dark at room temperature for 24 h. The $ABTS^{\bullet+}$ solution was diluted with distilled water to an absorbance of 1.00 ± 0.03 at 734 nm. Then, 100 μL of plant extracts were added to 3.8 mL of diluted $ABTS^{\bullet+}$ solution and the solutions were kept in the dark for 10 min. After 10 min, the absorbance was read at 734 nm by a UV-VIS spectrophotometer (UV-1800, Shimadzu, Kyoto, Japan) against the blank (distilled water). The trolox solution of various concentrations (0–15 μmol/L) was used to prepare the standard curve, and the results were expressed as μmol trolox per gram of dry weight extract (μmol trolox/g DW).

2.9. α-Amylase Inhibitory Assay

The α-amylase inhibitory activity of the plant samples was performed by the modified starch iodine method described by Hossain et al. [26]. Briefly, 100 μL of plant extracts (0.25 mg/mL) were taken in test tubes. To each test tube, an aliquot of 2 μL of α-amylase was added and incubated for 10 min at 37 °C. After incubating, 1% starch solution (20 μL) was added. Then, the mixture was incubated again for 60 min at 37 °C. After that, 20 μL of 1% iodine solution was added to the mixture. The absorbance of the mixture was taken at 565 nm, after the addition of 1 mL distilled water. Acarbose, a known α-amylase inhibitor, was used as a standard. The α-amylase inhibitory activity was calculated and expressed as percentage inhibition using the following formula:

$$(\%)\ \alpha\text{-amylase Inhibition} = (1 - (Abs_{sample}/Abs_{control})) \times 100 \quad (3)$$

where $Abs_{control}$ is the absorbance of the control reaction (containing all reagents minus plant extracts or acarbose) and Abs_{sample} is the absorbance of the plant extracts or acarbose.

2.10. Statistical Analysis

All experiments were carried out in three replicates and presented as mean ± standard deviation (SD) using Minitab version 18.0. (Minitab Inc., State College, PA, USA). One-way analysis of variance (ANOVA) and principal component analysis (PCA) were performed to check the differences between the nutrient contents, TPC, TFC, antioxidant activity, and α-amylase inhibition activity among the plant samples. The differences were declared significant at a level of $p < 0.05$. The Dunnett test to compare with control for α-amylase activity and Pearson correlation among variables were also calculated.

3. Results and Discussion

3.1. Proximate Composition

Table 1 summarizes the proximate composition, macro mineral (Ca, Na, K Mg) and micro mineral (Fe, Zn, Cu) content in the leaves and roots of the investigated wild plants.

The moisture content of the wild plants ranged from 82.78 ± 2.68 to 88.13 ± 1.55 g/100g fresh weight (FW) in the leaves and 55.44 ± 2.22 to 70.41 ± 2.11 g/100g FW in roots, which is in accordance with results reported by other authors [4,5]. On the other hand, Satter et al. [20] reported a higher content of moisture in some other wild plants of Bangladesh.

Protein contents were found between 18.13 ± 1.67 to 21.56 ± 1.10 g/100g DW in leaves and 7.08 ± 0.33 to 13.21 ± 0.93 g/100 g DW in roots, which is opposite to the findings of other studies [4,27]. However, our findings are similar to results observed by Satter et al. [20] and higher than those reported by Gupta et al. [5].

The contents of fat in this study were 1.88 ± 0.20 to 3.13 ± 0.51 g/100 g DW and 0.89 ± 0.07 to 1.12 ± 0.10 g/100 g DW in leaves and roots, respectively. These values are less than those reported by other authors [4,20,27].

The ash contents in the leaves and roots were observed from 19.78 ± 0.42 to 21.43 ± 0.33 g/100 g DW and 8.21 ± 0.61 to 17.37 ± 0.51 g/100 g DW, respectively. Afolayan & Jimoh [4] also reported similar values, whereas Gupta et al. [5] and Satter et al. [20] found lower contents than ours.

Dietary fiber content was in the range of 18.65 ± 1.23 to 20.28 ± 0.92 g/100 g DW in leaves and 13.78 ± 1.34 to 22.26 ± 0.56 g/100 g DW in roots, which is higher than that reported by previous studies [4,20]. Regional or other factors could be an explanation for these variations in proximate composition, as indicated elsewhere [22].

Table 1. Proximate composition (g/100g dry weight (DW)), and macro minerals and micro minerals (DW basis) of selected wild plants.

Wild Plants	*Achyranthes aspera* L. (Upat Lengra)		*Eclipta alba* L. (Kalokeshi)		*Vitex negundo* L. (Nirgundi)	
	Leaves	**Roots**	**Leaves**	**Roots**	**Leaves**	**Roots**
Proximate composition (g/100 g sample)						
Moisture	83.71 ± 1.33	61.23 ± 1.01	88.13 ± 1.55	55.44 ± 2.22	82.78 ± 2.68	70.41 ± 2.11
Protein	18.13 ± 1.67	7.08 ± 0.33	21.56 ± 1.10	13.21 ± 0.93	19.27 ± 0.85	11.35 ± 1.05
Fat	1.88 ± 0.20	0.89 ± 0.07	2.17 ± 0.11	0.94 ± 0.05	3.13 ± 0.51	1.12 ± 0.10
Fiber	18.65 ± 1.23	22.26 ± 0.56	20.28 ± 0.92	16.56 ± 0.52	19.70 ± 0.90	13.78 ± 1.34
Ash	21.43 ± 0.33	13.42 ± 0.88	19.78 ± 0.42	17.37 ± 0.51	20.15 ± 0.75	8.21 ± 0.61
Carbohydrate (CHO)	39.91 ± 1.85	56.35 ± 1.46	36.21 ± 0.63	51.92 ± 1.50	37.75 ± 0.75	65.54 ± 1.78
Mineral Composition						
Macro minerals (mg/100 g sample)						
Sodium (Na)	497.51 ± 3.66	135.20 ± 1.03	345.33 ± 1.25	100.50 ± 0.70	577.82 ± 2.23	202.72 ± 1.08
Potassium (K)	4866.45 ± 5.78	1185.37 ± 1.75	5174.82 ± 5.74	2359.90 ± 4.01	3345.20 ± 4.65	1058.39 ± 5.05
Magnesium (Mg)	333.51 ± 2.43	164.38 ± 0.96	274.20 ± 3.98	148.21 ± 2.22	315.15 ± 2.45	190.80 ± 0.70
Calcium (Ca)	1493.45 ± 3.73	842.16 ± 2.02	2221.33 ± 6.83	523.91 ± 1.13	1786.24 ± 7.88	1090.90 ± 1.10
Micro minerals (mg/100 g sample)						
Iron (Fe)	31.61 ± 0.70	19.83 ± 1.33	45.22 ± 1.12	16.13 ± 0.80	62.05 ± 1.01	23.40 ± 0.9
Zinc (Zn)	6.03 ± 0.09	3.51 ± 0.05	5.82 ± 0.96	2.80 ± 0.09	5.88 ± 0.44	4.35 ± 0.65
Copper (Cu)	1.13 ± 0.02	0.51 ± 0.01	2.33 ± 0.07	0.67 ± 0.03	1.08 ± 0.04	0.84 ± 0.02

Amounts of total available carbohydrate (CHO) were documented to be 36.21 ± 0.63 to 39.91 ± 1.85 g/100 g DW in leaves and 51.92 ± 1.50 to 65.54 ± 1.78 g/100 g DW in roots. Satter et al. [20] found higher values of total CHO in the leaves of other wild plants of Bangladesh, whereas Afolayan & Jimoh [4] and Seal [27] stated lower values than ours in South African and Indian wild edible plants. Thus, the proximate composition of wild plants grown in Bangladesh contain similar or even higher contents of specific nutrients to those of other wild plants growing in different global areas. These wild plants, as a source of high fiber, protein and ash, and with a low fat content, could be incorporated into weight control diet for obese people.

3.2. Mineral Composition

The mineral (macro and micro) composition of the studied wild plants is presented in Table 1, and the results revealed that these wild plants were rich in a wide variety of minerals including Na, K, Mg, Ca, Fe, Zn, and Cu. *Vitex negundo* leaf was found to contain higher amounts of Na and Fe, whereas Mg and Zn contents were higher in the leaf of *Achyranthes aspera*. *Eclipta alba* leaf was found to be a better source of potassium (5174.82 ± 5.74 mg/100 g DW), calcium (2221.33 ± 6.83 mg/100 g DW) and Cu (2.33 ± 0.07 mg/100 g DW) than others (Table 1). The roots of the plant had a relatively lower amount of minerals than their corresponding leaves ($p < 0.0001$). Mg and Cu values in the plant samples were found to be comparable to the values reported by several authors [4,20]. However, the concentrations of other minerals were much higher than those reported for other wild plant varieties [20,27], and some values were less than those stated by Afolayan & Jimoh [4]. These wild plants contain comparable or higher amounts of minerals than those documented in some commonly consumed vegetables such as spinach, cauliflower, cabbage, and lettuce and other cultivated vegetables [28]. Thus, the selected wild plants could potentially be utilized as a good source of major and trace minerals required for normal body function and maintenance.

3.3. Total Phenolic and Flavonoid Contents

The TPC of the leaf and root extracts of the studied plants was determined by the Folin–Ciocalteu method. The plant samples undertaken in this study showed the presence of TPC in ranges of 2.46 ± 0.06 to 8.45 ± 0.15 mg GAE/g DW, and 55.32 ± 0.47 to 72.11 ± 0.73 mg GAE/g DW for the roots and leaves, respectively (Table 2). The leaves of the samples were higher in TPC than their roots ($p < 0.0001$). The leaf and root of *Vitex negundo* showed the highest (72.11 ± 0.73 mg GAE/g DW) and lowest (2.46 ± 0.06 mg GAE/g DW) content of TPC, respectively.

The TFC of the root and leaf extracts of the samples was estimated by flavonoid–aluminum chloride conjugation method and is presented in Table 2. The estimation of TFC in different samples revealed values from 1.22 ± 0.09 to 4.88 ± 0.31 mg QE/g DW, and 31.55 ± 0.25 to 80.23 ± 0.55 mg QE/g DW for the roots and leaves, respectively (Table 2). The highest content (80.23 ± 0.55 mg QE/g DW) was observed in *Achyranthes aspera* leaf extract, while *Vitex negundo* root extract exhibited the lowest (1.22 ± 0.09 mg QE/g DW). Like TPC, TFC was also higher in leaves than roots ($p < 0.0001$). However, some studies reported higher contents of TPC and TFC in stems and roots than leaves, which is contradictory to our observation [29].

Table 2. Total phenolic, flavonoid, TEAC, and FRAP contents of the selected samples.

Scientific Name	Family	Local Name	TPC [1] (mg GAE/g DW)		TFC [2] (mg QE/g DW)		TEAC [3] (μmol trolox/g DW)		FRAP [4] (μmol Fe^{2+}/g DW)	
			Leaves	Roots	Leaves	Roots	Leaves	Roots	Leaves	Roots
Achyranthes aspera L.	Amaranthaceae	Upat Lengra	68.84 ± 0.61^a	4.55 ± 0.11^b	80.23 ± 0.55^a	2.23 ± 0.19^b	250.18 ± 1.08^a	12.13 ± 0.28^b	505.19 ± 1.56^a	65.22 ± 0.70^b
Eclipta alba L.	Asteraceae	Kalokeshi	55.32 ± 0.47^b	8.45 ± 0.15^a	31.55 ± 0.25^c	4.88 ± 0.31^a	184.31 ± 1.42^b	18.58 ± 0.20^a	474.35 ± 1.88^b	81.05 ± 0.55^a
Vitex negundo L.	Lamiaceae	Nirgundi	72.11 ± 0.73^a	2.46 ± 0.06^c	51.07 ± 0.88^b	1.22 ± 0.09^b	282.41 ± 1.25^a	7.50 ± 0.10^b	554.41 ± 2.38^a	53.78 ± 0.98^c

Values in the same column having different letters differ significantly ($p < 0.05$). [1] Total phenolic content; [2] Total flavonoid content; [3] Trolox equivalent antioxidant capacity; [4] Ferric reducing antioxidant power. FRAP: ferric-reducing antioxidant power; TEAC: trolox equivalent antioxidant capacity; TPC: total phenolic content: GAE: gallic acid equivalent; TFC: total flavonoid content; QE: quercetin equivalent.

To our knowledge, there are little or no available data in the literature about the TPC and TFC of the selected plants; thus, only a few papers could be found related to the TPC and TFC of other species of the same families. The leaf extract of *Vitex negundo* has recently been reported to contain 89.71 mg GAE/g of TPC and 63.11 mg QE/g of TFC [30], which are relatively higher values than our study. Shahat et al. [31] reported TPC in the range from 11 to 56 mg GAE/g of DW for different species of Asteraceae family, which is a lower value than our findings. Compared to a study by Nana et al. [32], using *Amaranthus cruentus* and *Amaranthus hybridus* of the Amaranthaceae family, we found lower TPC and TFC values in our sample of the Amaranthaceae family. We also found higher TPC in our study than some frequently consumed local vegetables [33]. On the other hand, our observed content was lower than some reported Asian vegetables [7,34]. However, it is well recognized that several factors, such as species, plant tissue, temperature, water stress and light conditions, as well as phenological development, can influence the TPC in the plants [22,24,35]. Thus, this explains the large differences observed between our and previous findings.

3.4. Antioxidant Capacities

The antioxidant capacities of the leaf and root extracts of the plants were investigated by various free radical scavenging assays, including DPPH, ABTS, and FRAP assays. In all of the antioxidant activity assays, the extract of leaves exhibited stronger antioxidant activity as compared to the extract of roots, which is in accordance with TPC and TFC.

Figure 2 represents the % inhibition of the DPPH free radical (a) and IC_{50} value (b) of the leaf and root extracts. The highest DPPH inhibition (86.65% inhibition) was observed in the *Vitex negundo* leaf, while the root of the same species showed the lowest inhibition (55.15% inhibition). The leaf extracts showed more potent scavenging activity than their root counterparts ($p < 0.001$). Adedapo et al. [29] also observed a similar trend but only at higher concentrations of the extract. Antioxidant activity measured by DPPH assay in different vegetables available in the local market ranged from 0.61 to 8328.80 µmol trolox equivalent per gram [33]. At 100 µg/mL, Adedapo et al. [29] noticed an 89.7% and 67.0% inhibition of DPPH by the leaf and stem of a South African medicinal plant, respectively, while Shahat et al. [31] recorded almost 100% inhibition of DPPH by Asteraceae family plants. Dasgupta & De [7] reported IC_{50} values of some leafy vegetables in India, measured by DPPH assay, ranging from 61.4 to 1946 µg/mL, which are higher values than what we observed in this study.

Table 2 represents the antioxidant activity of the plant extracts (leaf and root) based on ABTS and FRAP assays. The existence of antioxidant substances or reductants in the plant extracts directs the conversion of the ferric (Fe^{3+}) complex to the ferrous (Fe^{2+}) form, which is the principle of FRAP assay. The decrease in Fe^{3+} in the solution leads to the decrease in color, which implies the potent reducing power of the plant extracts. Among the selected plant extracts, *Vitex negundo* leaf and root extracts showed the highest (554.41 ± 2.38) and lowest (53.78 ± 0.98) FRAP values, respectively (Table 2).

In this study, the free radical scavenging power was also evaluated using the improved $ABTS^{\bullet+}$ method. The principle behind this assay is producing the ABTS radical cation ($ABTS^{\bullet+}$), a blue-green chromogen, by reacting ABTS and potassium persulphate. In the presence of antioxidant components, the color of the free radical is decreased, which has a characteristic absorbance at 734 nm [25]. Like DPPH inhibition and FRAP, *Vitex negundo* leaf and root extracts also showed the highest (282.41 ± 1.25) and lowest (7.50 ± 0.10) TEAC values, respectively (Table 2). Adedapo et al. [29] reported higher values of FRAP in stem than leaves, whereas in this study, we observed the opposite. However, a similar trend to our observation was reported by Adedapo et al. [29] in terms of ABTS value. The ABTS and FRAP values in some Indian leafy vegetables varied from 18.3 to 71.8 µmol trolox/g DW and 107.7 to 275.6 µmol Fe^{2+}/g DW, respectively [28]. The antioxidant competencies obtained from FRAP assay and those obtained from TEAC assay were highly correlated (R = 0.988) (Table 3), which signifies that antioxidants present in these plants were proficient in scavenging free radicals ($ABTS^{\bullet+}$) and reducing oxidants (ferric ions).

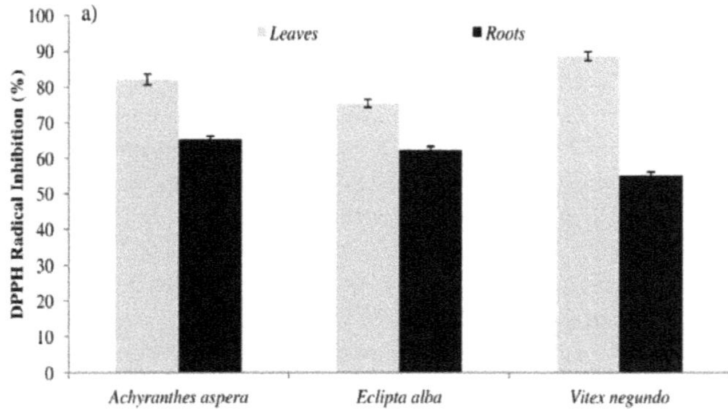

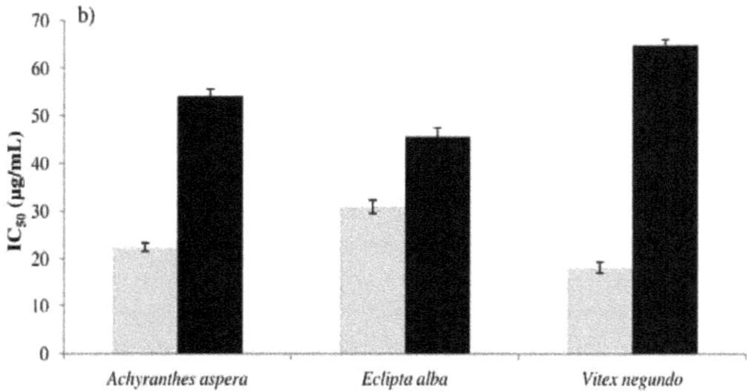

Figure 2. (%) inhibition of 2, 2-diphenyl-1-picrylhydrazyl radical (DPPH) free radical (**a**) and IC_{50} (**b**) value of the samples.

Table 3. Pearson correlation and corresponding *p*-values among variables.

			Correlation Matrix			
Variables	**TPC**	**TFC**	**TEAC**	**DPPH**	**FRAP**	**α-Amylase**
TPC		0.916	0.996	0.906	0.996	0.974
TFC	0.916		0.916	0.829	0.884	0.876
TEAC	0.996	0.916		0.914	0.988	0.966
DPPH	0.906	0.829	0.914		0.883	0.963
FRAP	0.996	0.884	0.988	0.883		0.967
α-amylase	0.974	0.876	0.966	0.963	0.967	
			***p*-Values**			
Variables	**TPC**	**TFC**	**TEAC**	**DPPH**	**FRAP**	**α-Amylase**
TPC		<0.0001	<0.0001	<0.0001	<0.0001	<0.0001
TFC	<0.0001		<0.0001	<0.0001	<0.0001	<0.0001
TEAC	<0.0001	<0.0001		<0.0001	<0.0001	<0.0001
DPPH	<0.0001	<0.0001	<0.0001		<0.0001	<0.0001
FRAP	<0.0001	<0.0001	<0.0001	<0.0001		<0.0001
α-amylase	<0.0001	<0.0001	<0.0001	<0.0001	<0.0001	

Plants rich in secondary metabolites including phenolics and flavonoids demonstrate powerful antioxidant properties, both in vitro and in vivo, which is attributed to their redox properties and chemical structures [35–37]. The results of the previous studies are also similar to ours [28,30–33,38]. Several studies reported a strong correlation between antioxidant activity and TPC, indicating the importance of polyphenols as a potent antioxidant component, which is emerging as a trend in numerous plant varieties [28,30–33,38]. In this study, we also observed that the higher TPC of the plant extracts resulted in higher antioxidant activity; moreover, we found that the relationship between the antioxidant capacity and phenolic compounds of the extracts was positively correlated (Table 3), accordingly signifying that phenolic compounds are major contributors to the antioxidant activity of the selected plant samples. Antioxidant molecules can neutralize the reactive free radicals and prevent the progression of chronic diseases, including diabetes, cancers, cardiovascular diseases, neurodegeneration, and inflammatory mediated diseases. Antioxidant activity from foods is normally generated from the combination of several compounds rather than a specific single compound, and hence it is difficult to relate the antioxidant activity to a specific compound [31]. Therefore, a diet supplemented with different wild plants can supply different antioxidant molecules and subsequently provides preventive measures. Thus, these wild edible plants, as a source of rich antioxidant compounds, should be brought to the attention of the general population as important health-promoting foods. Since this is the first study on the antioxidant activity of the selected plants, a detailed phytochemical analysis is an absolute necessity to isolate the active phenolic and flavonoid components.

3.5. α-Amylase Inhibitory Activity of the Selected Plants

The potential of the plant extracts to inhibit α-amylase activity was analyzed (Figure 3). The result revealed that the *Vitex negundo* leaf extract showed the highest inhibitory activity against α-amylase (70.95% inhibition) whereas the leaf extracts of *Achyranthes aspera* and *Eclipta alba* inhibited α-amylase by 64.49% and 56.16%, respectively. The lowest inhibitory activity (8.05% inhibition) was observed in the root extract of *Vitex negundo*. These plant extracts, especially the leaf extracts, showed appreciable α-amylase inhibitory effects when compared with acarbose and their root counterparts (Figure 3). Olubomehin et al. [39] also found higher α-amylase inhibition by the leaf compared to root of a traditional Nigerian plant. From Table 3, it can also be seen that the phenolic compounds exhibited significant correlation in inhibiting α-amylase activity. This result is also comparable to previous study findings using other Bangladeshi [40], Indian [41], and Egyptian wild plants [26]. On the other hand, some authors reported no significant inhibition on α-amylase activity by traditional plants [42]. Restraining or limiting the activity of α-amylase is one of the approaches in the prevention and/or management of type-2 diabetes. The inhibition of α-amylase delays carbohydrate absorption after food ingestion and thereby decreases the rate of glucose production and eventually lowers blood glucose levels [19]. Therefore, the leaves of wild plants used in this study could be used as functional food ingredients for regulating and maintaining carbohydrate metabolism and postprandial hyperglycemia.

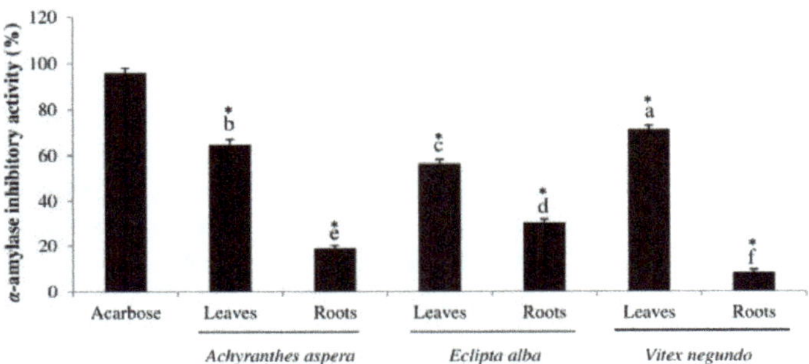

Figure 3. α-amylase inhibitory activity (%) of acarbose and the selected plant samples. Means that do not share a letter among samples are significantly different. * $p < 0.05$ compared to acarbose.

3.6. Principal Component Analysis

PCA analysis of pooled proximate variables, antioxidant activities, total phenolic content, total flavonoid content and α-amylase activity of the selected wild plants was carried out. In the PCA analysis, the first two principal components explained about 86.05% of the total variance (Figure 4): PC1 (77.09%) and PC2 (8.96%). The IC_{50} and CHO were negatively associated with PC1, whereas the loadings on PC2 specified high contributions from DPPH and Cu, with negative and positive values, respectively.

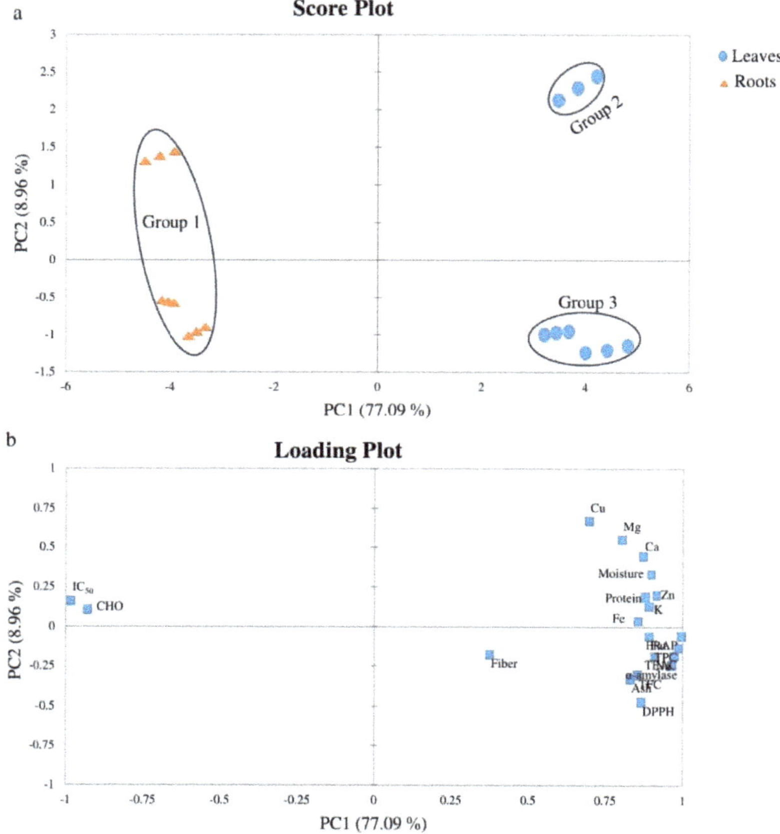

Figure 4. Score (**a**) and loading (**b**) plots of PCA analysis for the first and second components of selected plants.

Figure 4a,b represents the score and loading plots of analyzed variables of the plant samples, respectively. In Figure 4a, a clear separation between the leaves and roots of the analyzed sample was observed. Also, in Figure 4a, the formation of three groups can be seen. Group 1 consists of roots of *Vitex negundo*, *Achyranthes aspera* & *Eclipta alba*. Groups 2 & 3 consist of leaves of *Eclipta alba,* and *Achyranthes aspera* & *Vitex negundo*, respectively. IC_{50} and CHO variables contributed mostly to the separation of group 1, moisture, protein and all of the minerals except sodium for group 2, and TPC, TFC, α-amylase, antioxidant activities (DPPH, FRAP & TEAC), sodium, fat, protein, ash and fiber for group 3.

4. Conclusions

The results suggest that the leaves and roots of *Achyranthes aspera, Eclipta alba* and *Vitex negundo* are a promising source of fiber, protein, minerals, antioxidant molecules and could serve as material for dietary supplementation and functional food ingredients. Our investigation of the selected plants also provides in vitro evidence of α-amylase inhibition and justifies their use in the management of diabetes. However, the isolation of active compounds and the in vivo antidiabetic potential of these plants warrant further studies.

Author Contributions: Conceptualization, M.K.A. and M.A.; Methodology, Z.H.R., M.K.A. and M.A.; Software, Z.H.R. and M.K.A.; Validation, Z.H.R., M.K.A. and M.A.; Formal Analysis, Z.H.R., M.K.A. and M.A.; Investigation, Z.H.R. and M.K.A.; Resources, M.A.; Data Curation, Z.H.R. and M.K.A.; Writing—Original Draft Preparation, Z.H.R. and M.K.A.; Writing—Review & Editing, Z.H.R., M.K.A. and M.A.; Visualization, Z.H.R., M.K.A. and M.A.; Supervision, M.A.; Project Administration, M.K.A. and M.A.

Funding: The authors received no financial support for this research and publication of this article.

Acknowledgments: The authors would like to extend their gratitude to Taxonomist Maksuda Khatun, Department of Botany, University of Dhaka, for the identification of plant material.

Conflicts of Interest: The Authors declare that there is no conflict of interest.

References

1. Johnsen, S.P.; Overvad, K.; Stripp, C.; Tjønneland, A.; Husted, S.E.; Sørensen, H.T. Intake of fruit and vegetables and the risk of ischemic stroke in a cohort of danish men and women. *Am. J. Clin. Nutr.* **2003**, *78*, 57–64. [CrossRef] [PubMed]
2. Vauzour, D.; Vafeiadou, K.; Rendeiro, C.; Corona, G.; Spencer, J.P.E. The inhibitory effects of berry-derived flavonoids against neurodegenerative processes. *J. Berry Res.* **2010**, *1*, 45–52. [CrossRef]
3. Murimi, M.W.; Kanyi, M.G.; Mbogori, T.N.; Amin, M.R.; Rana, Z.H.; Nguyen, B.; Moyeda-Carabaza, A.F. Voices and perspectives of rural hispanic households on food insecurity in west texas: A qualitative study. *J. Hunger Environ. Nutr.* **2018**, 1–18. [CrossRef]
4. Afolayan, A.J.; Jimoh, F.O. Nutritional quality of some wild leafy vegetables in South Africa. *Int. J. Food Sci. Nutr.* **2009**, *60*, 424–431. [CrossRef] [PubMed]
5. Gupta, S.; Jyothi Lakshmi, A.; Manjunath, M.N.; Prakash, J. Analysis of nutrient and antinutrient content of underutilized green leafy vegetables. *LWT-Food Sci. Technol.* **2005**, *38*, 339–345. [CrossRef]
6. Nguyen, B.; Murimi, M.; Rana, Z.; Lee, H.; Halloran, R. Impact of a nutrition education intervention on nutrition knowledge and dietary intake of fruits, vegetables, and milk among fourth and fifth-grade elementary school children. *J. Nutr. Educ. Behav.* **2016**, *48*, S79. [CrossRef]
7. Dasgupta, N.; De, B. Antioxidant activity of some leafy vegetables of India: A comparative study. *Food Chem.* **2007**, *101*, 471–474. [CrossRef]
8. Alam, M.K.; Rana, Z.H.; Akhtaruzzaman, M. Chemical composition and fatty acid profile of Bangladeshi beef at retail. *Int. Food Res. J.* **2017**, *24*, 1897–1902.
9. Alam, M.K.; Rana, Z.H.; Akhtaruzzaman, M. Comparison of muscle and subcutaneous tissue fatty acid composition of Bangladeshi nondescript deshi bulls finished on pasture diet. *J. Chem.* **2017**, *2017*, 8579387. [CrossRef]
10. van Hecke, T.; van Camp, J.; de Smet, S. Oxidation during digestion of meat: Interactions with the diet and helicobacter pylori gastritis, and implications on human health. *Compr. Rev. Food Sci. Food Saf.* **2017**, *16*, 214–233. [CrossRef]
11. Ferguson, L.R. Chronic inflammation and mutagenesis. *Mutat. Res. Fundam. Mol. Mech. Mutagen.* **2010**, *690*, 3–11. [CrossRef] [PubMed]
12. Gülçin, I. Antioxidant activity of food constituents: An overview. *Arch. Toxicol.* **2012**, *86*, 345–391. [CrossRef] [PubMed]
13. Kumar, S.; Prasad, A.K.; Iyer, S.V.; Vaidya, S.K. Systematic pharmacognostical, phytochemical and pharmacological review on an ethno medicinal plant, *Basella alba* L. *J. Pharmacogn. Phyther.* **2013**, *5*, 53–58.

14. Aprile, A.; Negro, C.; Sabella, E.; Luvisi, A.; Nicolì, F.; Nutricati, E.; Vergine, M.; Miceli, A.; Blando, F.; De Bellis, L. Antioxidant activity and anthocyanin contents in olives (cv cellina di nardò) during ripening and after fermentation. *Antioxidants* **2019**, *8*, 138. [CrossRef] [PubMed]
15. Coelho, E.M.; de Souza, M.E.A.O.; Corrêa, L.C.; Viana, A.C.; de Azevêdo, L.C.; dos Santos Lima, M. Bioactive compounds and antioxidant activity of mango peel liqueurs (*Mangifera indica* L.) produced by different methods of maceration. *Antioxidants* **2019**, *8*, 102. [CrossRef] [PubMed]
16. Armendáriz-Fernández, K.; Herrera-Hernández, I.; Muñoz-Márquez, E.; Sánchez, E. Characterization of bioactive compounds, mineral content, and antioxidant activity in bean varieties grown with traditional methods in Oaxaca, Mexico. *Antioxidants* **2019**, *8*, 26. [CrossRef] [PubMed]
17. Nwosu, F.; Morris, J.; Lund, V.A.; Stewart, D.; Ross, H.A.; McDougall, G.J. Anti-proliferative and potential anti-diabetic effects of phenolic-rich extracts from edible marine algae. *Food Chem.* **2011**, *126*, 1006–1012. [CrossRef]
18. Mojica, L.; Meyer, A.; Berhow, M.A.; de Mejía, E.G. Bean cultivars (*Phaseolus vulgaris* L.) have similar high antioxidant capacity, in vitro inhibition of α-amylase and α-glucosidase while diverse phenolic composition and concentration. *Food Res. Int.* **2015**, *69*, 38–48. [CrossRef]
19. Hanhineva, K.; Törrönen, R.; Bondia-Pons, I.; Pekkinen, J.; Kolehmainen, M.; Mykkänen, H.; Poutanen, K. Impact of dietary polyphenols on carbohydrate metabolism. *Int. J. Mol. Sci.* **2010**, *11*, 1365–1402. [CrossRef]
20. Satter, M.M.A.; Khan, M.M.R.L.; Jabin, S.A.; Abedin, N.; Islam, M.F.; Shaha, B. Nutritional quality and safety aspects of wild vegetables consume in Bangladesh. *Asian-Pac. J. Trop. Biomed.* **2016**, *6*, 125–131. [CrossRef]
21. Ocvirk, S.; Kistler, M.; Khan, S.; Talukder, S.H.; Hauner, H. Traditional medicinal plants used for the treatment of diabetes in rural and urban areas of Dhaka, Bangladesh—An ethnobotanical survey. *J. Ethnobiol. Ethnomed.* **2013**, *9*, 43. [CrossRef] [PubMed]
22. Alam, M.; Rana, Z.; Islam, S. Comparison of the proximate composition, total carotenoids and total polyphenol content of nine orange-fleshed sweet potato varieties grown in Bangladesh. *Foods* **2016**, *5*, 64. [CrossRef] [PubMed]
23. Shajib, M.T.I.; Kawser, M.; Nuruddin Miah, M.; Begum, P.; Bhattacharjee, L.; Hossain, A.; Fomsgaard, I.S.; Islam, S.N. Nutritional composition of minor indigenous fruits: Cheapest nutritional source for the rural people of Bangladesh. *Food Chem.* **2013**, *140*, 466–470. [CrossRef] [PubMed]
24. Alam, M.K.; Rana, Z.H.; Islam, S.N.; Akhtaruzzaman, M. Total phenolic content and antioxidant activity of methanolic extract of selected wild leafy vegetables grown in Bangladesh: A cheapest source of antioxidants. *Potravin. Slovak J. Food Sci.* **2019**, *13*, 287–293. [CrossRef]
25. Miao, J.; Li, X.; Zhao, C.; Gao, X.; Wang, Y.; Gao, W. Active compounds, antioxidant activity and α-glucosidase inhibitory activity of different varieties of chaenomeles fruits. *Food Chem.* **2018**, *248*, 330–339. [CrossRef] [PubMed]
26. Hossain, S.; El-Sayed, M.; Aoshima, H. Antioxidative and anti-α-amylase activities of four wild plants consumed by pastoral nomads in Egypt. *Orient. Pharm. Exp. Med.* **2009**, *9*, 217–224. [CrossRef]
27. Seal, T. Wild edible plants of arunachal pradesh: Ethnomedicinal and nutritional importance. *Med. Plants* **2018**, *10*, 1–9. [CrossRef]
28. Saikia, P.; Deka, D.C. Mineral content of some wild green leafy vegetables of north-east India. *J. Chem. Pharm. Res.* **2013**, *5*, 117–121.
29. Adedapo, A.A.; Jimoh, F.O.; Koduru, S.; Afolayan, A.J.; Masika, P.J. Antibacterial and antioxidant properties of the methanol extracts of the leaves and stems of *Calpurnia aurea*. *BMC Complement. Altern. Med.* **2008**, *8*, 53. [CrossRef]
30. Saklani, S.; Mishra, A.; Chandra, H.; Atanassova, M.; Stankovic, M.; Sati, B.; Shariati, M.; Nigam, M.; Khan, M.; Plygun, S.; et al. Comparative evaluation of polyphenol contents and antioxidant activities between ethanol extracts of *Vitex negundo* and *Vitex trifolia* L. leaves by different methods. *Plants* **2017**, *6*, 45. [CrossRef]
31. Shahat, A.A.; Ibrahim, A.Y.; Elsaid, M.S. Polyphenolic content and antioxidant activity of some wild Saudi Arabian asteraceae plants. *Asian Pac. J. Trop. Med.* **2014**, *7*, 545–551. [CrossRef]
32. Nana, F.W.; Hilou, A.; Millogo, J.F.; Nacoulma, O.G. Phytochemical composition, antioxidant and xanthine oxidase inhibitory activities of *Amaranthus cruentus* L. and *Amaranthus hybridus* L. extracts. *Pharmaceuticals* **2012**, *5*, 613–628. [CrossRef] [PubMed]
33. Hossain, S.; Shaheen, N.; Mohiduzzaman, M.; Banu, C.P. Antioxidant capacity and total phenol content of commonly consumed selected vegetables of Bangladesh. *Malays. J. Nutr.* **2011**, *17*, 377–383.

34. Kaur, C.; Kapoor, H.C. Anti-oxidant activity and total phenolic content of some asian vegetables. *Int. J. Food Sci. Technol.* **2002**, *37*, 153–161. [CrossRef]
35. Mendoza-Wilson, A.M.; Castro-Arredondo, S.I.; Espinosa-Plascencia, A.; Del Refugio Robles-Burgueño, M.; Balandrán-Quintana, R.R.; Del Carmen Bermúdez-Almada, M. Chemical composition and antioxidant-prooxidant potential of a polyphenolic extract and a proanthocyanidin-rich fraction of apple skin. *Heliyon* **2016**, *2*. [CrossRef] [PubMed]
36. Geetha, S.; Ram, M.S.; Mongia, S.S.; Singh, V.; Ilavazhagan, G.; Sawhney, R.C. Evaluation of antioxidant activity of leaf extract of seabuckthorn (*Hippophae rhamnoides* L.) on chromium(vi) induced oxidative stress in albino rats. *J. Ethnopharmacol.* **2003**, *87*, 247–251. [CrossRef]
37. Woumbo, C.Y.; Kuate, D.; Womeni, H.M. Cooking methods affect phytochemical composition and anti-obesity potential of soybean (*Glycine max*) seeds in wistar rats. *Heliyon* **2017**, *3*. [CrossRef]
38. Petrus, A.J.A.; Kalpana, K.; Devi, A.B. Antioxidant capacity and lipophilic constitution of *Alternanthera bettzickiana* flower extract. *Orient. J. Chem.* **2014**, *30*, 491–499. [CrossRef]
39. Olubomehin, O.O.; Abo, K.A.; Ajaiyeoba, E.O. Alpha-amylase inhibitory activity of two anthocleista species and in vivo rat model anti-diabetic activities of *Anthocleista djalonensis* extracts and fractions. *J. Ethnopharmacol.* **2013**, *146*, 811–814. [CrossRef]
40. Uddin, N.; Hasan, M.R.; Hossain, M.M.; Sarker, A.; Hasan, A.H.M.N.; Islam, A.F.M.M.; Chowdhury, M.M.H.; Rana, M.S. In vitro α–amylase inhibitory activity and in vivo hypoglycemic effect of methanol extract of *Citrus macroptera* Montr. fruit. *Asian-Pac. J. Trop. Biomed.* **2014**, *4*, 473–479. [CrossRef]
41. Rao, P.S.; Mohan, G.K. In vitro alpha-amylase inhibition and in vivo antioxidant potential of *Momordica dioica* seeds in streptozotocin-induced oxidative stress in diabetic rats. *Saudi J. Biol. Sci.* **2017**, *24*, 1262–1267. [CrossRef] [PubMed]
42. Oyedemi, S.O.; Oyedemi, B.O.; Ijeh, I.I.; Ohanyerem, P.E.; Coopoosamy, R.M.; Aiyegoro, O.A. Alpha-amylase inhibition and antioxidative capacity of some antidiabetic plants used by the traditional healers in southeastern Nigeria. *Sci. World J.* **2017**, *2017*, 3592491. [CrossRef] [PubMed]

Article

Vitis vinifera L. Leaf Extract Inhibits In Vitro Mediators of Inflammation and Oxidative Stress Involved in Inflammatory-Based Skin Diseases

Enrico Sangiovanni [1,*], Chiara Di Lorenzo [1], Stefano Piazza [1], Yuri Manzoni [1], Cecilia Brunelli [1], Marco Fumagalli [1], Andrea Magnavacca [1], Giulia Martinelli [1], Francesca Colombo [1], Antonella Casiraghi [2], Gloria Melzi [1], Laura Marabini [3], Patrizia Restani [1] and Mario Dell'Agli [1,*]

1 Department of Pharmacological and Biomolecular Sciences, Università degli Studi di Milano, 20133 Milan, Italy; chiara.dilorenzo@unimi.it (C.D.L.); stefano.piazza@unimi.it (S.P.); yurimanzoni@alice.it (Y.M.); c.brunelli@biolifeitaliana.it (C.B.); marco.fumagalli3@unimi.it (M.F.); andrea.magnavacca@unimi.it (A.M.); giulia.martinelli@unimi.it (G.M.); francesca.colombo1@unimi.it (F.C.); gloria.melzi1@studenti.unimi.it (G.M.); patrizia.restani@unimi.it (P.R.)

2 Department of Pharmaceutical Sciences, Università degli Studi di Milano, 20133 Milan, Italy; antonella.casiraghi@unimi.it

3 Department Environmental Science and Policy, Università degli Studi di Milano, 20133 Milan, Italy; laura.marabini@unimi.it

* Correspondence: enrico.sangiovanni@unimi.it (E.S.); mario.dellagli@unimi.it (M.D.); Tel.: +39-02-503-183-98 (M.D.)

Received: 26 April 2019; Accepted: 13 May 2019; Published: 16 May 2019

Abstract: Psoriasis is a chronic cutaneous condition characterized by the release of pro-inflammatory mediators and oxidative stress. The reduction of these factors is currently the most effective strategy to inhibit the symptoms of pathology. Antioxidants from natural sources are increasingly used to improve skin conditions. Dried red leaves from grapevine (*Vitis vinifera* L., cv Teinturiers) showed anti-inflammatory and anti-bacterial activities, but their possible effects on keratinocytes have not been previously investigated. In this study we tested the ability of a water extract from grapevine leaves (VVWE) to inhibit inflammatory conditions in human keratinocytes (HaCaT cells), challenged with proinflammatory (tumor necrosis factor-α (TNF-α) or lipopolysaccharide (LPS)) or prooxidant (ultraviolet B radiation (UVB) or H_2O_2) mediators. VVWE inhibited interleukin-8 (IL-8) secretion induced by proinflammatory stimuli, acting on the IL-8 promoter activity, but the effect was lower when prooxidant mediators were used. The effect was partly explained by the reduction of nuclear factor-κB (NF-κB)-driven transcription and nuclear translocation. Furthermore, vascular endothelial growth factor (VEGF) secretion, a regulator of angiogenesis, was inhibited by VVWE, but not matrix metalloproteinase-9 (MMP-9), a protease involved in matrix remodeling. VVWE, assayed on Franz diffusion cell system, showed a marked reduction of High Performance Liquid Chromatography (HPLC)-identified compounds. Pure molecules individually failed to reduce TNF-α-induced IL-8 release, suggesting synergistic effects or the presence of other bioactive compounds still unknown.

Keywords: *Vitis vinifera* L.; grapevine leaves; keratinocytes; skin inflammation; oxidative stress; in vitro skin permeability; TNF-α; UVB; LPS; H_2O_2

1. Introduction

The epidermis, the outermost part of the skin, is the first barrier between our organism and the environment, and this layer is mainly constituted by keratinocytes [1]. This cell population possesses an active role in skin's defense but is also relevant to the pathogenesis of chronic inflammatory skin diseases, such as psoriasis and atopic dermatitis [2].

Psoriasis is a chronic skin disease affecting approximately the 2% of the worldwide population [3] and is characterized by inflammation, increased dermal angiogenesis, and hyperproliferation of keratinocytes. The pathology has a complex genetic inheritance [4] which causes dysregulation of the innate immune system [5], while an interplay between environmental and genetic factors is responsible for the disease-starting events, such as the release of pro-inflammatory mediators and oxidative stress [6]. Tumor necrosis factor-α (TNF-α) plays a central role in the complex cytokine network of psoriasis, as demonstrated by the clinical efficacy of monoclonal antibody therapy (anti-TNF-α) in psoriatic patients (i.e., with infliximab) [7]. In human keratinocytes, TNF-α induces the activation of pro-inflammatory mediators, including nuclear factor-κB (NF-κB) [8], which translocates from the cytoplasm into the nucleus. Ultraviolet B radiation (UVB) radiations activate the canonical NF-κB pathway [9] through the generation of reactive oxygen species (ROS), which in turn exacerbate oxidative stress [10]. Moreover, NF-κB regulates the expression of several genes involved in skin inflammatory conditions, such as interleukin-8 (IL-8) and vascular endothelial growth factor (VEGF) in different cell types, including keratinocytes [11]. IL-8 is a potent chemokine involved in the recruitment of leukocytes [12] whereas VEGF, which is increased in psoriatic lesions [13], promotes the formation of the typical psoriatic microvasculature [14].

Inhibition of VEGF has shown promising results improving symptoms of psoriasis [15], also in animal models [16]. Furthermore, TNF-α regulates extracellular matrix remodeling through matrix metalloproteinase (MMP) production in keratinocytes [17] including MMP-9 [18]. MMPs are deeply involved in cell migration, tissue remodeling, vasodilatation, and angiogenesis.

Monoclonal antibodies against TNF-α (i.e., infliximab, adalimumab, and golimumab) and circulating receptor fusion proteins (etanercept) are effective therapies for psoriasis, but the search for new strategies is mandatory due to the possible occurrence of serious side effects including lymphoma, infections, congestive heart failure, demyelinating disease, a lupus-like syndrome, induction of auto-antibodies, injection site reactions, and systemic adverse effects [19].

The reduction of cytokines and growth factors is currently the most effective strategy to inhibit symptoms of psoriasis, and different keratinocyte-based assays, using monolayer cultures, are widely employed to assess the effects of pharmacological treatments on proliferation and inflammation [20].

In this contest, antioxidants from natural source, including botanicals, are increasingly used to improve skin inflammatory and oxidative stress conditions [21]. In the last few years, a variety of botanicals have been tested for their antioxidant and anti-inflammatory activities, mostly due to the high content of polyphenols [22–25]. However, the support for the topical use in psoriasis is limited by the number of studies available in the literature [26].

Grapevine (*Vitis vinifera* L.) is a plant belonging to the genus of Vitaceae, originating in the Mediterranean area. Dried red leaves from the cultivar Teinturiers should contain at least 4.0 percent anthocyanins, expressed as cyanidin-3-*O*-glucoside according to the European Scientific Cooperative on Phytotherapy (ESCOP) Monograph [27]. Grapevine leaves contain a variety of phytoconstituents showing high antioxidant activity including condensed tannins, phenols, and anthocyanins. Moreover, grapevine leaves show other biological properties including anti-inflammatory, anti-bacterial, and vasorelaxant effects [28,29]. We have previously demonstrated that the water extract from *Vitis vinifera* leaves (VVWE) shows anti-inflammatory activity at the gastrointestinal level, acting on the NF-κB pathway [30]. Grapevine leaves are efficiently used in the treatment of varicose veins [31] as formulations for oral or topical use. However, their possible anti-inflammatory effects in keratinocytes have not been previously investigated.

The aim of the present study was to investigate the ability elicited by a water extract from grapevine leaves to inhibit inflammatory conditions induced by mediators of inflammation or oxidative stress in human keratinocytes. To reach this goal, cultured human keratinocytes (HaCaT) cells were used as reliable model of human keratinocytes, and the effect of VVWE on several markers of skin diseases was evaluated following activation with proinflammatory (TNF-α or lipopolysaccharide (LPS)) or prooxidant (UVB or H_2O_2) mediators.

2. Materials and Methods

2.1. Materials

HaCaT cells were purchased from Cell Line Service GmbH (Eppelheim, Germany). Dulbecco's Modified Eagle Medium (DMEM), 3-(4,5-dimethylthiazol-2-yl)-2,5-diphenyltetrazolium bromide (MTT), and 3,3′,5,5′-tetramethylbenzidine (TMB) were from Sigma Aldrich (Milan, Italy). Penicillin, streptomycin, L-glutamine, sodium pyruvate, trypsin-EDTA, and Lipofectamine® 2000 were from Life Technologies Italia (Monza, Italy).

Human TNF-α, the Human VEGF Elisa Development Kit, and the Human IL-8 Elisa Development Kit were from Peprotech Inc. (London, UK). Fetal bovine serum (FBS), and disposable materials for cell culture were purchased by Euroclone (Euroclone S.p.A., Pero-Milan, Italy). The plasmid NF-κB-LUC containing the luciferase gene under the control of three κB sites was a gift of Nikolaus Marx (Department of Internal Medicine-Cardiology, University of Ulm, Ulm, Germany). The native IL-8-LUC promoter was kindly provided by Takaaki Shimohata (Department of Preventive Environment and Nutrition, University of Tokushima Graduate School, Tokushima, Japan). The promoter contains sequences responsive to several transcription factors such as activator protein 1 (AP-1), CCAAT-enhancer-binding protein-β (C/EBPβ), and NF-κB. Britelite^{TM} plus was from Perkin Elmer (Monza, Italy).

2.2. Plant Material and Preparation of the Water Extract (VVWE)

Dried and powdered leaves from *Vitis vinifera* L. cv. Teinturiers were kindly donated by PhytoLab Company (Vestenbergsgreuth, Germany). Plant material was maintained at 4 °C until extraction. VVWE was prepared according to ESCOP monographs [27]. The plant material was extracted twice, at room temperature, in the dark, with deionized water in the ratio herb/water 10 g/100 mL, and lyophilized. The recovery (*w/w*) was 26% calculated on the dried drug weight. Samples were then stored at −20 °C until the assays. Before biological evaluations, the extract was dissolved in sterilized distilled water at a concentration of 50 mg/mL, and immediately stored in aliquots at −20 °C.

2.3. Cell Culture

HaCaT cells were grown at 37 °C in DMEM supplemented with 100 units penicillin per mL, 100 mg streptomycin per mL, 2 mM L-glutamine, and 10% heat-inactivated FBS, under a humidified atmosphere containing 5% CO_2.

2.4. Characterization of Grapevine Extract by High Performance Liquid Chromatography (HPLC)

VVWE was characterized by a validated HPLC-DAD method as previously described [26]. Briefly, two different HPLC methods coupled with a Diode Array Detector (DAD) have been used, the first for anthocyanin detection at 520 nm by the use of a Synergi 4u MAX-RP 80 Å column (250 × 4.60 mm × 4 μm) (Phenomenex, Torrance, CA, USA) and the second for flavonols and caffeic acid derivative detection at 360 nm by Synergi 4u MAX-RP 80 Å column (250 × 2 mm × 4 μm) (Phenomenex, Torrance, CA, USA). The anthocyanins standards, hyperoside and kaempferol-3-*O*-glucoside, were from Extrasynthese (Genay, France); all the other standards and solvents were bought from Sigma-Aldrich (St. Louis, MO, USA).

2.5. IL-8 Release

Cells were grown in 24-well plates for 48 h (30,000 cells per well) before the cytokine treatment. IL-8 was quantified using a Human Interleukin-8 ELISA Development Kit as previously described [30,32]. Preliminary time-course experiments were performed to set the best conditions for further experiments. HaCaT were treated with TNF-α (10 ng/mL), LPS (5 μg/mL), or H_2O_2 (100 μM), for 3, 6, 24, and 30 h. The IL-8 secretion induced by TNF-α or LPS was tested after 6 h of treatment, while the 24-h time point was chosen for H_2O_2. Epigallocatechin-3-gallate (EGCG) (20 μM) was used as the reference inhibitor of IL-8 secretion.

2.6. Transient Transfection Assays

HaCaT cells were plated in 24-well plates and transiently transfected with the plasmid NF-κB-LUC or native IL-8-LUC, both at 250 ng per well, using Lipofectamine® 2000, according to previous studies [32]. Sixteen hours later, the cells were treated for 6 h with increasing concentrations of VVWE in the presence of the pro-inflammatory mediators (TNF-α at 10 ng/mL or LPS at 5 μg/mL). After six hours, cells were harvested and the luciferase assay was performed using the BriteliteTM Plus reagent (PerkinElmer Inc., Massachusetts, USA) according to the manufacturer's instructions. Data were expressed considering 100% of the luciferase activity related to the cytokine-induced promoter activity.

2.7. NF-κB Nuclear Translocation

For the evaluation of the NF-κB (p65) translocation, HaCaT were plated at the concentration of 3×10^6 cells/mL in 100-mm plates with fresh complete medium. After 48 h, cells were treated for 1 h with the inflammatory mediators (TNF-α at 10 ng/mL or LPS at 5 μg/mL) and the extract (10–50 μg/mL) under study, using FBS-free medium. Nuclear extracts were prepared using the Nuclear Extraction Kit from the Cayman Chemical Company (Ann Arbor, MI, USA) as previously described [32].

Data were expressed considering 100% of the absorbance related to the cytokine-induced NF-κB nuclear translocation. EGCG (20 μM) was used as the reference inhibitor of the NF-κB nuclear translocation. The results are the mean ± SD of three experiments in triplicate.

2.8. UVB Radiation

HaCaT cells were grown in 24-well plates for 48 h (30,000 cells per well), washed with phosphate-buffered saline (PBS), and exposed to UVB (40 mJ/cm^2) light source (Triwood 31/36, W36, V230, Helios Italquartz, Milano, Italy) in a glass bath. Radiation time (about 50 s) was adjusted for each experimental day, measuring energy emission by the LP 471 UVB probe (Delta OHM, Padova, Italy). After radiation, serum free fresh culture medium was immediately added. For the evaluation of IL-8 release, cells were treated for additional 9 h, a time point selected in preliminary time-course experiments. For NF-κB (p65) translocation, HaCaT cells were treated for 1 h after UVB radiation. EGCG (20 μM) was used as reference inhibitor.

2.9. VEGF Release

Cells were grown in 24-well plates for 48 h (30,000 cells per well) before challenge with the pro-inflammatory mediator. VEGF was quantified by the Human VEGF ELISA Development Kit. Briefly, Corning 96-well EIA/RIA plates (Sigma-Aldrich, Milan, Italy) were coated with the antibody provided in the ELISA Kit (PeproTech Inc., London, UK) overnight at 4 °C. After blocking the reaction, 300 μL of samples were transferred into wells at room temperature for 2 h. The amount of VEGF in the samples was detected by spectroscopy (signal read: 450 nm, 0.1 s) as described above. The quantitative measurement of VEGF was done using an optimized standard curve supplied with the ELISA set (100–2000 pg/mL). The maximal release of VEGF was observed at 24 h for TNF-α (10 ng/mL) and H_2O_2 (100 μM) or 30 h for LPS (5 μg/mL). EGCG (20 μM) was used as the reference inhibitor of VEGF secretion.

2.10. MMP-9 Release

MMP-9 secretion was performed on HaCaT cells treated with TNF-α (10 ng/mL) and (LPS 5 μg/mL) for 3, 6, 24, and 48 h. Cells were grown in 24-well plates (30,000 cell/well) for 48 h, before the treatment. Human MMP-9 ELISA Kit from RayBio® (Norcross GA, USA) was used to quantify MMP-9 secretion according to manufacturer's instructions. 300 μL of samples in duplicate were transferred into a

96-well plate coated with anti-human MMP-9 and incubated overnight at 4 °C with gentle shaking. MMP-9 secreted was detected by the use of biotinylated and streptavidin–horseradish peroxidase (HRP) conjugate antibodies as previously described. The quantitative measurement of MMP-9 was done using an optimized standard curve supplied with the ELISA kit (8.23–6000 pg/mL). The MMP-9 secretion reached the maximum at 24 h for both TNF-α and LPS. VVWE was tested at 10–50 μg/mL in the presence of stimuli and between 2.5–200 μg/mL without the pro-inflammatory mediators.

2.11. Cytotoxicity Assay

The cell morphology before and after treatment was assessed by light microscope inspection. Cell viability was assessed by the MTT test and verified by lactate dehydrogenase (LDH) assay. No sign of cytotoxicity was observed in cells treated with VVWE at 5–500 μg/mL for 6 h.

2.12. In Vitro Skin Permeation Study

The human epidermis membrane used for in vitro permeation studies was obtained from the abdominal skin of a single donor. The epidermis was prepared according to the heat separation method, as previously reported [33].

Ex vivo skin permeation study: the study was performed using the Franz diffusion cell method. The human epidermis was mounted carefully on the receiver compartment of the Franz's cell with the stratum corneum side in contact with donor solution. The receiver compartment was filled with freshly prepared degassed HCl 0.1 M solution (receiver phase). VVWE (150 mg) solubilized in 0.1 M HCl:EtOH (50:50, *v*/*v*) was loaded in the donor compartment (0.5 mL). At predetermined intervals (1, 5, 7, 24, 32, 48 h), 0.2-mL samples were removed from the receiver compartment and immediately replaced with fresh receiver phase. Sink conditions were maintained throughout the experiment. The samples were assayed by HPLC analysis. Four parallel experiments were performed.

2.13. Statistical Analysis

All data are the mean ± SD of at least three experiments performed in duplicate (ELISA assays) or triplicate (transfection assays). Data were analyzed by unpaired one-way analysis of variance (ANOVA), or two-way analysis of variance (ANOVA) followed by Bonferroni's post hoc test. Statistical analyses were performed using GraphPad Prism 5.02 software (GraphPad Software Inc., San Diego, CA, USA). * $p < 0.05$ was considered statistically significant. The half maximal inhibitory concentration (IC_{50}) was calculated using GraphPad Prism 5.02.

3. Results and Discussion

3.1. Characterization of VVWE

The HPLC analysis of VVWE identified the following compounds: five flavonols (quercetin 3-*O*-glucoside, quercetin 3-*O*-glucuronide, kaempferol 3-*O*-glucoside, hyperoside, and rutin), two anthocyanosides (delphinidin 3-*O*-glucoside and cyanidin 3-*O*-glucoside), and caftaric acid (Table 1). Quercetin-3-*O*-glucuronide, quercetin-3-*O*-glucoside, and caftaric acid were, in order, the most abundant phenols in the extract; in particular, the quercetin-3-*O*-glucuronide value was 29.14 ± 1.92 mg/g and the quercetin-3-*O*-glucoside value was 21.68 ± 0.91 mg/g (mean ± SD).

Table 1. Data are expressed as mg([a]) or µg([b]) of pure compound per g water extract from *Vitis vinifera* leaves (VVWE) (mean ± SD). Recovery percentage (% recovery) was calculated on the availability in weight of pure compounds after permeation of 150 mg of VVWE in four replicates.

VVWE Composition: Identified Compounds	Donor Solution Content mg/g [a] ± SD	Permeated Amount µg/g [b] ± SD (% Recovery)	Retained Amount in the Epidermidis µg/g [b] ± SD (% Recovery)
Caftaric acid	9.99 ± 0.35	47.59 ± 27.64 (0.476%)	10.20 ± 9.00 (0.102%)
Rutin	1.31 ± 0.05	0.30 ± 0.35 (0.023%)	0.80 ± 0.91 (0.061%)
Hyperoside	2.30 ± 0.17	0.47 ± 0.68 (0.020%)	1.34 ± 1.43 (0.058%)
Quercetin 3-*O*-glucoside	21.68 ± 0.91	8.74 ± 9.50 (0.040%)	13.58 ± 15.88 (0.063%)
Quercetin 3-*O*-glucuronide	29.14 ± 1.92	51.66 ± 46.10 (0.177%)	28.08 ± 30.33 (0.096%)
Kaempferol 3-*O*-glucoside	3.77 ± 0.06	2.09 ± 1.70 (0.055%)	3.53 ± 3.90 (0.094%)
Delphinidin 3-*O*-glucoside	0.95 ± 0.03	0.48 ± 0.68 (0.050%)	0.04 ± 0.08 (0.004%)
Cyanidin 3-*O*-glucoside	2.29 ± 0.04	1.63 ± 2.08 (0.071%)	0.30 ± 0.60 (0.013%)
Petunidin 3-*O*-glucoside	0.66 ± 0.05	0.38 ± 0.53 (0.058%)	0.10 ± 0.20 (0.015%)
Peonidin 3-*O*-glucoside	1.91 ± 0.06	2.17 ± 2.72 (0.114%)	0.58 ± 1.17 (0.030%)
Malvidin 3-*O*-glucoside	1.27 ± 0.07	0.92 ± 1.22 (0.072%)	0.29 ± 0.58 (0.023%)

3.2. VVWE Reduces IL-8 Release and Promoter Activity Induced by Pro-Inflammatory Mediators

Skin inflammatory diseases are characterized by over-expression of a multitude of pro-inflammatory mediators which impact on the cells occurring in the epidermis, mostly keratinocytes.

IL-8 is one of the main chemokines released by keratinocytes during inflammatory processes, which can in turn recruit leukocytes at the site of inflammation. In the following experiments the ability of VVWE to inhibit IL-8 secretion in HaCaT cells, an immortalized cell line widely used as a model of human keratinocytes, overcoming the potential challenge of donor variation, was tested.

To test the effect of the extract on IL-8 release induced by different inflammatory conditions, cells were challenged with the endogenous stimulus TNF-α or with pure lipopolysaccharide (LPS) from *Escherichia coli* (*E. coli*), which mimics bacterial inflammation.

The extract was able to reduce TNF-α or LPS-induced IL-8 release in a concentration dependent fashion (Figure 1A,B). IL-8 secretion was more pronounced when TNF-α was used as stimulus compared to LPS; in both cases VVWE (50 µg/mL) reduced the IL-8 secretion close to the basal level (IC_{50} 2.60 and 14.04 µg/mL, respectively, for IL-8 induced by TNF-α or LPS).

To clarify if the effect of VVWE on the IL-8 release could be due to impairment of the corresponding promoter activity, HaCaT cells were transfected by IL-8-LUC plasmid as described in the materials and method section.

The extract was able to reduce IL-8 promoter activity although the effect was less pronounced when compared to the ability to inhibit IL-8 release induced by pro-inflammatory stimuli (Figure 2A,B). The IC_{50} was 22.73 µg/mL on the TNF-α-induced IL-8 promoter activity, whereas IC_{50} was 33.98 µg/mL on the LPS-induced IL-8 promoter activity.

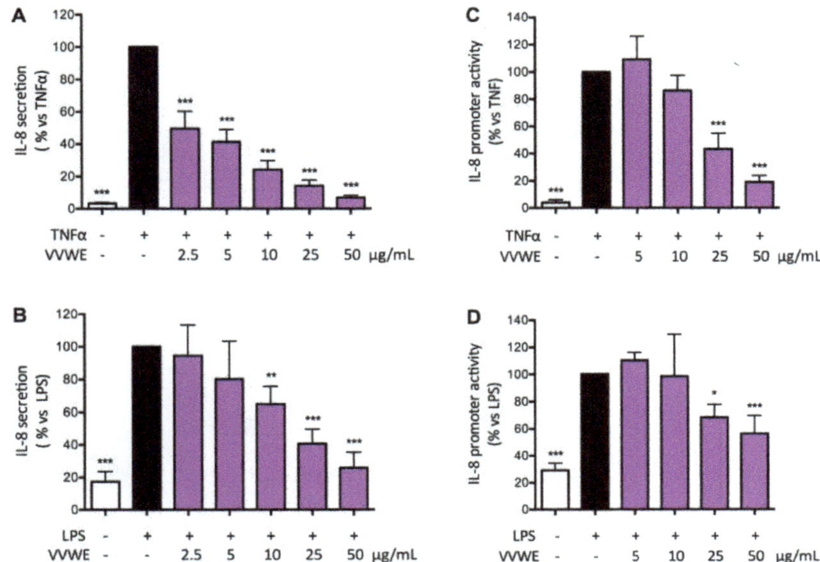

Figure 1. Effect of water extract from *Vitis vinifera* leaves (VVWE) on the tumor necrosis factor-α (TNF-α)-induced (**A**) or lipopolysaccharide (LPS)-induced (**B**) interleukin-8 (IL-8) secretion. HaCaT cells were treated for 6 h with TNF-α (10 ng/mL) or LPS (5 μg/mL) and VVWE (2.5–50 μg/mL). IL-8 secretion was evaluated by ELISA assay. Basal (without pro-inflammatory stimuli) and control (with TNF-α/LPS) levels of IL-8 were 17.4, 560.1 and 108.7 pg/mL, respectively. The effect of VVWE was evaluated on IL-8 promoter activity induced by TNF-α (**C**) and LPS (**D**). HaCaT cells were treated for 6 h with TNF-α (10 ng/mL) or LPS (5 μg/mL) and VVWE (5–50 μg/mL). IL-8 promoter activity was measured in transfected HaCaT cells by the luciferase assay. The graphs show the means ± SD of at least three experiments performed in triplicate. Statistical analysis: one-way ANOVA, followed by Bonferroni's post hoc test. * $p < 0.05$, ** $p < 0.01$, *** $p < 0.001$ vs. TNF-α alone. Epigallocatechin-3-gallate (EGCG) (20 μM) was used as the reference inhibitor of TNF-α or LPS-induced IL-8 secretion (62.1% and 84.9% inhibition, respectively) and IL-8 promoter activity (35.9% and 33.3% inhibition, respectively); the effect of the reference inhibitor is in agreement with that reported in the scientific literature.

The effect of VVWE on IL-8 secretion induced by H_2O_2, as a pro-oxidant, was also evaluated. H_2O_2 doubled the amount of IL-8 released by HaCaT cells, but VVWE reduced the chemokine release just at the highest concentration (50 μg/mL, data not shown) thus suggesting that the effect of the extract is higher when the chemokine is released by pro-inflammatory mediators.

3.3. VVWE Impairs the NF-κB Pathway Acting on Transcription and Nuclear Translocation

NF-κB represents a key factor in a variety of skin inflammatory conditions including psoriasis [34], and TNF-α strongly induces activation of the NF-κB pathway. NF-κB driven transcription and nuclear translocation were assessed to better clarify the involvement of this transcription factor in the mode of action elicited by the extract. Cells were transiently transfected by NF-κB-LUC plasmid and treated as previously described. TNF-α approximately doubled the NF-κB driven transcription, while LPS showed a slight minor effect. VVWE had an inhibition trend on the NF-κB driven transcription, but the effect was statistically significant only when cells were challenged with TNF-α as stimulus (Figure 3A,B).

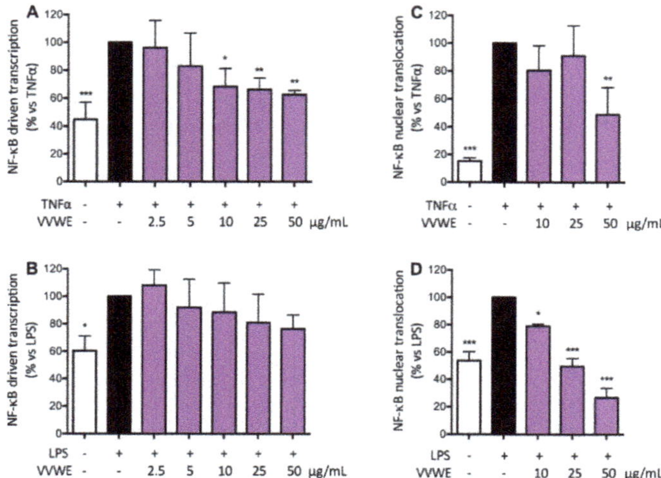

Figure 2. Effect of VVWE on the nuclear factor-κB (NF-κB)-driven transcription induced by TNF-α (**A**) or LPS (**B**). HaCaT cells were treated 6 h with TNF-α (10 ng/mL) or LPS (5 µg/mL) and VVWE. NF-κB nuclear translocation was evaluated in HaCaT cells treated for 1 h with TNF-α (**C**) or LPS (**D**) and VVWE (10–50 µg/mL). The graphs show the means ± SD of at least three experiments performed in triplicate and duplicate, for transcription and translocation respectively. Statistical analysis: one-way ANOVA, followed by Bonferroni's post hoc test. * $p < 0.05$, ** $p < 0.01$, *** $p < 0.001$ vs. TNF-α or LPS alone. Here, 20 µM EGCG were used as the reference inhibitor of TNF-α or LPS-induced NF-κB-driven transcription (59.9 and 29.9% inhibition, respectively) and NF-κB nuclear translocation (62.4% and 53.2% inhibition, respectively).

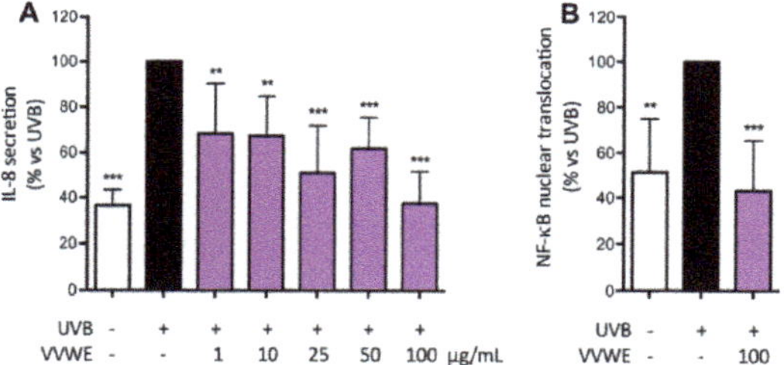

Figure 3. Effect of VVWE on IL-8 secretion (**A**) or NF-κB nuclear translocation (**B**) induced by ultraviolet B (UVB) irradiation. HaCaT cells were treated for 50 s with UVB (40 mJ/cm^2) and followed by VVWE for 9 h (IL-8 secretion) or 1 h (NF-κB nuclear translocation). The graphs show the means ± SD of at least three experiments performed in triplicate and duplicate, for IL-8 secretion and translocation, respectively. Statistical analysis: one-way ANOVA, followed by Bonferroni's post hoc test. ** $p < 0.01$, *** $p < 0.001$ vs. UVB alone. Here, 20 µM EGCG were used as the reference inhibitor for UVB-induced IL-8 secretion (61.7% inhibition) and NF-κB nuclear translocation (39.1% inhibition).

VVWE was also able to impair the NF-κB nuclear translocation induced by TNF-α, with a reduction of 50% at 50 µg/mL (Figure 4A). LPS showed weaker induction of translocation compared to the cytokine, whereas VVWE completely abolished LPS-induced nuclear translocation at 25 µg/mL (Figure 2D). Furthermore, VVWE (50 µg/mL) reduced the LPS-induced NF-κB nuclear translocation below the unstimulated control level by 50%.

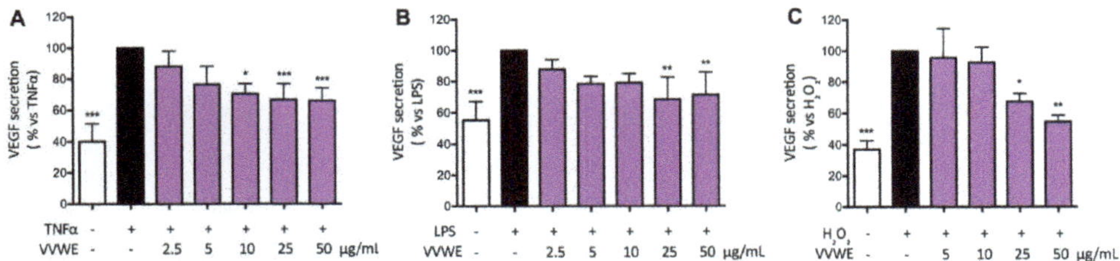

Figure 4. Effect of VVWE on vascular endothelial growth factor (VEGF) secretion induced by TNF-α (**A**), LPS (**B**), and H_2O_2 (**C**). HaCaT cells were treated for 24 h with TNF-α (10 ng/mL), H_2O_2 (100 μM) or 30 h with LPS (5 μg/mL) and VVWE (2.5–50 μg/mL). Secreted VEGF was evaluated by the ELISA assay. The graphs show the means ± SD of at least three experiments performed in triplicate. Statistical analysis: one-way ANOVA, followed by Bonferroni's post hoc test. * $p < 0.05$, ** $p < 0.01$, *** $p < 0.001$ vs. TNF-α alone. Here, 20 μM EGCG were used as the reference inhibitor of TNF-α, LPS, or H_2O_2-induced VEGF secretion (78.7%, 100%, and 100% inhibition, respectively).

Our groups previously demonstrated that VVWE is able to inhibit the NF-κB pathway through the impairment of the translocation from the cytoplasm into the nucleus in human gastric epithelial cells [30]. However, this is the first study reporting the effect of grapevine leaves as inhibitors of the NF-κB driven transcription and nuclear translocation induced by different pro-inflammatory stimuli.

3.4. VVWE Reduces IL-8 Release and NF-κB Nuclear Translocation Induced by UV-B Radiations

HaCaT cells and primary human keratinocytes display distinct keratinocyte morphology and undergo UVB-induced apoptosis [35]. UV-B induces oxidative stress in keratinocytes through the generation of reactive oxygen species (ROS) and activates several inflammatory pathways such as the mitogen-activated protein kinase (MAPK), NF-κB, and Janus kinase (JAK)/signal transduction and activation of transcription (STAT) signaling [10]. Keratinocytes are the major target of UV-B radiation and their response is predominantly regulated by the NF-κB.

VVWE reduced the UV-B induced IL-8 secretion (Figure 3A) at basal level at 100 μg/mL (IC_{50} 2.42 μg/mL) and the effect of the highest concentration paralleled the effect on the UV-B induced nuclear translocation (Figure 3B).

3.5. Effects of VVWE on VEGF and MMP-9 Release

VVWE was assayed on the ability to influence the release of markers widely involved in skin pathological processes including inflammatory-based conditions and wound injury. In particular, VEGF is a key regulator of the angiogenesis process whereas MMP9 is involved in the extracellular matrix remodeling. TNF-α and LPS increased the amount of VEGF approximately to 200% compared to the basal control level, and VVWE reduced the secretion significantly starting from 10 or 25 μg/mL (Figure 4A,B, respectively). H_2O_2 was able to induce VEGF release and the extent was comparable to that caused by TNF-α; the extract counteracted the release in a concentration-dependent fashion, with IC_{50} of 27.26 μg/mL (Figure 4C). The inhibition of MMP-9 secretion was evaluated in HaCaT cells by VVWE. Treatment for 24 h with TNF-α or LPS induced a 7.3- or 1.6-fold increase of MMP-9, respectively. VVWE (10–50 μg/mL) was not able to inhibit TNF-α-induced MMP-9 release (Figure 5A); surprisingly, VVWE showed further induction of MMP-9 released by LPS, with a three-fold at 50 μg/mL with respect to the unstimulated control (Figure 5B).

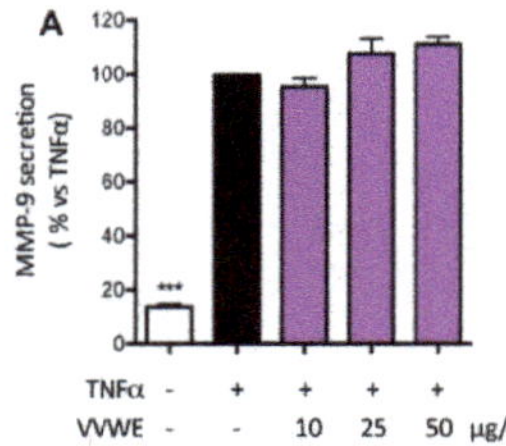

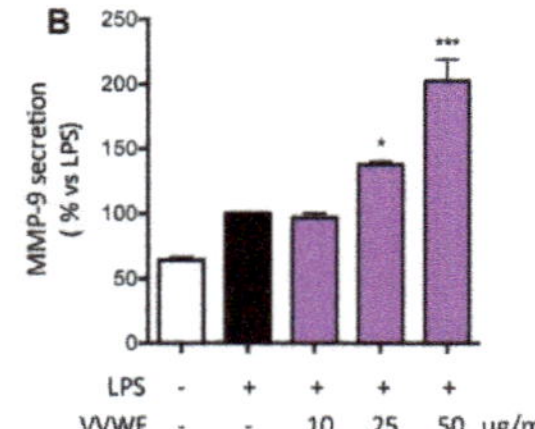

Figure 5. Effect of VVWE on MMP-9 secretion induced by TNF-α (**A**) or LPS (**B**). HaCaT cells were treated for 24 h with TNF-α (10 ng/mL) or 30 h with LPS (5 μg/mL) and VVWE (10–50 μg/mL). Matrix metalloproteinase-9 (MMP-9) release was evaluated by the ELISA assay. The graphs show the means ± SD of at least three experiments performed in triplicate. Statistical analysis: one-way ANOVA, followed by Bonferroni's post hoc test. * $p < 0.05$, *** $p < 0.001$ vs. TNF-α alone. EGCG (20 μM) was used as the reference compound of TNF-α or LPS-induced MMP-9 secretion (38.0% and no inhibition, respectively).

In contrast, administration of VVWE at 50 μg/mL in the absence of pro-inflammatory stimuli induced an increase of MMP-9 secretion of 1.4-fold compared to control, suggesting a synergistic effect with LPS. VVWE (200 μg/mL), without other pro-inflammatory stimuli, induced the release of MMP-9 in a concentration-dependent manner up to 6.2-fold with respect to control (data not shown).

3.6. Franz Diffusion Cell Method

In order to evaluate the ability of the extract components to permeate the epidermis and therefore their possible bioavailability, VVWE was assayed on Franz diffusion cells over a period of 48h. Among all compounds identified in the extract only a limited part of them pass through the epidermis (Table 1) during the permeation process. In very few cases and in small amounts these compounds were detected before 48 h in the receiver phase.

Starting from 150 mg of VVWE, the most abundant compounds able to cross the skin barrier were in the following order: quercetin 3-*O*-glucuronide (51.66 μg/g), caftaric acid (47.59 μg/g), and quercetin 3-*O*-glucoside (8.74 μg/g) (Table 1). The analysis of the compounds retained into the skin's portion revealed that the most abundant were quercetin 3-*O*-glucuronide (28.08 μg/g) > quercetin 3-*O*-glucoside (13.58 μg/g) > caftaric acid (10.20 μg/g). Being an aqueous extract, VVWE contains hydrophilic compounds that are less suitable for skin permeation. Ethanol, as a permeation enhancer, was added to improve solubility and to disorder skin lipids [36]. Antioxidants detected in the receiver phase and into the epidermis have log$p$ values below 1.15 (caftaric acid) and 1.21 (kaempferol 3-*O*-glucoside; from predicted properties, SciFinder). In the case of 3-*O*-glucoside derivatives, the predicted properties were not found other than for kaempferol and quercetin. Hyperoside and quercetin 3-*O*-glucoside are highly hydrophilic (log$p$ value < 0). Detection of the latter and quercetin 3-*O*-glucuronide (log$p$ = 0.62) is probably due to their abundance in the donor phase. To further improve these results, a proper vehicle should be tested as already reported in case of quercetin [37].

3.7. Effects of Pure Compounds on IL-8 Secretion

The most abundant compounds of VVWE that were able to cross the skin barrier (caftaric acid, quercetin-3-*O*-glucoside, and quercetin-3-*O*-glucuronide) were tested separately with respect to their ability to reduce IL-8 secretion in HaCaT cells. Caftaric acid, quercetin 3-*O*-glucoside, and quercetin 3-*O*-glucuronide (0.1–100 μM) failed to reduce IL-8 release, induced by TNF-α, up to the maximum concentration tested (100 μM), thus suggesting that the effect of the extract could be due to synergistic effects occurring among the constituents or to the presence of other bioactive compounds still unknown.

4. Conclusions

This is the first study showing grapevine leaves as inhibitors of the NF-κB pathway at the cutaneous level. VVWE reduced two typical markers of psoriatic lesions, IL-8 and VEGF. The effect of the extract was higher when pro-inflammatory stimuli were used; however it also showed anti-oxidant mechanisms of action against H_2O_2 and UVB radiation. In parallel, VVWE did not inhibit MMP-9 release, potentially promoting tissue remodeling while reducing other inflammatory markers. Analytical studies showed that diffusion of polar compounds through the skin layer is markedly reduced, but still able to reach potential bioactive concentrations. Unfortunately, the evaluation of individual compounds did not identify bioactive components, and further studies are required. Taken together, our results seem to suggest the possible use of grapevine leaves as anti-inflammatory agents for skin inflammatory conditions.

Author Contributions: E.S., C.D.L., A.C., L.M., P.R., and M.D. conceived and designed the experiments; S.P., Y.M., C.B., M.F., A.M., G.M. (Giulia Martinelli), F.C., and G.M. (Gloria Melzi) performed the experiments; E.S., S.P., C.D.L., and A.C. analyzed the data; and E.S. and M.D. wrote the paper.

Funding: The work was partially supported by Fondazione Cariplo (HyWoNNa project grant, 2015-0550 to Mario Dell'Agli). This research was supported by grants from MIUR "Progetto Eccellenza".

Acknowledgments: The authors thank Petra Boukamp and Norbert Fusenig from Deutsches Krebsforschungszentrum, Stiftung des öffentlichen Rechts (German Cancer Research Center) and Im Neuenheimer Feld 280, D-69120 Heidelberg, Germany, for providing the HaCaT cell line. Takaaki Shimohata and Akira Takahashi, Departments of Preventive Environment and Nutrition, University of Tokushima Graduate School, Japan, provided the plasmids carrying the native and mutated IL-8 promoters.

Conflicts of Interest: The authors declare no conflict of interest.

References

1. Nestle, F.O.; Di Meglio, P.; Qin, J.Z.; Nickoloff, B.J. Skin immune sentinels in health and disease. *Nat. Rev. Immunol.* **2009**, *9*, 679–691. [CrossRef]
2. de Bruin Weller, M.S.; Knulst, A.C.; Meijer, Y.; Bruijnzeel-Koomen, C.A.; Pasmans, S.G. Evaluation of the child with atopic dermatitis. *Clin. Exp. Allergy* **2012**, *42*, 352–362. [CrossRef]
3. Christophers, E. Psoriasis—Epidemiology and clinical spectrum. *Clin. Exp. Dermatol.* **2001**, *26*, 314–320. [CrossRef]
4. Farber, E.M.; Nall, M.L. The natural history of psoriasis in 5600 patients. *Dermatologica* **1974**, *148*, 1–18. [CrossRef] [PubMed]
5. Nickoloff, B.J. Skin innate immune system in psoriasis: Friend or foe? *J. Clin. Investig.* **1999**, *104*, 1161–1164. [CrossRef] [PubMed]
6. Bacchetti, T.; Campanati, A.; Ferretti, G.; Simonetti, O.; Liberati, G.; Offidani, A.M. Oxidative stress and psoriasis: The effect of antitumour necrosis factor-alpha inhibitor treatment. *Br. J. Dermatol.* **2013**, *168*, 984–989. [CrossRef] [PubMed]
7. Reich, K.; Nestle, F.O.; Papp, K.; Ortonne, J.P.; Evans, R.; Guzzo, C.; Li, S.; Dooley, L.T.; Griffiths, C.E. EXPRESS Study Investigators. Infliximab induction and maintenance therapy for moderate-to-severe psoriasis: A phase III, multicentre, double-blind trial. *Lancet* **2005**, *366*, 1367–1374. [CrossRef]
8. Wullaert, A.; Bonnet, M.C.; Pasparakis, M. Nf-kappab in the regulation of epithelial homeostasis and inflammation. *Cell Res.* **2011**, *21*, 146–158. [CrossRef]
9. Wang, W.; Mani, A.M.; Wu, Z.H. DNA damage-induced nuclear factor-kappa B activation and its roles in cancer progression. *J. Cancer Metastasis Treat.* **2017**, *3*, 45–59. [CrossRef]
10. Radhiga, T.; Agilan, B.; Muzaffer, U.; Karthikeyan, R.; Kanimozhi, G.; Paul, V.I.; Prasad, N.R. Phytochemicals as modulators of ultraviolet-b radiation induced cellular and molecular events: A review. *J. Radiat. Cancer Res.* **2016**, *7*, 2–12.
11. Cho, J.W.; Lee, K.S.; Kim, C.W. Curcumin attenuates the expression of IL-1beta, IL-6, and TNF-alpha as well as cyclin e in tnf-alpha-treated HaCat cells; NF-kappaB and MAPKs as potential upstream targets. *Int. J. Mol. Med.* **2007**, *19*, 469–474.

12. Colombo, I.; Sangiovanni, E.; Maggio, R.; Mattozzi, C.; Zava, S.; Corbett, Y.; Fumagalli, M.; Carlino, C.; Corsetto, P.A.; Scaccabarozzi, D.; et al. HaCat cells as a reliable in vitro differentiation model to dissect the inflammatory/repair response of human keratinocytes. *Mediat. Inflamm.* **2017**, *2017*, 7435621. [CrossRef] [PubMed]
13. Detmar, M.; Brown, L.F.; Claffey, K.P.; Yeo, K.T.; Kocher, O.; Jackman, R.W.; Berse, B.; Dvorak, H.F. Overexpression of vascular permeability factor/vascular endothelial growth factor and its receptors in psoriasis. *J. Exp. Med.* **1994**, *180*, 1141–1146. [CrossRef]
14. Nestle, F.O.; Kaplan, D.H.; Barker, J. Psoriasis. *N. Engl. J. Med.* **2009**, *361*, 496–509. [CrossRef]
15. Sauder, D.N.; Dekoven, J.; Champagne, P.; Croteau, D.; Dupont, E. Neovastat (AE-941), an inhibitor of angiogenesis: Randomized phase I/II clinical trial results in patients with plaque psoriasis. *J. Am. Acad. Dermatol.* **2002**, *47*, 535–541. [CrossRef]
16. Halin, C.; Fahrngruber, H.; Meingassner, J.G.; Bold, G.; Littlewood-Evans, A.; Stuetz, A.; Detmar, M. Inhibition of chronic and acute skin inflammation by treatment with a vascular endothelial growth factor receptor tyrosine kinase inhibitor. *Am. J. Pathol.* **2008**, *173*, 265–277. [CrossRef]
17. Cordiali-Fei, P.; Trento, E.; D'Agosto, G.; Bordignon, V.; Mussi, A.; Ardigo, M.; Mastroianni, A.; Vento, A.; Solivetti, F.; Berardesca, E.; et al. Effective Therapy with Anti-TNF-α in Patients with Psoriatic Arthritis Is Associated with Decreased Levels of Metalloproteinases and Angiogenic Cytokines in the Sera and Skin Lesions. *Ann. N. Y. Acad. Sci.* **2007**, *1110*, 578–589. [CrossRef] [PubMed]
18. Alenius, G.M.; Jonsson, S.; Wallberg Jonsson, S.; Ny, A.; Rantapaa Dahlqvist, S. Matrix metalloproteinase 9 (MMP-9) in patients with psoriatic arthritis and rheumatoid arthritis. *Clin. Exp. Rheumatol.* **2001**, *19*, 760.
19. Scheinfeld, N. A comprehensive review and evaluation of the side effects of the tumor necrosis factor alpha blockers etanercept, infliximab and adalimumab. *J. Dermatolog. Treat.* **2004**, *15*, 280–294. [CrossRef] [PubMed]
20. Svensson, L.; Ropke, M.A.; Norsgaard, H. Psoriasis drug discovery: Methods for evaluation of potential drug candidates. *Expert Opin. Drug Discov.* **2012**, *7*, 49–61. [CrossRef]
21. Smith, N.; Weymann, A.; Tausk, F.A.; Gelfand, J.M. Complementary and alternative medicine for psoriasis: A qualitative review of the clinical trial literature. *J. Am. Acad. Dermatol.* **2009**, *61*, 841–856. [CrossRef]
22. Fascella, G.; D'Angiolillo, F.; Mammano, M.M.; Amenta, M.; Romeo, F.V.; Rapisarda, P.; Ballistreri, G. Bioactive compounds and antioxidant activity of four rose hip species from spontaneous sicilian flora. *Food Chem.* **2019**, *289*, 56–64. [CrossRef] [PubMed]
23. D'Eliseo, D.; Pannucci, E.; Bernini, R.; Campo, M.; Romani, A.; Santi, L.; Velotti, F. In vitro studies on anti-inflammatory activities of kiwifruit peel extract in human thp-1 monocytes. *J. Ethnopharmacol.* **2019**, *233*, 41–46. [CrossRef]
24. Cefali, L.C.; Franco, J.G.; Nicolini, G.F.; Ataide, J.A.; Mazzola, P.G. In vitro antioxidant activity and solar protection factor of blackberry and raspberry extracts in topical formulation. *J. Cosmet. Dermatol.* **2019**, *18*, 539–544. [CrossRef] [PubMed]
25. Mastrogiovanni, F.; Mukhopadhya, A.; Lacetera, N.; Ryan, M.T.; Romani, A.; Bernini, R.; Sweeney, T. Anti-inflammatory effects of pomegranate peel extracts on in vitro human intestinal caco-2 cells and ex vivo porcine colonic tissue explants. *Nutrients* **2019**, *11*, 548. [CrossRef]
26. Deng, S.; May, B.H.; Zhang, A.L.; Lu, C.; Xue, C.C. Plant extracts for the topical management of psoriasis: A systematic review and meta-analysis. *Br. J. Dermatol.* **2013**, *169*, 769–782. [CrossRef]
27. ESCOP. *Escop Monographs: The Scientific Foundation for Herbal Medicinal Products*, 2nd ed.; Georg Thieme Verlag: New York, NY, USA, 2009.
28. Orhan, D.D.; Orhan, N.; Ozcelik, B.; Ergun, F. Biological activities of *Vitis vinifera* L. leaves. *Turk. J. Biol.* **2009**, *33*, 341–348.
29. Fernandes, F.; Ramalhosa, E.; Pires, P.; Verdial, J.; Valentao, P.; Andrade, P.; Bento, A.; Pereira, J.A. Vitis vinifera leaves towards bioactivity. *Ind. Crops Prod.* **2013**, *43*, 434–440. [CrossRef]
30. Sangiovanni, E.; Di Lorenzo, C.; Colombo, E.; Colombo, F.; Fumagalli, M.; Frigerio, G.; Restani, P.; Dell'Agli, M. The effect of in vitro gastrointestinal digestion on the anti-inflammatory activity of *Vitis vinifera* L. Leaves. *Food Funct.* **2015**, *6*, 2453–2463. [CrossRef]
31. Rabe, E.; Stucker, M.; Esperester, A.; Schafer, E.; Ottillinger, B. Efficacy and tolerability of a red-vine-leaf extract in patients suffering from chronic venous insufficiency-results of a double-blind placebo-controlled study. *Eur. J. Vasc. Endovasc. Surg.* **2011**, *41*, 540–547. [CrossRef]

32. Khalilpour, S.; Sangiovanni, E.; Piazza, S.; Fumagalli, M.; Beretta, G.; Dell'Agli, M. In vitro evidences of the traditional use of *Rhus coriaria* L. Fruits against skin inflammatory conditions. *J. Ethnopharmacol.* **2019**, *238*, 111829. [CrossRef] [PubMed]
33. Musazzi, U.M.; Franze, S.; Minghetti, P.; Casiraghi, A. Emulsion versus nanoemulsion: How much is the formulative shift critical for a cosmetic product? *Drug Deliv. Transl. Res.* **2018**, *8*, 414–421. [CrossRef]
34. Goldminz, A.M.; Au, S.C.; Kim, N.; Gottlieb, A.B.; Lizzul, P.F. Nf-kappab: An essential transcription factor in psoriasis. *J. Dermatol. Sci.* **2013**, *69*, 89–94. [CrossRef]
35. Chaturvedi, V.; Qin, J.Z.; Denning, M.F.; Choubey, D.; Diaz, M.O.; Nickoloff, B.J. Abnormal NF-kappab signaling pathway with enhanced susceptibility to apoptosis in immortalized keratinocytes. *J. Dermatol. Sci.* **2001**, *26*, 67–78. [CrossRef]
36. Richert, S.; Schrader, A.; Schrader, K. Transdermal delivery of two antioxidants from different cosmetic formulations. *Int. J. Cosmet. Sci.* **2003**, *25*, 5–13. [CrossRef]
37. Chen-yu, G.; Chun-fen, Y.; Qi-lu, L.; Qi, T.; Yan-wei, X.; Wei-na, L.; Guang-xi, Z. Development of a quercetin-loaded nanostructured lipid carrier formulation for topical delivery. *Int. J. Pharm.* **2012**, *430*, 292–298. [CrossRef]

Article

Natural Antioxidant Resveratrol Suppresses Uterine Fibroid Cell Growth and Extracellular Matrix Formation In Vitro and In Vivo

Hsin-Yuan Chen [1,†], Po-Han Lin [1], Yin-Hwa Shih [2], Kei-Lee Wang [3], Yong-Han Hong [4], Tzong-Ming Shieh [5,†], Tsui-Chin Huang [6,*] and Shih-Min Hsia [1,7,8,9,*]

1 School of Nutrition and Health Sciences, College of Nutrition, Taipei Medical University, Taipei 11031, Taiwan; hsin246@gmail.com (H.-Y.C.); phlin@tmu.edu.tw (P.-H.L.)
2 Department of Healthcare Administration, Asia University, Taichung 41354, Taiwan; evashih@asia.edu.tw
3 Department of Nursing, Ching Kuo Institute of Managemnet and Health, Keelung 20301, Taiwan; kellywang@tmu.edu.tw
4 Department of Nutrition, I-Shou University, Kaohsiung 84001, Taiwan; yonghan@isu.edu.tw
5 Department of Dental Hygiene, College of Health Care, China Medical University, Taichung 40402, Taiwan; tmshieh@mail.cmu.edu.tw
6 PhD Program for Cancer Biology and Drug Discovery, College of Medical Science and Technology, Taipei Medical University and Academia Sinica, Taipei 11031, Taiwan
7 Graduate Institute of Metabolism and Obesity Sciences, College of Nutrition, Taipei Medical University, Taipei 11031, Taiwan
8 School of Food and Safety, Taipei Medical University, Taipei 11031, Taiwan
9 Nutrition Research Center, Taipei Medical University Hospital, Taipei 11031, Taiwan
* Correspondence: tsuichin@tmu.edu.tw (T.-C.H.); bryanhsia@tmu.edu.tw (S.-M.H.); Tel.: +886-2-2736-1661 (ext. 6558) (S.-M.H.)
† These authors contributed equally to this paper.

Received: 9 March 2019; Accepted: 10 April 2019; Published: 12 April 2019

Abstract: Resveratrol (RSV) is a polyphenolic phytoalexin found in peanuts, grapes, and other plants. Uterine fibroids (UF) are benign growths that are enriched in extracellular matrix (ECM) proteins. In this study, we aimed to investigate the effects of RSV on UF using in vivo and in vitro approaches. In mouse xenograft models, tumors were implanted through the subcutaneous injection of Eker rat-derived uterine leiomyoma cells transfected with luciferase (ELT-3-LUC) in five-week-old female nude (Foxn1^{nu}) mice. When the tumors reached a size of 50–100 mm^3, the mice were randomly assigned to intraperitoneal treatment with RSV (10 mg·kg^{-1}) or vehicle control (dimethyl sulfoxide). Tumor tissues were assayed using an immunohistochemistry analysis. We also used primary human leiomyoma cells as in vitro models. Cell viability was determined using the sodium bicarbonate and 3-(4,5-dimethylthiazol-2-yl)-2,5-diphenyltetrazolium bromide (MTT) assay. The protein expression was assayed using Western blot analysis. The messenger ribonucleic acid (mRNA) expression was assayed using quantitative reverse transcription–polymerase chain reaction (qRT–PCR). Cell apoptosis was assayed using Annexin V-fluorescein isothiocyanate (FITC) and propidium iodide (PI) and Hoechst 33342 staining. RSV significantly suppressed tumor growth in vivo and decreased the proportion of cells showing expression of proliferating cell nuclear antigen (PCNA) and α-smooth muscle actin (α-SMA). In addition, RSV decreased the protein expression of PCNA, fibronectin, and upregulated the ratio of Bax (Bcl-2-associated X) and Bcl-2 (B-cell lymphoma/leukemia 2) in vivo. Furthermore, RSV reduced leiomyoma cell viability, and decreased the mRNA levels of fibronectin and the protein expression of collagen type 1 (COL1A1) and α-SMA (ECM protein marker), as well as reducing the levels of β-catenin protein. RSV induced apoptosis and cell cycle arrest at sub-G1 phase. Our findings indicated the inhibitory effects of RSV on the ELT-3-LUC xenograft model and indicated that RSV reduced ECM-related protein expression in primary human leiomyoma cells, demonstrating its potential as an anti-fibrotic therapy for UF.

Keywords: uterine fibroids; resveratrol; extracellular matrix; ELT-3-LUC xenograft model

1. Introduction

Benign uterine fibroids (UF), also known as myomas or leiomyomas, are the most common neoplasm of the uterus and occur in up to 77% of women by the onset of menopause in the United States [1,2]. Women with UF usually suffer from a reduced quality of life due to symptoms such as abnormal uterine bleeding, pelvic pain, frequent urination, and infertility [3,4]. Although the etiology remains unclear, genetic factors, cytokines, growth factors, steroid hormones (estrogens and progestogens) and/or their receptors, and excessive production of extracellular matrix (ECM) have been reported to play a pivotal role in the development of UF [4]. In general, the degradation of ECM is precisely regulated under normal physiological conditions, however, abnormal ECM metabolism is involved in pathogenesis of UF [5]. The major ECM components of UF include fibronectin, collagens, and proteoglycans such as biglycan and fibromodulin [6,7].

Most therapeutic treatments provide only temporary or partial relief from UF, and are not successful in every patient [8]. In comparison, hysterectomy is considered as the only option and the fastest treatment to reduce pain from UF, especially for women with uterine fibroids but a lack of cognition [9]. However, women who undergo hysterectomies encounter a number of problems, such as pelvic floor disorders, early menopause, and sexual dysfunction; these postoperative complications can be relieved using conventional medical treatment, but the cost related to uterine fibroids is considerable [9]. In recent years, more research has been undertaken to identify natural extracts as adjuncts to chemotherapy. In particular, dietary polyphenols, such as epigallocatechin gallate (EGCG) [10], green tea extract [11], and strawberry extract [12], have been shown to have anti-leiomyoma activities.

Resveratrol (RSV; trans-3,5,4′-trihydroxystilbene) is a natural polyphenolic compound belonging to the stilbene group. RSV is present in several plants [13], including blueberries [14], peanuts [15], and grapes [16], as well as in grape related products, such as wine [17]. In general, fresh grape skins contain 50–100 $mg{\cdot}g^{-1}$ resveratrol [18,19]. RSV is a potent antioxidant [20] with anti-inflammatory [21], anti-proliferative [22], and anti-adipogenic [23] effects on several cancer cells, including breast [24] and prostate cancers [25], and it might provide a potential treatment for dysmenorrhea [26]. Recently, some studies have showed that RSV can reduce tissue fibrogenesis in chronic kidney diseases [27]. In addition, we have previously shown that RSV inhibits leiomyoma cell proliferation, induces apoptosis, promotes cell cycle arrest, and regulates messenger ribonucleic acid (mRNA) and protein expression of ECM-associated proteins in vitro [28]. These data support the potential of RSV as an alternative therapeutic treatment for UF. However, its effects on leiomyoma growth in vivo remain unclear. Therefore, the present study investigated the effects of RSV on UF growth in a mouse xenograft model in vivo.

2. Materials and Methods

2.1. Reagents and Antibodies

Dulbecco's Modified Eagle Medium/Nutrient Mixture F-12 (DMEM/F12), antibiotic-antimycotic solution (100×), and 0.05% trypsin-ethylenediaminetetraacetic acid (EDTA, 1×) were purchased from CAISSON Labs (Smithfield, UT, USA). Fetal bovine serum (FBS), trypan blue, NucBlue™ Live ReadyProbes™ Reagent, a bicinchoninic acid protein assay kit, and enhanced chemiluminescence reagents were purchased from Thermo Fisher Scientific (Waltham, MA, USA). The Annexin V-fluorescein isothiocyanate (FITC) apoptosis detection kit I was purchased from Becton Dickinson (BD) Biosciences (San Jose, CA, USA). Bovine serum albumin (BSA) was purchased from BioShop (Burlington, Canada). Protease and phosphatase inhibitor cocktail tablets were purchased from Roche (Basel, Switzerland). Sodium bicarbonate, 3-(4,5-dimethylthiazol-2-yl)-2,5-diphenyltetrazolium bromide (MTT), propidium iodide (PI), and dimethyl sulfoxide (DMSO) were purchased from Sigma-Aldrich (Louis, MO, USA). VivoGlo™ Luciferin (in vivo grade) was purchased from Promega (Fitchburg, WI, USA). Matrigel® basement membrane matrix was purchased from Corning (Corning, NY, USA). Zoletil® 50 was purchased from Virbac (Carros, France). Rompun® 20 (xylazine hydrochloride) was purchased from Bayer (Pittsburgh, PA, USA). The following antibodies were used in this study: anti-proliferating cell nuclear antigen (PCNA), anti-Bax, anti-Bcl-2 and anti-β-catenin (Cell Signal Technology, Danvers, MA, USA), anti-β-actin and anti-α-SMA, anti-vimentin, anti-collagen type 1 (COL1A1) (GeneTex, Irvine, CA, USA), fibronectin, and goat anti-rabbit IgG and anti-mouse IgG antibodies (Abcam, Cambridge, MA, USA).

2.2. Preparation of RSV

RSV ($C_{14}H_{12}O_3$, chemical abstracts service number: 501-36-0) was purchased from ECHO CHEMICAL Co., Ltd. (purity >99%, Miaoli, Taiwan). A stock solution of 100 mM was prepared in DMSO, aliquoted, and then stored at −20 °C until use. For in vitro experiments, the final concentrations of RSV were prepared by diluting the stock with cell culture medium. The control cells were treated with vehicle (0.1% DMSO).

2.3. Cell Culture

The Eker rat-derived uterine leiomyoma (ELT-3) cell lines were kindly provided by Lin-Hung Wei (Department of Oncology, National Taiwan University Hospital, Taipei, Taiwan). ELT-3 cells transfected with luciferase reporter genes (ELT-3-LUC) were previously established in our laboratory. In addition, the primary cultures of human leiomyoma cells were isolated from uterine leiomyoma tumor tissue specimens, which were collected from women (30–40 years of age, $n = 6$) undergoing myomectomy at the Department of Oncology, National Taiwan University Hospital (Taipei, Taiwan). According to a previous study all the human tissue specimens were approved by the Institutional Review Board and Ethics Committee of the National Taiwan University Hospital (permit number: 201210072RIC). The process of purification of the leiomyoma cells was as described previously [29], and leiomyoma cells from passages 2–7 were used in this study. Both ELT-3-LUC and leiomyoma cells were cultured in DMEM/F12 containing 10% FBS, 1% antibiotics [10,000 units·mL^{-1} penicillin, 10,000 μg·mL^{-1} streptomycin, and 25 μg·mL^{-1} amphotericin with 8.5 g·L^{-1} NaCl], and 0.6 mg·mL^{-1} Geneticin® G418 Sulfate (Thermo Fisher Scientific, Waltham, MA, USA; ELT-3-LUC only); both cell lines were incubated at 37 °C with 5% CO_2.

2.4. Tumor Xenograft in Nude (Foxn1^{nu}) Mice

Five-week-old female nude (Foxn1^{nu}) mice (BioLASCO, Taipei, Taiwan) were housed under a 12 h light/12 h dark cycle in a pathogen-free environment, with ad libitum access to food and water. Tumors were implanted through the subcutaneous (s.c.) injection of ELT-3-LUC cells (1×10^6 cells suspended in 0.1 mL phosphate-buffered saline (PBS)/Matrigel solution for each mouse) into the right flank of the mice. After the tumors reached a size of 50–100 mm^3 (approximately 1 month), the mice were randomly assigned to two groups ($n = 5$): one group received an intraperitoneal (i.p.) injection of RSV (10 mg·kg^{-1}; treatment group), and the other group received a vehicle (DMSO; control group) twice a week for 4 weeks. The tumor volume was measured using calipers and calculated as $L \times W^2 \times 0.52$, where L is the length and W is the width. The tumor volumes and body weights were recorded until the animals were sacrificed by an i.p. injection of anesthetic mixtures [1 mL zoletil (Virbac, Carros, France) + 1 mL rompun (Bayer, Pittsburgh, PA, USA)]. Every week, the mice were administered an i.p. injection of luciferin (150 mg·kg^{-1} body weight) and detected using a non-invasive in vivo imaging system (IVIS). At the end of the experiment, the tumor tissues were stained with hematoxylin and eosin (H&E). All the animal studies were conducted according to the protocols approved by the Institutional Animal Care and Use Committee (IACUC) of Taipei Medical University (IACUC approval no. 2015–0115).

2.5. Immunohistochemistry Analysis

To observe the localization of specific proteins, immunohistochemistry analysis was assayed. Tumor tissues were embedded and sliced at a thickness of 2- or 6-μm by the animal experiment center of Taipei Medical University (Taipei, Taiwan). The tissue sections were stained by Bio-Check Laboratories Ltd (Taipei, Taiwan). To analyze the immunohistochemistry slides, five areas were photographed at 40× magnification (center, bottom, top, left, and right regions) using an EVOS® microscope (Thermo Fisher Scientific, Waltham, MA, USA), and the color of the PCNA and α-SMA staining in the tissue sections was observed.

2.6. Western Blot Analysis

The lysates of tumor tissues were prepared in ice-cold lysis buffer (50 mmol·L^{-1} Tris (pH 8.0), 100 mmol·L^{-1} NaCl, 0.1% sodium dodecyl sulfate (SDS), 1% NP-40, 0.5 mM EDTA) containing a protease inhibitor cocktail. The proteins (30 μg) were boiled for 5 min, separated using 12% SDS–polyacrylamide gel electrophoresis (SDS–PAGE), and then transferred electrophoretically to Immobilon-P polyvinylidene fluoride (PVDF) membranes (0.22 μm) for 150–180 min at 280 mA and 250 V. Then, the membranes were washed three times for 10 min/wash with Tris-buffered saline containing Tween 20 (TBST) buffer, blocked with blocking buffer (5% BSA) for 1 h at 25°C, and incubated for 8 h with primary antibodies (1:1000 in blocking buffer) at 4 °C. The next day, the membranes were washed three times for 10 min/wash with a TBST buffer, incubated for 1 h in a blocking buffer containing goat anti-rabbit or anti-mouse IgG (as appropriate) coupled to alkaline phosphatase (1:10,000), and washed three times with TBST (10 min/wash). Finally, the bands were detected using enhanced chemiluminescence. The densitometric values were normalized to the internal control (β-actin) using Image Lab™ Software Version 5.2 1. (Bio-Rad, Hercules, California, USA).

2.7. Cell Viability Assay

The effect of RSV treatment on cell viability was examined using the MTT assay. The human leiomyoma cells were seeded in 96-well plates (2 × 10^3 cells/well), cultured for 24 h, and treated with various concentrations of RSV in fresh medium containing 1% FBS. The MTT solution (1 mg·mL^{-1}) was then added directly to each well (100 μL/well) for 4 h. The absorbance was measured at 570 nm, with a reference wavelength of >630 nm, using a microplate reader (BioTek, Winooski, VT, USA).

2.8. Quantitative Real-Time RT–PCR (qRT–PCR)

The total cellular ribonucleic acid (RNA) was extracted from RSV-treated cells with TRIzol™ reagent (Thermo Fisher Scientific, Waltham, MA, USA) followed by Quick-RNATM MiniPrep Plus (Zymo Research, Irvine, CA, USA), and a total of 2 μg RNA was reverse transcribed using a RevertAid H minus first strand cDNA synthesis kit (Thermo Fisher Scientific, Waltham, MA, USA) according to the manufacturer's instructions. Amplification reactions were performed using the PowerUp™ SYBR™ Green master mix (Thermo Fisher Scientific, Waltham, MA, USA). qRT–PCR analyses were performed using the Applied Biosystems StepOnePlus™ real-time PCR system (Thermo Fisher Scientific, Waltham, MA, USA). Amplification of all genes was performed under the following cycling conditions: denaturation at 95 °C for 10 min, followed by 40 cycles for 15 s at 95 °C and 30 s at 60 °C. The synthesis of the DNA product of the expected size was confirmed using a melt curve analysis. The comparative threshold cycle (Ct) values of each gene were normalized to Ct values of glyceraldehyde 3-phosphate dehydrogenase (GAPDH, internal control). The primers used for qRT–PCR analysis are listed in Table 1.

Table 1. Sequences of quantitative reverse transcription–polymerase chain reaction (qRT–PCR) primers.

Gene	Forward (5′ to 3′)	Reverse (5′ to 3′)
FN1 [1]	GGCCAGTCCTACAACCAGTAT	TCGGGAATCTTCTCTGTCAGC
GAPDH [2]	TGCACCACCAACTGCTTAGC	GGCATGGACTGTGGTCATGAG

[1] *FN1*, fibronectin; [2] *GAPDH*, glyceraldehyde 3-phosphate dehydrogenase.

2.9. Hoechst 33342 Staining

To detect alterations of nuclei morphology of leiomyoma cells after RSV treatment, Hoechst 33342 staining was performed. The leiomyoma cells were seeded in 6 cm^2 culture dishes (5 × 10^4 cells) and treated with RSV (10, 50, 100 μM). After 48 h of treatment, the cells were directly stained with 2 drops/mL Hoechst 33342 by incubation for 20 min at room temperature. Images were acquired using a fluorescence microscope.

2.10. Apoptosis Analysis

The induction of apoptosis was determined using Annexin V-FITC/PI staining. The leiomyoma cells were seeded in 10 cm^2 culture dishes (1 × 10^6 cells) and treated with RSV (10, 50, 100 μM) for 48 h. The cells were stained with Annexin V-FITC and PI by incubation for 15 min at room temperature protected from light. The apoptotic cells were analyzed using BD Accuri™ C6 Plus Flow Cytometer (BD Biosciences, San Jose, CA, USA), and the results were analyzed using the BD Accuri™ C6 Plus software (BD Biosciences, San Jose, CA, USA).

2.11. Cell Cycle Analysis

To assess the cell cycle progression, the leiomyoma cells were seeded into 10 cm^2 culture dishes (1 × 10^6 cells) and then treated with RSV (10, 50, 100 μM) for 48 h. All the cells were collected, slowly added to 9 mL of 70% cold ethanol, and then stored at −20 °C overnight. The cells were washed twice with cold phosphate-buffered saline (PBS), resuspended in 500 μL propidium iodide (PI)/Triton X-100 staining solution (10 mL 0.1% (*v*/*v*) Triton X-100 in PBS containing 2 mg DNAse-free RNAse A and 0.40 mL of 500 μg·mL^{-1} PI), and incubated for 30 min at 20 °C. The fluorescence was measured using a fluorescence-activated cell-sorting (FACS) Calibur flow cytometer (BD Biosciences, San Jose, CA, USA) and the cell cycle distribution was analyzed using the CellQuest software program (BD Biosciences, San Jose, CA, USA).

2.12. Statistical Analysis

The data are presented as the mean ± standard deviation (SD), and the differences between the means were analyzed using Sigma Plot version 12.5 (SoftHome International, Taipei, Taiwan). For the comparison of the two groups, a Student's *t*-test was used. The group means were compared using the one-way analysis of variance and Duncan's multiple-range test. The difference between two means was considered as statistically significant when $p < 0.05$.

3. Results

3.1. The Inhibitory Effect of RSV on the Growth of UF in Vivo

As shown in Figure 1, the treatment group received RSV via i.p. injection twice per week for 4 weeks. During the treatment period, the mouse body weights were measured each time they were injected to investigate the effects of RSV on overall health. IVIS was used to track ELT-3-LUC tumor growth over time in this mouse xenograft model. Unfortunately, due to the individual differences and ELT-3-LUC cell instability, we did not show all the tracking results. From the appearance and size of the tumors, we can initially evaluate the effect of resveratrol. The tumor sizes and volumes were significantly reduced in the treatment group, as compared to the control group (Figure 2A,B). Notably, a significant difference in tumor volume was observed between the vehicle- and RSV-treated groups from day 56 of treatment (Figure 2C). No significant group difference in the mouse body weights was observed (Figure 2D). In addition, IVIS imaging identified a higher bioluminescent signal in the vehicle-treated group than in the RSV group at day 56, although large inter-individual differences were observed (Figure 2E,F). These data demonstrated the potent inhibitory effect of RSV on the growth of UF within a relatively short treatment period.

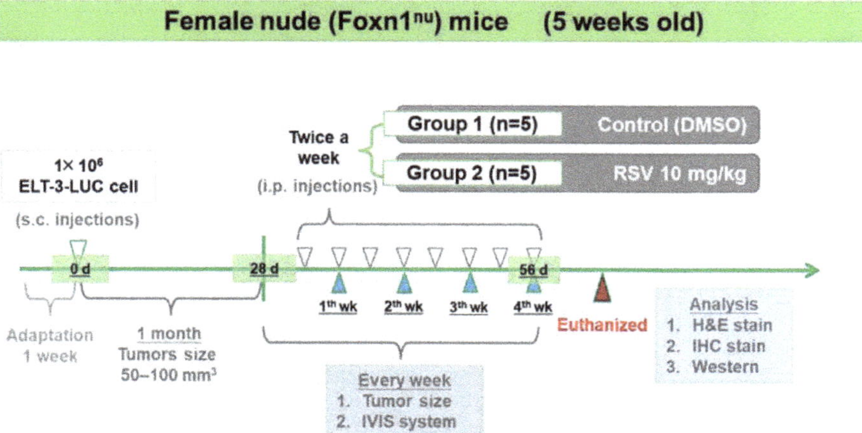

Figure 1. Schematic representation of the treatment plan for the xenograft mouse model. Cultured Eker rat-derived uterine leiomyoma cells transfected with luciferase (ELT-3-LUC) embedded in Dulbecco's Modified Eagle Medium/Nutrient Mixture F-12 (DMEM/F12)/Matrigel solution were transplanted into the right flank of female nude (Foxn1^{nu}) mice. When the tumors reached a size of 50–100 mm^3 (approximately 1 month), the mice received an intraperitoneal injection of either resveratrol (RSV; 10 mg·kg^{-1}) or vehicle control (dimethyl sulfoxide; DMSO) twice a week for 1 month. nude (Foxn1^{nu}) mice: nude mice with a spontaneous deletion in the *FOXN1* gene; IVIS: non-invasive in vivo imaging system; H&E: hematoxylin and eosin; IHC: immunohistochemical.

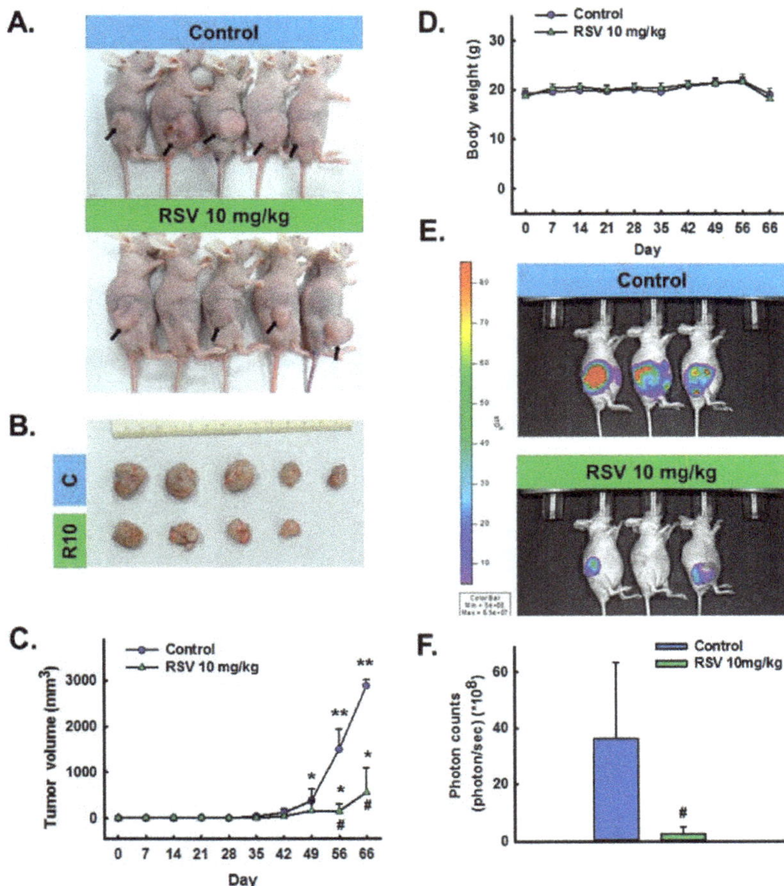

Figure 2. Effect of resveratrol (RSV) on tumor xenograft growth. (**A**) Morphology and (**B**) size of tumors isolated from the sacrificed mice in each group are shown at day 66, the end-point of the treatment. (**C**) Tumor volumes and (**D**) body weights of nude (Foxn1^{nu}) mice. (**E**) Total photon flux from imaging on day 56 after xenografting. (**F**) All luciferase images were normalized to the same photon saturation scale. Data are presented as the mean ± SD (n = 5 or 3); * $p < 0.05$ and ** $p < 0.001$ vs. day 0; # $p < 0.05$ vs. control.

To explore the effects of RSV further, immunohistochemical analyses were performed. Compared with the control group, the RSV-treated (10 mg·kg^{-1}) group showed a decrease in the proportions of cells that were positive for PCNA (a marker of cell growth, Figure 3A-b,-e) or α-SMA (a smooth muscle marker, Figure 3A-c,-f), as well as hematoxylin and eosin staining (H&E, Figure 3A-a,-d). In addition, Western blot analysis showed that mice treated with RSV showed reduced levels of PCNA and fibronectin in whole tissue extracts (Figure 3B), but enhanced levels of Bax/Bcl-2 (apoptosis-related markers).

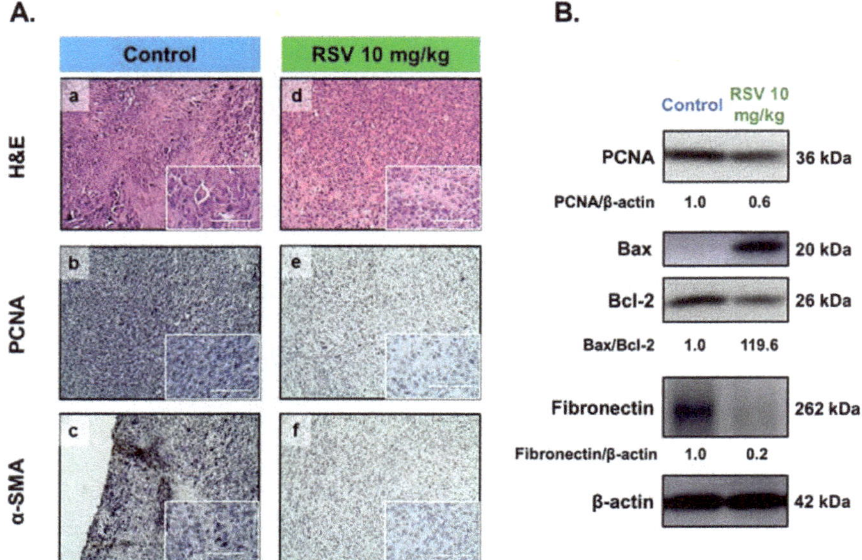

Figure 3. Effect of resveratrol (RSV) treatment on tumor xenografts. (**A**) Eker rat-derived uterine leiomyoma cells transfected with luciferase (ELT-3-LUC) tumors were excised and processed for hematoxylin and eosin (H&E) staining (**a**,**d**) and immunohistochemical (IHC) analysis of proliferating cell nuclear antigen (PCNA) (**b**,**e**) and α-smooth muscle actin (α-SMA) (**c**,**f**); the scale bars represent 100 μm. (**B**) Tumor lysates were separated by sodium dodecyl sulfate polyacrylamide gel electrophoresis and analyzed on Western blotting with an anti-PCNA, fibronectin, Bax and Bcl-2 antibody. β-actin was used as a loading control. The band intensities are expressed as a ratio, relative to the loading control.

3.2. Effects of RSV on Leiomyoma Cell Proliferation and Extracellular Matrix (ECM) Accumulation in Vitro

To evaluate whether RSV produced a cytotoxic effect, leiomyoma cells were treated with RSV (10, 50, or 100 μM) for 48 h or 72 h. Cell viability was measured using the MTT assay; the results showed that RSV has significantly reduced the viability of leiomyoma cells (Figure 4B), and narrow cells were observed at 100 μM RSV (Figure 4A). Numerous studies have shown that excessive ECM production is an important factor that cannot be ignored in relation to uterine fibroid growth. To examine the effect of RSV on the expression of ECM in leiomyoma cells, we chose more representative ECM proteins as markers, such as fibronectin, collagen type 1, vimentin, and α-SMA. As shown in Figure 4C, leiomyoma cells exposed to 100 μM RSV showed a significantly lower mRNA expression of *FN1*. In addition, Western blot analysis showed that 100 μM RSV significantly decreased the levels of COL1A1, α-SMA, and β-catenin compared to controls for 48 h (Figure 4D,F,G). These data demonstrate the potent inhibitory effect of RSV on tumor growth and ECM accumulation in leiomyoma cells in vitro.

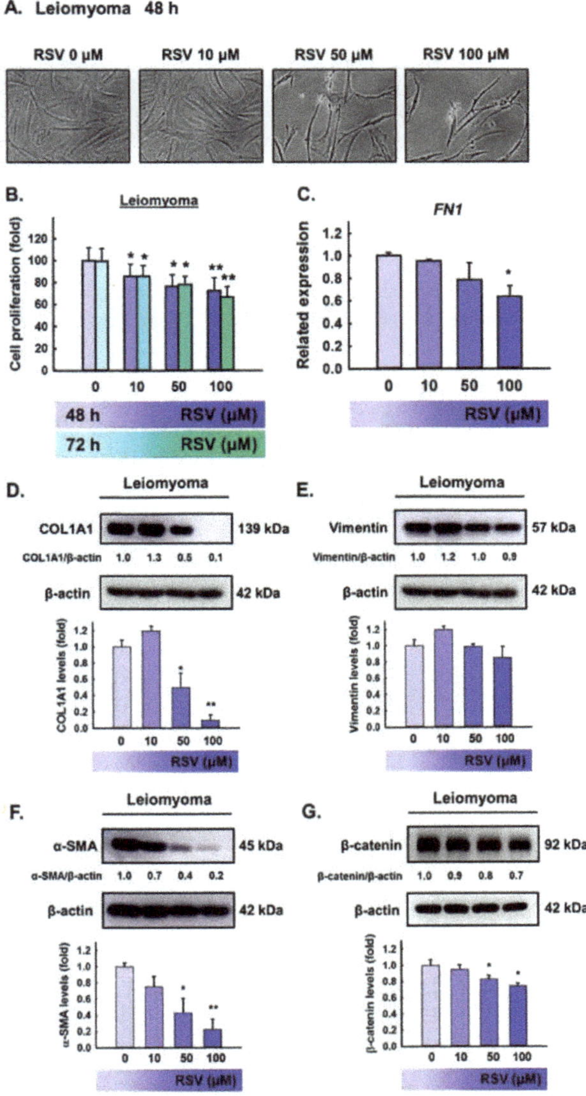

Figure 4. Cytotoxic effects of resveratrol (RSV) on primary human leiomyoma cells. Leiomyoma cells were exposed to either vehicle (dimethyl sulfoxide; DMSO) or RSV (10–100 μM) for 48 h or 72 h. (**A**) Morphology of leiomyoma cells after the indicated treatment (magnification, ×200). (**B**) Cell proliferation was measured using a 3-(4,5-dimethylthiazol-2-yl)-2,5-diphenyltetrazolium bromide (MTT) assay. (**C**) RNA samples were isolated from leiomyoma cells treated with RSV (0–100 μM) and subjected to quantitative reverse transcription–polymerase chain reaction (qRT–PCR) using primers specific for fibronectin (*FN1*). (**D–G**) Leiomyoma cell lysates were separated using sodium dodecyl sulfate polyacrylamide gel electrophoresis (SDS–PAGE) and analyzed using Western blot with anti-COL1A1, vimentin, α-SMA, and β-catenin. β-actin was used as a loading control. The values of the band intensity represent the densitometric estimation of each band normalized to β-actin. Protein quantification of COL1A1, vimentin, α-SMA, and β-catenin expression in leiomyoma cells is shown in the bar graph. The results are expressed as the means ± SD of three independent experiments; * $p < 0.05$, ** $p < 0.001$, as compared with the control.

3.3. Effects of RSV on Apoptosis and Cell Cycle Progression of Leiomyoma Cells in Vitro

Nuclear condensation and the nuclear morphology changes in leiomyoma cells were examined by using Hoechst 33342 staining at 48 h after RSV treatment. As shown in Figure 5A, leiomyoma cells exposed to 100 µM RSV showed stronger blue fluorescence and an increased number of cells with fragmented and condensed nuclei than the control group. To evaluate whether RSV induced apoptosis, Annexin V-FITC and PI staining were used. The apoptosis rate depended on the percentage of early apoptotic cells (FITC+/PI−) and late apoptotic cells (FITC+/PI+). As shown in Figure 5B, 100 µM RSV increased the percentage of apoptotic cells compared to the controls at 48 h. On the other hand, the fluorescence intensity of the sub-G1 cell fraction also represented an apoptotic cell population. As shown in Figure 5C, 100 µM RSV increased the percentage of sub-G1 cells compared to controls at 48 h. These data demonstrate that RSV has potent pro-apoptosis effects on tumor growth in leiomyoma cells in vitro.

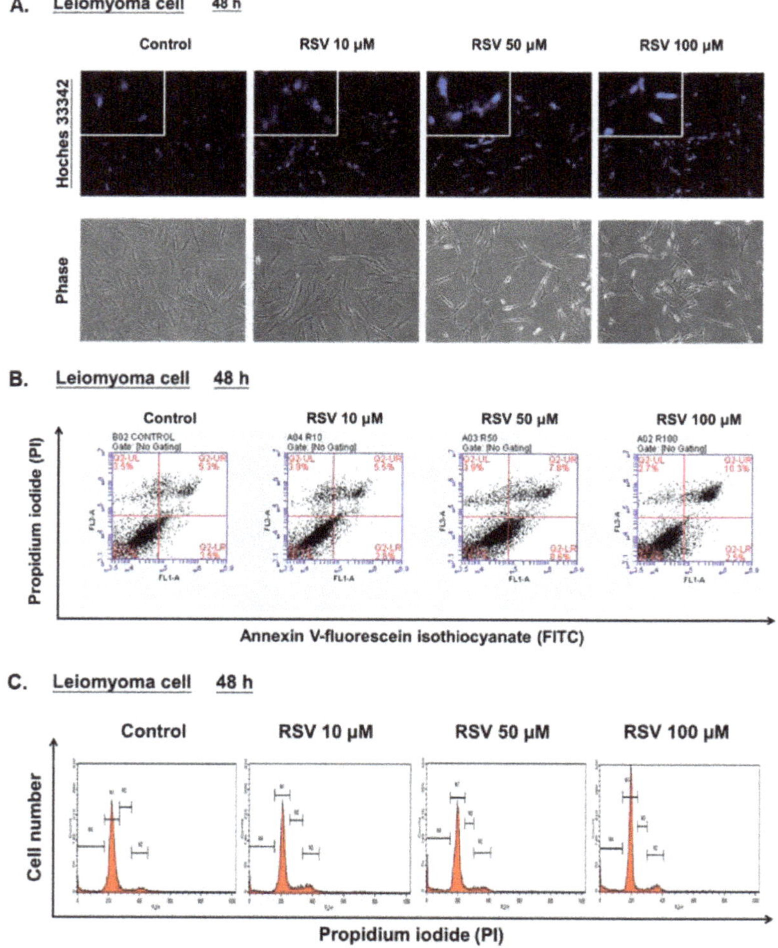

Figure 5. Resveratrol (RSV) induces apoptosis in primary human leiomyoma cells. Leiomyoma cells were exposed to either vehicle (dimethyl sulfoxide; DMSO) or RSV (10, 50, 100 µM) for 48 h. (**A**) Nuclear changes revealed by Hoechst 33342 (×200) and the morphology of leiomyoma cells. (**B**) The cells were harvested and stained with Annexin V-fluorescein isothiocyanate (FITC) and propidium iodide (PI), and cell apoptosis was analyzed using flow cytometry. (**C**) The cells were stained with propidium iodide (PI), and the histograms of cell cycle distribution was analyzed by flow cytometry.

4. Discussion

This study identified potentially beneficial inhibitory effects of RSV on UF growth in a mouse xenograft model in vivo, as well as on the proliferation of primary human leiomyoma cells in vitro. RSV exhibits pleiotropic activities in both in vivo and in vitro experimental models; these include anti-proliferation, pro-apoptosis, anti-carcinogenic, and anti-oxidant effects [20,22,24,25]. Each cell line has a different sensitivity to RSV and individual animal models also have different outcomes according to varying experimental conditions. For example, the intraperitoneal (i.p.) injection of RSV at a dose of 25 mg·kg^{-1} body weight reduced the tumor volume of MDA-MB-231 breast cancer cells in a xenograft mice model [30]. In addition, RSV (2.5 and 10 mg·kg^{-1}) administered intraperitoneally significantly reduced the tumor volume in mice bearing highly metastatic Lewis lung carcinoma (LLC) tumors [31]. In our study, we found that the i.p. injection of RSV at a dose of 10 mg·kg^{-1} body weight reduced the tumor volume of ELT-3 uterine leiomyoma cells.

In theory, a higher plasma level of RSV could be reached with a high dose of RSV. However, the consumption of a higher dose of RSV did not necessarily result in significantly higher plasma concentrations. According to a previous study, the plasma bioavailability of RSV was approximately 2% after a single-dose consumption [32], which is the result of rapid biotransformation to sulfate as well as the glucuronide conjugates. In addition, RSV was administered intraperitoneally at a concentration of 20 mg·kg^{-1} body weight; as a result, approximately 5 μM resveratrol glucuronide and 13 μM resveratrol sulfate were detected in the serum after 15 min, with concentrations reducing over the next 2 h [33]. Although the bioavailability of RSV is very low, many studies still use a higher than physiologically reasonable concentration for research purposes. For example, Garvin et al. found that 100 μM RSV induced significant morphological changes indicative of apoptosis in MDA-MB-231 breast cancer cells [30], and Wong et al. also found that 100 μM RSV promoted apoptosis by mediating caspase-3/7 activation and induced morphologic changes in cultured ovarian theca-interstitial (T-I) cells [22], these findings both based on the same concentration utilized in our study. However, in order to improve the bioavailability of RSV, Caddeo et al. changed the form of delivery of natural products and found the effectiveness of RSV can be potentiated by a polyphenol vesicular formulation [34].

The use of athymic nude mice is a commonly employed experimental model for cancer treatment [35]. In previous studies, scientific researchers have also used this mouse xenograft model to explore the therapy of leiomyoma [36,37]. For example, a previous in vivo study on nude mice injected subcutaneously with ELT-3 leiomyoma cells showed that EGCG treatment reduced tumor size, as compared to a control treatment (water). In addition, EGCG arrested the growth of ELT-3 cells and decreased leiomyoma size in Eker rat models as early as two weeks after treatment initiation [10]. Our study referred to the same mouse xenograft model and established a similar method, in which ELT-3 uterine leiomyoma cells were inoculated subcutaneously into the right flank of nude (Foxn1^{nu}) mice after anesthetizing. In agreement with the results of the study by Zhang et al. [10], our results showed that RSV significantly decreased tumor volume and arrested tumor growth. It is worth mentioning that Suzuk et al. [36] have established a novel and simple mouse xenograft model of human uterine leiomyomas according to this author's latest study, which will provide us with an ideal experimental model for the discovery of new compounds in the future.

PCNA is a DNA polymerase coenzyme that is closely related to cell proliferative activity because of its involvement in the synthesis of DNA in the nucleus [38]. In a previous study, immunohistochemical (IHC) staining revealed a decreased PCNA expression in Eker rat leiomyomas treated with 1, 25-dihydroxyvitamin D3, as compared to vehicle-treated control rats [39]. Zhang et al. demonstrated that the number of PCNA-positive cells decreased after 4- and 8-week treatments with EGCG, as compared to the number observed in water-treated control animals [10]. Similarly, the present study identified a decrease in the number of PCNA-positive cells and the protein expression of PCNA in mice treated with RSV, as compared to vehicle-treated animals.

Apoptosis is a process of programmed cell death; both Bax (Bcl-2-associated X) and Bcl-2 (B-cell lymphoma/leukemia 2) are markers of apoptosis-regulating proteins. The expression of Bcl-2 results in prolonged cell survival by restricting the activation of caspases. On the other hand, the overexpression of Bax results in accelerated programmed cell death. According to previous studies, the anti-apoptotic mechanism seems to be involved in the development of uterine leiomyoma; several studies have demonstrated that the anti-apoptotic *Bcl-2* gene was significantly over-expressed in uterine leiomyoma compared to homologous myometrium [40], and can even be influenced by the endocrine environment [41]. A previous study demonstrated that the inhibition of anti-apoptotic proteins Bcl-2/Bcl-xL promoted apoptotic cell death [42]. In the current study, we found that RSV enhanced the ratio of Bax and Bcl-2 and speculated that RSV may have induced apoptosis of UF growth in vivo. In addition, a previous study from Baarine et al. [43] found that RSV-treated cells exhibited apoptosis characteristics including nuclear fragmentation and condensation which were identified by Hoechst 33342. In agreement with the results of the study by Baarine et al. [43], our results showed that 100 μM RSV enhanced blue fluorescence and increased the number of cells with fragmented and condensed nuclei in primary human leiomyoma cells in vitro.

Because ECM accumulation is critical for the development of UF [44], it seems that the inhibitory effects of RSV could be mediated by ECM degradation. There are many factors related to ECM; first of all, increased deposition of ECM-associated proteins (fibronectin, collagens) and proteoglycans (biglycan, fibromodulin) is a typical characteristic of UF [6,7]. Myofibroblasts are the ECM-depositing cells active in wound healing, which are retained by UF when fibrotic responses are dysregulated [45]. The activation of myofibroblasts correlates with the expression of α-SMA, which is a key component supporting tissue contraction of ECM [46]. Previous studies have demonstrated that α-SMA is elevated in leiomyoma compared to myometrium [47]. In addition, several studies have reported that β-catenin expression was increased in UF compared to the adjacent myometrium tissue [48], which is associated with proliferation and ECM formation [49]. A recent study has shown that an increase in ECM stiffness triggers upregulation of β-catenin in UF cells [48].

In our previous study, we found that RSV reduced the levels of ECM-associated proteins (fibronectin and collagen type 1) and proteoglycans (fibromodulin and biglycan) in ELT3 cells in vitro [28]. In agreement with the results of these studies, our results showed that 100 μM RSV significantly decreased the protein expression of COL1A1, α-SMA, and β-catenin, as well as the mRNA level of *FN1* (fibronectin) in primary human leiomyoma cells compared to controls in vitro. Furthermore, RSV (10 $mg{\cdot}kg^{-1}$) reduced the proportion of α-SMA-positive cells and decreased the protein levels of fibronectin in vivo. However, the limitations of the present study are worth mentioning. The underlying mechanisms of RSV on ELT-3-LUC tumor xenografts still need to be elucidated in detail and further exploration of the molecular mechanisms and biological significance of RSV on ECM degradation is warranted.

5. Conclusions

The present study demonstrated that RSV suppressed tumor growth in vivo and inhibited primary human leiomyoma cells in vitro (Figure 6). In addition, RSV regulated ECM-associated protein expression. These findings indicate that RSV has the potential to reduce hyperplasia of leiomyoma cells. To the best of our knowledge, this is the first study to demonstrate the inhibitory potential of RSV on UF growth in vivo and may encourage further studies to highlight the molecular mechanisms involved in RSV and UF.

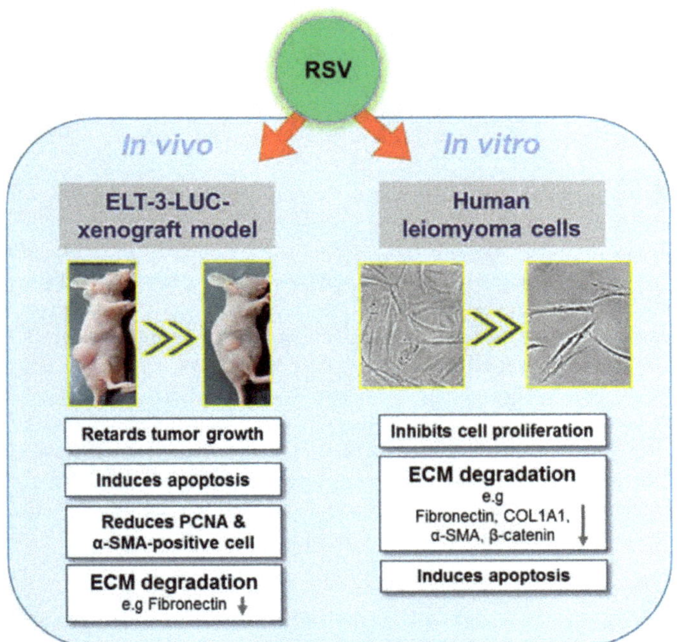

Figure 6. Schematic diagram of how the mechanism of RSV inhibits the growth of uterine fibroids. RSV significantly suppressed the tumor growth of ELT-3-LUC-xenografted mice and enhanced the Bax/Bcl-2 ratio, as well as reducing the proportion of PCNA and α-SMA-positive cells and the protein expression of fibronectin in an in vivo experiment. RSV also significantly inhibited the viability of primary human leiomyoma cells (magnification, ×200), induced apoptosis, and regulated the ECM-related gene (fibronectin) and proteins (COL1A1, vimentin, α-SMA, and β-catenin) in in vitro experiments. Abbreviations: RSV, resveratrol; ECM, extracellular matrix; PCNA, proliferating cell nuclear antigen; COL1A1, collagen type 1; α-SMA, alpha-smooth muscle actin.

Author Contributions: Conceptualization, H.-Y.C. and S.-M.H.; methodology, P.-H.L., K.-L.W. T.-M.S. and Y.-H.H.; investigation, H.-Y.C.; resources, Y.-H.S. and T.-M.S.; writing—review and editing, H.-Y.C. and S.-M.H.; supervision, S.-M.H. and T.-C.H.; writing—original draft preparation, H.-Y.C. and S.-M.H.; writing-review & editing, H.-Y.C., T.-M.S., T.-C.H. and S.-M.H.

Funding: This research was funded from the Ministry of Science and Technology (MOST), Taiwan, Republic of China, the grants number (MOST106-2320-B-038-064-MY3, MOST103-2313-B-038-003-MY3 and MOST106-2314-B-039-MY2); from Council of Agriculture, Taiwan, Republic of China, the grants number (106AS-16.4.1-ST-a4 and 107AS-13.4.1-ST-a6) from Council of Agriculture, Taiwan, Republic of China, and from China Medical University, Taiwan, Republic of China, the grants number (CMU107-S-37).

Conflicts of Interest: The authors declare no conflict of interest.

Abbreviations

RSV	Resveratrol
ELT-3	Eker uterine leiomyoma cells
UF	Uterine fibroids
ECM	Extracellular matrix
PCNA	Proliferating cell nuclear antigen
α-SMA	Alpha-smooth muscle actin
SDS-PAGE	Sodium dodecyl sulfate polyacrylamide gel electrophoresis
IVIS	In vivo imaging system (IVIS)

References

1. Cramer, S.F.; Patel, A. The frequency of uterine leiomyomas. *Am. J. Clin. Pathol.* **1990**, *94*, 435–438. [CrossRef] [PubMed]
2. Day Baird, D.; Dunson, D.B.; Hill, M.C.; Cousins, D.; Schectman, J.M. High cumulative incidence of uterine leiomyoma in black and white women: Ultrasound evidence. *Am. J. Obstet. Gynecol.* **2003**, *188*, 100–107. [CrossRef]
3. Okolo, S. Incidence, aetiology and epidemiology of uterine fibroids. *Best Pract. Res. Clin. Obstet. Gynaecol.* **2008**, *22*, 571–588. [CrossRef]
4. Islam, M.S.; Protic, O.; Giannubilo, S.R.; Toti, P.; Tranquilli, A.L.; Petraglia, F.; Castellucci, M.; Ciarmela, P. Uterine leiomyoma: Available medical treatments and new possible therapeutic options. *J. Clin. Endocrinol. Metab.* **2013**, *98*, 921–934. [CrossRef] [PubMed]
5. Islam, M.S.; Ciavattini, A.; Petraglia, F.; Castellucci, M.; Ciarmela, P. Extracellular matrix in uterine leiomyoma pathogenesis: A potential target for future therapeutics. *Hum. Reprod. Update* **2018**, *24*, 59–85. [CrossRef]
6. Arici, A.; Sozen, I. Transforming growth factor-beta3 is expressed at high levels in leiomyoma where it stimulates fibronectin expression and cell proliferation. *Fertil. Steril.* **2000**, *73*, 1006–1011. [CrossRef]
7. Leppert, P.C.; Baginski, T.; Prupas, C.; Catherino, W.H.; Pletcher, S.; Segars, J.H. Comparative ultrastructure of collagen fibrils in uterine leiomyomas and normal myometrium. *Fertil. Steril.* **2004**, *82*, 1182–1187. [CrossRef]
8. Commandeur, A.E.; Styer, A.K.; Teixeira, J.M. Epidemiological and genetic clues for molecular mechanisms involved in uterine leiomyoma development and growth. *Hum. Reprod. Update* **2015**, *21*, 593–615. [CrossRef] [PubMed]
9. Tinelli, A.; Mynbaev, O.A.; Sparić, R.; Kadija, S.; Stefanović, A.; Tinelli, R.; Malvasi, A. Physiology and Importance of the Myoma's Pseudocapsule. In *Hysteroscopy*; Tinelli, A., Alonso Pacheco, L., Haimovich, S., Eds.; Springer International Publishing: Cham, Switaerland, 2018; pp. 337–356.
10. Zhang, D.; Al-Hendy, M.; Richard-Davis, G.; Montgomery-Rice, V.; Sharan, C.; Rajaratnam, V.; Khurana, A.; Al-Hendy, A. Green tea extract inhibits proliferation of uterine leiomyoma cells in vitro and in nude mice. *Am. J. Obstet. Gynecol.* **2010**, *202*, e281–e289. [CrossRef]
11. Roshdy, E.; Rajaratnam, V.; Maitra, S.; Sabry, M.; Allah, A.S.; Al-Hendy, A. Treatment of symptomatic uterine fibroids with green tea extract: A pilot randomized controlled clinical study. *Int. J. Womens Health* **2013**, *5*, 477–486. [CrossRef]
12. Islam, M.S.; Giampieri, F.; Janjusevic, M.; Gasparrini, M.; Forbes-Hernandez, T.Y.; Mazzoni, L.; Greco, S.; Giannubilo, S.R.; Ciavattini, A.; Mezzetti, B.; et al. An anthocyanin rich strawberry extract induces apoptosis and ROS while decreases glycolysis and fibrosis in human uterine leiomyoma cells. *Oncotarget* **2017**, *8*, 23575–23587. [CrossRef]
13. Dei Cas, M.; Ghidoni, R. Cancer Prevention and Therapy with Polyphenols: Sphingolipid-Mediated Mechanisms. *Nutrients* **2018**, *10*, 940. [CrossRef] [PubMed]
14. Lyons, M.M.; Yu, C.; Toma, R.B.; Cho, S.Y.; Reiboldt, W.; Lee, J.; van Breemen, R.B. Resveratrol in Raw and Baked Blueberries and Bilberries. *J. Agric. Food Chem.* **2003**, *51*, 5867–5870. [CrossRef]
15. Sales, J.M.; Resurreccion, A.V.A. Resveratrol in Peanuts. *Crit. Rev. Food Sci. Nutr.* **2014**, *54*, 734–770. [CrossRef]
16. Jeandet, P.; Bessis, R.; Gautheron, B. The Production of Resveratrol (3,5,4'-trihydroxystilbene) by Grape Berries in Different Developmental Stages. *Am. J. Enol. Vitic.* **1991**, *42*, 41–46.
17. Jeandet, P.; Bessis, R.; Maume, B.F.; Meunier, P.; Peyron, D.; Trollat, P. Effect of Enological Practices on the Resveratrol Isomer Content of Wine. *J. Agric. Food Chem.* **1995**, *43*, 316–319. [CrossRef]
18. Jeandet, P.; Bessis, R.; Sbaghi, M.; Meunier, P. Production of the Phytoalexin Resveratrol by Grapes as a Response to Botrytis Attack Under Natural Conditions. *J. Phytopathol.* **1995**, *143*, 135–139. [CrossRef]
19. Aggarwal, B.B.; Bhardwaj, A.; Aggarwal, R.S.; Seeram, N.P.; Shishodia, S.; Takada, Y. Role of resveratrol in prevention and therapy of cancer: Preclinical and clinical studies. *Anticancer Res.* **2004**, *24*, 2783–2840. [PubMed]
20. Shahidi, F.; Ambigaipalan, P. Phenolics and polyphenolics in foods, beverages and spices: Antioxidant activity and health effects—A review. *J. Funct. Foods* **2015**, *18*, 820–897. [CrossRef]
21. De Sa Coutinho, D.; Pacheco, M.T.; Frozza, R.L.; Bernardi, A. Anti-Inflammatory Effects of Resveratrol: Mechanistic Insights. *Int. J. Mol. Sci.* **2018**, *19*, 1812. [CrossRef]

22. Wong, D.H.; Villanueva, J.A.; Cress, A.B.; Duleba, A.J. Effects of resveratrol on proliferation and apoptosis in rat ovarian theca-interstitial cells. *Mol. Hum. Reprod.* **2010**, *16*, 251–259. [CrossRef]
23. Carpene, C.; Les, F.; Casedas, G.; Peiro, C.; Fontaine, J.; Chaplin, A.; Mercader, J.; Lopez, V. Resveratrol Anti-Obesity Effects: Rapid Inhibition of Adipocyte Glucose Utilization. *Antioxidants (Basel)* **2019**, *8*, 74. [CrossRef] [PubMed]
24. Lin, C.Y.; Hsiao, W.C.; Wright, D.E.; Hsu, C.L.; Lo, Y.C.; Wang Hsu, G.S.; Kao, C.F. Resveratrol activates the histone H2B ubiquitin ligase, RNF20, in MDA-MB-231 breast cancer cells. *J. Funct. Foods* **2013**, *5*, 790–800. [CrossRef]
25. Hudson, T.S.; Hartle, D.K.; Hursting, S.D.; Nunez, N.P.; Wang, T.T.; Young, H.A.; Arany, P.; Green, J.E. Inhibition of prostate cancer growth by muscadine grape skin extract and resveratrol through distinct mechanisms. *Cancer Res.* **2007**, *67*, 8396–8405. [CrossRef] [PubMed]
26. Hsia, S.M.; Wang, K.L.; Wang, P.S. Effects of resveratrol, a grape polyphenol, on uterine contraction and Ca(2)+ mobilization in rats in vivo and in vitro. *Endocrinology* **2011**, *152*, 2090–2099. [CrossRef]
27. Bai, Y.; Lu, H.; Wu, C.; Liang, Y.; Wang, S.; Lin, C.; Chen, B.; Xia, P. Resveratrol inhibits epithelial-mesenchymal transition and renal fibrosis by antagonizing the hedgehog signaling pathway. *Biochem. Pharmacol.* **2014**, *92*, 484–493. [CrossRef]
28. Wu, C.H.; Shieh, T.M.; Wei, L.H.; Cheng, T.F.; Chen, H.Y.; Huang, T.C.; Wang, K.L.; Hsia, S.M. Resveratrol inhibits proliferation of myometrial and leiomyoma cells and decreases extracellular matrix-associated protein expression. *J. Funct. Foods* **2016**, *23*, 241–252. [CrossRef]
29. Hsia, S.M.; Lin, K.H.; Chiang, W.C.; Wu, C.H.; Shieh, T.M.; Huang, T.C.; Chen, H.Y.; Lin, L.C. Effects of Adlay Hull and Testa Ethanolic Extracts on the Growth of Uterine Leiomyoma Cells. *Adapt. Med.* **2017**, *9*, 85–96. [CrossRef]
30. Garvin, S.; Ollinger, K.; Dabrosin, C. Resveratrol induces apoptosis and inhibits angiogenesis in human breast cancer xenografts in vivo. *Cancer Lett.* **2006**, *231*, 113–122. [CrossRef]
31. Kimura, Y.; Okuda, H. Resveratrol Isolated from Polygonum cuspidatum Root Prevents Tumor Growth and Metastasis to Lung and Tumor-Induced Neovascularization in Lewis Lung Carcinoma-Bearing Mice. *J. Nutr.* **2001**, *131*, 1844–1849. [CrossRef]
32. Walle, T.; Hsieh, F.; DeLegge, M.H.; Oatis, J.E., Jr.; Walle, U.K. High absorption but very low bioavailability of oral resveratrol in humans. *Drug. Metab. Dispos.* **2004**, *32*, 1377–1382. [CrossRef] [PubMed]
33. Yu, C.; Shin, Y.G.; Chow, A.; Li, Y.; Kosmeder, J.W.; Lee, Y.S.; Hirschelman, W.H.; Pezzuto, J.M.; Mehta, R.G.; van Breemen, R.B. Human, rat, and mouse metabolism of resveratrol. *Pharm. Res.* **2002**, *19*, 1907–1914. [CrossRef]
34. Caddeo, C.; Nacher, A.; Vassallo, A.; Armentano, M.F.; Pons, R.; Fernandez-Busquets, X.; Carbone, C.; Valenti, D.; Fadda, A.M.; Manconi, M. Effect of quercetin and resveratrol co-incorporated in liposomes against inflammatory/oxidative response associated with skin cancer. *Int. J. Pharm.* **2016**, *513*, 153–163. [CrossRef] [PubMed]
35. Szadvari, I.; Krizanova, O.; Babula, P. Athymic nude mice as an experimental model for cancer treatment. *Physiol. Res.* **2016**, *65*, S441–S453. [PubMed]
36. Suzuki, Y.; Ii, M.; Saito, T.; Terai, Y.; Tabata, Y.; Ohmichi, M.; Asahi, M. Establishment of a novel mouse xenograft model of human uterine leiomyoma. *Sci. Rep.* **2018**, *8*, 8872. [CrossRef] [PubMed]
37. Vaezy, S.; Fujimoto, V.Y.; Walker, C.; Martin, R.W.; Chi, E.Y.; Crum, L.A. Treatment of uterine fibroid tumors in a nude mouse model using high-intensity focused ultrasound. *Am. J. Obstet. Gynecol.* **2000**, *183*, 6–11. [CrossRef]
38. Strzalka, W.; Ziemienowicz, A. Proliferating cell nuclear antigen (PCNA): A key factor in DNA replication and cell cycle regulation. *Ann. Bot.* **2011**, *107*, 1127–1140. [CrossRef]
39. Halder, S.K.; Sharan, C.; Al-Hendy, A. 1, 25-dihydroxyvitamin D3 treatment shrinks uterine leiomyoma tumors in the Eker rat model. *Biol. Reprod.* **2012**, *86*, 116. [CrossRef]
40. Csatlos, E.; Mate, S.; Laky, M.; Rigo, J., Jr.; Joo, J.G. Role of Apoptosis in the Development of Uterine Leiomyoma: Analysis of Expression Patterns of Bcl-2 and Bax in Human Leiomyoma Tissue with Clinical Correlations. *Int. J. Gynecol. Pathol.* **2015**, *34*, 334–339. [CrossRef]
41. Kovacs, K.A.; Lengyel, F.; Kornyei, J.L.; Vertes, Z.; Szabo, I.; Sumegi, B.; Vertes, M. Differential expression of Akt/protein kinase B, Bcl-2 and Bax proteins in human leiomyoma and myometrium. *J. Steroid. Biochem. Mol. Biol.* **2003**, *87*, 233–240. [CrossRef]

42. Rybka, V.; Suzuki, Y.J.; Shults, N.V. Effects of Bcl-2/Bcl-x(L) Inhibitors on Pulmonary Artery Smooth Muscle Cells. *Antioxidants (Basel)* **2018**, *7*, 150. [CrossRef] [PubMed]
43. Baarine, M.; Thandapilly, S.J.; Louis, X.L.; Mazué, F.; Yu, L.; Delmas, D.; Netticadan, T.; Lizard, G.; Latruffe, N. Pro-apoptotic versus anti-apoptotic properties of dietary resveratrol on tumoral and normal cardiac cells. *Genes Nutr.* **2011**, *6*, 161–169. [CrossRef] [PubMed]
44. Walker, C.L.; Stewart, E.A. Uterine fibroids: The elephant in the room. *Science* **2005**, *308*, 1589–1592. [CrossRef]
45. Darby, I.A.; Hewitson, T.D. Fibroblast differentiation in wound healing and fibrosis. *Int. Rev. Cytol.* **2007**, *257*, 143–179. [CrossRef] [PubMed]
46. Rao, K.B.; Malathi, N.; Narashiman, S.; Rajan, S.T. Evaluation of myofibroblasts by expression of alpha smooth muscle actin: A marker in fibrosis, dysplasia and carcinoma. *J. Clin. Diagn. Res.* **2014**, *8*, ZC14-17. [CrossRef]
47. Holdsworth-Carson, S.J.; Zaitseva, M.; Vollenhoven, B.J.; Rogers, P.A. Clonality of smooth muscle and fibroblast cell populations isolated from human fibroid and myometrial tissues. *Mol. Hum. Reprod.* **2014**, *20*, 250–259. [CrossRef]
48. Ko, Y.A.; Jamaluddin, M.F.B.; Adebayo, M.; Bajwa, P.; Scott, R.J.; Dharmarajan, A.M.; Nahar, P.; Tanwar, P.S. Extracellular matrix (ECM) activates beta-catenin signaling in uterine fibroids. *Reproduction* **2018**, *155*, 61–71. [CrossRef]
49. Tanwar, P.S.; Lee, H.J.; Zhang, L.; Zukerberg, L.R.; Taketo, M.M.; Rueda, B.R.; Teixeira, J.M. Constitutive activation of Beta-catenin in uterine stroma and smooth muscle leads to the development of mesenchymal tumors in mice. *Biol. Reprod.* **2009**, *81*, 545–552. [CrossRef]

Article

Inhibitory Effects of Auraptene and Naringin on Astroglial Activation, Tau Hyperphosphorylation, and Suppression of Neurogenesis in the Hippocampus of Streptozotocin-Induced Hyperglycemic Mice

Satoshi Okuyama *, Tatsumi Nakashima, Kumi Nakamura, Wakana Shinoka, Maho Kotani, Atsushi Sawamoto, Mitsunari Nakajima and Yoshiko Furukawa

Department of Pharmaceutical Pharmacology, College of Pharmaceutical Sciences, Matsuyama University, 4-2 Bunkyo-cho, Matsuyama, Ehime 790-8578, Japan; mu.yakuri.009@gmail.com (T.N.); mu.yakuri.008@gmail.com (K.N.); mu.yakuri.010@gmail.com (W.S.); mu.yakuri.011@gmail.com (M.K.); asawamot@g.matsuyama-u.ac.jp (A.S.); mnakajim@g.matsuyama-u.ac.jp (M.N.); furukawa@g.matsuyama-u.ac.jp (Y.F.)

* Correspondence: sokuyama@g.matsuyama-u.ac.jp; Tel.: +81-89-925-7111

Received: 24 July 2018; Accepted: 17 August 2018; Published: 19 August 2018

Abstract: Auraptene, a citrus-related compound, exerts anti-inflammatory effects in peripheral tissues, and we demonstrated these effects in the brains of a lipopolysaccharide-injected systemic inflammation animal model and a brain ischemic mouse model. Naringin, another citrus-related compound, has been shown to exert antioxidant effects in several animal models. Hyperglycemia induces oxidative stress and inflammation and causes extensive damage in the brain; therefore, we herein evaluated the anti-inflammatory and other effects of auraptene and naringin in streptozotocin-induced hyperglycemic mice. Both compounds inhibited astroglial activation and the hyperphosphorylation of tau at 231 of threonine in neurons, and also recovered the suppression of neurogenesis in the dentate gyrus of the hippocampus in hyperglycemic mice. These results suggested that auraptene and naringin have potential effects as neuroprotective agents in the brain.

Keywords: auraptene; naringin; hyperglycemia; neurogenesis; tau phosphorylation; anti-inflammation; anti-oxidation

---

1. Introduction

Recent research shows that hyperglycemia induces inflammation, reactive oxygen species production and neuronal dysfunction in the central nervous system, and those are linked to a number of disorders [1]. Diabetes was recently identified as a risk factor for Alzheimer's disease (AD) [2–4]. The AD brain is frequently associated with severe inflammation, oxidative stress, neuronal dysfunction, amyloid-beta accumulation, tau hyperphosphorylation, and memory impairment [5]. A well-established hyperglycemia diabetes model is induced by streptozotocin (STZ), a glucosamine-nitrosourea compound. STZ induces insulin-secreting pancreatic beta cell death through DNA methylation [6], resulting in chronic hyperglycemia and hypoinsulinemia. Severe oxidative stress, inflammation, tau hyperphosphorylation, and neuronal dysfunction have been observed in the brain of the STZ-induced hyperglycemia model [7,8].

We previously demonstrated that auraptene (AUR), a coumarin compound from citrus fruit, exerted anti-inflammatory and neuroprotective effects in the brain of a lipopolysaccharide (LPS)-injected inflammation model [9]. In this experiment, an intranigral injection of LPS induced microglial activation and dopaminergic neuronal death; on the other hand, an AUR treatment suppressed microglial activation and neuronal cell death. Furthermore, AUR exerted suppressive effects on astrocyte activation and neuronal cell death in the hippocampus of a transient global cerebral ischemic mouse model [10]. Naringin (NGI) is also a biologically active flavonoid substance from citrus that has been shown to exert antioxidant, anti-inflammatory, and neuroprotective effects in the brains of several brain disorder models [11–14]. Therefore, the aim of the present study was to investigate whether both AUR and NGI show anti-inflammatory and anti-tau hyperphosphorylation effects in the STZ-induced hyperglycemia mouse brain.

2. Materials and Methods

2.1. Experiment Schedule

C57BL/6 mice (nine-week-old, male) were purchased from Japan SLC, Inc. (Hamamatsu, Japan). All animal experiments were performed with the approved protocol by the Animal Care and Use Committee of Matsuyama University (#9-002; 2 September, 2009). Mice were housed in a room maintained at a constant temperature of 23 ± 1 °C with 12-h light/dark cycle (lights on 8:00–20:00). Mice were given food and water ad libitum for the duration of the study [15].

Randomly, mice were divided into the following four groups (n = 9 each). Intraperitoneally with vehicle (saline) treatment and the oral administration of vehicle (5% dimethyl sulfoxide/H_2O) were the control (CON) group. Intraperitoneal STZ treatment and the oral administration of the vehicle group were the STZ. STZ treatment and the oral administration of AUR (50 mg/kg; Ushio ChemiX Corp, Omaezaki, Japan) group, named AUR, and STZ treatment and the oral administration of NGI (100 mg/kg; LKT Laboratories, Inc., St. Paul, MN, USA) group, named NGI. Saline or 165 mg/kg of STZ (Wako, Osaka, Japan) was administered intraperitoneally depending on the groups on day 1. On day 8, test sample-administered groups were started to receive each sample orally once a day for 14 days. On day 22, mice were transcardially perfused with ice-cold phosphate-buffered saline (PBS) after the measurement of blood glucose, and the brains were removed.

2.2. Blood Glucose Measurement

To measure fasting glucose concentration on day 22, stock diets were removed from 8:00, and blood glucose was measured at 16:00. A Blood Glucose Monitoring System (Glucose Pilot; Iwai Chemicals Company, Tokyo, Japan) was used to measure blood glucose concentrations with a blood drop from the tail.

2.3. Immunohistochemistry and Immunofluorescence

Sagittal frozen brain sections were prepared at a thickness of 30 μm using a cryostat (CM3050S; Leica Microsystems, Heidelberger, Germany). Immunohistochemistry and immunofluorescence was performed as described in our previous study [15] with the specific primary antibodies (Table 1). Immunopositive signals in the micrographic images were quantified using ImageJ software (National Institute of Health, Bethesda, MD, USA) as described previously [15]. The positive signal densities were quantified using the "measure" tool in ImageJ software.

2.4. Western Blotting Analysis

Equal amounts of protein (25 μg) of hippocampal tissues were separated on 10% sodium dodecyl sulfate-polyacrylamide gels and electroblotted onto an Immun-Blot® PVDF Membrane (Bio-Rad, Hercules, CA, USA) as described in our previous study [16]. Western blotting analysis was performed with the specific primary antibodies (Table 1), and immunoreactive bands were visualized by Amersham ECL Prime Western Blotting Detection Reagent (GE Healthcare Life Sciences, Little Chalfont, UK). The band intensity was captured and measured using a LAS-3000 imaging system (Fujifilm, Tokyo, Japan).

2.5. Statistical Analysis

Data were analyzed by an unpaired *t*-test between two groups (CON vs STZ) and a one-factor ANOVA followed by Dunnett's multiple comparison test among three groups (STZ vs AUR or NGI) (Prism 6; GraphPad Software, La Jolla, CA, USA).

Table 1. Summary of primary antibodies used for immunohistochemiatry, immunofluorescence and western blotting analysis.

Antibody	Epitope Protein/Amino Acids	Host	Dilution	Resource
Iba1	ionized calcium-binding adaptor molecule 1	rabbit	1:1000	Wako, Osaka, Japan
GFAP	glial fibrillary acidic protein	mouse	1:200	Sigma-Aldrich, St. Louis, MO, USA
p-Thr231	phosphorylated-tau Threonine 231	rabbit	1:1000	AnaSpec, Fremont, CA, USA
p-Ser396	phosphorylated-tau Serine 396	rabbit	1:1000	AnaSpec
NeuN	neuronal nuclei	mouse	1:200	Millipore, Temecula, CA, USA
DCX	doublecortin	goat	1:50	Santa Cruz Biotechnology, Santa Cruz, CA, USA
PPARγ	peroxisome proliferator-activated receptor-gamma	rabbit	1:1000	Abcam, Cambridge, UK
Actin	actin	rabbit	1:1000	Sigma-Aldrich

3. Results

3.1. Suppressive Effects of AUR and NGI on Astrocyte Activation

Blood glucose concentrations were significantly elevated in the STZ group (Figure 1; *** $p < 0.001$); however, no significant changes were observed in the AUR and NGI groups.

Hyperglycemia induces inflammation and oxidative stress in the brain [7,8], and excess glial cell activation is known to be responsible for oxidative stress and inflammatory reactions in the brain [17]. Micoglia (Iba1-positive cells) activation were not confirmed in all groups (Figure 2a,b); however, the number of reactive astrocytes that is immunostained with the GFAP antibody significantly increased in the stratum lacunosum-moleculare in the hippocampus of the STZ group (Figure 3a,b; ** $p < 0.01$). In contrast, the GFAP-positive signals were significantly suppressed in the AUR and NGI groups (Figure 3a,b; ## $p < 0.01$, ### $p < 0.001$).

PPARγ regulates the cell signaling of the inflammation process and exerts anti-inflammatory effects [18], and the expression of PPARγ in the hippocampus was significantly suppressed in the STZ group (Figure 4; * $p < 0.05$); however, no significant changes were observed in the AUR and NGI groups.

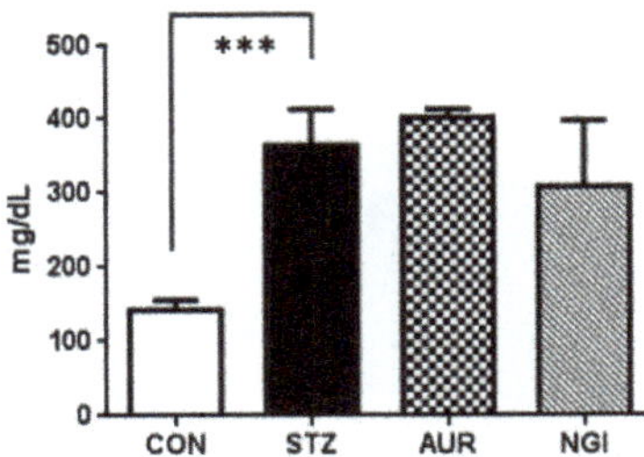

Figure 1. Blood glucose concentrations on Day 22. Values are means ± SEM. The symbol shows a significant difference between CON (control) and STZ (streptozotocin) (*** $p < 0.001$). AUR: auraptene; NGI: naringin.

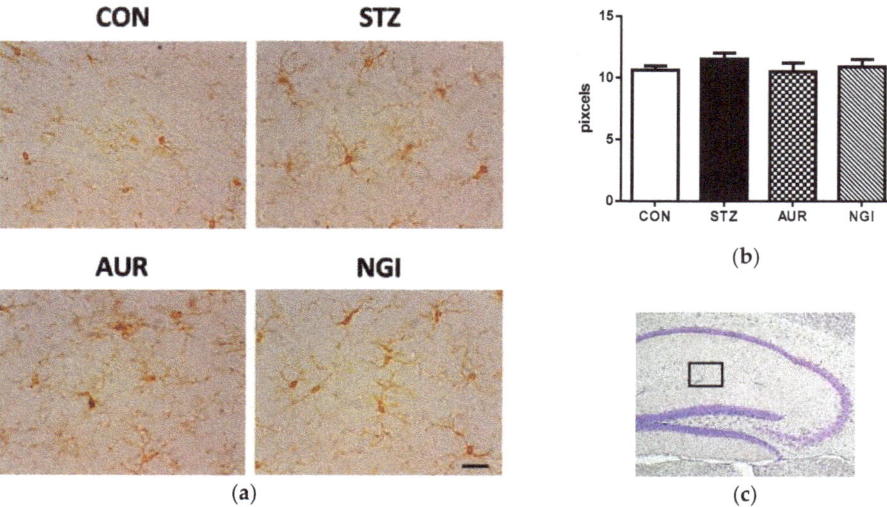

Figure 2. Effects of auraptene and naringin on Iba1 immunoreactivity in the hippocampus. (**a**) Sagittal sections were stained with an anti-Iba1 antibody. The scale bar shows 50 μm. (**b**) Quantitative analysis data of Iba1-positive signals using ImageJ software. (**c**) The location of the captured images. Values are means ± SEM.

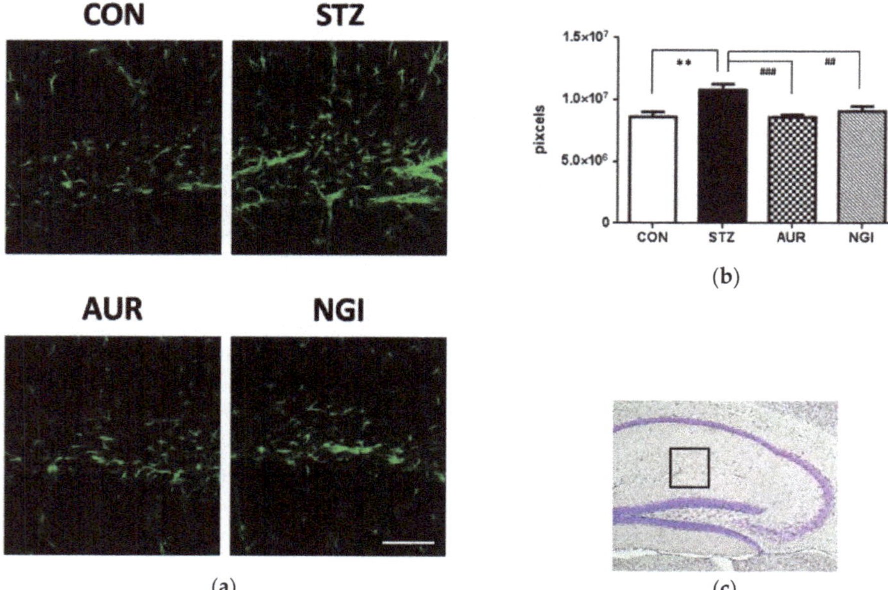

Figure 3. Effects of auraptene and naringin on GFAP immunoreactivity in the hippocampus. (**a**) Sagittal sections were stained with an anti-GFAP antibody. The scale bar shows 100 μm. (**b**) Quantitative analysis data of GFAP-positive signals using ImageJ software. (**c**) The location of the captured images. Values are means ± SEM. Symbols show significant differences between the following conditions: CON vs STZ (** $p < 0.01$), and STZ vs AUR or NGI (## $p < 0.01$, ### $p < 0.001$).

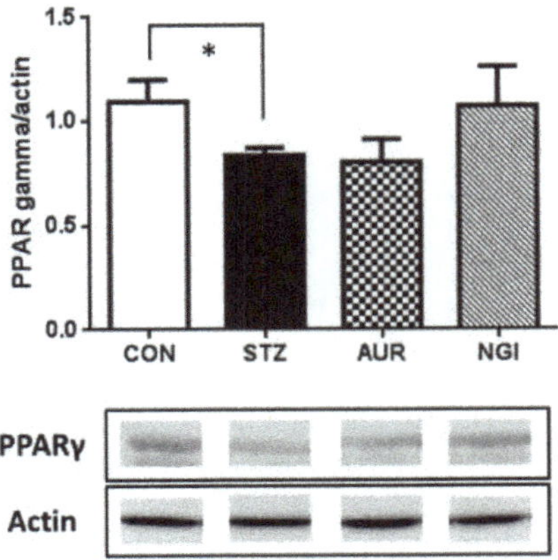

Figure 4. Effects of auraptene and naringin on the expression of PPARγ in the hippocampus. Values are means ± SEM. The symbol shows a significant difference between CON and STZ (* $p < 0.05$).

3.2. Effects of AUR and NGI on Tau Hyperphosphorylation

Tau, a microtubule-associated cytoskeletal protein, in the neuron has been understanding to relate to the molecular mechanisms for neurofibrillary tangle (NFT) formation through its multiple phosphorylative and conformational changes. It was previously shown that inflammation and oxidative stress induce tau hyperphosphorylation [19,20], and this was confirmed in STZ-injected mice brain [21]. We evaluated tau phosphorylation levels at 231 of threonine (p-Thr231) and 396 of serine (p-Ser396) in the hippocampus (Figures 5 and 6). Strong positive signals for p-Thr231 were confirmed in the CA3 region pyramidal cell layer and in the stratum lacunosum-moleculare in the hippocampus. The integrated densities of the immune-positive signals were evaluated, and that was significantly higher in the STZ group (Figure 5a,b; ** $p < 0.01$); however, significant suppressive effects on its phosphorylation was observed in the AUR and NGI treated groups (Figure 5a,b; # $p < 0.05$). Similar to p-Thr231, the levels of p-Ser396 in the hippocampus were significantly higher in the STZ group than in the CON group (Figure 6a,b; * $p < 0.05$). The signals were confirmed in hippocampal mossy fibers and the stratum lacunosum-moleculare. On the other hand, the AUR and NGI treatments exerted a tendency of suppressive effects (Figure 6a,b; $p = 0.056$ and $p = 0.087$, respectively).

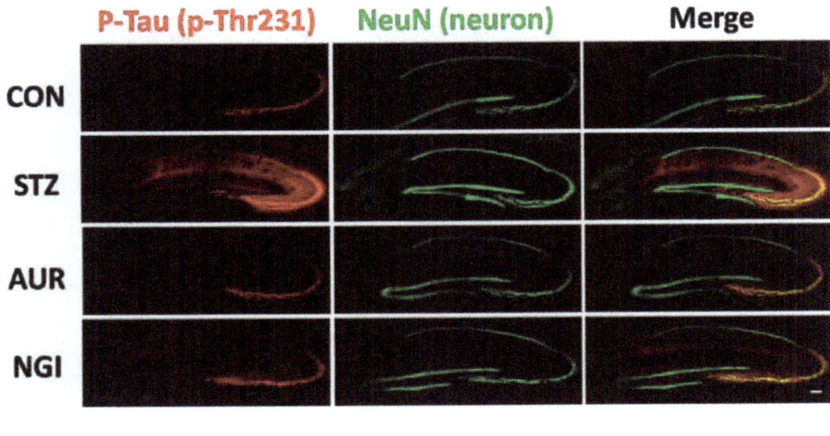

(a)

Figure 5. *Cont.*

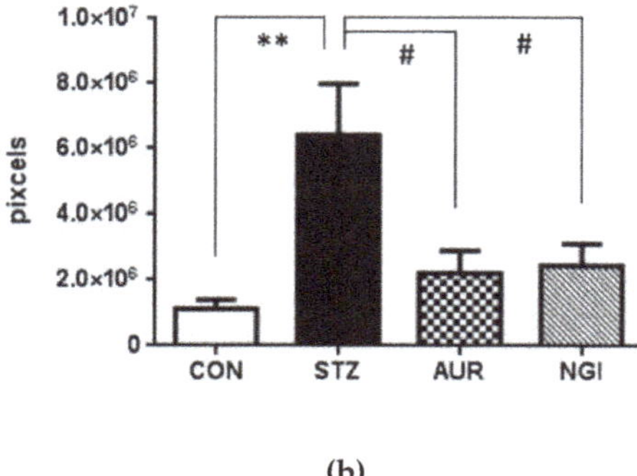

(b)

Figure 5. Effects of auraptene and naringin on the expression of phosphorylated Tau at 231 of threonine in the hippocampus. (**a**) Sagittal sections were stained with the anti-pThr231 (red) and NeuN (green) antibodies. The scale bar shows 100 μm. (**b**) Quantitative analysis data of pThr231-positive signals using ImageJ software. Values are means ± SEM. Symbols show significant differences between the following conditions: CON vs STZ (** $p < 0.01$), and STZ vs AUR or NGI (# $p < 0.05$).

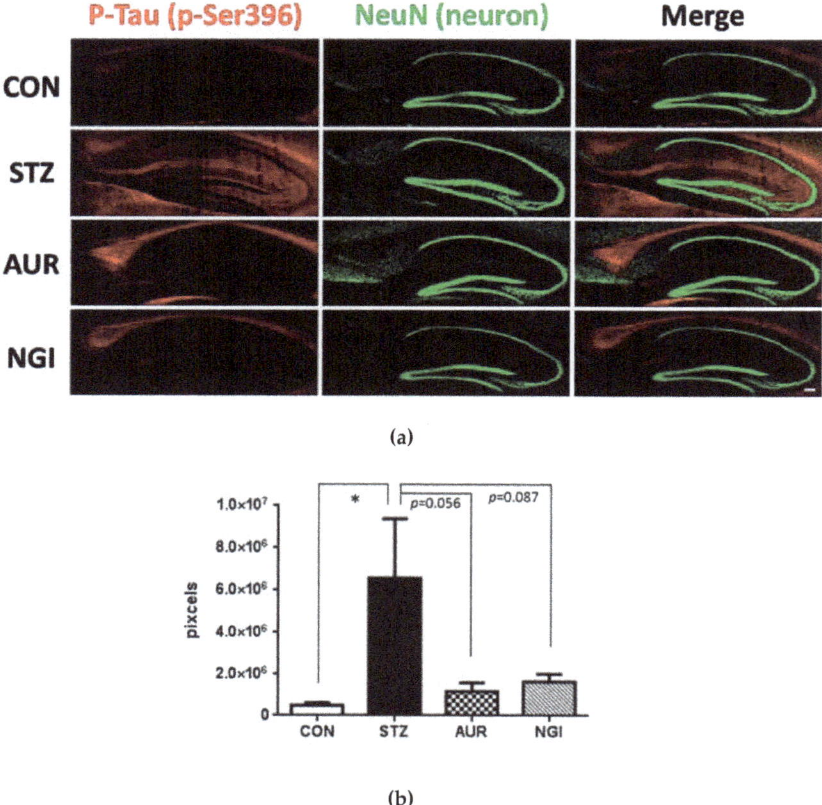

(a)

(b)

Figure 6. Effects of auraptene and naringin on the expression of phosphorylated Tau at 396 of serine in the hippocampus. (**a**) Sagittal sections were stained with the anti-pSer396 (red) and NeuN (green) antibodies. The scale bar shows 100 μm. (**b**) Quantitative analysis data of pSer396-positive signals using ImageJ software. Values are means ± SEM. Symbols show significant differences between CON vs STZ (* $p < 0.05$).

3.3. Enhancement of Neurogenesis by AUR and NGI in the Hippocampus

The subgranular zone (SGZ) of the dentate gyrus (DG) is one of the areas in which neurogenesis occurs in the hippocampus [22], and adult neurogenesis is known to play an important role in learning and memory. A previous report showed that suppression of the neurogenesis in the DG was confirmed, following a STZ administration [23]. Figure 6 shows the immunoreactivity of DCX, a marker for immature neurons, in the SGZ. DCX-positive cells were suppressibility observed in the STZ group, but markedly higher number in the AUR and NGI groups (Figure 7a). We manually counted the DCX-positive neurons, with checking the soma and nucleus, in the SGZ under the fluorescence microscopy and evaluated the number of DCX-positive newborn neurons in the SGZ. The numbers of the STZ group showed significantly suppressed expression (Figure 7b; *** $p < 0.001$); in contrast, the AUR and NGI groups ameliorated this suppression (Figure 7b; # $p < 0.05$, ## $p < 0.01$).

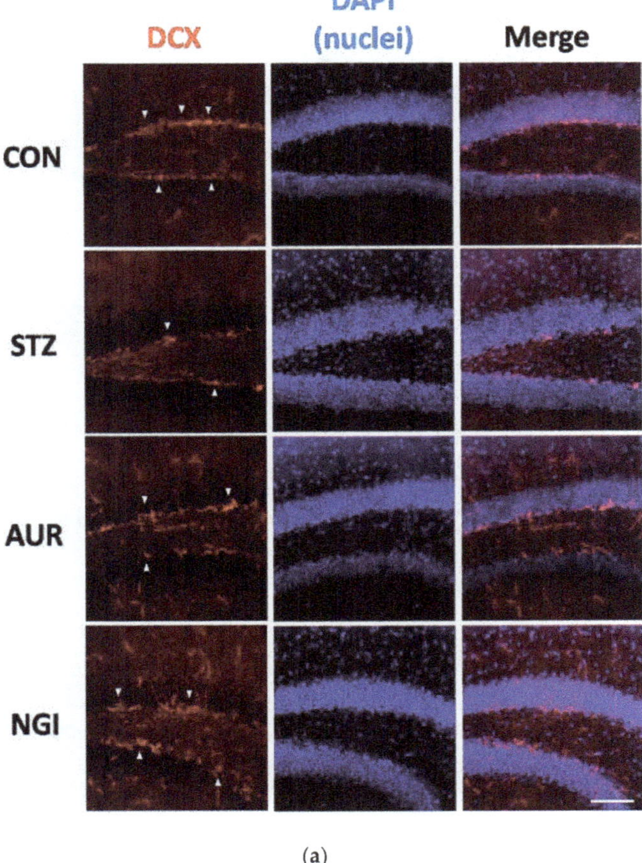

(a)

Figure 7. *Cont.*

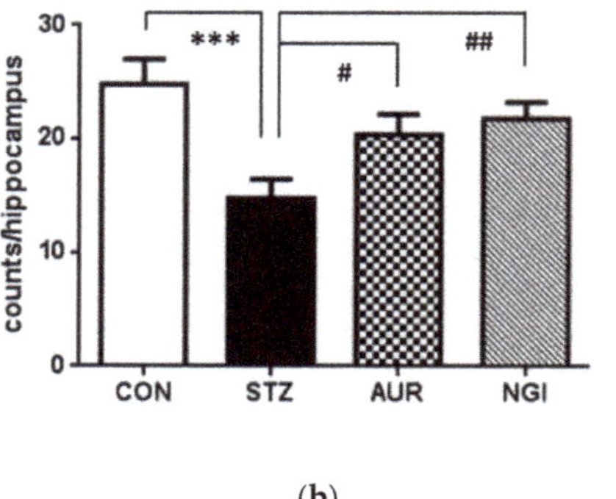

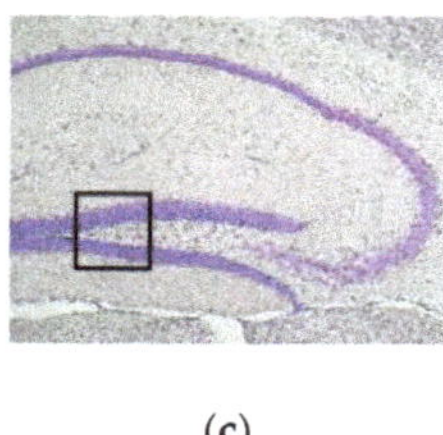

(**b**) (**c**)

Figure 7. Effects of auraptene and naringin on doublecortin immunoreactivity in the hippocampus. (**a**) Sagittal sections were stained with the anti-DCX antibody (red) and DAPI (blue). The white arrowheads indicate typical DCX-positive cells in the DG (dentate gyrus). The scale bar shows 100 μm. (**b**) Counting data of DCX-positive signals in the dentate gyrus. (**c**) The location of the captured images is shown with a square in the figure. Values are means ± SEM. Symbols show significant differences between the following conditions: CON vs STZ (*** $p < 0.001$), and STZ vs AUR or NGI (# $p < 0.05$, ## $p < 0.01$).

4. Discussion

Hyperglycemia induces inflammation and oxidative stress in the central nervous system [24], and diabetes is currently regarded as one of the risk factors for dementia, such as AD and vascular dementia [2–4]. The AD brain is frequently associated with severe inflammation, oxidative stress, neuronal dysfunction, amyloid-beta accumulation, tau hyperphosphorylation, and memory impairment [5]. Several studies indicate inflammation and oxidative stress increased tau hyperphosphorylation of neurons in the brain [19,20]; therefore, our primary focus was to clarify the anti-inflammatory and tau hyperphosphorylation suppression effects of AUR and NGI in a hyperglycemia model.

STZ-treated mice showed significantly elevated blood glucose concentrations, whereas suppressive effects were not observed in the NGI and AUR groups (Figure 1). Previous studies reported that the administration of 100 mg/kg of NGI suppressed blood glucose concentrations in STZ-treated rats, or 0.2% of AUR administration resulted in suppressive effects on high-fat diet-induced obese mice, respectively [25,26]. These findings suggested that NGI and AUR have the potential to reduce blood glucose concentrations in several diabetic models. However, treatments with NGI and AUR did not reduce blood glucose concentrations in our experiment.

In hyperglycemia models, astrocyte activation is related to immune response in the brain [27,28]. Treatments with AUR and NGI suppressed astrocyte activation in this experiment (Figure 3), and we previously demonstrated that the activation of astrocytes was inhibited by the AUR treatment in a transient global cerebral ischemic mouse model [10]. Strong microglial activation, also an immune response cell in the brain, was not observed in the STZ group (Figure 2) in this experiment, whereas astrocytes were detected. Microglia is activated earlier than astrocytes in a manner that is dependent on disease conditions [29]; therefore, we considered strong microglial activation to only occur in the early phase of the STZ treatment. In fact, we previously demonstrated that microglial activation was confirmed one week after the administration of STZ [30]; however, the time point of sacrifice in this experiment was two weeks after the STZ treatment. PPARγ is an important target in diabetes therapy, and regulates the cell signaling of the inflammation process [18]. AUR and NGI did not significantly affect the protein expression of PPARγ in the hippocampus in this experiment (Figure 4); on the other hand, a previous study reported that NGI ameliorated cognitive deficits via oxidative stress and proinflammatory factor suppression, and activated the protein expression of PPAR in the hippocampus of an STZ-injected rat model [25]. AUR also activates PPARγ in adipocytes [31]. Collectively, these findings in our experiment indicated that AUR and NGI exerted anti-inflammatory effects by suppressing the activation of astrocytes in the hippocampus, though we still have to do further experiments to see the detail mechanism.

Increases in inflammation and oxidative stress induce the hyperphosphorylated tau protein in neurons and this is enhanced in the hyperglycemic brain [19,20]. The microtubule function of neurons is maintained by the phosphorylation of the tau protein, and the regulation of kinases (including CDK-5 and GSK-3β) and phosphatases (such as PP2A) are very important; however, the hyperphosphorylation of tau induces microtubule dysfunction, leading to the formation of NFT, which is often observed in the AD brain [32]. In tau protein, several strong phosphorylation sites have been identified, such as Thr231 and Ser396, in hyperglycemic animal brains [21,33], and oxidative stress and inflammation may induce a kinase and phosphatase imbalance [34]. We focused on phosphorylation sites in Thr231 and Ser396 in the present study, and AUR and NGI treatment exerted suppressive effects on tau phosphorylation in the hippocampus in STZ-treated mice (Figures 5 and 6). Neurogenesis in the SGZ of the DG in the hippocampus is of particular importance for hippocampal-dependent memory function [35,36]. It is suppressed by a number of conditions, including depression, AD, and aging; in addition, hyperglycemia has also been shown to suppress neurogenesis [37,38]. Staining for neurogenesis with DCX, the AUR and NGI treatments enhanced its expression (Figure 7). Inhibitory effects of AUR and NGI on tau hyperphosphorylation and suppression of neurogenesis were our new findings.

5. Conclusions

When AUR and NGI were administered to STZ-injected hyperglycemia mice, they (1) suppressed astroglial activation; (2) diminished tau phosphorylation; and (3) stimulated neurogenesis in the SGZ of the DG in the hippocampus. These results suggest that AUR and NGI, citrus-related compounds, exert anti-inflammatory and antioxidative effect against hyperglycemia-induced changes in the brain, and have potential as novel neuroprotective agents obtained from food materials.

Author Contributions: S.O. designed and performed the research, and wrote the manuscript; T.N., K.N., W.S., M.K., and A.S. performed the research; M.N. and Y.F. supervised this research project.

Funding: This work was funded by The Tojuro Iijima Foundation for Food Science and Technology, year of 2015 and 2017, and JSPS KAKENHI Grant Number JP26750058.

Acknowledgments: The authors wish to thank Ushio ChemiX Corporation for kindly supplying AUR.

Conflicts of Interest: The authors declare no conflict of interest.

References

1. Dinel, A.L.; André, C.; Aubert, A.; Ferreira, G.; Layé, S.; Castanon, N. Cognitive and emotional alterations are related to hippocampal inflammation in a mouse model of metabolic syndrome. *PLoS ONE* **2011**, *6*, e24325. [CrossRef] [PubMed]
2. Kim, B.; Backus, C.; Oh, S.; Feldman, E.L. Hyperglycemia-induced tau cleavage in vitro and in vivo: A possible link between diabetes and Alzheimer's disease. *J. Alzheimer's Dis.* **2013**, *34*, 727–739. [CrossRef] [PubMed]
3. Kimura, N. Diabetes Mellitus Induces Alzheimer's Disease Pathology: Histopathological Evidence from Animal Models. *Int. J. Mol. Sci.* **2016**, *17*, 503. [CrossRef] [PubMed]
4. Bosco, D.; Fava, A.; Plastino, M.; Montalcini, T.; Pujia, A. Possible implications of insulin resistance and glucose metabolism in Alzheimer's disease pathogenesis. *J. Cell Mol. Med.* **2011**, *15*, 1807–1821. [CrossRef] [PubMed]
5. McKee, A.C.; Carreras, I.; Hossain, L.; Ryu, H.; Klein, W.L.; Oddo, S.; LaFerla, F.M.; Jenkins, B.G.; Kowall, N.W.; Dedeoglu, A. Ibuprofen reduces Abeta, hyperphosphorylated tau and memory deficits in Alzheimer mice. *Brain Res.* **2008**, *1207*, 225–236. [CrossRef] [PubMed]
6. Ho, N.; Balu, D.T.; Hilario, M.R.; Blendy, J.A.; Lucki, I. Depressive phenotypes evoked by experimental diabetes are reversed by insulin. *Physiol. Behav.* **2012**, *105*, 702–708. [CrossRef] [PubMed]

7. Cai, Z.; Yan, Y.; Wang, Y. Minocycline alleviates beta-amyloid protein and tau pathology via restraining neuroinflammation induced by diabetic metabolic disorder. *Clin. Interv. Aging* **2013**, *8*, 1089–1095. [CrossRef] [PubMed]
8. Sharma, R.; Buras, E.; Terashima, T.; Serrano, F.; Massaad, C.A.; Hu, L.; Bitner, B.; Inoue, T.; Chan, L.; Pautler, R.G. Hyperglycemia induces oxidative stress and impairs axonal transport rates in mice. *PLoS ONE* **2010**, *5*, e13463. [CrossRef] [PubMed]
9. Okuyama, S.; Semba, T.; Toyoda, N.; Epifano, F.; Genovese, S.; Fiorito, S.; Taddeo, V.A.; Sawamoto, A.; Nakajima, M.; Furukawa, Y. Auraptene and Other Prenyloxyphenylpropanoids Suppress Microglial Activation and Dopaminergic Neuronal Cell Death in a Lipopolysaccharide-Induced Model of Parkinson's Disease. *Int. J. Mol. Sci.* **2016**, *17*, 1716. [CrossRef] [PubMed]
10. Okuyama, S.; Morita, M.; Kaji, M.; Amakura, Y.; Yoshimura, M.; Shimamoto, K.; Ookido, Y.; Nakajima, M.; Furukawa, Y. Auraptene Acts as an Anti-Inflammatory Agent in the Mouse Brain. *Molecules* **2015**, *20*, 20230–20239. [CrossRef] [PubMed]
11. Gaur, V.; Aggarwal, A.; Kumar, A. Protective effect of naringin against ischemic reperfusion cerebral injury: Possible neurobehavioral, biochemical and cellular alterations in rat brain. *Eur. J. Pharmacol.* **2009**, *616*, 147–154. [CrossRef] [PubMed]
12. Leem, E.; Nam, J.H.; Jeon, M.T.; Shin, W.H.; Won, S.Y.; Park, S.J.; Choi, M.S.; Jin, B.K.; Jung, U.J.; Kim, S.R. Naringin protects the nigrostriatal dopaminergic projection through induction of GDNF in a neurotoxin model of Parkinson's disease. *J. Nutr. Biochem.* **2014**, *25*, 801–806. [CrossRef] [PubMed]
13. Golechha, M.; Chaudhry, U.; Bhatia, J.; Saluja, D.; Arya, D.S. Naringin protects against kainic acid-induced status epilepticus in rats: Evidence for an antioxidant, anti-inflammatory and neuroprotective intervention. *Biol. Pharm. Bull.* **2011**, *34*, 360–365. [CrossRef] [PubMed]
14. Okuyama, S.; Yamamoto, K.; Mori, H.; Sawamoto, A.; Amakura, Y.; Yoshimura, M.; Tamanaha, A.; Ohkubo, Y.; Sugawara, K.; Sudo, M.; et al. Neuroprotective effect of *Citrus kawachiensis* (*Kawachi Bankan*) peels, a rich source of naringin, against global cerebral ischemia/reperfusion injury in mice. *Biosci. Biotechnol. Biochem.* **2018**, *82*, 1216–1224. [CrossRef] [PubMed]
15. Okuyama, S.; Morita, M.; Miyoshi, K.; Nishigawa, Y.; Kaji, M.; Sawamoto, A.; Terugo, T.; Toyoda, N.; Makihata, N.; Amakura, Y.; et al. 3,5,6,7,8,3′,4′-Heptamethoxyflavone, a citrus flavonoid, on protection against memory impairment and neuronal cell death in a global cerebral ischemia mouse model. *Neurochem. Int.* **2014**, *70*, 30–38. [CrossRef] [PubMed]
16. Okuyama, S.; Shimada, N.; Kaji, M.; Morita, M.; Miyoshi, K.; Minami, S.; Amakura, Y.; Yoshimura, M.; Yoshida, T.; Watanabe, S.; et al. Heptamethoxyflavone, a citrus flavonoid, enhances brain-derived neurotrophic factor production and neurogenesis in the hippocampus following cerebral global ischemia in mice. *Neurosci. Lett.* **2012**, *528*, 190–195. [CrossRef] [PubMed]
17. Wilms, H.; Sievers, J.; Rickert, U.; Rostami-Yazdi, M.; Mrowietz, U.; Lucius, R. Dimethylfumarate inhibits microglial and astrocytic inflammation by suppressing the synthesis of nitric oxide, IL-1beta, TNF-alpha and IL-6 in an in vitro model of brain inflammation. *J. Neuroinflamm.* **2010**, *7*, 30. [CrossRef] [PubMed]
18. Collino, M.; Aragno, M.; Mastrocola, R.; Gallicchio, M.; Rosa, A.C.; Dianzani, C.; Danni, O.; Thiemermann, C.; Fantozzi, R. Modulation of the oxidative stress and inflammatory response by PPAR-gamma agonists in the hippocampus of rats exposed to cerebral ischemia/reperfusion. *Eur. J. Pharmacol.* **2006**, *530*, 70–80. [CrossRef] [PubMed]
19. Barron, M.; Gartlon, J.; Dawson, L.A.; Atkinson, P.J.; Pardon, M.C. A state of delirium: Deciphering the effect of inflammation on tau pathology in Alzheimer's disease. *Exp. Gerontol.* **2017**, *94*, 103–107. [CrossRef] [PubMed]
20. Wu, Z.; Nakanishi, H. Connection between periodontitis and Alzheimer's disease: Possible roles of microglia and leptomeningeal cells. *J. Pharmacol. Sci.* **2014**, *126*, 8–13. [CrossRef] [PubMed]
21. Clodfelder-Miller, B.J.; Zmijewska, A.A.; Johnson, G.V.; Jope, R.S. Tau is hyperphosphorylated at multiple sites in mouse brain in vivo after streptozotocin-induced insulin deficiency. *Diabetes* **2006**, *55*, 3320–3325. [CrossRef] [PubMed]
22. Aimone, J.B.; Li, Y.; Lee, S.W.; Clemenson, G.D.; Deng, W.; Gage, F.H. Regulation and function of adult neurogenesis: From genes to cognition. *Physiol. Rev.* **2014**, *94*, 991–1026. [CrossRef] [PubMed]

23. Guo, J.; Yu, C.; Li, H.; Liu, F.; Feng, R.; Wang, H.; Meng, Y.; Li, Z.; Ju, G.; Wang, J. Impaired neural stem/progenitor cell proliferation in streptozotocin-induced and spontaneous diabetic mice. *Neurosci. Res.* **2010**, *68*, 329–336. [CrossRef] [PubMed]
24. Maher, P.; Dargusch, R.; Ehren, J.L.; Okada, S.; Sharma, K.; Schubert, D. Fisetin lowers methylglyoxal dependent protein glycation and limits the complications of diabetes. *PLoS ONE* **2011**, *6*, e21226. [CrossRef] [PubMed]
25. Qi, Z.; Xu, Y.; Liang, Z.; Li, S.; Wang, J.; Wei, Y.; Dong, B. Naringin ameliorates cognitive deficits via oxidative stress, proinflammatory factors and the PPARγ signaling pathway in a type 2 diabetic rat model. *Mol. Med. Rep.* **2015**, *12*, 7093–7101. [CrossRef] [PubMed]
26. Takahashi, N.; Senda, M.; Lin, S.; Goto, T.; Yano, M.; Sasaki, T.; Murakami, S.; Kawada, T. Auraptene regulates gene expression involved in lipid metabolism through PPARα activation in diabetic obese mice. *Mol. Nutr. Food Res.* **2011**, *55*, 1791–1797. [CrossRef] [PubMed]
27. Infante-Garcia, C.; Jose Ramos-Rodriguez, J.; Marin-Zambrana, Y.; Teresa Fernandez-Ponce, M.; Casas, L.; Mantell, C.; Garcia-Alloza, M. Mango leaf extract improves central pathology and cognitive impairment in a type 2 diabetes mouse model. *Brain Pathol.* **2017**, *27*, 499–507. [CrossRef] [PubMed]
28. Zheng, Y.; Yang, Y.; Dong, B.; Zheng, H.; Lin, X.; Du, Y.; Li, X.; Zhao, L.; Gao, H. Metabonomic profiles delineate potential role of glutamate-glutamine cycle in db/db mice with diabetes-associated cognitive decline. *Mol. Brain* **2016**, *9*, 40. [CrossRef] [PubMed]
29. Webster, K.M.; Sun, M.; Crack, P.; O'Brien, T.J.; Shultz, S.R.; Semple, B.D. Inflammation in epileptogenesis after traumatic brain injury. *J. Neuroinflamm.* **2017**, *14*, 10. [CrossRef] [PubMed]
30. Okuyama, S.; Shinoka, W.; Nakamura, K.; Kotani, M.; Sawamoto, A.; Sugawara, K.; Sudo, M.; Nakajima, M.; Furukawa, Y. Suppressive effects of the peel of *Citrus kawachiensis* (*Kawachi Bankan*) on astroglial activation, tau phosphorylation, and inhibition of neurogenesis in the hippocampus of type 2 diabetic db/db mice. *Biosci. Biotechnol. Biochem.* **2018**, *82*, 1216–1224. [CrossRef] [PubMed]
31. Kuroyanagi, K.; Kang, M.S.; Goto, T.; Hirai, S.; Ohyama, K.; Kusudo, T.; Yu, R.; Yano, M.; Sasaki, T.; Takahashi, N.; et al. Citrus auraptene acts as an agonist for PPARs and enhances adiponectin production and MCP-1 reduction in 3T3-L1 adipocytes. *Biochem. Biophys. Res. Commun.* **2008**, *366*, 219–225. [CrossRef] [PubMed]
32. Medeiros, R.; Baglietto-Vargas, D.; LaFerla, F.M. The role of tau in Alzheimer's disease and related disorders. *CNS Neurosci. Ther.* **2011**, *17*, 514–524. [CrossRef] [PubMed]
33. Kim, B.; Backus, C.; Oh, S.; Hayes, J.M.; Feldman, E.L. Increased tau phosphorylation and cleavage in mouse models of type 1 and type 2 diabetes. *Endocrinology* **2009**, *150*, 5294–5301. [CrossRef] [PubMed]
34. Ballatore, C.; Lee, V.M.; Trojanowski, J.Q. Tau-mediated neurodegeneration in Alzheimer's disease and related disorders. *Nat. Rev. Neurosci.* **2007**, *8*, 663–672. [CrossRef] [PubMed]
35. Winocur, G.; Wojtowicz, J.M.; Sekeres, M.; Snyder, J.S.; Wang, S. Inhibition of neurogenesis interferes with hippocampus-dependent memory function. *Hippocampus* **2006**, *16*, 296–304. [CrossRef] [PubMed]
36. Shors, T.J.; Miesegaes, G.; Beylin, A.; Zhao, M.; Rydel, T.; Gould, E. Neurogenesis in the adult is involved in the formation of trace memories. *Nature* **2001**, *410*, 372–376. [CrossRef] [PubMed]
37. Yi, S.S.; Hwang, I.K.; Choi, J.W.; Won, M.H.; Seong, J.K.; Yoon, Y.S. Effects of hypothyroidism on cell proliferation and neuroblasts in the hippocampal dentate gyrus in a rat model of type 2 diabetes. *Anat. Cell Biol.* **2010**, *43*, 185–193. [CrossRef] [PubMed]
38. Hamilton, A.; Patterson, S.; Porter, D.; Gault, V.A.; Holscher, C. Novel GLP-1 mimetics developed to treat type 2 diabetes promote progenitor cell proliferation in the brain. *J. Neurosci. Res.* **2011**, *89*, 481–489. [CrossRef] [PubMed]

Article

Protective Effect of Aqueous Extract from the Leaves of *Justicia tranquebariesis* against Thioacetamide-Induced Oxidative Stress and Hepatic Fibrosis in Rats

Kumeshini Sukalingam [1], Kumar Ganesan [1,2] and Baojun Xu [2,*]

[1] Faculty of Medicine, International Medical School, Management and Science University, Shah Alam 40100, Malaysia; meshni_anat@yahoo.com.my (K.S.); drbiokumar@yahoo.com (K.G.)

[2] Food Science and Technology Program, Beijing Normal University–Hong Kong Baptist University United International College, Zhuhai 519087, China

* Correspondence: baojunxu@uic.edu.hk; Tel.: +86-756-3620-636; Fax: +86-756-3620-882

Received: 6 May 2018; Accepted: 17 June 2018; Published: 22 June 2018

Abstract: The present study aims to examine the protective effect of *Justicia tranquebariesis* on thioacetamide (TAA)-induced oxidative stress and hepatic fibrosis. Male Wister albino rats (150–200 g) were divided into five groups. Group 1 was normal control. Group 2 was *J. tranquebariensis* (400 mg/kg bw/p.o.)-treated control. Group 3 was TAA (100 mg/kg bw/s.c.)-treated control. Groups 4 and 5 were orally administered with the leaf extract of *J. tranquebariensis* (400 mg/kg bw) and silymarin (50 mg/kg bw) daily for 10 days with a subsequent administration of a single dose of TAA (100 mg/kg/s.c.). Blood and livers were collected and assayed for various antioxidant enzymes (SOD, CAT, GPx, GST, GSH, and GR). Treatment with *J. tranquebariensis* significantly reduced liver TBARS and enhanced the activities of antioxidant enzymes in TAA-induced fibrosis rats. Concurrently, pretreatment with *J. tranquebariensis* significantly reduced the elevated liver markers (AST, ALT, ALP, GGT, and TB) in the blood. In addition, *J. tranquebariensis*- and silymarin- administered rats demonstrated the restoration of normal liver histology and reduction in fibronectin and collagen deposition. Based on these findings, *J. tranquebariensis* has potent liver protective functions and can alleviate thioacetamide-induced oxidative stress, hepatic fibrosis and possible engross mechanisms connected to antioxidant potential.

Keywords: *Justicia tranquebariesis*; TAA; oxidative stress; hepatic fibrosis; hepatoprotection

1. Introduction

Liver fibrosis is a scarring mechanism of the liver related to prominent accretion of an extracellular matrix. Without efficient treatment at an early stage, reversible liver fibrosis develops into irreversible cirrhosis leading to liver failure, portal hypertension and the need for liver transplantation in many cases [1]. These progressive scarring insults result in liver cirrhosis, which is the foremost health burden leading to death worldwide. At present, no known therapeutic approach is widely available, except for the removal of the fibrogenic stimuli. However, in vivo and clinical studies have established that liver fibrosis and cirrhosis are reversible to heal, but still inadequate for widely used known treatments [2–5]. The liver fibrosis is usually caused by diverse chronic insults including, chemicals, parasitic infections, alcohol, viral hepatitis B and C, and autoimmune hepatitis. Due to the worldwide occurrence of these insults, liver fibrosis is widespread and is related to liver linked morbidity and mortality [6]. Chronic liver injury normally causes progressive liver fibrosis distinguished by alterations of both quality and quantity of hepatic extracellular matrix proteins including collagen, which occurs in most types of end-stage liver diseases [7]. Furthermore, this chronic liver injury is triggered by an inappropriate balance between the generation and destruction of ROS, which results in hepatocyte damage and abnormal tissue injury [8]. Thioacetamide (TAA) is an organosulphur. It is a chemical, which is extensively used as a fungicide in various industries including textile dyes [7]. Presently, it is considered as a carcinogen. It is rapidly metabolized into free radical derivatives such as TAA sulfoxide and TAA-S-S-dioxide, which leads to lipid peroxidation, and eventually culminates in centrilobular damage and liver injury [9]. Earlier studies have also demonstrated that the exposure to TAA caused liver injury, fibrosis, steatosis, and cirrhosis in experimental animals [1,9]. Hence, TAA is recognized as a model of liver fibrosis in rats. Currently, the widely used treatment of liver fibrosis and cirrhosis is inadequate; and there is no effectively widely used therapy that can prevent the development of hepatic diseases. Although recently developed drugs have been used to heal liver diseases, often these drugs have numerous side-effects. There is, thus, an urgently requirement for alternative remedies or drugs for the treatment of chronic liver disorders to replace present drugs of uncertain safety and effectiveness [10]. For this intention, herbal constituents and dietary supplements have potential as alternative medicines for the treatment of chronic liver diseases and related metabolic derailments [11].

Justicia tranquebariensis L. (Family: *Acanthaceae*), a common shrub is broadly scattered in various regions of India, Malaysia and Sri Lanka. The fresh leaves are coolant and aperient, generally used in jaundice, liver diseases, and smallpox [12]. Bruised leaves are applied to contusions, diaphoretic, diuretic, and rheumatism and used as antidotes for snake bite [13]. The juice of the leaves is used as an expectorant in cold, cough, nasal disorders, whereas, the paste of the leaves is applied externally to treat skin diseases, swelling and pain. The root could also be made into a paste to treat toothaches. Phytochemical studies of *J. tranquebariensis* revealed that the leaves contain adequate quantities of phytosterols, flavonoids, glycosides, triterpenoids, alkaloids, saponins, and tannins [14,15]. Aerial parts of the plant contain lignans including aryl tetralin, β-cubebin, lariciresinol, isolariciresinol, lyoniresinol and medioresinol [16]. In addition, the alcoholic extract yielded 7,22-ergostadienol, 28-isofucosterol, β-sitosterol-3-*O*-glucoside, brassicasterol, campesterol, stigmasterol, sitosterol, and spinasterol [12]. Furthermore, the plant has various pharmacological potentials including free radical scavenging, anti-inflammatory [14], antipyretic [15], antimicrobial [17], and antihepatotoxicity [18] potentials. Based on previous literature, no information was available on the hepatoprotective and antioxidant effect of the *J. tranquebariensis* leaf extract on liver fibrosis. Thus, this current investigation aimed to observe the antifibrotic and hepatoprotective effect of *J. tranquebariesis* on TAA-intoxicated rats.

2. Materials and Methods

2.1. Plant Material and Preparation

The fresh leaves of *Justicia tranquebariensis* L. (Figure 1) were collected during February–March 2014 from Shah Alam, Selangor, Malaysia and were authenticated by a taxonomist, Sujit Sarker, Department of Pharmacognosy and comparison with reference materials conserved in the Herbarium and voucher specimens were kept in the institution. Coarse powdered leaves (1000 g) were subjected to 2 L of distilled water, and extraction was maintained with regular stirring for 8 h. The extract was centrifuged, and the supernatants were evaporated using a vacuum rotary evaporator and residues were kept in refrigeration for further use (yield: 180 g/1000 g).

2.2. Animals

Male Wister albino rats (150–200 g) were acquired from the central animal house in Management and Science University, Shah Alam, Selangor, Malaysia. Animals were kept in animal cages under standard lab conditions (12 h alternating day and night cycles, rooms were air-conditioning at 25–28 °C). Rats were adapted to the laboratory settings for a week prior to the initiation of experimental treatments. Rats were supplied with free access to standard pellet food and water. The experimental studies were carried out based on the ethical approval and the protocol was permitted by the Institutional Ethical Committee of Management and Science University, Malaysia (Reg no: 12/2011/CPCSEA, proposal no: 75).

Figure 1. Leaves of *Justicia tranquebariensis* L.

2.3. Chemicals

TAA, thiobarbituric acid (TBA), 1-chloro-2,4-dinitrobenzene (CDNB), and nicotinamide adenine dinucleotide hydrogen phosphate (NADPH) were procured from Sigma-Aldrich Co., St. Louis, MO, USA. Reagents and chemicals used in the studies were of the analytical grade.

2.4. Dose Determination

A preliminary study was carried out to validate the optimal dose of plant extract by examining serum hepatic marker enzymes in TAA-intoxicated rats. Administration of aqueous extract from the leaves of *J. tranquebariensis* was given at various doses of 100, 200, 400, 800 mg/kg bw to different groups of rats. Among the doses, the 400 mg/kg bw showed more effectiveness than the other doses. Hence 400 mg/kg bw was used in this investigation. The dosage of TAA (100 mg/kg bw) and standard drug silymarin (50 mg/kg bw) used in the present study were chosen according to a previous study [19].

2.5. Experimental Design

Following laboratory adaptation, animals were randomly divided into five groups, each group comprising of six rats. All rats were kept fasting for 24 h before the experiment. Group 1 (normal control): Rats of this group received 5 mL of distilled water/kg bw. Group 2 (*J. tranquebariensis* control): Rats of this group were pretreated with *J. tranquebariensis* (400 mg/kg bw p.o./day) for 10 days. Group 3 (toxic control): Rats of this group were treated with a single dose of TAA (100 mg/kg bw/s.c.) on the 10th day. Group 4 (*J. tranquebariensis* plus TAA): Rats of this group were pretreated with *J. tranquebariensis* (400 mg/kg bw p.o./day) for 10 days and subsequent administration of a single dosage of TAA (100 mg/kg/s.c.). Group 5 (Silymarin plus TAA): Rats of this group were pretreated with silymarin (50 mg/kg bw/p.o./day) for 10 days followed by a single dose subcutaneous injection of TAA (100 mg/kg). TAA hepatotoxicity induction was followed our previous investigation [20]. All animals were deprived of food overnight and sacrificed by using anesthesia followed by cervical dislocation. Blood was collected from each respective group for the assay of biochemical parameters. The liver was immediately isolated, immersed in cold saline and weighed. A piece of one gm of liver from each rat was taken and homogenized to make liver homogenate, which were then centrifuged and the supernatant obtained was subjected to tissue biochemical parameters.

2.6. Biochemical Parameters

The activities of serum ALP, ALT, AST, GGT and TB were quantified by using commercial kits (Premier Diagnostics Sdn Bhd, Shah Alam, Malaysia). In the liver tissue homogenates, LPO was calculated using TBARS based on the method of Ohkawa et al. [21]. SOD and CAT were measured in liver tissue homogenate based the method of Marklund et al. [22] and Sinha [23] respectively. Activities of the respective enzymes were assayed by the methods as GR by Bellomo et al. [24], GPx by

Rotruck et al. [25], GST by Habig et al. [26], and total GSH by Moron et al. [27]. The quantification of protein was done by the method of Lowry et al. [28].

2.7. Histological Investigations

After sacrificing the animals, the livers were quickly removed and preserved in 10% formosaline and processed for paraffin embedding following the standard micro techniques. Sections (5 µm thick) of liver tissues stained with hematoxylin and eosin (H&E) were evaluated for histopathological under a light microscope.

2.8. Statistical Analysis

All data obtained in the study were expressed as mean ± standard deviation (S.D.). The data of the groups were statistically done using one-way analysis of variance (ANOVA) and the individual comparison was obtained by Duncan's Multiple Range Test (DMRT) by the SPSS software for Windows Version 20.0 (IBM Corp. Armonk, New York, NY, USA). A value of $p < 0.05$ was considered to indicate a significant difference between groups.

3. Results

3.1. Effects of *J. tranquebariensis* on Hepatic TBARS and Antioxidant Enzymes

The levels of tissue TBARS formation and the activities of antioxidant enzymes SOD, CAT, GPx, GR, GST and total reduced GSH in the liver of normal control and experimental rats are demonstrated in Figure 2A–G. In TAA (100 mg/kg bw)-treated rats, all antioxidant enzymes SOD, CAT, GPx, GR, GST and total GSH were found to be significantly decreased ($p < 0.05$), whereas tissue TBARS level significantly increased ($p < 0.05$) when compared with the normal control group. However, pretreatment administration of *J. tranquebariensis* (400 mg/kg bw) and silymarin (50 mg/kg bw) in TAA-induced rats have significantly altered to the above changes by regulating the TBARS level which subsequently increased those antioxidant enzymes.

3.2. Effects of *J. tranquebariensis* on Liver Marker Enzymes and Total Bilirubin

The activities of liver marker enzymes such as ALT, AST, ALP, GGT and the content of TB in the serum of control and experimental groups are shown in Figure 3A,B. The activities of ALT, AST, ALP, GGT and the content of total bilirubin in serum has significantly increased ($p < 0.05$) in TAA-intoxicated rats when compared with the normal control group. However, pretreatment of *J. tranquebariensis* (400 mg/kg bw) significantly decreased ($p < 0.05$) the above liver marker enzymes and total bilirubin levels observed in TAA-intoxicated rats.

3.3. Histological Observation

The histology of normal control, TAA control, *J. tranquebariensis* and silymarin-treated rats are exhibited in Figure 4. The liver sections of normal control groups showed the normal architecture of hepatocytes with prominent nucleus, well-preserved cytoplasm and visible central vein. TAA-administered animals showed hepatocytes with severe toxicity described by extensive necrosis, collagen and fibronectin deposition with sinusoidal spaces, and a central venule. Tissue showed severe cell swelling, vacuolar degeneration, loss of cell boundaries and the replacement of the cytoplasm with fluid and a centrally positioned nucleus. The livers of the rats pretreated with *J. tranquebariensis* (400 mg/kg bw) and silymarin (50 mg/kg bw) showed a normal lobular pattern with sinusoidal spaces, moderate swelling, and restoration of normal liver histology and thereby a decrease in collagen and fibronectin deposition.

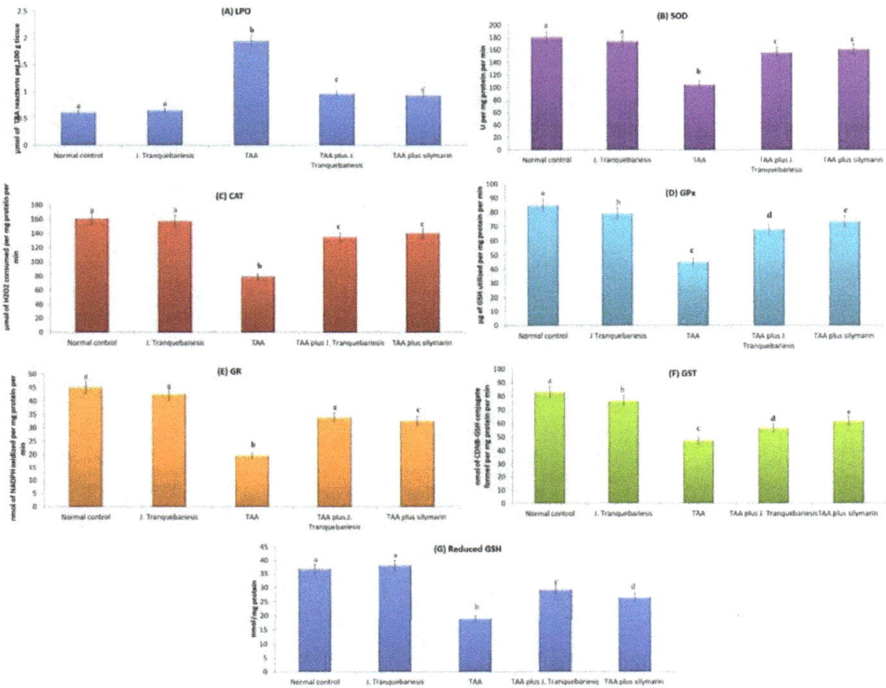

Figure 2. (**A–G**) Effects of *Justicia tranquebariesis* L. on liver lipid peroxidation and antioxidant enzyme activities in TAA intoxicated rats. Values are expressed as mean ± S.D. for six animals in each group. Values not sharing a common superscript (a–e) differ significantly. Letter "a" is significant to b, c, d, and e; likewise the letter "b" is significant to a, c, d and e. LPO—lipid peroxidation; TBA—thiobarbituric acid; SOD—superoxide dismutase; CAT—catalase; GPx—glutathione peroxidase; GR—glutathione reductase; NADPH—nicotinamide dinucleotide phosphatase; GST—glutathione transferase; CDNB—1-chloro-2,4-dinitrobenzene; GSH—glutathione.

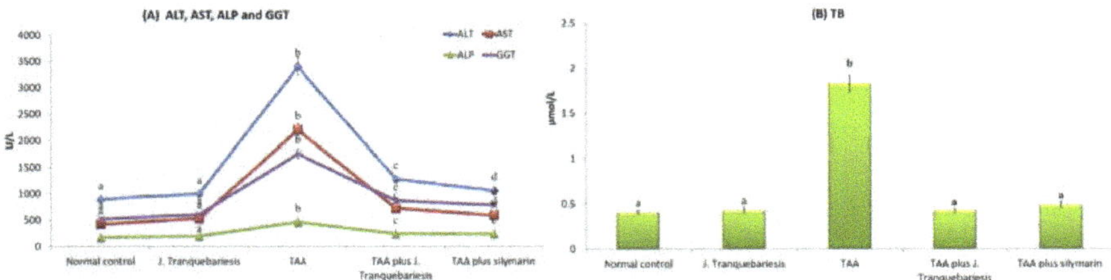

Figure 3. (**A**,**B**) Effects of *Justicia tranquebariesis* L. on hepatic markers and TB in TAA intoxicated rats. Values are expressed as mean ± S.D. for six animals in each group. Values not sharing a common superscript (a–d) differ significantly. Letter "a" is significant to b, c, and d; likewise a letter "b" is significant to a, c, and d. TB—total bilirubin; ALT—alanine aminotransferase; AST—aspartate aminotransferase; ALP—alkaline phosphatase; GGT—gamma-glutamyltransferase.

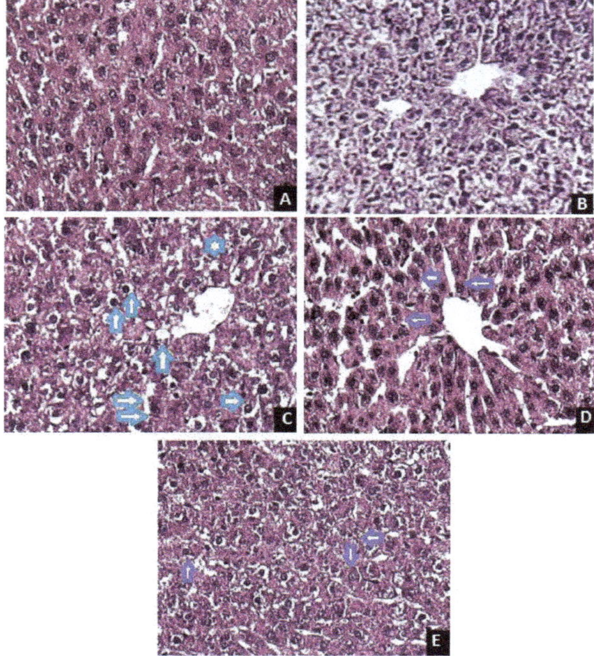

Figure 4. Micrographs showing the effect of *Justicia tranquebariensis* on TAA-induced hepatic fibrosis in rats. (**A**) Section of normal control rats showing the histological appearance of hepatocytes with prominent nuclei and cytoplasm (H&E. 400×); (**B**) Section of *J. tranquebariensis* (400 mg/kg bw/p.o.) treated control rats showing the histological appearance of hepatocytes with prominent nuclei and cytoplasm (H&E. 400×); (**C**) Section of TAA (100 mg/kg bw/s.c.) treated control showing fatty degeneration of some hepatocytes, Kupfer cells characterized by cell swelling, the replacement of the cytoplasm with a clear fluid and a centrally located nucleus (↑↑ blue), loss of cell boundaries, hepatic necrosis, collagen and fibronectin deposition and inflammatory cell infiltration (H&E. 400×); (**D**) Section of TAA (100 mg/kg bw/s.c.) plus *J. tranquebariensis* (400 mg/kg bw/p.o.) treated rats showing regenerated cells and the almost normal architecture of the liver (↑ violet) with a decrease in collagen and fibronectin deposition (H&E. 400×); (**E**) Section of TAA (100 mg/kg bw/s.c.) plus silymarin (50 mg/kg bw)-treated rats showing regenerated cells and the almost normal architecture of the liver (↑ violet) with decrease in collagen and fibronectin deposition (H&E. 400×).

4. Discussion

In the present study, we investigated the effect of an aqueous extract from the leaves of *J. tranquebariensis* on TAA-induced liver fibrosis in rats. A noticeable indication of liver injury is the release of cytoplasmic cellular enzymes into the blood due to the obstructions caused by chemicals in the transport functions of liver cells [8]. The increases in serum enzymes are markers for liver injury/damage. A significant increase in ALT, AST, and GGT can be considered as marker enzymes of liver damage. Serum transaminases were reduced and GGT returned to normal by pretreatment of *J. tranquebariensis* with a healing of hepatic parenchyma and regeneration of hepatocytes [29,30]. The concentrations of the ALP and TB have also been used as hepatic markers in chemically induced hepatic injury. In the toxicity studies, about 80% of serum ALP levels have been found to be elevated in rodents [31,32]. Pretreatment of *J. tranquebariensis* prevented the TAA toxicity effect on ALP activity in serum. In addition, the normalization of serum TB levels is done by the administration of *J. tranquebariensis*, which shows an indication of the normal functions of the hepatocyte.

GSH in the cytosolic pool consists of 85% hepatocellular GSH and 15% mitochondrial GSH. The reductions of hepatic GSH provide valuable evidence of the defensive role of GSH against noxious foreign materials [33]. Therefore, GSH is considered as an endogenous defensive mediator against various drugs [34]. A large dose of TAA causes hepatic GSH depletion because the excess TAA derivative reacts rapidly with GSH, which exacerbates oxidative stress in conjunction with mitochondrial dysfunction [8,35]. In the present study, pretreatment of *J. tranquebariensis* clearly enhanced GSH levels and facilitated the rapid and efficient consumption of ROS generation in TAA-intoxicated rats. GST is a soluble enzyme located in the cytosol, which plays a significant function in the detoxification of xenobiotics [8]. It increases the solubility of hydrophobic substances and metabolizes toxic compounds to non-toxic ones, which means they have an increasing protective activity of the liver [36]. The increased hepatic GST activity induced by *J. tranquebariensis* can, therefore, reduce TAA hepatotoxicity. There was a decrease in GPx activity in animals administered with TAA, which could be due to the higher production of toxicity. In the presence of *J. tranquebariensis*, GPx levels were restored back to control levels. The increase in hepatic GR activities was shown in *J. tranquebariensis* administered rats, as compared with the liver of TAA-induced rats. Elevated levels of SOD and CAT are insightful indexes into liver damage as they eradicate ROS, thereby reducing the harmful effects. On the contrary, an earlier report showed that TAA induction raises the TBARS levels, which instigated to reduce the levels of SOD and CAT [1]. However, pretreatment with *J. tranquebariensis* and silymarin significantly decreased TBARS and elevated antioxidant enzymes in TAA-induced rats. Earlier studies have also suggested that *J. tranquebariensis* has strong antioxidant potential [18] and prevents ROS and/or RNS-mediated tissue damage [17].

The degree of protection was maximally observed in silymarin (50 mg/kg bw/p.o)-treated groups when compared with *J. tranquebariensis* (400 mg/kg bw). This degree of hepatoprotection could be based on its various properties of anti-oxidation, inhibition of lipid peroxidation, regeneration of intracellular glutathione content, protein synthesis and improvement of liver regeneration from hepatocellular necrosis [37–43]. Furthermore, it stabilizes cellular membranes and regulates membrane permeability that inhibits toxins entry into liver cells. Silymarin inhibits fibrogenesis in the liver by inhibition of stellate cell proliferation and its further transformation into myofibroblasts [44–47].

In the present study, we induced hepatic fibrosis by subcutaneous injection of a single dose of TAA. TAA is normally metabolized by CYP2E1 releasing ROS, which attacks DNA, lipid, and protein [48]. This may cause centrilobular necrosis, dilated sinusoidal spaces, collagen and fibronectin deposition and diffuse hyaline necrosis with blood pooling in sinusoidal spaces, accompanied by a marked attenuation of a normal liver function. The results of the TAA group were in agreement with other studies [1,2]. However, the histopathological patterns of the livers of rats pretreated with *J. tranquebariensis* showed a normal lobular pattern with minimal pooling of blood in the sinusoidal spaces, moderate swelling, condense collagen and fibronectin deposition and re-establishment of liver architecture to normal. In conclusion, the result of this study revealed that pretreatment of *J. tranquebariensis* has a strong hepato-protective effect on TAA-induced liver oxidative stress and liver fibrosis in rats owing to its strong antioxidant properties. However, further evidence is needed to establish the possible mechanisms of the hepatoprotection in order to widely use *J. tranquebariensis* as an antihepatotoxic agent.

Author Contributions: K.S. and K.G. conceived, designed, and performed the experiments; K.G. analyzed the data and wrote the paper. B.X. critically read the manuscript.

Funding: This research received no external funding.

Acknowledgments: The authors sincerely acknowledge to the faculty members and technical staffs in Faculty of Medicine, International Medical School, Management and Science University, Shah Alam, Malaysia who provided a great deal of support and cooperation in the study.

Conflicts of Interest: The authors declare no conflict of interest.

Abbreviations

ALP	alkaline phosphatase
ALT	alanine transaminase
ANOVA	one-way analysis of variance
AST	aspartate transaminase
bw	body weight
CAT	catalase
CDNB	1-chloro-2,4-dinitrobenzene
DMART	Duncan's Multiple Range Test
GGT	gamma glutamyl transpeptidase
GPx	glutathione peroxidase
GR	glutathione reductase
GSH	reduced glutathione
GST	glutathione s-transferase
H&E	hematoxylin and eosin
Kg	kilogram
LPO	lipid peroxidation
mg	milligram
NADPH	nicotinamide adenine dinucleotide hydrogen phosphate
ROS	reactive oxygen species
s.c.	subcutaneous
S.D.	standard deviation
SOD	superoxide dismutase
SPSS	Statistical Package for the Social Sciences
TAA	thioacetamide
TB	total bilirubin
TBA	thiobarbituric acid
TBARS	thiobarbituric acid reactive substances
RNS	reactive nitrogen species
CYP2E1	cytochrome P450 2E1

References

1. Kaur, V.; Kumar, M.; Kaur, P.; Kaur, S.; Singh, A.P.; Kaur, S. Hepatoprotective activity of *Butea monosperma* bark against thioacetamide-induced liver injury in rats. *Biomed. Pharmacother.* **2017**, *89*, 332–341. [CrossRef] [PubMed]
2. El-Mihi, K.A.; Kenawy, H.I.; El-Karef, A.; Elsherbiny, N.M.; Eissa, L.A. Naringin attenuates thioacetamide-induced liver fibrosis in rats through modulation of the PI3K/Akt pathway. *Life Sci.* **2017**, *187*, 50–57. [CrossRef] [PubMed]
3. Fallowfield, J.A.; Kendall, T.J.; Iredale, J.P. Reversal of fibrosis: No longer a pipe dream? *Clin. Liver Dis.* **2006**, *10*, 481–497. [CrossRef] [PubMed]
4. Ismail, M.H.; Pinzani, M. Reversal of liver fibrosis. *Saudi J. Gastroenterol.* **2009**, *15*, 72–79. [CrossRef] [PubMed]
5. Schuppan, D.; Ashfaq-Khan, M.; Yang, A.T.; Kim, Y.O. Liver fibrosis: Direct antifibrotic agents and targeted therapies. *Matrix Biol.* **2018**. [CrossRef] [PubMed]
6. Sebastiani, G.; Gkouvatsos, K.; Pantopoulos, K. Chronic hepatitis C and liver fibrosis. *World J. Gastroenterol.* **2014**, *20*, 11033–11053. [CrossRef] [PubMed]
7. Al-Attar, A.M.; Al-Rethea, H.A. Chemoprotective effect of omega-3 fatty acids on thioacetamide-induced hepatic fibrosis in male rats. *Saudi J. Biol. Sci.* **2017**, *24*, 956–965. [CrossRef] [PubMed]
8. Ganesan, K.; Sukalingam, K.; Xu, B. *Solanum trilobatum* L. ameliorate thioacetamide-induced oxidative stress and hepatic damage in albino rats. *Antioxidants* **2017**, *6*, 68. [CrossRef] [PubMed]
9. Bashandy, S.A.; Alaamer, A.; Moussa, S.A.; Omara, E. Role of zinc oxide nanoparticles in alleviating hepatic fibrosis and nephrotoxicity induced by thioacetamide in rats. *Can. J. Physiol. Pharmacol.* **2017**, *16*, 1–8. [CrossRef] [PubMed]

10. Feng, Y.M.; Wang, X.; Wang, L.; Ma, X.W.; Wu, H.; Bu, H.R.; Xie, X.Y.; Qi, J.N.; Zhu, Q. Efficacy and safety of combination therapy of chemoembolization and radiofrequency ablation with different time intervals for hepatocellular carcinoma patients. *Surg. Oncol.* **2017**, *26*, 236–241. [CrossRef] [PubMed]
11. Rehman, J.; Akhtar, N.; Asif, H.M.; Sultana, S.; Ahmad, M. Hepatoprotective evaluation of aqueous-ethanolic extract of *Capparis decidua* (Stems) in paracetamol-induced hepatotoxicity in experimental rabbits. *Pak. J. Pharm. Sci.* **2017**, *30*, 507–511. [PubMed]
12. Prajapathi, N.D.; Purohit, S.S.; Sharma, A.K.; Kumar, T. *A Hand Book of Medicinal Plants: A Complete Source of Book*, 1st ed.; Agrobios Publisher: Jodhpur, India, 2003; p. 554. ISBN 13: 978-8177541342.
13. Sekhar, J.; Penchala Pratap, G.; Sudarsanam, G.; Prasad, G.P. Ethnic information on treatments for snake bites in Kadapa district of Andhra Pradesh. *Life Sci. Leaflet* **2011**, *12*, 368–375.
14. Akilandeswari, S.; Kumarasundari, S.K.; Valarmath, R.; Manimaran, S.; Sivakumar, M. Studies on anti-inflammatory activity of leaf extract of *Justicia tranquebariensis* L. *Indian J. Nat. Prod.* **2001**, *17*, 14–16.
15. Begum, M.S.; Ilyas, M.H.M.; Burkanudeen, A. Antipyretic activity of extract of leaves of *Justicia tranquebariensis* (Linn) in albino mice. *Pharmacist* **2009**, *4*, 49–51.
16. Raju, G.V.S.; Pillai, K.R. Lignans from *Justicia tranquebariensis* Linn. *Indian J. Chem.* **1989**, *28*, 558–561.
17. Balamurugan, G.; Arunkumar, M.P.; Muthusamy, P.; Anbazhagan, S. Preliminary phytochemical screening, free radical scavenging and antimicrobial activities of *Justicia tranquebariensis* L. *Res. J. Pharm. Technol.* **2008**, *1*, 116–118.
18. Begum, M.S.; Ilyas, M.H.M.; Burkanudeen, A. Protective and curative effects of *Justicia tranquebariensis* (Linn) leaves in acetaminophen-induced hepatotoxicity. *Int. J. Pharm. Biol. Arch.* **2011**, *2*, 989–995.
19. Kantah, M.K.; Kobayashi, R.; Sollano, J.; Naito, Y.; Solimene, U.; Jains, S.; Catanzaro, R.; Minelli, E.; Polimeni, A.; Marotta, F. Hepatoprotective activity of a phytotherapeutic formula on thioacetamide-induced liver fibrosis model. *Acta Biomed.* **2011**, *82*, 82–89. [PubMed]
20. Kumar, G.; Banu, G.S.; Pappa, P.V.; Sundararajan, M.; Pandian, M.R. Hepatoprotective activity of *Trianthema portulacastrum* L. against paracetamol and thioacetamide intoxication in albino rats. *J. Ethnopharmacol.* **2004**, *92*, 37–40. [CrossRef] [PubMed]
21. Ohkawa, H.; Ohishi, N.; Yagi, K. Assay for lipid peroxides in animal tissues by thiobarbituric acid reaction. *Anal. Biochem.* **1979**, *95*, 351–358. [CrossRef]
22. Marklund, S.; Marklund, G. Involvement of the superoxide anion radical in the autoxidation of pyrogallol and a convenient assay for superoxide dismutase. *Eur. J. Biochem.* **1974**, *47*, 469–474. [CrossRef] [PubMed]
23. Sinha, A.K. Colorimetric assay of catalase. *Anal. Biochem.* **1972**, *47*, 389–394. [CrossRef]
24. Bellomo, G.; Mirabelli, F.; DiMonte, D.; Richelmi, P.; Thor, H.; Orrenius, C.; Orrenius, S. Formation and reduction of glutathione-protein mixed disulfides during oxidative stress: A study with isolated hepatocytes and menadione (2-methyl-1,4-naphthoquinone). *Biochem. Pharmacol.* **1987**, *36*, 1313–1320. [CrossRef]
25. Rotruck, J.T.; Pope, A.L.; Ganther, H.E.; Swanson, A.B.; Hafeman, D.G.; Hoekstra, W.G. Selenium: Biochemical role as a component of glutathione peroxidase. *Science* **1973**, *179*, 588–590. [CrossRef] [PubMed]
26. Habig, W.H.; Pabst, M.J.; Jakoby, W.B. Glutathione S-transferases: The first enzymatic step in mercapturic acid formation. *J. Biol. Chem.* **1974**, *249*, 7130–7139. [PubMed]
27. Moron, M.S.; Depierre, J.W.; Mannervik, B.; Moron, M.S.; Depierre, J.W.; Mannervik, B. Levels of glutathione, glutathione reductase and glutathione S-transferase activities in rat lung and liver. *Biochim. Biophys. Acta* **1979**, *582*, 67–78. [CrossRef]
28. Lowry, O.H.; Rosebrough, N.J.; Farr, A.L.; Randall, R.J. Protein measurement with the Folin phenol reagent. *J. Biol. Chem.* **1951**, *193*, 265–275. [PubMed]
29. Kumar, G.; Banu, G.S.; Pandian, M.R. Biochemical activity of selenium and glutathione on country made liquor (CML) induced hepatic damage in rats. *Indian J. Clin. Biochem.* **2007**, *22*, 105–108. [CrossRef] [PubMed]
30. Lin, Y.L.; Lin, H.W.; Chen, Y.C.; Yang, D.J.; Li, C.C.; Chang, Y.Y. Hepatoprotective effects of naturally fermented noni juice against thioacetamide-induced liver fibrosis in rats. *J. Chin. Med. Assoc.* **2017**, *80*, 212–221. [CrossRef] [PubMed]
31. Kumar, G.; Banu, G.S.; Kannan, V. Effects of Arrack on liver protein of *Mus musculus*. *J. Ecobiol.* **2006**, *18*, 321–324.
32. Vakiloddin, S.; Fuloria, N.; Fuloria, S.; Dhanaraj, S.A.; Balaji, K.; Karupiah, S. Evidences of hepatoprotective and antioxidant effect of *Citrullus colocynthis* fruits in paracetamol-induced hepatotoxicity. *Pak. J. Pharm. Sci.* **2015**, *28*, 951–957. [PubMed]

33. Simeonova, R.; Bratkov, V.M.; Kondeva-Burdina, M.; Vitcheva, V.; Manov, V.; Krasteva, I. Experimental liver protection of *n*-butanolic extract of *Astragalus monspessulanus* L. on carbon tetrachloride model of toxicity in rat. *Redox Rep.* **2015**, *20*, 145–153. [CrossRef] [PubMed]
34. Kumar, G.; Banu, G.S.; Kannan, V.; Pandian, M.R. Antihepatotoxic effect of β-carotene on paracetamol induced hepatic damage in rats. *Indian J. Exp. Biol.* **2005**, *43*, 351–355. [PubMed]
35. Kumar, G.; Banu, G.S.; Balapala, K.R. Ameliorate the effect of *Solanum trilobatum* L. on hepatic enzymes in experimental diabetes. *Nat. Prod. Indian J.* **2011**, *7*, 315–319.
36. Miguel, F.M.; Schemitt, E.G.; Colares, J.R.; Hartmann, R.M.; Morgan-Martins, M.I.; Marroni, N.P. Action of vitamin E on experimental severe acute liver failure. *Arq. Gastroenterol.* **2017**, *54*, 123–129. [CrossRef] [PubMed]
37. Thakare, V.N.; Aswar, M.K.; Kulkarni, Y.P.; Patil, R.R.; Patel, B.M. Silymarin ameliorates experimentally induced depressive-like behavior in rats: Involvement of hippocampal BDNF signaling, inflammatory cytokines, and oxidative stress response. *Physiol. Behav.* **2017**, *179*, 401–410. [CrossRef] [PubMed]
38. Vahabzadeh, M.; Amiri, N.; Karimi, G. Effects of Silymarin on the Metabolic Syndrome; a Review. *J. Sci. Food Agric.* **2018**. [CrossRef] [PubMed]
39. Mazhari, S.; Razi, M.; Sadrkhanlou, R. Silymarin and celecoxib ameliorate experimental varicocele-induced pathogenesis: Evidences for oxidative stress and inflammation inhibition. *Int. Urol. Nephrol.* **2018**. [CrossRef] [PubMed]
40. Taleb, A.; Ahmad, K.A.; Ihsan, A.U.; Qu, J.; Lin, N.; Hezam, K.; Koju, N.; Hui, L.; Qilong, D. Antioxidant effects and mechanism of silymarin in oxidative stress-induced cardiovascular diseases. *Biomed. Pharmacother.* **2018**, *102*, 689–698. [CrossRef] [PubMed]
41. Rahimi, R.; Karimi, J.; Khodadadi, I.; Tayebinia, H.; Kheiripour, N.; Hashemnia, M.; Goli, F. Silymarin ameliorates expression of urotensin II (U-II) and its receptor (UTR) and attenuates toxic oxidative stress in the heart of rats with type 2 diabetes. *Biomed. Pharmacother.* **2018**, *101*, 244–250. [CrossRef] [PubMed]
42. Eraky, S.M.; El-Mesery, M.; El-Karef, A.; Eissa, L.A.; El-Gayar, A.M. Silymarin and caffeine combination ameliorates experimentally-induced hepatic fibrosis through down-regulation of LPAR1 expression. *Biomed. Pharmacother.* **2018**, *101*, 49–57. [CrossRef] [PubMed]
43. El-Newary, S.A.; Shaffie, N.M.; Omer, E.A. The protection of Thymus vulgaris leaves alcoholic extract against hepatotoxicity of alcohol in rats. *Asian Pac. J. Trop. Med.* **2017**, *10*, 361–371. [CrossRef] [PubMed]
44. Fraschini, F.; Demartini, G.; Esposti, D. Pharmacology of silymarin. *Clin. Drug Investig.* **2002**, *22*, 51–65. [CrossRef]
45. Crocenzi, F.A.; Roma, M.G. Silymarin as a new hepatoprotective agent in experimental cholestasis: New possibilities for an ancient medication. *Curr. Med. Chem.* **2006**, *13*, 1055–1074. [CrossRef] [PubMed]
46. Gazák, R.; Walterová, D.; Kren, V. Silybin and silymarin-New and emerging applications in medicine. *Curr. Med. Chem.* **2007**, *14*, 315–338. [CrossRef] [PubMed]
47. Harati, K.; Behr, B.; Wallner, C.; Daigeler, A.; Hirsch, T.; Jacobsen, F.; Renner, M.; Harati, A.; Lehnhardt, M.; Becerikli, M. Anti-proliferative activity of epigallocatechin-3-gallate and silibinin on soft tissue sarcoma cells. *Mol. Med. Rep.* **2017**, *15*, 103–110. [CrossRef] [PubMed]
48. Salama, S.M.; Abdulla, M.A.; Alrashdi, A.S.; Hadi, A.H. Mechanism of hepatoprotective effect of *Boesenbergia rotunda* in thioacetamide-induced liver damage in rats. *Evid. Based Complement. Altern. Med.* **2013**, *2013*, 157456. [CrossRef] [PubMed]

 antioxidants

Review

Vitamin C in Plants: From Functions to Biofortification

Costantino Paciolla, Stefania Fortunato, Nunzio Dipierro, Annalisa Paradiso, Silvana De Leonardis, Linda Mastropasqua and Maria Concetta de Pinto *

Department of Biology, University of Bari "Aldo Moro", Via E. Orabona 4, 70125 Bari, Italy; costantino.paciolla@uniba.it (C.P.); stefania.fortunato@uniba.it (S.F.); nunzio.dipierro@uniba.it (N.D.); annalisa.paradiso@uniba.it (A.P.); silvana.deleonardis@uniba.it (S.D.L.); linda.mastropasqua@uniba.it (L.M.)
* Correspondence: mariaconcetta.depinto@uniba.it; Tel.: +39-080-544-2156

Received: 2 October 2019; Accepted: 26 October 2019; Published: 29 October 2019

Abstract: Vitamin C (L-ascorbic acid) is an excellent free radical scavenger, not only for its capability to donate reducing equivalents but also for the relative stability of the derived monodehydroascorbate radical. However, vitamin C is not only an antioxidant, since it is also a cofactor for numerous enzymes involved in plant and human metabolism. In humans, vitamin C takes part in various physiological processes, such as iron absorption, collagen synthesis, immune stimulation, and epigenetic regulation. Due to the functional loss of the gene coding for L-gulonolactone oxidase, humans cannot synthesize vitamin C; thus, they principally utilize plant-based foods for their needs. For this reason, increasing the vitamin C content of crops could have helpful effects on human health. To achieve this objective, exhaustive knowledge of the metabolism and functions of vitamin C in plants is needed. In this review, the multiple roles of vitamin C in plant physiology as well as the regulation of its content, through biosynthetic or recycling pathways, are analyzed. Finally, attention is paid to the strategies that have been used to increase the content of vitamin C in crops, emphasizing not only the improvement of nutritional value of the crops but also the acquisition of plant stress resistance.

Keywords: ascorbate; antioxidant; biofortification; light; plant growth; reactive oxygen species; vitamin C

1. Introduction

Vitamin C (L-ascorbic acid) was isolated from the adrenal cortex by Albert Szent-Györgyi in 1928. Szent-Györgyi demonstrated that this compound, which can act as a powerful reducing agent, indicated with the empirical formula of $C_6H_8O_6$, had a molecular mass of 178 ± 3 and was a lactone with an acidic hydrogen atom. Due to its similarity to simple sugars and its acidic properties, Szent-Györgyi called this compound "hexuronic acid" [1]. In 1932, Charles Glen King isolated an antiscorbutic compound from lemon juice that was recognized as the hexuronic acid found by Szent-Györgyi [2,3]. At the same time, Szent-Györgyi showed that 1 mg/day of hexuronic acid provided ample protection against scurvy [4]. The definitive structure of vitamin C, which is a hexonic acid aldono-1,4-lactone with an enediol group on C2 and C3, was achieved by Norman Haworth in 1933 [5]. The evidence that this compound was able to prevent scurvy led to it being renamed from hexuronic acid to ascorbic acid [6].

Vitamin C is the most abundant water-soluble compound working in one-electron reactions, and it is an essential micronutrient and a key element for the metabolism of almost all living organisms. In humans, vitamin C has numerous functions, mainly acting as an antioxidant and a cofactor for mono-oxygenases and dioxygenases [7].

The roles of vitamin C as an antioxidant in humans has been established based on a large body of scientific evidence. Vitamin C, by scavenging free radicals, protects DNA, proteins, and lipids from oxidative damages [8]. Vitamin C is used as an antioxidant throughout the body but may have specific roles in some organs. For instance, vitamin C is required in the eyes at a millimolar concentration to guarantee protection from oxidative damage due to solar radiation [9]. Vitamin C inhibits the synthesis of the carcinogenic nitrosamines, which can be synthesized in the intestine or absorbed with food [10], and reduces tetrahydrobiopterin, the cofactor of nitric oxide synthase, which catalyzes the synthesis of nitric oxide [11]. Vitamin C is also important for iron bioavailability, reducing non-heme iron from the ferric (Fe^{3+}) to the ferrous (Fe^{2+}) form, which is more easily absorbed in the intestine; for this reason, this vitamin is indirectly needed to protect against anemia [12]. Vitamin C also influences iron metabolism through the stimulation of ferritin synthesis and the inhibition of ferritin degradation [13].

Vitamin C, as a cofactor of peptidyl-glycine alpha-amidating monooxygenase, is involved in the biosynthesis of many signaling peptides, such as oxytocin, vasopressin, cholecystokinin, and calcitonin [14–16]. Vitamin C functions as a cofactor for many dioxygenases, reducing the iron in the active site of these enzymes to Fe^{2+}. Vitamin C contributes to the correct formation of collagen through post-translational modifications of procollagen. In particular, this vitamin acts as a cofactor for the reaction catalyzed by prolyl 3-hydroxylase, prolyl 4-hydroxylase, and lysyl hydroxylase, which are involved in the hydroxylation of lysine and proline and permit the formation of the stable structure of collagen [17–19]. Vitamin C, as a cofactor of hydroxylase, is used for the synthesis of norepinephrine and carnitine [20–22]. Vitamin C intervenes in many cytochrome-P450-dependent hydroxylation reactions, such as the transformation of cholesterol into bile acids, the degradation of exogenous substances such as pollutants and drugs, and the synthesis of steroid hormones [23].

Recently, vitamin C has been identified as a cofactor for the methylcytosine dioxygenases ten-eleven translocation (TET), which is involved in DNA demethylation and JmjC-domain-containing proteins, which catalyze the demethylation of histones [24]. As a result of vitamin C deficiency, especially in the nucleus, the requirements of TETs or some JmjC-domain-containing histone demethylases may not be met, leading to alterations in the methylation–demethylation dynamics of DNA and histones, which can subsequently contribute to phenotypic alterations or even diseases. By regulating the epigenome, vitamin C can be involved in embryonic development, postnatal development and aging, and cancer and other diseases [24]. Being able to modulate the epigenome, vitamin C has been proposed as an effective molecule in anticancer therapies [25].

Vitamin C is considered a vitamin only for a few vertebrate species, including humans, that are unable to synthesize it [26]. Indeed, vitamin C can be synthesized by plants and most animals [27]. Primates, guinea pigs, bats, some species of birds, insects, invertebrates, and fish are examples of species not able to synthesize this vitamin [26,27]. The inability of humans to synthesize vitamin C lies in the functional loss of the gene coding L-gulonolactone oxidase, the last enzyme involved in the animal biosynthetic pathway of this vitamin [26].

The best way for humans to obtain vitamin C is by diet, and plant foods represent the primary source of this vitamin. Although synthetic vitamin C is chemically indistinguishable from the plant-derived vitamin, fruits and vegetables have different micronutrients and phytochemicals that can affect its bioavailability [28]. Many studies conducted on vitamin-C-deficient animals have shown that vitamin C in plant foods has greater bioavailability than that found in drugs or supplements [28]. For instance, in homozygote Gulo mice, the uptake and tissue distribution of vitamin C were higher when the vitamin was furnished by kiwifruit gel than when it was added as a synthetic supplement in drinking water [29]. Nevertheless, studies conducted on humans have not shown significant differences in bioavailability between synthetic and plant-derived vitamin C [28]. Despite the comparable dosage and bioavailability, it has been shown that orange juice and not synthetic vitamin C drink protects leukocytes from oxidative DNA damage [30]. Probably, the improved effects of vitamin C dispensed with fruits and vegetables is due to the interaction with other micronutrients, such as vitamin E [31] and iron [32]. Consistently, plant-derived vitamin C is related to reduced occurrence of different chronic diseases [33].

The loss of the capability to synthesize vitamin C in our ancestors would not have been a disadvantage with a diet rich in vegetables and fruits, which could have provided enough vitamin

C [34]. On the contrary, since L-gulonolactone oxidase produces the potentially toxic H_2O_2, this loss could have been an evolutionary improvement in the control of redox homeostasis [35].

Nowadays, very low levels of vitamin C are present in main crops, with the consequence that diet does not provides enough intake of this vitamin. Thus, obtaining plant foods with enhanced vitamin C content represents an important goal for human health. In-depth knowledge of the metabolism and functions of vitamin C in plants is needed to achieve biofortification.

2. Vitamin C as an Antioxidant

Vitamin C is an essential element of plant and animal antioxidant systems, which can be defined as complex redox networks, including metabolites and enzymes, with mutual interactions and synergistic effects [36]. Antioxidants can spontaneously provide electrons to free radicals, alleviating the oxidative cellular environments caused by aerobic metabolism.

Chemically, vitamin C is a dibasic acid with an enediol group on C2 and C3 of a heterocyclic lactone ring. At physiological pH, the hydroxyl group at C3 is deprotonated, giving a monovalent anion, which is indicated as ascorbate (ASC) [37]. The enediol group permits the donation of one or two electrons, forming monodehydroascorbate (MDHA) and dehydroascorbate (DHA), respectively [36]. The ASC redox potential ranges from +0.40 to +0.50 V [38,39]; thus, the molecule can directly donate electrons to reactive oxygen species (ROS), such as singlet oxygen, superoxide anions, and hydroxyl radicals, as well as to tocopheroxyl radicals (Figure 1) [36]. Being able to reduce tocopheroxyl radicals, ASC is indirectly involved in the scavenging of lipid peroxides and radicals, contributing to the decrease of lipid peroxidation and, consequently, to the protection of membranes [40]. Due to the fast reduction of ROS by ASC, the damage of biomolecules can be prevented before the activation of antioxidant enzymes.

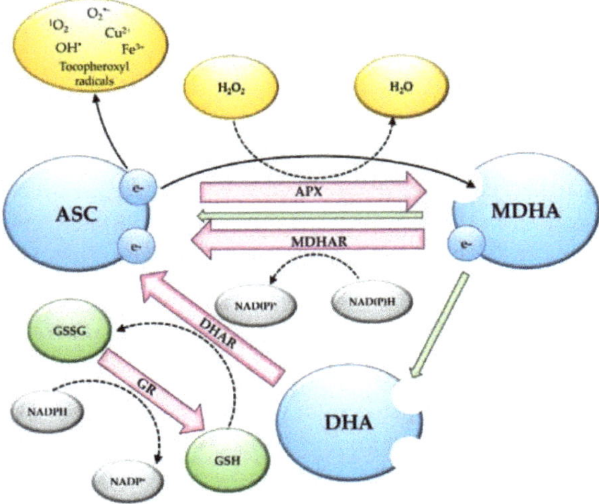

Figure 1. Redox-dependent reactions of vitamin C. ASC can donate electrons directly to reactive oxygen species, metals, and tocopheroxyl radicals. The reduction of H_2O_2 by ASC occurs via APX. MDHA can undergo dismutation (green arrows), providing ASC and DHA. The reduction of oxidized forms occurs through the ASC-GSH cycle. MDHA and DHA are reduced by MDHAR and DHAR, respectively, whereas the reduced glutathione is recovered by GR (more details are provided in the text). Abbreviations: ASC, ascorbate; APX, ascorbate peroxidase; DHA, dehydroascorbate; DHAR, dehydroascorbate reductase; MDHA, monodehydroascorbate; MDHAR, monodehydroascorbate reductase; GR, glutathione reductase; GSH, glutathione; GSSG, glutathione disulphide.

ASC can also reduce metals such as copper and iron, leading to the formation of ROS through the Haber-Weiss and Fenton reactions [41]. Thus, in some cases, ASC, acting as a reducing agent, will generate oxidants. This can occur in cell culture media, within the physiologic concentration of ASC, in presence of metals or in vivo in humans only when plasma and extracellular fluids contain millimolar concentrations of ASC [42].

Due to resonance stabilization of unpaired electrons, MDHA, derived from the loss of one electron, has very low reactivity with other radicals and, consequently, has very low toxicity. Two molecules of MDHA can spontaneously dismutate in ASC and DHA (Figure 1) [43]. DHA, if not rapidly reduced to ASC, will be permanently hydrolyzed to threonate, oxalate, oxalyl threonate, or tartrate [44].

ASC has low direct reactivity with H_2O_2, but in plants, ASC works as a specific electron donor for ascorbate peroxidase (APX), a heme peroxidase which catalyzes the conversion of H_2O_2 to H_2O and O_2, giving MDHA (Figure 1). APX has a high affinity for H_2O_2 and removes this ROS, even at low concentrations [36]. In plants, different APX isoenzymes have been identified in cytosol, mitochondria, peroxisomes, and chloroplasts, but all are coded by nuclear genes. The plant model system *Arabidopsis thaliana* possesses six APX genes coding for six isoenzymes, two of which are targeted to the cytosol (APX1 and 2), two to peroxisomes (APX3 and 5), one to the thylakoid membrane (tAPX), and one that is dual-targeted to chloroplast stroma and mitochondrial matrix (sAPX) [45]. The crop species rice and tomato possess seven and eight APX isozymes, respectively [46].

APX is part of the ASC-glutathione (GSH) cycle (Figure 1), which is involved in ASC regeneration [47]. MDHA is reduced to ASC by MDHA reductase (MDHAR), which is a flavin enzyme that utilizes NAD(P)H as electron donors [43]. Many cell compartments possess MDHAR activity. *Arabidopsis* has five genes coding for MDHAR2 and 3, localized in the cytosol; MDHAR1 and 4, in peroxisomes and membranes; and MDHAR6 in chloroplasts and mitochondria [48]. DHA can be reduced to ASC by DHA reductase (DHAR), which utilizes GSH as an electron donor, leading to the formation of glutathione disulphide (GSSG). DHAR has an important role in maintaining the reduced ASC in order to avoid DHA degradation [49]. DHAR activity has been identified in cytosol, chloroplasts, mitochondria, and peroxisomes [48]. *Arabidopsis* has three genes coding for DHAR localized differently in the cells: *DHAR1* localized in cytosol and peroxisomes, *DHAR2* localized only in the cytosol, and *DHAR3* targeted to chloroplasts [50]. In the ASC-GSH cycle, GSSG is reduced to GSH by the NADPH-dependent glutathione reductase (GR). GR plays a pivotal role in maintaining the correct balance between reduced GSH and ASC pools [51]. GR activity has been detected in chloroplasts, cytosol, mitochondria, and peroxisomes [52]. In *Arabidopsis*, two genes encode for GR in plants: GR1 is predicted to code a cytosolic isoenzyme and GR2 encodes for a dual-targeted plastidic/mitochondrial protein [53].

Owing to its high antioxidant properties and to the presence of an effective system for redox regeneration, in plants, vitamin C plays a significant role in the defense against oxidative stress, which arises in response to biotic or abiotic stresses [54]. The important role of vitamin C in the tolerance to several stresses is underlined by the increase in the enzymes involved in biosynthesis and recycling, observed in the presence of adverse environmental conditions [49,55]. Interestingly, feeding with L-galactono-1,4-lactone, which enhances the vitamin C content, can increase resistance to various kinds of stress [56–59].

3. Multiple Roles of Vitamin C in Plants

A significant part of accessible glucose (about 1%) is used for vitamin C production, which is present at high concentration in plants [60]. Vitamin C was found in all cell compartments, including the apoplast (the cell wall and extracellular space), reaching a concentration of 20 mM in chloroplasts [61]. However, the vitamin C content significantly differs among plant species and in the same species between diverse cultivars [62]. Moreover, the vitamin C content varies among different tissues and organs, usually being high in leaves, meristematic tissues, flowers, or young fruits and low in non-photosynthetic organs such as stems and roots [54,62]. Only seeds that reach maturity in a stage of strong dehydration (orthodox seeds) contain little vitamin C, which is essentially in the oxidized form [63,64]. In the same organ or tissue, vitamin C content is influenced by the plant developmental stage and environmental changes [65–68]. Light is one of the most significant environmental signals involved in the regulation of vitamin C levels [69,70].

As in humans, vitamin C favors iron uptake in plants. The ASC efflux in the apoplast contributes to the reduction of Fe^{3+}, catalyzed by the ferric chelate reductase plasma membrane enzyme. *Arabidopsis* mutants having low vitamin C content (*vtc* mutants) show a decrease in Fe^{3+} reducing capability and a consequent reduction of iron accumulation in the seeds [71].

Vitamin C, having different functions in chloroplasts, is essential for the correct functionality of photosynthesis. Firstly, ASC has a key role in the direct scavenging of ROS and in the removal of H_2O_2 through the water-water cycle [72,73]. ASC also participates in the xanthophyll cycle, which is needed to protect photosystem II (PSII) from photoinhibition. In this cycle, ASC is the cofactor of violaxanthin de-epoxidase, which converts violaxanthin in zeaxanthin, the xanthophyll responsible for dissipating excess excitation energy in the light harvesting complexes of PSII [74]. Finally, ASC can donate electrons to both photosystems, especially when they are damaged by stress conditions [75,76]. Changes in vitamin C content significantly modify the expression of genes linked to photosynthesis [77]. The lowering of the ASC content, through the suppression of DHAR expression, leads to the loss of chlorophyll a, the reduction of the RUBISCO large subunit, and a decrease in CO_2 assimilation [78]. Consistently, vitamin C-deficient *Arabidopsis* mutants enter senescence earlier than wild-type [79]. Thus, vitamin C, by preserving photosynthetic functioning and limiting ROS-mediated damage, slows down leaf senescence [78–81].

Vitamin C is involved in the synthesis of the plant hormone ethylene, acting as a cofactor of 1-aminocyclopropane-1-carboxylic acid (ACC) oxidase, the enzyme that catalyzes the last biosynthetic step. Indeed, ASC contributes to the ring opening of ACC by supplying the electron to the active site of the enzyme [62,82]. Being a cofactor of dioxygenases, vitamin C could also be involved in the synthesis of abscisic acid and gibberellins, as well as in the catabolism of auxins [62].

A complex interplay between vitamin C and hormone signaling intervenes in different phases of plant growth and development, as well as in plant response to the environment and pathogens [83,84]. In particular, vitamin C involvement in the defense response against pathogens is strictly dependent on pathogen lifestyles [84]. It is known that defense against biotrophic pathogens is mediated by salicylic acid signaling, whereas defense against necrotrophic pathogens is mediated by jasmonic acid and ethylene signaling [85]. *Arabidopsis* mutants with low vitamin C levels show an increase in salicylic acid, pathogenesis-related proteins, and camalexin and are more resistant to *Pseudomonas syringae* and *Peronospora parasitica* [79,86–88]. On the contrary, the same mutants are more susceptible to the necrotrophic ascomycete *Alternaria brassicicola* [89]. Nevertheless, exogenous addition of ASC acts as an inducer of disease resistance in different plant-pathogen interactions [89–92].

In plants, vitamin C can control the division, elongation, and differentiation of cells, as well as programmed cell death (PCD). Vitamin C plays a significant role in the control of cell division. This metabolite in the meristematic cells of root meristems can shorten the G1 phase and stimulate entry into the S phase [93,94]. In the quiescent center of the root meristem, where cells are not dividing, low levels of ASC, linked to a significant increase of the ASC-consuming enzyme ASC oxidase (AOX), are responsible for the arrest of the cell cycle in the G1 phase [95]. In tobacco BY-2 cells, a peak in ASC, as well as L-galactono-1,4-lactone dehydrogenase (GLDH), activity overlaps with the peak in the mitotic index [96,97]. Moreover, cells enriched with ASC show stimulation of cell division, whereas enrichment with DHA leads to a reduction in cell division, suggesting that the ASC redox state is fundamental to cell cycle progression [96]. Low ASC levels and an altered redox state negatively affect cell cycle progression in the root meristem of *Arabidopsis*, with a consequent decrease in the number of cells in the proliferation zone [98]. An ASC increase has also been shown during cell divisions in developing embryos [99]. The stimulation of division in the apical meristem by ASC seems to be principally due to the inhibition of peroxidase involved in the crosslink of cell wall components. Inhibition of vitamin C biosynthesis leads to abortion of the meristem [100].

The stimulation of cell elongation is due to the expression of AOX, the activity of which determines an increase in oxidized forms of vitamin C in the apoplast [101,102]. Indeed, apoplastic MDHA participates in transmembrane electron transfer, accepting electrons from cytochrome b. This process induces plasma membrane hyperpolarization and activation of H^+-ATPase, with an acidification of the apoplast that favors cell wall relaxing [103]. The parallel oxidation of NADH acidifies the cytoplasm and activates vacuolar H^+-ATPase, increasing vacuolization and cell expansion [103–105]. In the apoplast, ASC in the presence of Cu^{2+} can exert its pro-oxidant role, producing H_2O_2, which induces degradation of polysaccharides [106,107]. Moreover, by reducing lignin precursors utilized by peroxidases, ASC delays cell wall lignification [108]. With the transition from meristematic to differentiated cells, ASC levels significantly decrease, permitting the activity of secretory peroxidases and, consequently, cell wall stiffening and lignification, occurring during differentiation [109].

Vitamin C is also involved in the control of PCD. *A. thaliana* mutants, with low vitamin C content, spontaneously trigger localized cell death like that which occurs during hypersensitive response, a plant-defense mechanism activated to block pathogen invasion [87]. An ASC decrease is necessary for PCD induced by H_2O_2 and heat shock (HS) in tobacco BY-2 cells [110–113]. The decrease in ASC during HS-induced PCD is due to inactivation of the last enzyme of the vitamin C biosynthetic pathway [114]. Low levels of ASC during PCD are parallel to a decrease in the level of transcript, protein, and activity of APX [111,112,115]. The impairment in ASC and APX is needed to increase ROS, which are essential for PCD induction [116,117]. Interestingly, the increase in vitamin C biosynthesis by the supply of L-galactono-1,4-lactone delays PCD occurring during kernel maturation in durum wheat, with a consequent postponement of dehydration and improvement in kernel filling [118].

Vitamin C, regulating the abovementioned processes at molecular and cellular levels, is therefore involved in different phases of plant growth and development, such as seed maturation and germination, flowering, fruit ripening, and senescence [119].

4. Vitamin C Biosynthesis in Plants

Vitamin C biosynthesis in higher plants has been the subject of dispute for many years. The first investigations related to vitamin C biosynthesis in plants date back to 1950 [120]. A definitive mechanism was formulated only 40 years later [60]. Unlike the animal pathway, in plants, no carbon inversion occurs in the biosynthesis of vitamin C, and the C1 in the D-glucose molecule remains as C1 after conversion. The Smirnoff–Wheeler pathway, in which vitamin C is synthesized from D-mannose and L-galactose (D-mannose/L-galactose pathway), represents the major route of vitamin C biosynthesis in plants [62,121]. Three other routes have been proposed for vitamin C biosynthesis: the gulose pathway, the myoinositol pathway, and the galacturonate pathway (Figure 2) [122–124].

4.1. D-Mannose/L-Galactose Pathway

In the Smirnoff-Wheeler pathway, D-glucose-6-phosphate is transformed into D-fructose-6-phosphate by phosphoglucose isomerase (PGI) and then is directed into D-mannose metabolism by phosphomannose isomerase (PMI), which produces D-mannose-6-phosphate. In *Arabidopsis*, PMI1 expression increases concomitantly with vitamin C levels under continuous light, and knockdown *pmi1* plants showed decreased levels of this metabolite [125]. D-Mannose-6-phosphate is then converted into D-mannose-1-phosphate by phosphomannose mutase (PMM). Genetic evidence for the involvement of PMM in vitamin C biosynthesis has been obtained in *Nicotiana benthamiana* and *Arabidopsis* [126,127].

GDP-D-mannose pyrophosphorylase (GMP) transfers guanosine monophosphate from GTP to give GDP-D-mannose. In *Arabidopsis*, GMP is coded by VTC1; the *vtc1* mutants accumulate ~25–30% of vitamin C levels of wild type and are hypersensitive to ozone [128,129]. *vtc1* mutants also have altered sensitivity to other ROS-generating conditions, including H_2O_2, UV-B, SO_2, and combined high light and salt stress [61]. Additional support for the involvement of GMP in vitamin C biosynthesis was obtained in potato plants constitutively expressing the antisense GMP gene. These plants showed a significant decrease in the activity of the enzyme and a significant reduction of vitamin C in leaves and tubers [130].

GDP-L-galactose is produced directly by GDP-D-mannose through a 3′5′ epimerization catalyzed by GDP-D-mannose epimerase (GME). GME has been characterized in *Chlorella*, flax, and *Arabidopsis* [131,132]. The enzyme belongs to the extended short-chain dehydratase/reductase protein family, with a modified NAD^+ binding Rossman fold domain [133]. GME is also able to catalyze the 5′ epimerization of GDP-mannose, giving GDP-L-gulose, which is the precursor of a possible side-branch biosynthetic pathway (the gulose pathway) for vitamin C synthesis [122,132,134].

GDP-D-mannose and GDP-L-galactose are substrates for the synthesis of glycoproteins and polysaccharides of cell walls [135,136]. Thus, the first dedicated step for vitamin C synthesis in the D-mannose/L-galactose pathway is the conversion of GDP-L-galactose into L-galactose-1-phosphate, catalyzed by GDP-L-galactose-phosphorylase (GGP) [137].

In *Arabidopsis*, GGP is encoded by the VTC2 and VTC5 genes [121]. The VTC2 expression levels are significantly higher (100–1000 times) than that of VTC5; moreover, T-DNA insertion mutants of VTC2 and VTC5 have 20% and 80% of the vitamin C content of wild-type plants, respectively. The double *vtc2 vtc5* mutants are unable to grow after cotyledon expansion, unless there is feeding of galactose or ASC, suggesting that at least in *Arabidopsis*, the D-mannose/L-galactose pathway is the substantial font of vitamin C [121]. The key role of GGP as a control point in vitamin C biosynthesis has been shown not only in *Arabidopsis* but also in tobacco, tomato, kiwifruit, strawberry, potato, citrus, and blueberry [138–143]. The transcript levels of VTC2 and VTC5 strongly correlate with the vitamin C content and the increase with light irradiation [121,144]. Moreover, VTC2, as well as other orthologue genes, can be controlled at the translation level by a noncanonical upstream open reading frame (uORF). In the presence of a high amount of ASC, the uORF encodes for a peptide, which acts as an inhibitor of translation, whereas under a low amount of ASC, the uORF is bypassed and GGP is translated [145].

L-Galactose-1-phosphate is converted into L-galactose by L-galactose-1-phosphate phosphatase (GPP), which is encoded by VTC4 in *Arabidopsis* [146,147]. However, *vtc4* mutants have a partial decrease of GPP activity and vitamin C content [147,148]. Accordingly, in *Arabidopsis*, the reaction can also be catalyzed by the purple acid phosphatase AtPAP15 [149]. L-galactose dehydrogenase (GDH) is the NAD-dependent enzyme catalyzing the conversion of L-galactose into L-galactono-1,4-lactone [60]. This step of ASC biosynthesis is not limiting. Indeed, in *Arabidopsis*, the overexpression of GDH does not change the vitamin C content, and antisense plants show vitamin C decrease only under high light [150].

The last step of the Smirnoff-Wheeler pathway is catalyzed by GLDH, a flavoprotein which converts L-galactono-1,4-lactone into L-ascorbate, transferring electrons to cytochrome c [27,120,151]. An important observation is that GLDH does not release H_2O_2, as happens with L-gulonolactone oxidase in animals, and therefore the production of vitamin C in plants does not affect the redox state of the cell [152]. Unlike the other enzymes of the D-mannose/L-galactose pathway, which are all localized in the cytosol, GLDH is an integral protein of the mitochondrial inner membrane [97,153,154]. Specifically, GLDH has been detected in complex I and acts as an essential plant-specific factor for complex I assembly [155–158]. Due to the GLDH localization, vitamin C biosynthesis is very sensitive to stresses that cause impairments of electron flux [111,114,159]. On the other hand, an increase in respiration, like that observed during tomato ripening, is associated with an enhancement of vitamin C content [160]. The different steps of the Smirnoff-Wheeler pathway are schematized in Figures 2 and 3.

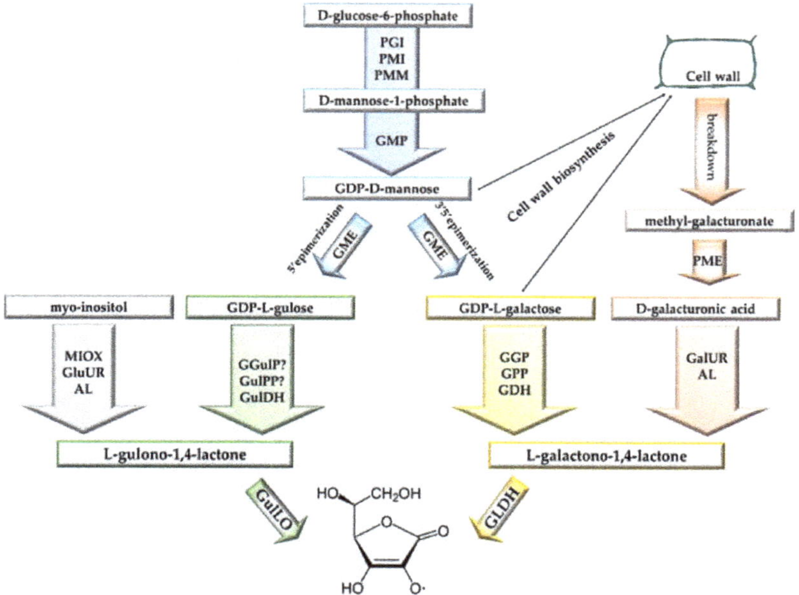

Figure 2. Schematic representation of vitamin C biosynthetic pathways. Different colors indicate different pathways. In grey, the myoinositol pathway; in green, the gulose pathway; in yellow, the D-mannose/L-galactose pathway; in orange, the galacturonate pathway. Represented in blue are the initial steps leading to GDP-D-mannose, which is a common precursor to the D-mannose/L-galactose and gulose pathways. A question mark indicates enzymes not identified in plants (more details are provided in the text). Abbreviations: AL, aldono lactonase; GalUR, D-galacturonate reductase; GDH, L-galactose dehydrogenase; GGP, GDP-L-galactose-phosphorylase; GGulP, GDP-L-gulose pyrophosphatase; GLDH, L-galactono-1,4-lactone dehydrogenase; GluUR, glucuronate reductase; GME, GDP-D-mannose epimerase; GMP, GDP-D-mannose pyrophosphorylase; GPP, L-galactose-1-phosphate phosphatase; GulDH, L-gulose dehydrogenase; GulPP, L-gulose-1-phosphate phosphatase; GulLO, gulonolactone oxidase; MIOX, myoinositol oxygenase; PGI, phosphoglucose isomerase; PME, pectin methyl esterase; PMI, phosphomannose isomerase; PMM, phosphomannose mutase.

4.2. Other Vitamin C Biosynthetic Pathways

As reported above, the gulose pathway starts from the 3′ epimerization of the GDP-D-mannose catalyzed by GME, with the formation of GDP-L-gulose (Figure 2) [122]. In this pathway, GDP-L-gulose is successively converted into L-guolse-1P, L-gulose, and L-gulono-1,4-lactone. L-gulono-1,4-lactone has been detected in plant extracts [161]. Moreover, external supplementation of L-gulono-1,4-lactone causes an increase in vitamin C content in *Arabidopsis* cells and tobacco leaves but is less efficient than feeding with L-galactono-1,4-lactone [162–164]. Activity of gulonolactone oxidase (GulLO) has been found in potato [163], and more recently, two genes coding for GulLOs have been identified in *Arabidopsis*. These enzymes are dehydrogenases with specificity for L-gulono-1,4-lactone and differ from plant GLDHs and mammalian GulLOs. The *Arabidopsis* GulLOs seem to be regulated post-transcriptionally and the limited enzyme availability could explain the slow utilization of the substrate [165].

L-Gulono-1,4-lactone is also the last precursor of vitamin C biosynthesis in the myoinositol pathway (Figure 2). In this pathway, myoinositol is converted into D-glucuronate by myoinositol oxygenase (MIOX). The other two steps producing L-gulonic acid and L-gulono-1,4-lactone are respectively catalyzed by glucuronate reductase and aldono lactonase [166]. A myoinositol oxygenase (MIOX4) has been identified in *Arabidopsis* [123]. However, the effective contribution of myoinositol to vitamin C synthesis in vivo is strongly debated [167–169].

The galacturonate pathway is also known as the salvage pathway, since it utilizes sugars provided by the breakdown of cell walls [124]. The degradation of pectin releases methyl-galacturonate, which is converted into D-galacturonate by methyl esterase and successively into L-galactonate by D-galacturonate reductase (GalUR). An aldono lactonase converts L-galactonate into L-galactono-1,4-lactone, which is the last precursor of vitamin C in the Smirnoff–Wheeler pathway (Figure 2) [27]. A gene encoding GalUR was initially identified in strawberry [124,170]. This pathway seems to be active mainly during fruit ripening in some species [124,170–173]. In tomato, vitamin C synthesis in immature green fruit is enhanced only by the supply of L-galactose, whereas in red ripened fruits by feeding with both L-galactose and D-galacturonate [170]. Moreover, the high vitamin C content found in tomato introgression lines IL12–4 compared with the parental M82 seems to be due to a higher expression of a pectinesterase and two polygalacturonases [171].

5. Light-Dependent Vitamin C Accumulation in Plants

Several papers have reported that plant exposure to light significantly increases vitamin C content [139,174,175]. Consistently, probably due to a reduction of irradiance levels, plants grown in greenhouses show lower levels of vitamin C compared with plants cultivated in the field [176]. However, the enhancement of vitamin C in plants seems to be dependent on the total amount of incident global radiation, which can be regulated by modulating light intensity or the duration of plant's exposure to light. The use of continuous light for 48 and 72 h after a period of darkness causes a great increase of vitamin C levels in lettuce and *Arabidopsis*, respectively [70,177]. However, the continuous exposure of lettuce plants to very high irradiation causes a loss of vitamin C, whereas continuous low irradiation improves the content of this metabolite [178]. Shen et al. [179] found that in lettuce, continuous illumination with red-blue light emitting diodes (LEDs) increases the vitamin C content in relation to the exposure time. Interestingly, in the same plants cultivated for 15 days under continuous red-blue LEDs exposure, the maximum peak of vitamin C content was found after nine days from the beginning of light exposure [180].

The quality of light can also influence the vitamin C pool. The regulatory effects of monochromatic lights of the UV-Vis spectrum on the modulation of vitamin C content have been studied in different species, but the obtained results are often contradictory, suggesting that different species can respond differently to specific wavelengths. High-red/far-red ratios enhanced vitamin C levels in the leaves of *Phaseolus vulgaris* [67]. Lettuce plants illuminated with single or combined blue and red lights showed higher vitamin C contents than plants grown in white light [181]. In Chinese kale, different lights, except for blue, used during sprout growth improved the vitamin C content, and the highest concentration of ASC was found in shoots exposed to white LEDs [182]. The influence of light on the vitamin C content was also evaluated in numerous fruits, such as apple [183], tomato [175,184], Satsuma mandarin, Valencia orange, and Lisbon lemon [185]. In these three citrus varieties, the enhancement of vitamin C content in the fruits was greater with the increasing intensity of blue LEDs. Furthermore, continuous irradiation with blue LEDs is more effective than pulsed irradiation and is related to increased expression of genes involved in the modulation of the vitamin C pool, suggesting a control at the transcriptional level [185]. On the contrary, accumulation of vitamin C in lettuce grown under continuous light is mainly due to changes in activity, rather than in expression, of the enzymes involved in vitamin C biosynthesis and oxidation [180].

Light treatments have also been tested in the postharvest, but also in this case, contradictory results have been reported. In cabbages stored for 15 days at low temperatures, the content of vitamin C was higher in the presence of blue light [186], unlike what was observed in asparagus stored at 4 °C, in which the vitamin C content after six days of blue light did not differ from the control in the dark [187]. In broccoli as well, blue light had no positive effects on the vitamin C content, whereas green, red, and yellow lights, probably stimulating metabolic and physiological activity, permitted de novo vitamin C synthesis [188].

In addition to visible light radiation, plants are also exposed in nature to UV radiation, which constitutes about 7% of solar radiation [189]. UV was shown to improve vitamin C content in soybean sprouts [190]. Increased levels of vitamin C in cucumber plants illuminated with UV-B have been linked to a significant increase in the activity of MIOX, GLDH, and enzymes of the ASC-GSH cycle [191].

Light Regulation of Vitamin C Accumulation

The pivotal role of light-dependent vitamin C accumulation in green tissues is due to photosynthesis. In *Arabidopsis* leaves and tomato green fruits, the photosynthetic inhibitor DCMU blocks vitamin C accumulation by reducing the expression of genes involved in its biosynthesis [70,170]. Moreover, photosynthesis increases the amount of soluble carbohydrates, which are biosynthetic precursors of vitamin C. Three genes involved in carbohydrate accumulation and translocation have been related to a quantitative trait locus (QTL) associated with a 1.4-fold vitamin C increase in tomato [192]. On the other hand, Ntagkas et al. [193] found no correlation between vitamin C content and levels of soluble carbohydrates.

Light can control vitamin C accumulation through different types of regulation (Figure 3).

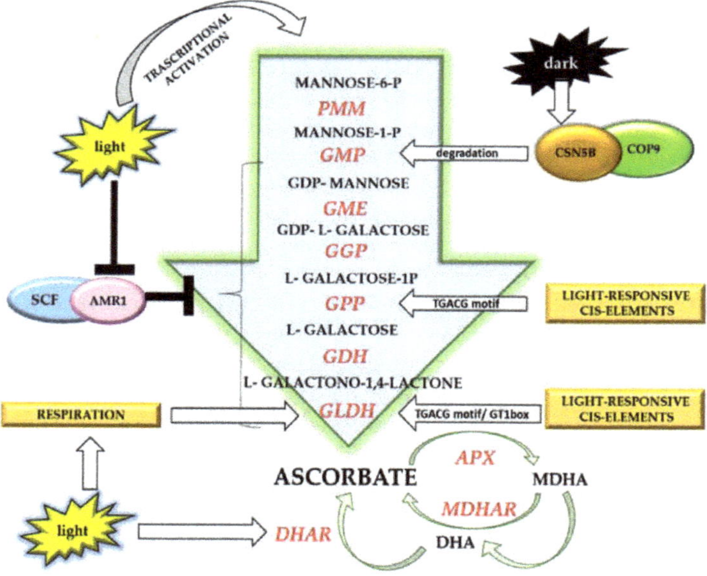

Figure 3. Light-dependent mechanisms involved in vitamin C accumulation (details are provided in the text). Abbreviations: PMM, phosphomannose mutase; GMP, GDP-D-mannose pyrophosphorylase; GME, GDP-D-mannose epimerase; GGP, GDP-L-galactose-phosphorylase; GPP, L-galactose-1-phosphate phosphatase; GDH, L-galactose dehydrogenase; GLDH, L-galactono-1,4-lactone dehydrogenase; APX, ascorbate peroxidase; MDHA, monodehydroascorbate; MDHAR monodehydroascorbate reductase; DHA, dehydroascorbate; DHAR, dehydroascorbate reductase.

Light-dependent accumulation of vitamin C in plants seems to be principally due to the enhanced expression of different genes involved in the D-mannose/L-galactose biosynthetic pathway [70,121,142,175]. It has been shown that in rice, GLDH and GPP contain light-responsive cis-elements (GT1 box and TGACG motif) in their promoters [194]. Concerning GLDH, light beyond regulating its expression can induce changes in respiration, indirectly modulating enzymatic activity [69,159]. Light also influences the expression of genes involved in ASC recycling. During germination, corn seeds exposed to high light exhibited a higher DHAR expression along with an increased vitamin C content (Figure 3) [195].

In tomato, light induces the expression of the transcription factor HZ24, which in turn activates GMP transcription in leaves and immature fruits [196]. High light also induces a decrease in the expression of AMR1, a transcription factor acting as a negative regulator of the last six genes of the D-mannose/L-galactose biosynthetic pathway, consequently increasing vitamin C biosynthesis (Figure 3) [197]. Finally, in the dark, CSN5B promotes GMP degradation via 26S proteasome, reducing vitamin C levels [198]. Additionally, darkness promotes vitamin C catabolism [44].

The complexity of light-dependent vitamin C regulation (Figure 3) highlights that obtaining vitamin-C-biofortified plants by light treatments requires in-depth knowledge of the metabolic and physiological processes involved [199].

6. Vitamin C Biofortification

Increasing the vitamin C content in plants can have a triple-positive effect: producing food with a high content of vitamin C for human health, increasing the postharvest shelf life, and, not less important, increasing the resistance of plants to various kinds of stress. The different strategies adopted for vitamin C biofortification (Figure 4) are discussed below.

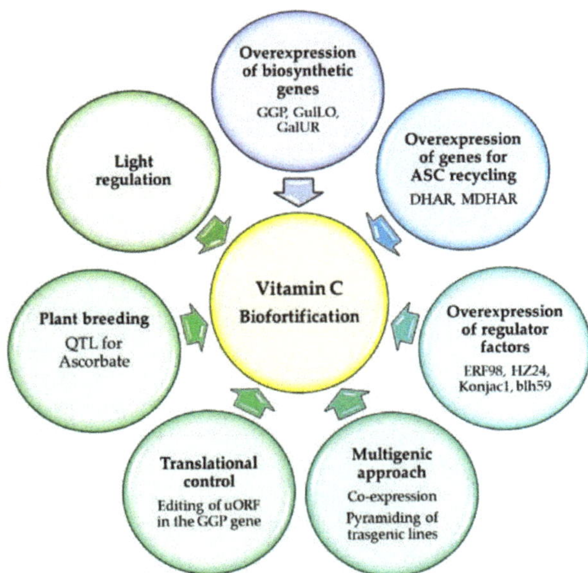

Figure 4. Strategies for vitamin C biofortification. Modulation of light intensity and quality can be used to obtain vitamin C enrichment in crops. The overexpression of single or multiple genes belonging to the biosynthetic and recycling pathways, as well as to the regulatory network, represents a good tool for vitamin C biofortification. Vitamin C accumulation can be controlled by modulating translation by the editing of the uORF in the GGP gene. Finally, vitamin C biofortification can be obtained by plant breeding that exploits candidate genes in the quantitative trait locus (QTL) associated with high vitamin C content (more details are provided in the text).

6.1. Manipulation of D-Mannose/L-Galactose Pathway

Several genes of the D-mannose/L-galactose pathway have been overexpressed in different crops in order to enhance vitamin C levels, but not all have given good results [34,35]. It has been proved that overexpression of GGP, which represents the bottleneck of vitamin C biosynthesis [137], is a good strategy for biofortification [200]. For instance, in *Arabidopsis*, transient overexpression of GGP leads to a 2.5-fold increase in vitamin C content, whereas the overexpression of the other genes involved in the same pathway does not cause relevant differences in terms of vitamin C [142]. Similar results were obtained in rice where, among different transgenic lines overexpressing six *Arabidopsis* genes involved in vitamin C biosynthesis, the highest vitamin C content was found in the line overexpressing GGP [201]. A kiwi gene coding for GGP, initially tested with good results in *Arabidopsis* [138], has been overexpressed in three crops, leading to a vitamin C increase of six-fold in tomato, three-fold in potato, and two-fold in strawberry; however, in tomato, GGP overexpression has led to some morphological fruit alterations, such as seed loss [140]. The GGP gene of acerola, a well-known crop with high vitamin C content, under the control of a leaf-specific promoter, has been overexpressed in rice, increasing the foliar content up to 2.5-fold, which did not cause morphological changes and conferred multistress tolerance [202]. Interestingly, also, the editing of the uORF, which controls the translation of the GGP2 gene in lettuce and tomato, increases the vitamin C content by 150% and confers tolerance to oxidative stress, providing a good strategy to obtain transgene-free lines with improved vitamin C [203,204]. Moreover, in apple, three paralogs of GGP, colocated in ASC-QTL clusters and, specifically, the GGP1 allele, play a key role in the regulation of vitamin C content in fruits. This suggests that a single-nucleotide polymorphism of this allele is an excellent candidate for breeding in order to improve vitamin C levels in fruits [205].

The multigenic approach, based on the coexpression of genes of the D-mannose/L-galactose pathway, represents an interesting strategy to obtain high levels of vitamin C in crops. The transient coexpression of GGP and GME in tobacco leaves caused a seven-fold increase in vitamin C content [138]. In *Arabidopsis*, GGP overexpressing lines had a 2.9-fold enhancement of vitamin C, whereas the double-gene transformation with GGP-GPP and GGP-GLDH led to an up to 4.1-fold vitamin C increase [206]. The contemporary overexpression of acerola GGP, GMP, and GME genes in tomato protoplasts caused an increase in vitamin C content, which was approximately four-fold higher than in wild type [207]. A stable transformation with GME, GMP, GGP, and GPP was obtained in tomato through pyramiding, which is a conventional hybridization that is technically achievable and generates stable inherited target genes [208,209]. Pyramiding transgenic lines GME × GMP and GME × GMP × GGP × GPP showed a substantial increase in vitamin C content in leaves and fruits. Moreover, in these lines, vitamin C transport capability, fruit shape and size, as well as stress tolerance were significantly ameliorated [209].

6.2. Manipulation of Other Biosynthetic Pathways

The overexpression of genes of the alternative biosynthetic pathways have also given good results in terms of vitamin C content in different crops. Regarding the gulose pathway, positive results have been obtained with the expression of rat cDNA encoding GulLO, the enzyme involved in the final step of the animal vitamin C biosynthetic pathway [163,210,211]. Lettuce and tobacco plants constitutively expressing this gene showed four- and seven-fold increases in vitamin C levels, respectively [163]. Transgenic potato plants, overexpressing the same gene, show improved vitamin C accumulation in tubers and increased tolerance to several abiotic stresses [210]. In the same way, *Arabidopsis* lines overexpressing this GulLO contained high vitamin C contents and exhibited improved growth and enhanced biomass of shoots and roots, as well as higher tolerance to diverse abiotic stresses [211].

Interesting results have also been obtained by manipulating the galacturonate pathway. Overexpression of the strawberry FaGalUR led to a two-fold increase in vitamin C content in potato, and this enhancement allowed for an increase in tolerance to abiotic stresses in the transgenic lines [212]. Similar positive results have been reported for tomato, where, although there was a moderate increase in vitamin C content, an increase in total antioxidants occurred that was linked to redox state regulation [213]; moreover, tomato plants overexpressing FaGalUR were found to be more tolerant to abiotic stresses [214]. Interestingly, in the tomato introgression line IL12-4-SL, the genes encoding for pectin methylesterase, polygalacturonase, and UDP-D-glucuronic-acid-4-epimerase, which are involved in pectin degradation, have been identified as candidate genes for a QTL associated with high vitamin C content, suggesting that marker-assisted selection could be a good strategy to enhance vitamin C accumulation [215].

6.3. Manipulation of Recycling Genes

Vitamin C enhancement in crops can also be achieved by manipulating the genes coding for MDHAR and DHAR, which are the enzymes involved in the reduction of MDHA and DHA, respectively. Several papers have reported that ASC regeneration by DHAR overexpression could represent an efficient method of vitamin C biofortification in different species, such as corn [216], tomato [217], and blueberry [143]. Recently, a cytosolic DHAR identified in the woody plant *Liriodendron chinense* was overexpressed in *Arabidopsis*, which led not only to vitamin C enhancement but also to an improvement of growth under stress conditions [218]. In apple, colocation between *DHAR3-3* and a QTL for browning has been found, showing a relationship between ASC redox state and fruit vulnerability to browning [205].

Research conducted on MDHAR shows discordant results depending on the species. The overexpression in tobacco of the *Arabidopsis* cytosolic isoform of MDHAR enhances vitamin C content [219]. Similarly, the acerola MDHAR, overexpressed in tobacco, led to a two-fold increase in vitamin C content and a better tolerance to salt stress [220]. On the other hand, overexpression of the cytosolic-targeted tomato MDHAR caused a 0.7-fold reduction in vitamin C content in tomato fruits [217]. Another study conducted on tomato indicates that transgenic lines overexpressing MDHAR display a reduction in vitamin C content in leaves, while lines with silencing of MDHAR show an increase of vitamin C in both fruits and leaves [221]. The enhancement of vitamin C in silenced MDHAR lines could be due to a decrease in degradation [44]. In cherry tomato, the suppression of AOX was also found to increase vitamin C, lycopene, and carotene contents of the fruits and to confer tolerance to salt stress [222].

6.4. Manipulation of Regulatory Networks

Despite the many results obtained by overexpressing vitamin C-related genes, limited success has been reported in most species. In light of this, attention has shifted to the manipulation of components of the regulatory network, such as transcription and regulator factors. The overexpression of ERF98, which is a positive regulator of GMP, GGP, and GLDH genes, enhanced vitamin C content and increased tolerance to salt stress in *Arabidopsis* [223]. Similarly, tomato plants overexpressing HZ24, a transcriptional factor that binds the promoters of GMP, GME2, and GGP, showed increased vitamin C levels and reduced sensitivity to oxidative stress [196]. Overexpression of the regulator factors KONJAC1 and 2, which are two nucleotide sugar pyrophosphorylase-like proteins that modulate GMP activity, led to an increase in vitamin C content in *Arabidopsis* [224]. *Arabidopsis* and tomato plants, overexpressing the regulator factor SlZF3, showed inhibition of GMP degradation by COP9 signalosome, with a consequent enhancement of vitamin C content and tolerance to salt stress [225]. A new transcription factor bHLH59, which can activate the transcription of PMI, PMM, and GMP 2–4 and colocalize with the vitamin C QTL TFA9, has been identified in tomato. The overexpression of bHLH59 causes vitamin C accumulation and increases oxidative stress tolerance. The differences in vitamin C accumulation within different tomato accessions is ascribed to nucleotide differences in the promoter region of HLH59. This finding could be used to plan breeding strategies for vitamin C improvement [226].

7. Conclusions

In humans, different physiological processes require vitamin C as an antioxidant or a cofactor of mono-oxygenases and dioxygenases, whereby low vitamin C levels prevent optimal functioning. An increase in vitamin C intake through food is surely beneficial for human physiology [227,228]. The recommended daily intake (RDI) of vitamin C is 75–90 mg/day [229]. Nevertheless, 100 g of potatoes and tomatoes have about one-fourth of the RDI and cereal grains contain very low, almost undetectable, quantities of vitamin C [230]. Assuming their potential to make available adequate vitamin C levels, biofortified crops could be decisive in the elimination of vitamin C deficiency on a worldwide scale. Apart from having beneficial effects on human health, vitamin C biofortification also has the potential to improve plant tolerance to various stresses, which is a prominent target to guarantee crop productivity in an era of global climate change.

However, vitamin C accumulation in different plant organs is dependent on multiple metabolic processes, such as biosynthesis, recycling, degradation, and transport. Moreover, the vitamin C content is influenced by endogenous stimuli and environmental factors, among which light is of primary importance. Thus, as for other micronutrients, comprehensive knowledge of the genetic, biochemical, and molecular networks that govern vitamin C levels is mandatory to obtain vitamin-C-biofortified crops [231].

Considerable progresses on the understanding of the multiple roles of vitamin C and its interaction with other antioxidants, as well as with signal transduction pathways of hormones and ROS have been made. However, little is known about the influence that increased vitamin C levels may have on the different physiological processes of plants. Since changes in vitamin C levels greatly modify gene expression, and in particular, the transcript levels of genes involved in photosynthesis and the defense response to pathogens [77], the possibility of undesirable consequences resulting from the altered vitamin C content has to be carefully considered. Thus, efforts to obtain vitamin-C-biofortified plants necessitate an in-depth investigation into how these changes can alter plant growth, development, and responses to biotic and abiotic stresses under field conditions. To limit the possible unplanned consequences, targeted approaches altering vitamin C levels in specific tissues or organs are required.

Particular attention must be also paid to the choice of methodology utilized for vitamin C biofortification. The multigenic approach, obtained with co-expression or pyramiding, has led to a significant increase in vitamin C content. However, considering that genetically modified organisms are not always easily accepted by public opinion, methodologies avoiding the use of transgenes must be taken in consideration. For instance, the editing of the uORF on the promoter of genes coding for GGP is a good method to obtain non transgenic plants enriched with vitamin C. The identification of candidate genes in QTL associated with high vitamin C content could also allow for obtaining vitamin-C-biofortified plants by marker-assisted selection, thus avoiding the use of transgenes.

Thoroughly understanding the regulatory mechanisms involved in vitamin C accumulation, which can differ between crops and growth phases, together with the choice of better approaches to be utilized to improve vitamin C levels are important goals for developing more efficient strategies for vitamin C biofortification.

Author Contributions: Writing—original draft preparation, C.P., S.F., N.D., A.P., S.D.L., L.M., and M.C.d.P.; writing—review and editing, C.P. and M.C.d.P.; visualization, A.P. and S.D.L.; supervision, M.C.d.P.

Funding: This research was funded by University of Bari Aldo Moro, grant number H95E10000710005.

Conflicts of Interest: The authors declare no conflict of interest.

References

1. Szent-Gyorgyi, A. Observations on the function of peroxidase systems and the chemistry of the adrenal cortex: Description of a new carbohydrate derivative. *Biochem. J.* **1928**, *22*, 1387–1409. [CrossRef] [PubMed]
2. King, C.G.; Waugh, W.A. The chemical nature of vitamin C. *Science* **1932**, *75*, 357–358. [CrossRef] [PubMed]
3. Svirbely, J.L.; Szent-Gyorgyi, A. The chemical nature of vitamin C. *Biochem. J.* **1932**, *26*, 865–870. [CrossRef] [PubMed]
4. Svirbely, J.L.; Szent-Gyorgyi, A. Hexuronic acid as the antiscorbutic factor. *Nature* **1932**, *129*, 690. [CrossRef]
5. Haworth, W.H.; Hirst, E.L. Synthesis of ascorbic acid. *J. Soc. Chem. Ind.* **1933**, *52*, 645–646. [CrossRef]
6. Szent-Györgyi, A.; Haworth, W.H. Hexuronic acid (ascorbic acid) as the antiscorbutic factor. *Nature* **1933**, *131*, 24. [CrossRef]
7. Padayatty, S.J.; Levine, M. Vitamin C: The known and the unknown and Goldilocks. *Oral Dis.* **2016**, *22*, 463–493. [CrossRef]

8. Turck, D.; Bresson, J.L.; Burlingame, B.; Dean, T.; Fairweather-Tait, S.; Heinonen, M.; Hirsch-Ernst, K.I.; Mangelsdorf, I.; McArdle, H.J.; Naska, A.; et al. Scientific opinion on Vitamin C and protection of DNA, proteins and lipids from oxidative damage: Evaluation of a health claim pursuant to Article 14 of Regulation (EC) No 1924/2006. EFSA NDA Panel (EFSA Panel on Dietetic Products, Nutrition and Allergies). *EFSA J.* **2017**, *15*, 4685. [CrossRef]
9. Brubaker, R.F.; Bourne, W.M.; Bachman, L.A.; McLaren, J.W. Ascorbic acid content of human corneal epithelium. *Invest. Ophthalmol. Vis. Sci.* **2000**, *41*, 1681–1683.
10. Tannenbaum, S.R.; Wishnok, J.S.; Leaf, C.D. Inhibition of nitrosamine formation by ascorbic acid. *Am. J. Clin. Nutr.* **1991**, *53*, 247S–250S. [CrossRef]
11. Huang, A.; Vita, J.A.; Venema, R.C.; Keaney, J.F. Ascorbic acid enhances endothelial nitric-oxide synthase activity by increasing intracellular tetrahydrobiopterin. *J. Biol. Chem.* **2000**, *275*, 17399–17406. [CrossRef] [PubMed]
12. Lane, D.J.; Richardson, D.R. The active role of vitamin C in mammalian iron metabolism: Much more than just enhanced iron absorption! *Free Radic. Biol. Med.* **2014**, *75*, 69–83. [CrossRef] [PubMed]
13. Lane, D.J.; Bae, D.H.; Merlot, A.M.; Sahni, S.; Richardson, D.R. Duodenal cytochrome b (DCYTB) in iron metabolism: An update on function and regulation. *Nutrients* **2015**, *7*, 2274–2296. [CrossRef] [PubMed]
14. Eipper, B.A.; Mains, R.E. The role of ascorbate in the biosynthesis of neuroendocrine peptides. *Am. J. Clin. Nutr.* **1991**, *54*, 1153S–1156S. [CrossRef] [PubMed]
15. Prigge, S.T.; Mains, R.E.; Eipper, B.A.; Amzel, L.M. New insights into copper monooxygenases and peptide amidation: Structure, mechanism and function. *Cell. Mol. Life Sci.* **2000**, *57*, 1236–1259. [CrossRef] [PubMed]
16. Kumar, D.; Mains, R.E.; Eipper, B.A. 60 YEARS OF POMC: From POMC and alpha-MSH to PAM, molecular oxygen, copper, and vitamin C. *J. Mol. Endocrinol.* **2016**, *56*, T63–T76. [CrossRef]
17. Peterkofsky, B. Ascorbate requirement for hydroxylation and secretion of procollagen: Relationship to inhibition of collagen synthesis in scurvy. *Am. J. Clin. Nutr.* **1991**, *54*, 1135S–1140S. [CrossRef]
18. Pekkala, M.; Hieta, R.; Kursula, P.; Kivirikko, K.I.; Wierenga, R.K.; Myllyharju, J. Crystallization of the proline-rich-peptide binding domain of human type I collagen prolyl 4-hydroxylase. *Acta Crystallogr. D Biol. Crystallogr.* **2003**, *59*, 940–942. [CrossRef]
19. Mandl, J.; Szarka, A.; Banhegyi, G. Vitamin C: Update on physiology and pharmacology. *Br. J. Pharmacol.* **2009**, *157*, 1097–1110. [CrossRef]
20. Fleming, P.J.; Kent, U.M. Cytochrome b561, ascorbic acid, and transmembrane electron transfer. *Am. J. Clin. Nutr.* **1991**, *54*, 1173S–1178S. [CrossRef]
21. Dunn, W.A.; Rettura, G.; Seifter, E.; Englard, S. Carnitine biosynthesis from gamma-butyrobetaine and from exogenous protein-bound 6-N-trimethyl-L-lysine by the perfused guinea pig liver. Effect of ascorbate deficiency on the in situ activity of gamma-butyrobetaine hydroxylase. *J. Biol. Chem.* **1984**, *259*, 10764–10770. [PubMed]
22. Rebouche, C.J. Ascorbic acid and carnitine biosynthesis. *Am. J. Clin. Nutr.* **1991**, *54*, 1147S–1152S. [CrossRef] [PubMed]
23. Kobayashi, M.; Hoshinaga, Y.; Miura, N.; Tokuda, Y.; Shigeoka, S.; Murai, A.; Horio, F. Ascorbic acid deficiency decreases hepatic cytochrome P-450, especially CYP2B1/2B2, and simultaneously induces heme oxygenase-1 gene expression in scurvy-prone ODS rats. *Biosci. Biotechnol. Biochem.* **2014**, *78*, 1060–1066. [CrossRef] [PubMed]
24. Camarena, V.; Wang, G. The epigenetic role of vitamin C in health and disease. *Cell. Mol. Life Sci.* **2016**, *73*, 1645–1658. [CrossRef] [PubMed]
25. Blaszczak, W.; Barczak, W.; Masternak, J.; Kopczynski, P.; Zhitkovich, A.; Rubis, B. Vitamin C as a modulator of the response to cancer therapy. *Molecules* **2019**, *24*, 453. [CrossRef] [PubMed]
26. Linster, C.L.; Van Schaftingen, E.; Vitamin, C. Biosynthesis, recycling and degradation in mammals. *FEBS J.* **2007**, *274*, 1–22. [CrossRef]
27. Smirnoff, N.; Conklin, P.L.; Loewus, F.A. Biosynthesis of ascorbic acid in plants: A renaissance. *Ann. Rev. Plant. Physiol. Plant. Mol. Biol.* **2001**, *52*, 437–467. [CrossRef]
28. Carr, A.C.; Vissers, M.C. Synthetic or food-derived vitamin C—Are they equally bioavailable? *Nutrients* **2013**, *5*, 4284–4304. [CrossRef]

29. Vissers, M.C.; Bozonet, S.M.; Pearson, J.F.; Braithwaite, L.J. Dietary ascorbate intake affects steady state tissue concentrations in vitamin C-deficient mice: Tissue deficiency after suboptimal intake and superior bioavailability from a food source (kiwifruit). *Am. J. Clin. Nutr.* **2011**, *93*, 292–301. [CrossRef]
30. Tanaka, K.; Hashimoto, T.; Tokumaru, S.; Iguchi, H.; Kojo, S. Interactions between vitamin C and vitamin E are observed in tissues of inherently scorbutic rats. *J. Nutr.* **1997**, *127*, 2060–2064. [CrossRef]
31. Guarnieri, S.; Riso, P.; Porrini, M. Orange juice vs vitamin C: Effect on hydrogen peroxide-induced DNA damage in mononuclear blood cells. *Br. J. Nutr.* **2007**, *97*, 639–643. [CrossRef] [PubMed]
32. Beck, K.; Conlon, C.A.; Kruger, R.; Coad, J.; Stonehouse, W. Gold kiwifruit consumed with an iron-fortified breakfast cereal meal improves iron status in women with low iron stores: A 16-week randomised controlled trial. *Br. J. Nutr.* **2011**, *105*, 101–109. [CrossRef] [PubMed]
33. Carr, A.C.; Frei, B. Toward a new recommended dietary allowance for vitamin C based on antioxidant and health effects in humans. *Am. J. Clin. Nutr.* **1999**, *69*, 1086–1107. [CrossRef] [PubMed]
34. Macknight, R.C.; Laing, W.A.; Bulley, S.M.; Broad, R.C.; Johnson, A.A.; Hellens, R.P. Increasing ascorbate levels in crops to enhance human nutrition and plant abiotic stress tolerance. *Curr. Opin. Biotechnol.* **2017**, *44*, 153–160. [CrossRef] [PubMed]
35. Locato, V.; Cimini, S.; Gara, L.D. Strategies to increase vitamin C in plants: From plant defense perspective to food biofortification. *Front. Plant. Sci.* **2013**, *4*, 152. [CrossRef] [PubMed]
36. Paciolla, C.; Paradiso, A.; de Pinto, M.C. Cellular redox homeostasis as central modulator in plant stress. In *Redox State as a Central Regulator of Plant-Cell Stress Responses*; Gupta, D.K., Palma, J.M., Corpas, F.J., Eds.; Springer: Cham, Switzerland, 2016; pp. 1–23. [CrossRef]
37. Tripathi, R.P.; Singh, B.; Bisht, S.S.; Pandey, J. L-Ascorbic acid in organic synthesis: An overview. *Curr. Org. Chem.* **2009**, *13*, 99–122. [CrossRef]
38. Zhang, L.; Dong, S.J. The electrocatalytic oxidation of ascorbic acid on polyaniline film synthesized in the presence of camphorsulfonic acid. *J. Electroanal. Chem.* **2004**, *568*, 189–194. [CrossRef]
39. Matsui, T.; Kitagawa, Y.; Okumura, M.; Shigeta, Y. Accurate standard hydrogen electrode potential and applications to the redox potentials of vitamin C and NAD/NADH. *J. Phys. Chem. A* **2015**, *119*, 369–376. [CrossRef]
40. Szarka, A.; Tomasskovics, B.; Banhegyi, G. The ascorbate-glutathione-alpha-tocopherol triad in abiotic stress response. *Int. J. Mol. Sci.* **2012**, *13*, 4458–4483. [CrossRef]
41. Buettner, G.R.; Jurkiewicz, B.A. Catalytic metals, ascorbate and free radicals: Combinations to avoid. *Radiat. Res.* **1996**, *145*, 532–541. [CrossRef]
42. Parrow, N.L.; Leshin, J.A.; Levine, M. Parenteral ascorbate as a cancer therapeutic: A reassessment based on pharmacokinetics. *Antioxid. Redox Signal.* **2013**, *19*, 2141–2156. [CrossRef] [PubMed]
43. Hossain, M.A.; Asada, K. Monodehydroascorbate reductase from cucumber is a flavin adenine dinucleotide enzyme. *J. Biol. Chem.* **1985**, *260*, 12920–12926. [PubMed]
44. Truffault, V.; Fry, S.C.; Stevens, R.G.; Gautier, H. Ascorbate degradation in tomato leads to accumulation of oxalate, threonate and oxalyl threonate. *Plant J.* **2017**, *89*, 996–1008. [CrossRef] [PubMed]
45. Maruta, T.; Sawa, Y.; Shigeoka, S.; Ishikawa, T. Diversity and evolution of ascorbate peroxidase functions in chloroplasts: More than just a classical antioxidant enzyme? *Plant Cell Physiol.* **2016**, *57*, 1377–1386. [CrossRef]
46. Teixeira, F.K.; Menezes-Benavente, L.; Margis, R.; Margis-Pinheiro, M. Analysis of the molecular evolutionary history of the ascorbate peroxidase gene family: Inferences from the rice genome. *J. Mol. Evol.* **2004**, *59*, 761–770. [CrossRef]
47. Foyer, C.H.; Noctor, G. Ascorbate and glutathione: The heart of the redox hub. *Plant Physiol.* **2011**, *155*, 2–18. [CrossRef]
48. Sano, S. Molecular and functional characterization of monodehydroascorbate and dehydroascorbate reductases. In *Ascorbic Acid in Plant Growth, Development and Stress Tolerance*; Hossain, M.A., Munnè-Bosch, S., Burritt, D.J., Diaz-Vivancos, P., Fujita, M., Lorence, A., Eds.; Springer: Cham, Switzerland, 2017; pp. 129–156. [CrossRef]
49. Gallie, D.R. The role of L-ascorbic acid recycling in responding to environmental stress and in promoting plant growth. *J. Exp. Bot.* **2013**, *64*, 433–443. [CrossRef]

50. Noshi, M.; Yamada, H.; Hatanaka, R.; Tanabe, N.; Tamoi, M.; Shigeoka, S. Arabidopsis dehydroascorbate reductase 1 and 2 modulate redox states of ascorbate-glutathione cycle in the cytosol in response to photooxidative stress. *Biosci. Biotechnol. Biochem.* **2017**, *81*, 523–533. [CrossRef]
51. Ding, S.; Lu, Q.; Zhang, Y.; Yang, Z.; Wen, X.; Zhang, L.; Lu, C. Enhanced sensitivity to oxidative stress in transgenic tobacco plants with decreased glutathione reductase activity leads to a decrease in ascorbate pool and ascorbate redox state. *Plant Mol. Biol.* **2009**, *69*, 577–592. [CrossRef]
52. Harshavardhan, V.T.; Wu, T.M.; Hong, C.Y. Glutathione reductase and abiotic stress tolerance in plants. In *Glutathione in Plant Growth, Development, and Stress Tolerance*; Hossain, M.A., Mostofa, M.G., Diaz-Vivancos, P., Burritt, D.J., Fujita, M., Tram, L.S.P., Eds.; Springer: Cham, Switzerland, 2017; pp. 265–286. [CrossRef]
53. Chew, O.; Rudhe, C.; Glaser, E.; Whelan, J. Characterization of the targeting signal of dual-targeted pea glutathione reductase. *Plant Mol. Biol* **2003**, *53*, 341–356. [CrossRef]
54. Gest, N.; Gautier, H.; Stevens, R. Ascorbate as seen through plant evolution: The rise of a successful molecule? *J. Exp. Bot.* **2013**, *64*, 33–53. [CrossRef] [PubMed]
55. Holler, S.; Ueda, Y.; Wu, L.; Wang, Y.; Hajirezaei, M.R.; Ghaffari, M.R.; von Wiren, N.; Frei, M. Ascorbate biosynthesis and its involvement in stress tolerance and plant development in rice (*Oryza sativa* L.). *Plant Mol. Biol.* **2015**, *88*, 545–560. [CrossRef] [PubMed]
56. Maddison, J.; Lyons, T.; Plochl, M.; Barnes, J. Hydroponically cultivated radish fed L-galactono-1,4-lactone exhibit increased tolerance to ozone. *Planta* **2002**, *214*, 383–391. [CrossRef] [PubMed]
57. Paradiso, A.; Berardino, R.; de Pinto, M.C.; Sanita di Toppi, L.; Storelli, M.M.; Tommasi, F.; De Gara, L. Increase in ascorbate-glutathione metabolism as local and precocious systemic responses induced by cadmium in durum wheat plants. *Plant Cell Physiol.* **2008**, *49*, 362–374. [CrossRef]
58. Wang, S.D.; Zhu, F.; Yuan, S.; Yang, H.; Xu, F.; Shang, J.; Xu, M.Y.; Jia, S.D.; Zhang, Z.W.; Wang, J.H.; et al. The roles of ascorbic acid and glutathione in symptom alleviation to SA-deficient plants infected with RNA viruses. *Planta* **2011**, *234*, 171–181. [CrossRef]
59. Sgobba, A.; Paradiso, A.; Dipierro, S.; De Gara, L.; de Pinto, M.C. Changes in antioxidants are critical in determining cell responses to short- and long-term heat stress. *Physiol. Plant.* **2015**, *153*, 68–78. [CrossRef]
60. Wheeler, G.L.; Jones, M.A.; Smirnoff, N. The biosynthetic pathway of vitamin C in higher plants. *Nature* **1998**, *393*, 365–369. [CrossRef]
61. Smirnoff, N. Ascorbic acid: Metabolism and functions of a multi-facetted molecule. *Curr. Opin. Plant Biol.* **2000**, *3*, 229–235. [CrossRef]
62. Smirnoff, N. Ascorbic acid metabolism and functions: A comparison of plants and mammals. *Free Radic. Biol. Med.* **2018**, *122*, 116–129. [CrossRef]
63. De Gara, L.; de Pinto, M.C.; Arrigoni, O. Ascorbate synthesis and ascorbate peroxidase activity during the early stage of wheat germination. *Physiol. Plant.* **1997**, *100*, 894–900. [CrossRef]
64. Tommasi, F.; Paciolla, C.; de Pinto, M.C.; De Gara, L. A comparative study of glutathione and ascorbate metabolism during germination of *Pinus pinea* L. seeds. *J. Exp. Bot.* **2001**, *52*, 1647–1654. [CrossRef] [PubMed]
65. Veljovic-Jovanovic, S.D.; Pignocchi, C.; Noctor, G.; Foyer, C.H. Low ascorbic acid in the vtc-1 mutant of Arabidopsis is associated with decreased growth and intracellular redistribution of the antioxidant system. *Plant Physiol.* **2001**, *127*, 426–435. [CrossRef] [PubMed]
66. Kukavica, B.; Jovanovic, S.V. Senescence-related changes in the antioxidant status of ginkgo and birch leaves during autumn yellowing. *Physiol. Plant.* **2004**, *122*, 321–327. [CrossRef]
67. Bartoli, C.G.; Tambussi, E.A.; Diego, F.; Foyer, C.H. Control of ascorbic acid synthesis and accumulation and glutathione by the incident light red/far red ratio in *Phaseolus vulgaris* leaves. *FEBS Lett.* **2009**, *583*, 118–122. [CrossRef]
68. Heyneke, E.; Luschin-Ebengreuth, N.; Krajcer, I.; Wolkinger, V.; Muller, M.; Zechmann, B. Dynamic compartment specific changes in glutathione and ascorbate levels in Arabidopsis plants exposed to different light intensities. *BMC Plant Biol.* **2013**, *13*, 104. [CrossRef]
69. Bartoli, C.G.; Yu, J.; Gomez, F.; Fernandez, L.; McIntosh, L.; Foyer, C.H. Inter-relationships between light and respiration in the control of ascorbic acid synthesis and accumulation in *Arabidopsis thaliana* leaves. *J. Exp. Bot.* **2006**, *57*, 1621–1631. [CrossRef]

70. Yabuta, Y.; Mieda, T.; Rapolu, M.; Nakamura, A.; Motoki, T.; Maruta, T.; Yoshimura, K.; Ishikawa, T.; Shigeoka, S. Light regulation of ascorbate biosynthesis is dependent on the photosynthetic electron transport chain but independent of sugars in Arabidopsis. *J. Exp. Bot.* **2007**, *58*, 2661–2671. [CrossRef]
71. Grillet, L.; Ouerdane, L.; Flis, P.; Hoang, M.T.; Isaure, M.P.; Lobinski, R.; Curie, C.; Mari, S. Ascorbate efflux as a new strategy for iron reduction and transport in plants. *J. Biol. Chem.* **2014**, *289*, 2515–2525. [CrossRef]
72. Asada, K. THE WATER-WATER CYCLE IN CHLOROPLASTS: Scavenging of active oxygens and dissipation of excess photons. *Annu. Rev. Plant Physiol. Plant Mol. Biol.* **1999**, *50*, 601–639. [CrossRef]
73. Awad, J.; Stotz, H.U.; Fekete, A.; Krischke, M.; Engert, C.; Havaux, M.; Berger, S.; Mueller, M.J. 2-Cysteine peroxiredoxins and thylakoid ascorbate peroxidase create a water-water cycle that is essential to protect the photosynthetic apparatus under high light stress conditions. *Plant Physiol.* **2015**, *167*, 1592–1603. [CrossRef]
74. Saga, G.; Giorgetti, A.; Fufezan, C.; Giacometti, G.M.; Bassi, R.; Morosinotto, T. Mutation analysis of violaxanthin de-epoxidase identifies substrate-binding sites and residues involved in catalysis. *J. Biol. Chem.* **2010**, *285*, 23763–23770. [CrossRef] [PubMed]
75. Toth, S.Z.; Nagy, V.; Puthur, J.T.; Kovacs, L.; Garab, G. The physiological role of ascorbate as photosystem II electron donor: Protection against photoinactivation in heat-stressed leaves. *Plant Physiol.* **2011**, *156*, 382–392. [CrossRef] [PubMed]
76. Ivanov, B.N. Role of ascorbic acid in photosynthesis. *Biochemistry (Moscow)* **2014**, *79*, 282–289. [CrossRef] [PubMed]
77. Kiddle, G.; Pastori, G.M.; Bernard, S.; Pignocchi, C.; Antoniw, J.; Verrier, P.J.; Foyer, C.H. Effects of leaf ascorbate content on defense and photosynthesis gene expression in Arabidopsis thaliana. *Antioxid. Redox Signal.* **2003**, *5*, 23–32. [CrossRef] [PubMed]
78. Chen, Z.; Gallie, D.R. Dehydroascorbate reductase affects leaf growth, development, and function. *Plant Physiol.* **2006**, *142*, 775–787. [CrossRef] [PubMed]
79. Barth, C.; Moeder, W.; Klessig, D.F.; Conklin, P.L. The timing of senescence and response to pathogens is altered in the ascorbate-deficient Arabidopsis mutant *vitamin c-1*. *Plant Physiol.* **2004**, *134*, 1784–1792. [CrossRef] [PubMed]
80. Barth, C.; De Tullio, M.; Conklin, P.L. The role of ascorbic acid in the control of flowering time and the onset of senescence. *J. Exp. Bot.* **2006**, *57*, 1657–1665. [CrossRef] [PubMed]
81. Gallie, D.R. Increasing Vitamin C content in plant foods to improve their nutritional value-successes and challenges. *Nutrients* **2013**, *5*, 3424–3446. [CrossRef]
82. Murphy, L.J.; Robertson, K.N.; Harroun, S.G.; Brosseau, C.L.; Werner-Zwanziger, U.; Moilanen, J.; Tuononen, H.M.; Clyburne, J.A.C. A simple complex on the verge of breakdown: Isolation of the elusive cyanoformate ion. *Science* **2014**, *344*, 75–78. [CrossRef]
83. Akram, N.A.; Shafiq, F.; Ashraf, M. Ascorbic acid-a potential oxidant scavenger and its role in plant development and abiotic stress tolerance. *Front. Plant Sci.* **2017**, *8*, 613. [CrossRef]
84. Boubakri, H. The role of ascorbic acid in plant-pathogen interactions. In *Ascorbic Acid in Plant Growth, DEVELOPMENT and stress Tolerance*; Hossain, M.A., Munnè-Bosch, S., Burritt, D.J., Diaz-Vivancos, P., Fujita, M., Lorence, A., Eds.; Springer: Cham, Switzerland, 2017; pp. 255–268. [CrossRef]
85. Glazebrook, J. Contrasting mechanisms of defense against biotrophic and necrotrophic pathogens. *Annu. Rev. Phytopathol.* **2005**, *43*, 205–227. [CrossRef] [PubMed]
86. Pastori, G.M.; Kiddle, G.; Antoniw, J.; Bernard, S.; Veljovic-Jovanovic, S.; Verrier, P.J.; Noctor, G.; Foyer, C.H. Leaf vitamin C contents modulate plant defense transcripts and regulate genes that control development through hormone signaling. *Plant Cell* **2003**, *15*, 939–951. [CrossRef] [PubMed]
87. Pavet, V.; Olmos, E.; Kiddle, G.; Mowla, S.; Kumar, S.; Antoniw, J.; Alvarez, M.E.; Foyer, C.H. Ascorbic acid deficiency activates cell death and disease resistance responses in Arabidopsis. *Plant Physiol.* **2005**, *139*, 1291–1303. [CrossRef] [PubMed]
88. Mukherjee, M.; Larrimore, K.E.; Ahmed, N.J.; Bedick, T.S.; Barghouthi, N.T.; Traw, M.B.; Barth, C. Ascorbic acid deficiency in Arabidopsis induces constitutive priming that is dependent on hydrogen peroxide, salicylic acid, and the NPR1 gene. *Mol. Plant Microbe* **2010**, *23*, 340–351. [CrossRef]
89. Botanga, C.J.; Bethke, G.; Chen, Z.; Gallie, D.R.; Fiehn, O.; Glazebrook, J. Metabolite Profiling of Arabidopsis Inoculated with *Alternaria brassicicola* reveals that ascorbate reduces disease severity. *Mol. Plant Microbe* **2012**, *25*, 1628–1638. [CrossRef] [PubMed]

90. Egan, M.J.; Wang, Z.Y.; Jones, M.A.; Smirnoff, N.; Talbot, N.J. Generation of reactive oxygen species by fungal NADPH oxidases is required for rice blast disease. *Proc. Natl. Acad. Sci. USA* **2007**, *104*, 11772–11777. [CrossRef]
91. Fujiwara, A.; Shimura, H.; Masuta, C.; Sano, S.; Inukai, T. Exogenous ascorbic acid derivatives and dehydroascorbic acid are effective antiviral agents against *Turnip mosaic virus* in *Brassica rapa*. *J. Gen. Plant Pathol.* **2013**, *79*, 198–204. [CrossRef]
92. Li, J.Y.; Trivedi, P.; Wang, N. Field evaluation of plant defense inducers for the control of citrus Huanglongbing. *Phytopathology* **2016**, *106*, 37–46. [CrossRef]
93. Liso, R.; Calabrese, G.; Bitonti, M.B.; Arrigoni, O. Relationship between ascorbic acid and cell division. *Exp. Cell Res.* **1984**, *150*, 314–320. [CrossRef]
94. Liso, R.; Innocenti, A.M.; Bitonti, M.B.; Arrigoni, O. Ascorbic acid-induced progression of quiescent center cells from G1-phase to S-phase. *New Phytol.* **1988**, *110*, 469–471. [CrossRef]
95. Kerk, N.M.; Feldman, L.J. A Biochemical-model for the initiation and maintenance of the quiescent center—Implications for organization of root-meristems. *Development* **1995**, *121*, 2825–2833.
96. de Pinto, M.C.; Francis, D.; De Gara, L. The redox state of the ascorbate-dehydroascorbate pair as a specific sensor of cell division in tobacco BY-2 cells. *Protoplasma* **1999**, *209*, 90–97. [CrossRef] [PubMed]
97. de Pinto, M.C.; Tommasi, F.; De Gara, L. Enzymes of the ascorbate biosynthesis and ascorbate-glutathione cycle in cultured cells of Tobacco Bright Yellow-2 s. *Plant Physiol. Biochem.* **2000**, *38*, 541–550. [CrossRef]
98. de Simone, A.; Hubbard, R.; de la Torre, N.V.; Velappan, Y.; Wilson, M.; Considine, M.J.; Soppe, W.J.J.; Foyer, C.H. redox changes during the cell cycle in the embryonic root meristem of *Arabidopsis thaliana*. *Antioxid. Redox Sign.* **2017**, *27*, 1505–1519. [CrossRef]
99. Stasolla, C.; Yeung, E.C. Ascorbic acid metabolism during white spruce somatic embryo maturation and germination. *Physiol. Plant.* **2001**, *111*, 196–205. [CrossRef]
100. Stasolla, C.; Yeung, E.C. Cellular ascorbic acid regulates the activity of major peroxidases in the apical poles of germinating white spruce (*Picea glauca*) somatic embryos. *Plant Physiol. Biochem.* **2007**, *45*, 188–198. [CrossRef]
101. Pignocchi, C.; Fletcher, J.M.; Wilkinson, J.E.; Barnes, J.D.; Foyer, C.H. The function of ascorbate oxidase in tobacco. *Plant Physiol.* **2003**, *132*, 1631–1641. [CrossRef]
102. Li, R.; Xin, S.; Tao, C.C.; Jin, X.; Li, H.B. Cotton ascorbate oxidase promotes cell growth in cultured tobacco bright yellow-2 cells through generation of apoplast oxidation. *Int. J. Mol. Sci.* **2017**, *18*, 1346. [CrossRef]
103. Gonzalez-Reyes, J.A.; Alcain, F.J.; Caler, J.A.; Serrano, A.; Cordoba, F.; Navas, P. Relationship between apoplastic ascorbate regeneration and the stimulation of root-growth in *Allium-Cepa* L. *Plant Sci.* **1994**, *100*, 23–29. [CrossRef]
104. Horemans, N.; Foyer, C.H.; Asard, H. Transport and action of ascorbate at the plant plasma membrane. *Trends Plant Sci.* **2000**, *5*, 263–267. [CrossRef]
105. Horemans, N.; Foyer, C.H.; Potters, G.; Asard, H. Ascorbate function and associated transport systems in plants. *Plant Physiol. Biochem.* **2000**, *38*, 531–540. [CrossRef]
106. Fry, S.C. Oxidative scission of plant cell wall polysaccharides by ascorbate-induced hydroxyl radicals. *Biochem. J.* **1998**, *332*, 507–515. [CrossRef] [PubMed]
107. Schopfer, P. Hydroxyl radical-induced cell-wall loosening in vitro and in vivo: Implications for the control of elongation growth. *Plant J.* **2001**, *28*, 679–688. [CrossRef] [PubMed]
108. Padu, E. Apoplastic peroxidases, ascorbate and lignification in relation to nitrate supply in wheat stem. *J. Plant Physiol.* **1999**, *154*, 576–583. [CrossRef]
109. de Pinto, M.C.; De Gara, L. Changes in the ascorbate metabolism of apoplastic and symplastic spaces are associated with cell differentiation. *J. Exp. Bot.* **2004**, *55*, 2559–2569. [CrossRef]
110. de Pinto, M.C.; Tommasi, F.; De Gara, L. Changes in the antioxidant systems as part of the signaling pathway responsible for the programmed cell death activated by nitric oxide and reactive oxygen species in tobacco Bright-Yellow 2 cells. *Plant Physiol.* **2002**, *130*, 698–708. [CrossRef]
111. Vacca, R.A.; de Pinto, M.C.; Valenti, D.; Passarella, S.; Marra, E.; De Gara, L. Production of reactive oxygen species, alteration of cytosolic ascorbate peroxidase, and impairment of mitochondrial metabolism are early events in heat shock-induced programmed cell death in tobacco bright-yellow 2 cells. *Plant Physiol.* **2004**, *134*, 1100–1112. [CrossRef]

112. de Pinto, M.C.; Paradiso, A.; Leonetti, P.; De Gara, L. Hydrogen peroxide, nitric oxide and cytosolic ascorbate peroxidase at the crossroad between defence and cell death. *Plant J.* **2006**, *48*, 784–795. [CrossRef]
113. Locato, V.; Gadaleta, C.; De Gara, L.; De Pinto, M.C. Production of reactive species and modulation of antioxidant network in response to heat shock: A critical balance for cell fate. *Plant Cell Environ.* **2008**, *31*, 1606–1619. [CrossRef]
114. Valenti, D.; Vacca, R.A.; de Pinto, M.C.; De Gara, L.; Marra, E.; Passarella, S. In the early phase of programmed cell death in Tobacco Bright Yellow 2 cells the mitochondrial adenine nucleotide translocator, adenylate kinase and nucleoside diphosphate kinase are impaired in a reactive oxygen species-dependent manner. *BBA Biochim. Biophys. Acta* **2007**, *1767*, 66–78. [CrossRef]
115. de Pinto, M.C.; Locato, V.; Sgobba, A.; Romero-Puertas, M.D.; Gadaleta, C.; Delledonne, M.; De Gara, L. S-Nitrosylation of ascorbate peroxidase is part of programmed cell death signaling in Tobacco Bright Yellow-2 cells. *Plant Physiol.* **2013**, *163*, 1766–1775. [CrossRef] [PubMed]
116. de Pinto, M.C.; Locato, V.; De Gara, L. Redox regulation in plant programmed cell death. *Plant Cell Environ.* **2012**, *35*, 234–244. [CrossRef] [PubMed]
117. Locato, V.; Paradiso, A.; Sabetta, W.; De Gara, L.; de Pinto, M.C. Nitric Oxide and Reactive Oxygen Species in PCD Signaling. *Adv. Bot. Res.* **2016**, *77*, 165–192. [CrossRef]
118. Paradiso, A.; de Pinto, M.C.; Locato, V.; De Gara, L. Galactone-gamma-lactone-dependent ascorbate biosynthesis alters wheat kernel maturation. *Plant Biol.* **2012**, *14*, 652–658. [CrossRef] [PubMed]
119. Ortiz-Espín, A.; Sánchez-Guerrero, A.; Sevilla, F.; Jiménez, A. The role of ascorbate in plant growth and development. In *Ascorbic Acid in Plant Growth, Development and Stress Tolerance*; Hossain, M.A., Munnè-Bosch, S., Burritt, D.J., Diaz-Vivancos, P., Fujita, M., Lorence, A., Eds.; Springer: Cham, Switzerland, 2017; pp. 25–45. [CrossRef]
120. Hancock, R.D.; Viola, R. Biosynthesis and catabolism of L-ascorbic acid in plants. *Crit. Rev. Plant Sci.* **2005**, *24*, 167–188. [CrossRef]
121. Dowdle, J.; Ishikawa, T.; Gatzek, S.; Rolinski, S.; Smirnoff, N. Two genes in *Arabidopsis thaliana* encoding GDP-L-galactose phosphorylase are required for ascorbate biosynthesis and seedling viability. *Plant J.* **2007**, *52*, 673–689. [CrossRef] [PubMed]
122. Wolucka, B.A.; Van Montagu, M. GDP-mannose 3′,5′-epimerase forms GDP-L-gulose, a putative intermediate for the de novo biosynthesis of vitamin C in plants. *J. Biol. Chem.* **2003**, *278*, 47483–47490. [CrossRef]
123. Lorence, A.; Chevone, B.I.; Mendes, P.; Nessler, C.L. Myo-inositol oxygenase offers a possible entry point into plant ascorbate biosynthesis. *Plant Physiol.* **2004**, *134*, 1200–1205. [CrossRef]
124. Agius, F.; Gonzalez-Lamothe, R.; Caballero, J.L.; Munoz-Blanco, J.; Botella, M.A.; Valpuesta, V. Engineering increased vitamin C levels in plants by overexpression of a D-galacturonic acid reductase. *Nat. Biotechnol.* **2003**, *21*, 177–181. [CrossRef]
125. Maruta, T.; Yonemitsu, M.; Yabuta, Y.; Tamoi, M.; Ishikawa, T.; Shigeoka, S. Arabidopsis phosphomannose isomerase 1, but not phosphomannose isomerase 2, is essential for ascorbic acid biosynthesis. *J. Biol. Chem.* **2008**, *283*, 28842–28851. [CrossRef]
126. Qian, W.; Yu, C.; Qin, H.; Liu, X.; Zhang, A.; Johansen, I.E.; Wang, D. Molecular and functional analysis of phosphomannomutase (PMM) from higher plants and genetic evidence for the involvement of PMM in ascorbic acid biosynthesis in Arabidopsis and *Nicotiana benthamiana*. *Plant J.* **2007**, *49*, 399–413. [CrossRef] [PubMed]
127. Hoeberichts, F.A.; Vaeck, E.; Kiddle, G.; Coppens, E.; van de Cotte, B.; Adamantidis, A.; Ormenese, S.; Foyer, C.H.; Zabeau, M.; Inze, D.; et al. A Temperature-sensitive mutation in the *Arabidopsis thaliana* phosphomannomutase gene disrupts protein glycosylation and triggers cell death. *J. Biol. Chem.* **2008**, *283*, 5708–5718. [CrossRef] [PubMed]
128. Conklin, P.L.; Williams, E.H.; Last, R.L. Environmental stress sensitivity of an ascorbic acid-deficient Arabidopsis mutant. *Proc. Natl. Acad. Sci. USA* **1996**, *93*, 9970–9974. [CrossRef] [PubMed]
129. Conklin, P.L.; Norris, S.R.; Wheeler, G.L.; Williams, E.H.; Smirnoff, N.; Last, R.L. Genetic evidence for the role of GDP-mannose in plant ascorbic acid (vitamin C) biosynthesis. *Proc. Natl. Acad. Sci. USA* **1999**, *96*, 4198–4203. [CrossRef]
130. Keller, R.; Renz, F.S.; Kossmann, J. Antisense inhibition of the GDP-mannose pyrophosphorylase reduces the ascorbate content in transgenic plants leading to developmental changes during senescence. *Plant J.* **1999**, *19*, 131–141. [CrossRef]

131. Barber, G.A. Observations on the mechanism of the reversible epimerization of GDP-D-mannose to GDP-L-galactose by an enzyme from Chlorella pyrenoidosa. *J. Biol. Chem.* **1979**, *254*, 7600–7603.
132. Wolucka, B.A.; Persiau, G.; Van Doorsselaere, J.; Davey, M.W.; Demol, H.; Vandekerckhove, J.; Van Montagu, M.; Zabeau, M.; Boerjan, W. Partial purification and identification of GDP-mannose 3″,5″-epimerase of Arabidopsis thaliana, a key enzyme of the plant vitamin C pathway. *Proc. Natl. Acad. Sci. USA* **2001**, *98*, 14843–14848. [CrossRef]
133. Fenech, M.; Amaya, I.; Valpuesta, V.; Botella, M.A. Vitamin C content in fruits: Biosynthesis and regulation. *Front. Plant Sci.* **2019**, *9*, 2006. [CrossRef]
134. Maruta, T.; Ichikawa, Y.; Mieda, T.; Takeda, T.; Tamoi, M.; Yabuta, Y.; Ishikawa, T.; Shigeoka, S. The contribution of Arabidopsis homologs of L-gulono-1,4-lactone oxidase to the biosynthesis of ascorbic acid. *Biosci. Biotechnol. Biochem.* **2010**, *74*, 1494–1497. [CrossRef]
135. Lukowitz, W.; Nickle, T.C.; Meinke, D.W.; Last, R.L.; Conklin, P.L.; Somerville, C.R. Arabidopsis cyt1 mutants are deficient in a mannose-1-phosphate guanylyltransferase and point to a requirement of N-linked glycosylation for cellulose biosynthesis. *Proc. Natl. Acad. Sci. USA* **2001**, *98*, 2262–2267. [CrossRef]
136. Reiter, W.D.; Vanzin, G.F. Molecular genetics of nucleotide sugar interconversion pathways in plants. *Plant Mol. Biol.* **2001**, *47*, 95–113. [CrossRef] [PubMed]
137. Bulley, S.; Laing, W. The regulation of ascorbate biosynthesis. *Curr. Opin. Plant Biol.* **2016**, *33*, 15–22. [CrossRef] [PubMed]
138. Bulley, S.M.; Rassam, M.; Hoser, D.; Otto, W.; Schunemann, N.; Wright, M.; MacRae, E.; Gleave, A.; Laing, W. Gene expression studies in kiwifruit and gene over-expression in Arabidopsis indicates that GDP-L-galactose guanyltransferase is a major control point of vitamin C biosynthesis. *J. Exp. Bot.* **2009**, *60*, 765–778. [CrossRef] [PubMed]
139. Li, M.; Ma, F.; Liang, D.; Li, J.; Wang, Y. Ascorbate biosynthesis during early fruit development is the main reason for its accumulation in kiwi. *PLoS ONE* **2010**, *5*, e14281. [CrossRef]
140. Bulley, S.; Wright, M.; Rommens, C.; Yan, H.; Rassam, M.; Lin-Wang, K.; Andre, C.; Brewster, D.; Karunairetnam, S.; Allan, A.C.; et al. Enhancing ascorbate in fruits and tubers through over-expression of the L-galactose pathway gene GDP-L-galactose phosphorylase. *Plant Biotechnol. J.* **2012**, *10*, 390–397. [CrossRef]
141. Alos, E.; Rodrigo, M.J.; Zacarias, L. Differential transcriptional regulation of L-ascorbic acid content in peel and pulp of citrus fruits during development and maturation. *Planta* **2014**, *239*, 1113–1128. [CrossRef]
142. Yoshimura, K.; Nakane, T.; Kume, S.; Shiomi, Y.; Maruta, T.; Ishikawa, T.; Shigeoka, S. Transient expression analysis revealed the importance of VTC2 expression level in light/dark regulation of ascorbate biosynthesis in Arabidopsis. *Biosci. Biotechnol. Biochem.* **2014**, *78*, 60–66. [CrossRef]
143. Liu, F.; Wang, L.; Gu, L.; Zhao, W.; Su, H.; Cheng, X. Higher transcription levels in ascorbic acid biosynthetic and recycling genes were associated with higher ascorbic acid accumulation in blueberry. *Food Chem.* **2015**, *188*, 399–405. [CrossRef]
144. Gao, Y.; Badejo, A.A.; Shibata, H.; Sawa, Y.; Maruta, T.; Shigeoka, S.; Page, M.; Smirnoff, N.; Ishikawa, T. Expression analysis of the *VTC2* and *VTC5* genes encoding GDP-L-galactose phosphorylase, an enzyme involved in ascorbate biosynthesis, in *Arabidopsis thaliana*. *Biosci. Biotechnol. Biochem.* **2011**, *75*, 1783–1788. [CrossRef]
145. Laing, W.A.; Martinez-Sanchez, M.; Wright, M.A.; Bulley, S.M.; Brewster, D.; Dare, A.P.; Rassam, M.; Wang, D.; Storey, R.; Macknight, R.C.; et al. An upstream open reading frame is essential for feedback regulation of ascorbate biosynthesis in Arabidopsis. *Plant Cell* **2015**, *27*, 772–786. [CrossRef]
146. Laing, W.A.; Bulley, S.; Wright, M.; Cooney, J.; Jensen, D.; Barraclough, D.; MacRae, E. A highly specific L-galactose-1-phosphate phosphatase on the path to ascorbate biosynthesis. *Proc. Natl. Acad. Sci. USA* **2004**, *101*, 16976–16981. [CrossRef] [PubMed]
147. Conklin, P.L.; Gatzek, S.; Wheeler, G.L.; Dowdle, J.; Raymond, M.J.; Rolinski, S.; Isupov, M.; Littlechild, J.A.; Smirnoff, N. *Arabidopsis thaliana VTC4* encodes L-galactose-1-P phosphatase, a plant ascorbic acid biosynthetic enzyme. *J. Biol. Chem.* **2006**, *281*, 15662–15670. [CrossRef] [PubMed]
148. Torabinejad, J.; Donahue, J.L.; Gunesekera, B.N.; Allen-Daniels, M.J.; Gillaspy, G.E. VTC4 is a bifunctional enzyme that affects myoinositol and ascorbate biosynthesis in plants. *Plant Physiol.* **2009**, *150*, 951–961. [CrossRef] [PubMed]
149. Zhang, W.; Gruszewski, H.A.; Chevone, B.I.; Nessler, C.L. An Arabidopsis purple acid phosphatase with phytase activity increases foliar ascorbate. *Plant Physiol.* **2008**, *146*, 431–440. [CrossRef] [PubMed]

150. Gatzek, S.; Wheeler, G.L.; Smirnoff, N. Antisense suppression of L-galactose dehydrogenase in *Arabidopsis thaliana* provides evidence for its role in ascorbate synthesis and reveals light modulated L-galactose synthesis. *Plant J.* **2002**, *30*, 541–553. [CrossRef]
151. Leferink, N.G.H.; van den Berg, W.A.M.; van Berkel, W.J.H. L-Galactono-gamma-lactone dehydrogenase from *Arabidopsis thaliana*, a flavoprotein involved in vitamin C biosynthesis. *FEBS J.* **2008**, *275*, 713–726. [CrossRef]
152. Wheeler, G.; Ishikawa, T.; Pornsaksit, V.; Smirnoff, N. Evolution of alternative biosynthetic pathways for vitamin C following plastid acquisition in photosynthetic eukaryotes. *Elife* **2015**, *4*, e06369. [CrossRef]
153. Siendones, E.; Gonzalez-Reyes, J.A.; Santos-Ocana, C.; Navas, P.; Cordoba, F. Biosynthesis of ascorbic acid in kidney bean. L-galactono-gamma-lactone dehydrogenase is an intrinsic protein located at the mitochondrial inner membrane. *Plant Physiol.* **1999**, *120*, 907–912. [CrossRef]
154. Bartoli, C.G.; Pastori, G.M.; Foyer, C.H. Ascorbate biosynthesis in mitochondria is linked to the electron transport chain between complexes III and IV. *Plant Physiol.* **2000**, *123*, 335–343. [CrossRef]
155. Heazlewood, J.L.; Howell, K.A.; Millar, A.H. Mitochondrial complex I from Arabidopsis and rice: Orthologs of mammalian and fungal components coupled with plant-specific subunits. *BBA Biochim. Biophys. Acta* **2003**, *1604*, 159–169. [CrossRef]
156. Pineau, B.; Layoune, O.; Danon, A.; De Paepe, R. L-Galactono-1,4-lactone dehydrogenase is required for the accumulation of plant respiratory complex I. *J. Biol. Chem.* **2008**, *283*, 32500–32505. [CrossRef] [PubMed]
157. Schertl, P.; Sunderhaus, S.; Klodmann, J.; Grozeff, G.E.G.; Bartoli, C.G.; Braun, H.P. L-Galactono-1,4-lactone dehydrogenase (GLDH) forms part of three subcomplexes of mitochondrial complex I in *Arabidopsis thaliana*. *J. Biol. Chem.* **2012**, *287*, 14412–14419. [CrossRef] [PubMed]
158. Schimmeyer, J.; Bock, R.; Meyer, E.H. L-Galactono-1,4-lactone dehydrogenase is an assembly factor of the membrane arm of mitochondrial complex I in Arabidopsis. *Plant Mol. Biol.* **2016**, *90*, 117–126. [CrossRef] [PubMed]
159. Millar, A.H.; Mittova, V.; Kiddle, G.; Heazlewood, J.L.; Bartoli, C.G.; Theodoulou, F.L.; Foyer, C.H. Control of ascorbate synthesis by respiration and its implications for stress responses. *Plant Physiol.* **2003**, *133*, 443–447. [CrossRef]
160. Ioannidi, E.; Kalamaki, M.S.; Engineer, C.; Pateraki, I.; Alexandrou, D.; Mellidou, I.; Giovannonni, J.; Kanellis, A.K. Expression profiling of ascorbic acid-related genes during tomato fruit development and ripening and in response to stress conditions. *J. Exp. Bot.* **2009**, *60*, 663–678. [CrossRef]
161. Wagner, C.; Sefkow, M.; Kopka, J. Construction and application of a mass spectral and retention time index database generated from plant GC/EI-TOF-MS metabolite profiles. *Phytochemistry* **2003**, *62*, 887–900. [CrossRef]
162. Davey, M.W.; Gilot, C.; Persiau, G.; Ostergaard, J.; Han, Y.; Bauw, G.C.; Van Montagu, M.C. Ascorbate biosynthesis in Arabidopsis cell suspension culture. *Plant Physiol.* **1999**, *121*, 535–543. [CrossRef]
163. Jain, A.K.; Nessler, C.L. Metabolic engineering of an alternative pathway for ascorbic acid biosynthesis in plants. *Mol. Breed.* **2000**, *6*, 73–78. [CrossRef]
164. Imai, T.; Niwa, M.; Ban, Y.; Hirai, M.; Oba, K.; Moriguchi, T. Importance of the L-galactonolactone pool for enhancing the ascorbate content revealed by L-galactonolactone dehydrogenase-overexpressing tobacco plants. *Plant Cell Tissue Organ Cult.* **2009**, *96*, 105–112. [CrossRef]
165. Aboobucker, S.I.; Suza, W.P.; Lorence, A. Characterization of two Arabidopsis l-gulono-1,4-lactone oxidases, AtGulLO3 and AtGulLO5, involved in ascorbate biosynthesis. *React. Oxyg. Spec. (Apex)* **2017**, *4*, 389–417. [CrossRef]
166. Valpuesta, V.; Botella, M.A. Biosynthesis of L-ascorbic acid in plants: New pathways for an old antioxidant. *Trends Plant Sci.* **2004**, *9*, 573–577. [CrossRef] [PubMed]
167. Endres, S.; Tenhaken, R. Myoinositol oxygenase controls the level of myoinositol in Arabidopsis, but does not increase ascorbic acid. *Plant Physiol.* **2009**, *149*, 1042–1049. [CrossRef] [PubMed]
168. Endres, S.; Tenhaken, R. Down-regulation of the myo-inositol oxygenase gene family has no effect on cell wall composition in Arabidopsis. *Planta* **2011**, *234*, 157–169. [CrossRef] [PubMed]
169. Ivanov Kavkova, E.; Blöchl, C.; Tenhaken, R. The Myo -inositol pathway does not contribute to ascorbic acid synthesis. *Plant Biol.* **2018**, *21*, 95–102. [CrossRef] [PubMed]

170. Badejo, A.A.; Wada, K.; Gao, Y.; Maruta, T.; Sawa, Y.; Shigeoka, S.; Ishikawa, T. Translocation and the alternative D-galacturonate pathway contribute to increasing the ascorbate level in ripening tomato fruits together with the D-mannose/L-galactose pathway. *J. Exp. Bot.* **2012**, *63*, 229–239. [CrossRef]
171. Di Matteo, A.; Sacco, A.; Anacleria, M.; Pezzotti, M.; Delledonne, M.; Ferrarini, A.; Frusciante, L.; Barone, A. The ascorbic acid content of tomato fruits is associated with the expression of genes involved in pectin degradation. *BMC Plant Biol.* **2010**, 10. [CrossRef]
172. Melino, V.J.; Soole, K.L.; Ford, C.M. Ascorbate metabolism and the developmental demand for tartaric and oxalic acids in ripening grape berries. *BMC Plant Biol.* **2009**, *9*. [CrossRef]
173. Cruz-Rus, E.; Amaya, I.; Sanchez-Sevilla, J.F.; Botella, M.A.; Valpuesta, V. Regulation of L-ascorbic acid content in strawberry fruits. *J. Exp. Bot.* **2011**, *62*, 4191–4201. [CrossRef]
174. Tabata, K.; Takaoka, T.; Esaka, M. Gene expression of ascorbic acid-related enzymes in tobacco. *Phytochemistry* **2002**, *61*, 631–635. [CrossRef]
175. Massot, C.; Stevens, R.; Genard, M.; Longuenesse, J.J.; Gautier, H. Light affects ascorbate content and ascorbate-related gene expression in tomato leaves more than in fruits. *Planta* **2012**, *235*, 153–163. [CrossRef]
176. Massot, C.; Genard, M.; Stevens, R.; Gautier, H. Fluctuations in sugar content are not determinant in explaining variations in vitamin C in tomato fruit. *Plant Physiol. Biochem.* **2010**, *48*, 751–757. [CrossRef] [PubMed]
177. Zhou, W.L.; Liu, W.K.; Yang, Q.C. Quality changes in hydroponic lettuce grown under pre-harvest short-duration continuous light of different intensities. *J. Hortic. Sci. Biotechem.* **2012**, *87*, 429–434. [CrossRef]
178. Riga, P.; Benedicto, L.; Gil-Izquierdo, A.; Collado-Gonzalez, J.; Ferreres, F.; Medina, S. Diffuse light affects the contents of vitamin C, phenolic compounds and free amino acids in lettuce plants. *Food Chem.* **2019**, *272*, 227–234. [CrossRef] [PubMed]
179. Shen, Y.Z.; Guo, S.S.; Ai, W.D.; Tang, Y.K. Effects of illuminants and illumination time on lettuce growth, yield and nutritional quality in a controlled environment. *Life Sci. Space Res.* **2014**, *2*, 38–42. [CrossRef]
180. Zha, L.Y.; Zhang, Y.B.; Liu, W.K. Dynamic responses of ascorbate pool and metabolism in lettuce to long-term continuous light provided by red and blue LEDs. *Environ. Exp. Bot.* **2019**, *163*, 15–23. [CrossRef]
181. Ohashi-Kaneko, K.; Takase, M.; Kon, N.; Fujiwara, K.; Kurata, K. Effect of light quality on growth and vegetable quality in leaf lettuce, spinach and komatsuna. *Environ. Control Biol.* **2007**, *45*, 189–198. [CrossRef]
182. Qian, H.; Liu, T.; Deng, M.; Miao, H.; Cai, C.; Shen, W.; Wang, Q. Effects of light quality on main health-promoting compounds and antioxidant capacity of Chinese kale sprouts. *Food Chem.* **2016**, *196*, 1232–1238. [CrossRef]
183. Li, M.J.; Ma, F.W.; Shang, P.F.; Zhang, M.; Hou, C.M.; Liang, D. Influence of light on ascorbate formation and metabolism in apple fruits. *Planta* **2009**, *230*, 39–51. [CrossRef]
184. Ntagkas, N.; Woltering, E.; Nicole, C.; Labrie, C.; Marcelis, L.F.M. Light regulation of vitamin C in tomato fruit is mediated through photosynthesis. *Environ. Exp. Bot.* **2019**, *158*, 180–188. [CrossRef]
185. Zhang, L.C.; Ma, G.; Yamawaki, K.; Ikoma, Y.; Matsumoto, H.; Yoshioka, T.; Ohta, S.; Kato, M. Regulation of ascorbic acid metabolism by blue LED light irradiation in citrus juice sacs. *Plant Sci.* **2015**, *233*, 134–142. [CrossRef]
186. Lee, Y.J.; Ha, J.Y.; Oh, J.E.; Cho, M.S. The effect of LED irradiation on the quality of cabbage stored at a low temperature. *Food Sci. Biotechnol.* **2014**, *23*, 1087–1093. [CrossRef]
187. Mastropasqua, L.; Tanzarella, P.; Paciolla, C. Effects of postharvest light spectra on quality and health-related parameters in green *Asparagus officinalis* L. *Postharvest Biol. Technol.* **2016**, *112*, 143–151. [CrossRef]
188. Loi, M.; Liuzzi, V.C.; Fanelli, F.; De Leonardis, S.; Maria Creanza, T.; Ancona, N.; Paciolla, C.; Mule, G. Effect of different light-emitting diode (LED) irradiation on the shelf life and phytonutrient content of broccoli (*Brassica oleracea* L. var. italica). *Food Chem.* **2019**, *283*, 206–214. [CrossRef]
189. Frohnmeyer, H.; Staiger, D. Ultraviolet-B radiation-mediated responses in plants. Balancing damage and protection. *Plant Physiol.* **2003**, *133*, 1420–1428. [CrossRef] [PubMed]
190. Xu, M.J.; Dong, J.F.; Zhu, M.Y. Effects of germination conditions on ascorbic acid level and yield of soybean sprouts. *J. Sci. Food Agric.* **2005**, *85*, 943–947. [CrossRef]
191. Liu, P.; Li, Q.; Gao, Y.; Wang, H.; Chai, L.; Yu, H.; Jiang, W. A New perspective on the effect of UV-B on L-ascorbic acid metabolism in cucumber seedlings. *J. Agric. Food Chem.* **2019**, *67*, 4444–4452. [CrossRef] [PubMed]

192. Calafiore, R.; Aliberti, A.; Ruggieri, V.; Olivieri, F.; Rigano, M.M.; Barone, A. Phenotypic and molecular selection of a superior *Solanum pennellii* introgression sub-line suitable for improving quality traits of cultivated tomatoes. *Front. Plant Sci.* **2019**, *10*, 190. [CrossRef]
193. Ntagkas, N.; Woltering, E.; Bouras, S.; de Vos, R.C.; Dieleman, J.A.; Nicole, C.C.; Labrie, C.; Marcelis, L.F. Light-induced vitamin c accumulation in tomato fruits is independent of carbohydrate availability. *Plants* **2019**, *8*, 86. [CrossRef]
194. Fukunaga, K.; Fujikawa, Y.; Esaka, M. Light regulation of ascorbic acid biosynthesis in rice via light responsive cis-elements in genes encoding ascorbic acid biosynthetic enzymes. *Biosci. Biotechem. Biochem.* **2010**, *74*, 888–891. [CrossRef]
195. Liu, F.Y.; Nan, X.; Jian, G.H.; Yan, S.J.; Xie, L.H.; Brennan, C.S.; Huang, W.J.; Guo, X.B. The manipulation of gene expression and the biosynthesis of Vitamin C, E and folate in light-and dark-germination of sweet corn seeds. *Sci. Rep.-UK* **2017**, *7*, 7484. [CrossRef]
196. Hu, T.; Ye, J.; Tao, P.; Li, H.; Zhang, J.; Zhang, Y.; Ye, Z. The tomato HD-Zip I transcription factor SlHZ24 modulates ascorbate accumulation through positive regulation of the D-mannose/L-galactose pathway. *Plant J.* **2016**, *85*, 16–29. [CrossRef] [PubMed]
197. Zhang, W.Y.; Lorence, A.; Gruszewski, H.A.; Chevone, B.I.; Nessler, C.L. *AMR1*, an Arabidopsis gene that coordinately and negatively regulates the mannose/L-galactose ascorbic acid biosynthetic pathway. *Plant Physiol.* **2009**, *150*, 942–950. [CrossRef]
198. Wang, J.; Yu, Y.; Zhang, Z.; Quan, R.; Zhang, H.; Ma, L.; Deng, X.W.; Huang, R. *Arabidopsis* CSN5B interacts with VTC1 and modulates ascorbic acid synthesis. *Plant Cell* **2013**, *25*, 625–636. [CrossRef] [PubMed]
199. Ntagkas, N.; Woltering, E.J.; Marcelis, L.F.M. Light regulates ascorbate in plants: An integrated view on physiology and biochemistry. *Environ. Exp. Bot.* **2018**, *147*, 271–280. [CrossRef]
200. Mellidou, I.; Kanellis, A.K. Genetic control of ascorbic acid biosynthesis and recycling in horticultural crops. *Front. Chem.* **2017**, *5*, 50. [CrossRef] [PubMed]
201. Zhang, G.Y.; Liu, R.R.; Zhang, C.Q.; Tang, K.X.; Sun, M.F.; Yan, G.H.; Liu, Q.Q. Manipulation of the rice L-galactose pathway: Evaluation of the effects of transgene overexpression on ascorbate accumulation and abiotic stress tolerance. *PLoS ONE* **2015**, *10*, e0125870. [CrossRef]
202. Ali, B.; Pantha, S.; Acharya, R.; Ueda, Y.; Wu, L.B.; Ashrafuzzaman, M.; Ishizaki, T.; Wissuwa, M.; Bulley, S.; Frei, M. Enhanced ascorbate level improves multi-stress tolerance in a widely grown indica rice variety without compromising its agronomic characteristics. *J. Plant Physiol.* **2019**, *240*, 152998. [CrossRef]
203. Zhang, H.; Si, X.; Ji, X.; Fan, R.; Liu, J.; Chen, K.; Wang, D.; Gao, C. Genome editing of upstream open reading frames enables translational control in plants. *Nat. Biotechnol.* **2018**, *36*, 894–898. [CrossRef]
204. Li, T.D.; Yang, X.P.; Yu, Y.; Si, X.M.; Zhai, X.W.; Zhang, H.W.; Dong, W.X.; Gao, C.X.; Xu, C. Domestication of wild tomato is accelerated by genome editing. *Nat. Biotechnol.* **2018**, *36*, 1160–1163. [CrossRef]
205. Mellidou, I.; Chagne, D.; Laing, W.A.; Keulemans, J.; Davey, M.W. Allelic variation in paralogs of GDP-l-galactose phosphorylase is a major determinant of vitamin C concentrations in apple fruit. *Plant Physiol.* **2012**, *160*, 1613–1629. [CrossRef]
206. Zhou, Y.; Tao, Q.C.; Wang, Z.N.; Fan, R.; Li, Y.; Sun, X.F.; Tang, K.X. Engineering ascorbic acid biosynthetic pathway in Arabidopsis leaves by single and double gene transformation. *Biol. Plant.* **2012**, *56*, 451–457. [CrossRef]
207. Suekawa, M.; Fujikawa, Y.; Inoue, A.; Kondo, T.; Uchida, E.; Koizumi, T.; Esaka, M. High levels of expression of multiple enzymes in the Smirnoff-Wheeler pathway are important for high accumulation of ascorbic acid in acerola fruits. *Biosci. Biotechnol. Biochem.* **2019**, *83*, 1713–1716. [CrossRef] [PubMed]
208. Saltzman, A.; Birol, E.; Bouis, H.E.; Boy, E.; De Moura, F.F.; Islam, Y.; Pfeiffer, W.H. Biofortification: Progress toward a more nourishing future. *Glob. Food Secur.* **2013**, *2*, 9–17. [CrossRef]
209. Li, X.; Ye, J.; Munir, S.; Yang, T.; Chen, W.; Liu, G.; Zheng, W.; Zhang, Y. Biosynthetic gene pyramiding leads to ascorbate accumulation with enhanced oxidative stress tolerance in tomato. *Int. J. Mol. Sci.* **2019**, *20*, 1558. [CrossRef] [PubMed]
210. Hemavathi, *!!! REPLACE !!!*; Upadhyaya, C.P.; Akula, N.; Young, K.E.; Chun, S.C.; Kim, D.H.; Park, S.W. Enhanced ascorbic acid accumulation in transgenic potato confers tolerance to various abiotic stresses. *Biotechnol. Lett.* **2010**, *32*, 321–330. [CrossRef] [PubMed]

211. Lisko, K.A.; Torres, R.; Harris, R.S.; Belisle, M.; Vaughan, M.M.; Jullian, B.; Chevone, B.I.; Mendes, P.; Nessler, C.L.; Lorence, A. Elevating vitamin C content via overexpression of myo-inositol oxygenase and L-gulono-1,4-lactone oxidase in Arabidopsis leads to enhanced biomass and tolerance to abiotic stresses. *In Vitro Cell. Dev. Biol. Plant* **2013**, *49*, 643–655. [CrossRef]
212. Hemavathi; Upadhyaya, C.P.; Young, K.E.; Akula, N.; Kim, H.S.; Heung, J.J.; Oh, O.M.; Aswath, C.R.; Chun, S.C.; Kim, D.H.; et al. Over-expression of strawberry D-galacturonic acid reductase in potato leads to accumulation of vitamin C with enhanced abiotic stress tolerance. *Plant Sci.* **2009**, *177*, 659–667. [CrossRef]
213. Amaya, I.; Osorio, S.; Martinez-Ferri, E.; Lima-Silva, V.; Doblas, V.G.; Fernandez-Munoz, R.; Fernie, A.R.; Botella, M.A.; Valpuesta, V. Increased antioxidant capacity in tomato by ectopic expression of the strawberry D-galacturonate reductase gene. *Biotechnol. J.* **2015**, *10*, 490–500. [CrossRef]
214. Lim, M.Y.; Jeong, B.R.; Jung, M.; Harn, C.H. Transgenic tomato plants expressing strawberry D-galacturonic acid reductase gene display enhanced tolerance to abiotic stresses. *Plant Biotechnol. Rep.* **2016**, *10*, 105–116. [CrossRef]
215. Rigano, M.M.; Lionetti, V.; Raiola, A.; Bellincampi, D.; Barone, A. Pectic enzymes as potential enhancers of ascorbic acid production through the D-galacturonate pathway in Solanaceae. *Plant Sci.* **2018**, *266*, 55–63. [CrossRef]
216. Naqvi, S.; Zhu, C.; Farre, G.; Ramessar, K.; Bassie, L.; Breitenbach, J.; Perez Conesa, D.; Ros, G.; Sandmann, G.; Capell, T.; et al. Transgenic multivitamin corn through biofortification of endosperm with three vitamins representing three distinct metabolic pathways. *Proc. Natl. Acad. Sci. USA* **2009**, *106*, 7762–7767. [CrossRef] [PubMed]
217. Haroldsen, V.M.; Chi-Ham, C.L.; Kulkarni, S.; Lorence, A.; Bennett, A.B. Constitutively expressed DHAR and MDHAR influence fruit, but not foliar ascorbate levels in tomato. *Plant Physiol. Biochem.* **2011**, *49*, 1244–1249. [CrossRef] [PubMed]
218. Hao, Z.; Wang, X.; Zong, Y.; Wen, S.; Cheng, Y.; Li, H. Enzymatic activity and functional analysis under multiple abiotic stress conditions of a dehydroascorbate reductase gene derived from *Liriodendron Chinense*. *Environ. Exp. Bot* **2019**, *167*, 103850. [CrossRef]
219. Yin, L.N.; Wang, S.W.; Eltayeb, A.E.; Uddin, M.I.; Yamamoto, Y.; Tsuji, W.; Takeuchi, Y.; Tanaka, K. Overexpression of dehydroascorbate reductase, but not monodehydroascorbate reductase, confers tolerance to aluminum stress in transgenic tobacco. *Planta* **2010**, *231*, 609–621. [CrossRef] [PubMed]
220. Eltelib, H.A.; Fujikawa, Y.; Esaka, M. Overexpression of the acerola (*Malpighia glabra*) monodehydroascorbate reductase gene in transgenic tobacco plants results in increased ascorbate levels and enhanced tolerance to salt stress. *S. Afr. J. Bot.* **2012**, *78*, 295–301. [CrossRef]
221. Gest, N.; Garchery, C.; Gautier, H.; Jimenez, A.; Stevens, R. Light-dependent regulation of ascorbate in tomato by a monodehydroascorbate reductase localized in peroxisomes and the cytosol. *Plant Biotechnol. J.* **2013**, *11*, 344–354. [CrossRef]
222. Abdelgawad, K.F.; El-Mogy, M.M.; Mohamed, M.I.A.; Garchery, C.; Stevens, R.G. Increasing ascorbic acid content and salinity tolerance of cherry tomato plants by suppressed expression of the ascorbate oxidase genes. *Agronomy* **2019**, *9*, 51. [CrossRef]
223. Zhang, Z.J.; Wang, J.; Zhang, R.X.; Huang, R.F. The ethylene response factor AtERF98 enhances tolerance to salt through the transcriptional activation of ascorbic acid synthesis in Arabidopsis. *Plant J.* **2012**, *71*, 273–287. [CrossRef]
224. Sawake, S.; Tajima, N.; Mortimer, J.C.; Lao, J.; Ishikawa, T.; Yu, X.; Yamanashi, Y.; Yoshimi, Y.; Kawai-Yamada, M.; Dupree, P.; et al. KONJAC1 and 2 are key factors for GDP-Mannose generation and affect L-ascorbic acid and glucomannan biosynthesis in Arabidopsis. *Plant Cell* **2015**, *27*, 3397–3409. [CrossRef]
225. Li, Y.; Chu, Z.N.; Luo, J.Y.; Zhou, Y.H.; Cai, Y.J.; Lu, Y.G.; Xia, J.H.; Kuang, H.H.; Ye, Z.B.; Ouyang, B. The C2H2 zinc-finger protein SlZF3 regulates AsA synthesis and salt tolerance by interacting with CSN5B. *Plant Biotechnol. J.* **2018**, *16*, 1201–1213. [CrossRef]
226. Ye, J.; Li, W.F.; Ai, G.; Li, C.X.; Liu, G.Z.; Chen, W.F.; Wang, B.; Wang, W.Q.; Lu, Y.G.; Zhang, J.H.; et al. Genome-wide association analysis identifies a natural variation in basic helix-loop-helix transcription factor regulating ascorbate biosynthesis via D-mannose/L-galactose pathway in tomato. *PLoS Genet.* **2019**, *15*, e1008149. [CrossRef] [PubMed]

227. Johnston, C.S.; Corte, C.; Swan, P.D. Marginal vitamin C status is associated with reduced fat oxidation during submaximal exercise in young adults. *Nutr. Metab.* **2006**, *3*, 35. [CrossRef] [PubMed]
228. Johnston, C.S.; Barkyoumb, G.M.; Schumacher, S. Vitamin C supplementation slightly improves physical activity levels and reduces cold incidence in men with marginal vitamin C status: A randomized controlled trial. *Nutrients* **2014**, *6*, 2572–2583. [CrossRef] [PubMed]
229. Monsen, E.R. Dietary reference intakes for the antioxidant nutrients: Vitamin C, vitamin E, selenium, and carotenoids. *J. Am. Diet. Assoc.* **2000**, *100*, 637–640. [CrossRef]
230. Food Data Central. Available online: Fdc.nal.usda.gov (accessed on 16 September 2019).
231. Bhullar, N.K.; Gruissem, W. Nutritional enhancement of rice for human health: The contribution of biotechnology. *Biotechnol. Adv.* **2013**, *31*, 50–57. [CrossRef] [PubMed]

Review

Effects and Mechanisms of Tea and Its Bioactive Compounds for the Prevention and Treatment of Cardiovascular Diseases: An Updated Review

Shi-Yu Cao [1], Cai-Ning Zhao [1], Ren-You Gan [2,*], Xiao-Yu Xu [1], Xin-Lin Wei [2], Harold Corke [2], Atanas G. Atanasov [3,4,5] and Hua-Bin Li [1,*]

1 Guangdong Provincial Key Laboratory of Food, Nutrition and Health, Department of Nutrition, School of Public Health, Sun Yat-Sen University, Guangzhou 510080, China; caoshy3@mail2.sysu.edu.cn (S.-Y.C.); zhaocn@mail2.sysu.edu.cn (C.-N.Z.); xuxy53@mail2.sysu.edu.cn (X.-Y.X.)

2 Department of Food Science & Technology, School of Agriculture and Biology, Shanghai Jiao Tong University, Shanghai 200240, China; weixinlin@sjtu.edu.cn (X.-L.W.); hcorke@sjtu.edu.cn (H.C.)

3 The Institute of Genetics and Animal Breeding, Polish Academy of Sciences, Jastrzębiec, 05-552 Magdalenka, Poland; atanas.atanasov@univie.ac.at

4 Department of Pharmacognosy, University of Vienna, 1090 Vienna, Austria

5 Institute of Neurobiology, Bulgarian Academy of Sciences, 23 Acad. G. Bonchev str., 1113 Sofia, Bulgaria

* Correspondence: renyougan@sjtu.edu.cn (R.-Y.G.); lihuabin@mail.sysu.edu.cn (H.-B.L.); Tel.: +86-21-3420-8517 (R.-Y.G.); +86-20-873-323-91 (H.-B.L.)

Received: 18 May 2019; Accepted: 4 June 2019; Published: 6 June 2019

Abstract: Cardiovascular diseases (CVDs) are critical global public health issues with high morbidity and mortality. Epidemiological studies have revealed that regular tea drinking is inversely associated with the risk of CVDs. Additionally, substantial in vitro and in vivo experimental studies have shown that tea and its bioactive compounds are effective in protecting against CVDs. The relevant mechanisms include reducing blood lipid, alleviating ischemia/reperfusion injury, inhibiting oxidative stress, enhancing endothelial function, attenuating inflammation, and protecting cardiomyocyte function. Moreover, some clinical trials also proved the protective role of tea against CVDs. In order to provide a better understanding of the relationship between tea and CVDs, this review summarizes the effects of tea and its bioactive compounds against CVDs and discusses potential mechanisms of action based on evidence from epidemiological, experimental, and clinical studies.

Keywords: tea; bioactive compounds; polyphenols; EGCG; cardiovascular diseases; mechanisms

1. Introduction

Cardiovascular diseases (CVDs), a group of disorders of the heart and blood vessels, mainly include coronary heart disease (CHD), stroke, heart failure, hypertensive heart disease, rheumatic heart disease, etc. As reported by the World Health Organization (WHO), CVDs are the leading causes of death globally and were responsible for 17.9 million deaths in 2016, accounting for 31% of all global deaths [1]. The proven risk factors of CVDs include unhealthy diet, tobacco consumption, physical inactivity, and harmful use of alcohol [2]. Among these risk factors, diet is suggested to be the most adjustable factor in preventing CVDs. Many studies have shown that fruits, vegetables, cereals, spices, nuts, and mushrooms can prevent CVDs [3–11]. Moreover, several studies have indicated that tea and its bioactive components can prevent and treat CVDs as well as improve cardio-metabolic health [12,13].

Tea is the second most consumed beverage worldwide and has a long drinking history of over 2000 years [14]. Tea contains abundant bioactive compounds, which possess favorable effects against many diseases, such as CVDs, obesity, diabetes, liver diseases, and cancers [15–21]. Numerous epidemiological studies have demonstrated that tea consumption is reversely associated with CVD risk [12,13,22,23]. In addition, in in vitro and in vivo experimental studies, tea and its bioactive components, mainly epicatechin, catechin, and epigallocatechin-3-gallate (EGCG) (Figure 1), have been found to be effective in preventing CVDs, with the mechanisms mainly including lowering blood lipid, ameliorating ischemia/reperfusion injury, attenuating oxidative stress, enhancing endothelial function, relieving inflammation, and protecting cardiomyocyte function [14,24]. Furthermore, clinical trials have also revealed the beneficial effects of tea and its bioactive compounds against CVDs [25,26].

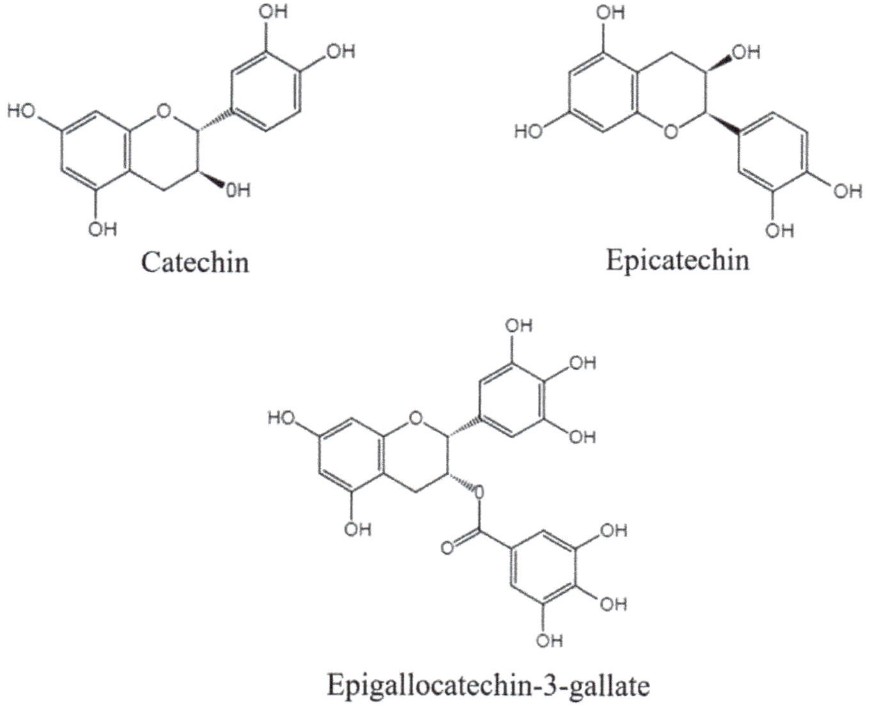

Catechin

Epicatechin

Epigallocatechin-3-gallate

Figure 1. The chemical structures of main catechins in tea associated with cardiovascular disease (CVD) protection.

In order to provide a better understanding of the relationship between tea and CVDs, we therefore searched the recent epidemiological, in vitro and in vivo experimental, and clinical studies from the last five years from the Web of Science Core Collection and PubMed databases based on keywords in the title and abstract, including tea, cardiovascular diseases, heart diseases, heart failure, hypertensive heart disease, rheumatic heart disease, and myocardial infarction. The literature types were mainly article and review papers, while meeting abstracts were excluded. This paper provides a comprehensive and updated review on the effects of tea and its bioactive compounds against CVDs, with special attention paid to the relevant mechanisms.

2. Epidemiological Studies

Several epidemiological studies have reported that tea consumption has a protective effect against CVDs. A meta-analysis indicated that green tea consumption could significantly reduce the risk of CVDs, and the odds ratio (OR) of myocardial infarction for those drinking 1–3 cups/day of green tea was 0.81 (95% CI: 0.67–0.98) compared with those drinking less than 1 cup/day [12]. Data from the Japan Public Health Center-based prospective study found that consumption of green tea could reduce the risk of heart disease in both women and men, and specifically, that the effect is marginal for decreasing the risk of death from heart disease in non-smoking men [27]. The results from a Chinese cohort study including 165,000 adult men also revealed that habitual green tea drinking was inversely associated with CVD death risk, and the hazard ratios (HR) were 0.93 (95% confidence interval (CI): 0.85–1.01) for ≤5 g/day, 0.91 (95% CI: 0.85–0.98) for 5–10 g/day, and 0.86 (95% CI: 0.79–0.93) for >10 g/day [22]. Additionally, two prospective cohort studies found that drinking green tea could reduce the risk of CVD death with HR 0.86 (95% CI: 0.77–0.97) in middle-aged and elderly Chinese adults [28]. In a Netherlands cohort study, tea consumption was found to be remarkably and nonlinearly associated with the decreased CVD risk in men, with those drinking 2–3 cups/day possessing the lowest HR (0.72, 95% CI: 0.57–0.91) [29]. In addition, a Dongfeng-Tongji cohort study found that green tea consumption could reduce the risk of CHD (HR = 0.89, 95% CI: 0.81–0.98) in the middle-aged and older Chinese population [30]. Furthermore, evidence from the Multi-Ethnic Study of Atherosclerosis conducted on white, Chinese-American, black, and Hispanic populations showed that habitual tea drinking (≥1 cup/day) could inhibit the progression of coronary artery calcification, which led to a decreased cardiovascular event incidence, with HR 0.71 (95% CI: 0.53–0.95), and compared to other race/ethnicity groups, the Chinese-American group had a higher tea consumption and lower incidence of cardiovascular events [31].

Notably, the bioactive compounds in tea also exhibited cardiovascular protective effects in some epidemiological studies. A dose-response meta-analysis regarding flavonoids mainly from tea revealed that flavonoids, such as flavonols, flavones, and flavanones, showed strong effects on reducing CVD risks in a dose-dependent manner, and an increase of 100 mg/day exerted a linear reduced risk of 4% CVD mortality [20]. In addition, the intake of flavonoids from tea and other food was found to be inversely related to CVD mortality and the relevant HRs were 0.34 (95% CI: 0.17–0.69) for data from United States Department of Agriculture (USDA) and 0.32 (95% CI: 0.16–0.61) for data from Phenol-Explorer database [32]. Besides, it has been demonstrated that the Polish population is characterized by a high polyphenol intake, and interestingly, most of the polyphenols are derived from tea and coffee [33]. Subsequently, it has been demonstrated that a higher intake of tea in this population was inversely associated with the risk of cardio-metabolic events [13,34]. In a prospective cohort study with 774 Dutch men aged 65–84 years, epicatechin was found to be associated with a reduced CVD mortality in men with CVDs (HR = 0.54, 95% CI: 0.31–0.96) [23]. Moreover, a high intake of catechins was inversely associated with the risk of CVDs and the HR for a 1-point increment of 10 mg/day was 0.98 (95%CI: 0.96–0.99) in the Nutrinet-Santé French cohort [35]. Furthermore, a prospective, nested case-control study conducted on middle-aged Japanese men found that high serum levels of EGCG could decrease the risk of stroke in non-smoking men, with adjusted OR 0.53 (95% CI: 0.29–0.98) for the highest EGCG level compared with the non-detectable one [36].

Epidemiological studies have indicated that tea consumption could ameliorate cardiovascular risk factors. Hypertension is a major risk factor in CHD and total stroke [37]. The results from the Observation of Cardiovascular Risk Factors in Luxembourg study showed that daily consumption of 100 mL of tea decreased the systolic blood pressure (SBP) by 0.6 mmHg and pulse pressure by 0.5 mmHg [38]. In another study, a cross-sectional study conducted on a rural elderly population in Jiangsu, China, found that tea consumption was significantly and inversely associated with diastolic blood pressure (DBP) (coefficient = −0.74, p = 0.003), and frequent tea drinking could reduce the risk of hypertension with OR 0.79 (95% CI: 0.65–0.95), p = 0.011 [39]. Moreover, in a longitudinal study conducted on 80,182 Chinese individuals (49 ± 12 years of age), regular tea drinking was found to inhibit the decrease of the serum high-density lipoprotein cholesterol (HDL-C) level in men aged 60 or older, which could reduce the risk of CVDs because a low concentration of HDL-C was suggested to be responsible for high risk of CVDs [40,41]. However, a case-control study using data from INTERHEART China found that habitual tea drinking would increase the risk of acute myocardial infarction, with OR 1.29 (95% CI: 1.03–1.61) for 4 cups/day tea drinkers compared with tea nondrinkers [42]. This inconsistent result may be due to the racial/ethnic factor or the different tea bioactive compound profiles.

Overall, epidemiological studies from Japan, China, the Netherlands, Luxembourg, France, America, and Poland have suggested a favorable role of tea and its bioactive compounds in reducing the risk of CVD incidence and mortality, although a few studies reported that tea could not protect against CVDs. The results of the epidemiological studies are summarized in Table 1.

Table 1. The effects of tea on CVDs based on epidemiological studies.

Subjects	Study Type	Effects	Risk Estimates (95%CI)	Ref.
90,914 Japanese participants aged 40–69 y	cohort study	Reducing the risk of heart disease and cerebrovascular disease	heart disease: 0.70 (0.56–0.87) for 3–4 cups/day; cerebrovascular disease: 0.73 (0.56–0.94) for 3–4 cups/day	[27]
165,000 Chinese adult men without pre-existing disease	cohort study	Reducing the risk of CVDs	0.93(0.85–1.01) for ≤5 g/day; 0.91 (0.85–0.98) for 5–10 g/day; 0.86 (0.79–0.93) for >10 g/day	[22]
74,941 women aged 40–70 y and 61,491 men aged 40–74 y in China	cohort study	Reducing the risk of CVDs	0.86 (0.77–0.97)	[28]
120,852 men and women in the Netherlands aged 55–69 y	cohort study	Reducing the risk of CVDs	0.72 (0.57–0.91) for 2–3 cups/day in men	[29]
19,471 participants free of CHD, stroke or cancer	cohort study	Reducing the risk of CHD	0.89 (0.81–0.98)	[30]
6508 participants from Multi-Ethnic Study of Atherosclerosis	cohort study	Slowing the progression of coronary artery calcium	0.71 (0.53–0.95) for ≥1 cup/day	[31]
1063 women aged >75 y in Australia	cohort study	Reducing the mortality of CVDs	0.34 (0.17–0.69) for data from USDA; 0.32 (0.16–0.61) for data from Phenol-Explorer databases	[32]
774 Dutch men aged 65–84 y	cohort study	Reducing the risk of CVDs	0.54 (0.31–0.96)	[23]
80,182 Chinese participants aged 37–61 y free of CVDs, cancers, and cholesterol-lowering agent use	cohort study	Increasing blood HDL-C	NA	[40]
29,876 participants aged 40–69 y free of heart disease, stroke, or cancer in Japan	case-control study	Lowering the risk of stroke in non-smoking men	0.53 (0.29–0.98)	[36]
1352 participants aged 18–69 y in Luxembourg	cross-sectional study	Decreasing the SBP and pulse pressure	NA	[38]
4579 participants aged ≥60 y in China	cross-sectional study	Lowering DBP and the risk of hypertension	0.79 (0.65–0.95)	[39]
5856 participants (case 2909, control 2947) in China	case-control study	Increasing the risk of acute myocardial infarction	1.29 (1.03–1.61) for 4 cups/d	[42]

Abbreviations: y, year; NA, Not available; CVD, cardiovascular disease; CHD, coronary heart disease; SBP, systolic blood pressure; DBP, diastolic blood pressure.

3. Experimental Studies

Increasing in vitro and in vivo experimental studies indicate that tea and its bioactive compounds possess cardiovascular protective effects, including lowering blood lipids, ameliorating ischemia/reperfusion injury, protecting endothelial function, protecting cardiomyocyte function, reducing oxidative stress, and alleviating inflammation, which are discussed below highlighting the relevant mechanisms.

3.1. Lowering Blood Lipids

Hyperlipidemia plays a key role in the development of atherosclerosis, and is a vital risk factor for cardiovascular diseases, which can be characterized by changes in the profile of serum lipids, including high triglyceride (TG) level, high cholesterol level, and low HDL-C level [43]. Treatment of male hamsters with mixed extracts of green tea, cocoa, coffee, and garcinia for 6 weeks was found to be effective in lowering serum TG, total cholesterol (TC), low-density lipoprotein-cholesterol (LDL-C), and hepatic TG and cholesterol in a dose-dependent manner [44]. Furthermore, combining the use of green tea extract and eriodictyol could suppress the mRNA level of 3-hydroxy-3-methylglutary-coenzyme A reductase (HMGCR) and 3-hydroxy-3-methylglutary-coenzyme A synthase (HMGCS), and increase the level of LDL receptor, leading to a lowered cholesterol level in male C57BL/6J mice fed with high-fat and high-sucrose diets [45]. Matcha, a kind of powdered green tea, could down-regulate TC, TG, and LDL-C levels, increase HDL level, decrease the serum glucose level, and elevate the superoxide dismutase (SOD) activity and malondialdehyde (MAD) content [24]. Moreover, a study found that a novel *Bacillus*-fermented green tea could suppress pancreatic lipase activity in vitro (IC_{50} = 0.48 mg/mL) and reduce postprandial lipaemia by 26% with 500 mg/kg tea in rats [46].

EGCG, the most abundant catechin in green tea, could attenuate the endothelial dysfunction induced by oxidized-LDL through the Jagged-1/Notch signaling pathway in human umbilical vein endothelial cells. Meanwhile, in apolipoprotein E (ApoE) knockout mice, EGCG was also confirmed to be effective in alleviating high-fat diet (HFD)-induced atherosclerosis through the Jagged-1/Notch signaling pathway [47]. Furthermore, the administration of EGCG to HFD-fed $ApoE^{-/-}$ mice significantly inhibited atherosclerotic plaque formation with upregulation of interleukin-10 (IL-10) levels and downregulation of plasma IL-6 and tumor necrosis factor-α (TNF-α) levels and attenuated HFD-induced dyslipidemia through modulating the liver X receptor (LXR)/sterol regulatory element binding transcription factor-1(SREBP-1) pathway [48]. In addition, green tea catechins were found to be effective in inhibiting LDL oxidation through incorporating themselves into LDL particles in nonconjugated forms in vitro [49]. Epicatechin could lower TC, LDL-C, and TG, mitigate liver fat accumulation, and increase HDL-C in hyperlipidemic rats induced by high-fat, high-cholesterol diets. These alterations were achieved through regulating the Insig-1-SREBP-SCAP pathway and other lipid metabolic-associated genes, including LXR-α, fatty acid synthase (FAS) and sirtuin 1 (SIRT1) [50]. Furthermore, a study revealed that green tea polyphenol could improve lipid metabolism disorders, inhibit atherogenesis, and elevate the expression of hepatic PPAR α and autophagy markers (LC3, Beclin1, and p62) in the vessel wall of ApoE-knockout mice [51].

3.2. Ameliorating Ischemia/Reperfusion Injury

Studies on ischemia/reperfusion have indicated the protective role of tea extract against ischemia/reperfusion related injuries [52,53]. A polyphenol trimer from green tea, cinnamon, and resveratrol were found effective in decreasing mitochondrial reactive oxygen species (ROS) and cell swelling in endothelial cells suffering from ischemic injury [52]. In addition, green tea showed a stronger effect than other teas against ischemia/reperfusion in male Wistar rats [53]. In another study, EGCG in combination with zinc could inhibit hypoxia/reoxygenation-induced cell apoptosis through activating the phosphatidylinositol-3-kinase (PI3K)/RAC-α serine/threonine-protein kinase (Akt) signaling pathway in H9c2 rat cardiac myoblast cells [54]. Additionally, pretreatment of EGCG to H9c2 cells could reduce the apoptosis induced by hypoxia/reoxygenation through stabilizing mitochondrial membrane potential and decreasing the expression of mitochondrial damage-related proteins [55]. Pretreatment of EGCG to albino Westar rats could protect against myocardial infarction induced by isoproterenol through reducing myocardial apoptosis. The related cardio-protective effects of EGCG was achieved by sustaining the balance of anti-apoptotic/pro-apoptotic proteins involved in the mitochondrial apoptotic pathway, restricting oxidative stress, and maintaining the integrity of DNA [56]. Besides, EGCG post-conditioning could attenuate ischemia/reperfusion injury and inhibit myocardial apoptosis via the PI3K/Akt signaling pathway in rats [57]. Additionally, EGCG showed cardioprotective effects, including alleviating myocardial injury and preventing ventricular arrhythmia in a rat ischemia/reperfusion model, with the mechanisms inhibiting the release of mitochondrial DNA (a potent pro-inflammatory mediator) and regulating the PI3K/Akt signaling pathway [58]. EGCG could improve hemodynamic recovery during reperfusion, elevate the adenosine triphosphate (ATP)-level, and relieve oxidative stress in excised perfused rabbit hearts [59]. Epicatechin could protect from cardiac injury induced by ischemia and inhibit myocardial apoptosis, cardiac fibrosis, and myocardial hypertrophy, which was achieved by the phosphatase and tensin homolog (PTEN)/PI3K/Akt signaling pathway [60]. Catechin could alleviate hypoxia-induced injuries through decreasing microRNA-92a and modulating the JNK signaling pathway in H9c2 cells [61].

3.3. Protecting Endothelial Function

Many studies have revealed that tea and its bioactive compounds could improve endothelial function. Black tea administration, rich in theaflavins, could prevent endothelial dysfunction, reduce the reduced form of nicotinamide adenine dinucleotide phosphate (NADPH) oxidase, serum total cholesterol and ROS levels, and restore the level of phospho endothelial nitric oxide synthase (eNOS) in ovariectomized rats [62]. Another study reported that black tea could alleviate the endothelial injury caused by hypertension via reducing the serum homocysteine level and endothelial cell endoplasmic reticulum stress in male Sprague Dawley rats [63]. Hyperhomocysteinemia would cause vascular endothelial dysfunction and promote the development of atherosclerosis. EGCG was effective in inhibiting homocysteine-induced apoptosis via regulating the mitochondrial apoptotic and PI3K/Akt/eNOS signaling pathways in human umbilical vein endothelial cells [64]. Besides, EGCG could inhibit the proliferation of vascular smooth muscle cell induced by homocysteine with ERK1/2 and p38MAPK signaling pathways involved [65]. Moreover, EGCG could stimulate the proliferation, migration, and tube formation of endothelial cells and promote angiogenesis in mice through transient receptor potential vanilloid type 1(TRPV1) activation [66].

3.4. Protecting Cardiomyocyte Function

Oolong tea could dose-dependently alleviate 24 h hypoxia-induced cardiomyocyte loss and hypertrophy through inhibiting caspase-3-cleavage and apoptosis and enhancing *p*-Akt-associated survival [67]. Additionally, EGCG could normalize the increased Ca^{2+} sensitivity of myofilaments caused by a mutation in human cardiac troponin i (k206i), which is related to hypertrophic cardiomyopathy [68]. Moreover, EGCG was found to be effective in reducing cardiac hypertrophy and fibrosis through increasing the diameter and volume of cardiomyocytes and decreasing the generation of ROS in aged rats [69]. Furthermore, EGCG could protect the heart development of zebrafish embryos from injuries caused by bisphenol A, an emerging contaminant associated with CVDs [70].

3.5. Reducing Oxidative Stress

Oxidative stress is closely associated with many chronic diseases, such as cardiovascular diseases [71]. Green tea and γ-amino butyric acid (GABA) green tea are rich in polyphenol, theanine, glutamine, and caffeine, which were found to be effective in reducing oxidative stress, modulating antioxidant endogenous defenses, and improving post-stroke depression in mice [71]. Besides, a study reported that treatment of white tea could improve cardiac glycolytic and heart antioxidant capacity in prediabetic rats [72]. Moreover, EGCG could prevent human umbilical vein endothelial cells from oxidative stress injury induced by $PM_{2.5}$, an ambient fine particulate matter which could cause certain CVDs. These antioxidant effects of EGCG were achieved by activating the p38 mitogen-activated protein kinase (MAPK) and extracellular signal regulated kinase (ERK)1/2 signaling pathways and subsequently upregulating the nuclear factor E2-related factor 2 (Nrf2)/heme oxygenase-1 (HO-1) pathway [73]. Furthermore, EGCG could decrease myocardial oxidative stress and free fatty-acid levels, thus inhibiting the development of heart failure induced by the heart/muscle-specific deletion of manganese superoxide dismutase in mice [74]. Theanine could protect H9c2 cells against hydrogen peroxide-induced apoptosis via enhancing antioxidant capacity, such as elevating the activities of glutathione peroxidase and SOD, and reducing the levels of ROS, nitric oxide, and oxidized glutathione [75].

3.6. Alleviating Inflammation

Inflammation is involved in the development of many metabolic diseases, such as CVDs, obesity, and cancers [76]. A study reported that EGCG could suppress the production of blood angiotensin II-associated C-reactive protein, which plays a vital role in the progression of atherosclerosis and inflammatory hepatic diseases, through the angiotensin II type 1 receptor-ROS-ERK1/2 signaling pathway [77]. EGCG could also inhibit the inflammatory response via regulating the Notch pathway in human macrophages [78]. Moreover, EGCG alleviated inflammation through the increase of E3 ubiquitin ligase RNF 216, followed by downregulation of toll-like receptor 4 [79]. Additionally, green tea extract treatment could attenuate cardiac macrophage infiltration and improve insulin secretion function through activating the adenosine monophosphate-activated protein kinase (AMPK) signaling pathway in weanling rats [80].

It's noteworthy that the bioavailability of tea polyphenols is usually very poor, for example, the peak plasma concentration of EGCG was only 0.15 μM after consuming two cups of green tea in humans [81,82]. Tea catechins are deeply modified by gut bacteria. Phenyl-γ-valerolactones and phenylvaleric acids are the major gut metabolic products of catechins and have been found to be effective in preventing chronic diseases [83]. Besides, black tea theaflavin and its galloyl derivatives are barely absorbed in human digestive tracts, but the galloyl part of theaflavin released by the microbiota has been reported to have many bioactivities [84]. Moreover, it has been reported that black tea has similar effects to green tea in protecting cardiovascular diseases [9]. Although green tea is more abundant in catechins than black tea, black tea is rich in theaflavins and thearubigins, which could compensate its functions due to the lack of catechins.

Collectively, tea, especially green tea, black tea, and white tea, and their bioactive compounds, such as EGCG, catechin, and theanine, possess remarkable protective effects against CVDs. The effects of tea on CVDs by in vitro and in vivo experimental studies are summarized in Table 2. The main cardiovascular protective mechanisms of tea include the reduction of blood lipid, alleviation of ischemia/reperfusion injury, enhancement of endothelial function, protection of cardiomyocytes, attenuation of oxidative stress, and relief of inflammation (Figure 2).

Table 2. The effects and mechanisms of tea on CVDs based on in vitro and in vivo experimental studies.

Substances	Subjects	Study Type	Dose	Effects and Mechanisms	Ref.
Green tea extract	Male C57BL/6J mice	In vivo	0.2% (*w/w*)	HMGCR↓, HMGCS↓, cholesterol↓	[45]
Matcha	Male ICR mice	In vivo	0.025%, 0.05%, 0.075% (*w/w*)	TC↓, TG↓, LDL-C↓, serum glucose↓; HDL↑, SOD↑, MAD↑	[24]
Bacillus-fermented green tea	Pancreatic lipase; Sprague-Dawley male Rats	In vitro and in vivo	IC_{50} 0.48 mg/mL; 500mg/kg	TG↓, pancreatic lipase activity↓	[46]
Green tea infusion	Male Wistar rats	In vivo	400 mg/kg	Hippocampal oxidative stress↓, necrosis↓	[53]
GABA green tea	Male balb/c mice	In vivo	50 and 100 mg/kg	Oxidative stress↓; Antioxidant endogenous defenses↑	[71]
Green tea extract	Pregnant Wistar rats	In vivo	0.12%, 0.24%	Cardiac macrophage infiltration↓; Insulin↑	[80]
Black tea	Female Sprague-Dawley rats	In vivo	15 mg/kg/day	NADPH oxidases↓, ROS↓; Flow-mediated dilatation↑	[62]
Black tea	Rat aortic endothelial cells; Male Sprague Dawley rats	In vitro and in vivo	0.3–5 μM; 15 mg/kg/day	Endothelial injury↓, serum homocysteine↓, endoplasmic reticulum stress↓	[63]
Oolong tea	H9c2 cardiac myoblast cells; Neonatal rat ventricular cardiomyocytes	In vitro	100, 200, 400 mg/mL	Cardiomyocyte loss ↓, hypertrophy ↓	[67]
White tea	Male Wistar rats	In vivo	1 g/100 mL	Cardiac glycolytic↑, antioxidant capacity↑	[72]
EGCG	Human umbilical vein endothelial cells; $ApoE^{-/-}$ mice	In vitro and in vivo	50 μM; 0.8 g/L	Endothelial dysfunction↓; Jagged-1/Notch activated	[47]
EGCG	$ApoE^{-/-}$ mice	In vivo	40 mg/kg/d	IL-6↓, TNF-α↓, TG↓, TC↓, LDL↓; IL-10↑, HDL↑, LXR/SREBP-1 pathways modulated	[48]
EGCG	H9c2 cardiac myoblast cells	In vitro	5, 10, 15, and 20 μM	Hypoxia/reoxygenation induced apoptosis↓	[54]
EGCG	H9c2 cardiac myoblast cells	In vitro	10 μM	Apoptosis↓; Stabilizing mitochondrial membrane potential	[55]
EGCG	Albino Westar rats	In vivo	15 mg/kg.	Myocardial infarction↓	[56]
EGCG	Male Sprague-Dawley rats	In vivo	10 mg/kg	Myocardial apoptosis↓	[57]
EGCG	Chinchilla rabbit heart	In vitro	20 μM/L	Oxidative stress↓; ATP↑	[59]
EGCG	Male Wistar rats	In vivo	10 mg/kg	Plasma mtDNA↓, TNF↓, IL-6 ↓, IL-8↓, ventricular arrhythmia↓	[58]
EGCG	Human umbilical vein endothelial cells	In vitro	10, 20, 30 μM	Apoptosis↓	[64]
EGCG	Human aortic smooth muscle cells	In vitro	20 μM	Homocysteine-induced proliferation↓	[65]
EGCG	Bovine aortic endothelial cells; WT C57BL mice and $TRPV1^{-/-}$ mice	In vitro and in vivo	0, 1.25, 2.5, 10, 20 μM; 10 μM	Angiogenesis↑	[66]
EGCG	Wistar albino rats	In vivo	200 mg/kg	Cardiac hypertrophy↓, fibrosis↓, LDL↓, VLDL↓, TG↓, TC↓; HDL↑, TGFβ↑, TNFα↑, NF-κB↑	[69]
EGCG	Zebrafish embryos	In vivo	50, 100 μM	Damage caused by bisphenol A↓	[70]
EGCG	Human umbilical vein endothelial cells	In vitro	50, 100, 200, 300, 400 μM	Oxidative stress↓; Nrf2↑, HO-1↑	[73]
EGCG	Heart/muscle-specific MnSOD-deficient mice	In vivo	10 mg/L, 100 mg/L	Myocardial oxidative stress↓, free fatty acid↓	[74]
EGCG	Male Sprague-Dawley rats	In vivo	25, 50 mg/kg/day	Ang II type 1 receptor↓, ERK1/2↓; PPARγ ↑	[77]
EGCG	Human monocyte cell line	In vitro	50 μg/mL	Inflammatory response↓	[78]
EGCG	Male C57/BL6 mice	In vivo	0, 2.5, 5, 10 μM	TLR4 expression↓	[79]
Epicatechin	Male Sprague-Dawley rats	In vivo	10, 20, 40 mg/kg	TC↓, LDL-C↓, TG↓; HDL-C↑	[50]
Theanine	H9c2 cardiac myoblast cells	In vitro	0, 4, 8, 16 μM	Peroxide-induced apoptosis↓, ROS↓; SOD↑	[75]

Up arrows mean increase, down arrows mean decrease. HMGCR, 3-hydroxy-3-methylglutary-coenzyme A reductase; HMGCS, 3-hydroxy-3-methylglutary-coenzyme A synthase; TC, total cholesterol; TG, triglyceride; LDL-C, low-density lipoprotein-cholesterol; HDL, high-density lipoprotein; SOD, superoxide dismutase; MAD, malondialdehyde; NADPH, nicotinamide adenine dinucleotide phosphate; ROS, reactive oxygen species; IL, interleukin; TNF, tumor necrosis factor; VLDL, very low-density lipoprotein; TGF, transforming growth factor; NF, nuclear factor; Nrf2, nuclear factor E2-related factor 2; HO-1, heme oxygenase-1; ERK, extracellular signal-regulated kinases; PPAR, peroxisome proliferator-activated receptor; TLR, toll-like receptor.

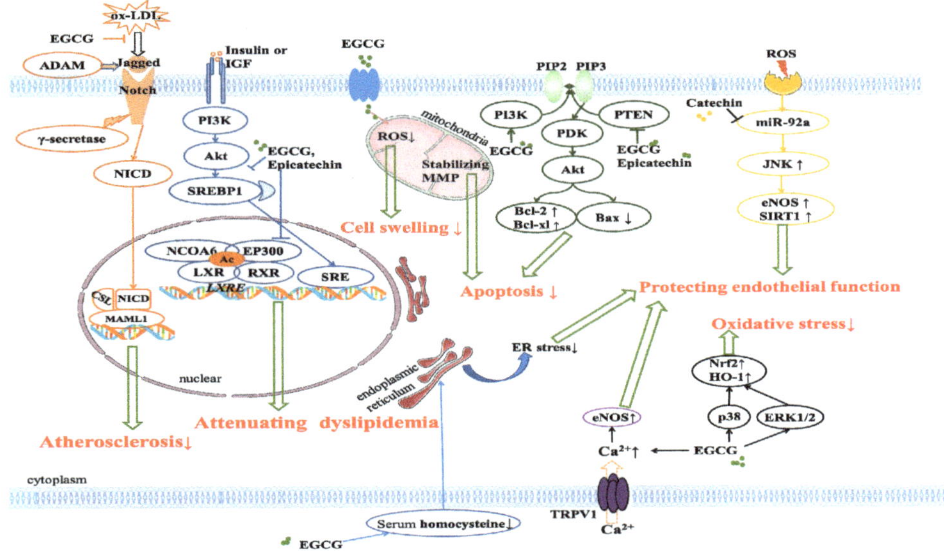

Figure 2. Signaling pathways involved in the protective effects of tea bioactive compounds against cardiovascular diseases. Epigallocatechin-3-gallate (EGCG) reduced atherosclerosis by inhibiting the activation of the Notch receptor induced by oxidized-LDL. EGCG and epicatechin could attenuate dyslipidemia through regulating the SREBP1 pathway. EGCG could reduce the reactive oxygen species level in mitochondria and stabilize the mitochondrial membrane potential, thus attenuating cell swelling and apoptosis of endothelial cells. EGCG and epicatechin could reduce the apoptosis of cardiac cells through regulating the PI3K pathway. EGCG could protect endothelial function through alleviating endoplasmic reticulum stress. EGCG and catechin could elevate the endothelial nitric oxide synthase (eNOS), thus protecting endothelial function. EGCG could reduce oxidative stress by regulating the p38 MAPK and ERK1/2 pathways. Abbreviations: ADAM, A-Disintegrin-And-Metalloprotease; NICD, Notch intracellular domain; PI3K, phosphatidylinositol-3-kinase; Akt, α serine/threonine-protein kinase; SREBP, sterol regulatory element binding transcription factor; LXR, liver X receptor; RXR, retinoid X receptor; NCOA6, nuclear receptor coactivator 6; PTEN, phosphatase and tensin homolog; PDK, phosphoinositide dependent kinase; Nrf, nuclear factor E2-related factor; HO-1, heme oxygenase-1; TRPV, transient receptor potential vanilloid type.

4. Clinical Trails

Dyslipidemia is a major risk factor of CVD development [43]. Benifuuki green tea, containing O-methylated catechin, was shown to remarkably reduce serum concentrations of LDL-C and lectin-like oxidized low-density lipoprotein receptor-1 ligands containing apolipoprotein B (LAB) in a randomized controlled trial (RCT) conducted on 155 volunteers [85]. Another RCT including 151 volunteer subjects also showed that Benifuuki and Yabukita green teas could decrease the LDL-C level [86]. In addition, consumption of green tea catechin extract for one year could lower serum TC ($p = 0.0004$), LDL-C ($p < 0.0001$), and non-HDL cholesterol ($p = 0.0032$) in healthy postmenopausal women [26]. A study investigated the hypolipidemic and antioxidant activities of catechin-enriched green and oolong teas in mild hypercholesterolemic individuals and found that the two teas showed similar antioxidant capacity and catechin-enriched oolong tea exerted a stronger hypolipidemic activity [87]. In addition, a meta-analysis concerned with ten RCTs of black tea drinking and serum cholesterol level found that black tea greatly reduced the serum LDL-C level, especially in those with higher risk of CVDs [25]. In another RCT involving 99 participants aged 25–60 years old with mild hypercholesterolemia, phytosterol-enriched instant black tea was found to be effective in lowering the levels of total cholesterol ($p < 0.001$), LDL-C ($p < 0.001$), apolipoprotein B ($p < 0.05$), and oxidative stress ($p < 0.05$), and elevating the levels of adiponectin, total antioxidant capacity, and tissue-plasminogen activator ($p < 0.05$), which were beneficial to cardiovascular function [88].

Hypertension is an independent predictor of cardiovascular-related death [89]. A study investigated the antihypertensive effects of short-term green tea consumption in 15 young volunteers aged 18–35 years old and 15 older volunteers aged 55–75 years old, and found that green tea could improve SBP, skin microvascular function, and oxygen tension [89]. In another study, the antihypertensive property of short-term consumption of green tea was revealed by a crossover RCT in 20 obese and prehypertensive women aged 41.1 ± 8.4 years old. Compared with a placebo group, women who had been drinking green tea for four weeks showed a decrease ($p < 0.05$) in 24-hour SBP (−3.61 ± −1.23 mmHg), daytime SBP (−3.61 ± −1.26 mmHg), and nighttime SBP (−3.94 ± −1.70 mmHg), with no significant changes in DBP [90]. Moreover, dietary flavonoids obtained from green tea, dehydrated red apple, and dark chocolate, at a dose of 425.8 ± 13.9 mg epicatechin equivalents combining with antihypertensive treatments (telmisartan or captopril), were found to significantly lower SBP and DBP in a RCT with 79 hypertension patients aged 20–55 years old [91]. Furthermore, in a RCT conducted on 19 hypertensive patients, consumption of black tea for eight days was shown to reduce SBP by 3.2 mmHg ($p < 0.005$) and DBP by 2.6 mmHg ($p < 0.0001$), inhibit the increase of BP within a fat load ($p < 0.0001$), and lower the index of reflection and stiffness [92].

Black tea also showed remarkable endothelial protective effects in a RCT conducted on 19 patients with hypertension. The participants of this study consumed black tea (containing 150 mg polyphenols) or a placebo twice a day for eight days, and the results indicated that black tea could stimulate the circulating amount of angiogenic cells and improve acute oral fat load-induced dysfunction of endothelial cells [93]. Additionally, a RCT involving 50 healthy men compared the endothelial protective effects of EGCG in three formulas including a green tea beverage, green tea extract, and pure EGCG, and found that only the green tea beverage could improve flow-mediated dilation [94]. Another RCT conducted with 14 healthy participants found that the intake of green tea polyphenol-enriched ice cream could immediately enhance vascular function and reduce oxidative stress [95]. Moreover, the administration of epicatechin (100 mg/d) for four weeks was found to improve the endothelial function and attenuate inflammation in 37 (pre)hypertensive participants [96].

In summary, numerous clinical trials (Table 3) support the cardiovascular-protective properties of tea and its bioactive compounds, with main mechanisms including reducing blood lipids, lowering BP, and protecting endothelial function.

Table 3. The effects of tea against CVDs based on clinical studies.

Subjects	Substances	Treatments	Effects and Mechanisms	Ref.
155 healthy participants	A green tea containing *O*-methylated catechin	12 g/d for 12 weeks	LDL-C↓, LAB↓	[85]
151 participants aged 30–70 y	Green tea	1.8 g/d for 12 weeks	LDL-C↓	[86]
15 participants aged 18–35 y and 15 participants aged 55–75 y	Green tea	2 cups/d for 14 days	Improving SBP and skin microvascular function	[89]
20 women aged 32.7–49.5 y	Green tea extract	500 mg for 4 weeks	SBP↓	[90]
50 healthy men	Green tea	A single dose of 200 mg EGCG	Improving flow-mediated dilation	[94]
14 healthy individuals	Green tea polyphenol-enriched ice cream	A single dose of 100 g	Oxidative stress↓,Vascular function↑	[95]
79 hypertension patients aged 20–55 y	Flavonoids from green tea	425.8 ± 13.9 mg epicatechin equivalents for 6 months	SBP ↓, DBP↓	[91]
60 individuals with mild hypercholesterolemia	Catechin-enriched green or oolong tea	780.6 mg/d or 640.4 mg/d catechin for 12 weeks	TC↓, LDL-C↓, TG↓	[87]
1075 healthy postmenopausal women	Catechins	1315 mg for 1 year	TC↓, LDL-C↓, non-HDL-C levels↓	[26]
99 participants aged 25–60 y with mild hypercholesterolemia	Phytosterol-enriched instant black tea	2.5 g/d for 4 weeks	Blood lipids↓	[88]
19 hypertensive patients	Black tea	129 mg/d flavonoids for 8 days	SBP↓	[92]
19 hypertension patients	Black tea	150 mg polyphenols for 8 days	Endothelial function↑	[93]
37 (Pre)hypertensive participants aged 40–80 y	Epicatechin or quercetin-3-glucoside	100 mg/d or 160 mg/d, respectively, for 4 weeks	Inflammation↓,Endothelial function ↑	[96]

Up arrows mean increase, down arrows mean decrease; LDL-C, low-density lipoprotein-cholesterol; LAB, apolipoprotein B; SBP, systolic blood pressure; DBP, diastolic blood pressure; TC, total cholesterol; TG, triglyceride; HDL, high-density lipoprotein.

5. Conclusions

Results from substantial epidemiological research has indicated that tea consumption is reversely associated with CVD risks, especially in those drinking tea habitually. In addition, a number of in vitro and in vivo experimental studies supported the protective effects of tea and its bioactive compounds against CVDs. The underlying mechanisms of action mainly include reducing blood lipid, alleviating ischemia/reperfusion injury, protecting cardiomyocyte function, enhancing endothelial function, lowering oxidative stress, and attenuating inflammation. Furthermore, clinical trials have also revealed that tea consumption could protect against CVDs. Therefore, it is valuable to recommend tea consumption for the public to protect cardiovascular health. Except for catechins and theaflavins, tea also contained procyanidin, phenolic acids, and so on. For example, the total proanthocyanidins in nine Mauritian black teas varied from 25 ± 2 to 74 ± 10 mg cyanidin chloride/g dry weight [97]. But few studies have focused on the effects of tea procyanidin and phenolic acids against cardiovascular diseases. In the future, different teas should be further evaluated considering their cardiovascular protective effects from the bench to bed, and more effective compounds should be separated and identified. More importantly, the mechanisms of action of teas, such as the molecular targets and mediated signaling pathways, should be further clarified to provide a better understanding for the action of tea. Last but not least, the safety of tea should be paid attention to, since its health benefits must be established on its safety.

Author Contributions: Conceptualization, S.-Y.C., R.-Y.G., and H.-B.L.; software, S.-Y.C.; writing—original draft preparation, S.-Y.C., C.-N.Z. and X.-Y.X.; writing—review and editing, R.-Y.G., X.-L.W., H.C., A.G.A., and H.-B.L.; visualization, S.-Y.C.; supervision, R.-Y.G. and H.-B.L.; funding acquisition, R.-Y.G. and H.-B.L.

Funding: This study was supported by the National Key R&D Program of China (No. 2018YFC1604400), Shanghai Basic and Key Program (No. 18JC1410800), Shanghai Pujiang Talent Plan (No. 18PJ1404600), the Agri-X Interdisciplinary Fund of Shanghai Jiao Tong University (No. Agri-X2017004), and the Key Project of Guangdong Provincial Science and Technology Program (No. 2014B020205002).

Conflicts of Interest: The authors declare no conflict of interest.

References

1. Cardiovascular Disease. Available online: https://www.who.int/cardiovascular_diseases/en/ (accessed on 20 April 2019).
2. Yusuf, S.; Reddy, S.; Ounpuu, S.; Anand, S. Global burden of cardiovascular diseases: part I: general considerations, the epidemiologic transition, risk factors, and impact of urbanization. *Circulation* **2001**, *104*, 2746–2753. [CrossRef] [PubMed]
3. Zhao, C.N.; Meng, X.; Li, Y.; Li, S.; Liu, Q.; Tang, G.Y.; Li, H.B. Fruits for prevention and treatment of cardiovascular diseases. *Nutrients* **2017**, *9*, 598. [CrossRef] [PubMed]
4. Tang, G.Y.; Meng, X.; Li, Y.; Zhao, C.N.; Liu, Q.; Li, H.B. Effects of vegetables on cardiovascular diseases and related mechanisms. *Nutrients* **2017**, *9*, 857. [CrossRef]
5. Zheng, J.; Zhou, Y.; Li, S.; Zhang, P.; Zhou, T.; Xu, D.P.; Li, H.B. Effects and mechanisms of fruit and vegetable juices on cardiovascular diseases. *Int. J. Mol. Sci.* **2017**, *18*, 555. [CrossRef] [PubMed]
6. Deng, G.F.; Xu, X.R.; Guo, Y.J.; Xia, E.Q.; Li, S.; Wu, S.; Chen, F.; Ling, W.H.; Li, H.B. Determination of antioxidant property and their lipophilic and hydrophilic phenolic contents in cereal grains. *J. Funct. Foods* **2012**, *4*, 906–914. [CrossRef]
7. Zheng, J.; Zhou, Y.; Li, Y.; Xu, D.P.; Li, S.; Li, H.B. Spices for prevention and treatment of cancers. *Nutrients* **2016**, *8*, 495. [CrossRef] [PubMed]
8. Guo, Y.J.; Deng, G.F.; Xu, X.R.; Wu, S.; Li, S.; Xia, E.Q.; Li, F.; Chen, F.; Ling, W.H.; Li, H.B. Antioxidant capacities, phenolic compounds and polysaccharide contents of 49 edible macro-fungi. *Food Funct.* **2012**, *3*, 1195–1205. [CrossRef] [PubMed]
9. Angelino, D.; Godos, J.; Ghelfi, F.; Tieri, M.; Titta, L.; Lafranconi, A.; Marventano, S.; Alonzo, E.; Gambera, A.; Sciacca, S.; et al. Fruit and vegetable consumption and health outcomes: an umbrella review of observational studies. *Int. J. Food Sci. Nutr.* **2019**, *15*, 1–16. [CrossRef] [PubMed]

10. Aune, D.; Keum, N.; Giovannucci, E.; Fadnes, L.T.; Boffetta, P.; Greenwood, D.C.; Tonstad, S.; Vatten, L.J.; Riboli, E.; Norat, T. Whole grain consumption and risk of cardiovascular disease, cancer, and all cause and cause specific mortality: Systematic review and dose-response meta-analysis of prospective studies. *BMJ* **2016**, *353*, i2716. [CrossRef] [PubMed]
11. Kim, Y.; Keogh, J.; Clifton, P.M. Nuts and cardio-metabolic disease: A review of meta-analyses. *Nutrients* **2018**, *10*, 1935. [CrossRef]
12. Pang, J.; Zhang, Z.; Zheng, T.Z.; Bassig, B.A.; Mao, C.; Liu, X.; Zhu, Y.; Shi, K.; Ge, J.; Yang, Y.J.; et al. Green tea consumption and risk of cardiovascular and ischemic related diseases: A meta-analysis. *Int. J. Cardiol.* **2016**, *202*, 967–974. [CrossRef] [PubMed]
13. Grosso, G.; Stepaniak, U.; Micek, A.; Topor-Mądry, R.; Pikhart, H.; Szafraniec, K.; Pająk, A. Association of daily coffee and tea consumption and metabolic syndrome: results from the Polish arm of the HAPIEE study. *Eur. J. Nutr.* **2015**, *54*, 1129–1137. [CrossRef] [PubMed]
14. Hodgson, J.M.; Croft, K.D. Tea flavonoids and cardiovascular health. *Mol. Aspects Med.* **2010**, *31*, 495–502. [CrossRef] [PubMed]
15. Higdon, J.V.; Frei, B. Tea catechins and polyphenols: Health effects, metabolism, and antioxidant functions. *Crit. Rev. Food Sci. Nutr.* **2003**, *43*, 89–143. [CrossRef] [PubMed]
16. Li, S.; Gan, L.Q.; Li, S.K.; Zheng, J.C.; Xu, D.P.; Li, H.B. Effects of herbal infusions, tea and carbonated beverages on alcohol dehydrogenase and aldehyde dehydrogenase activity. *Food Funct.* **2014**, *5*, 42–49. [CrossRef] [PubMed]
17. Li, F.; Li, S.; Li, H.B.; Deng, G.F.; Ling, W.H.; Xu, X.R. Antiproliferative activities of tea and herbal infusions. *Food Funct.* **2013**, *4*, 530–538. [CrossRef] [PubMed]
18. Xu, X.Y.; Zhao, C.N.; Cao, S.Y.; Tang, G.Y.; Gan, R.Y.; Li, H.B. Effects and mechanisms of tea for the prevention and management of cancers: An updated review. *Crit. Rev. Food Sci. Nutr.* **2019**, in press. [CrossRef]
19. Wang, X.; Ouyang, Y.Y.; Liu, J.; Zhao, G. Flavonoid intake and risk of CVD: A systematic review and meta-analysis of prospective cohort studies. *Br. J. Nutr.* **2014**, *111*, 1–11. [CrossRef]
20. Grosso, G.; Micek, A.; Godos, J.; Pajak, A.; Sciacca, S.; Galvano, F.; Giovannucci, E.L. Dietary flavonoid and lignan intake and mortality in prospective cohort studies: systematic review and dose-response meta-analysis. *Am. J. Epidemiol.* **2017**, *185*, 1304–1316. [CrossRef]
21. Grosso, G.; Godos, J.; Lamuela-Raventos, R.; Ray, S.; Micek, A.; Pajak, A.; Sciacca, S.; D'Orazio, N.; Del Rio, D.; Galvano, F. A comprehensive meta-analysis on dietary flavonoid and lignan intake and cancer risk: Level of evidence and limitations. *Mol. Nutr. Food Res.* **2017**, *61*, 1600930. [CrossRef]
22. Liu, J.X.; Liu, S.W.; Zhou, H.M.; Hanson, T.; Yang, L.; Chen, Z.M.; Zhou, M.G. Association of green tea consumption with mortality from all-cause, cardiovascular disease and cancer in a Chinese cohort of 165,000 adult men. *Eur. J. Epidemiol.* **2016**, *31*, 853–865. [CrossRef] [PubMed]
23. Dower, J.I.; Geleijnse, J.M.; Hollman, P.C.H.; Soedamah-Muthu, S.S.; Kromhout, D. Dietary epicatechin intake and 25-y risk of cardiovascular mortality: The Zutphen Elderly Study. *Am. J. Clin. Nutr.* **2016**, *104*, 58–64. [CrossRef] [PubMed]
24. Xu, P.; Ying, L.; Hong, G.J.; Wang, Y.F. The effects of the aqueous extract and residue of Matcha on the antioxidant status and lipid and glucose levels in mice fed a high-fat diet. *Food Funct.* **2016**, *7*, 294–300. [CrossRef] [PubMed]
25. Zhao, Y.M.; Asimi, S.; Wu, K.J.; Zheng, J.S.; Li, D. Black tea consumption and serum cholesterol concentration: Systematic review and meta-analysis of randomized controlled trials. *Clin. Nutr.* **2015**, *34*, 612–619. [CrossRef] [PubMed]
26. Samavat, H.; Newman, A.R.; Wang, R.; Yuan, J.; Wu, A.H.; Kurzer, M.S. Effects of green tea catechin extract on serum lipids in postmenopausal women: a randomized, placebo-controlled clinical trial. *Am. J. Clin. Nutr.* **2016**, *104*, 1671–1682. [CrossRef] [PubMed]
27. Saito, E.; Inoue, M.; Sawada, N.; Shimazu, T.; Yamaji, T.; Iwasaki, M.; Sasazuki, S.; Noda, M.; Iso, H.; Tsugane, S. Association of green tea consumption with mortality due to all causes and major causes of death in a Japanese population: The Japan Public Health Center-based Prospective Study (JPHC Study). *Ann. Epidemiol.* **2015**, *25*, 512–518. [CrossRef] [PubMed]
28. Zhao, L.G.; Li, H.L.; Sun, J.W.; Yang, Y.; Ma, X.; Shu, X.O.; Zheng, W.; Xiang, Y.B. Green tea consumption and cause-specific mortality: Results from two prospective cohort studies in China. *J. Epidemiol.* **2017**, *27*, 36–41. [CrossRef] [PubMed]

29. Van den Brandt, P.A. Coffee or Tea? A prospective cohort study on the associations of coffee and tea intake with overall and cause-specific mortality in men versus women. *Eur. J. Epidemiol.* **2018**, *33*, 183–200. [CrossRef] [PubMed]
30. Tian, C.; Huang, Q.; Yang, L.L.; Légaré, S.; Angileri, F.; Yang, H.D.; Li, X.L.; Min, X.W.; Zhang, C.; Xu, C.W.; et al. Green tea consumption is associated with reduced incident CHD and improved CHD-related biomarkers in the Dongfeng-Tongji cohort. *Sci. Rep.* **2016**, *6*, 24353. [CrossRef]
31. Miller, P.E.; Zhao, D.; Frazier-Wood, A.C.; Michos, E.D.; Averill, M.; Sandfort, V.; Burke, G.L.; Polak, J.F.; Lima, J.A.C.; Post, W.S.; et al. Associations of coffee, tea, and caffeine intake with coronary artery calcification and cardiovascular events. *Am. J. Med.* **2017**, *130*, 188–197. [CrossRef]
32. Ivey, K.L.; Hodgson, J.M.; Croft, K.D.; Lewis, J.R.; Prince, R.L. Flavonoid intake and all-cause mortality. *Am. J. Clin. Nutr.* **2015**, *101*, 1012–1020. [CrossRef] [PubMed]
33. Grosso, G.; Stepaniak, U.; Topor-Mądry, R.; Szafraniec, K.; Pająk, A. Estimated dietary intake and major food sources of polyphenols in the Polish arm of the HAPIEE study. *Nutrition* **2014**, *30*, 1398–1403. [CrossRef] [PubMed]
34. Micek, A.; Grosso, G.; Polak, M.; Kozakiewicz, K.; Tykarski, A.; Puch Walczak, A.; Drygas, W.; Kwaśniewska, M.; Pająk, A.; On behalf of WOBASZ II Investigators. Association between tea and coffee consumption and prevalence of metabolic syndrome in Poland—Results from the WOBASZ II study (2013–2014). *Int. J. Food Sci. Nutr.* **2018**, *69*, 358–368. [CrossRef] [PubMed]
35. Adriouch, S.; Lampuré, A.; Nechba, A.; Baudry, J.; Assmann, K.; Kesse-Guyot, E.; Hercberg, S.; Scalbert, A.; Touvier, M.; Fezeu, L.K. Prospective association between total and specific dietary polyphenol intakes and cardiovascular disease risk in the Nutrinet-Santé French cohort. *Nutrients* **2018**, *10*, 1587. [CrossRef] [PubMed]
36. Ikeda, A.; Iso, H.; Yamagishi, K.; Iwasaki, M.; Yamaji, T.; Miura, T.; Sawada, N.; Inoue, M.; Tsugane, S. Plasma tea catechins and risk of cardiovascular disease in middle-aged Japanese subjects: The JPHC study. *Atherosclerosis* **2018**, *277*, 90–97. [CrossRef] [PubMed]
37. Sowers, J.R.; Epstein, M.; Frohlich, E.D. Diabetes, hypertension, and cardiovascular disease—An update. *Hypertension* **2001**, *37*, 1053–1059. [CrossRef] [PubMed]
38. Alkerwi, A.; Sauvageot, N.; Crichton, G.E.; Elias, M.F. Tea, but not coffee consumption, is associated with components of arterial pressure. The observation of cardiovascular risk factors study in Luxembourg. *Nutr. Res.* **2015**, *35*, 557–565. [CrossRef]
39. Yin, J.Y.; Duan, S.Y.; Liu, F.C.; Yao, Q.K.; Tu, S.; Xu, Y.; Pan, C.W. Blood pressure is associated with tea consumption: A cross-sectional study in a rural, elderly population of Jiangsu China. *J. Nutr. Health Aging* **2017**, *21*, 1151–1159. [CrossRef] [PubMed]
40. Huang, S.E.; Li, J.J.; Wu, Y.T.; Ranjbar, S.; Xing, A.J.; Zhao, H.Y.; Wang, Y.X.; Shearer, G.C.; Bao, L.; Lichtenstein, A.H.; et al. Tea consumption and longitudinal change in high-density lipoprotein cholesterol concentration in Chinese adults. *J. Am. Heart Assoc.* **2018**, *7*, e008814. [CrossRef]
41. Chapman, M.J.; Ginsberg, H.N.; Amarenco, P.; Andreotti, F.; Boren, J.; Catapano, A.L.; Descamps, O.S.; Fisher, E.; Kovanen, P.T.; Kuivenhoven, J.A.; et al. Triglyceride-rich lipoproteins and high-density lipoprotein cholesterol in patients at high risk of cardiovascular disease: Evidence and guidance for management. *Eur. Heart J.* **2011**, *32*, 1345–1361. [CrossRef]
42. Hao, G.; Li, W.; Teo, K.; Wang, X.Y.; Yang, J.G.; Wang, Y.; Liu, L.S.; Yusuf, S. Influence of tea consumption on acute myocardial infarction in China population: The INTERHEART China Study. *Angiology* **2015**, *66*, 265–270. [CrossRef] [PubMed]
43. Bertolotti, M.; Maurantonio, M.; Gabbi, C.; Anzivino, C.; Carulli, N. Review article: hyperlipidaemia and cardiovascular risk. *Aliment Pharm. Ther.* **2005**, *222*, 28–30. [CrossRef] [PubMed]
44. Chang, C.; Hsu, Y.; Chen, Y.M.; Huang, W.C.; Huang, C.C.; Hsu, M.C. Effects of combined extract of cocoa, coffee, green tea and garcinia on lipid profiles, glycaemic markers and inflammatory responses in hamsters. *BMC Complement. Altern. Med.* **2015**, *15*, 269. [CrossRef] [PubMed]
45. Yamashita, M.; Kumazoe, M.; Nakamura, Y.; Won, Y.S.; Bae, J.; Yamashita, S.; Tachibana, H. The combination of green tea extract and eriodictyol inhibited high-fat/high-sucrose diet-induced cholesterol upregulation is accompanied by suppression of cholesterol synthesis enzymes. *J. Nutr. Sci. Vitaminol.* **2016**, *62*, 249–256. [CrossRef]

46. Seo, D.B.; Jeong, H.W.; Kim, Y.J.; Kim, S.; Kim, J.; Lee, J.H.; Joo, K.; Choi, J.K.; Shin, S.S.; Lee, S.J. Fermented green tea extract exhibits hypolipidaemic effects through the inhibition of pancreatic lipase and promotion of energy expenditure. *Brit. J. Nutr.* **2017**, *117*, 177–186. [CrossRef] [PubMed]
47. Yin, J.; Huang, F.; Yi, Y.; Yin, L.; Peng, D. EGCG attenuates atherosclerosis through the Jagged-1/Notch pathway. *Int. J. Mol. Med.* **2016**, *37*, 398–406. [CrossRef] [PubMed]
48. Pan, L.L.; Wu, Y.; Wang, R.Q.; Chen, J.W.; Chen, J.; Zhang, Y.; Chen, Y.; Geng, M.; Xu, Z.D.; Dai, M.; et al. (−)-Epigallocatechin-3-Gallate ameliorates atherosclerosis and modulates hepatic lipid metabolic gene expression in apolipoprotein E knockout mice: Involvement of TTC39B. *Front. Pharmacol.* **2018**, *9*, 195.
49. Suzuki-Sugihara, N.; Kishimoto, Y.; Saita, E.; Taguchi, C.; Kobayashi, M.; Ichitani, M.; Ukawa, Y.; Sagesaka, Y.M.; Suzuki, E.; Kondo, K. Green tea catechins prevent low-density lipoprotein oxidation via their accumulation in low-density lipoprotein particles in humans. *Nutr. Res.* **2016**, *36*, 16–23. [CrossRef]
50. Cheng, H.; Xu, N.; Zhao, W.; Su, J.J.; Liang, M.R.; Xie, Z.W.; Wu, X.L.; Li, Q.L. (−)-Epicatechin regulates blood lipids and attenuates hepatic steatosis in rats fed high-fat diet. *Mol. Nutr. Food Res.* **2017**, *61*, 1700303. [CrossRef]
51. Ding, S.B.; Jiang, J.J.; Yu, P.X.; Zhang, G.F.; Zhang, G.H.; Liu, X.T. Green tea polyphenol treatment attenuates atherosclerosis in high-fat diet-fed apolipoprotein E-knockout mice via alleviating dyslipidemia and upregulating autophagy. *PLoS ONE* **2017**, *12*, e0181666. [CrossRef]
52. Panickar, K.S.; Qin, B.; Anderson, R.A. Ischemia-induced endothelial cell swelling and mitochondrial dysfunction are attenuated by cinnamtannin D1, green tea extract, and resveratrol in vitro. *Nutr. Neurosci.* **2015**, *18*, 297–306. [CrossRef] [PubMed]
53. Martins, A.; Schimidt, H.L.; Garcia, A.; Altermann, C.D.C.; Santos, F.W.; Carpes, F.P.; Da Silva, W.C.; Mello-Carpes, P.B. Supplementation with different teas from *Camellia sinensis* prevents memory deficits and hippocampus oxidative stress in ischemia-reperfusion. *Neurochem. Int.* **2017**, *108*, 287–295. [CrossRef] [PubMed]
54. Zeng, X.; Tan, X. Epigallocatechin-3-gallate and zinc provide anti-apoptotic protection against hypoxia/ reoxygenation injury in H9c2 rat cardiac myoblast cells. *Mol. Med. Rep.* **2015**, *12*, 1850–1856. [CrossRef] [PubMed]
55. Wang, W.; Huang, X.; Shen, D.; Ming, Z.; Zheng, M.; Zhang, J. Polyphenol epigallocatechin-3-gallate inhibits hypoxia/reoxygenation-induced H9c2 cell apoptosis. *Minerva Med.* **2018**, *109*, 95–102. [PubMed]
56. Othman, A.I.; Elkomy, M.M.; El-Missiry, M.A.; Dardor, M. Epigallocatechin-3-gallate prevents cardiac apoptosis by modulating the intrinsic apoptotic pathway in isoproterenol-induced myocardial infarction. *Eur. J. Pharmacol.* **2017**, *794*, 27–36. [CrossRef]
57. Xuan, F.; Jian, J. Epigallocatechin gallate exerts protective effects against myocardial ischemia/reperfusion injury through the PI3K/Akt pathway-mediated inhibition of apoptosis and the restoration of the autophagic flux. *Int. J. Mol. Med.* **2016**, *38*, 328–336. [CrossRef]
58. Qin, C.Y.; Zhang, H.W.; Gu, J.; Xu, F.; Liang, H.M.; Fan, K.J.; Shen, J.Y.; Xiao, Z.H.; Zhang, E.Y.; Hu, J. Mitochondrial DNA-induced inflammatory damage contributes to myocardial ischemia reperfusion injury in rats: Cardioprotective role of epigallocatechin. *Mol. Med. Rep.* **2017**, *16*, 7569–7576. [CrossRef]
59. Salameh, A.; Schuster, R.; Daehnert, I.; Seeger, J.; Dhein, S. Epigallocatechin gallate reduces ischemia/reperfusion injury in isolated perfused rabbit hearts. *Int. J. Mol. Sci.* **2018**, *19*, 628. [CrossRef]
60. Li, J.W.; Wang, X.Y.; Zhang, X.; Gao, L.; Wang, L.F.; Yin, X.H. (−)-Epicatechin protects against myocardial ischemia-induced cardiac injury via activation of the PTEN/PI3K/AKT pathway. *Mol. Med. Rep.* **2018**, *17*, 8300–8308. [CrossRef]
61. Fang, J.F.; Dai, J.H.; Ni, M.; Cai, Z.Y.; Liao, Y.F. Catechin protects rat cardiomyocytes from hypoxia-induced injury by regulating microRNA-92a. *Int. J. Clin. Exp. Pathol.* **2018**, *11*, 3257–3266.
62. Leung, F.P.; Yung, L.M.; Ngai, C.Y.; Cheang, W.S.; Tian, X.Y.; Lau, C.W.; Zhang, Y.; Liu, J.; Chen, Z.Y.; Bian, Z.X.; et al. Chronic black tea extract consumption improves endothelial function in ovariectomized rats. *Eur. J. Nutr.* **2016**, *55*, 1963–1972. [CrossRef] [PubMed]
63. Cheang, W.S.; Ngai, C.Y.; Tam, Y.Y.; Tian, X.Y.; Wong, W.T.; Zhang, Y.; Lau, C.W.; Chen, Z.Y.; Bian, Z.X.; Huang, Y.; et al. Black tea protects against hypertension-associated endothelial dysfunction through alleviation of endoplasmic reticulum stress. *Sci. Rep.* **2015**, *5*, 10340. [CrossRef] [PubMed]
64. Liu, S.M.; Sun, Z.W.; Chu, P.; Li, H.L.; Ahsan, A.; Zhou, Z.R.; Zhang, Z.H.; Sun, B.; Wu, J.J.; Xi, Y.L.; et al. EGCG protects against homocysteine-induced human umbilical vein endothelial cells apoptosis by modulating mitochondrial-dependent apoptotic signaling and PI3K/Akt/eNOS signaling pathways. *Apoptosis* **2017**, *22*, 672–680. [CrossRef] [PubMed]

65. Zhan, X.L.; Yang, X.H.; Gu, Y.H.; Guo, L.L.; Jin, H.M. Epigallocatechin gallate protects against homocysteine-induced vascular smooth muscle cell proliferation. *Mol. Cell Biochem.* **2018**, *439*, 131–140. [CrossRef] [PubMed]
66. Guo, B.C.; Wei, J.; Su, K.H.; Chiang, A.N.; Zhao, J.F.; Chen, H.Y.; Shyue, S.K.; Lee, T.S. Transient receptor potential vanilloid type 1 is vital for (−)-epigallocatechin-3-gallate mediated activation of endothelial nitric oxide synthase. *Mol. Nutr. Food Res.* **2015**, *59*, 646–657. [CrossRef] [PubMed]
67. Shibu, M.A.; Kuo, C.; Chen, B.; Ju, D.; Chen, R.; Lai, C.; Huang, P.; Viswanadha, V.P.; Kuo, W.; Huang, C. Oolong tea prevents cardiomyocyte loss against hypoxia by attenuating p-JNK mediated hypertrophy and enhancing P-IGF1R, p-Akt, and p-Bad^{ser136} activity and by fortifying Nrf2 antioxidation system. *Environ. Toxicol.* **2018**, *33*, 220–233. [CrossRef] [PubMed]
68. Warren, C.M.; Karam, C.N.; Wolska, B.M.; Kobayashi, T.; de Tombe, P.P.; Arteaga, G.M.; Bos, J.M.; Ackerman, M.J.; Solaro, R.J. Green tea catechin normalizes the enhanced Ca^{2+} sensitivity of myofilaments regulated by a hypertrophic cardiomyopathy-associated mutation in human cardiac troponin I (K206I). *Circ-Cardiovasc. Genet.* **2015**, *8*, 765–773. [CrossRef]
69. Muhammed, I.; Sankar, S.; Govindaraj, S. Ameliorative effect of epigallocatechin gallate on cardiac hypertrophy and fibrosis in aged rats. *J. Cardiovasc. Pharm.* **2018**, *71*, 65–75.
70. Lombo, M.; Gonzalez-Rojo, S.; Fernandez-Diez, C.; Paz Herraez, M. Cardiogenesis impairment promoted by bisphenol A exposure is successfully counteracted by epigallocatechin gallate. *Environ. Pollut.* **2019**, *246*, 1008–1019. [CrossRef]
71. Di Lorenzo, A.; Nabavi, S.F.; Sureda, A.; Moghaddam, A.H.; Khanjani, S.; Arcidiaco, P.; Nabavi, S.M.; Daglia, M. Antidepressive-like effects and antioxidant activity of green tea and GABA green tea in a mouse model of post-stroke depression. *Mol. Nutr. Food Res.* **2016**, *60*, 566–579. [CrossRef]
72. Alves, M.G.; Martins, A.D.; Teixeira, N.F.; Rato, L.; Oliveira, P.F.; Silva, B.M. White tea consumption improves cardiac glycolytic and oxidative profile of prediabetic rats. *J. Funct. Foods* **2015**, *14*, 102–110. [CrossRef]
73. Yang, G.Z.; Wang, Z.J.; Bai, F.; Qin, X.J.; Cao, J.; Lv, J.Y.; Zhang, M.S. Epigallocatechin-3-gallate protects HUVECs from $PM_{2.5}$-induced oxidative stress injury by activating critical antioxidant pathways. *Molecules* **2015**, *20*, 6626–6639. [CrossRef] [PubMed]
74. Oyama, J.I.; Shiraki, A.; Nishikido, T.; Maeda, T.; Komoda, H.; Shimizu, T.; Makino, N.; Node, K. EGCG, a green tea catechin, attenuates the progression of heart failure induced by the heart/muscle-specific deletion of MnSOD in mice. *J. Cardiol.* **2017**, *69*, 417–427. [CrossRef] [PubMed]
75. Li, C.; Yan, Q.; Tang, S.; Xiao, W.; Tan, Z. L-Theanine protects H9c2 cells from hydrogen peroxide-induced apoptosis by enhancing antioxidant capability. *Med. Sci. Monit.* **2018**, *24*, 2109–2118. [CrossRef] [PubMed]
76. Hotamisligil, G.S. Inflammation and metabolic disorders. *Nature* **2006**, *444*, 860–867. [CrossRef] [PubMed]
77. Zhao, J.; Liu, J.; Pang, X.; Zhang, X.; Wang, S.; Wu, D. Epigallocatechin-3-gallate inhibits angiotensin II-induced C-reactive protein generation through interfering with the AT_1-ROS-ERK1/2 signaling pathway in hepatocytes. *Naunyn Schmiedebergs Arch. Pharmacol.* **2016**, *389*, 1225–1234. [CrossRef] [PubMed]
78. Wang, T.F.; Xiang, Z.M.; Wang, Y.; Li, X.; Fang, C.Y.; Song, S.; Li, C.L.; Yu, H.S.; Wang, H.; Yan, L.; et al. (−)-Epigallocatechin gallate targets Notch to attenuate the inflammatory response in the immediate early stage in human macrophages. *Front. Immunol.* **2017**, *8*, 433. [CrossRef] [PubMed]
79. Kumazoe, M.; Nakamura, Y.; Yamashita, M.; Suzuki, T.; Takamatsu, K.; Huang, Y.; Bae, J.; Yamashita, S.; Murata, M.; Yamada, S.; et al. Green tea polyphenol epigallocatechin-3-gallate suppresses toll-like receptor 4 expression via upregulation of E3 ubiquitin-protein ligase RNF216. *J. Biol. Chem.* **2017**, *292*, 4077–4088. [CrossRef] [PubMed]
80. Matsumoto, E.; Kataoka, S.; Mukai, Y.; Sato, M.; Sato, S. Green tea extract intake during lactation modified cardiac macrophage infiltration and AMP-activated protein kinase phosphorylation in weanling rats from undernourished mother during gestation and lactation. *J. Dev. Orig. Health Dis.* **2017**, *8*, 178–187. [CrossRef]
81. Clifford, M.N.; van der Hooft, J.J.; Crozier, A. Human studies on the absorption, distribution, metabolism, and excretion of tea polyphenols. *Am. J. Clin. Nutr.* **2013**, *98*, 1619S–1630S. [CrossRef]
82. Zhang, J.; Nie, S.; Wang, S. Nanoencapsulation enhances epigallocatechin-3-gallate stability and its antiatherogenic bioactivities in macrophages. *J. Agric. Food Chem.* **2013**, *61*, 9200–9209. [CrossRef] [PubMed]
83. Mena, P.; Bresciani, L.; Brindani, N.; Ludwig, I.A.; Pereira-Caro, G.; Angelino, D.; Llorach, R.; Calani, L.; Brighenti, F.; Clifford, M.N.; et al. Phenyl-γ-valerolactones and phenylvaleric acids, the main colonic metabolites of flavan-3-ols: synthesis, analysis, bioavailability, and bioactivity. *Nat. Prod. Rep.* **2019**, *36*, 714–752. [CrossRef] [PubMed]

84. Pereira-Caro, G.; Moreno-Rojas, J.M.; Brindani, N.; Del Rio, D.; Lean, M.E.J.; Hara, Y.; Crozier, A. Bioavailability of black tea theaflavins: Absorption, metabolism, and colonic catabolism. *J. Agric. Food Chem.* **2017**, *65*, 5365–5374. [CrossRef] [PubMed]
85. Imbe, H.; Sano, H.; Miyawaki, M.; Fujisawa, R.; Miyasato, M.; Nakatsuji, F.; Haseda, F.; Tanimoto, K.; Terasaki, J.; Maeda-Yamamoto, M.; et al. "Benifuuki" green tea, containing O-methylated EGCG reduces serum low-density lipoprotein cholesterol and lectin-like oxidized low-density lipoprotein receptor-1 ligands containing apolipoprotein B: A double-blind, placebo-controlled randomized trial. *J. Funct. Foods* **2016**, *25*, 25–37. [CrossRef]
86. Igarashi, Y.; Obara, T.; Ishikuro, M.; Matsubara, H.; Shigihara, M.; Metoki, H.; Kikuya, M.; Sameshima, Y.; Tachibana, H.; Maeda-Yamamotog, M.; et al. Randomized controlled trial of the effects of consumption of 'Yabukita' or 'Benifuuki' encapsulated tea-powder on low-density lipoprotein cholesterol level and body weight. *Food Nutr. Res.* **2017**, *61*, 1334484. [CrossRef]
87. Venkatakrishnan, K.; Chiu, H.F.; Cheng, J.C.; Chang, Y.; Lu, Y.Y.; Han, Y.C.; Shen, Y.C.; Tsai, K.S.; Wang, C.K. Comparative studies on the hypolipidemic, antioxidant and hepatoprotective activities of catechin-enriched green and oolong tea in a double-blind clinical trial. *Food Funct.* **2018**, *9*, 1205–1213. [CrossRef]
88. Orem, A.; Alasalvar, C.; Kural, B.V.; Yaman, S.; Orem, C.; Karadag, A.; Pelvan, E.; Zawistowski, J. Cardio-protective effects of phytosterol-enriched functional black tea in mild hypercholesterolemia subjects. *J. Funct. Foods* **2017**, *31*, 311–319. [CrossRef]
89. Wasilewski, R.; Ubara, E.O.; Klonizakis, M. Assessing the effects of a short-term green tea intervention in skin microvascular function and oxygen tension in older and younger adults. *Microvasc. Res.* **2016**, *107*, 65–71. [CrossRef]
90. Nogueira, L.D.P.; Nogueira Neto, J.F.; Klein, M.R.S.T.; Sanjuliani, A.F. Short-term effects of green tea on blood pressure, endothelial function, and metabolic profile in obese prehypertensive women: A crossover randomized clinical trial. *J. Am. Coll. Nutr.* **2017**, *36*, 108–115. [CrossRef]
91. De Jesús Romero-Prado, M.M.; Curiel-Beltrán, J.A.; Miramontes-Espino, M.V.; Cardona-Muñoz, E.G.; Rios-Arellano, A.; Balam-Salazar, L.B. Dietary flavonoids added to pharmacological antihypertensive therapy are effective in improving blood pressure. *Basic Clin. Pharmacol.* **2015**, *117*, 57–64. [CrossRef]
92. Grassi, D.; Draijer, R.; Desideri, G.; Mulder, T.; Ferri, C. Black tea lowers blood pressure and wave reflections in fasted and postprandial conditions in hypertensive patients: A randomised study. *Nutrients* **2015**, *7*, 1037–1051. [CrossRef]
93. Grassi, D.; Draijer, R.; Schalkwijk, C.; Desideri, G.; D'Angeli, A.; Francavilla, S.; Mulder, T.; Ferri, C. Black tea increases circulating endothelial progenitor cells and improves flow mediated dilatation counteracting deleterious effects from a fat load in hypertensive patients: A randomized controlled study. *Nutrients* **2016**, *8*, 727. [CrossRef] [PubMed]
94. Lorenz, M.; Rauhut, F.; Hofer, C.; Gwosc, S.; Mueller, E.; Praeger, D.; Zimmermann, B.F.; Wernecke, K.; Baumann, G.; Stangl, K.; et al. Tea-induced improvement of endothelial function in humans: No role for epigallocatechin gallate (EGCG). *Sci. Rep.* **2017**, *7*, 2279. [CrossRef] [PubMed]
95. Sanguigni, V.; Manco, M.; Sorge, R.; Gnessi, L.; Francomano, D. Natural antioxidant ice cream acutely reduces oxidative stress and improves vascular function and physical performance in healthy individuals. *Nutrition* **2017**, *33*, 225–233. [CrossRef] [PubMed]
96. Dower, J.I.; Geleijnse, J.M.; Gijsbers, L.; Schalkwijk, C.; Kromhout, D.; Hollman, P.C. Supplementation of the pure flavonoids epicatechin and quercetin affects some biomarkers of endothelial dysfunction and inflammation in (pre)hypertensive adults: A randomized double-blind, placebo-controlled, crossover trial. *J. Nutr.* **2015**, *145*, 1459–1463. [CrossRef] [PubMed]
97. Luximon-Ramma, A.; Bahorun, T.; Crozier, A.; Zbarsky, V.; Datla, K.P.; Dexter, D.T.; Aruoma, O.I. Characterization of the antioxidant functions of flavonoids and proanthocyanidins in Mauritian black teas. *Food Res. Int.* **2005**, *38*, 357–367. [CrossRef]

Review

Dietary Flavonoids as Cancer Chemopreventive Agents: An Updated Review of Human Studies

Carmen Rodríguez-García [1,2], Cristina Sánchez-Quesada [1,2,3] and José J. Gaforio [1,2,3,4,*]

1 Center for Advanced Studies in Olive Grove and Olive Oils, University of Jaen, Campus las Lagunillas s/n, 23071 Jaén, Spain; crgarcia@ujaen.es (C.R.-G.); csquesad@ujaen.es (C.S.-Q.)
2 Department of Health Sciences, Faculty of Experimental Sciences, University of Jaén, 23071 Jaén, Spain
3 Agri-Food Campus of International Excellence (ceiA3), 14005 Córdoba, Spain
4 CIBER Epidemiología y Salud Pública (CIBER-ESP), Instituto de Salud Carlos III, 28029 Madrid, Spain
* Correspondence: jgaforio@ujaen.es; Tel.: +34-953-212-002

Received: 25 April 2019; Accepted: 16 May 2019; Published: 18 May 2019

Abstract: Over the past few years, interest in health research has increased, making improved health a global goal for 2030. The purpose of such research is to ensure healthy lives and promote wellbeing across individuals of all ages. It has been shown that nutrition plays a key role in the prevention of some chronic diseases such as obesity, cardiovascular disease, diabetes, and cancer. One of the aspects that characterises a healthy diet is a high intake of vegetables and fruits, as both are flavonoid-rich foods. Flavonoids are one of the main subclasses of dietary polyphenols and possess strong antioxidant activity and anti-carcinogenic properties. Moreover, some population-based studies have described a relationship between cancer risk and dietary flavonoid intake. In this context, the goal of this review was to provide an updated evaluation of the association between the risk of different types of cancers and dietary flavonoid intake. We analysed all relevant epidemiological studies from January 2008 to March 2019 using the PUBMED and Web of Science databases. In summary, this review concludes that dietary flavonoid intake is associated with a reduced risk of different types of cancer, such as gastric, breast, prostate, and colorectal cancers.

Keywords: flavonoids; diet; antioxidants; cancer

1. Introduction

Cancer is among the diseases that have the greatest impact on society [1]. Even though its incidence has increased over the years, its mortality has decreased because of advances in treatment [2]. However, efforts to improve cancer prevention are needed. The aetiology of cancer is multifactorial, involving both environmental and genetic factors [3]. Diet is one of the lifestyle factors that affect cancer incidence and mortality [4]. Recently, several studies have reported that diets based on high levels of vegetables and fruits are strongly associated with a significant reduction in cancer risk [5,6].

Furthermore, there are some bioactive compounds in foods that have potential health benefits, such as flavonoids, carotenoids, stilbenes, lignans, and phenolic acids [7,8]. Flavonoids are a large group of phenolic compounds and are usually involved in protection against harsh environmental conditions, UV radiation, and microorganism attacks in plants [9,10]. Because of their potent antioxidant activity against oxidative stress, the interest in flavonoids has recently increased [11]. In vitro and in vivo studies have demonstrated that they have anti-carcinogenic properties against different types of cancers [5,12]. Moreover, many population-based studies have described an association between dietary flavonoids and cancer risk [13,14]. Hence, the goal of this review is to perform an updated evaluation of the association between the risk of different types of cancers and dietary flavonoids, as well as the intake of each flavonoid subclass.

2. Methodology

Recently, interest in flavonoids has increased because their strong antioxidant and anti-carcinogenic activities may have possible beneficial effects on cancer. Thus, in this review, we analysed all relevant cancer epidemiological studies from January 2008 to March 2019 using the PUBMED and Web of Science databases [15,16]. Since different reviews have already been published on flavonoids and cancer before 2008. Search entries included [flavonoids and cancer], [flavonoids and "breast cancer"], [flavonoids and "lung cancer"], [flavonoids and "prostate cancer"], [flavonoids and "gastric cancer"], [flavonoids and "pancreatic cancer"] and [flavonoids and "colorectal cancer"]. Selection criteria applied were the following: human studies, randomized controlled trials, cross-sectional, cohort and case-control studies and information about dietary intake. Reviews studies and Meta-Analyses were excluded. Besides, population studies were catalogued based on type of study: case-control or cohort study and the type of cancer.

3. Biosynthesis and Subclasses of Flavonoids

Flavonoids are secondary metabolites synthesised mainly by plants [9]. To date, more than 6000 different flavonoids have been identified, and they are distributed in a wide range of plants [17]. The general structure of flavonoids is composed of a 15-carbon skeleton, containing 2 benzene rings connected by a 3-carbon linking chain (Figure 1) [9]. Therefore, they are depicted as C6-C3-C6 compounds. Their biosynthesis involves two different biosynthetic pathways: the shikimic acid pathway and the acetate pathway (Figure 1) [9].

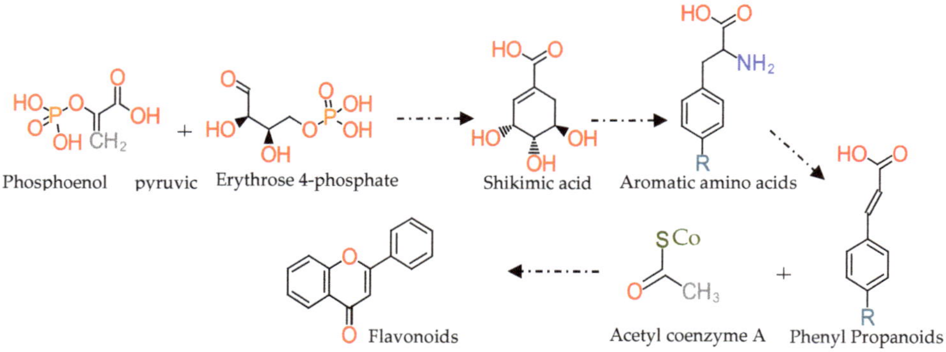

Phosphoenol pyruvic Erythrose 4-phosphate Shikimic acid Aromatic amino acids Flavonoids Acetyl coenzyme A Phenyl Propanoids

Figure 1. Flavonoid biosynthesis pathways (general structure of flavonoids).

Depending on the chemical structure, degree of oxidation, and unsaturation of the linking chain (C3), flavonoids can be classified into different groups, such as anthocyanidins, chalcones, flavonols, flavanones, flavan-3-ols, flavanonols, flavones, and isoflavonoids (Figure 2). Furthermore, flavonoids can be found in plants in glycoside-bound and free aglycone forms [9]. The glycoside-bound form is the most common flavone and flavonol form consumed in the diet [9].

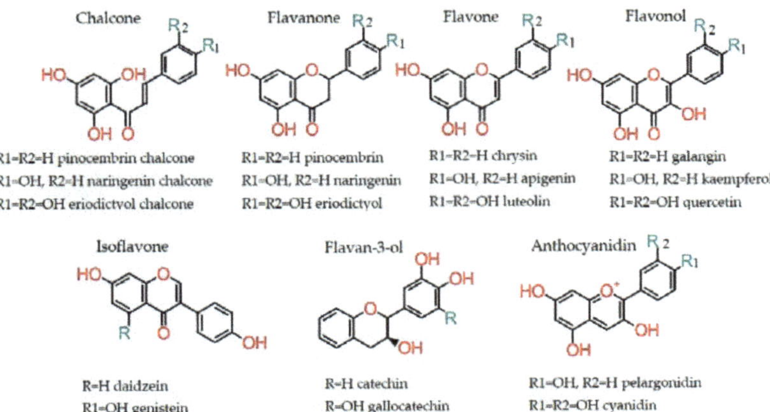

Figure 2. Flavonoid subclasses.

4. Dietary Flavonoids

Flavonoids are widely spread in different foods and beverages (such wine and tea), but the sources with the highest levels are fruits and vegetables [10]. Among the fruits (Table 1), the highest levels of flavonoids are found in berries, such as black elderberry (1358.66 mg/100 g) and black chokeberry (1012.98 mg/100 g) [18,19]. In the drupes group, some fruits such as plum and sweet cherry have higher levels of flavonoids than the rest of the group, 101.67 mg/100 g and 185.05 mg/100 g, respectively [20,21]. In the pomes group, apple has the level (56.35 mg/100 g) [21,22]. Furthermore, tropical fruits have a very low flavonoid content [23]. Depending on the type of fruit, the main flavonoid subclass groups vary: anthocyanins predominate in berries, and flavanols predominate in pomes, tropical fruits, and drupes (except in sweet cherry).

Table 1. Flavonoid contents of fruits (mg/100 g food). Data collected from Phenol Explorer [24].

Fruits	ANT	DYC	FVA	FVO	Total
		Berries			
Aestivalis grape	79.74	-	-	1.7	81.44
American cranberry	49.89	-	-	43.84	93.73
Black chokeberry	878.11	-	-	134.87	1012.98
Black elderberry	1316.66	-	-	42	1358.66
Black raspberry	-	-	-	19	19
Blackberry	172.59	-	13.87	16.87	203.33
Blackcurrant	593.58	-	1.17	13.68	608.43
Black grape	72.15	-	14.03	4.01	90.19
Green grape	-	-	3.78	2.49	6.27
Green currant	-	-	-	11.07	11.07
Highbush blueberry	156.6	-	1.11	54.77	212.48
Lingonberry	60.21	-	-	48.98	109.19
Lowbush blueberry	204.56	-	-	-	204.56
Red raspberry	72.47	-	5.73	16.26	94.46
Redcurrant	33.13	-	4.68	0.77	38.58
Strawberry	26.87	-	9.1075	2.32	38.29
		Drupes			
Nectarine	0.86	-	17.65	1.35	19.86
Peach	0.28	-	45.18	1.42	46.88
Plum	47.79	-	46.9	6.98	101.67
Sour cherry	54.43	-	0.2	-	54.63
Sweet cherry	170.18	-	14.87	-	185.05

Table 1. *Cont.*

Fruits	ANT	DYC	FVA	FVO	Total
			Pomes		
Apple	0.93	5.38	39.42	10.62	56.35
Pear	-	-	4.98	0.84	5.82
Quince	-	-	7.49	0.67	8.16
			Tropical Fruits		
Banana	-	-	1.55	-	1.55
Kiwi	-	-	0.7	-	0.7
Mango	-	-	1.72	-	1.72
Persimmon	-	-	1.28	-	1.28
Pomegranate	-	-	1.1	-	1.1

ANT: Anthocyanins, DYC: Dihydrochalcones, FVA: Flavan-3-ols, FVO: Flavonols.

Regarding vegetables (Table 2), the foods with the highest levels of flavonoids are broad bean pod (189.54 mg/100 g) [25], black olive (159.83 mg/100 g) [26], red onion (131.51 mg/100 g) [27], spinach (119.27 mg/100 g), and shallot (112.22 mg/100 g) [28,29]. Except for broad bean pod, the predominate flavonoid subclass in vegetables is flavanols.

Table 2. Flavonoid contents of vegetables (mg/100 g food) [24].

Vegetables	ANT	CHA	FVA	FNE	FVE	FVO	Total
			Cabbages				
Broccoli	-	-	-	-	-	27.8	27.8
			Fruit Vegetables				
Avocado	-	-	0.55	-	-	-	0.55
Black olive	82.97	-	-	-	27.43	49.43	159.83
Green olive	-	-	-	-	0.56	-	0.56
Green sweet pepper	-	-	-	-	2.11	5.49	7.6
Red sweet pepper	-	-	-	-	0.05	0.24	0.29
Tomato	-	-	-	0.14	-	0.014	0.15
			Leaf Vegetables				
Curly	-	-	-	-	-	24.06	24.06
Escarole	-	-	-	-	-	18.23	18.23
Green lettuce	-	-	-	-	0.4	3.99	4.39
Red lettuce	3.53		-	-	2.51	16.74	22.78
Spinach	-	-	-	-	-	119.27	119.27
			Onion-Family Vegetables				
Red onion	9	-	-	-	-	122.51	131.51
White onion	-	-	-	-	-	5.4	5.4
Yellow onion	-	-	-	-	-	59.1	59.1
Shallot	-	-	-	-	-	112.22	112.22
			Pod Vegetables				
Broad bean pod	-	0.08	154.45	-	0.37	34.64	189.54
Green bean	-	-	2.42	-	-	5.55	7.97
			Shoot Vegetables				
Asparagus	-	-	-	-	-	23.19	23.19
Globe artichoke, heads	-	-	-	-	57.8	-	57.8

ANT: Anthocyanins, CHA: Chalcones, FVA: Flavan-3-ols, FNE: Flavanones, FVE: Flavones, FVO: Flavonols.

Regarding seeds (Table 3), although common bean has high levels of flavonoids (from anthocyanins and flavonols), the foods with the highest levels are those derived from soy, and soy products have been suggested to play a key role in the prevention of different diseases [30].

Table 3. Flavonoid contents of seeds (mg/100 g food) [24].

Seeds	ANT	FVA	FNE	FVE	FVO	IFA	Total
			Nuts				
Almond	-	4.93	0.5	-	3.81	0.06	9.3
Cashew nut	-	1.1	-	-	-	-	1.1
Chestnut	-	0.05	-	-	-	-	0.05
Hazelnut	-	5.7	-	-	-	-	5.7
Peanut	-	-	-	-	-	0.51	0.51
Pecan nut	-	16.7	-	-	-	-	16.7
Pistachio		6.9	0.12	0.103	0.07		7.193
			Common Bean				
Black common bean	41.05	-	-	-	10	1.4	52.45
Others common bean	7.42	-	-	-	69.58	0.2	77.2
White common bean	0.13	-	-	-	49.96	0.5	50.59
			Other Beans				
Broad bean seed whole	-	49.37	-	-	-	-	49.37
Sunflower seed meal	-	-	-	-	-	0.02	0.02
			LENTILS				
Lentils	-	5.17		0.95	1.09	-	7.21
			Soy Products				
Soy paste miso	-	-	-	-	-	63.09	63.09
Soy tempeh	-	-	-	-	-	147.74	147.74
Soy tofu	-	-	-	-	-	39.24	39.24
Soybean roasted	-	-	-	-	-	253.11	253.11

ANT: Anthocyanins, FVA: Flavan-3-ols, FNE: Flavanones, FVE: Flavones, FVO: Flavonols, IFA: Isoflavonoids.

Regarding cereals (Table 4), some such as barley, buckwheat, and common wheat contain average levels of flavonoids (35.2 mg/100 g, 37.04 mg/100 g, and 77.4 mg/100 g, respectively). However, it is important to note that the highest levels are found in whole grains, and levels are greatly reduced when grains are heat treated or refined [30,31].

Table 4. Flavonoid contents of cereals (mg/100 g food) [24].

Cereals	FVA	FVE	FVO	Total
	Cereals			
Barley, whole grain flour	35.2	-	-	35.2
Buckwheat groats, thermally treated	-	-	8.96	8.96
Buckwheat, refined flour	-	-	5.86	5.86
Buckwheat, whole grain flour	-	0.9	36.14	37.04
Common wheat, refined flour	-	18.4	0.08	18.48
Common wheat, whole grain flour	-	77.29	0.11	77.4

FVA: Flavan-3-ols, FVE: Flavones, FVO: Flavonols.

Cocoa and its products, such as dark and milk chocolate, are flavonoid-rich foods (Table 5). In these foods, the main flavonoids are flavanols, with cocoa containing 511.63 mg/100 g [32,33].

Table 5. Flavonoid contents of cocoa (mg/100 g food) [24].

Cocoa	FVA	FVO	Total
	Cocoa		
Chocolate dark	212.36	25	237.36
Chocolate milk	19.22	-	19.22
Cocoa powder	511.62	-	511.62

FVA: Flavan-3-ols, FVO: Flavonols.

Regarding oils, the data collected from the Phenol Explorer database refer only to oils made from olives (Table 6). In ascending order, refined, virgin, and extra virgin olive oil contain 0.15 mg, 0.23 mg, and 1.53 mg of flavones in 100 g, respectively [34,35].

Table 6. Flavonoid contents of oils (mg/100 g oil) [24].

Oils	FVE	Total
	Oils	
Extra virgin olive oil	1.53	1.53
Virgin olive oil	0.23	0.23
Refined olive oil	0.15	0.15

FVE: Flavones.

For beverages, a distinction can be made between non-alcoholic (Table 7) and alcoholic drinks (Table 8). The non-alcoholic drinks with the highest levels of flavonoids are tea infusions, particularly black (83.35 mg/100 g) and green tea (77.44 mg/100 g), and these are mainly flavanols [36,37]. The second most flavonoid-rich beverages are fruit juices, notably pure apple juice (54.99 mg/100 g), pure orange juice (48.02 mg/100 g), pure grapefruit juice (47.12 mg/100 g), and pure lemon juice (37.43 mg/100 g) [38]. The main flavonoids in citrus juices and grapefruit juice are flavanones [39]. However, the main flavonoids in pome juices are flavanols. Regarding alcoholic beverages, wine red contains the highest flavonoid level (83.96 mg/100 mL) [40,41].

Table 7. Flavonoid contents of non-alcoholic beverages (mg/100 g drink) [24].

Non-Alcoholic Beverages	ANT	DYC	FVA	FNE	FVE	FVO	IFA	Total
				Cocoa Beverage				
Chocolate, milk	-	-	20.33	-	-	-	-	20.33
				Fruit Juices				
				Berry Juices				
Fox grape juice	-	-	5.9	-	-	-	-	5.9
Green grape juice	-	-	3.88	-	-	-	-	3.88
Grapefruit juice	-	-	-	46.44	-	0.68	-	47.12
				Citrus Juices				
Lemon juice	-	-	-	32.66	4.77	-	-	37.43
Lime juice	-	-	-	19.61	-	-	-	19.61
Orange juice	3.17	-	-	37.63	6.14	1.08	-	48.02
Pummelo juice	-	-	-	8.48	-	-	-	8.48
Red raspberry juice	-	-	-	-	-	9.58	-	9.58
Rowanberry	-	-	-	-	-	7.04	-	7.04

Table 7. *Cont.*

Non-Alcoholic Beverages	ANT	DYC	FVA	FNE	FVE	FVO	IFA	Total
			Drupe Juices					
Plum juice	-	5.85	24.7	-	-	-	-	30.55
			Pome Juices					
Apple juice	-	4.39	48.45	-	-	2.15	-	54.99
Apple (cider) juice	-	4.78	22.66	-	-	-	-	27.44
Pear juice	-	-	3.24	-	-	-	-	3.24
			Tropical Juices					
Kiwi juice	-	-	0.38	-	-	0.09	-	0.47
Pomegranate juice	10.13	0.1	-	-	-	0.25	-	10.48
			Herb Infusions					
German chamomile, tea	-	-	2.07	-	-	-	-	2.07
Lemon verbena	-	-	10.6	-	-	-	-	10.6
Peppermint, tea	-	-	10.23	-	-	-	-	10.23
			Tea Infusion					
Fennel tea	-	-	-	-	-	3.26	-	3.26
Black tea	-	-	73.29	-	-	10.06	-	83.35
Green tea	-	-	71.18	-	-	6.26	-	77.44
Oolong tea	-	-	35.72	-	-	-	-	35.72
			Soy Products					
Soy milk	-	-	-	-	-	-	18	18

ANT: Anthocyanins, DYC: Dihydrochalcones, FVA: Flavanols, FNE: Flavanones, FVE: Flavones, FVO: Flavonols, IFA: Isoflavonoids.

Table 8. Flavonoid contents of alcoholic beverages (mg/100 g drink and mg/100 mL wine) [24].

Alcoholic Beverages	ANT	DYC	DYF	FVA	FNE	FVE	FVO	IFA	Total
				Beer					
Beer (alcohol free)	-	0.0003	-	0.11	0.01	-	-	-	0.12
Beer (ale)	-	0.01	-	0.38	0.24	-	-	0.02	0.65
Beer (dark)	-	0.03	-	0.03	0.15	-	-	-	0.21
Beer (regular)	-	0.001	-	0.61	0.04	0.004	0.09	0.02	0.77
				Wines					
Red wine	23.3	-	5.44	47.02	0.85	-	7.35	-	83.96
Rosé wine	-	-	0.38	2	-	-	-	-	2.38
White wine	0.04	-	0.57	2.07	0.23	-	0.695	-	3.61

ANT: Anthocyanins, DYC: Dihydrochalcones, DYF: Dihydroflavonols, FVA: Flavan-3-ols, FNE: Flavanones, FVE: Flavones, FVO: Flavonols, IFA: Isoflavonoids.

Therefore, a diet rich in fruits, vegetables, seeds, and cereals will provide large amounts of flavonoids. However, it is important to know that there are some foods which contain high quantities of flavonoids, including berries, black olives, spinach, onions, soy products, cocoa, whole grain cereals, tea infusions, and red wine.

5. Pharmacokinetics

In order to determine the biological activity and physiological functions of flavonoids in vivo, their bioavailability must be known. Hence, it is necessary to understand the processes of absorption, digestion, metabolism, and excretion in the digestive tract.

Although dietary flavonoids are mostly found in their glucoside form (Figure 3), they are not found in plasma [42,43] because, once flavonoids enter the oral cavity, they begin to be hydrolysed [42].

In addition, their absorption throughout the digestive tract is associated with the hydrolysing activity of different enzymes [44]. In the small intestine, deglycosylation occurs in which two enzymes that act as β-glucosidases are involved: lactase-phlorizin hydrolase (LPH) and cytosolic β-glucosidase (CBG), which are located in the brush border of epithelial cells and enterocytes, respectively [42,45]. Flavonoid-O-β-D-glucosides, for which LPH has high specificity, can enter into cells by passive diffusion. However, glucosides enter enterocytes via sodium-glucose co-transporter type 1 (SGLT1)) [42,44,46]. Although β-glucosidases cannot hydrolyse non-monoglucosidic glycosides, gut microbiota compensate for this through the production of absorbable aglycon in the large intestine and cecum (Figure 3) [42].

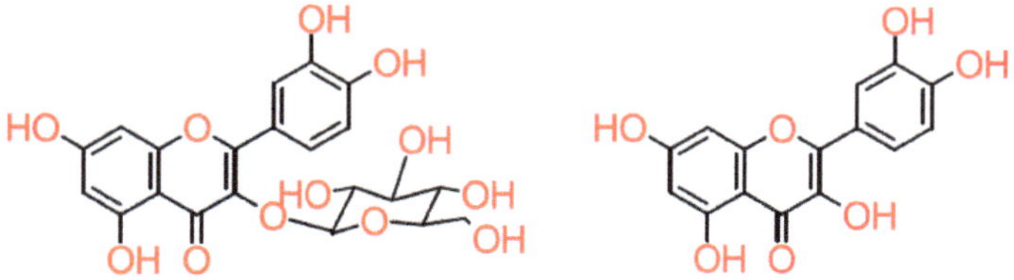

Figure 3. Structure of glycoside and aglycone flavonoids.

Once flavonoids and aglycons are absorbed via the small and the large intestine, respectively, the second phase of enzymatic metabolism begins [42,44]. In this stage, three types of enzymes are involved (uridine-5′-diphosphate-glucuronosyltransferases, sulfotransferases, and catechol-O-methyltransferases) that can conjugate flavonoids with glucuronic acid, sulphate, and methyl groups, making them more water-soluble [13,47]. This phase begins in the wall of the small intestine where metabolites pass to the portal vein and are transported to the liver. In the liver, metabolites are conjugated by sulphation and methylation processes [42]. In the systemic circulation and urine, there are different chemical forms of flavonoids. However, in human plasma, aglycons are rarely detected [42,48–50]. Certain plasmatic metabolites are usually excreted into the intestine through bile, and here, they are deconjugated by microbiota and reabsorbed [42,51]. Thus, enterohepatic circulation increases the half-life of flavonoids in human plasma [40].

The gut microbiome plays a main role in the metabolism and absorption of flavonoids. However, these processes could be modified due to flavonoids interaction with other nutrients [52,53]. Among them, flavonoids could alter glucose absorption after high carbohydrate food intake, because inhibit carbohydrate-hydrolyzing enzymes (α-amylase and α-glucosidase) [54]. Besides, flavonoids inhibit glucose transporter in the brush border [54]. However, flavonoid bioavailability is modified with fats intake that improves flavonoid intestinal absorption due to the increment of bile salts secretion which enhances micellar incorporation of flavonoids [54]. However, regarding proteins intake, flavonoid bioavailability became worse [55]. It has been demonstrated that the interaction of phenolic acids with proteins affects antioxidant efficacy and protein digestibility [56].

Depending on the type of flavonoid and its source, bioavailability may differ. Quercetin is one of the most frequently consumed flavonoids (the main sources of quercetin are onions, apples, tea, and wine), being mainly found in its glycosylated form [13]. For example, quercetin glycosides from apples have lower bioavailability than those from onions [13,57]. The plasma levels of quercetin metabolites range from 0.7 to 7.6 μM [13].

Other studies have analysed the levels of flavonoids in human plasma after the intake of flavonoid-rich foods [13]. They could be grouped according to the flavonoid subclass. Flavonols present in apples, onions, and buckwheat tea are found after intake at plasma levels of 0.30 μM, 0.74–7.60 μM, and 2.10 μM, respectively [13,57]. For flavanols in red wine, black tea, green tea, and cocoa, the plasmatic concentration after intake is around 0.08 μM, 0.09–0.34 μM, 1.00–1.80 μM, and 4.92–5.92 μM, respectively [13,58–60]. The base plasma levels of flavanones in orange juice and grapefruit juice are around 0.06–0.64 μM and 5.99 μM, respectively, after intake [61]. Finally, the plasma levels of anthocyanidins after consuming red wine, elderberry extract, and blackcurrant juice are around 0.01 μM, 0.10 μM, and 0.11 μM, respectively [13,61,62].

The highest concentration of plasma flavonoids in humans usually occurs 1 to 2 h after the consumption of flavonoid-rich foods [36]. However, the level depends on the type of flavonoid, as anthocyanins and catechins have a half-life elimination that is 5 to 10 times less than that of flavonols [55]. Although data on the concentration of flavonoids in human tissue are scarce, flavonoids have been shown to play an important role in antioxidant defence in both cells and tissues [13].

6. Worldwide Flavonoid Intake

The intake of flavonoids depends not only on the food itself and its bioavailability but also on geography, agricultural practices, climate stress, and cultural factors. Diets may differ in different locations [63]. Therefore, based on food frequency questionnaires (FFQs) administered in different studies, we extracted the following distribution of flavonoid consumption around the world (Figure 4).

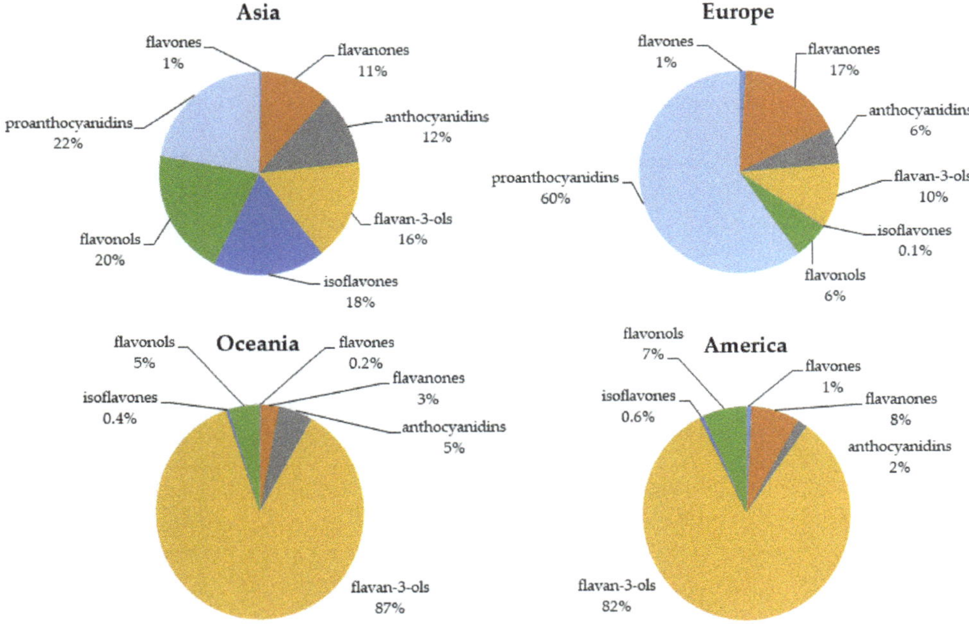

Figure 4. Worldwide intake of flavonoid subclasses.

6.1. Asia

A study performed by Ying Zhang et al. (2010) in China investigated the main sources of flavonoids in adults [64]. The mean intakes of total flavonoids, flavones, and flavonols were 19.13, 4.19, and 13.38 mg/day, respectively. The total intake of flavones and flavonols was attributable to fruits and vegetables. The main sources of flavonoids were *Actinidia* (5%), eggplant (7%), celery (7%), potato (8%), and apple (12%) [64]. Similarly, another study on female adolescents of northern China observed that the mean total flavonoid intake was 20.60 mg/day, with flavone and flavonol intakes of 4.31 and 16.29 mg/day, respectively. From lowest to highest, the food sources of flavonoids were aubergine (3.9%), leeks (3.9%), soybean sprouts (4.2%), celery (4.2%), tomatoes (4.2%), Chinese cabbage (4.7%), oranges (7%), lettuce (7.3%), potatoes (9.9%), and apple (11.7%) [65].

The major sources of dietary flavonoids in Korean adults were identified in a study performed by You Kin Kim et al. (2015) [66]. In this study, they observed that the total daily flavonoid intake was 107 mg/day and that the anthocyanidin, flavan-3-ol, flavanone, flavone, flavonol, and isoflavone intakes were 24.3, 21.8, 8.81, 0.97, 27.8, and 24.3 mg/day, respectively. The main food sources of flavonoids were, in ascending order, tangerine, radish, tofu, onion, soybeans, persimmons, green tea, and Kimchi (traditional fermented vegetable product) [66]. Another study performed by Shinyoung Jun et al. (2015) evaluated the dietary flavonoid intake in Korean adults (33,581 subjects, aged 19 years and older) [67]. The mean total flavonoid intake was 318 mg/day. The intakes of flavonoid subclasses were, in ascending order, flavones (0.4%), flavanones (11.3%), anthocyanidins (11.6%), flavan-3-ols (16.2%), isoflavones (18.1%), flavonols (20.3%), and proanthocyanidins (22.3%) [67].

6.2. Europe

Diets can differ according to region. Hence, Europe can be divided into southern and northern diets, known as the Mediterranean and Non-Mediterranean (UK and Nordic) diets, respectively.

The Mediterranean diet has been closely studied. Therefore, there is a large amount of data on Mediterranean nutrition. Among these studies, the European Prospective Investigation into Cancer and Nutrition Study (EPIC) is one of the most important. This study included 477,312 subjects from different countries in Europe, aged 35 to 70 years. For the Spanish cohort (40,683 subjects) in 2010, the mean total flavonoid intake was 313.26 mg/day [68]. From lowest to highest, the flavonoid subclass intakes were isoflavones (<0.01%), flavones (1.1%), anthocyanidins (5.8%), flavonols (5.9%), flavan-3-ols (10.3%), flavanones (16.9%), and proanthocyanidins (60.1%). The main dietary food sources of flavonoids were tea (2.2%), chocolate (2.6%), peaches (3.3%), pears (4%), beans (4.9%), oranges (9.3%), red wine (21%), and apples (23%) [68]. The main sources of proanthocyanidins were apples, red wines, and beans. Similarly, the most abundant sources of flavan-3-ols were, in ascending order, some fruits (such as plums, grapes, apricots, pears, and peaches), chocolate, apples, tea, and red wine. However, the most abundant sources of flavanones were citrus fruits and their derived products (such as juices) [69]. In the European region, the main food sources of anthocyanidin were fruits (such as pears, apples, and grapes), seeds, and nuts. These were followed by wine, isotonic drinks (in the Northern region), juices (Central region), and vegetables [70].

Another study performed by EPIC (36037 subjects aged 35–74 years) demonstrated that there are differences in the in flavonoid intake between European countries (Norway, Sweden, Denmark, the UK, the Netherlands, Germany, France, Italy, Spain, and Greece) [71]. The daily proanthocyanidin intake was the lowest in Greece and the highest in Spain. In contrast, the lowest intake of flavan-3-ols was observed in Greek women and men (124.8 and 160.5 mg/day, respectively), and the highest total intake was observed in women of the UK General population cohort (377.6 mg/day) and health-conscious men (453.6 mg/day). Likewise, flavan-3-ol monomer intake was the lowest in Greece (20.7 and 26.6 mg/day in women and men, respectively) and the highest in the UK general population (178.6 and 213.5 mg/day in women and men, respectively) [71]. The most important sources of flavan-3-ols in Mediterranean countries, non-Mediterranean countries, and the UK are non-citrus fruit, mainly apples, followed by wine and tea. Tea is responsible for the high flavan-3-ol intake in the UK [71]. For proanthocyanidins, the most important sources in Mediterranean countries are non-citrus fruits, and those in the UK are tea, wine, puddings, and pulses. However, in non-Mediterranean countries, the most important sources are non-citrus fruits, wine, and chocolate [71].

A study performed by Anna Vogiatzoglou et al. (2015) identified the main sources of flavonoids in the European Union [64]. The mean intake of total flavonoids was 428 mg/day, with the lowest intake in the Southern Region (301 mg/day), followed by the Northern Region (348 mg/day), and the highest intake was in the Central Region (506 mg/day), with flavan-3-ols the main flavonoid subclass consumed. Except for flavones and anthocyanidins (which had the highest intakes in the Northern Region), the highest intakes of all other flavonoid subclasses were in the Central Region. Regarding flavonoid sources, in the Southern region, the main sources of flavonoids were fruits and fruit products (mainly pome fruits and berries), but in the Northern and Central regions, tea was the main source of total flavonoids [72]. There were many regional differences, and in the Northern region, the intakes of flavanones and anthocyanidins were the highest, mainly in Finland where the primary sources are citrus fruits and berries, respectively. Nevertheless, in the Southern region, France had the highest intake of anthocyanidins and flavan-3-ols. This study also reported that Germany and Belgium had very low intakes of flavonoid-rich foods [72].

6.3. Oceania

A study that estimated the flavonoid intake in the Australian population (13,858 participants) obtained an average total flavonoid intake of 351 mg/day (of which 75% was flavan-3-ols and 15% was flavanones). In ascending order, the most important flavonoids sources were stalk vegetables, leaf, apples, wine, grapes, oranges, and black tea (which provided 76% of the flavonoid intake) [73]. In the Australian diet, the predominant sources of flavonols and flavon-3-ols were green and black tea as well as pears, apples, and wine for the latter [74]. Other significant sources of flavonols were beans, grapes, apple, broccoli, and onion. Wine was the main source of anthocyanidin. The main sources of flavone and flavanone were spinach and oranges, respectively [74]. However, the most recent Australian population study performed by Murphy KJ et al. (2019) reported an average total flavonoid intake of 660 and 566 mg/day for women and men, respectively [75]. In ascending order, the contributions to total flavonoids intake by subclass were flavones (0.2%), isoflavones (0.4%), flavanones (2.9%), flavonols (4.8%), anthocyanidins (5.3%), and flavan-3-ols (86.5%) [75]. Regarding the dietary sources of flavonoids, tea was responsible for 85% of the total flavonoid intake, followed by fruit juice (2.4%), apple (2.2%), wine (1.7%), berries (1.6%), banana (1.1%), cocoa (0.6%), citrus fruit (0.6%), plum (0.4%), grapes (0.4%), and nuts (0.4%) [75].

6.4. North America

In America, diets differ depending on the region. In North America, processed foods predominate in diets, whereas in South America, fruits and vegetables are the main components of the diet [76]. A study performed by Monica L Bertoia et al. (2016) analysed the dietary flavonoid intake of three prospective cohorts in United States, finding estimated averages of 236 mg/day and 224 mg/day for women and men, respectively [77]. Another prospective study in the United States estimated the intake of flavonoid subclasses, in increasing order, as isoflavones (0.6%), flavones (0.8%), anthocyanidins (1.6%), flavonols (6.8%), flavanones (7.6%), and flavan-3-ols (82.5%) [78]. Thus, the major dietary flavonoid sources were citrus fruits, wine, citrus fruit juices, and tea. In fact, tea was the main source for flavan-3-ols and flavonols [78]. Another study performed by Kim K et al. (2016) estimated the intake and major food sources of flavonoids in adults in the United States [79]. The major dietary sources were apples, wine, citrus fruit, berries, citrus fruit juices, and tea, with tea as the major contributor of flavan-3-ols and flavonols, at 155.9 and 164.4 mg/day, respectively, of the total flavonoids [79].

For Europe and the United States, numerous descriptive studies on flavonoid intake have been published, but for Latin-American countries, there are insufficient data available. However, *Raul* Zamora-Ros et al. (2018) analysed polyphenol dietary intake in the Mexican Teachers' Cohort, reporting an average total flavonoid intake of 235 mg/day. In this population, the main food sources of total polyphenols were orange juice (4.8%), mandarins (5.1%), apples (7.2%), and coffee (47.4%) [80].

There are some differences between the intake of flavonoid subclasses around the world (Table 9). However, it is unclear if these differences are related to differences in cancer incidence. To clarify this issue, for this review, the latest epidemiological studies and GLOBOCAN data (2018) [2] were collected.

Table 9. Flavonoids intake and main food sources worldwide.

Country	Intake (mg/d)	Subclass	Food Sources
Asia	107	Protanthocyanidins > flavonols > isoflavons > flavan-3-ols > anthocyanidins > flavanones > flavones	Kimchi, green tea, persimmons, soybeans, onions
Southern europe	313	Proanthocyanidins > flavanones > flavan-3-ols > flavonols > anthocyanidins > flavones > isoflavones	Apples, red wine, oranges, beans, pears, peaches
Northern europe	348	Flavan-3-ols > flavones > anthocyanidins > flavonols > flavanones > isoflavones	Tea, citrus fruits, berries
Central europe	506	Flavan-3-ols > anthocyanidins > proanthocyanidins > flavanones > flavonols > flavones > isoflavones	Tea, non-citrus fruits, wine
Oceania	351	Flavan-3-ols > anthocyanidins > flavonols > flavanones > isoflavones > flavones	Black tea, oranges, grapes, wine, apples
North America	230	Flavan-3-ols > flavanones > flavonols > anthocyanidins > flavones > isoflavones	Apples, wine, citrus fruit juices and tea

7. Antioxidant Activity of Dietary Flavonoids and Cancer Incidence

All biological processes in an organism must remain in homeostasis. When the pro-oxidant load and antioxidant defence are unbalanced, reactive oxygen species (ROS) are produced, and free radicals are generated [81]. Oxidative stress is characterised by the amount of ROS produced and is closely related to development of some diseases such as cancer caused by oxidative lesions in DNA. However, there are other mechanisms that protect organisms against oxidation, including good nutrition [81]. Thus, the interest in finding compounds with antioxidant activity such as flavonoids has increased. Among them, apigenin (a plant-derived food polyphenol, with sources such as chamomile tea and celery) seems to have strong antioxidant activity in neurological disorders [82]. Myricitrin has been isolated from Daebong persimmon peel, and this flavonoid has strong antioxidant activity through its ferric ion reducing antioxidant ability [83]. Another flavonoid, hesperetin, was shown to ameliorate oxidative stress in disease conditions such as dyslipidaemia and hyperglycaemia in a murine model [84]. In addition, in diabetic rats, galangin reduced hyperglycaemia-mediated oxidative stress and improved the antioxidant status [85]. Under abnormal conditions such as hyperammonemia in rats, quercetin was found to protect against oxidative stress and exert anti-inflammatory activity [86]. During induced oxidative stress in rats, rutin was found to act as a strong antioxidant protecting against oxidative effects [87]. Moreover, in another in vitro study, it was demonstrated that kaempferol has moderate oxygen radical absorption capacity and strong radical-scavenging activity [80]. In murine tissues, quercetin protects against induced oxidative damage [88]. Several studies have investigated the antioxidant activity of flavonoids in humans [89]. Among them, a study performed by Alipour B. et al. (2016) suggested an association between serum total antioxidant capacity and total flavonoid consumption [90]. However, they attributed antioxidant activity to anthocyanins [90]. Thus, there is evidence indicating the strong antioxidant activity of flavonoids in vitro and in vivo, and many epidemiological studies have shown that dietary flavonoids are associated with a lower incidence of cancer. Therefore, because cancer is a major health problem worldwide, it would be of value to determine if its incidence is associated with dietary flavonoid intake and what intake amount would reduce cancer risk.

The latest data collected from GLOBOCAN [2] indicate differences in total cancer incidence around the world. Asia is responsible for 48% of the total cancer incidence (Figure 5).

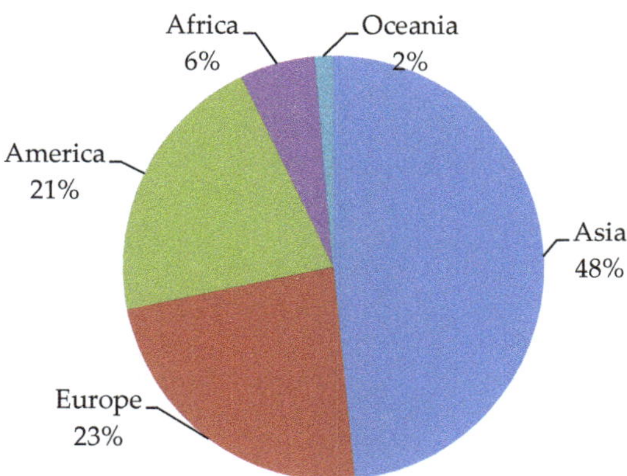

Figure 5. Worldwide cancer incidence. Data collected from GLOBOCAN (2018) [2].

However, incidence can vary according to gender for different types of cancer around the world. According to GLOBOCAN [2] data, the cancer types with the highest incidence in males are, in decreasing order, lung, prostate, stomach, liver, and colorectal cancers (Figure 6). For females, breast cancer accounts for the highest number of cancer cases, followed by lung cancer (Figure 7).

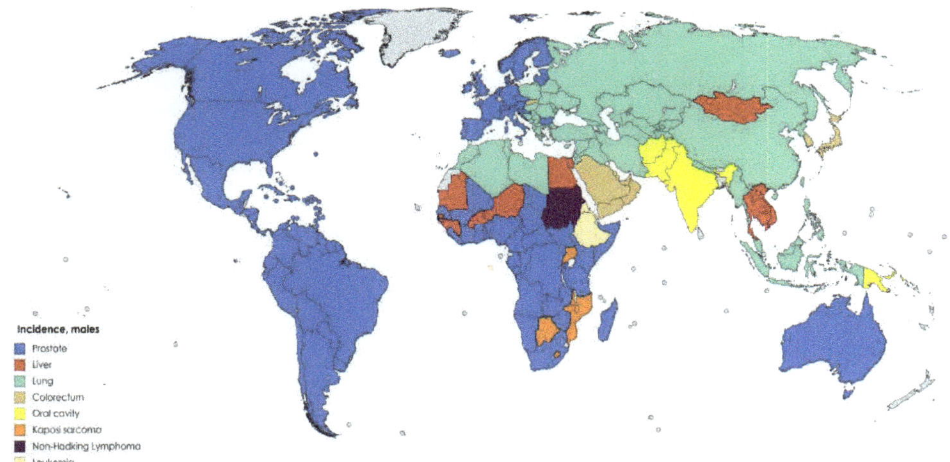

Figure 6. Worldwide cancer incidence by type in males [2]. Created with mapchart.net.

Regarding breast cancer (Table 10), in the European Prospective Investigation into Cancer and Nutrition (EPIC) study, flavonoid dietary intake and breast cancer risk were analysed in a cohort of 334,850 women with an 11.5 year follow up [91]. Within this cohort were 11,576 breast cancer cases. However, there was no statistically significant association between total flavonoid (Hazard Ratio (HR) 0.97, 95% Confidence Interval (CI): 0.90–1.07) and isoflavone (HR 1, 95 %CI: 0.91–1.10) intakes and breast cancer risk [91]. Another prospective study evaluated coffee and tea intake and its relationship with breast cancer risk in black women [92]. The results showed that among the 52.062 participants, there were 1268 incident cases of breast cancer during 12 years of follow up. The data showed that that the intake of coffee (Internal Rate of Return (IRR): 1.03, 95% CI: 0.77–1.39) or tea (IRR: 1.13, 95% CI: 0.78–1.63) was not associated with the risk of breast cancer [92]. Regarding tea and coffee intake, a study performed in Sweden suggested that tea intake is positively associated with oestrogen and progesterone receptor-positive breast cancer, but that coffee consumption is negatively associated with the risk of oestrogen receptor-positive, progesterone receptor-negative breast cancer [93]. Another study performed in Shanghai attempted to associate urinary polyphenols with breast cancer risk [94]. They measured tea flavonols (kaempferol and quercetin) and polyphenols as epicatechin in a cohort with 353 cases and 701 controls. They observed an inverse association between breast cancer risk and urinary excretion of epicatechin (Odds Ratio (OR) 0.59, 95% CI: 0.39–0.88) [94]. Thus, it was concluded that epicatechin-rich foods could reduce breast cancer risk.

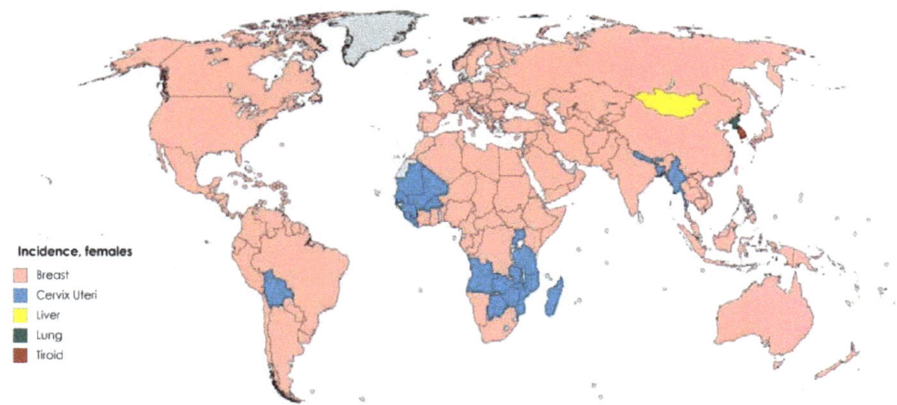

Figure 7. Worldwide cancer incidence by type in females [2]. Created with mapchart.net.

There is controversy regarding the association between breast cancer and isoflavone intake because of its possible role in oestrogen metabolism. Thus, a case-control study in south-western China investigated the relationship between oestrogen metabolism, soy isoflavones, and breast cancer risk [95]. The findings suggested a protective effect of a high soy isoflavone intake on breast cancer risk based on the relation of oestrogen metabolites and breast cancer [95]. Furthermore, regarding other flavonoid subclasses, a meta-analysis of epidemiologic studies performed in 2013 demonstrated that breast cancer risk had a direct association with flavone (Relative Risk (RR): 0.83, 95% CI: 0.76–0.91) and flavonol intake in women (RR: 0.88, 95% IC: 0.80–0.98) [96]. Likewise, in another study performed by Cutler et al. (2008) that analysed cancer risk in postmenopausal women in relation with dietary flavonoid intake, an inverse association was obtained between isoflavone intake and cancer incidence (HR: 0.93, 95% CI: 0.86–1.00), and an inverse association was found between proanthocyanidin (HR: 0.75, 95% CI: 0.57–0.97) and flavanone (HR: 0.68, 95% CI: 0.53–0.86) intake and lung cancer incidence [14].

Regarding total polyphenol intake and breast cancer risk, a study performed by Gardeazabal et al. (2018) that included more than 22,000 Spanish university graduates showed that menopausal status is an important factor in breast cancer risk [97]. Thus, they found no significant association between breast cancer risk and total polyphenol intake. However, in postmenopausal women, they observed an inverse association between breast cancer risk and total polyphenol intake (HR: 0.31, 95% CI: 0.13–0.77) [97]. Because they have antioxidant activity and similar chemical compositions as oestrogens, flavonoids are able to reduce menopause symptoms [98]. However, further research is needed to demonstrate the effect of flavonoid intake on pre- and post-menopause breast cancer risk.

Table 10. Association between flavonoid intake and risk of breast, lung, and prostate cancers.

Authors	Methods	Results
	Breast	
Itziar Gardeazabal et al. (2018) [97]	Prospective Cohort Study 10,713 Spanish Women Food Frequency Questionnaire (FFQ) Phenol-Explorer database HPLC	There was not a statistically significant association between total flavonoids and breast cancer risk. However, in postmenopausal women, the data indicate an inverse association between breast risk cancer and total polyphenol intake (HR: 0.31, 95% CI: 0.13–0.77).
Oh, J.K et al. (2015) [93]	Prospective Cohort Study 42,099 Swedish Women 30–49 years FFQ	Data showed that compared with no consumption, women who consumed >1 cup tea/day had an increased breast cancer risk (Relative Risk (RR): 1.19, 95% Confidence Interval (CI): 1.00–1.42), but women with a high intake of coffee (3–4 cups/day) had a decreased breast cancer risk (RR: 0.87, 95% CI: 0.76–1.00).
Raul Zamora-Ros et al. (2013) [91]	Prospective Cohort Study 334,850 women, 35–70 years 11.5 years follow up FFQ Phenol-Explorer database	There was no statistically significant association between total flavonoid (Hazard Ratio (HR) 0.97, 95% CI: 0.90–1.07) and isoflavone (HR 1, 95%CI: 0.91–1.10) intake and breast cancer risk.
Wang et al. (2011) [95]	Case-Control Study. 400 cases and 400 controls. Daily intake of soy isoflavones data Gene sequencing	They suggested a protective role of high soy isoflavone intake against breast cancer risk based on the relation of oestrogen metabolites, breast cancer, and isoflavone metabolism.
Boggs et al. (2010) [92]	Prospective Cohort Study 52,062 women, 21–69 years 12 years follow up FFQ	Data showed that the intake of coffee (Internal Rate of Return (IRR): 1.03, 95% CI: 0.77–1.39) or tea (IRR: 1.13, 95% CI: 0.78–1.63) was not associated with risk of breast cancer in participants.
Luo JF et al. (2010) [94]	Case-Control Study 353 cases, 701 controls, 40–70 years FFQ Liquid chromatography	There was an inverse association between breast cancer risk and urinary excretion of epicatechin (Odds Ratio (OR) 0.59, 95% CI: 0.39–0.88).
	Lung	
Christensen KY et al. (2012) [99]	Case-Control Study 1061 cases and 1425 controls FFQ	A low intake of flavonoids was related to an increased risk of lung cancer. OR: 0.63, 95% CI: 0.47–0.85.
Cutler et al. (2008) [14]	Prospective Cohort Study USDA database FFQ	There was an inverse association between isoflavone intake and cancer incidence (HR: 0.93, 95% CI: 0.86–1.00) and an inverse association between proanthocyanidin (HR: 0.75, 95% CI: 0.57–0.97) and flavanone (HR: 0.68, 95% CI: 0.53–0.86) intake with lung cancer incidence.
	Prostate	
Giulio Reale et al. (2018) [100]	Case-Control study 118 cases and 222 controls FFQ Prostate Specific Antigen	High intake of some subclasses of flavonoids (catechin (OR: 0.12, 95% CI: 0.04–0.36) and flavonol (OR: 0.19, 95% CI: 0.06–0.56) significantly reduces the risk of prostate cancer.

A case-control study performed by Christensen et al. (2012) (Table 10), which analysed the association of lung cancer risk with flavonoid intake, did not find an association between flavonoid intake and risk reduction. However, a low intake of total and different subclasses of flavonoids was related to an increased risk of lung cancer. The ORs (95%CI) were 0.63 (0.47–0.85) for total flavonoids, 0.70 (0.53–0.94) for flavanones, 0.62 (0.45–0.84) for flavonols, 0.68 (0.50–0.93) for flavones, 0.67 (0.50–0.90) for flavan-3-ols, and 0.82 (0.61–1.11) for anthocyanidins [99].

A population-based case-control study carried out on a population of Sicilian men analysed the association between dietary factors, such as flavonoids, and prostate cancer risk (Table 10). The results suggested that prostate cancer risk could be reduced by a high intake of catechins (OR: 0.12, 95% CI: 0.04–0.36) and flavonol (OR: 0.19, 95% CI: 0.06–0.56). However, the risk seemed to increase with a high intake of flavanones [100].

Gastric cancer is the second main cause of cancer deaths and the fourth most common cancer worldwide (Table 11) [68]. In a case-control study performed in Korea, a significant association was found between total flavonoid intake and gastric cancer risk reduction in women (OR 0.33, 95% CI 0.15–0.73) but not in men [101]. Furthermore, the EPIC study investigated the association between gastric adenocarcinoma risk and flavonoid intake [68]. They observed an inverse association between gastric adenocarcinoma risk and total flavonoid intake in women (HR 0.81, 95% CI 0.70, 0.94). This association was observed for some flavonoid subclasses such as flavanols, flavones, flavonols, and anthocyanidins [68]. However, in a prospective study carried out in the United States that analysed all cancers, researchers observed that flavonoid intake was associated with protection against neck and head cancer risk but not gastric cancer risk [102].

Pancreatic cancer has the worst prognosis of all cancers (Table 11), and its mortality/incidence ratio is 0.98 [96]. However, a study performed on the EPIC cohort examined the association between pancreatic cancer risk and flavonoid intake and found an inverse association between them, although it was not statistically significant [103].

Regarding colorectal cancer (Table 11), it has been demonstrated that flavonoids are able to inhibit the growth of colon cancer cells in vitro [104]. However, in human-based studies, the results are different. In a prospective study that examined daily flavonoid intake and its relationship with colorectal cancer, the data showed there was no association between the risk of colorectal cancer and flavonoid intake [104]. Important results were obtained in a study performed by Xu M. et al. (2016). Their data showed that there was an inverse association between flavonoid intake as anthocyanidins, flavanones, and flavones and colorectal cancer risk. However, this only occurred when the sources of flavonoids were fruits and vegetables [105]. A case-control study performed on a Spanish population found an inverse association between the risk of colorectal cancer and intake of total flavonoids (OR: 0.59, 95% CI, 0.35–0.99) and some flavonoid subclasses (such as proanthocyanidins and flavones) [106]. The same researchers performed a case-control study to analyse the relationship between flavonoid intake and colorectal cancer recurrence and survival. However, their results did not support the beneficial effects of flavonoids on colorectal cancer prognosis [107].

Table 11. Association between flavonoid intake and risk of gastric, pancreatic, and colorectal cancers.

Authors, Year	Methods	Results
	Gastric Adenocarcinoma	
Sun L et al. (2017) [102]	Prospective cohort study 469,008 participants 12 years follow up FFQ	Data suggested that total flavonoid intake was associated with a reduced risk of neck and head cancer (HR: 0.76, 95% CI: 0.66–0.86).
Hae Donw Woo et al. (2014) [101]	Case-Control study. 334 cases and 334 controls, aged 35–75 years, from Korea. FFQ USDA database	Total flavonoids and their subclasses were significantly associated with a reduced risk of gastric cancer in women (OR 0.33, 95% CI 0.15–0.73) but not in men.
Zamora-Ros et al. (2012) [68]	Observational study 477,312 subjects, aged 35–70 years, from 10 European countries. Average Follow-up of 11 years FFQ USDA and Phenol Explorer Databases.	Total dietary intake was associated with a significant reduction in the risk of gastric adenocarcinoma in women (HR 0.81, 95% CI 0.70, 0.94).
	Pancreatic Cancer	
Molina-Montes E. et al. (2016) [103]	Prospective cohort 477,309 participants Average Follow-up of 11 years FFQ USDA and Phenol Explorer Databases.	There was no association with pancreatic cancer risk and dietary flavonoid intake (HR: 1.09, 95% CI: 0.95–1.11); however, there was an inverse association, but not statistically significant, between prostate cancer risk and flavanone intake.
	Colorectal Cancer	
Nimptsch K et al. (2016) [104]	Prospective cohort 42,478 male 76,364 female 26 years follow up FFQ	Data did not show an association between colorectal cancer risk and flavonoid subclass intake. RR (95% IC) values were 0.98 (0.81, 1.19) for anthocyanins, 1.07 (0.95, 1.21) for flavan-3-ols, 0.96 (0.84, 1.10) for flavanones, 1.01 (0.89, 1.15) for flavones, and 1.04 (0.91, 1.18) for flavonols.
Xu M. et al. (2016) [105]	Case-Control study 1632 cases and 1632 controls FFQ	There was an inverse association between colorectal cancer risk and flavone (OR: 0.54, 95% CI 0.43, 0·67), flavanone (OR: 0.28, 95% CI: 0.22–0.36), and anthocyanidin (OR: 0.80, 95% CI: 0.64–1.00) intake.
Zamora Ros et al. (2015) [107]	Case-control study 523 participants FFQ	Flavonoid intake was not associated with colorectal cancer survival or recurrence.
Zamora Ros et al. (2013) [108]	Case-Control study 424 cases and 401 hospital-based controls FFQ Phenol Explorer Database	Data showed an inverse correlation between risk of colorectal cancer and intake of total flavonoids (OR: 0.59, 95% CI, 0.35–0.99) and some flavonoid subclasses (such as proanthocyanidins and flavones).

8. Conclusions

In summary, there remains controversy regarding the possible protective effect of flavonoids on cancer in epidemiological studies. However, this association could vary depending on many factors such as geographical location and diet. It appears that some flavonoid subclasses suggest a decrease of the risk of different types of cancer, such as catechin and flavonols for prostate cancer, epicatechin for breast cancer, proanthocyanidins for lung cancer, flavones for colorectal cancer, and total flavonoids for gastric cancer. Thus, because the main sources of these flavonoids are different, the risk of cancer could be reduced by including them in a healthy diet, which would be mainly based on vegetables and fruits, whole grain cereals, legumes, seeds, and nuts, as well as cocoa, coffee, fruit juices, and tea. However, further studies are needed to investigate and confirm this hypothesis that a healthy diet can help decrease the incidence of different types of cancer.

Author Contributions: All authors participated actively in the design and conception of this review. All authors assessed the present form of the review and have approved it for publication.

Funding: Carmen Rodríguez-García receives only a pre-doctoral research grant from the University of Jaén (Ayudas predoctorales para la formación del personal investigador. Acción 4_El_CTS_1_2017).

Conflicts of Interest: The authors declare no conflict of interest.

References

1. Imran, M.; Rauf, A.; Abu-Izneid, T.; Nadeem, M.; Shariati, M.A.; Khan, I.A.; Imran, A.; Orhan, I.E.; Rizwan, M.; Atif, M.; et al. Luteolin, a flavonoid, as an anticancer agent: A review. *Biomed. Pharmacother.* **2019**, *112*, 108612. [CrossRef]
2. Bray, F.; Ferlay, J.; Soerjomataram, I.; Siegel, R.L.; Torre, L.A.; Jemal, A. Global cancer statistics 2018: GLOBOCAN estimates of incidence and mortality worldwide for 36 cancers in 185 countries. *CA Cancer J. Clin.* **2018**, *68*, 394–424. [CrossRef] [PubMed]
3. Theodoratou, E.; Timofeeva, M.; Li, X.; Meng, X.; Ioannidis, J.P.A. Nature, nurture, and cancer risks: Genetic and nutritional contributions to cancer. *Annu. Rev. Nutr.* **2017**, *37*, 293–320. [CrossRef] [PubMed]
4. Afshin, A.; Sur, P.J.; Fay, K.A.; Cornaby, L.; Ferrara, G.; Salama, J.S.; Mullany, E.C.; Abate, K.H.; Abbafati, C.; Abebe, Z.; et al. Health effects of dietary risks in 195 countries, 1990–2017: A systematic analysis for the global burden of disease study 2017. *Lancet* **2019**, *393*, 1958–1972. [CrossRef]
5. Darband, S.G.; Kaviani, M.; Yousefi, B.; Sadighparvar, S.; Pakdel, F.G.; Attari, J.A.; Mohebbi, I.; Naderi, S.; Majidinia, M. Quercetin: A functional dietary flavonoid with potential chemo-preventive properties in colorectal cancer. *J. Cell Physiol.* **2018**, *233*, 6544–6560. [CrossRef] [PubMed]
6. Giovannucci, E. Nutritional epidemiology and cancer: A tale of two cities. *Cancer Causes Control* **2018**, *29*, 1007–1014. [CrossRef] [PubMed]
7. Xu, D.P.; Li, Y.; Meng, X.; Zhou, T.; Zhou, Y.; Zheng, J.; Zhang, J.J.; Li, H.B. Natural antioxidants in foods and medicinal plants: extraction, assessment and resources. *Int. J. Mol. Sci.* **2017**, *18*, 96. [CrossRef] [PubMed]
8. Shashirekha, M.N.; Mallikarjuna, S.E.; Rajarathnam, S. Status of bioactive compounds in foods, with focus on fruits and vegetables. *Crit. Rev. Food Sci. Nutr.* **2015**, *55*, 1324–1339. [CrossRef] [PubMed]
9. Nabavi, S.M.; Samec, D.; Tomczyk, M.; Milella, L.; Russo, D.; Habtemariam, S.; Suntar, I.; Rastrelli, L.; Daglia, M.; Xiao, J.; et al. Flavonoid biosynthetic pathways in plants: Versatile targets for metabolic engineering. *Biotechnol. Adv.* **2018**. [CrossRef] [PubMed]
10. Liu, J.; Wang, X.; Yong, H.; Kan, J.; Jin, C. Recent advances in flavonoid-grafted polysaccharides: Synthesis, structural characterization, bioactivities and potential applications. *Int. J. Biol. Macromol.* **2018**, *116*, 1011–1025. [CrossRef] [PubMed]
11. Chen, H.Y.; Lin, P.H.; Shih, Y.H.; Wang, K.L.; Hong, Y.H.; Shieh, T.M.; Huang, T.C.; Hsia, S.M. Natural antioxidant resveratrol suppresses uterine fibroid cell growth and extracellular matrix formation in vitro and in vivo. *Antioxidants* **2019**, *8*, 99. [CrossRef]
12. Sudhakaran, M.; Sardesai, S.; Doseff, A.I. Flavonoids: new frontier for immuno-regulation and breast cancer control. *Antioxidants* **2019**, *8*, 103. [CrossRef]

13. Lotito, S.B.; Frei, B. Consumption of flavonoid-rich foods and increased plasma antioxidant capacity in humans: Cause, consequence, or epiphenomenon? *Free Radic. Biol. Med.* **2006**, *41*, 1727–1746. [CrossRef]
14. Cutler, G.J.; Nettleton, J.A.; Ross, J.A.; Harnack, L.J.; Jacobs, D.R., Jr.; Scrafford, C.G.; Barraj, L.M.; Mink, P.J.; Robien, K. Dietary flavonoid intake and risk of cancer in postmenopausal women: The Iowa Women's Health Study. *Int. J. Cancer* **2008**, *123*, 664–671. [CrossRef]
15. PubMed. Available online: https://www.ncbi.nlm.nih.gov/pubmed/ (accessed on 9 May 2019).
16. Web of Science Database. Available online: https://apps.webofknowledge.com/WOS_GeneralSearch_input.do?product=WOS&search_mode=GeneralSearch&SID=D6uwPoN96kn8K5NwIjw&preferencesSaved= (accessed on 9 May 2019).
17. Scarano, A.; Chieppa, M.; Santino, A. Looking at flavonoid biodiversity in horticultural crops: A colored mine with nutritional benefits. *Plants* **2018**, *7*, 98. [CrossRef]
18. Maatta-Riihinen, K.R.; Kamal-Eldin, A.; Torronen, A.R. Identification and quantification of phenolic compounds in berries of *Fragaria* and *Rubus* species (family Rosaceae). *J. Agric. Food Chem.* **2004**, *52*, 6178–6187. [CrossRef] [PubMed]
19. Wu, X.; Gu, L.; Prior, R.L.; McKay, S. Characterization of anthocyanins and proanthocyanidins in some cultivars of Ribes, Aronia, and Sambucus and their antioxidant capacity. *J. Agric. Food Chem.* **2004**, *52*, 7846–7856. [CrossRef]
20. de Pascual-Teresa, S.; Santos-Buelga, C.; Rivas-Gonzalo, J.C. Quantitative analysis of flavan-3-ols in spanish foodstuffs and beverages. *J. Agric. Food Chem.* **2000**, *48*, 5331–5337. [CrossRef] [PubMed]
21. Arts, I.C.; van de Putte, B.; Hollman, P.C. Catechin contents of foods commonly consumed in The Netherlands. 1. Fruits, vegetables, staple foods, and processed foods. *J. Agric. Food Chem.* **2000**, *48*, 1746–1751. [CrossRef] [PubMed]
22. Vrhovsek, U.; Rigo, A.; Tonon, D.; Mattivi, F. Quantitation of polyphenols in different apple varieties. *J. Agric. Food Chem.* **2004**, *52*, 6532–6538. [CrossRef]
23. Landberg, R.; Naidoo, N.; van Dam, R.M. Diet and endothelial function: From individual components to dietary patterns. *Curr. Opin. Lipidol.* **2012**, *23*, 147–155. [CrossRef]
24. Medina-Remón, A.; Manach, C.; Knox, C.; Wishart, D.S.; Perez-Jimenez, J.; Rothwell, J.A.; M'Hiri, N.; García-Lobato, P.; Eisner, R.; Neveu, V.; et al. Phenol-Explorer 3.0: A major update of the Phenol-Explorer database to incorporate data on the effects of food processing on polyphenol content. *Database* **2013**, *2013*. [CrossRef]
25. Tomas-Barberan, F.A.; Garcia-Grau, M.M.; Tomas-Lorente, F. Flavonoid concentration changes in maturing broad bean pods. *J. Agric. Food Chem.* **1991**, *39*, 255–258. [CrossRef]
26. Romani, A.; Mulinacci, N.; Pinelli, P.; Vincieri, F.F.; Cimato, A. Polyphenolic content in five tuscany cultivars of Olea europaea L. *J. Agric. Food Chem.* **1999**, *47*, 964–967. [CrossRef]
27. Rodriguez Galdon, B.; Pena-Mendez, E.; Havel, J.; Rodriguez, E.M.; Romero, C.D. Cluster analysis and artificial neural networks multivariate classification of onion varieties. *J. Agric. Food Chem.* **2010**, *58*, 11435–11440. [CrossRef]
28. Howard, L.R.; Pandjaitan, N.; Morelock, T.; Gil, M.I. Antioxidant capacity and phenolic content of spinach as affected by genetics and growing season. *J. Agric. Food Chem.* **2002**, *50*, 5891–5896. [CrossRef]
29. Fattorusso, E.; Iorizzi, M.; Lanzotti, V.; Taglialatela-Scafati, O. Chemical composition of shallot (*Allium ascalonicum* Hort.). *J. Agric. Food Chem.* **2002**, *50*, 5686–5690. [CrossRef]
30. Rienks, J.; Barbaresko, J.; Nothlings, U. Association of isoflavone biomarkers with risk of chronic disease and mortality: A systematic review and meta-analysis of observational studies. *Nutr. Rev.* **2017**, *75*, 616–641. [CrossRef]
31. Steadman, K.J.; Burgoon, M.S.; Lewis, B.A.; Edwardson, S.E.; Obendorf, R.L. Minerals, phytic acid, tannin and rutin in buckwheat seed milling fractions. *J. Sci. Food Agric.* **2001**, *81*, 1094–1100. [CrossRef]
32. Tomas-Barberan, F.A.; Cienfuegos-Jovellanos, E.; Marin, A.; Muguerza, B.; Gil-Izquierdo, A.; Cerda, B.; Zafrilla, P.; Morillas, J.; Mulero, J.; Ibarra, A.; et al. A new process to develop a cocoa powder with higher flavonoid monomer content and enhanced bioavailability in healthy humans. *J. Agric. Food Chem.* **2007**, *55*, 3926–3935. [CrossRef]
33. Miller, K.B.; Stuart, D.A.; Smith, N.L.; Lee, C.Y.; McHale, N.L.; Flanagan, J.A.; Ou, B.; Hurst, W.J. Antioxidant activity and polyphenol and procyanidin contents of selected commercially available cocoa-containing and chocolate products in the United States. *J. Agric. Food Chem.* **2006**, *54*, 4062–4068. [CrossRef]

34. Medina, E.; de Castro, A.; Romero, C.; Brenes, M. Comparison of the concentrations of phenolic compounds in olive oils and other plant oils: Correlation with antimicrobial activity. *J. Agric. Food Chem.* **2006**, *54*, 4954–4961. [CrossRef]
35. Carrasco-Pancorbo, A.; Gomez-Caravaca, A.M.; Cerretani, L.; Bendini, A.; Segura-Carretero, A.; Fernandez-Gutierrez, A. Rapid quantification of the phenolic fraction of Spanish virgin olive oils by capillary electrophoresis with UV detection. *J. Agric. Food Chem.* **2006**, *54*, 7984–7991. [CrossRef]
36. Uysal, U.D.; Aturki, Z.; Raggi, M.A.; Fanali, S. Separation of catechins and methylxanthines in tea samples by capillary electrochromatography. *J. Sep. Sci.* **2009**, *32*, 1002–1010. [CrossRef]
37. Khokhar, S.; Magnusdottir, S.G. Total phenol, catechin, and caffeine contents of teas commonly consumed in the United Kingdom. *J. Agric. Food Chem.* **2002**, *50*, 565–570. [CrossRef]
38. Guo, J.; Yue, T.; Yuan, Y.; Wang, Y. Chemometric classification of apple juices according to variety and geographical origin based on polyphenolic profiles. *J. Agric. Food Chem.* **2013**, *61*, 6949–6963. [CrossRef]
39. Belajová, E.; Suhaj, M. Determination of phenolic constituents in citrus juices: Method of high performance liquid chromatography. *Food Chem.* **2004**, *86*, 339–343. [CrossRef]
40. Jandera, P.; Skeifikova, V.; Rehova, L.; Hajek, T.; Baldrianova, L.; Skopova, G.; Kellner, V.; Horna, A. RP-HPLC analysis of phenolic compounds and flavonoids in beverages and plant extracts using a CoulArray detector. *J. Sep. Sci.* **2005**, *28*, 1005–1022. [CrossRef]
41. La Torre, G.L.; Saitta, M.; Vilasi, F.; Pellicanò, T.; Dugo, G. Direct determination of phenolic compounds in Sicilian wines by liquid chromatography with PDA and MS detection. *Food Chem.* **2006**, *94*, 640–650. [CrossRef]
42. Murota, K.; Nakamura, Y.; Uehara, M. Flavonoid metabolism: The interaction of metabolites and gut microbiota. *Biosci. Biotechnol. Biochem.* **2018**, *82*, 600–610. [CrossRef]
43. Sesink, A.L.; O'Leary, K.A.; Hollman, P.C. Quercetin glucuronides but not glucosides are present in human plasma after consumption of quercetin-3-glucoside or quercetin-4'-glucoside. *J. Nutr.* **2001**, *131*, 1938–1941. [CrossRef]
44. Rodriguez-Mateos, A.; Vauzour, D.; Krueger, C.G.; Shanmuganayagam, D.; Reed, J.; Calani, L.; Mena, P.; del Rio, D.; Crozier, A. Bioavailability, bioactivity and impact on health of dietary flavonoids and related compounds: An update. *Arch. Toxicol.* **2014**, *88*, 1803–1853. [CrossRef]
45. Day, A.J.; Gee, J.M.; DuPont, M.S.; Johnson, I.T.; Williamson, G. Absorption of quercetin-3-glucoside and quercetin-4'-glucoside in the rat small intestine: The role of lactase phlorizin hydrolase and the sodium-dependent glucose transporter. *Biochem. Pharmacol.* **2003**, *65*, 1199–1206. [CrossRef]
46. Walgren, R.A.; Lin, J.T.; Kinne, R.K.; Walle, T. Cellular uptake of dietary flavonoid quercetin 4'-beta-glucoside by sodium-dependent glucose transporter SGLT1. *J. Pharmacol. Exp. Ther.* **2000**, *294*, 837–843.
47. Erlund, I.; Kosonen, T.; Alfthan, G.; Maenpaa, J.; Perttunen, K.; Kenraali, J.; Parantainen, J.; Aro, A. Pharmacokinetics of quercetin from quercetin aglycone and rutin in healthy volunteers. *Eur. J. Clin. Pharmacol.* **2000**, *56*, 545–553. [CrossRef]
48. Mullen, W.; Edwards, C.A.; Crozier, A. Absorption, excretion and metabolite profiling of methyl-, glucuronyl-, glucosyl- and sulpho-conjugates of quercetin in human plasma and urine after ingestion of onions. *Br. J. Nutr.* **2006**, *96*, 107–116. [CrossRef]
49. Clarke, D.B.; Lloyd, A.S.; Botting, N.P.; Oldfield, M.F.; Needs, P.W.; Wiseman, H. Measurement of intact sulfate and glucuronide phytoestrogen conjugates in human urine using isotope dilution liquid chromatography-tandem mass spectrometry with [13C(3)]isoflavone internal standards. *Anal. Biochem.* **2002**, *309*, 158–172. [CrossRef]
50. Nakamura, T.; Murota, K.; Kumamoto, S.; Misumi, K.; Bando, N.; Ikushiro, S.; Takahashi, N.; Sekido, K.; Kato, Y.; Terao, J. Plasma metabolites of dietary flavonoids after combination meal consumption with onion and tofu in humans. *Mol. Nutr. Food Res.* **2014**, *58*, 310–317. [CrossRef]
51. Arts, I.C.; Sesink, A.L.; Faassen-Peters, M.; Hollman, P.C. The type of sugar moiety is a major determinant of the small intestinal uptake and subsequent biliary excretion of dietary quercetin glycosides. *Br. J. Nutr.* **2004**, *91*, 841–847. [CrossRef]
52. Williamson, G.; Kay, C.D.; Crozier, A. The bioavailability, transport, and bioactivity of dietary flavonoids: A review from a historical perspective. *Compr. Rev. Food Sci. Food Saf.* **2018**, *17*, 1054–1112. [CrossRef]
53. Jakobek, L. Interactions of polyphenols with carbohydrates, lipids and proteins. *Food Chem.* **2015**, *175*, 556–567. [CrossRef]

54. Gonzales, G.B.; Smagghe, G.; Grootaert, C.; Zotti, M.; Raes, K.; van Camp, J. Flavonoid interactions during digestion, absorption, distribution and metabolism: A sequential structure-activity/property relationship-based approach in the study of bioavailability and bioactivity. *Drug Metab. Rev.* **2015**, *47*, 175–190. [CrossRef]
55. Hollman, P.C.H. Absorption, bioavailability, and metabolism of flavonoids. *Pharm. Biol.* **2004**, *42*, 74–83. [CrossRef]
56. Swieca, M.; Gawlik-Dziki, U.; Dziki, D.; Baraniak, B.; Czyz, J. The influence of protein-flavonoid interactions on protein digestibility in vitro and the antioxidant quality of breads enriched with onion skin. *Food Chem.* **2013**, *141*, 451–458. [CrossRef]
57. Graefe, E.U.; Wittig, J.; Mueller, S.; Riethling, A.K.; Uehleke, B.; Drewelow, B.; Pforte, H.; Jacobasch, G.; Derendorf, H.; Veit, M. Pharmacokinetics and bioavailability of quercetin glycosides in humans. *J. Clin. Pharmacol.* **2001**, *41*, 492–499. [CrossRef]
58. Donovan, J.L.; Bell, J.R.; Kasim-Karakas, S.; German, J.B.; Walzem, R.L.; Hansen, R.J.; Waterhouse, A.L. Catechin is present as metabolites in human plasma after consumption of red wine. *J. Nutr.* **1999**, *129*, 1662–1668. [CrossRef]
59. Van het Hof, K.H.; Wiseman, S.A.; Yang, C.S.; Tijburg, L.B. Plasma and lipoprotein levels of tea catechins following repeated tea consumption. *Proc. Soc. Exp. Biol. Med.* **1999**, *220*, 203–209.
60. Holt, R.R.; Lazarus, S.A.; Sullards, M.C.; Zhu, Q.Y.; Schramm, D.D.; Hammerstone, J.F.; Fraga, C.G.; Schmitz, H.H.; Keen, C.L. Procyanidin dimer B2 [epicatechin-(4beta-8)-epicatechin] in human plasma after the consumption of a flavanol-rich cocoa. *Am. J. Clin. Nutr.* **2002**, *76*, 798–804. [CrossRef]
61. Erlund, I.; Meririnne, E.; Alfthan, G.; Aro, A. Plasma kinetics and urinary excretion of the flavanones naringenin and hesperetin in humans after ingestion of orange juice and grapefruit juice. *J. Nutr.* **2001**, *131*, 235–241. [CrossRef]
62. Manach, C.; Morand, C.; Gil-Izquierdo, A.; Bouteloup-Demange, C.; Remesy, C. Bioavailability in humans of the flavanones hesperidin and narirutin after the ingestion of two doses of orange juice. *Eur. J. Clin. Nutr.* **2003**, *57*, 235–242. [CrossRef]
63. Nielsen, I.L.; Dragsted, L.O.; Ravn-Haren, G.; Freese, R.; Rasmussen, S.E. Absorption and excretion of black currant anthocyanins in humans and watanabe heritable hyperlipidemic rabbits. *J. Agric. Food Chem.* **2003**, *51*, 2813–2820. [CrossRef]
64. Haytowitz, D.B.; Bhagwat, S.; Holden, J.M. Sources of variability in the flavonoid content of foods. *Procedia Food Sci.* **2013**, *2*, 46–51. [CrossRef]
65. Zhang, Y.; Li, Y.; Cao, C.; Cao, J.; Chen, W.; Zhang, Y.; Wang, C.; Wang, J.; Zhang, X.; Zhao, X. Dietary flavonol and flavone intakes and their major food sources in Chinese adults. *Nutr. Cancer* **2010**, *62*, 1120–1127. [CrossRef] [PubMed]
66. Sun, C.; Wang, H.; Wang, D.; Chen, Y.; Zhao, Y.; Xia, W. Using an FFQ to assess intakes of dietary flavonols and flavones among female adolescents in the Suihua area of northern China. *Public Health Nutr.* **2015**, *18*, 632–639. [CrossRef]
67. Kim, Y.J.; Park, M.Y.; Chang, N.; Kwon, O. Intake and major sources of dietary flavonoid in Korean adults: Korean National Health and Nutrition Examination Survey 2010–2012. *Asia Pac. J. Clin. Nutr.* **2015**, *24*, 456–463. [CrossRef] [PubMed]
68. Jun, S.; Shin, S.; Joung, H. Estimation of dietary flavonoid intake and major food sources of Korean adults. *Br. J. Nutr.* **2016**, *115*, 480–489. [CrossRef]
69. Zamora-Ros, R.; Agudo, A.; Lujan-Barroso, L.; Romieu, I.; Ferrari, P.; Knaze, V.; Bueno-de-Mesquita, H.B.; Leenders, M.; Travis, R.C.; Navarro, C.; et al. Dietary flavonoid and lignan intake and gastric adenocarcinoma risk in the European Prospective Investigation into Cancer and Nutrition (EPIC) study. *Am. J. Clin. Nutr.* **2012**, *96*, 1398–1408. [CrossRef]
70. Zamora-Ros, R.; Andres-Lacueva, C.; Lamuela-Raventos, R.M.; Berenguer, T.; Jakszyn, P.; Barricarte, A.; Ardanaz, E.; Amiano, P.; Dorronsoro, M.; Larranaga, N.; et al. Estimation of dietary sources and flavonoid intake in a Spanish adult population (EPIC-Spain). *J. Am. Diet. Assoc.* **2010**, *110*, 390–398. [CrossRef]
71. Zamora-Ros, R.; Knaze, V.; Lujan-Barroso, L.; Slimani, N.; Romieu, I.; Touillaud, M.; Kaaks, R.; Teucher, B.; Mattiello, A.; Grioni, S.; et al. Estimation of the intake of anthocyanidins and their food sources in the European Prospective Investigation into Cancer and Nutrition (EPIC) study. *Br. J. Nutr.* **2011**, *106*, 1090–1099. [CrossRef]

72. Knaze, V.; Zamora-Ros, R.; Lujan-Barroso, L.; Romieu, I.; Scalbert, A.; Slimani, N.; Riboli, E.; van Rossum, C.T.; Bueno-de-Mesquita, H.B.; Trichopoulou, A.; et al. Intake estimation of total and individual flavan-3-ols, proanthocyanidins and theaflavins, their food sources and determinants in the European Prospective Investigation into Cancer and Nutrition (EPIC) study. *Br. J. Nutr.* **2012**, *108*, 1095–1108. [CrossRef]
73. Vogiatzoglou, A.; Mulligan, A.A.; Lentjes, M.A.; Luben, R.N.; Spencer, J.P.; Schroeter, H.; Khaw, K.T.; Kuhnle, G.G. Flavonoid intake in European adults (18 to 64 years). *PLoS ONE* **2015**, *10*, e0128132. [CrossRef] [PubMed]
74. Johannot, L.; Somerset, S.M. Age-related variations in flavonoid intake and sources in the Australian population. *Public Health Nutr.* **2006**, *9*, 1045–1054. [CrossRef] [PubMed]
75. Somerset, S.M.; Johannot, L. Dietary flavonoid sources in Australian adults. *Nutr. Cancer* **2008**, *60*, 442–449. [CrossRef] [PubMed]
76. Murphy, K.J.; Walker, K.M.; Dyer, K.A.; Bryan, J. Estimation of daily intake of flavonoids and major food sources in middle-aged Australian men and women. *Nutr. Res.* **2019**, *61*, 64–81. [CrossRef] [PubMed]
77. Grotto, D.; Zied, E. The Standard American Diet and its relationship to the health status of Americans. *Nutr. Clin. Pract.* **2010**, *25*, 603–612. [CrossRef]
78. Bertoia, M.L.; Rimm, E.B.; Mukamal, K.J.; Hu, F.B.; Willett, W.C.; Cassidy, A. Dietary flavonoid intake and weight maintenance: Three prospective cohorts of 124,086 US men and women followed for up to 24 years. *BMJ* **2016**, *352*, i17. [CrossRef]
79. Chun, O.K.; Song, W.O.; Chung, S.J. Estimated dietary flavonoid intake and major food sources of U.S. adults. *J. Nutr.* **2007**, *137*, 1244–1252. [CrossRef]
80. Kim, K.; Vance, T.M.; Chun, O.K. Estimated intake and major food sources of flavonoids among US adults: Changes between 1999–2002 and 2007–2010 in NHANES. *Eur. J. Nutr.* **2016**, *55*, 833–843. [CrossRef]
81. Zamora-Ros, R.; Biessy, C.; Rothwell, J.A.; Monge, A.; Lajous, M.; Scalbert, A.; Lopez-Ridaura, R.; Romieu, I. Dietary polyphenol intake and their major food sources in the Mexican Teachers' Cohort. *Br. J. Nutr.* **2018**, *120*, 353–360. [CrossRef]
82. Saha, S.K.; Lee, S.B.; Won, J.; Choi, H.Y.; Kim, K.; Yang, G.-M.; Dayem, A.A.; Cho, S.-G. Correlation between oxidative stress, nutrition, and cancer initiation. *Int. J. Mol. Sci.* **2017**, *18*, 1544. [CrossRef]
83. Kim, M.; Jung, J. The natural plant flavonoid apigenin is a strong antioxidant that effectively delays peripheral neurodegenerative processes. *Anat. Sci. Int.* **2019**. [CrossRef]
84. Hwang, I.W.; Chung, S.K. Isolation and identification of myricitrin, an antioxidant flavonoid, from daebong persimmon peel. *Prev. Nutr. Food Sci.* **2018**, *23*, 341–346. [CrossRef]
85. Jayaraman, R.; Subramani, S.; Abdullah, S.H.S.; Udaiyar, M. Antihyperglycemic effect of hesperetin, a citrus flavonoid, extenuates hyperglycemia and exploring the potential role in antioxidant and antihyperlipidemic in streptozotocin-induced diabetic rats. *Biomed. Pharmacother.* **2018**, *97*, 98–106. [CrossRef]
86. Aloud, A.A.; Veeramani, C.; Govindasamy, C.; Alsaif, M.A.; el Newehy, A.S.; Al-Numair, K.S. Galangin, a dietary flavonoid, improves antioxidant status and reduces hyperglycemia-mediated oxidative stress in streptozotocin-induced diabetic rats. *Redox Rep.* **2017**, *22*, 290–300. [CrossRef]
87. Kanimozhi, S.; Bhavani, P.; Subramanian, P. Influence of the flavonoid, quercetin on antioxidant status, lipid peroxidation and histopathological changes in hyperammonemic Rats. *Indian J. Clin. Biochem.* **2017**, *32*, 275–284. [CrossRef]
88. Abarikwu, S.O.; Olufemi, P.D.; Lawrence, C.J.; Wekere, F.C.; Ochulor, A.C.; Barikuma, A.M. Rutin, an antioxidant flavonoid, induces glutathione and glutathione peroxidase activities to protect against ethanol effects in cadmium-induced oxidative stress in the testis of adult rats. *Andrologia* **2017**, *49*. [CrossRef]
89. Torres Castaneda, G.H.; Dulcey, A.J.C.; Martinez, J.H.I. Flavonoid glycosides from *Siparuna gigantotepala* leaves and their antioxidant activity. *Chem. Pharm. Bull.* **2016**, *64*, 502–506. [CrossRef]
90. Alipour, B.; Rashidkhani, B.; Edalati, S. Dietary flavonoid intake, total antioxidant capacity and lipid oxidative damage: A cross-sectional study of Iranian women. *Nutrition* **2016**, *32*, 566–572. [CrossRef]
91. Olayinka, E.T.; Ore, A.; Adeyemo, O.A.; Ola, O.S.; Olotu, O.O.; Echebiri, R.C. Quercetin, a flavonoid antioxidant, ameliorated procarbazine-induced oxidative damage to murine tissues. *Antioxidants* **2015**, *4*, 304–321. [CrossRef]

92. Zamora-Ros, R.; Ferrari, P.; Gonzalez, C.A.; Tjonneland, A.; Olsen, A.; Bredsdorff, L.; Overvad, K.; Touillaud, M.; Perquier, F.; Fagherazzi, G.; et al. Dietary flavonoid and lignan intake and breast cancer risk according to menopause and hormone receptor status in the European Prospective Investigation into Cancer and Nutrition (EPIC) Study. *Breast Cancer Res. Treat.* **2013**, *139*, 163–176. [CrossRef]
93. Boggs, D.A.; Palmer, J.R.; Stampfer, M.J.; Spiegelman, D.; Adams-Campbell, L.L.; Rosenberg, L. Tea and coffee intake in relation to risk of breast cancer in the Black Women's Health Study. *Cancer Causes Control* **2010**, *21*, 1941–1948. [CrossRef]
94. Oh, J.K.; Sandin, S.; Strom, P.; Lof, M.; Adami, H.O.; Weiderpass, E. Prospective study of breast cancer in relation to coffee, tea and caffeine in Sweden. *Int. J. Cancer* **2015**, *137*, 1979–1989. [CrossRef]
95. Luo, J.; Gao, Y.T.; Chow, W.H.; Shu, X.O.; Li, H.; Yang, G.; Cai, Q.; Rothman, N.; Cai, H.; Shrubsole, M.J.; et al. Urinary polyphenols and breast cancer risk: Results from the Shanghai Women's Health Study. *Breast Cancer Res. Treat.* **2010**, *120*, 693–702. [CrossRef]
96. Wang, Q.; Li, H.; Tao, P.; Wang, Y.P.; Yuan, P.; Yang, C.X.; Li, J.Y.; Yang, F.; Lee, H.; Huang, Y. Soy isoflavones, CYP1A1, CYP1B1, and COMT polymorphisms, and breast cancer: A case-control study in southwestern China. *DNA Cell Biol.* **2011**, *30*, 585–595. [CrossRef]
97. Hui, C.; Qi, X.; Qianyong, Z.; Xiaoli, P.; Jundong, Z.; Mantian, M. Flavonoids, flavonoid subclasses and breast cancer risk: A meta-analysis of epidemiologic studies. *PLoS ONE* **2013**, *8*, e54318. [CrossRef]
98. Gardeazabal, I.; Romanos-Nanclares, A.; Martinez-Gonzalez, M.A.; Sanchez-Bayona, R.; Vitelli-Storelli, F.; Gaforio, J.J.; Aramendia-Beitia, J.M.; Toledo, E. Total polyphenol intake and breast cancer risk in the SUN cohort. *Br. J. Nutr.* **2018**, 1–23. [CrossRef]
99. Baeza, I.; de la Fuente, M. The Role of Polyphenols in Menopause. In *Nutrition and Diet in Menopause*; Martin, C.J.H., Watson, R.R., Preedy, V.R., Eds.; Humana Press: Totowa, NJ, USA, 2013; pp. 51–63. [CrossRef]
100. Christensen, K.Y.; Naidu, A.; Parent, M.E.; Pintos, J.; Abrahamowicz, M.; Siemiatycki, J.; Koushik, A. The risk of lung cancer related to dietary intake of flavonoids. *Nutr. Cancer* **2012**, *64*, 964–974. [CrossRef]
101. Reale, G.; Russo, G.I.; di Mauro, M.; Regis, F.; Campisi, D.; Giudice, A.L.; Marranzano, M.; Ragusa, R.; Castelli, T.; Cimino, S.; et al. Association between dietary flavonoids intake and prostate cancer risk: A case-control study in Sicily. *Complement. Ther. Med.* **2018**, *39*, 14–18. [CrossRef]
102. Woo, H.D.; Lee, J.; Choi, I.J.; Kim, C.G.; Lee, J.Y.; Kwon, O.; Kim, J. Dietary flavonoids and gastric cancer risk in a Korean population. *Nutrients* **2014**, *6*, 4961–4973. [CrossRef]
103. Sun, L.; Subar, A.F.; Bosire, C.; Dawsey, S.M.; Kahle, L.L.; Zimmerman, T.P.; Abnet, C.C.; Heller, R.; Graubard, B.I.; Cook, M.B.; et al. Dietary flavonoid intake reduces the risk of head and neck but not esophageal or gastric cancer in US men and women. *J. Nutr.* **2017**, *147*, 1729–1738. [CrossRef]
104. Molina-Montes, E.; Sanchez, M.J.; Zamora-Ros, R.; Bueno-de-Mesquita, H.B.; Wark, P.A.; Obon-Santacana, M.; Kuhn, T.; Katzke, V.; Travis, R.C.; Ye, W.; et al. Flavonoid and lignan intake and pancreatic cancer risk in the European prospective investigation into cancer and nutrition cohort. *Int. J. Cancer* **2016**, *139*, 1480–1492. [CrossRef]
105. Nimptsch, K.; Zhang, X.; Cassidy, A.; Song, M.; O'Reilly, E.J.; Lin, J.H.; Pischon, T.; Rimm, E.B.; Willett, W.C.; Fuchs, C.S.; et al. Habitual intake of flavonoid subclasses and risk of colorectal cancer in 2 large prospective cohorts. *Am. J. Clin. Nutr.* **2016**, *103*, 184–191. [CrossRef]
106. Xu, M.; Chen, Y.M.; Huang, J.; Fang, Y.J.; Huang, W.Q.; Yan, B.; Lu, M.S.; Pan, Z.Z.; Zhang, C.X. Flavonoid intake from vegetables and fruits is inversely associated with colorectal cancer risk: A case-control study in China. *Br. J. Nutr.* **2016**, *116*, 1275–1287. [CrossRef]
107. Zamora-Ros, R.; Not, C.; Guino, E.; Lujan-Barroso, L.; Garcia, R.M.; Biondo, S.; Salazar, R.; Moreno, V. Association between habitual dietary flavonoid and lignan intake and colorectal cancer in a Spanish case-control study (the Bellvitge Colorectal Cancer Study). *Cancer Causes Control* **2013**, *24*, 549–557. [CrossRef]
108. Zamora-Ros, R.; Guinó, E.; Alonso, M.H.; Vidal, C.; Barenys, M.; Soriano, A.; Moreno, V. Dietary flavonoids, lignans and colorectal cancer prognosis. *Sci. Rep.* **2015**, *5*, 14148. [CrossRef]

Review

Inula L. Secondary Metabolites against Oxidative Stress-Related Human Diseases

Wilson R. Tavares [1] and Ana M. L. Seca [2,3,*]

1 Faculty of Sciences and Technology, University of Azores, 9501-801 Ponta Delgada, Portugal; wrt-94@hotmail.com
2 cE3c—Centre for Ecology, Evolution and Environmental Changes/Azorean Biodiversity Group & University of Azores, Rua Mãe de Deus, 9501-801 Ponta Delgada, Portugal
3 QOPNA & LAQV-REQUIMTE, University of Aveiro, 3810-193 Aveiro, Portugal
* Correspondence: ana.ml.seca@uac.pt; Tel.: +351-296-650-172

Received: 29 March 2019; Accepted: 2 May 2019; Published: 6 May 2019

Abstract: An imbalance in the production of reactive oxygen species in the body can cause an increase of oxidative stress that leads to oxidative damage to cells and tissues, which culminates in the development or aggravation of some chronic diseases, such as inflammation, diabetes mellitus, cancer, cardiovascular disease, and obesity. Secondary metabolites from *Inula* species can play an important role in the prevention and treatment of the oxidative stress-related diseases mentioned above. The databases Scopus, PubMed, and Web of Science and the combining terms *Inula*, antioxidant and secondary metabolites were used in the research for this review. More than 120 articles are reviewed, highlighting the most active compounds with special emphasis on the elucidation of their antioxidative-stress mechanism of action, which increases the knowledge about their potential in the fight against inflammation, cancer, neurodegeneration, and diabetes. Alantolactone is the most polyvalent compound, reporting interesting EC_{50} values for several bioactivities, while 1-*O*-acetylbritannilactone can be pointed out as a promising lead compound for the development of analogues with interesting properties. The *Inula* genus is a good bet as source of structurally diverse compounds with antioxidant activity that can act via different mechanisms to fight several oxidative stress-related human diseases, being useful for development of new drugs.

Keywords: *Inula*; oxidative stress; ROS; secondary metabolites; inflammation; diabetes; neurologicaldamage; cancer; sesquiterpene lactones

1. Introduction

Oxygen metabolism, which involves mainly redox reactions, is fundamental for human life, but it leads to the production of reactive oxygen species (ROS) and reactive nitrogen species (RNS) [1,2], affecting regulation of several biological processes and cell functions [3]. ROS and RNS include not only radical species such as hydroxyl radical ($^{\bullet}OH$), superoxide radical anion ($O_2^{\bullet-}$), and nitric oxide radical ($^{\bullet}NO$), having unpaired electrons and exhibiting short biological half-lives, but also labile nonradicals species like singlet oxygen ($^1O_2$), peroxynitrite ($ONOO^-$), and hydrogen peroxide (H_2O_2), which can also be transformed into some of the radical species mentioned above [4,5]. All these species, due their irreversible and nonselective reactivity, are associated with oxidative-stress related damage [4]. In fact, when cellular production of ROS and RNS overwhelms the antioxidant capacity of cells, it leads to a state of oxidative stress, which in turn can cause oxidative damage to large biomolecules such as proteins, lipids, and deoxyribonucleic acid (DNA) [6]. The consequent degradation of cellular integrity and tissue functions culminates in the development or aggravation of some disorders such as inflammation, ageing, diabetes, cancer, cardiovascular, neurodegenerative disease, and obesity [6–9].

A recent topic of increasing interest and investigation in the scientific community is the use of plants and their secondary metabolites as therapeutic agents [10–13]. Plants are an excellent source of compounds with pharmacological potential and/or possessing leading chemical structures in the development of new drugs [10–12], and they have always been used effectively as medicine for treatment of human diseases. The *Inula* species (more than 100 species [14]) from the Asteraceae family (also known as Compositae) are widely distributed in Africa, Asia, and Europe and have been reported to possess more than 400 compounds, mainly terpenoids (sesquiterpene lactones and dimers, diterpenes, and triterpenoids) and flavonoids, with many of them exhibiting interesting pharmacological activities [12,13], and are of great scientific and medicinal interest, as evidenced by the two ongoing clinical studies involving herbal preparations containing *Inula* species (ClinicalTrials.gov Identifier: NCT03256708 and NCT02918487). Furthermore, many studies continue to be published showing the potential of *Inula* species in the treatment and prevention of diseases related to oxidative stress, showing traditional medicine applications of plant, in vitro, and in vivo biological activities of *Inula* extracts. In the Kashmir Himalayas, the roots and seeds of *Inula racemosa* Hook. f. are used to treat various health conditions including inflammation and rheumatism [15], while in Pakistan, to treat rheumatism, they use *Inula orientalis* Lam. (syn. *Inula grandiflora* Willd) [16]. The ethanol extract of *Inula helenium* L. exhibits antioxidant and anti-neuroinflammatory activities in lipopolysaccharide (LPS)-stimulated BV-2 microglia cells, suggesting that the extract could act by inhibiting NO production and inducible nitric oxide synthase (iNOS) expression levels through suppression of the expression of interleukin-6 (IL-6) levels [17]. Qun et al. [18] revealed that the hydroethanolic extract of *Inula helenium* presented anti-inflammatory activity in a mouse model, acting by inhibition of tumor necrosis factor-α (TNF-α)-induced activation of nuclear factor kappa-B (NF-κB) and the expression of IL-1, IL-4 and TNF-α, as shown by the test in human keratinocyte HaCat cell line. Another study [19], revealed that ethanol extract from flowers of *Inula japonica* Thunb. inhibited lipid accumulation in 3T3-L1 adipocytes in vitro and reported also that C57BL/6J mice models fed with high-fat diet with 2.5 g of the extract showed a decrease in body fat mass, hepatic lipid accumulation, and body weight gain, while increasing muscle weight.

The taxonomy of some *Inula* species, as in many other genera, has been altered in recent years, and in this review, only the published works involving species whose binominal Latin name is considered by the "The Plant List" database [14] as an *Inula* accepted name are considered. The abovementioned studies are only a few examples of the great interest in *Inula* anti oxidative-stress related disorders research, which led to an increase in the investigation of the metabolites responsible for the activities exhibited, providing support for *Inula*'s use in traditional medicine, as well as establishing the *Inula* genus as a source of antioxidant compounds. This paper intends to provide a critical bibliographic review that demonstrates this, showing a selection of *Inula* compounds with the highest pharmacological potential for the treatment of oxidative-stress related pathological problems as well as to discuss the mechanisms of action involved in their pharmacological action.

2. Radical Scavenging Activity of Secondary Metabolites from *Inula* Species Determined Using DPPH and ABTS Methods

There are many methods available to allow a first approach for evaluating the antioxidant potential of a compound or extract [20]. Among them, the 1,1-diphenyl-2-picrylhydrazyl (DPPH) and 2,2′-azino-bis(3-ethylbenzothiazoline-6-sulphonic acid) (ABTS) free radical scavenging colorimetric methods are the most popular, since they offer advantages of being rapid, simple, and inexpensive and provide first-hand information on the overall antioxidant capacity of the tested sample [21,22]. However, the two methods are not equivalent: The DPPH scavenging test measures the ability of a compound to neutralize the DPPH radical by a mechanism involving single-electron transfer (SET), while in ABTS assay, the radical neutralization mechanism is mainly hydrogen-atom transfer (HAT), although in some cases, it could also be electron transfer, resulting in a more sensitive method [23,24]. As already mentioned, more than 400 secondary metabolites isolated from *Inula* species are known, and many of them exhibit radical scavenging properties by DPPH and/or ABTS methods. A critical non-exhaustive selection of the most representative *Inula* secondary metabolites, which exhibit an activity identical or superior to that of a reference compound, are presented in Table 1, and the respective chemical structures are shown in Figure 1. In addition, in this selection, we preferentially consider the published works in which the authors present an associated statistical parameter, thus guaranteeing the reliability of the result, and a low associated error (*c.a.* 10% of the mean).

Table 1. Scavenging effects of *Inula* secondary metabolites **1–10** and reference compound on 1,1-diphenyl-2-picrylhydrazyl (DPPH) and 2,2′-azino-bis(3-ethylbenzothiazoline-6-sulphonic acid) (ABTS) radicals (EC_{50}, μM).

Compound	DPPH (Reference Compound)	ABTS (Reference Compound)	*Inula* Source
1,3-dicaffeoylquinic acid (**1**)		12 ± 0.4 (Ascorbic acid: 15 ± 0.01) [25]	*Inula helenium* [26]
β-caryophyllene (**2**)	1.25 ± 0.06 (Ascorbic acid: 1.5 ± 0.03) [27]		*Inula cappa* (Buch.-Ham. ex D.Don) DC. * [28]
Caffeic acid (**3**)	25.0 ± 1.7 (Ascorbic acid: 20.7 ± 1.31) ** [29]	8.82 ± 0.33 (Ascorbic acid: 15.05 ± 2.61) ** [29]	*Inula helenium* [30]
Chlorogenic acid (**4**)	36.83 ± 0.76 (Caffeic acid: 35.02 ± 2.11) ** [31]		*Inula ensifolia* L. [32], *Inula cappa* [33], *Inula helenium* [34]
Isoquercitrin (**5**)	12.68 ± 0.54 (Trolox: 18.10 ± 0.44) ** [35]		*Inula japonica* [36], *Inula ensifolia* [32], *Inula helenium* [34]
Kaempferol (**6**)	27.18 ± 1.05 (Ascorbic acid: 20.72 ± 1.31) ** [29] 47.97 ± 0.03 (Ascorbic acid: 20.27 ± 0.11) ** [37]	12.93 ± 0.52 (Ascorbic acid: 15.05 ± 2.61) ** [29]	*Inula salsoloides* (Turcz.) Ostenf. [38]
Luteolin (**7**)	6.69 ± 0.15 (Ascorbic acid: 16.88 ± 0.02) [39]		*Inula japonica* [36], *Inula salsoloides* [38], *Inula britannica* L. [40]
Quercetin (**8**)	8.80 ± 0.79 (Ascorbic acid: 20.72 ± 1.31) ** [29] 19.75 ± 1.06 (Caffeic acid: 35.02 ± 2.11) ** [31]	6.25 ± 1.09 (Ascorbic acid: 15.05 ± 2.61) ** [29]*	*Inula japonica* [36], *Inula britannica* [41], *Inula helenium* [34]
Quercitrin (**9**)	9.93 ± 0.38 (Trolox: 18.10 ± 0.44) [35]		*Inula japonica* [36], *Inula ensifolia* [32], *Inula helenium* [34]
Rutin (**10**)	19.31 ± 0.39 (Caffeic acid: 35.02 ± 2.11) ** [31]		*Inula helenium* [34]

* According to "The plant list" database [14], this is an unresolved name. ** After unit conversion from μg/mL to μM.

In some assigned cases (see Table 1 note), there was the necessity to convert the EC_{50} values from the original bibliographic source from μg/mL to μM, to allow a comparison of antioxidant activity between the compounds.

According to the DPPH assay values in Table 1, β-caryophyllene (**2**), with an EC_{50} of 1.25 ± 0.06 μM, is by far the most active compound, followed by quercetin (**8**) and quercitrin (**9**), also with interesting EC_{50} values (EC_{50} < 10 μM). It should be noticed that all these compounds showed better EC_{50} values than the reference compound used in their studies, i.e., ascorbic acid or trolox.

As it is possible to see in Table 1, regarding the ABTS assay, a lot fewer published results are available in the literature. Quercetin (**8**) and caffeic acid (**3**) are the compounds with the lowest EC_{50} values, i.e., 6.25 ± 1.09 μM and 8.82 ± 0.33 μM, respectively. Both compounds presented better radical scavenging activity than the reference compound ascorbic acid.

The higher sensitivity of the ABTS method is reflected in lower EC_{50} values when compared to those obtained by the DPPH method for the same compound tested.

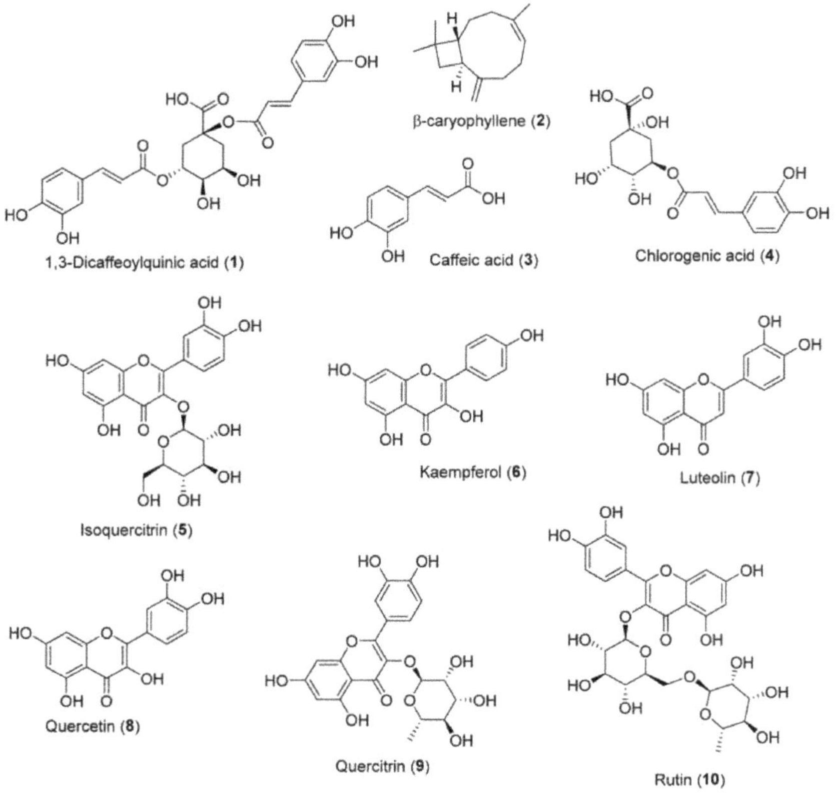

Figure 1. Chemical structure of *Inula* secondary metabolites (**1–10**) with DPPH and/or ABTS antioxidant activity.

It should be emphasized that the results of DPPH and ABTS are somewhat dependent on the used experimental conditions, and therefore, different works may report different DPPH and ABTS EC_{50} values for the same compound (see example: Kaempferol (**6**), Table 1). To mitigate this, it is very important to present the EC_{50} value of an appropriate reference, thus allowing a more reliable comparison of the level of activity in the different publications. Surprisingly, even in recent publications, a significant number of published papers continue to be found that do not meet this requirement. This is a point at which researchers and the peer review process should be more demanding and rigorous, contributing greatly to making the published data more comparable and therefore more useful and of greater impact.

The data in Table 1 show that *Inula* species have relevant compounds with great antioxidant activity, many of them more active than some of the reference compounds, such as ascorbic acid, already used by industry as antioxidants.

Although the antioxidant activity assays by the DPPH and ABTS methods are simple, rapid, and very useful as a first approach, the extrapolation of their results to the antioxidant effect at a cellular level in a biological environment is impossible, and they do not give any information about the cellular mechanisms in which the compounds tested act. This information is very relevant and is obtained using methods and approaches very different from those discussed so far.

3. Secondary Metabolites from *Inula* Species against Oxidative-Stress Related Diseases

As noted above, compounds isolated from *Inula* species exhibit a wide range of biological activities against oxidative stress diseases such as inflammation, diabetes, cancer, and neurodegenerative diseases. Thus, much research has been developed to understand how *Inula* compounds act, using models more complex than the model of radical scavenging referred in point 2, and therefore closer to real biological systems. In this section, we present not an exhaustive compilation but rather a critical analysis of the more in-depth studies and the most relevant aspects of the action mechanisms exhibited by the *Inula* compounds that have, as a final consequence, the reduction of the oxidative stress nature inherent to the mentioned diseases.

3.1. Inflammation

Since overproduction of ROS leads to cellular and tissue damage, inflammation is intrinsically linked to oxidative stress [42]. Inflammation is a complex defense mechanism that is vital to health since it is the immune system's response to harmful stimuli, such as damaged cells, toxic compounds, pathogens or irradiation [43]. Cellular and molecular events are triggered in an acute inflammatory response in order to mitigate the impact of an injury or infection, allowing restoration of tissue homeostasis [44]. However, uncontrolled acute inflammation may become chronic, leading to the development of a variety of chronic inflammatory diseases [45]. Intracellular inflammatory signaling pathways include NF-κB, the mitogen-activated protein kinase (MAPK) and Janus kinase/signal transducer and activator of transcription 3 (JAK/STAT3) pathways. All of them are activated by inflammatory stimuli such as TNF-α, interleukin-1β (IL-1β), and IL-6 that interact with the Toll-like receptors (TLR), TNF receptor (TNFR), IL-1 receptor (IL-1R), and IL-6 receptor (IL-6R), mediating inflammation through the production of more inflammatory stimuli [46]. NO is also fundamental in the cellular defense mechanism of inflammation, since NO synthase is induced by pro-inflammatory cytokines; however, it can cause adverse effects such as autoimmune reactions and neurodegenerative syndromes when overproduction of NOs occurs [47]. Cyclooxygenase 2 (COX-2) is a prostaglandin–endoperoxide synthase 2 enzyme that is responsible for generation of prostanoids like prostaglandin E2 (PGE2) that act in the modulation of multiple inflammation and pro-carcinogenic processes [48,49]. The overexpression of COX-2 has been associated with carcinogenesis, resistance to apoptosis, and inflammatory diseases [50,51]. COX-2 expression is controlled by the binding of many trans-factors to the corresponding sites on its promoters, like NF- κB, which in turn, depends on the degradation of IκB proteins by an IκB kinase (IKK) complex [52].

Direct myocardial injury can be caused by inflammatory cytokines response, microcirculation dysfunction, and insufficient energy [53]. The work of Huang et al. [54] clarifies the mechanism by which isoquercitrin (**5**) (Figure 1) attenuates the inflammatory response on LPS-induced cardiac dysfunction on C57BL/6 mice or H9c2 cardiomyoblasts. After LPS stimulation, production of large amounts of TNF-α, monocyte chemoattractant protein 1 (MCP1), and IL6 (all pro-inflammatory cytokines) starts, regulated via the NF-κB signaling pathway, leading to cardiac injury. According to this study, pretreatment with isoquercitrin (**5**) (40 μM) attenuates LPS-induced cardiac dysfunction as well as decreases the levels of TNF-α, IL6, MCP1, and iNOS in vivo and in vitro by blocking the MAPK and NF-κB pathways.

Alantolactone (**11**) (Figure 2) is a eudesmanolide sesquiterpene lactone with an α-methylene–γ-lactone moiety that is considered the active principle of *Inula helenium* [55]. Alantolactone (**11**) is found in several *Inula* species besides *Inula helenium*, e.g., *Inula japonica*, *Inula racemosa*, *Inula royleana* DC., and *Inula falconeri* Hook.f. [12]. Zhang et al. [56] showed that alantolactone (**11**) inhibits LPS-induced NO production in RAW 264.7 macrophages, presenting an IC_{50} value of 7.39 ± 0.36 μM, being better than the positive control aminoguanidine (IC_{50} = 9.12 ± 0.35 μM). These results are in accordance with the ones presented by Chun et al. [57], where compound **11** at 10 μM inhibited the production of NO, PGE2, and TNF-α, as well as COX-2 and iNOS protein and mRNA transcription in LPS-stimulated RAW 264.7 cells. The same study showed that alantolactone (**11**) disrupted the NF-κB signaling pathway through inhibition of the phosphorylation of inhibitory κB-α (IκB-α) and IKK, as well as the MAPK pathway. A recent study [18] with HaCat cell line revealed that alantolactone

(**11**) presented anti-inflammatory activity, since it also could inhibit the expression of IL-1, IL-4, and TNF-α and TNF-α-induced activation of NF-κB, in a dose-dependent manner.

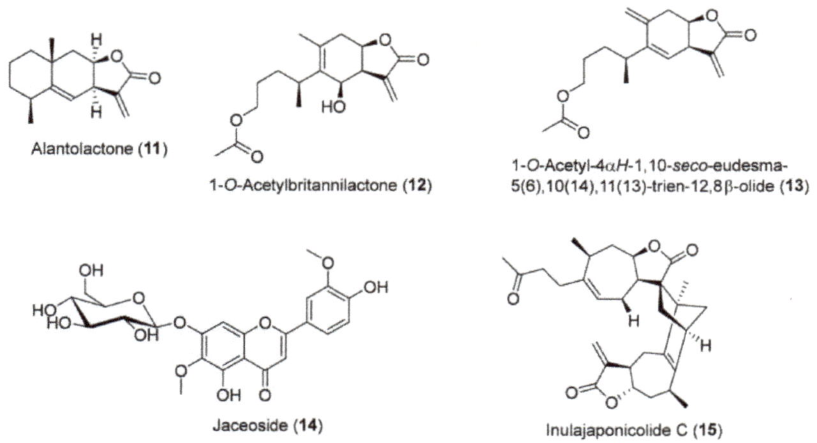

Alantolactone (**11**)

1-*O*-Acetylbritannilactone (**12**)

1-*O*-Acetyl-4α*H*-1,10-*seco*-eudesma-5(6),10(14),11(13)-trien-12,8β-olide (**13**)

Jaceoside (**14**)

Inulajaponicolide C (**15**)

Figure 2. Chemical structure of *Inula* secondary metabolites (**11, 12, 14, 15**) and the semisynthetic derivative (**13**) with reported activity against oxidative-stress inflammatory process.

1-*O*-acetylbritannilactone (**12**) (Figure 2) is a 1,10-*seco*-eudesmanolide sesquiterpene that, like compound **11**, has an α-methylene–γ-lactone skeleton, found in *Inula britannica var. chinensis* and *Inula japonica* [12], and that possesses cytotoxic potential [58,59] and anti-inflammatory properties [60,61]. A recent study by Wei et al. [62], found that the 6-deoxy1-*O*-acetylbritannilactone with a methylene at C-14 position, an analogue of 1-*O*-acetylbritannilactone (**12**) labelled as 1-*O*-acetyl-4α*H*-1,10-*seco*-eudesma-5(6),10(14),11(13)-trien-12,8β-olide (**13**) (Figure 2), exhibits an anti-inflammatory effect. In fact, compound **13** decreased NO production and iNOS expression in RAW 264.7 macrophage normal cell line with IC_{50} value of 1.3 μM.

Several compounds from *Inula montana* L. possessed promising anti-inflammatory activity through inhibition of NO production in murine macrophages RAW 264.7 cell line, jaceoside (**14**) (Figure 2) being the compound most active with IC_{50} of 0.34 ± 0.01 μM, being several times better than the positive control drug dexamethasone (IC_{50} of 3.89 ± 0.94 μM) [63].

Several dimeric- and trimeric-sesquiterpenes isolated from *Inula japonica* exhibit anti-inflammatory properties [64]. One of them, the 2,4-linked sesquiterpene lactone dimer named inulajaponicolide C (**15**) (Figure 2), presented the most potent inhibitory effect over NO production in LPS-stimulated RAW 264.7 cells with IC_{50} value of 1.0 ± 0.1 μM, being much better than the indomethacin (IC_{50} = 14.6 ± 0.5 μM) used as positive control.

3.2. Diabetes

Diabetes mellitus is characterized by chronic hyperglycemia resulting from flaws in insulin action, insulin secretion, or both [65]. Hyperglycemia induces the increase of ROS production, which in turns causes damages in cells and activation of inflammation processes [66] and triggers apoptosis in the β-cells, worsening the lack of insulin [67]. Thus, acquired insulin resistance and glucose intolerance are associated with chronic inflammation [68,69], the pro-inflammatory cytokine being IL-6 the main link between both processes [70].

A randomized double-blind clinical trial placebo-controlled performed in 30 patients suffering from impaired glucose tolerance showed that the administration of 400 mg of chlorogenic acid (**4**) (Figure 1) three times a day for 12 weeks decreased fasting plasma glucose and increased insulin sensitivity, despite the fact that insulin secretion decreased [71]. The authors suggest that the antidiabetic effect of chlorogenic acid (**4**) could be due to its action on hepatic peroxisome proliferation-activated receptor α (PPARα), which plays a role as a facilitator in clearing lipids from the liver and enhancing insulin sensitivity [72].

The most significant component of the regulating post-prandial insulin secretion mechanism is glucagon-like peptide-1 (GLP-1) that is secreted from cells in the gastrointestinal tract in response to nutrient absorption [73]. GLP-1 is rapidly inactivated in vivo by circulating dipeptidyl peptidase 4 (DPP-IV) [74]. A recent study [75], using colorectal adenocarcinoma NCI-H716 cells as an in vitro model of gastrointestinal cells, showed that isoquercitrin (**5**) is a promising compound to treat type 2 diabetes since it was identified as a DPP-IV inhibitor, with an IC_{50} of 96.8 μM. Furthermore, the levels of GLP-1 increased, suggesting that isoquercitrin (**5**) may also stimulate GLP-1 secretion and bioavailability in a dose-dependent manner. In addition, the same work [75] using in vivo assays with type 2 diabetic Chinese Kunming mice showed that isoquercitrin (**5**) treatment for 8 weeks (80 mg/kg b.w. per day), significantly increased GLP-1 and insulin levels in plasma while lowering the fasting blood glucose levels. These results are in accordance with the ones obtained by Huang et al. [76] that reported hepatoprotective potential of isoquercitrin (**5**) (10 and 30 mg/kg b.w. per day) against type 2 diabetes-induced hepatic injury in rats after 21 days of treatment with significant suppression of DPP-IV mRNA level expression.

Kim et al. [77] demonstrated that alantolactone (**11**) (Figure 2) could increase glucose uptake levels, suggesting it as a great candidate for the treatment of insulin resistance and glucose intolerance. In fact, the 4 h pretreatment of L6 rat myoblast cell line with alantolactone (**11**) (at 0.5 μM), followed by 24 h exposure to IL-6, caused a decrease in the IL-6 induced insulin resistance and allowed the increase of glucose uptake levels to the levels of the control group (without exposure to IL-6). Therefore, alantolactone (**11**) possess antidiabetic potential resulting from its effect against IL-6 induced inflammatory process.

3.3. Neurological Damages

Formation and deposition of amyloid beta (Aβ) plaques in the brain in excess, a characteristic of Alzheimer disease (AD), can generate oxidative stress, which triggers inflammatory processes and exacerbates the destruction of hippocampal and neighboring tissues [78]. Therefore, in order to ameliorate or prevent the progression of ROS-mediated neurological damages, antioxidants are considered as promising candidates for therapeutics not only in AD but also in other neurodegenerative diseases like Huntington or Parkinson's [79,80].

There are indications in the literature that alantolactone (**11**) (Figure 2) exhibits relevant properties to combat oxidative stress, not only in inflammatory processes, as noted above, but also in neurological system. In fact, Seo et al. [81] showed that alantolactone (**11**) at 0.1 to 1 μM has neuroprotective effects on mouse cortical neurons since cell viability was little affected by exposure to $A\beta_{25-35}$ (10 μM), preventing also the shortening of dendrite length, in contrast to what happened in the control group exposed only to $A\beta_{25-35}$ (10 μM). In addition, alantolactone (**11**) treatment decreased acetylcholinesterase (AChE) activity and decreased intracellular ROS production in a dose-dependent manner [81]. However, the authors alert that alantolactone (**11**) at high doses (i.e., >5 μM) could act as prooxidant promoting ROS production. Moreover, the administration of alantolactone (**11**) (1 mg/kg b.w.) reverts scopolamine-induced cognitive impairments in male C57BL/6J and C57BL/6J/Nrf2 knockout mouse, indicating that alantolactone (**11**) improves working memory, probably mediated by activation of the nuclear factor erythroid 2-related factor 2 (Nrf2) signaling pathway [81], a factor that modulates the antioxidant response to an oxidant exposure by a increasing the expression of genes encoding antioxidant enzymes, like the glutathione reductase (GSR), γ-glutamylcysteine ligase (GCL), heme oxygenase-1 (HO-1) and NAD(P)H:quinone oxidoreductase-1 (NQO1) [82,83].

Oxidative stress following traumatic brain injury (TBI) can have devastating effects on brain tissues, since it causes oxidase enzymes activation, mitochondrial functions become impaired, membrane phospholipids are destroyed, and several cellular components, such as DNAs, RNA, carbohydrates, lipids, and proteins, are harmed, which ultimately leads to irreversible damage to neuronal cells and brain tissue [84]. A very recent study [85] reported that treatment of TBI in male Sprague–Dawley rats with alantolactone (**11**) (Figure 2) at 10 and 20 mg/kg b.w. alleviated cerebral edema and improved neurological function via anti-apoptosis, anti-inflammatory, and antioxidative pathways. Furthermore, the same study [85] reported that alantolactone (**11**) significantly suppressed COX-2 expression by inhibiting the activation of the NF-κB pathway, diminishing the levels of glutathione disulphide (GSSG) and malondialdehyde (MDA) (products of lipid peroxidation and an important marker of oxidative damage level [86]) while causing in brain tissues after TBI an increase in the level of glutathione (GSH) and in the activity of superoxide dismutase (SOD), the antioxidant first line defense [87].

Neurodegenerative diseases like AD are closely related with neuroinflammation [88]. In fact, excessive amount of NO accumulates in the central nervous system (CNS) as a result of inflammatory response over damaged microglia cells, which in turns exacerbates neuroinflammation and aggravates neurodegenerative diseases [89]. Liu et al. [90] isolated various compounds from *Inula japonica*, in an attempt to find potentially useful compounds with NO inhibitory effects for the treatment of neuroinflammation. Inujaponin F (**16**) (Figure 3) and 1-oxo-4αH-eudesma-5(6),11(13)-dien-12,8β-olide (**17**) (Figure 3) presented higher NO inhibitory activity in LPS-induced murine microglial BV-2 cells with IC_{50} values of 1.3 ± 0.1 μM and 1.5 ± 0.2 μM, respectively, higher activity than the one reported by the positive control 2-methyl-2-thiopseudourea sulphate (SMT) that presented an IC_{50} value of 2.9 ± 0.5 μM [90]. This anti-neuroinflammatory effect of compounds **16** and **17**, according to the molecular docking studies, could be due to their ability to interact with residues of the active cavities of iNOS protein, blocking it [90]. The iNOS protein is the most critical component in charge of the amount of NO in inflammatory response [91].

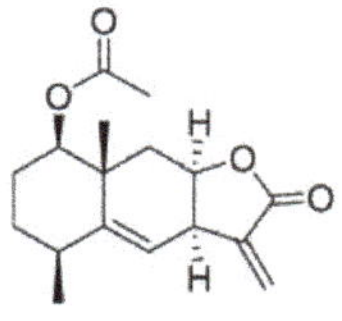

Inujaponin F (**16**)

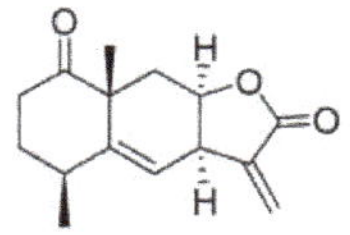

1-Oxo-4α*H*-eudesma-5(6),11(13)-dien-12,8β-olide (**17**)

Figure 3. Chemical structure of *Inula* secondary metabolites (**16** and **17**) with reported activity against neurological oxidative-stress damages.

3.4. *Carcinogenesis*

Carcinogenesis is a complex process through which cancer develops, but putting it simple, it basically involves genetic modification of genomic DNA (creation of a mutated cell) followed by growth and division of the aberrant cell with accumulation of additional genetic and epigenetic changes [92]. A recurrent characteristic of cancer progression and resistance to treatment is deregulated redox signaling, which means alteration in redox balance and culminates in elevated levels of ROS [93]. ROS production causes more DNA damage and triggers signaling pathways that activate pro-carcinogenic factors and anti-apoptotic responses, favoring cancer survival and progression [94,95].

Dahham et al. [27] found that β-caryophyllene (**2**) (Figure 1) demonstrated a selective anti-proliferative effect against colon cancer HCT 116 cells (IC_{50} = 19 μM,) and pancreatic cancer PANC-1 cells (IC_{50} = 27 μM), with selectivity index (SI) values from 5.8 to 27.9. It should be pointed out that β-caryophyllene (**2**) presented IC_{50} values not too far from the positive controls 5-fluorouracil (IC_{50} = 12.7 μM) and betulinic acid (IC_{50} = 19.4 μM, SI = 2.7-5). Additionally, β-caryophyllene (**2**) demonstrated apoptotic properties in the HCT 116 cells, by caspase-3 enzyme activation, loss of mitochondrial membrane potential, and DNA fragmentation pathways [27].

An interesting in vitro and in vivo study [96] investigated the effects of alantolactone (**11**) (Figure 2) on several glioblastoma multiforme cells (GBM) (i.e., U87, U251, U118, and SH-SY5Y cell lines) and determined that it suppresses the growth of GBM cells. According to the results, alantolactone (**11**) reduced in a dose- and time-dependent manner the survival rate of the tested cell lines exhibiting the highest cytotoxic activity against U251 cell line (IC_{50} = 16.33 ± 1.93 μM), without displaying cytotoxicity against normal human glial cell line, SVG, at concentrations below 25 μM. Furthermore, against U251 and U87 cell lines, alantolactone (**11**) reported IC_{50} values significantly lower than those of celecoxib (CB), a classical and potent commercial COX-2 inhibitor, which reported IC_{50} values of 120.32 μM and 135.27 μM, respectively [96]. In addition, this study [96] also found that the antitumor effect of alantolactone (**11**) in the GBM cells could be in part via NF-κB/COX-2-mediated signaling cascades through inhibition of IKKβ kinase activity. As referred above, the overexpression of COX-2 has been associated with inflammatory processes and also related with carcinogenesis and resistance to apoptosis [50,51]. Since IKKβ is the major subunit of this complex, its inhibition by alantolactone (**11**) ultimately leads to a decrease in the COX-2 expression and consequent intensification of the cytotoxic effect in the cells. Taking into account the results of the in vitro studies, the authors [96] also investigated the possible therapeutic effect of alantolactone (**11**) against tumor growth in BALB/c male nude mice. They noticed that toxic effects were not detected in the mice treated only with alantolactone (**11**) (10 and 20 mg/kg b.w.), and tumor weights and volumes decreased in the study group when compared with the control group (tumor inhibition rates of 47.73 ± 9.32% and 70.45 ± 13.33%, respectively).

Alantolactone (**11**) seems to be a very versatile compound. Not only due to its activities referred to in the previous points, but also because it exhibits cytotoxic activity against solid tumors, as referred to in the previous paragraph, and also against nonsolid tumors, as shown by Ding et al. [97]. In this work [97], alantolactone (**11**) shows selective (SI > 8) antitumor activity against several acute myeloid leukemia stem cell lines (AML), such as THP-1 (IC_{50} = 2.17 ± 0.72 μM), KG1a (IC_{50} = 2.75 ± 0.65 μM), K562 (IC_{50} = 2.75 ± 0.64 μM), and HL60 (IC_{50} = 3.26 ± 0.88 μM), as well as in the multidrug-resistant cell lines K562/A02 (IC_{50} = 2.73 ± 0.83 μM) and HL60/ADR (IC_{50} = 3.28 ± 0.80 μM), where alantolactone (**11**) is more cytotoxic than the clinically used drug adriamycin (ADR) (IC_{50} = 8.94 ± 3.79 μM against K562/A02 and IC_{50} = 5.54 ± 1.21 μM against HL60/ADR). Unfortunately, the results of this work should be considered under reserve, since the associated standard deviation is very high (about 20% of the mean). Above all, this applies to the cytotoxicity of the clinical drug against the K562/A02 multiresistant cell line, where the standard deviation reaches 42% of the mean value, which means a high dispersion of the results obtained in different replicates and, therefore, a low confidence in the result. The authors [97] also noticed that treatment with alantolactone (**11**) on HL60 and KG1a cell lines caused induction of cellular apoptosis by suppression of the NF-κB pathway, an important pathway involved in oxidative-stress related complications. An overexpression of the pro-apoptotic protein Bax was observed, while the expression of Bcl-2, an apoptosis inhibitor, and of NF-κB p65 subunit were reduced significantly. The alantolactone also caused the reduction of the downstream target proteins of the NF-κB pathway, the X-linked inhibitor of apoptosis protein (XIAP) and the FLICE-inhibitory protein (FLIP) that play important roles in cell apoptosis [97].

1-*O*-Acetylbritannilactone (**12**) (Figure 2), like alantolactone (**11**) (Figure 2), is a sesquiterpene lactone very common in *Inula* species [12] that elicits apoptosis in cancer cell lines through partially targeting the NF-κB pathway [98]. In fact, Wang et al. [98] showed that the combination of 1-*O*-acetylbritannilactone (**12**) (10 μM) and the approved chemotherapy drug gemcitabine (10 μg/mL) had a synergistic effect on the suppression of A549 cells proliferation, by inducing apoptosis in a 72 h treatment. The mixture decreases significantly the cell survival rates (mix of the two compounds cell survival = 30.2%) when compared with the control (100%), and with the compounds alone (1-*O*-acetylbritannilactone = 59.1%; gemcitabine alone = 49.7%). The authors also found that 1-*O*-acetylbritannilactone (**12**) and the combination treatment significantly decreased the expression of NF-κB and Bcl-2, while upregulating Bax expression [98].

Angiogenesis is a complex and normal process that allows the formation of new blood vessels (capillary formation) from the pre-existing ones, being crucial during wound healing or embryo development; however, it is abnormally present in cancer [99]. As a critical component of tumor angiogenesis, glycoprotein vascular endothelial growth factor (VEGF) is widely expressed in many cancers [100,101], while the vascular endothelial growth factors receptor-2 (VEGFR2) increased signaling is also characteristic of angiogenesis in tumors [102–104]. Alantolactone (**11**) (Figure 2) exhibits anti-angiogenesis property, since it shows anti-proliferative activity against human umbilical vascular endothelial cells (HUVEC) (IC_{50} = 14.2 μM), a model cell line used to study angiogenesis processes [105]. The alantolactone (**11**) anti-angiogenesis property could be related with its capacity to decrease capillary formation, by suppressing VEGFR2 signaling and decreasing the expression of its multiple downstream protein kinases, e.g., focal adhesion kinase (FAK) [105].

Anti-angiogenic activity is also exhibited by 1-*O*-acetylbritannilactone (**12**) (Figure 2) [106]. In the in vitro assay, 1-*O*-acetylbritannilactone (**12**) at 5 μM and 10 μM dose-dependently inhibits VEGF (25 ng/mL)-stimulated HUVEC migration, proliferation, and capillary structure formation [106]. Regarding the in vivo assay, administration for 20 consecutive days of 1-*O*-acetylbritannilactone (**12**) (12 mg/kg b.w. per day) to A549 tumor xenografts male nude BALB/c mice cause a significant decrease in tumor cell angiogenesis and tumor growth when compared to the control group, without significant toxicity or adverse effects to the experimental animals [106]. The 1-*O*-acetylbritannilactone (**12**) seems to have the ability to suppress the VEGFR2 downstream Src-FAK signaling pathway, by remarkable inhibition of steroid receptor coactivator (Src) and FAK phosphorylation [106]. This last two are crucial signaling kinases in VEGF-mediated angiogenesis, by working together, or separately, to promote growth, migration, and survival of endothelial cells as well as capillary tube formation [100,101].

Another study [107] found that the 5α-epoxyalantolactone (**18**) (Figure 4), a sesquiterpene lactone isolated from the roots of *Inula helenium* and with a chemical structure very similar to alantolactone (**11**), had antiproliferative activity against human leukemia stem-like cell line KG1a. It presents an IC_{50} value of 3.36 ± 0.18 μM and was found to reduce the expression of anti-apoptotic protein Bcl-2 and increased the expression of pro-apoptotic protein Bax in a dose-dependent manner, while increasing the release of cytochrome into the cytoplasm, culminating in apoptosis of the cells [107].

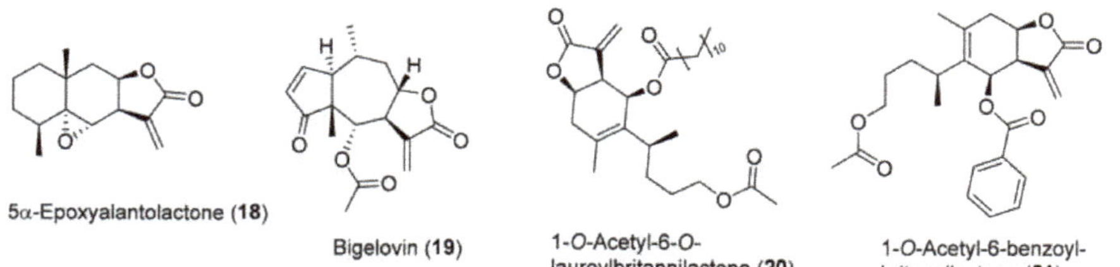

5α-Epoxyalantolactone (**18**)

Bigelovin (**19**)

1-*O*-Acetyl-6-*O*-lauroylbritannilactone (**20**)

1-*O*-Acetyl-6-benzoyl-britannilactone (**21**)

Figure 4. Chemical structure of *Inula* secondary metabolites (**18–19**) and semisynthetic derivatives (**20–21**) with reported activity against oxidative-stress carcinogenesis.

Several important physiological functions in inflammation, cell differentiation, proliferation, and cell survival, as well as apoptosis and immune modulation are mediated by many cytokines [108,109]. The activation of the cytokines signals transduction of the Janus kinase (JAK) and STAT pathway, where JAKs phosphorylate STATs, causing their activation, associated with cancer and other proliferative diseases [110,111]. A study [112] showed that bigelovin (**19**) (Figure 4), a very abundant sesquiterpene lactone found in several *Inula* species [12,13], is a potent inhibitor of the JAK2/STAT3 signaling pathway. It directly inactivates JAK2 and blocks the downstream signaling transduction pathway, blocking IL-6-induced activation of STAT3. This explains the bigelovin (**19**) remarkable antitumor activity against several cancer cell lines from different tissues [112,113], e.g., human lung carcinoma cell lines (A549 $IC_{50} \cong 4.5$ μM and H460 $IC_{50} \cong 8.5$ μM), human cervical carcinoma cell line (HeLa $IC_{50} \cong 3.3$ μM), human hepatocellular carcinoma cell line (HepG2 $IC_{50} \cong 7.1$ μM), human breast adenocarcinoma cell line (MDA-MB-231 $IC_{50} \cong 1.3$ μM, MDAMB-453 $IC_{50} \cong 2.5$ μM and MDA-MB-468 $IC_{50} \cong 1.1$ μM), and human leukemia cell lines (HL-60 $IC_{50} \cong 0.5$ μM, Jurkat $IC_{50} \cong 0.9$ μM and U937 $IC_{50} \cong 0.6$ μM) [112]. Li et al. [114] showed that bigelovin (**19**) also acts mainly via the IL6/STAT3 pathway, significantly and effectively exerting anti-inflammatory and antitumor effects on colorectal cancer cells (CRC). In in vitro assay, cell viability, proliferation and colony formation of colon cancer cells colon-26 and its most aggressive version colon-26-M01 cells are inhibited in time- and dose-dependent manners, by bigelovin (**19**), with IC_{50} values of 0.99 ± 0.3 μM and 1.12 ± 0.33 μM, respectively [114]. In in vivo assay, the male BALB/c mice inoculated with human colon adenocarcinoma cell line HCT 116 and murine colon cancer cell line 26-M01 were subjected to treatment with bigelovin (**19**), at 0.3, 1, and 3 mg/kg b.w., applied every three days for 6 times. All doses significantly suppressed tumor growth and inhibited metastasis without decrease of body weight in both CRC mouse models [114].

As confirmed by all the above, several compounds isolated from *Inula* species exhibit relevant properties in the fight against oxidative-stress related diseases, with 1-*O*-acetylbritannilactone (**12**) (Figure 2) being one of the most studied compounds. The interest in this compound led to the publication of several studies on the synthesis of derivatives and evaluation of their biological activity. In some cases, the results obtained are very interesting. For example, the semisynthetic derivative 1-*O*-acetyl-6-*O*-lauroylbritannilactone (**20**) (Figure 4) is one of the most promising 1-*O*-acetyl-britannilactone derivatives (it bearing a lauroyl group at C-6 position) and exhibits cytotoxic activity against several cell lines (HCT 116, HEp-2 and HeLa), with IC_{50} values of 2.91 ± 0.61 μM, 5.85 ± 0.45 μM, and 6.78 ± 0.23 μM, respectively [115]. It is not so effective as etoposide (IC_{50} values of 2.13 ± 0.23 μM, 4.79 ± 0.54 μM, and 2.97 ± 0.25 μM, respectively) but a lot better than 1-*O*-acetylbritannilactone (**12**), (IC_{50} values of 36.1 ± 3.1 μM, 19.3 ± 1.5 μM, and 32.6 ± 2.5 μM, respectively) [115]. It should be noticed that, at least in the case of the HCT 116 cell line, 1-O-Acetyl-6-O-lauroylbritannilactone (**20**) could rival etoposide while being less toxic to the CHO normal cell line (IC_{50} = 5.97 ± 0.12 μM) than the reference compound etoposide (IC_{50} = 2.60 ± 0.15 μM). In addition, 1-O-Acetyl-6-O-lauroylbritannilactone (**20**) was also found to cause cell-cycle arrest in the G2/M phase in HCT 116 cell line [115].

In a similar work [116], the 6-OH position of 1-*O*-acetylbritannilactone (**12**) was modified with a variety of substituents, being the semisynthetic derivative, 1-*O*-acetyl-6-benzoyl-britannilactone (**21**) (Figure 4), the most promising antitumor derivative with IC_{50} values of 5.19 ± 0.10 μM and 9.93 ± 0.06 μM against HeLa and SGC-7901 cell lines, respectively, an activity level not much different from those of reference drug etoposide (HeLa IC_{50} = 2.97 ± 0.25 μM and SGC-7901 IC_{50} = 6.56 ± 0.68 μM), but it does not rival with a 5-fluorouracil drug against SGC-7901 cell line (IC_{50} = 0.86 ± 0.05 μM) [116]. In addition to this, it is worth mentioning that this type of approach is very interesting and worth investing in, because the adequate structural modification of the natural compounds enables the development of new affordable, efficient, and safe antineoplastic drugs [117].

As referred above, under impaired antioxidant pathways, critical cellular gene mutations can be induced by oxidative stress, which can be the major carcinogenic inductor [7,118]. However, in some cases, the increase in oxidative stress levels could also contribute to antitumor activity [119]. In fact, alantolactone (**11**) (Figure 2) [120] and bigelovin (**19**) (Figure 4) [121], two compounds described above as cytotoxic agents by antioxidant pathways, can have cytotoxic activity also through pro-oxidant pathways. These two studies [120,121], among several in the literature [119,121–123], are presented here as examples of a new perspective on the role of ROS, showing that in some cases, the production of ROS may be beneficial. In fact, the cytotoxic activity by pro-oxidant action opens new perspectives in research on the role of ROS species in biological systems as well as on new ways of fighting cancer. However, understanding the factors related to the cytotoxic effect by pro-oxidant mechanism and its effects in an integrated perspective require much more in-depth studies. Its discussion in more detail, although interesting, falls outside the scope of this review.

4. Conclusions

Taking into account the recent literature presented on this review regarding compounds with antioxidant properties and action mechanisms that target the reduction of the oxidative stress nature inherent to the various mentioned diseases, it should be mentioned that many aspects still require clarification and further studies. Knowledge about the interactions of the mentioned compounds with others, as well as the precise pathways through which some compounds exert their therapeutic activities remains scarce. The *Inula* species showed to be a good source of interesting and active compounds that act against oxidative-stress related diseases, through antioxidant mechanisms and/or other nonspecific antioxidant pathways, culminating in a melioration of the oxidative-stress induced problems. From all compounds, β-caryophyllene (**2**) is one of the most promising ones, since it presented higher antioxidant activity in the DPPH assay (IC_{50} of 1.25 ± 0.06 μM), more active than the reference ascorbic acid. Jaceoside (**14**) exhibits the best anti-inflammatory activity from all compounds (IC_{50} of 0.34 ± 0.01 μM), through inhibition of NO production. Jaceoside (**14**) should be taken in consideration as another promising compound for future studies regarding different bioactivities and its mechanisms of action. Alantolactone (**11**) is the most polyvalent compound, reporting interesting IC_{50} values for several bioactivities (i.e., anti-inflammatory, anti-diabetic, neuroprotective, and antitumoral). 1-*O*-acetylbritannilactone (**12**) can be also pointed out as a promising compound, since it can be used as a blueprint for the development of analogues with interesting properties. This work expects to highlight the relevance of *Inula* species as a source of compounds with relevant bioactivities against stress-oxidative related diseases.

Author Contributions: W.R.T. and A.M.L.S. conceived and wrote the paper.

Funding: This research was funded by FCT/MCT, by financial support to the cE3c centre (FCT Unit, UID/BIA/00329/2013, 2015-2018, and UID/BIA/00329/2019), and to the QOPNA research Unit (FCT UID/QUI/00062/2019) through national founds and, where applicable, co-financed by the FEDER, within the PT2020 Partnership Agreement, and to the Portuguese NMR Network.

Acknowledgments: Thanks are due to the University of Azores and University of Aveiro.

Conflicts of Interest: The authors declare no conflict of interest.

Abbreviations

3T3-L1	Mouse adipocytes cells
26-M01	Murine aggressive colorectal cancer
A549	Human lung carcinoma
ABTS	2,2′-azino-bis(3-ethylbenzothiazoline-6-sulphonic acid
AChE	Acetylcholinesterase
AD	Alzheimer disease
ADR	Adriamycin
AML	Acute myeloid leukemia
Aβ	Amyloid-β
BALB/c	Strain of laboratory mouse
Bax	Bcl-2-associated X
Bcl-2	B-cell lymphoma 2
BV-2	Mouse microglia cells
b.w.	Body weight
C57BL/6J	Strain of laboratory mouse
CB	Celecoxib
CHO	Normal hamster cell line
CNS	Central nervous system
COX-2	Cyclooxygenase 2
CRC	Colorectal cancer
DNA	Deoxyribonucleic acid
DPPH	1,1-Diphenyl-2-picrylhydrazyl
DPP-IV	Dipeptidyl peptidase 4
FAK	Focal adhesion kinase
FLIP	FLICE-inhibitory protein
GBM	Glioblastoma multiforme
GLP-1	Glucagon-like peptide 1
GSH	Glutathione
GSR	Glutathione reductase

GSSG	Glutathione disulphide
H460	Human lung carcinoma
H9c2	Rat cardiomyoblasts
HaCaT	Nontumorigenic human epidermal cells
HAT	Hydrogen-atom transfer
HCT 116	Human colon cancer
HeLa	Human cervical carcinoma
HEp-2	Human larynx epidermal carcinoma
HepG2	Human hepatocellular carcinoma
HL-60	Human acute promyelocytic leukemia
HO-1	Heme oxygenase-1
HUVEC	Human umbilical vascular endothelial cells
IC_{50}	Half maximal inhibitory concentration
IKK	IκB kinase
IκB-α	Inhibitory κB-α
IL-1	Interleukin 1
IL-1β	Interleukin-1β
IL-1R	Interleukin-1 receptor
IL-4	Interleukin 4
IL-6	Interleukin 6
IL-6R	Interleukin 6 receptor
iNOS	Inducible nitric oxide synthase
JAK	Janus kinase
Jurkat	Human acute T cell leukemia
K562	Human bone marrow chronic myelogenous leukemia
K562/A02	Human chronic myelogenous leukemia multidrug-resistant
KG1a	Human acute monocytic leukemia
LPS	Lipopolysaccharide
MAPK	Mitogen-activated protein kinase
MCP1	Monocyte chemoattractant protein 1
MDA	Malondialdehyde
MDA-MB-231	Human breast adenocarcinoma
MDA-MB-453	Human breast metastatic carcinoma
MDA-MB-468	Human breast adenocarcinoma (ethnicity: black)
MMP	Mitochondrial membrane potential
mRNA	Messenger ribonucleic acid
NCI-H716	Human colorectal adenocarcinoma
NF-κB	Nuclear factor kappa-B
NO	Nitric oxide
NQO1	NAD(P)H:quinone oxidoreductase-1
Nrf2	Nuclear factor erythroid 2-related factor 2
PANC-1	Human pancreatic epithelioid carcinoma
PGE2	Prostaglandin E2
PPARα	Peroxisome proliferation-activated receptor α
RAW 264.7	Macrophage normal cell line
RNA	Ribonucleic acid
ROS	Reactive oxygen species
RNS	Reactive nitrogen species
SET	Single-electron transfer
SGC-7901	Gastric carcinoma
SH-SY5Y	Human neuroblastoma
SI	Selectivity index
SMT	2-methyl-2-thiopseudourea sulphate
SOD	Superoxide dismutase

Src	Steroid receptor coactivator
STAT	Signal transducer and activator of transcription
STAT3	Signal transducer and activator of transcription 3
SVG	Normal human glial cell
TBI	Traumatic brain injury
THP-1	Human acute monocytic leukemia
TLR	Toll-like receptor
TNF-α	Tumor necrosis factor α
TNFR	Tumor necrosis factor receptor
TRAIL	TNF-related apoptosis inducing ligand
U87	Human primary glioblastoma
U118	Human glioblastoma
U251	Human glioblastoma
U937	Human histiocytic lymohoma
VEGF	Vascular endothelial growth factor
VEGFR2	Vascular endothelial growth factors receptor-2
XIAP	X-linked inhibitor of apoptosis protein

References

1. Chandel, N.S.; Budinger, G.R.S. The cellular basis for diverse responses to oxygen. *Free. Radic. Biol. Med.* **2007**, *42*, 165–174. [CrossRef] [PubMed]
2. Szymanska, R.; Pospíšil, P.; Kruk, J. Plant-derived antioxidants in disease prevention 2018. *Oxid. Med. Cell Longev.* **2018**, *2018*, e2068370. [CrossRef] [PubMed]
3. Vara, D.; Pula, G. Reactive oxygen species: Physiological roles in the regulation of vascular cells. *Curr. Mol. Med.* **2014**, *14*, 1103–1125. [CrossRef] [PubMed]
4. Thomas, C.; Mackey, M.M.; Diaz, A.A.; Cox, D.P. Hydroxyl radical is produced via the Fenton reaction in submitochondrial particles under oxidative stress: Implications for diseases associated with iron accumulation. *Redox Rep.* **2009**, *14*, 102–108. [CrossRef] [PubMed]
5. Chen, Q.; Wang, Q.; Zhu, J.; Xiao, Q.; Zhang, L. Reactive oxygen species: Key regulators in vascular health and diseases. *Br. J. Pharmacol.* **2018**, *175*, 1279–1292. [CrossRef]
6. Sies, H.; Berndt, C.; Jones, D.P. Oxidative stress. *Annu. Rev. Biochem.* **2017**, *86*, 715–748. [CrossRef]
7. Klaunig, J.E.; Wang, Z. Oxidative stress in carcinogenesis. *Curr. Opin. Toxicol.* **2018**, *7*, 116–121. [CrossRef]
8. McGarry, T.; Biniecka, M.; Veale, D.J.; Fearon, U. Hypoxia, oxidative stress and inflammation. *Free Radic. Biol. Med.* **2018**, *125*, 15–24. [CrossRef] [PubMed]
9. Ighodaro, O.M. Molecular pathways associated with oxidative stress in diabetes mellitus. *Biomed. Pharmacother.* **2018**, *108*, 656–662. [CrossRef]
10. Seca, A.M.L.; Pinto, D.C.G.A. Plant secondary metabolites as anticancer agents: Successes in clinical trials and therapeutic application. *Int. J. Mol. Sci.* **2018**, *19*, 263. [CrossRef]
11. Wright, G.D. Unlocking the potential of natural products in drug discovery. *Microb. Biotechnol.* **2019**, *12*, 55–57. [CrossRef] [PubMed]
12. Seca, A.M.L.; Grigore, A.; Pinto, D.C.G.A.; Silva, A.M.S. The genus *Inula* and their metabolites: from ethnopharmacological to medicinal uses. *J. Ethnopharmacol.* **2014**, *154*, 286–310. [CrossRef]
13. Seca, A.M.L.; Pinto, D.C.G.A.; Silva, A.M.S. Metabolomic profile of the genus *Inula*. *Chem. Biodivers.* **2015**, *12*, 859–906. [CrossRef]
14. The Plant List. Available online: http://www.theplantlist.org/1.1/browse/A/Compositae/Inula/ (accessed on 16 February 2019).
15. Jeelani, S.M.; Rather, G.A.; Sharma, A.; Lattoo, S.K. In perspective: Potential medicinal plant resources of Kashmir Himalayas, their domestication and cultivation for commercial exploitation. *J. Appl. Res. Med. Aromat. Plants* **2018**, *8*, 10–25. [CrossRef]
16. Alamgeer; Uttra, A.M.; Ahsan, H.; Hasan, U.H.; Chaudhary, M.A. Traditional medicines of plant origin used for the treatment of inflammatory disorders in Pakistan: A review. *J. Tradit. Chin. Med.* **2018**, *38*, 636–656. [CrossRef]

17. Lee, S.G.; Kang, H. Anti-neuroinflammatory effects of ethanol extract of *Inula helenium* L (Compositae). *Trop. J. Pharm. Res.* **2016**, *15*, 521–526. [CrossRef]
18. Wang, Q.; Gao, S.; Wu, G.Z.; Yang, N.; Zu, X.P.; Li, W.C.; Xie, N.; Zhang, R.R.; Li, C.W.; Hu, Z.L.; et al. Total sesquiterpene lactones isolated from *Inula helenium* L. attenuates 2,4-dinitrochlorobenzene-induced atopic dermatitis-like skin lesions in mice. *Phytomedicine* **2018**, *46*, 78–84. [CrossRef]
19. Park, S.H.; Lee, D.H.; Kim, M.J.; Ahn, J.; Jang, Y.J.; Ha, T.Y.; Jung, C.H. *Inula japonica* Thunb. flower ethanol extract improves obesity and exercise endurance in mice fed a high-fat diet. *Nutrients* **2019**, *11*, e17. [CrossRef] [PubMed]
20. Alam, M.N.; Bristi, N.J.; Rafiquzzaman, M. Review on in vivo and in vitro methods evaluation of antioxidant activity. *Saudi Pharm. J.* **2013**, *21*, 143–152. [CrossRef]
21. Kedare, S.B.; Singh, R.P. Genesis and development of DPPH method of antioxidant assay. *J. Food Sci. Technol.* **2011**, *48*, 412–422. [CrossRef]
22. Olszowy, M.; Dawidowicz, A.L. Is it possible to use the DPPH and ABTS methods for reliable estimation of antioxidant power of colored compounds? *Chem. Pap.* **2018**, *72*, 393–400. [CrossRef]
23. Badarinath, A.V.; Rao, K.M.; Chetty, C.M.S.; Ramkanth, S.; Rajan, T.V.S.; Gnanaprakash, K. A review of in vitro antioxidant methods: Comparisons, correlations and considerations. *Int. J. PharmTech. Res.* **2010**, *2*, 1276–1285.
24. Nimse, S.B.; Pal, D. Free radicals, natural antioxidants and their reaction mechanisms. *RSC Adv.* **2015**, *5*, 27986–28006. [CrossRef]
25. Danino, O.; Gottlieb, H.E.; Grossman, S.; Bergman, M. Antioxidant activity of 1,3-dicaffeoylquinic acid isolated from *Inula viscosa*. *Food Res. Int.* **2009**, *42*, 1273–1280. [CrossRef]
26. Stojakowska, A.; Malarz, J.; Kiss, A.K. Hydroxycinnamates from elecampane (*Inula helenium* L.) callus culture. *Acta Physiol. Plant.* **2016**, *38*, 1–5. [CrossRef]
27. Dahham, S.S.; Tabana, Y.M.; Iqbal, M.A.; Ahamed, M.B.K.; Ezzat, M.O.; Majid, A.S.A.; Majid, A.M.S.A. The anticancer, antioxidant and antimicrobial properties of the sesquiterpene β-caryophyllene from the essential oil of *Aquilaria crassna*. *Molecules* **2015**, *20*, 11808. [CrossRef]
28. Priydarshi, R.; Melkani, A.B.; Mohan, L.; Pant, C.C. Terpenoid composition and antibacterial activity of the essential oil from *Inula cappa* (Buch-Ham. ex. D. Don) DC. *J. Essent. Oil Res.* **2015**, *28*, 172–176. [CrossRef]
29. Lee, K.J.; Oh, Y.C.; Cho, W.K.; Ma, J.Y. Antioxidant and anti-inflammatory activity determination of one hundred kinds of pure chemical compounds using offline and online screening HPLC assay. *Evid. Based Complement. Alternat. Med.* **2015**, *2015*, e165457. [CrossRef] [PubMed]
30. Wang, J.; Zhao, Y.M.; Zhang, M.L.; Shi, Q.W. Simultaneous determination of chlorogenic acid, caffeic acid, alantolactone and isoalantolactone in *Inula helenium* by HPLC. *J. Chromatogr. Sci.* **2015**, *53*, 526–530. [CrossRef] [PubMed]
31. Palmieri, M.G.S.; Cruz, L.T.; Bertges, F.S.; Húngaro, H.M.; Batista, L.R.; da Silva, S.S.; Fonseca, M.J.V.; Rodarte, M.P.; Vilela, F.M.P.; do Amaral, M.P.H. Enhancement of antioxidant properties from green coffee as promising ingredient for food and cosmetic industries. *Biocatal. Agric. Biotechnol.* **2018**, *16*, 43–48. [CrossRef]
32. Stojakowska, A.; Malarz, J.; Zubek, S.; Turnau, K.; Kisiel, W. Terpenoids and phenolics from *Inula ensifolia*. *Biochem. Syst. Ecol.* **2010**, *38*, 232–235. [CrossRef]
33. Wu, Z.J.; Shan, L.; Lu, M.; Shen, Y.H.; Tang, J.; Zhang, W.D. Chemical constituents from *Inula cappa*. *Chem. Nat. Compd.* **2010**, *46*, 298–300. [CrossRef]
34. Nan, M.; Vlase, L.; Eșianu, S.; Tămaș, M. The analysis of flavonoids from *Inula helenium* L. flowers and leaves. *Acta Med. Marisiensis* **2011**, *57*, 319–323.
35. Li, X.; Jiang, Q.; Wang, T.; Liu, J.; Chen, D. Comparison of the antioxidant effects of quercitrin and isoquercitrin: Understanding the role of the 6″-OH group. *Molecules* **2016**, *21*, e1246. [CrossRef] [PubMed]
36. Yu, N.J.; Zhao, Y.M.; Zhang, Y.Z.; Li, Y.F. Japonicins A and B from the flowers of *Inula japonica*. *J. Asian Nat. Prod. Res.* **2006**, *8*, 385–390. [CrossRef] [PubMed]
37. Al-Rifai, A. Identification and evaluation of *in-vitro* antioxidant phenolic compounds from the *Calendula tripterocarpa* Rupr. *S. Afr. J. Bot.* **2018**, *116*, 238–244. [CrossRef]
38. Hu, X.J.; Jin, H.Z.; Liu, X.H.; Zhang, W.D. Two new sesquiterpenes from *Inula salsoloides* and their inhibitory activities against NO production. *Helv. Chim. Acta* **2011**, *94*, 306–312. [CrossRef]

39. Choi, J.S.; Islam, M.N.; Ali, M.Y.; Kim, Y.M.; Park, H.J.; Sohn, H.S.; Jung, H.A. The effects of C-glycosylation of luteolin on its antioxidant, anti-Alzheimer's disease, anti-diabetic, and anti-inflammatory activities. *Arch. Pharm. Res.* **2014**, *37*, 1354–1363. [CrossRef] [PubMed]
40. Ivanova, V.; Trendafilova, A.; Todorova, M.; Danova, K.; Dimitrov, D. Phytochemical profile of *Inula britannica* from Bulgaria. *Nat. Prod. Commun.* **2017**, *12*, 153–154. [CrossRef]
41. Geng, H.M.; Zhang, D.Q.; Zha, J.P.; Qi, J.L. Simultaneous HPLC determination of five flavonoids in *Flos Inulae*. *Chromatographia* **2007**, *66*, 271–275. [CrossRef]
42. Forrester, S.J.; Kikuchi, D.S.; Hernandes, M.S.; Xu, Q.; Griendling, K.K. Reactive oxygen species in metabolic and inflammatory signaling. *Circ. Res.* **2018**, *122*, 877–902. [CrossRef] [PubMed]
43. Medzhitov, R. Inflammation 2010: new adventures of an old flame. *Cell* **2010**, *140*, 771–776. [CrossRef]
44. Sugimoto, M.A.; Sousa, L.P.; Pinho, V.; Perretti, M.; Teixeira, M.M. Resolution of inflammation: What controls its onset? *Front. Immunol.* **2016**, *7*, e160. [CrossRef] [PubMed]
45. Straub, R.H.; Schradin, C. Chronic inflammatory systemic diseases: An evolutionary trade-off between acutely beneficial but chronically harmful programs. *Evol. Med. Public Health* **2016**, *2016*, 37–51. [CrossRef]
46. Chen, L.; Deng, H.; Cui, H.; Fang, J.; Zuo, Z.; Deng, J.; Li, Y.; Wang, X.; Zhao, L. Inflammatory responses and inflammation-associated diseases in organs. *Oncotarget* **2018**, *9*, 7204–7218. [CrossRef] [PubMed]
47. Sharma, J.N.; Al-Omran, A.; Parvathy, S.S. Role of nitric oxide in inflammatory diseases. *Inflammopharmacology* **2007**, *15*, 252–259. [CrossRef] [PubMed]
48. Ricciotti, E.; FitzGerald, G.A. Prostaglandins and inflammation. *Arterioscler. Thromb. Vasc. Biol.* **2011**, *31*, 986–1000. [CrossRef]
49. Goradel, N.H.; Najafi, M.; Salehi, E.; Farhood, B.; Mortezaee, K. Cyclooxygenase-2 in cancer: a review. *J. Cell. Physiol.* **2018**, *234*, 5683–5699. [CrossRef]
50. Todoric, J.; Antonucci, L.; Karin, M. Targeting inflammation in cancer prevention and therapy. *Cancer Prev. Res.* **2016**, *9*, 895–905. [CrossRef]
51. Chauhan, G.; Roy, K.; Kumar, G.; Kumari, P.; Alam, S.; Kishore, K.; Panjwani, U.; Ray, K. Distinct influence of COX-1 and COX-2 on neuroinflammatory response and associated cognitive deficits during high altitude hypoxia. *Neuropharmacology* **2019**, *146*, 138–148. [CrossRef]
52. Kim, H.N.; Kim, D.H.; Kim, E.H.; Lee, M.H.; Kundu, J.K.; Na, H.K.; Cha, Y.N.; Surh, Y.J. Sulforaphane inhibits phorbol ester-stimulated IKK-NF-κB signaling and COX-2 expression in human mammary epithelial cells by targeting NF-κB activating kinase and ERK. *Cancer Lett.* **2014**, *351*, 41–49. [CrossRef] [PubMed]
53. Suzuki, T.; Suzuki, Y.; Okuda, J.; Kurazumi, T.; Suhara, T.; Ueda, T.; Nagata, H.; Morisaki, H. Sepsis-induced cardiac dysfunction and β-adrenergic blockade therapy for sepsis. *J. Intensive Care* **2017**, *5*, e22. [CrossRef]
54. Huang, S.-H.; Xu, M.; Wu, H.-M.; Wan, C.-X.; Wang, H.-B.; Wu, Q.-Q.; Liao, H.-H.; Deng, W.; Tang, Q.-Z. Isoquercitrin attenuated cardiac dysfunction via AMPKα-dependent pathways in LPS-treated mice. *Mol. Nutr. Food Res.* **2018**, *62*, e1800955. [CrossRef] [PubMed]
55. Wang, G.-W.; Qin, J.-J.; Cheng, X.-R.; Shen, Y.-H.; Shan, L.; Jin, H.-Z.; Zhang, W.-D. *Inula* sesquiterpenoids: structural diversity, cytotoxicity and anti-tumor activity. *Expert Opin. Investig. Drugs* **2014**, *23*, 317–345. [CrossRef]
56. Zhang, S.-D.; Qin, J.-J.; Jin, H.-Z.; Yin, Y.-H.; Li, H.-L.; Yang, X.-W.; Li, X.; Shan, L.; Zhang, W.-D. Sesquiterpenoids from *Inula racemosa* Hook. f. inhibit nitric oxide production. *Planta Med.* **2012**, *78*, 166–171. [CrossRef]
57. Chun, J.; Choi, R.J.; Khan, S.; Lee, D.-S.; Kim, Y.-C.; Nam, Y.-J.; Lee, D.-U.; Kim, Y.S. Alantolactone suppresses inducible nitric oxide synthase and cyclooxygenase-2 expression by down-regulating NF-κB, MAPK and AP-1 via the MyD88 signaling pathway in LPS-activated RAW 264.7 cells. *Int. Immunopharmacol.* **2012**, *14*, 375–383. [CrossRef]
58. Liu, B.; Han, M.; Sun, R.-H.; Wang, J.-J.; Zhang, Y.-P.; Zhang, D.-Q.; Wen, J.-K. ABL-N-induced apoptosis in human breast cancer cells is partially mediated by c-Jun NH_2-terminal kinase activation. *Breast Cancer Res.* **2010**, *12*, R9. [CrossRef]
59. Fang, X.-M.; Liu, B.; Liu, Y.-B.; Wang, J.-J.; Wen, J.-K.; Li, B.-H.; Han, M. Acetylbritannilactone suppresses growth via upregulation of krüppel-like transcription factor 4 expression in HT-29 colorectal cancer cells. *Oncol. Rep.* **2011**, *26*, 1181–1187. [CrossRef]

60. Khan, A.L.; Hussain, J.; Hamayun, M.; Gilani, S.A.; Ahmad, S.; Rehman, G.; Kim, Y.-H.; Kang, S.-M.; Lee, I.-J. Secondary metabolites from *Inula britannica* L. and their biological activities. *Molecules* **2010**, *15*, 1562. [CrossRef] [PubMed]
61. Liu, B.; Wen, J.K.; Li, B.H.; Fang, X.M.; Wang, J.J.; Zhang, Y.P.; Shi, C.J.; Zhang, D.Q.; Han, M. Celecoxib and acetylbritannilactone interact synergistically to suppress breast cancer cell growth via COX-2-dependent and -independent mechanisms. *Cell Death Dis.* **2011**, *2*, e185. [CrossRef]
62. Wei, X.-P.; Chen, Y.-F.; Zhu, H.; Wu, X.-R.; Yu, Y.; Kong, D.-X.; Duan, H.-Q.; Jin, M.-H.; Qin, N. Synthesis and anti-inflammatory activities of 1-*O*-acetylbritannilactone analogues. *Phytochem. Lett.* **2017**, *19*, 248–253. [CrossRef]
63. Garayev, E.; Di Giorgio, C.; Herbette, G.; Mabrouki, F.; Chiffolleau, P.; Roux, D.; Sallanon, H.; Ollivier, E.; Elias, R.; Baghdikian, B. Bioassay-guided isolation and UHPLC-DAD-ESI-MS/MS quantification of potential anti-inflammatory phenolic compounds from flowers of *Inula montana* L. *J. Ethnopharmacol.* **2018**, *226*, 176–184. [CrossRef]
64. Jin, Q.; Lee, J.W.; Jang, H.; Lee, H.L.; Kim, J.G.; Wu, W.; Lee, D.; Kim, E.-H.; Kim, Y.; Hong, J.T.; Lee, M.K.; Hwang, B.Y. Dimeric- and trimeric sesquiterpenes from the flower of *Inula japonica*. *Phytochemistry* **2018**, *155*, 107–113. [CrossRef]
65. Kharroubi, A.T.; Darwish, H.M. Diabetes mellitus: The epidemic of the century. *World J. Diabetes* **2015**, *6*, 850–867. [CrossRef]
66. Volpe, C.M.O.; Villar-Delfino, P.H.; dos Anjos, P.M.F.; Nogueira-Machado, J.A. Cellular death, reactive oxygen species (ROS) and diabetic complications. *Cell Death Dis.* **2018**, *9*, 119. [CrossRef]
67. Kohnert, K.-D.; Freyse, E.-J.; Salzsieder, E. Glycaemic variability and pancreatic β-cell dysfunction. *Curr. Diabetes Rev.* **2012**, *8*, 345–354. [CrossRef]
68. Kim, J.J.; Sears, D.D. TLR4 and insulin resistance. *Gastroenterol. Res. Pract.* **2010**, *2010*, 212563. [CrossRef] [PubMed]
69. Chen, L.; Chen, R.; Wang, H.; Liang, F. Mechanisms linking inflammation to insulin resistance. *Int. J. Endocrinol.* **2015**, *2015*, 508409. [CrossRef] [PubMed]
70. Kim, T.H.; Choi, S.E.; Ha, E.S.; Jung, J.G.; Han, S.J.; Kim, H.J.; Kim, D.J.; Kang, Y.; Lee, K.W. IL-6 induction of TLR-4 gene expression via STAT3 has an effect on insulin resistance in human skeletal muscle. *Acta Diabetol.* **2013**, *50*, 189–200. [CrossRef]
71. Zuñiga, L.Y.; Aceves-de la Mora, M.C.; González-Ortiz, M.; Ramos-Núñez, J.L.; Martínez-Abundis, E. Effect of chlorogenic acid administration on glycemic control, insulin secretion, and insulin sensitivity in patients with impaired glucose tolerance. *J. Med. Food* **2018**, *21*, 469–473. [CrossRef] [PubMed]
72. Chan, S.M.H.; Sun, R.-Q.; Zeng, X.-Y.; Choong, Z.-H.; Wang, H.; Watt, M.J.; Ye, J.-M. Activation of PPARα ameliorates hepatic insulin resistance and steatosis in high fructose-fed mice despite increased endoplasmic reticulum stress. *Diabetes* **2013**, *62*, 2095–2105. [CrossRef] [PubMed]
73. Wang, P.; Yan, Z.; Zhong, J.; Chen, J.; Ni, Y.; Li, L.; Ma, L.; Zhao, Z.; Liu, D.; Zhu, Z. Transient receptor potential vanilloid 1 activation enhances gut glucagon-like peptide-1 secretion and improves glucose homeostasis. *Diabetes* **2012**, *61*, 2155–2165. [CrossRef]
74. Deacon, C.F.; Mannucci, E.; Ahrén, B. Glycaemic efficacy of glucagon-like peptide-1 receptor agonists and dipeptidyl peptidase-4 inhibitors as add-on therapy to metformin in subjects with type 2 diabetes-a review and meta analysis. *Diabetes Obes. Metab.* **2012**, *14*, 762–767. [CrossRef] [PubMed]
75. Zhang, L.; Zhang, S.-T.; Yin, Y.-C.; Xing, S.; Li, W.-N.; Fu, X.-Q. Hypoglycemic effect and mechanism of isoquercitrin as an inhibitor of dipeptidyl peptidase-4 in type 2 diabetic mice. *RSC Adv.* **2018**, *8*, 14967. [CrossRef]
76. Huang, X.-L.; He, Y.; Ji, L.-L.; Wang, K.-Y.; Wang, Y.-L.; Chen, D.-F.; Geng, Y.; OuYang, P.; Lai, W.-M. Hepatoprotective potential of isoquercitrin against type 2 diabetes-induced hepatic injury in rats. *Oncotarget.* **2017**, *8*, 101545–101559. [CrossRef]
77. Kim, M.; Song, K.; Kim, Y.S. Alantolactone improves prolonged exposure of interleukin-6-induced skeletal muscle inflammation associated glucose intolerance and insulin resistance. *Front. Pharmacol.* **2017**, *8*, 405. [CrossRef]
78. Wang, J.; Gu, B.J.; Masters, C.L.; Wang, Y.-J. A systemic view of Alzheimer disease - insights from amyloid-β metabolism beyond the brain. *Nat. Rev. Neurol.* **2017**, *13*, 612–623. [CrossRef] [PubMed]

79. Chaturvedi, R.K.; Beal, M.F. Mitochondrial diseases of the brain. *Free Radic. Biol. Med.* **2013**, *63*, 1–29. [CrossRef]
80. Manoharan, S.; Guillemin, G.J.; Abiramasundari, R.S.; Essa, M.M.; Akbar, M.; Akbar, M.D. The role of reactive oxygen species in the pathogenesis of Alzheimer's disease, Parkinson's disease, and Huntington's disease: a mini review. *Oxid. Med. Cell. Longev.* **2016**, *2016*, e8590578. [CrossRef] [PubMed]
81. Seo, J.Y.; Lim, S.S.; Kim, J.; Lee, K.W.; Kim, J.-S. Alantolactone and isoalantolactone prevent amyloid $\beta_{25\text{-}35}$-induced toxicity in mouse cortical neurons and scopolamine-induced cognitive impairment in mice. *Phytother. Res.* **2017**, *31*, 801–811. [CrossRef]
82. Ma, Q. Role of Nrf2 in oxidative stress and toxicity. *Annu. Rev. Pharmacol. Toxicol.* **2013**, *53*, 401–426. [CrossRef] [PubMed]
83. Zhang, M.; An, C.; Gao, Y.; Leak, R.K.; Chen, J.; Zhang, F. Emerging roles of Nrf2 and phase II antioxidant enzymes in neuroprotection. *Prog. Neurobiol.* **2013**, *100*, 30–47. [CrossRef]
84. Zhang, Q.-G.; Laird, M.D.; Han, D.; Nguyen, K.; Scott, E.; Dong, Y.; Dhandapani, K.M.; Brann, D.W. Critical role of NADPH oxidase in neuronal oxidative damage and microglia activation following traumatic brain injury. *PLoS ONE* **2012**, *7*, e34504. [CrossRef]
85. Wang, X.; Lan, Y.-L.; Xing, J.-S.; Lan, X.-Q.; Wang, L.-T.; Zhang, B. Alantolactone plays neuroprotective roles in traumatic brain injury in rats via anti-inflammatory, anti-oxidative and anti-apoptosis pathways. *Am. J. Transl. Res.* **2018**, *10*, 368–380.
86. Wang, Q.-S.; Xie, K.-Q.; Zhang, C.-L.; Zhu, Y.-J.; Zhang, L.-P.; Guo, X.; Yu, S.-F. Allyl chloride-induced time dependent changes of lipid peroxidation in rat nerve tissue. *Neurochem. Res.* **2005**, *30*, 1387–1395. [CrossRef] [PubMed]
87. Ighodaro, O.M.; Akinloye, O.A. First line defence antioxidants-superoxide dismutase (SOD), catalase (CAT) and glutathione peroxidase (GPX): Their fundamental role in the entire antioxidant defence grid. *Alexandria J. Med.* **2018**, *54*, 287–293. [CrossRef]
88. Heneka, M.T.; Carson, M.J.; El Khoury, J.; Landreth, G.E.; Brosseron, F.; Feinstein, D.L.; Jacobs, A.H.; Wyss-Coray, T.; Vitorica, J.; Ransohoff, R.M.; et al. Neuroinflammation in Alzheimer's disease. *Lancet Neurol.* **2015**, *14*, 388–405. [CrossRef]
89. Shadfar, S.; Hwang, C.J.; Lim, M.-S.; Choi, D.-Y.; Hong, J.T. Involvement of inflammation in Alzheimer's disease pathogenesis and therapeutic potential of anti-inflammatory agents. *Arch. Pharm. Res.* **2015**, *38*, 2106–2119. [CrossRef]
90. Liu, F.; Dong, B.; Yang, X.; Yang, Y.; Zhang, J.; Jin, D.-Q.; Ohizumi, Y.; Lee, D.; Xu, J.; Guo, Y. NO inhibitors function as potential anti-neuroinflammatory agents for AD from the flowers of *Inula japonica*. *Bioorg. Chem.* **2018**, *77*, 168–175. [CrossRef]
91. Lind, M.; Hayes, A.; Caprnda, M.; Petrovic, D.; Rodrigo, L.; Kruzliak, P.; Zulli, A. Inducible nitric oxide synthase: Good or bad? *Biomed. Pharmacother.* **2017**, *93*, 370–375. [CrossRef]
92. Tanaka, T.; Shimizu, M.; Kochi, T.; Moriwaki, H. Chemical-induced carcinogenesis. *J. Exp. Clin. Med.* **2013**, *5*, 203–209. [CrossRef]
93. Kumari, S.; Badana, A.K.; Gavara, M.M.; Gugalavath, S.; Malla, R. Reactive oxygen species: A key constituent in cancer survival. *Biomark. Insights* **2018**, *13*, 1–9. [CrossRef]
94. Kim, J.; Kim, J.; Bae, J.-S. ROS homeostasis and metabolism: a critical liaison for cancer therapy. *Exp. Mol. Med.* **2016**, *48*, e269. [CrossRef] [PubMed]
95. Tafani, M.; Sansone, L.; Limana, F.; Arcangeli, T.; De Santis, E.; Polese, M.; Fini, M.; Russo, M.A. The interplay of reactive oxygen species, hypoxia, inflammation, and sirtuins in cancer initiation and progression. *Oxid. Med. Cell Longev.* **2016**, *2016*, e3907147. [CrossRef] [PubMed]
96. Wang, X.; Yu, Z.; Wang, C.; Cheng, W.; Tian, X.; Huo, X.; Wang, Y.; Sun, C.; Feng, L.; Xing, J.; et al. Alantolactone, a natural sesquiterpene lactone, has potent antitumor activity against glioblastoma by targeting IKKβ kinase activity and interrupting NF-κB/COX-2-mediated signaling cascades. *J. Exp. Clin. Cancer Res.* **2017**, *36*, 93. [CrossRef]
97. Ding, Y.; Gao, H.; Zhang, Y.; Li, Y.; Vasdev, N.; Gao, Y.; Chen, Y.; Zhang, Q. Alantolactone selectively ablates acute myeloid leukemia stem and progenitor cells. *J. Hematol. Oncol.* **2016**, *9*, e93. [CrossRef]
98. Wang, F.; Li, H.; Qiao, J.-O. 1-*O*-Acetylbritannilactone combined with gemcitabine elicits growth inhibition and apoptosis in A549 human non-small cell lung cancer cells. *Mol. Med. Rep.* **2015**, *12*, 5568–5572. [CrossRef] [PubMed]

99. Rajabi, M.; Mousa, S.A. The role of angiogenesis in cancer treatment. *Biomedicines* **2017**, *5*, e34. [CrossRef] [PubMed]
100. Albini, A.; Tosetti, F.; Li, V.W.; Noonan, D.M.; Li, W.W. Cancer prevention by targeting angiogenesis. *Nat. Rev. Clin. Oncol.* **2012**, *9*, 498–509. [CrossRef] [PubMed]
101. Sennino, B.; McDonald, D.M. Controlling escape from angiogenesis inhibitors. *Nat. Rev. Cancer* **2012**, *12*, 699–709. [CrossRef]
102. Cébe-Suarez, S.; Zehnder-Fjällman, A.; Ballmer-Hofer, K. The role of VEGF receptors in angiogenesis; complex partnerships. *Cell. Mol. Life Sci.* **2006**, *63*, 601–615. [CrossRef]
103. Fontanella, C.; Ongaro, E.; Bolzonello, S.; Guardascione, M.; Fasola, G.; Aprile, G. Clinical advances in the development of novel VEGFR2 inhibitors. *Ann. Transl. Med.* **2014**, *2*, e123. [CrossRef]
104. Jayasinghe, C.; Simiantonaki, N.; Habedank, S.; Kirkpatrick, C.J. The relevance of cell type- and tumor zone-specific VEGFR-2 activation in locally advanced colon cancer. *J. Exp. Clin. Cancer Res.* **2015**, *34*, e42. [CrossRef]
105. Liu, Y.-R.; Cai, Q.-Y.; Gao, Y.-G.; Luan, X.; Guan, Y.-Y.; Lu, Q.; Sun, P.; Zhao, M.; Fang, C. Alantolactone, a sesquiterpene lactone, inhibits breast cancer growth by antiangiogenic activity via blocking VEGFR2 signaling. *Phytother. Res.* **2018**, *32*, 643–650. [CrossRef] [PubMed]
106. Zhengfu, H.; Hu, Z.; Huiwen, M.; Zhijun, L.; Jiaojie, Z.; Xiaoyi, Y.; Xiujun, C. 1-o-acetylbritannilactone (ABL) inhibits angiogenesis and lung cancer cell growth through regulating VEGF-Src-FAK signaling. *Biochem. Biophys. Res. Commun.* **2015**, *464*, 422–427. [CrossRef] [PubMed]
107. Ding, Y.; Pan, W.; Xu, J.; Wang, T.; Chen, T.; Liu, Z.; Xie, C.; Zhang, Q. Sesquiterpenoids from the roots of *Inula helenium* inhibit acute myelogenous leukemia progenitor cells. *Bioorg. Chem.* **2019**, *86*, 363–367. [CrossRef] [PubMed]
108. Schindler, C.; Levy, D.E.; Decker, T. JAK-STAT signaling: from interferons to cytokines. *J. Biol. Chem.* **2007**, *282*, 20059–20063. [CrossRef]
109. Knoops, L.; Hornakova, T.; Royer, Y.; Constantinescu, S.N.; Renauld, J.-C. JAK kinases overexpression promotes in vitro cell transformation. *Oncogene.* **2008**, *27*, 1511–1519. [CrossRef] [PubMed]
110. Wang, T.; Niu, G.; Kortylewski, M.; Burdelya, L.; Shain, K.; Zhang, S.; Bhattacharya, R.; Gabrilovich, D.; Heller, R.; Coppola, D.; Dalton, W.; Jove, R.; Pardoll, D.; Yu, H. Regulation of the innate and adaptive immune responses by Stat-3 signaling in tumor cells. *Nat. Med.* **2004**, *10*, 48–54. [CrossRef] [PubMed]
111. Levine, R.L.; Pardanani, A.; Tefferi, A.; Gilliland, D.G. Role of JAK2 in the pathogenesis and therapy of myeloproliferative disorders. *Nat. Rev. Cancer.* **2007**, *7*, 673–683. [CrossRef] [PubMed]
112. Zhang, H.-H.; Kuang, S.; Wang, Y.; Sun, X.-X.; Gu, Y.; Hu, L.-H.; Yu, Q. Bigelovin inhibits STAT3 signaling by inactivating JAK2 and induces apoptosis in human cancer cells. *Acta Pharmacol. Sin.* **2015**, *36*, 507–516. [CrossRef] [PubMed]
113. Zeng, G.-Z.; Tan, N.-H.; Ji, C.-J.; Fan, J.-T.; Huang, H.-Q.; Han, H.-J.; Zhou, G.-B. Apoptosis inducement of bigelovin from *Inula helianthus-aquatica* on human leukemia U937 cells. *Phytother. Res.* **2009**, *23*, 885–891. [CrossRef] [PubMed]
114. Li, M.; Yue, G.G.-L.; Song, L.-H.; Huang, M.-B.; Lee, J.K.-M.; Tsui, S.K.-W.; Fung, K.-P.; Tan, N.-H.; Lau, C.B.-S. Natural small molecule bigelovin suppresses orthotopic colorectal tumor growth and inhibits colorectal cancer metastasis via IL6/STAT3 pathway. *Biochem. Pharmacol.* **2018**, *150*, 191–201. [CrossRef] [PubMed]
115. Dong, S.; Tang, J.-J.; Zhang, C.-C.; Tian, J.-M.; Guo, J.-T.; Zhang, Q.; Li, H.; Gao, J.-M. Semisynthesis and in vitro cytotoxic evaluation of new analogues of 1-*O*-acetylbritannilactone, a sesquiterpene from *Inula britannica*. *Eur. J. Med. Chem.* **2014**, *80*, 71–82. [CrossRef] [PubMed]
116. Tang, J.-J.; Dong, S.; Han, Y.-Y.; Lei, M.; Gao, J.-M. Synthesis of 1-*O*-acetylbritannilactone analogues from *Inula britannica* and in vitro evaluation of their anticancer potential. *Med. Chem. Commun.* **2014**, *5*, 1584. [CrossRef]
117. Maier, M.E. Design and synthesis of analogues of natural products. *Org. Biomol. Chem.* **2015**, *13*, 5302–5343. [CrossRef]
118. Klaunig, J.E.; Wang, Z.; Pu, X.; Zhou, S. Oxidative stress and oxidative damage in chemical carcinogenesis. *Toxicol. Appl. Pharmacol.* **2011**, *254*, 86–99. [CrossRef]
119. Tai, Y.; Cao, F.; Li, M.; Li, P.; Xu, T.; Wang, X.; Yu, Y.; Gu, B.; Yu, X.; Cai, X.; et al. Enhanced mitochondrial pyruvate transport elicits a robust ROS production to sensitize the antitumor efficacy of interferon-γ in colon cancer. *Redox Biol.* **2019**, *20*, 451–457. [CrossRef] [PubMed]

120. Cui, L.; Bu, W.; Song, J.; Feng, L.; Xu, T.; Liu, D.; Ding, W.; Wang, J.; Li, C.; Ma, B.; et al. Apoptosis induction by alantolactone in breast cancer MDA-MB-231 cells through reactive oxygen species-mediated mitochondrion-dependent pathway. *Arch. Pharm. Res.* **2018**, *41*, 299–313. [CrossRef]
121. Li, M.; Song, L.H.; Yue, G.G.L.; Lee, J.K.M.; Zhao, L.M.; Li, L.; Zhou, X.; Tsui, S.K.W.; Ng, S.S.-M.; Fung, K.-P.; et al. Bigelovin triggered apoptosis in colorectal cancer in vitro and in vivo via upregulating death receptor 5 and reactive oxidative species. *Sci. Rep.* **2017**, *7*, 42176. [CrossRef] [PubMed]
122. Jiang, Y.; Xu, H.; Wang, J. Alantolactone induces apoptosis of human cervical cancer cells via reactive oxygen species generation, glutathione depletion and inhibition of the Bcl-2/Bax signaling pathway. *Oncol. Lett.* **2016**, *11*, 4203–4207. [CrossRef] [PubMed]
123. Zhang, J.; Li, Y.; Duan, D.; Yao, J.; Gao, K.; Fang, J. Inhibition of thioredoxin reductase by alantolactone prompts oxidative stress-mediated apoptosis of HeLa cells. *Biochem. Pharmacol.* **2016**, *102*, 34–44. [CrossRef] [PubMed]

MDPI
St. Alban-Anlage 66
4052 Basel
Switzerland
www.mdpi.com

Antioxidants Editorial Office
E-mail: antioxidants@mdpi.com
www.mdpi.com/journal/antioxidants

www.ingramcontent.com/pod-product-compliance
Lightning Source LLC
LaVergne TN
LVHW071613170726
843515LV00009B/2384
9783036587646